Engineering mathematics
through
applications

Second edition

KULDEEP SINGH

Senior Lecturer in Mathematics
School of Physics, Astronomy and Mathematics
University of Hertfordshire

palgrave
macmillan

First edition 2003
This edition first published 2011 by
PALGRAVE MACMILLAN

Palgrave Macmillan in the UK is an imprint of Macmillan Publishers Limited,
registered in England, company number 785998, of Houndmills, Basingstoke,
Hampshire RG21 6XS.

Palgrave Macmillan in the US is a division of St Martin's Press LLC,
175 Fifth Avenue, New York, NY 10010.

Palgrave Macmillan is the global academic imprint of the above companies
and has companies and representatives throughout the world.

Palgrave® and Macmillan® are registered trademarks in the United States,
the United Kingdom, Europe and other countries.

ISBN 978–0–230–27479–2 paperback

This book is printed on paper suitable for recycling and made from fully
managed and sustained forest sources. Logging, pulping and manufacturing
processes are expected to conform to the environmental regulations of the
country of origin.

A catalogue record for this book is available from the British Library.

A catalog record for this book is available from the Library of Congress.

10   9   8   7   6   5   4   3
20   19   18   17   16   15   14

Printed and bound in China

Acknowledgements

The author and publishers would like to thank the following for the use of copyright material:

Maplesoft Inc.

Cambridge University Press for the table on page 918, taken from J. C. P. Miller & F. C. Powell, *The Cambridge Elementary Mathematical Tables*, 2nd edn, 1979. Reproduced with permission.

The Engineering Council.

The author and publishers are grateful to the following for permission to reproduce questions from past examination papers:

Jay Abramson, Arizona State University, USA; Eric Bahuaud, Stanford University, USA; Fabrice Baudoin, Purdue University, USA; David Bayer, Barnard College, Columbia University, New York, USA; Maretta Brennan, Cork Institute of Technology, Ireland; Dietrich Burbulla, University of Toronto, Canada; Michael Chung, University of Aberdeen, UK; Francis Coghlan, University of Manchester, UK; Stephen Gourley, University of Surrey, UK; Rhian Green, University of Loughborough, UK; Alexander Hulpke, Colorado State University, USA; Jennifer Kloke, Stanford University, USA; Duane Kouba, University of California, Davis, USA; Roger Luther, University of Sussex, UK; Fiona Message, University of Portsmouth, UK; Wayne Nagata, University of British Columbia, Canada; Erhard Neher, University of Ottawa, Canada; Pascal O'Connor, Cork Institute of Technology, Ireland; Sean O'Rourke, University of California, Davis, USA; John Parkinson, University of Manchester, UK; Nicolai Reshetikhin, University of California, Berkeley, USA; Alyssa Sankey, University of New Brunswick, Canada; Marshall Slemrod, University of Wisconsin, USA; Peter Sollich, King's College London, UK; Toby Stafford, University of Manchester, UK; Colin Steele, University of Manchester, UK; Shannon Sullivan, Memorial University of Newfoundland, Canada; Yi Sun, North Carolina State University, USA; Joachim Vogt, Jacobs University, Germany; Louise Walker, University of Manchester, UK; Joe Ward, University of Loughborough, UK; Jack Williams, University of Manchester, UK; Francis Wright, Queen Mary, University of London, UK.

Dedication

— To Bibi Chanan Kaur

Summary of Contents

Contents

Note to the Student

This book is ideally suited for anyone doing mathematics as part of their engineering or science undergraduate studies. Very little mathematical knowledge is assumed.

The heart of the book is the wealth of engineering examples drawn from a wide range of disciplines such as aerospace, building services, civil, control, electrical, manufacturing, mechanical, etc. In this respect the book is unique. The inclusion of these engineering disciplines will show from the outset the many applications of the mathematics you are studying. Understanding why you need to learn the mathematics and using these mathematical principles in real engineering examples are a real motivating factor.

These examples do not assume prior engineering knowledge. However you will be asked to reflect on your final answer, i.e. how does your result relate to the problem? Reflection is a good way of learning mathematics and therefore frequent use of a question and answer format is made throughout the book. A question mark icon [?] represents the questions. The question mark icon will help you to stop and think rather than pressing on and reading the answer on the next line.

The key to learning mathematics is to do exercises. If you want to become familiar with mathematical language and understand the theory there is no substitute for practice. There are exercises associated with each section and a miscellaneous exercise at the end of each chapter. The exercises are an integral part of the book and not just a bolt on. Moreover every section begins with a list of objectives and ends with a summary. The chapter also begins with a list of objectives. Doing the exercises and checking your answers are the only way you can meet these objectives.

Complete solutions to *all* exercises are provided on the book's website at **www.palgrave.com/engineering/singh**. This makes the book suitable for students working by themselves or on a distance learning course, as well as for lecture-based courses. These complete solutions provide a thorough step-by-step guide through each problem. Also on the website is a link through to dozens of test questions which offer immediate feedback and scores, an online glossary of key terms, some additional sections and an e-index to help you find the right page in the book immediately.

The proof of mathematical results is kept to a minimum. Instead the emphases of the book are on you learning by investigating results, observing patterns, visualizing graphs, answering questions using technology, etc.

You need to understand the first two chapters thoroughly, because basic arithmetic and algebra are vital ingredients for the remaining chapters. For example, you cannot do any calculus unless you have a firm footing in basic algebra. The content of these chapters should eventually become second nature to you. Recent news articles have stated that there are fewer and fewer students applying for engineering courses at university, which is a serious problem for industry. One of the reasons is that the mathematical nature of engineering is perceived as difficult. This book addresses this problem by using an everyday language step-by-step guide through each example. For example, if an expansion of brackets is carried out at a particular step in the manipulation, then this is stated as [Expanding] adjacent to the step.

Every formula has been given a reference number. For example, reference number **2.1** means that it is the first formula in Chapter 2. If a particular formula is used, then it is either stated in the main text or placed in the footnote on the page so that you do not need to search for it.

Major figures in the history of engineering mathematics are discussed. Of course these are not central to the mathematics, but they do provide motivation for that particular topic.

There are occasionally some examples that are more difficult to follow. These are highlighted with an *.

We assume you have access to a basic scientific calculator. Use is also made of a computer algebra system called MAPLE[†] and a graphical calculator. However you can use any computer algebra system. The book does not try to teach MAPLE but uses it to illuminate examples. The use of this software is not integral to understanding the mathematics.

Finally, good luck with your work through this book, and enjoy it!

Kuldeep Singh
University of Hertfordshire
July 2011

[†] MAPLE is a registered trademark of Maplesoft Inc., www.maplesoft.com

Preface to Second Edition

Mathematics is a subject that is applied within fields as diverse as business, economics, engineering, physics, computer science, ecology, sociology, demography and genetics.

Historically, one of my primary concerns has been in finding tangible and accessible text books to recommend to my students. Based on the popularity of the first edition of *Engineering Mathematics Through Applications*, I have felt compelled to construct a second edition that bridges the considerable divide between A-level and undergraduate engineering mathematics.

I am somewhat fortunate in that I have had several hundred students to assist me in evaluating each chapter and, in response to their reactions, I have consequently modified, expanded and added sections to ensure that the content entirely encompasses the ability of students with a limited mathematical background, as well as the more advanced scholars. I believe this has allowed me to create a book that is unparalleled in the simplicity of its explanation, yet comprehensive in its approach to even the most challenging aspects.

Level

This book is intended for first- and second-year undergraduates with low to average mathematics A-level grades or equivalent. Many students find the transition between A-level and undergraduate engineering mathematics difficult, and this book specifically addresses that gap and allows seamless progression. It assumes very limited prior mathematical knowledge, yet also covers difficult material and answers tough questions through the use of clear explanations and a wealth of illustrations. The emphasis of the book is on students learning for themselves by gradually absorbing clearly presented text, supported by patterns, graphs and associated questions.

Pedagogical issues

The strength of the text is in the large number of examples and the step-by-step explanation of each topic as it is introduced. It is compiled in a way that allows distance learning, with explicit solutions to all the set problems freely available online. The exercises at the end of each chapter comprise questions from past examination papers from various universities, helping to reinforce the reader's confidence. Also included are short historical biographies of the leading players in the field of engineering mathematics. These are generally placed at the beginning of a section to engage the interest of the student from the outset.

This is a book that allows the student to gradually develop an understanding of a topic, without the need for constant additional support from a tutor.

Online content

We have a big and growing companion website at **www.palgrave.com/engineering/singh** which includes:

- The fully worked solutions to **all** exercises in the book – a fantastic extra resource for students
- Additional teaching content:

  Chapter 2, Section G Applications of equations to electrical circuits

                  G1   Modelling electrical circuits

  Chapter 5, Section F Hyperbolic properties

                  F1   Other hyperbolic functions

                  F2   Hyperbolic identities

                  F3   Inverse hyperbolic functions

  Chapter 7, Section J Power series

  Chapter 10, Section F Functions of complex numbers

                  F1   Identities

  Plus additional notes on various other topics

- Additional exercises on various topics
- Excellent introduction to MATLAB
- Interactive online questions
- A searchable glossary of key concepts
- An e-index to allow quick look-up of the right page in the book

Background

My interest in mathematics began at school. I am originally of Indian descent, and as a young child often found English difficult to comprehend, but I discovered an affinity with mathematics, a universal language that I could begin to learn from the same starting point as my peers.

My passion has always been to teach, and I have taught engineering mathematics at the University of Hertfordshire since 1992. The first edition of *Engineering Mathematics Through Applications*, I am proud to say, has been used widely as the basis for undergraduate studies in many different countries. I also host and regularly update a website dedicated to mathematics.

My family and career leave little room for outside interests, but I am a keen football fan and occasional cyclist.

I would particularly like to thank Timothy Peacock who has helped in the significant improvement of this second edition. Thanks too to the Reprographics Unit, in particular the late Lesley Corney at the University of Hertfordshire.

Kuldeep Singh

Important Formulae and Methods

Basic Formulae

Reference Number	Formulae	Notes
(1.13)	$(a + b)^2 = a^2 + 2ab + b^2$	
(1.14)	$(a - b)^2 = a^2 - 2ab + b^2$	
(1.15)	$(a + b)(a - b) = a^2 - b^2$	Difference of two squares
(1.16)	$x = \dfrac{-b \pm \sqrt{b^2 - 4ac}}{2a}$	Quadratic formula
(2.1)	$y = mx + c$	A straight line with gradient m and y intercept c
(4.1)	$\sin(\theta) = opp/hyp$	
(4.2)	$\cos(\theta) = adj/hyp$	
(4.3)	$\tan(\theta) = opp/adj$	
(4.35)	$\tan(A) = \dfrac{\sin(A)}{\cos(A)}$	
(4.53)	$\sin(2A) = 2\sin(A)\cos(A)$	
(4.54)	$\cos(2A) = \cos^2(A) - \sin^2(A)$ $= 2\cos^2(A) - 1 = 1 - 2\sin^2(A)$	
(4.64)	$\sin^2(A) + \cos^2(A) = 1$	
(4.65)	$1 + \tan^2(A) = \sec^2(A)$	
(5.12)	$\ln(A) + \ln(B) = \ln(A \times B)$	

Reference Number	Formulae	Notes
(5.13)	$\ln(A) - \ln(B) = \ln(A/B)$	
(5.14)	$\ln(A^n) = n \ln(A)$	
(6.2)	$\dfrac{d}{dx}[x^n] = nx^{n-1}$	
(6.11)	$\dfrac{d}{dx}[e^{kx}] = ke^{kx}$	
(6.12)	$\dfrac{d}{dx}[\sin(kx)] = k\cos(kx)$	
(6.13)	$\dfrac{d}{dx}[\cos(kx)] = -k\sin(kx)$	
(8.38)	$\displaystyle\int \cos(kx+m)\,dx = \dfrac{1}{k}\sin(kx+m) + C$	
(8.39)	$\displaystyle\int \sin(kx+m)\,dx = -\dfrac{1}{k}\cos(kx+m) + C$	
(8.41)	$\displaystyle\int e^{kx+m}\,dx = \dfrac{1}{k}e^{kx+m} + C$	
(13.4)	$y(IF) = \displaystyle\int \left[(IF)\,Q(x)\right] dx$	Integrating Factor
(14.4)	$y = Ae^{m_1 x} + Be^{m_2 x}$	Real and different, m_1, m_2
(14.5)	$y = (A + Bx)e^{mx}$	Real and equal, m
(14.6)	$y = e^{\alpha x}\left[A\cos(\beta x) + B\sin(\beta x)\right]$	Complex, $m = \alpha \pm j\beta$
(16.1)	$\bar{x} = \dfrac{x_1 + x_2 + x_3 + \cdots + x_n}{n}$	Mean of data
(16.3)	$s = \sqrt{\dfrac{(x_1 - \bar{x})^2 + (x_2 - \bar{x})^2 + (x_3 - \bar{x})^3 + \cdots + (x_n - \bar{x})^2}{n}}$	Standard deviation of data

Methods for Differentiating and Integrating

In Chapters 6–9 we will cover the fundamental methods employed in working with calculus. Here is a reference list of the different methods used for differentiating and integrating given functions.

Differentiation	Method	Integration	Method
Sum/Difference of 2 functions Eg: $2x + 3x^2$ $\dfrac{dy}{dx} = 2 + 6x$	Simply add or subtract differentiated functions	Sum/Difference of 2 functions Eg: $2x + 3x^2 = f(x)$ $\int f(x)dx = x^2 + x^3 + C$	Simply add or subtract integrated functions
Product of 2 functions Eg: $y = x\cos(x)$ $\dfrac{dy}{dx} = \cos(x) - x\sin(x)$	Use the product rule: $\dfrac{dy}{dx} = u'v + uv'$	Product of 2 functions Eg: $\int xe^x$	Use integration by parts: $\int uv'dx = uv - \int uv'dx$ Take u to be the function in the following order: 1) In x, 2) x^n 3) e^{kx} If integration by parts fails try integration by substitution.
Differentiating quotients Eg: $y = \dfrac{\sin(x)}{x^2}$	Use the quotient rule: $\dfrac{dy}{dx} = \dfrac{u'v - uv'}{v^2}$	Integrating quotients Eg: $\int \dfrac{x^2}{x^3 - 1}\,dx$	Use: $\int \dfrac{f'(x)}{f(x)}\,dx = \ln|f(x)| + c$ And partial fractions if required.
Chain rule: y is a function of u and u is a function of x Eg: $y = u = (x^2 + 2)^3$ $\dfrac{dy}{dx} = 6x(x^2 + 2)^2$	Refer to table 3 or use chain rule: $\dfrac{dy}{dx} = \dfrac{dy}{du} \cdot \dfrac{du}{dx}$	Integration by substitution: Eg: $\int \sin(2x)dx$	Use integration by substitution. Let $u = 2x$ $\dfrac{du}{dx} = 2 \quad dx = \dfrac{du}{2}$

Reference Number	Formulae	Notes
	My Important Formulae	

INTRODUCTION
Arithmetic for Engineers

SECTION A **Whole numbers**

By the end of this section you will be able to:
▶ understand the symbols $<$, $\leq$, $>$ and $\geq$
▶ evaluate the arithmetic of positive and negative numbers

A1 **Directed numbers**

Figure 1 shows the number line. The negative numbers are to the left of zero and have a negative symbol, $-$,
in front of them.
The positive
numbers are to

Fig. 1

the right of zero but do **not** have an associated $+$ symbol. For example, $+3 = 3$.

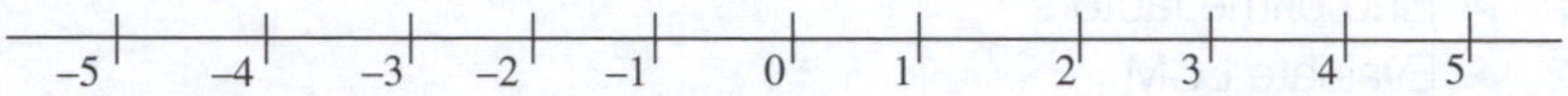

Numbers to the right are greater than numbers to the left, and greater than is denoted by the symbol $>$.

5 is greater than 2 is written $5 > 2$.

Similarly -1 is greater than -3 because -1 is to the right of -3, see Fig. 1.

? **How can we write this in symbol format?**

$$-1 > -3$$

The symbol $\geq$ means greater than or equal to.

Conversely numbers to the left are less than numbers to the right and less than is denoted by the symbol $<$.

? **We know 1 is less than 4, so how can we write this in terms of symbols?**

$$1 < 4$$

? **In the same manner, -4 lies to the left of -1, so how can we write this?**

$$-4 < -1$$

The symbol $\leq$ represents less than or equal to.

Example 1

Fill in the most appropriate symbol $>$, $\geq$, $<$, $\leq$ for $\square$ in the following:

a $2 \ \square \ 3$ **b** $-1 \ \square \ -3$ **c** $-1 \ \square \ -1$ **d** $11 \ \square \ -11$

e $-11 \ \square \ 11$ **f** $-5 \ \square \ -100$ **g** $-7 \ \square \ 0$

Example 1 *continued*

Solution

a Since 2 is less than 3 we have $2 < 3$.
b On the number line -1 is to the right of -3, so -1 is greater than -3:

$$-1 > -3$$

c -1 is equal to -1, so either $\leq$ or $\geq$ will do. Thus

$$-1 \leq -1 \text{ or } -1 \geq -1$$

Both of these are correct. However we normally write this as

$$-1 = -1$$

d 11 is to the right of -11, so 11 is greater than -11. Hence $11 > -11$.
e This is the reverse of **d**, so $-11 < 11$.
f -5 is to the right of -100 on the number line (Fig. 2).
-5 is greater
than -100,
therefore
$-5 > -100$.

Fig. 2

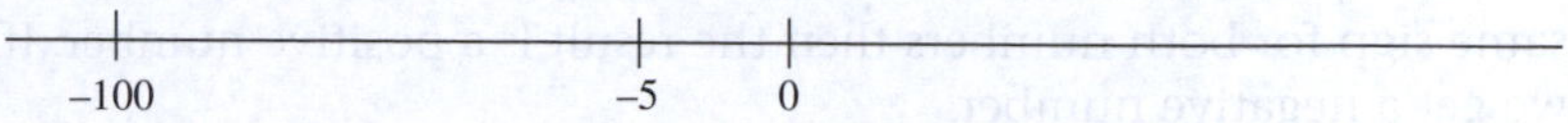

| -100 | | -5 | 0 |

g -7 is to the left of 0 so it is less than 0. Thus $-7 < 0$.

We can add (+), subtract (−), multiply (×) and divide (÷) numbers. To add and subtract negative numbers (numbers left of zero), we need to apply the following rules:

Rule **1**. Adding a negative number is the same as subtracting a positive number.

Rule **2**. Subtracting a negative number is the same as adding a positive number.

Example 2

Evaluate the following: **a** $5 + (-3)$ **b** $-3 + (-4)$ **c** $8 - (-4)$
     d $-7 - (-5)$ **e** $6 + (-6)$

Solution

a Using Rule **1** gives

$$5 + (-3) = 5 - 3 = 2$$

b Using Rule **1** again gives

$$-3 + (-4) = -3 - 4$$

From -3 we move 4 steps to the left (Fig. 3).

$$-3 - 4 = -7$$

Fig. 3
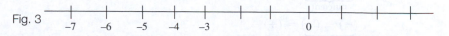

$-7 \quad -6 \quad -5 \quad -4 \quad -3 \qquad\qquad\qquad 0$

Example 2 *continued*

c Applying Rule **2** gives

$$8 - (-4) = 8 + 4 = 12$$

d Applying Rule **2** again gives

$$-7 - (-5) = -7 + 5 = -2$$

(From -7 you are moving 5 steps to the right.)

e Applying Rule **1** gives

$$6 + (-6) = 6 - 6 = 0$$

Next we consider the multiplication and division of negative numbers. It's worth remembering

'Minus times minus makes a plus.'

For multiplication and division of two non-zero numbers, the rule is that if we have the same sign for both numbers then the result is a positive number. If the signs are different, we get a negative number.

For example,

$$9 \times (-3) = -27$$
$$(-9) \times (-3) = 27$$
$$9 \div (-3) = -3$$
$$(-9) \div (-3) = 3$$

For multiplication:

$$\underbrace{(+) \times (+)}_{\text{same}} = (+)$$

$$\underbrace{(+) \times (-)}_{\text{different}} = (-)$$

$$\underbrace{(-) \times (+)}_{\text{different}} = (-)$$

$$\underbrace{(-) \times (-)}_{\text{same}} = (+)$$

Similarly for division:

$$\underbrace{(+) \div (+)}_{\text{same}} = (+)$$

$$\underbrace{(+) \div (-)}_{\text{different}} = (-)$$

$$\underbrace{(-) \div (+)}_{\text{different}} = (-)$$

$$\underbrace{(-) \div (-)}_{\text{same}} = (+)$$

Example 3

Evaluate the following:

a $(-6) \times 2$ **b** $(-3) \times (-4)$ **c** $8 \div (-2)$

d $(-15) \div (-3)$ **e** $(-2) \times (-3) \times (-4)$

Solution

a Since -6 is a negative number multiplied by a positive number 2, the result is a negative number:

$$(-6) \times 2 = -12$$

b Minus times minus makes a plus. So $(-3) \times (-4) = 12$.

c Positive number, 8, divided by a negative number, -2, gives a negative number:

$$8 \div (-2) = -4$$

d Both numbers, -15 and -3, have the same sign so the result is a positive number:

$$(-15) \div (-3) = 5$$

e Minus times minus gives a plus, so $(-2) \times (-3) = 6$.
Then $6 \times (-4) = -24$. Hence

$$\underbrace{(-2) \times (-3)}_{=6} \times (-4) = 6 \times (-4) = -24$$

It's worth remembering that if the number of negative numbers in the multiplication is odd then the final result is a negative number. Conversely if the number of negative numbers is even, then the final result is a positive number:

$$\underbrace{(-) \times (-) \times \cdots \times (-)}_{\text{odd number of negative numbers}} = (-)$$

$$\underbrace{(-) \times \cdots \times (-)}_{\text{even number of negative numbers}} = (+)$$

We can represent multiplication in several ways, such as $\times$, $\cdot$ or no space between two bracketed numbers. For example

$$3 \times 2 = 3 \cdot 2 = (3)(2)$$

Division can also be represented in several ways. The number $4 \div 2$ can also be written as $\frac{4}{2}$ or 4/2. Thus

$$4 \div 2 = \frac{4}{2} = 4/2$$

The evaluations with negative numbers can be implemented on a calculator by using the $\boxed{(-)}$ button or the $\boxed{+/-}$ button.

To compute $(-1729) \times (343)$ on a calculator, PRESS

| (−) | | 1729 | | × | | 343 | | = | which should show -593047.

Whole numbers are called **integers**. There are numbers between whole numbers which will be discussed in subsequent sections.

Fig. 4

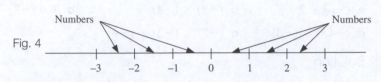

Figure 4 shows a portion of the real number line indicating some of the whole numbers. Any number on this line is a **real number**.

SUMMARY

The symbols $<$ and $\leq$ mean less than and less than or equal to respectively. Similarly $>$ and $\geq$ mean greater than and greater than or equal to respectively.

For adding and subtracting, remember the following rules:

Adding a negative number is the same as subtracting a positive number.

Subtracting a negative number is the same as adding a positive number.

For division and multiplication:

If both numbers have the same sign, then the result is positive, otherwise it is negative.

Exercise **Intro(a)**

Solutions at end of book. Complete solutions available at www.palgrave.com/engineering/singh

1 Fill in the most appropriate symbol $<$, $\leq$, $>$ or $\geq$ in place of $\square$ for the following:

  a $17 \square -17$ **b** $-17 \square 17$
  c $-2 \square -1$ **d** $-5 \square -5$
  e $-2 \square 0$

2 Complete the correct symbol, $+$, $-$, $\div$, $\times$, in place of $\square$ for the following:

  a $3 \square 2 = 5$ **b** $3 \square -2 = 5$
  c $9 \square 3 = 3$ **d** $-7 \square (-11) = 77$
  e $-6 \square (-5) = -1$

3 Evaluate the following:

  a $7 - 12$ **b** $-3 + 1$
  c $-3 - (-1)$ **d** $-11 + (-11)$
  e $-11 - (-11)$

4 Calculate

  a $(-6) \div (-2)$ **b** $(-6) \times 2$
  c $(-6) \times (-2)$ **d** $(-6) \div 2$
  e $(-6) \div (-2)$ **f** $6 \div (-2)$
  g $(-1) \times (-2) \times (-8)$

5 By using your calculator, or otherwise, compute the following:

  a $(-343) \times 343$
  b $(-343) \times (-343)$
  c $(-729) \div 81$
  d $\dfrac{(-729)}{81}$
  e $(-666) - (-1945)$
  f $(-2) \times (-5) \times (-7) \times (-10)$

SECTION B **Indices**

By the end of this section you will be able to:
▶ understand the terms **index**, **power** and **indices**
▶ evaluate indices
▶ evaluate square roots, cube roots, etc.

B1 **Powers and roots**

? Instead of writing $\underbrace{2 \times 2 \times 2 \times 2 \times 2}_{\text{5 copies}}$, can you remember a shorter way of writing this number?

It can be written as 2 with superscript 5, 2^5. The superscript is called the index or the power, so in 2^5 the 5 is the **power** or **index**. The plural of index is **indices**. We can write 7×7 as 7^2 pronounced '7 squared', that is

$$7 \times 7 = 7^2$$

? How can we write $7 \times 7 \times 7$ in this index format?

$$\underbrace{7 \times 7 \times 7}_{\text{3 copies}} = 7^3$$

7^3 is pronounced '7 cubed'. We can write repetitive multiplication of the same number with an index (or power).

? What is 9^7 (9 to the index 7) equal to?

$$\underbrace{9 \times 9 \times 9 \times 9 \times 9 \times 9 \times 9}_{\text{7 copies}}$$

To evaluate this we use a calculator. PRESS

| 9 | x^y | 7 | = |, which should show 4782969. Therefore

$$9^7 = 4782969$$

On some calculators you might have to press the $\boxed{\wedge}$ key instead of the $\boxed{x^y}$ key.
See the handbook of your calculator.

Powers of numbers can be easily evaluated on a calculator. Let's do an example.

Example 4

By using your calculator, or otherwise, compute the following:

a 3^4 **b** 88^2 **c** 20^3 **d** 20^1 **e** 7^0 **f** $(-2)^3$

Solution

Use the x^y or $\wedge$ button on your calculator.
a $3^4 = \underbrace{3 \times 3 \times 3 \times 3}_{\text{4 copies}} = 81$

Example 4 *continued*

b $88^2 = 88 \times 88 = 7744$
c $20^3 = 20 \times 20 \times 20 = 8000$

[?] **d How can we write 20^1?**

$$20^1 = 20$$

Any number to the index 1 is the number itself because it is **not** multiplied by itself again.
e Use your calculator to find 7^0:

$$7^0 = 1$$

There is a general result which says that any number, apart from zero, to the index 0 gives 1.
f Odd number of negatives multiplied gives a negative number:

$$(-2)^3 = \underbrace{(-2) \times (-2)}_{=4} \times (-2)$$
$$= 4 \times (-2)$$
$$= -8$$

[?] **Since $9^2 = 81$, can we somehow extract 9 from the number 81?**

We move in the opposite direction, from 81 to 9. The 9 is the square root of 81. The square root is represented by the symbol $\sqrt{\ }$. Thus we write it as

$$\sqrt{81} = 9$$

[?] **Can you find another square root of 81 or is 9 the only one?**

[?] **What is $(-9) \times (-9) = (-9)^2$ equal to?**

$$(-9) \times (-9) = 81$$

because minus times minus gives a plus. So -9 is also a square root of 81.
$\sqrt{81} = 9$ or -9, sometimes written as

$$\sqrt{81} = \pm 9$$

where $\pm$ is the plus or minus symbol. The square root of a positive number gives you two numbers, one positive and the other negative.

[?] **What is $\sqrt{64}$ equal to?**

$$\sqrt{64} = \pm 8 \ (+8 \text{ or } -8)$$

We will generally use the following notation:

$$\sqrt{64} = 8 \text{ (only positive root)}$$
$$\pm\sqrt{64} = \pm 8$$

Most calculators will only give the positive square root. For the other square root we just place a minus sign in front.

In this Introduction chapter when we refer to roots we mean **only** the **real** roots.

We also have the cube root of a number. For example, $12^3 = 1728$, so a cube root of 1728 is 12. The symbol for the cube root is $\sqrt[3]{}$. Hence

$$\sqrt[3]{1728} = 12$$

There is **only one** real cube root.

? **What is the cube root of 27?**

A number multiplied by itself three times gives 27:

$$(\text{number}) \times (\text{number}) \times (\text{number}) = 27$$

Use a calculator. There should be a $\sqrt[3]{}$ button on your calculator.

? **What is the answer?**

It's 3, thus

$$\sqrt[3]{27} = 3$$

Remember there is only one cube root. Similarly we can define other roots, for example

$$3^4 = 81$$

? **What is the 4th root of 81, $\sqrt[4]{81}$, equal to?**

$$\sqrt[4]{81} = 3 \text{ because } 3^4 = 81$$

In this case -3 is also a root. So

$$\pm\sqrt[4]{81} = \pm 3$$

In general, the **even** root of a number results in two **numbers**, one positive and the other negative. For **odd** roots we only obtain **one** root.

Example 5

Calculate

a $\pm\sqrt{25}$ **b** $\pm\sqrt{1444}$ **c** $\sqrt{19600}$ **d** $\sqrt[3]{8}$ **e** $\sqrt[3]{-8}$ **f** $\pm\sqrt[4]{625}$

Solution

a Since $5 \times 5 = 25$ and $(-5) \times (-5) = 25$ we have

$$\pm\sqrt{25} = \pm 5 \ (5 \text{ or } -5)$$

b Use the $\sqrt{}$ button on your calculator:

$$\pm\sqrt{1444} = \pm 38$$

c Similarly $\sqrt{19600} = 140$ (only the positive root)

? **d** **What is the cube root of 8, $\sqrt[3]{8}$, equal to?**

We know $2 \times 2 \times 2 = 8$, so $\sqrt[3]{8} = 2$ (only one real root).

e Also $(-2) \times (-2) \times (-2) = -8$. Thus

$$\sqrt[3]{-8} = -2$$

Try this on your calculator by using the $\sqrt[3]{}$ button.

Example 5 *continued*

f To find $\sqrt[4]{625}$ we use a calculator. You will need to consult the handbook of your calculator for instructions to evaluate $\sqrt[4]{625}$. Hence

$$\sqrt[4]{625} = 5$$

Since we want to find the 4th (even) roots of 625, there is another root, -5. So

$$\pm\sqrt[4]{625} = \pm 5$$

The square root is most useful in various fields of engineering such as electronics, structures, vibrations, control, etc. In the next example we examine the properties of the square root, $\sqrt{\ }$.

Example 6

Compute the following:

a $\sqrt{25 - 16}$ **b** $\sqrt{64 + 36}$ **c** $\sqrt{36 \times 81}$ **d** $\sqrt{\dfrac{256}{16}}$

Solution

a We first carry out the subtraction and then take the square root of the result:

$$\sqrt{25 - 16} = \sqrt{9} = 3$$

It is **critical** that you know

$$\sqrt{25 - 16} \neq \sqrt{25} - \sqrt{16} \qquad \text{[Not equal]}$$

b Similarly $\sqrt{64 + 36} = \sqrt{100} = 10$ [Addition first and then root]

c With multiplication under the square root, it doesn't matter about the order:

$$\sqrt{36 \times 81} = \underset{=6}{\sqrt{36}} \times \underset{=9}{\sqrt{81}}$$
$$= 6 \times 9 = 54$$

Also by using a calculator we could have

$$\sqrt{36 \times 81} = \sqrt{2916} = 54$$

However for the first approach we didn't need to use a calculator.

d Division is similar to multiplication, so it doesn't matter about the order:

$$\sqrt{\frac{256}{16}} = \frac{\sqrt{256}}{\sqrt{16}} = \frac{16}{4} = 4$$

Note that for addition (and subtraction) the arithmetic under the square root is executed first. For multiplication (and division) you can apply the square root first and then multiply or vice versa.

The same properties apply to the cube root, 4th roots, etc.

SUMMARY

A number squared is a number with superscript 2 and means it is multiplied by itself:

$$(\text{number})^2 = (\text{number}) \times (\text{number})$$

The superscript is called the index or power. We also have indices of 3, 4, 5, etc.

Square root, $\sqrt{\ }$, is the reverse of squaring:

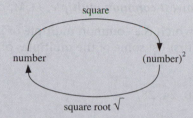

The cube root $\left(\sqrt[3]{\ }\right)$ and the 4th root $\left(\sqrt[4]{\ }\right)$ are defined in a similar way. The even real root of a given positive number results in two numbers, one positive and the other negative (± number).

If we have addition (or subtraction) under the square root sign, then we have to implement the addition (or subtraction) first. For multiplication (and division), order does **not** matter.

Exercise **intro(b)**

Solutions at end of book. Complete solutions available at www.palgrave.com/engineering/singh

1 Evaluate the following:
  a 8^2 **b** 8^3
  c 8^8 **d** 45^2
  e 264^2 **f** 20001^0
  g $(-3)^5$

2 Compute

  a $\pm\sqrt{49}$ **b** $\pm\sqrt{1681}$

  c $\sqrt{183184}$ **d** $\sqrt[3]{2197}$

  e $\sqrt[3]{729}$ **f** $\pm\sqrt[4]{16}$

  g $\sqrt[4]{2401}$ **h** $\pm\sqrt[8]{256}$

  i $\sqrt[5]{243}$ **j** $\sqrt[5]{(-243)}$

3 Calculate the following:

  a $\sqrt{100-36}$ **b** $\sqrt{129+40}$

  c $\sqrt{7225-5929}$ **d** $\sqrt[4]{7225-5929}$

  e $\sqrt[3]{500+500}$ **f** $\sqrt[3]{(-500-500)}$

4 Evaluate

  a $\sqrt{16\times25}$ **b** $\sqrt{144\times81}$

  c $\sqrt{\dfrac{144}{36}}$ **d** $\sqrt{\dfrac{225}{25}}$

  e $\sqrt[5]{\dfrac{7776}{243}}$ **f** $\sqrt[5]{-\left(\dfrac{7776}{243}\right)}$

SECTION C **Numbers**

By the end of this section you will be able to:
▶ evaluate the lowest common multiple, LCM, of numbers
▶ understand what is meant by a prime number
▶ find the prime factors of a whole number

C1 **Lowest common multiple – LCM**

[?] **What do you think the term** *lowest common multiple, LCM,* **of two numbers means?**

It's the lowest positive number which is a common multiple of both numbers. For example, to find the LCM of 4 and 6, we first list some of the multiples of these numbers.

[?] **What are the multiples of 4?**

Multiples of 4 are 4, 8, 12, 16, 20, 24, 28, . . .

[?] **What are the multiples of 6?**

Multiples of 6 are 6, 12, 18, 24, 30, . . .

Amongst these two lists we look for numbers which are **common** to both these lists.

[?] **Can you spot a number in both these lists?**

There are two numbers, 12 and 24, which lie in both lists. Both these numbers, 12 and 24, are common multiples of 4 and 6.

[?] **Which one of these is the lowest common multiple (LCM) of 4 and 6?**

It's 12 because it is the smaller number. We say the LCM of 4 and 6 is 12. Let's try another example.

Example 7

Find the lowest common multiple (LCM) of 7 and 8.

Solution

[?] **What do we need to do first?**

List the multiples of 7 and 8.

The multiples of 7 are 7, 14, 21, 28, 35, 42, 49, 56, 63, 70, . . .

The multiples of 8 are 8, 16, 24, 32, 40, 48, 56, 64, . . .

[?] **What is the smallest number common to both lists?**

56. Hence the lowest common multiple (LCM) of 7 and 8 is 56.

C2 Prime factors

This method of finding the lowest common multiple (LCM) can be laborious, even for simple numbers. There is another technique used to find the LCM which involves finding the prime factors.

? **What do we mean by prime factors?**

In **Example 7** we say 7 and 8 are factors of 56 – whole numbers which divide into 56.

? **Can you think of any factors of 12?**

> 3 and 4

Of course there are other factors of 12:

> 1, 2, 3, 4, 6 and 12

Factors are whole numbers that divide exactly into a given number.

? **What are the factors of 20?**

> 1, 2, 4, 5, 10 and 20. That is, all the whole numbers that divide into 20 exactly.

? **What are the factors of 7?**

> 7 and 1

? 7 and 1 are the only factors of 7. **Can you remember the specific name given to a number whose only factors are the number itself and 1?**

It's called a **prime number**. The only factors of a prime number are 1 and itself. For example, prime numbers less than 20 are 2, 3, 5, 7, 11, 13, 17 and 19. Note that 1 is not a prime number.

? **What's so special about prime numbers?**

Every whole number can be written as a prime or a multiple of its prime factors.

? **What do we mean by prime factors?**

These are factors of a number which are prime. For example, 10 can be written as

> $10 = 2 \times 5$

where 2 and 5 are prime. Thus the prime factors of 10 are 2 and 5.

? **How can we write 20 in terms of prime factors?**

> $20 = 2 \times 10$

and 10 can be written as 2×5, so we have

> $20 = 2 \times 2 \times 5$

Also we can write 2×2 as 2^2:

> $20 = 2^2 \times 5$

We say $2^2 \times 5$ is the **prime decomposition** of 20.

> **Example 8**
>
> Find the prime decomposition of the numbers:
>
> **a** 55 **b** 1000 **c** 31
>
> Solution
>
> **a** We know that $5 \times 11 = 55$ and since 5 and 11 are primes, so the prime decomposition of 55 is 5×11.
>
> **b** We know $100 \times 10 = 1000$ and we can write 100 as 10×10. We have
>
> $$1000 = \underbrace{100}_{=10 \times 10} \times 10 = 10 \times 10 \times 10$$
>
> **What are the prime factors of 10?**
>
> $$10 = 5 \times 2$$
>
> Hence:
>
> $$1000 = 10 \times 10 \times 10$$
> $$= (5 \times 2) \times (5 \times 2) \times (5 \times 2)$$
> $$= \underbrace{(5 \times 5 \times 5)}_{3 \text{ copies}} \times \underbrace{(2 \times 2 \times 2)}_{3 \text{ copies}}$$
> $$1000 = 5^3 \times 2^3$$
>
> **c** 31 is a prime number, so $31 = 31$.

We can use this idea of writing a number in its prime factors to find the lowest common multiple (LCM). If you write down the prime factors of each number, then the LCM is the multiplication of the highest power of each prime factor.

> **Example 9**
>
> Obtain the lowest common multiple (LCM) of 9, 15 and 40.
>
> Solution
>
> Writing the prime factors of each number gives:
>
> $$9 = 3 \times 3 = 3^2$$
> $$15 = 5 \times 3$$
> $$40 = 8 \times 5$$
> $$= (4 \times 2) \times 5$$
> $$= \underbrace{(2 \times 2 \times 2)}_{3 \text{ copies}} \times 5$$
> $$40 = 2^3 \times 5$$
>
> So all the prime factors are 2, 3 and 5.

Example 9 *continued*

? **What is the highest power of each prime factor 2, 3 and 5?**

Since $40 = 2^3 \times 5$, $15 = 5 \times 3$ and $9 = 3^2$ therefore we have 2^3, 3^2 and 5.

So the LCM $= 2^3 \times 3^2 \times 5$

$\qquad\qquad = 8 \times 9 \times 5$

$\qquad\qquad = 360$

The LCM of 9, 15 and 40 is 360.

Note that we can find the LCM for more than two numbers. The initial method for finding the LCM was to write down the multiples of 9, 15 and 40 and then spot the lowest number common among the list of all three multiples. This would have been tedious. Try it!

The LCM is used in the addition and subtraction of fractions covered later in this Introduction chapter.

SUMMARY

The lowest common multiple (LCM) of a collection of numbers is the smallest positive number which is common to all the multiples of the numbers.

A factor is a whole number which divides exactly into another number.

A prime number is a whole number with only 1 and itself as factors.

Every whole number can be written as a prime or a multiple of prime factors. The LCM of numbers can be evaluated by multiplying the highest power of each prime factor.

Exercise **Intro(c)**

Solutions at end of book. Complete solutions available at www.palgrave.com/engineering/singh

1 By listing the multiples, find the LCM of

  a 5, 6

  b 6, 14

  c 6, 27

  d 3, 7, 42

2 Write the following numbers in terms of its prime factors:

  a 90

  b 144

  c 94

  d 495

3 Find the LCM of

  a 24, 54

  b 8, 27, 64

  c 121, 125, 144

SECTION D Fractions

By the end of this section you will be able to:
► understand the different types of fractions
► write a fraction in its simplest form

D1 Fractions

? **What does the word fraction mean?**

$$\frac{\text{whole number}}{\text{whole number}}$$

A fraction is the division of one number by another. Examples are $\frac{1}{2}, \frac{1}{4}, \frac{17}{18}$, etc.

These numbers, $\frac{1}{2}, \frac{1}{4}, \frac{17}{18}$, are examples of **proper fractions**, that is the top number is smaller than the bottom number. A fraction is normally denoted by

$$\frac{\text{top number}}{\text{bottom number}}$$

The top number is called the **numerator** and the bottom number is called the **denominator**:

$$\frac{\text{numerator}}{\text{denominator}}$$

A proper fraction, $\frac{\text{numerator}}{\text{denominator}}$, is where the numerator is always less than the denominator.

? **What is an *improper fraction*?**

This is when the numerator is greater than the denominator, such as $\frac{8}{7}, \frac{4}{3}, \frac{377}{120}$, etc.

An improper fraction is also called a **top-heavy** fraction.

? **What does the term *mixed fraction* mean?**

This is a whole number with a proper fraction such as $2\frac{1}{2}, 3\frac{7}{11}, 7\frac{1}{8}$, etc.

Considering a fraction as a division, for example $\frac{22}{7}$, is the same as $22 \div 7$. In this way we can express top-heavy fractions as mixed fractions by dividing the numerator by the denominator. For example,

$$\frac{22}{7} = 22 \div 7 \text{ which gives 3 remainder 1}$$

So we can write $\frac{22}{7}$ as 3 whole ones and $\frac{1}{7}$ left over, thus

$$\frac{22}{7} = 3\frac{1}{7}$$

Also $\frac{21}{7}$ is equal to 3.

? Going the other way, a mixed fraction can be expressed as a top-heavy fraction. **How do we write $2\frac{7}{9}$ as a top-heavy fraction?**

$$2 = \frac{18}{9} = \frac{2 \times 9}{9}$$

Thus we have

$$2\frac{7}{9} = \frac{(2 \times 9) + 7}{9}$$

$$= \frac{18 + 7}{9} = \frac{25}{9}$$

We can also reduce a fraction to its simplest terms if the numerator and denominator have a common factor between them. Otherwise the fraction is already in its simplest form. For example, $\frac{8}{21}$ is in its simplest form because there is **no** whole number which goes into both 8 and 21. However $\frac{8}{18}$ is not in its simplest form because 2 divides into 8 and 2 divides into 18. Hence

$$\frac{8}{18} = \frac{4 \times \cancel{2}}{9 \times \cancel{2}} = \frac{4}{9} \qquad \text{[Cancelling 2's]}$$

If the same number appears in the numerator and denominator, then we can cancel this number. In this example we had $\frac{2}{2} = 1$.

Example 10

Express $\frac{36}{48}$ in its simplest form.

Solution

? **Which whole number goes into 36 and 48?**

12, that is $3 \times 12 = 36$ and $4 \times 12 = 48$. Thus

$$\frac{36}{48} = \frac{3 \times \cancel{12}}{4 \times \cancel{12}} = \frac{3}{4} \qquad \text{[Cancelling 12's]}$$

? **Conversely we can move from $\frac{3}{4}$ to $\frac{36}{48}$, how?**

By multiplying the numerator and denominator by 12.

$\frac{3}{4}$ and $\frac{36}{48}$ are called **equivalent fractions**. You can also write a fraction in its simplest terms by using a calculator. To write $\frac{512}{640}$ in its simplest form on a calculator, PRESS

| 512 | | a b/c | | 640 | | = |, which should show | 4 | | ⌐ | | 5 |. If you do not have an

| a b/c | button on your calculator, then try looking at the handbook for instructions. Thus

$$\frac{512}{640} = \frac{4}{5}$$

SUMMARY

There are different types of fractions – proper fractions, top-heavy fractions, mixed fractions and equivalent fractions.

We can write a top-heavy fraction as a mixed fraction and vice versa.

We can express a fraction in its simplest form by cancelling a number common to the numerator and the denominator.

Exercise **Intro(d)**

Solutions at end of book. Complete solutions available at www.palgrave.com/engineering/singh

1 Express the following numbers as mixed fractions:

$$\frac{355}{113}, \frac{213}{71}, \frac{878}{323}, \frac{577}{408}, \frac{64}{7}$$

2 Write the following fractions in their simplest form:

a $\dfrac{7}{21}$ **b** $\dfrac{8}{20}$ **c** $\dfrac{72}{100}$

d $\dfrac{56}{75}$ **e** $\dfrac{272}{272}$

3 Which of the following pairs are equivalent fractions?

a $\dfrac{16}{64}, \dfrac{1}{4}$ **b** $\dfrac{26}{65}, \dfrac{1}{5}$ **c** $\dfrac{49}{89}, \dfrac{4}{8}$

4 Write the following mixed fractions as top-heavy fractions:

a $3\dfrac{1}{3}$ **b** $1\dfrac{70}{69}$ **c** $9\dfrac{87}{100}$

SECTION E **Arithmetic of fractions**

By the end of this section you will be able to:
▶ add and subtract fractions
▶ multiply fractions
▶ divide fractions

E1 **Addition and subtraction of fractions**

We can only add and subtract the numerators of fractions if they have the same denominator. For example

$$\frac{2}{7} + \frac{3}{7} = \frac{2+3}{7} = \frac{5}{7}$$

$$\frac{2}{7} - \frac{3}{7} = \frac{2-3}{7} = \frac{-1}{7} = -\frac{1}{7}$$

Note that −1 over 7 is the same as putting the negative sign in front, $-\dfrac{1}{7}$.

? **How can we add fractions with different denominators?**

? **We have to rewrite the fractions with a common denominator first, how?**

We find the lowest common multiple (LCM) of the denominators of the fractions concerned. The procedure for adding or subtracting fractions is

1 Find the LCM of the denominators.
2 Express each fraction with the LCM as the denominator.
3 Add or subtract the numerators.
Let's try an example.

Example 11

Find: **a** $\dfrac{1}{4} + \dfrac{3}{10}$ **b** $\dfrac{22}{7} - \dfrac{25}{8}$

Solution

? **a How do we find the LCM of 4 and 10?**

We first find the multiples of 4 and 10:
Multiples of 4 are 4, 8, 12, 16, 20, 24, 28, ...
Multiples of 10 are 10, 20, 30, 40, ...

? **Among these two lists, is there a number in both lists?**

Yes, 20. This is the lowest common multiple (LCM) of 4 and 10. We need to place each fraction with 20 as the denominator. Since 4×5 makes 20, we multiply the numerator and denominator of $\dfrac{1}{4}$ by 5:

$$\frac{1}{4} = \frac{1 \times 5}{4 \times 5} = \frac{5}{20}$$

Similarly 2×10 makes 20, so we multiply the numerator and denominator of $\dfrac{3}{10}$ by 2:

$$\frac{3}{10} = \frac{3 \times 2}{10 \times 2} = \frac{6}{20}$$

We can now add the numerators:

$$\frac{1}{4} + \frac{3}{10} = \frac{5}{20} + \frac{6}{20} \qquad \left[\text{Writing } \frac{1}{4} \text{ and } \frac{3}{10} \text{ with denominator 20} \right]$$

$$= \frac{5 + 6}{20}$$

$$= \frac{11}{20}$$

b The multiples of 7 and 8 are as follows:

$$7, 14, 21, 28, 35, 42, 49, \underline{56}, 63, 70, \ldots$$

$$8, 16, 24, 32, 40, 48, \underline{56}, 64, \ldots$$

The LCM is 56, so we need to write each fraction under the common denominator of 56. Remember it is 7×8 which makes 56, so we multiply the numerator and denominator of $\dfrac{22}{7}$ by 8:

$$\frac{22}{7} = \frac{22 \times 8}{7 \times 8} = \frac{176}{56}$$

Example 11 *continued*

Similarly multiply the numerator and denominator of $\dfrac{25}{8}$ by 7:

$$\frac{25}{8} = \frac{25 \times 7}{8 \times 7} = \frac{175}{56}$$

Thus subtracting the numerators:

$$\frac{22}{7} - \frac{25}{8} = \frac{176}{56} - \frac{175}{56}$$

$$= \frac{176 - 175}{56}$$

$$= \frac{1}{56}$$

Note that both fractions in **b** are top-heavy fractions. The evaluation in **Example 11b** of $\dfrac{22}{7} - \dfrac{25}{8}$ can also be carried out by writing each fraction as a mixed number:

$$\frac{22}{7} = 22 \div 7 = 3 \text{ remainder } 1$$

So $\dfrac{22}{7} = 3\dfrac{1}{7}$. Also $\dfrac{25}{8} = 3$ remainder 1, thus $\dfrac{25}{8} = 3\dfrac{1}{8}$. Hence

$$\frac{22}{7} - \frac{25}{8} = 3\frac{1}{7} - 3\frac{1}{8}$$

$$= (3 - 3) + \left(\frac{1}{7} - \frac{1}{8}\right)$$

$$= 0 + \underbrace{\left(\frac{1}{7} - \frac{1}{8}\right)}_{= \,^{1}/_{56}}$$

$$= 0 + \frac{1}{56}$$

$$\frac{22}{7} - \frac{25}{8} = \frac{1}{56}$$

(Both $\dfrac{22}{7}$ and $\dfrac{25}{8}$ are approximations for π.)

The evaluations with fractions can also be executed on a calculator. For example, to evaluate $\dfrac{22}{7} - \dfrac{25}{8}$ on a calculator, PRESS

| 22 | | a b/c | | 7 | | (−) | | 25 | | a b/c | | 8 | | = |, which should show | 1 ⌐ 56 |

E2 Multiplication and division of fractions

Multiplication and division of fractions is effortless compared to addition and subtraction. For multiplication, multiply the numerators and multiply the denominators.

> ### Example 12
>
> Evaluate **a** $\dfrac{2}{9} \times \dfrac{4}{5}$ **b** $1\dfrac{1}{5} \times 1\dfrac{5}{12}$
>
> Solution
>
> **a** This is straightforward:
>
> $$\frac{2}{9} \times \frac{4}{5} = \frac{2 \times 4}{9 \times 5} = \frac{8}{45}$$
>
> **b** $1\dfrac{1}{5}$ and $1\dfrac{5}{12}$ are mixed fractions, so we can write them as top-heavy fractions:
>
> $$1\frac{1}{5} = \frac{(1 \times 5) + 1}{5} = \frac{6}{5}$$
> $$1\frac{5}{12} = \frac{(1 \times 12) + 5}{12} = \frac{17}{12}$$
>
> **How do we multiply these fractions?**
>
> As part **a**, multiply the numerators and multiply the denominators:
>
> $$1\frac{1}{5} \times 1\frac{5}{12} = \frac{6}{5} \times \frac{17}{12} \qquad \text{[Using the above conversion to top-heavy fractions]}$$
> $$= \frac{6 \times 17}{5 \times 12}$$
> $$= \frac{102}{60}$$
>
> **The fraction $\dfrac{102}{60}$ is not in its simplest form, how can we place this in its simplest form?**
>
> The factor 6 is common between 102 and 60 so $\dfrac{102}{60} = \dfrac{17 \times 6}{10 \times 6} = \dfrac{17}{10} = 1\dfrac{7}{10}$.
>
> Hence $1\dfrac{1}{5} \times 1\dfrac{5}{12} = 1\dfrac{7}{10}$.
>
> The evaluation **b** can be made somewhat simpler as follows:
>
> $$1\frac{1}{5} \times 1\frac{5}{12} = \frac{6}{5} \times \frac{17}{12} = \frac{\cancel{6}}{5} \times \frac{17}{2 \times \cancel{6}} \underbrace{=}_{\substack{\text{cancelling} \\ \text{the 6's}}} \frac{1 \times 17}{5 \times 2} = \frac{17}{10} = 1\frac{7}{10}$$
>
> We cancel the 6's before multiplication.

For division of fractions we turn the second fraction upside down and multiply. Turning a fraction upside down is generally referred to as **inverting** the fraction.

> ### Example 13
>
> Determine **a** $\dfrac{5}{8} \div \dfrac{2}{9}$ **b** $2\dfrac{18}{25} \div 2\dfrac{71}{100}$
>
> Solution
>
> **a** $\dfrac{5}{8} \div \dfrac{2}{9} = \dfrac{5}{8} \times \dfrac{9}{2} = \dfrac{5 \times 9}{8 \times 2} = \dfrac{45}{16}$
>
> turning the second
fraction upside down
>
> **?** **We can write $\dfrac{45}{16}$ as a mixed fraction, how?**
>
> $45 \div 16 = 2$ remainder 13, so $\dfrac{45}{16} = 2\dfrac{13}{16}$. Hence $\dfrac{5}{8} \div \dfrac{2}{9} = 2\dfrac{13}{16}$.
>
> **b** We can write both, $2\dfrac{18}{25}$ and $2\dfrac{71}{100}$, as top-heavy fractions:
>
> $$2\dfrac{18}{25} = \dfrac{(2 \times 25) + 18}{25} = \dfrac{68}{25}$$
>
> $$2\dfrac{71}{100} = \dfrac{(2 \times 100) + 71}{100} = \dfrac{271}{100}$$
>
> Substituting these gives
>
> $$2\dfrac{18}{25} \div 2\dfrac{71}{100} = \dfrac{68}{25} \div \dfrac{271}{100} \quad \text{[Using the above conversion to top-heavy fractions]}$$
>
> $$= \dfrac{68}{25} \times \dfrac{100}{271}$$
>
> **?** **How can we evaluate $\dfrac{68}{25} \times \dfrac{100}{271}$?**
>
> Since 25 is common to 25 and 100, so we can cancel:
>
> $$\dfrac{68}{25} \times \dfrac{100}{271} = \dfrac{68}{25} \times \dfrac{4 \times 25}{271} \underset{\substack{\text{cancelling}\\ \text{25's}}}{=} \dfrac{68}{1} \times \dfrac{4}{271} = \dfrac{68 \times 4}{1 \times 271} = \dfrac{272}{271} = 1\dfrac{1}{271}$$
>
> Hence $2\dfrac{18}{25} \div 2\dfrac{71}{100} = 1\dfrac{1}{271}$.

We can also divide and multiply fractions on a calculator. Get familiar with your calculator to compute fractions.

SUMMARY

To add and subtract fractions we first have to find the lowest common multiple (LCM) of the denominators and put each fraction with this new denominator. It is then just a matter of adding or subtracting the numerators.

For multiplication of fractions we simply multiply the numerators and denominators separately. Cancelling a common factor first simplifies the multiplication.

For division, invert the second fraction and then multiply the fractions.

Exercise Intro(e)

Solutions at end of book. Complete solutions available at www.palgrave.com/engineering/singh

1 Determine the following:

a $\dfrac{1}{2} + \dfrac{1}{3}$ b $\dfrac{1}{4} + \dfrac{1}{9}$ c $\dfrac{1}{6} + \dfrac{1}{60}$

d $\dfrac{1}{5} + \dfrac{1}{7}$ e $\dfrac{2}{3} + \dfrac{6}{8}$

2 Evaluate

a $\dfrac{1}{2} - \dfrac{1}{3}$ b $\dfrac{22}{7} - \dfrac{16}{9}$

c $\dfrac{41}{29} - \dfrac{204}{145}$ d $\dfrac{3}{31} - \dfrac{2}{41}$

3 Calculate

a $\dfrac{1}{2} + \dfrac{1}{3} + \dfrac{1}{6}$ b $\dfrac{1}{2} + \dfrac{1}{3} - \dfrac{1}{4} + \dfrac{1}{5}$

c $\dfrac{1}{7} - \dfrac{1}{3} + \dfrac{2}{8}$

4 Compute the following and write your final answers in their simplest form:

a $\dfrac{1}{2} \times \dfrac{1}{3}$ b $\dfrac{17}{12} \times \dfrac{84}{60}$

c $\dfrac{235}{19} \times \dfrac{38}{5}$

5 Write the following as a single fraction in its simplest form:

a $\dfrac{1}{2} \div \dfrac{1}{3}$ b $\dfrac{99}{70} \div \dfrac{22}{7}$

c $\dfrac{235}{19} \div \dfrac{5}{38}$

For question 6 use your calculator.

6 Evaluate the following:

a $1 + \dfrac{24}{60} + \dfrac{51}{3600}$

b $\dfrac{1}{56} + \dfrac{1}{679} + \dfrac{1}{776}$

c $\dfrac{1351}{780} - \dfrac{265}{153}$ d $\dfrac{377}{120} - \dfrac{333}{106}$

e $\dfrac{577}{408} \times \dfrac{17}{12}$ f $\left(\dfrac{17}{12}\right)^2$

g $\dfrac{41}{29} \div \dfrac{99}{70}$ h $\dfrac{377}{120} \div \dfrac{22}{7}$

SECTION F **Decimals**

By the end of this section you will be able to:
▶ understand the decimal number system
▶ round off numbers correct to a number of decimal places or significant figures
▶ understand the term irrational numbers

F1 **Decimal number system**

You must have seen numbers in decimal format such as 253.679.

? **What does the point (.) in 253.679 represent?**

It's called the decimal point and is the cut-off position between whole numbers and fractions of whole numbers.

? **What does the 253 signify in 253.679?**

253 is made up from 2 hundreds, 5 tens and 3 units:

Hundreds (100)	Tens (10)	Units (1)
2	5	3

Thus

$$253 = (2 \times 100) + (5 \times 10) + (3 \times 1)$$

? **What is the value of .679 in 253.679?**

.679 is 6 tenths, 7 hundredths and 9 thousandths:

tenths $\left(\dfrac{1}{10}\right)$	hundredths $\left(\dfrac{1}{100}\right)$	thousandths $\left(\dfrac{1}{1000}\right)$
6	7	9

Thus

$$.679 = \left(6 \times \frac{1}{10}\right) + \left(7 \times \frac{1}{100}\right) + \left(9 \times \frac{1}{1000}\right)$$

Table 1 shows the place value of certain numbers with respect to the decimal point.

Number	Hundreds (100)	Tens (10)	Units (1)	Decimal point	Tenths $\left(\dfrac{1}{10}\right)$	Hundredths $\left(\dfrac{1}{100}\right)$	Thousandths $\left(\dfrac{1}{1000}\right)$
53.76		5	3	.	7	6	
0.213			0	.	2	1	3
25.001		2	5	.	0	0	1
829.705	8	2	9	.	7	0	5

TABLE 1

? **What is the place value of 1 in 25.001?**

Since it is in the third position after the decimal point, its value is 1 thousandth, or $\frac{1}{1000}$.

Let's investigate decimal places.

F2 Decimal place

In countless numbers of cases the calculator does not give you an accurate answer but is only correct to the number of digits on your display. We can approximate numbers by stating them accurate to a number of decimal places (d.p.). For example

$$25.3691 = 25.369 \text{ (correct to 3 d.p.)}$$

$$= 25.37 \text{ (correct to 2 d.p.)}$$

$$= 25.4 \text{ (correct to 1 d.p.)}$$

For 3 decimal places you examine the fourth number after the decimal point:

$$25.369\ \underset{\substack{\text{fourth}\\\text{number}}}{1}$$

If this number is 5 or larger we increase the third number after the decimal point by 1, otherwise we leave the third number intact:

$$25.36\ \underset{\substack{\text{third}\\\text{number}}}{9}\ 1$$

Since we are interested in writing the number correct to 3 d.p. we erase the fourth number after the decimal point. In this case

$$25.3691 = 25.369 \text{ (correct to 3 d.p.)}$$

Similarly, for 2 decimal places, since 9 is greater than 5 we increase the second number after the decimal point to 7 and erase the numbers to the right:

$$25.3691 = 25.37 \text{ (correct to 2 d.p.)}$$

For 1 decimal place we examine the second position after the decimal point:

$$25.3\ \underset{\substack{\text{second}\\\text{number}}}{6}\ 91$$

and since this is bigger than 5 we increase the first number after the decimal point to 4 and erase the remaining numbers to the right:

$$25.3691 = 25.4 \text{ (correct to 1 d.p.)}$$

Example 14

Write the following numbers correct to 2 d.p.:

a 0.534   **b** 7.697

c 0.0461   **d** 0.005

Solution

a Since the 4 in 0.534 is less than 5 we leave the second number after the decimal place intact:

$$0.534 = 0.53 \ (2 \text{ d.p.})$$

b Since the last 7 on the right in 7.697 is bigger than 5 we increase the 9 to 10. Thus

$$7.697 = 7.70 \ (2 \text{ d.p.})$$

Notice that 69 becomes 70 after the decimal point.

c Similarly $0.0461 = 0.05$ (2 d.p.).

d Since we have 5 in the third position after the decimal point in 0.005 we increase the 0 before 5 by 1. Thus

$$0.005 = 0.01 \ (2 \text{ d.p.})$$

(2 d.p.) means the answer is correct to 2 decimal places.

F3 Significant figures

We can also approximate numbers by using significant figures. The following illustrates the significance of each of the digits in 386:

3 8 6

The 3 in 386 has a value of 300, the 8 a value of 80 and the 6 a value of 6. Note that the 3 is the most significant figure although it is the smallest digit in 386. Rounding 386 to 1 significant figure (1 s.f.) gives 400 because the 8 next to the 3 is bigger than 5, so we increase 3 to 4 and place zeros after that. Writing a number to 1 significant figure means that at most one digit is not zero, hence 386 rounds to 400. **What is 386 to 2 significant figures?**

Since the 6 is larger than 5 we increase the 8 to 9 and place a zero in the remaining right position:

$$386 = 390 \ (2 \text{ s.f.})$$

(2 s.f.) means that the answer is correct to 2 significant figures.

Example 15

Write the following correct to the number of significant figures stated in the brackets:

a 7.526 (2 s.f.)   **b** 186 895 (3 s.f.)   **c** 0.000 617 (1 s.f.)
d 9.99 (1 s.f.)   **e** 11.399 (2 s.f.)

Solution

a In 7.526, the 2 in the 3rd position is less than 5 so
$$7.526 = 7.5 \text{ (2 s.f.)}$$

b Similarly 186 895 = 187 000 (3 s.f.)
c In this case we examine the second digit after the zeros:
$$0.000\ 6 \quad \underset{\text{this digit}}{1} \quad 7$$

Since 1 is less than 5 we leave the 6 intact, thus
$$0.000\ 617 = 0.000\ 6 \text{ (1 s.f.)}$$

d Similarly 9.99 = 10.00 (1 s.f.)
e Since 3 is less than 5 in 11.399, therefore
$$11.399 = 11.000 \text{ (2 s.f.)}$$

F4 Irrational numbers

There are some fundamental numbers in mathematics and engineering such as π (pi), $\sqrt{2}$ and e. You might remember π from the area of a circle (πr^2).

What is the exact value of π?

If we evaluate π on a 10-digit calculator it shows
$$3.141592654$$

Of course this is **only** correct to 10 s.f., π goes on for an infinite number of decimal places. Today π has been approximated to more than a trillion decimal places. Yes, I said approximated because it is **not** the exact value of π. We cannot write down the exact fraction. The only way to write π exactly is to leave it as π.

There is another number called e which like π cannot be expressed exactly. The number e correct to 12 d.p. is
$$e = 2.718281828459$$

The number e crops up in many fields of engineering such as mechanics, aerodynamics, electronics, control, etc. We will examine this number further in **Chapter 5**. Another notable number is $\sqrt{2}$. On a 10-digit calculator
$$\sqrt{2} = 1.414213562 \text{ (correct to 9 d.p.)}$$

Again we cannot evaluate the correct value of $\sqrt{2}$ in decimal or fraction format. The only way to write $\sqrt{2}$ exactly is to leave it as $\sqrt{2}$.

A number which we cannot write as a whole number or a fraction is called an **irrational number.**

These numbers, π, e and $\sqrt{2}$, are all examples of irrational numbers.

There are an infinite number of irrational numbers on the real line, such as $\sqrt{3}$, $\sqrt{5}$, 3π, etc.

SUMMARY

We can round off numbers to so many decimal places (d.p.) or significant figures (s.f.).

Irrational numbers often used in engineering are π, e and $\sqrt{2}$.

Exercise **Intro(f)**

Solutions at end of book. Complete solutions available at www.palgrave.com/engineering/singh

1 Write the numbers correct to the number of decimal places stated:

  a 1.618 034   **i** 3 d.p.   **ii** 2 d.p.
  b 4.6692   **i** 4 d.p.   **ii** 2 d.p.
  c 2.502 907 875   **i** 3 d.p.   **ii** 1 d.p.
  d 0.373 96   **i** 3 d.p.   **ii** 2 d.p.

2 Write the following numbers correct to the number of significant figures stated:

  a 1.618 034   **i** 3 s.f.   **ii** 2 s.f.

  b 2.973 214   **i** 5 s.f.   **ii** 2 s.f.
  c 0.110 001   **i** 4 s.f.   **ii** 1 s.f.
  d 9.869   **i** 2 s.f.   **ii** 1 s.f.

3 Write the following numbers to 2 s.f.:

  a 1729   **b** 99 954
  c 107 928 278 317

4 Write π, e and $\sqrt{2}$ to

  a 2 d.p.   **b** 2 s.f.

SECTION G **Powers of 10**

By the end of this section you will be able to:
▶ write numbers in standard form
▶ examine the SI units and their symbols
▶ write physical quantities in the most appropriate symbol format
▶ estimate basic arithmetical results

G1 **Standard form**

? What is the value of 10^2, 10^3 and 10^4?

These are all powers of 10 and can be expanded as

$$10^2 = 10 \times 10 = 100$$

$$10^3 = 10 \times 10 \times 10 = 1000$$

$$10^4 = 10 \times 10 \times 10 \times 10 = 10\ 000$$

? We have $10^2 = 100$, $10^3 = 1000$ and $10^4 = 10\,000$. **What do you notice?**

The index number (superscript number) is the same as the number of zeros after 1.

We will use this concept in this section:

$$10^{\text{index number}} = 1\ \underbrace{0\cdots0}_{\substack{\text{number of zeros} \\ = \text{ index number}}}$$

? **However, can you remember how you multiplied numbers by 10?**

We moved all the figures one place to the left of the decimal point, for example

$$7.530 \times 10 = 75.30$$

? **How did you multiply by 100?**

We moved all the figures two places to the left of the decimal point, for example

$$7.530 \times 100 = 753.0$$

? **What about multiplying by 1000?**

We moved all the figures three places to the left of the decimal point, for example

$$7.530 \times 1000 = 7530$$

Writing a number in standard form involves going the other way, that is we are given a number and we write this as a number between 1 and 10 multiplied by a power of 10:

$$\text{number} = (\text{number between 1 and 10}) \times 10^{\text{index}}$$

For example, 9801 can be written in standard form as

$$9801 = 9.801 \times 1000 = 9.801 \times 10^3$$

? **What's the point? Why can't we leave it as 9801?**

Well we can, but this particular notation (standard form) allows us to express very large numbers in a more compact form. For example, the distance of the earth from the sun is 93 million miles, or 93 000 000 miles, and writing this number in standard form gives **?** a more concise representation. **How can we write 93 000 000 in standard form?** 93 times a million, and since a million has 6 zeros:

$$93\,000\,000 = 93.0 \times 10^6$$

The number is still not in standard form. Remember the first number should be between 1 and 10 but we have 93.0.

We have to move the decimal point in 93.0 one place to the left which means multiplying by another 10.

The index of 10 is 7 because it is $10^6 \times 10$. So we have

$$93\,000\,000 = 9.3 \times 10^7$$

Let's do some more examples.

Example 16

Write the following numbers in standard form:

a 333 000

b 133 000 000 000 000 000 000

c 1054.35

Solution

We need to write each number between 1 and 10 multiplied by a power of 10.

a **The number between 1 and 10 is 3.33, so how many places does the decimal point have to move?**

$$333\ 000 = \underbrace{333\ 000.0}_{5\ places}$$

5 places to the left. Thus $333\ 000 = 3.33 \times 10^5$.

b **The number of zeros in this large number is 18 but then we have 133 which we need to write as 1.33×100. Hence there are 2 more places to move, which makes 20 altogether. We have**

$$\underbrace{133\ 000\ 000\ 000\ 000\ 000\ 000}_{20\ places} = 1.33 \times 10^{20}$$

Notice how such a large number can be written in compact form.

c **What is the number between 1 and 10 for 1054.35?**

It's 1.05435, so how many places do we shift the decimal point?

$$\underbrace{1054}_{3\ places}.35$$

3 places to the left. Hence

$$1054.35 = 1.05435 \times 10^3$$

We need to use the following rule for the next part:

$$\frac{1}{10^{index}} = 10^{-index}$$

Small numbers can also be represented in standard form. By small numbers we mean numbers between 0 and 1. For small numbers we multiply by $\frac{1}{10}, \frac{1}{100}, \frac{1}{1000}$, etc. which is the same as dividing by 10, 100, 1000, etc. respectively.

? **What is the result of multiplying 753.67 by $\dfrac{1}{10}$ (or dividing by 10)?**

All the figures move to the right by 1 place of the decimal point:

$$753.67 \times \frac{1}{10} = 75.367$$

Similarly, multiplying by $\dfrac{1}{100}$ moves all the figures to the right by 2 places of the decimal point:

$$753.67 \times \frac{1}{100} = 7.5367$$

? **What is the consequence of multiplying by $\dfrac{1}{1000}$?**

All the figures move to the right by 3 places of the decimal point:

$$753.67 \times \frac{1}{1000} = 0.75367$$

Using $10^1 = 10$, $10^2 = 100$ and $10^3 = 1000$, we can write $\dfrac{1}{10}$, $\dfrac{1}{100}$ and $\dfrac{1}{1000}$ in terms of negative indices because $\dfrac{1}{10^{\text{index}}} = 10^{-\text{index}}$:

$$\frac{1}{10} = \frac{1}{10^1} = 10^{-1}$$

$$\frac{1}{100} = \frac{1}{10^2} = 10^{-2}$$

$$\frac{1}{1000} = \frac{1}{10^3} = 10^{-3}$$

? **How can we write 0.007 78 in standard form?**

Remember numbers in standard form can be written as

$$(\text{number between 1 and 10}) \times 10^{\text{index}}$$

? **For 0.007 78 the number between 1 and 10 is 7.78, so how many places do we have to move the decimal point?**

$$0.\underset{\text{3 places}}{\underline{007}}\,78, \text{ we move the decimal point 3 places to the right}$$

Hence $0.007\,78 = 7.78 \times 10^{-3}\,(= 7.78 \div 10^3)$.

Notice the negative index, that's because we are shifting the decimal point to the right.

Example 17

Write each of the following numbers in standard form:

a 0.000 072 92   **b** 0.000 000 000 066 73   **c** $2 \div 10^5$

Solution

a **The number between 1 and 10 is 7.292, so how many places do we shift the decimal point?**
0.000 072 92, 5 places to the right. Hence
$$\underbrace{0.000\ 072}_{5\ \text{places}}\ 92$$

$$0.000\ 072\ 92 = 7.292 \times 10^{-5}$$

b **What number do we write down between 1 and 10 for 0.000 000 000 066 73?**

6.673

We shift the decimal point by 11 places to the right:

$$\underbrace{0.000\ 000\ 000\ 066}_{11\ \text{places}}\ 73$$

Hence $0.000\ 000\ 000\ 066\ 73 = 6.673 \times 10^{-11}$.

c We can write 10^5 as $\dfrac{10^5}{1}$ because dividing by 1 gives the same number:

$$2 \div 10^5 = 2 \div \frac{10^5}{1} = 2 \times \frac{1}{10^5} = 2 \times 10^{-5} \text{ (turning the second fraction upside down)}$$

We can use a calculator to enter numbers in standard form. Use the $\boxed{\text{EXP}}$ (or $\boxed{\text{E}}$ or $\boxed{\text{EE}}$) button on your calculator. For example, to enter 1.6×10^{-19} on a calculator, PRESS $\boxed{1.6}\ \boxed{\text{EXP}}\ \boxed{(-)}\ \boxed{19}\ \boxed{=}$, which should show 1.6^{-19}.

Standard form is also known as exponential notation, or scientific notation, that's the reason for the $\boxed{\text{EXP}}$ button. Most calculators can place numbers in standard form. See the handbook of your calculator.

G2 **SI units**

In engineering, some indices of 10 are more popular than others, particularly the multiples of 3: 10^3, 10^6, 10^{-3}, etc. Moreover these powers of 10 have a particular name, for example 10^3 is kilo. When we say Milton Keynes is 85 kilometres from London we mean

$$85 \times 10^3 \text{ metres or } 85\ 000 \text{ metres}$$

The kilo is a prefix. Table 2 shows the prefix for some powers of 10 and their associated symbols.

Power of 10	Prefix	Symbol for prefix
10^{18}	exa	E
10^{15}	peta	P
10^{12}	tera	T
10^9	giga	G
10^6	mega	M
10^3	kilo	k
10^{-3}	milli	m
10^{-6}	micro	μ
10^{-9}	nano	n
10^{-12}	pico	p
10^{-15}	femto	f
10^{-18}	atto	a

TABLE 2

Generally we will use the SI (*Système International*) units of measurement. Table 3 shows some primary physical quantities with SI units and their associated symbol for the SI unit.

Quantity	SI unit	Symbol for SI unit
Mass	kilogram	kg
Length	metre	m
Time	second	s
Resistance	ohm	Ω
Capacitance	farad	F
Current	ampere	A
Inductance	henry	H
Potential	volt	V

TABLE 3

(Table 3 continued on next page)

	Quantity	SI unit	Symbol for SI unit
TABLE 3 CONTINUED	Power	watt	W
	Frequency	hertz	Hz
	Pressure	pascal	Pa
	Energy	joule	J
	Temperature	kelvin	K
	Force	newton	N

The SI units also allow the use of some secondary units such as hour, day, etc. We can use Table 2 in conjunction with Table 3 to express a physical quantity. For example, the distance 85 thousand metres is better expressed as 85 kilometres, or in symbol format 85 km.

Example 18 *electrical principles*

Write the following electrical quantities in the most appropriate symbol format:

a 0.005 amp   **b** 0.000 003 5 farad   **c** 15 000 000 ohms

Solution

a 0.005 amp = 5×10^{-3} amp = 5 milli amps = 5 mA
by Table 2

b 0.000 003 5 farad = 3.5×10^{-6} farad = 3.5 micro farads = 3.5 μF
by Table 2

c 15 000 000 ohms = 15×10^{6} ohms = 15 mega ohms = 15 MΩ
by Table 2

Note that the number does not have to be in standard form.

Example 19

Convert the following into the appropriate units of Table 3:

a 85 km   **b** 15 mH   **c** 15 GHz   **d** 3 ns   **e** 100 pF   **f** 2.54 mm

Solution

a 85 km is 85 kilometres, so we need to convert this into metres. **What does kilo represent?**

Kilo = 10^{3}, hence 85 km = 85×10^{3} m

Example 19 *continued*

b 15 mH is 15 millihenrys. We need to convert 15 mH into henrys because that is the unit of inductance in Table 3. Remember that milli = 10^{-3}, so we have

$$15 \text{ mH} = 15 \times 10^{-3} \text{ H}$$

?
?
c What is 15 GHz?

15 gigahertz. **What does giga represent?**

Giga is 10^9, so

$$15 \text{ GHz} = 15 \times 10^9 \text{ Hz}$$

d 3 ns is 3 nanoseconds. By Table 2, nano = 10^{-9}, hence

$$3 \text{ ns} = 3 \times 10^{-9} \text{ s}$$

?
e 100 pF is 100 picofarads. **What does pico represent?**

By Table 2, pico is 10^{-12}:

$$100 \text{ pF} = 100 \times 10^{-12} \text{ F}$$

f 2.54 mm is 2.54 millimetres. Hence

$$2.54 \text{ mm} = 2.54 \times 10^{-3} \text{ m}$$

G3 Estimation and accuracy

You often find that people go to extreme lengths to find an accurate answer to a numerical engineering problem. However in many cases this is not required. For example, in some electronics industries you can only obtain certain resistance values such as 10, 12, 15 . . . . Moreover each resistor has a tolerance limit and a 10 Ω resistor may vary between say 9 Ω and 11 Ω. Thus the resistor value lies between two limits. So evaluating a resistance value correct to 10 d.p. in an electrical problem is laborious and superfluous. Generally there is widespread use of tolerance limits in the field of engineering.

? **So why do we go to these extraordinary measures to evaluate answers accurate to the number of digits on our calculator?**

Most students copy down the numbers from the display of their calculator. It's very straightforward to plug numbers into your calculator and then reproduce the results on paper. Some students have adopted this habit from their primary school years and find difficulty in abandoning this procedure.

It's valuable practice to estimate your answer first and then evaluate it on a calculator. Let's investigate by an example.

Example 20

Estimate the following:

a 705×0.83 **b** $70.4 \div 0.73$ **c** $\dfrac{510 \times 0.62}{0.33}$

Solution

a We can round 705 to 700 and 0.83 to 0.8. Thus

$$705 \times 0.83 \approx 700 \times 0.8 = 70 \times 8 = 560$$

So 705×0.83 is approximately 560. The notation $\approx$ is used for approximation.

b We can round 70.4 to 70 and 0.73 to 0.7. We have

$$70.4 \div 0.73 \approx 70 \div 0.7 = 700 \div 7 = 100$$

c Similarly rounding 510 to 500, 0.62 to 0.6 and 0.33 to 0.3 we have

$$\frac{510 \times 0.62}{0.33} \approx \frac{500 \times 0.6}{0.3}$$

$$= \frac{50 \times 6}{0.3}$$

$$= \frac{300}{0.3}$$

$$= \frac{3000}{3}$$

$$= 1000$$

So $\dfrac{510 \times 0.62}{0.33}$ is approximately 1000. Compare these results with the evaluation on your calculator.

Estimation gives you a rough guide to your final result and can be used as a check on your calculator output.

The calculator and computer are very useful tools but we need to be cautious. There is a grave danger that you may bypass the thinking phase and carry out evaluations on autopilot. I have heard of students implementing division by 1 on a calculator.

SUMMARY

A number in standard form is written as

(number between 1 and 10) $\times 10^{\text{index}}$

Large numbers have a positive index and small numbers have a negative index.

In engineering, indices of multiples of 3 are popular and are given a particular name.

The SI units in Table 3 are the units for physical quantities.

In many engineering problems you do not require exact evaluations but a close approximation will do.

Exercise Intro(g)

Solutions at end of book. Complete solutions available at www.palgrave.com/engineering/singh

1 Write the following numbers in standard form:

   a 186 000   **b** 1 392 000
   c 136 000   **d** 0.000 000 034 39
   e 0.000 000 095 1   **f** 0.009 29
   g 0.000 025 8   **h** 14.96×10^6
   i 273.15   **j** 7.6×10^2

2 The following numbers are in standard form. Write them in conventional form:

   a 6.4×10^6 **b** 3.3×10^{-9}
   c 7.292×10^{-5} **d** 3×10^8

3 Arrange the following numbers in order of size, smallest first:

   a 12 750, 12.75×10^2, 12.75×10^{-3} and 12.75
   b 3.14×10^3, $3.14 \div 10^3$ and 3.14×10^{-2}

4 Evaluate the following on a calculator and give your final answer correct to 2 d.p.:

   a $\dfrac{1.25 \times 10^3 \times 0.15 \times 348}{15 \times 10^5}$

   b $0.5 \times 9.11 \times 10^{-31} \times (1.86 \times 10^5)^2 \times 10^{20}$

   c $2\pi\sqrt{\dfrac{6 \times 10^{-2}}{9.81}}$

5 [electrical principles] Write the following electrical quantities in the most appropriate symbols:

   a 100×10^{-12} farads
   b 30 000 ohms
   c 0.000 3 amps

6 [mechanics] Write the following mechanical quantities in the most appropriate symbol format:

   a 8536 N   **b** 75 000 000 W
   c 0.2×10^{12} Pa

7 Rewrite the following quantities in terms of the SI units in Table 2 (in standard form):

   a 3000 mm   **b** 573 kN
   c 25 MJ   **d** 12 ps
   e 25 mW

8 Estimate the following:

   a $\dfrac{22}{7}$ **b** $\dfrac{333}{106}$

   c 99×99 **d** $\dfrac{714 \times 0.63}{14.45}$

SECTION H **Conversion**

By the end of this section you will be able to:
▶ convert SI units
▶ convert imperial to SI units

H1 **Conversion of units**

Generally we use SI units throughout the book. Conversion of units is best understood by examples.

Example 21 *mechanics*

An object has a velocity of 5 m/s. Find the velocity in km/h, where h = hour.

Solution

What does 5 m/s mean?

5 m/s means the object has covered a distance of 5 m in 1 second.

How many metres will it cover in 1 hour?

Since there are 60×60 seconds (s) in 1 hour, the object covers

$$5 \times 60 \times 60 = 18\ 000$$

18 000 metres in 1 hour.

How can we write this in km/h (km per hour)?

Divide the result by 1000 because 1000 m = 1 km:

$$\frac{18\cancel{000}}{1\cancel{000}} = 18$$

Thus 5 m/s = 18 km/h. This means 5 m per second is equivalent to 18 km per hour.

Example 22 *fluid mechanics*

The density of mercury is 13.55 g/cm^3. Write this in units of kg/m^3 (g = gram, cm = centimetre and 100 cm = 1 m).

Solution

13.55 g/cm^3 means 1 cm^3 of mercury has a mass of 13.55 g. We need to find the mass for 1 m^3.

Example 22 *continued*

What is 1 m³ in terms of (centimetre)³?

$$1 \text{ m}^3 = 1 \text{ m} \times 1 \text{ m} \times 1 \text{ m} = 100 \text{ cm} \times 100 \text{ cm} \times 100 \text{ cm}$$

How many grams is 1 m³ of mercury?

$$13.55 \times 100 \times 100 \times 100 = 13\,550\,000$$

So 1 m³ of mercury has a mass of 13 550 000 g. **Since we need kg/m³, we have to write 13 550 000 g in kg. How?**

Divide by 1000 because 1000 g = 1 kg:

$$\frac{13\,550\,\cancel{000}}{1000} = 13\,550$$

So 1 m³ has a mass of 13 550 kg, hence 13.55 g/cm³ = 13 550 kg/m³.

Example 23

Given that 1 mile = 1.6 km, convert 70 miles per hour to m/s.

Solution

We have

$$70 \text{ miles} = 70 \times 1.6 \text{ km} = 112 \text{ km}$$

What is 112 km in metres (m)?

$$112 \text{ km} = 112 \times \underbrace{1000}_{\text{kilo} = 1000} \text{ m} = 112\,000 \text{ m}$$

70 miles per hour = 112 000 metres per hour, but we need to write this in metres per second.

How many seconds are there in 1 hour?

$$1 \text{ hour} = 60 \text{ minutes} = 60 \times 60 \text{ s} = 3600 \text{ s} \text{ (s = seconds)}$$

How do we convert 112 000 metres per hour into m/s?

Divide by 3600:

$$70 \text{ miles per hour} = \frac{112\,000}{3600} \text{ m/s} = 31.11 \text{ m/s (2 d.p.)}$$

The answer is correct to 2 d.p.

The unit of miles per hour (mph) is an example of imperial units. There are other imperial units such as feet per second, inches, ft³, etc.

SUMMARY

For SI units we look at the prefix of the unit and then multiply or divide accordingly. For converting imperial units into SI units we multiply (or divide) by the relevant factor.

Exercise Intro(h)

Solutions at end of book. Complete solutions available at www.palgrave.com/engineering/singh

Use the following data (correct to 3 d.p.) for this exercise:

1 foot = 0.305 m, 1 km = 0.621 mile and 1 mile = 1.609 km

1 [mechanics]   An object has a velocity of 31 m/s. Convert this into km/h.

2 [mechanics]   Acceleration due to gravity is 32.174 feet/s². Write this in m/s².

3 [mechanics]   The velocity of sound is 342 m/s. Express this in miles per hour.

4 [fluid mechanics]   The density of air is 1.206 kg/m³. Convert this into g/cm³.

5 [materials]   The second moment of area of a body is 0.63 cm⁴. Convert this into m⁴.

6 Convert 5 feet² to m².

7 Convert the following into m/s:
a 30 mph   **b** 80 km/h
c 186 000 miles per second

8 [materials]   Modulus of elasticity is 0.2×10^{11} N/m². Write this in N/cm².

SECTION I Arithmetical operations

By the end of this section you will be able to:
▶ use BROIDMAS to understand the priority of arithmetical operations
▶ evaluate arithmetical problems using BROIDMAS

I1 Sequence of operations

The sequence of operations such as $+, \times, \sqrt{\ }$, etc. is critical in evaluating arithmetical problems. **For example, what is the value of 5 + 2 − 3 × 6?**

Which operation is carried out first? The addition, multiplication or subtraction?

We use the mnemonic BROIDMAS to remember the priority of operations:

$\underline{B}$rackets   } First

$\underline{RO}$ots
$\underline{I}$ndices   } Second

Division	}	
Multiplication		Third

Addition	}	
Subtraction		Fourth

Combining the underlined letters makes the word **BROIDMAS**.

To evaluate $5 + 2 - 3 \times 6$, which operation do we implement first?

Multiplication, because it has priority over addition and subtraction:

$$5 + 2 - 3 \times 6 = 5 + 2 - 18$$

At this point, $5 + 2 - 18$, we work from left to right because addition and subtraction have the same priority. Hence

$$\underbrace{5+2}_{=7} - 18 = 7 - 18 = -11$$

Let's proceed with another example.

Example 24

Compute

† $\left(5 + \sqrt{25 - 16} \right) \div 2$

Solution

What do we evaluate first in $\left(5 + \sqrt{25 - 16} \right) \div 2$?

Brackets have first priority, so we initially need to calculate

$$5 + \sqrt{25 - 16}$$

Within this, the square root has priority over addition (B<u>RO</u>ID<u>M</u>A<u>S</u>). We have to first carry out the subtraction and then take the square root:

$$\sqrt{25 - 16} = \sqrt{9} = 3$$

Hence

$$5 + \sqrt{25 - 16} = 5 + 3 = 8$$

Replacing $5 + \sqrt{25 - 16}$ with 8 in † gives

$$8 \div 2 = 4$$

Therefore the final result is 4. Note that we could write the above as

* $\left(5 + \sqrt{25 - 16} \div 2 \right) = \dfrac{5\sqrt{25 - 16}}{2} = 4$

Even if there are no brackets around the numerator (the middle expression in *) we still divide the whole numerator by 2. We will encounter these types of arithmetical problems in subsequent chapters.

Example 25

Compute the following:

$$6^3 - 5 \times 5^3 + 10 \times 4^3 - 10 \times 3^3 + 5 \times 2^3 - 1$$

Solution

Which operation do we employ first?

BRO<u>ID</u>MAS discloses that the indices have preference over multiplication, so we first evaluate the indices by using a calculator. Hence

$$6^3 = 216, \quad 5^3 = 125, \quad 4^3 = 64, \quad 3^3 = 27 \text{ and } 2^3 = 8$$

Substituting these into the original problem gives

$$216 - \underbrace{5 \times 125}_{=625} + \underbrace{10 \times 64}_{=640} - \underbrace{10 \times 27}_{=270} + \underbrace{5 \times 8}_{=40} - 1$$

$$= 216 - 625 + 640 - 270 + 40 - 1$$

$$= 0$$

Hence

$$6^3 - 5 \times 5^3 + 10 \times 4^3 - 10 \times 3^3 + 5 \times 2^3 - 1 = 0$$

Note that 5×5^3 can written as 5^4 because it is $5 \times \underbrace{5 \times 5 \times 5}_{3 \text{ copies}}$.

Example 26

Compute $\dfrac{4}{3} \pi \times 5^3$ correct to 2 d.p.

Solution

If we want our final answer correct to 2 d.p., then we have to work to 3 d.p. and round off at the end.

How do we calculate $\dfrac{4}{3} \pi \times 5^3$?

Employing BRO<u>ID</u>MAS tells us that indices have preference over multiplication, so we first calculate 5^3:

$$5^3 = 125$$

We have

$$\frac{4}{3} \pi \times 5^3 = \frac{4}{3} \pi \times 125$$

Multiplication and division have the same priority so we can multiply $4 \times \pi \times 125$, and then divide the result by 3; or we can divide first, $4 \div 3$, and then multiply the result by π and 125:

$$\frac{4}{3} \pi \times 125 = (4 \times \pi \times 125) \div 3 = 523.60 \text{ (2 d.p.)}$$

SUMMARY

BROIDMAS is used as a priority list to evaluate arithmetical problems.

Exercise **Intro(i)**

Solutions at end of book. Complete solutions available at www.palgrave.com/engineering/singh

1 Evaluate the following:
 a $(1 + 3 + 5) \times (1 \times 3 \times 5)$
 b $(1 + 4 + 4) \times (1 \times 4 \times 4)$
 c $(49 \times 1) + (49 \times 3) + (49 \times 5)$
 d $1 + 7^2 + 5^3$
 e $(2 \times 4^3) + (4 \times 3^3) + (3 \times 1)$
 f $(2 \times 3 \times 5 \times 7 \times 11) + 1$

2 Calculate

 a $2^{11} - 1$ **b** $\dfrac{1 - 5}{2}$

 c $\dfrac{1}{2} - \dfrac{1}{3 \times 2^3} + \dfrac{1}{5 \times 2^5}$

3 Compute

 a $\dfrac{1 + \sqrt{5}}{2} + \dfrac{1 - \sqrt{5}}{2}$

 b $\dfrac{5 + \sqrt{25 - 24}}{2}$

 c $\dfrac{5 \pm \sqrt{(-5)^2 - (4 \times 1 \times 6)}}{2}$

 d $\dfrac{-4 \pm \sqrt{(4)^2 - 4 \times (-5)}}{2}$

 e $\dfrac{-65 \pm \sqrt{(-65)^2 + (4 \times 14 \times 25)}}{28}$

4 Evaluate the following correct to 2 d.p.:

 a $\pi \times 7^2 \times 3$

 b $\dfrac{\pi}{3} \times 4^2 \times 5$ **c** $\dfrac{4\pi \times 5^3}{3}$

 d $2\pi \times 7(5 + 7)$

 e $2\pi \sqrt{\dfrac{3^2 + 4^2}{2}}$

5 Evaluate

 a $2^4(2^5 - 1)$ **b** $2^{2^3} + 1$

SECTION J **Percentages**

By the end of this section you will be able to:
▶ convert percentages to a fraction or a decimal
▶ convert a fraction or a decimal to a percentage
▶ apply percentages to engineering problems

J1 **Percentages, fractions and decimals**

? What does the word *percentage* mean?

It means out of 100 and is denoted by the symbol %. For example, 10% means 10 out of 100 or $\dfrac{10}{100}$ $(= 10 \div 100)$:

$$10\% = \frac{10}{100} = \frac{1}{10} = 0.1$$

These are **all** equivalent. Hence we can write a percentage as a fraction or a decimal. Let's investigate percentages by doing some examples.

Example 27

Convert the following percentages to their simplest fractions:

a 15%   **b** $8\frac{1}{3}$ %

Solution

a What does 15% mean?

15 out of 100:

$$15\% = \frac{15}{100} = \frac{\cancel{5} \times 3}{\cancel{5} \times 20} = \frac{3}{20} \qquad \text{[Cancelling 5's]}$$

Hence $15\% = \frac{3}{20}$.

b Similarly $8\frac{1}{3}$ % $= \frac{8^{1}/_{3}}{100}$. **How else can we write $8\frac{1}{3}$?**

As a top-heavy fraction: $8\frac{1}{3} = \frac{(8 \times 3) + 1}{3} = \frac{25}{3}$

$$\frac{8^{1}/_{3}}{100} = \frac{25/3}{100} = \frac{25}{3} \div 100$$

$$= \frac{25}{3} \div \frac{100}{1} \qquad \left[\text{Because } 100 = \frac{100}{1} \right]$$

$$= \frac{25}{3} \times \frac{1}{100} \qquad \left[\begin{array}{l} \text{Inverting the} \\ \text{second fraction} \end{array} \right]$$

$$= \frac{\cancel{25}}{3} \times \frac{1}{4 \times \cancel{25}} \qquad \text{[Cancelling 25's]}$$

$$= \frac{1}{3 \times 4} = \frac{1}{12}$$

Hence $8\frac{1}{3}$ % $= \frac{1}{12}$.

To convert a percentage into a decimal number is straightforward, for example 16.8% means 16.8 out of 100 which can also be represented by

$$\frac{16.8}{100} = 16.8 \div 100 = 0.168$$

(We move the decimal point 2 places to the left.) Hence

$$16.8\% = 0.168$$

Conversely we can write decimals and fractions as a percentage. To convert a decimal or a fraction into a percentage, multiply by 100.

Example 28

Transform the following into percentages:

a $\dfrac{9}{20}$ **b** $\dfrac{7}{15}$ **c** 0.6735

Solution

Multiply each of these by 100:

a $\dfrac{9}{20} \times 100 = \dfrac{9 \times 10}{2} = \dfrac{90}{2} = 45$. Thus $\dfrac{9}{20} = 45\%$.

b Similarly

$$\dfrac{7}{15} \times 100 = \dfrac{7}{5 \times 3} \times (5 \times 20) = \dfrac{7 \times 20}{3} = \dfrac{140}{3} = 140 \div 3 = 46.67$$

Hence $\dfrac{7}{15} = 46.67\%$ (2 d.p.)

c $0.6735 \times 100 = 67.35$. We have $0.6735 = 67.35\%$.

J2 Evaluation of percentages

It is often easier to use a calculator to evaluate percentages of quantities. There should be a percentage symbol, %, on your calculator. For percentage calculations, see the handbook of your calculator.

Let's attempt some percentages with engineering examples.

Example 29 *electrical principles*

A resistor has a value of 1300 Ω. If the tolerance limits of the resistor are 3%, find the smallest and largest value of the resistor.

Solution

We need to evaluate 3% of 1300.

What does 3% mean?

3 out of 100, $\dfrac{3}{100}$. Hence

$$3\% \text{ of } 1300 = \dfrac{3}{100} \times 1300$$

$$= \dfrac{3 \times 13}{1} = 39$$

Example 29 *continued*

The resistor value lies between $1300 - 39$ and $1300 + 39$ (or 1300 ± 39). The smallest and largest values are given by 1261 Ω and 1339 Ω respectively. We sometimes say the range of the resistor is 1261–1339 Ω.

Example 30 *heat transfer*

A rod has a length of 0.583 m. After heating, it is expanded by 0.23%. Find, correct to 3 d.p., the new length.

Solution

First we want to find 0.23% of 0.583. How?

$$0.23\% = \frac{0.23}{100}, \text{ hence}$$

$$0.23\% \text{ of } 0.583 = \frac{0.23}{100} \times 0.583 = 0.0023 \times 0.583$$

$$= 1.34 \times 10^{-3}$$

The new length $= 0.583 + (1.34 \times 10^{-3}) = 0.584$ m (correct to 3 d.p.).

The percentage error, % error, for an experiment is defined as

I.1 $$\% \text{ error} = \frac{\text{experimental value} - \text{exact value}}{\text{exact value}} \times 100$$

I.1 is a reference number which we refer to later on.

Example 31 *heat transfer*

The experimental temperature value at which water freezes is 275.16 K; the exact temperature is 273.15 K.

Find the percentage error in the experimental value.

Solution

Substitute 275.16 for the experimental value and 273.15 for the exact value into I.1 :

$$\% \text{ error} = \frac{\text{experimental value} - \text{exact value}}{\text{exact value}} \times 100$$

We have

$$\% \text{ error} = \left(\frac{275.16 - 273.15}{273.15}\right) \times 100$$

$$= \left(\frac{2.01}{273.15}\right) \times 100$$

$$= 0.74\% \ (2 \text{ d.p.})$$

SUMMARY

Percentage, %, means out of a 100.

To transform a percentage to a fraction, write the percentage figure over 100 and simplify.

To express a fraction or decimal in terms of a percentage, multiply by 100.

Use your calculator for percentage calculations.

I.1 $\% \text{ error} = \dfrac{\text{experimental value} - \text{exact value}}{\text{exact value}} \times 100$

Exercise Intro(j)

Solutions at end of book. Complete solutions available at www.palgrave.com/engineering/singh

1 Convert the following to percentages:

  a $\dfrac{1}{10}$ **b** $\dfrac{1}{12}$

  c $\dfrac{7}{8}$ **d** $\dfrac{3}{13}$

  e 0.167   **f** 2.583

2 Transform the following percentages to fractions in their simplest form:

  a 4%   **b** 9%

  c 17.5%   **d** 2.5%

3 [electrical principles] The following are tolerance limits of various resistors. Calculate the range of each resistor:

  a 900 Ω ± 10%

  b 1200 Ω ± 5%

  c 27 kΩ ± 1%

  d 19 kΩ ± 3%

  e 5 MΩ ± 0.15%

4 A speedometer has an error of 4.5%. If the speed indicated is 120 km/h, find the range of the speed.

5 [heat transfer] The length of a bar is 3.567 m. If the bar expands by 0.15%, what is the new length correct to 4 s.f.?

6 The exact value of sound velocity is 342 m/s. If the experimental value produces 345 m/s, find the percentage error.

7 [thermodynamics] The volume of a gas in a piston is 0.005 00 m^3. If pressure is applied, then the volume of the gas is compressed by 0.57%. Find the new volume.

8 [heat transfer] A rod of length 0.35 m is expanded to a length of 0.38 m. Find the expansion as a percentage of the original length.

9 [electrical principles] An ammeter is specified to have a tolerance limit of ± 4.3%. Find the highest value of a current reading of 25 mA.

SECTION K **Ratios**

By the end of this section you will be able to:
▶ understand the term **ratio**
▶ simplify and evaluate ratios

K1 **Ratio**

? What does the word *ratio* mean?

It's a relationship that tells us the amount to which one set of objects is compared to another. For example, the ratio of boys to girls in a classroom. Consider another example. Mortar is made by mixing 3 bags of sand for every bag of cement and then adding water to the mixture. We say that the sand to cement ratio is 3 to 1 and it is denoted by 3:1. This notation 3:1 is equivalent to $3 \div 1$ or a fraction $\dfrac{3}{1}$. The ratio 3:1 is also the same as 9:3, that is

$$3:1 = 9:3$$

In the case of making mortar it means we need 9 bags of sand for every 3 bags of cement. Multiplying and dividing (not by zero) by the same number leaves the ratio unaltered. Thus

$$3:1 = 6:2 = 9:3 = 27:9 = 1:\frac{1}{3} \text{ etc.}$$

We say 3:1 is the simplified ratio because it is placed into the simplest whole numbers.

Example 32

Simplify the following ratios:

a 7:14   **b** 100:5   **c** $\dfrac{1}{5}$:20   **d** $\dfrac{1}{5} : \dfrac{1}{4}$ **e** 0.36:0.06

Solution

? a What do 7 and 14 have in common?

Both have a common factor of 7. Dividing both, 7 and 14, by 7 gives 1 and 2 respectively. Hence

$$7:14 = 1:2$$

? b What is a common factor of 5 and 100?

5, so dividing both, 100 and 5, by 5 gives 20 and 1 respectively. Hence

$$100:5 = 20:1$$

For **c**, **d** and **e** we need to use whole numbers rather than fractions or decimals.

Example 32 *continued*

? **c** **Since we have** $\dfrac{1}{5}$**, it is better to multiply by 5. Why?**

Because $5 \times \dfrac{1}{5} = 1$. We also need to multiply 20 by 5, thus

$$\dfrac{1}{5}:20 = 1:100$$

? **d** **How can we simplify the ratio** $\dfrac{1}{5}:\dfrac{1}{4}$**?**

Find the lowest common multiple (LCM) of 4 and 5 and then multiply by this number. The LCM of 4 and 5 is 20, thus

$$20 \times \dfrac{1}{5} = 4$$

$$20 \times \dfrac{1}{4} = 5$$

We have

$$\dfrac{1}{5}:\dfrac{1}{4} = 4:5$$

? **e** **First we convert the decimals, 0.36 and 0.06, into whole numbers. How?**

Multiply by 100:

$$0.36 \times 100 = 36$$
$$0.06 \times 100 = 6$$

We have

$$0.36:0.06 = 36:6$$

? **Can we simplify 36:6?**

Yes, both 36 and 6 have a common factor of 6:

$$36 \div 6 = 6$$
$$6 \div 6 = 1$$

So 36:6 = 6:1, hence

$$0.36:0.06 = 6:1$$

We can write 0.36:0.06 directly as 6:1 because $6 \times 0.06 = 0.36$.

We can have a ratio of more than two quantities. For example, say we want to divide £500 in the ratio 2:3:5. We need to find the amount of money contained in each part.

? **How many parts are there altogether?**

$$\text{Total number of parts} = 2 + 3 + 5 = 10$$
$$\text{Money for each part} = £500 \div 10 = £50$$

The first part is 2 out of 10, so it contains $2 \times £50 = £100$. Similarly the second part has $3 \times £50 = £150$, and the third part has $5 \times £50 = £250$.

Example 33

Engineering students at a university are divided into civil, mechanical, electrical and aerospace in the ratio 2:3:5:1 respectively. If there are 1089 engineering students, find the number of students in each of the disciplines.

Solution

Students are divided into $2 + 3 + 5 + 1 = 11$ parts.

How many students are there in one part?

$$1089 \div 11 = 99$$

Number of students in civil $= 99 \times 2 = 198$

Number of students in mechanical $= 99 \times 3 = 297$

Number of students in electrical $= 99 \times 5 = 495$

Number of students in aerospace $= 99 \times 1 = 99$

As a check, you can total the number of students:

$$198 + 297 + 495 + 99 = 1089$$

SUMMARY

A ratio between two objects makes a comparison of their relative sizes. Ratio is denoted by ':' and we can compare more than two objects.

Exercise **Intro(k)**

Solutions at end of book. Complete solutions available at www.palgrave.com/engineering/singh

1 Simplify the following ratios:
   a 100:10   **b** 69:23
   c 5:25:45   **d** 6:42:54:72
   e $\dfrac{2}{3}:\dfrac{5}{6}$

2 By using a calculator, or otherwise, simplify the following:
   a $1\dfrac{1}{5}:2\dfrac{1}{4}$ **b** $3\dfrac{1}{2}:2\dfrac{5}{12}$

3 Simplify the following ratios:
   a 0.4:0.5   **b** 0.52:0.72
   c $\sqrt{2}:\sqrt{8}$

4 A piece of metal of length 0.64 m is cut into three pieces in the ratio 2:3:5. Find the length of each piece.

5 A metal brass is a mixture of copper and zinc where the copper to zinc ratio is 8:3. If the total mass of the brass is 66 kg, find the mass of copper and zinc.

6 A metal alloy is made from copper, zinc and nickel in the ratio 3:4:6. If the alloy has a mass of 45.5 kg, find the mass of each metal.

7 Engineering students at a university are divided into manufacturing, building services, vehicle and control in the ratio 2:3:5:4 respectively. If there are 1260 engineering students, find the number of students in each discipline.

Miscellaneous exercise **Intro**

1 Convert 1 mm^3 to m^3.

2 Evaluate

   a 5^3 **b** $\sqrt{100}$ **c** $\sqrt[3]{-8}$

   d $\sqrt{\dfrac{196}{49}}$ **e** $\sqrt{12^2 + 5^2}$

3 Calculate the following and write your answer in its simplest form:

   a $\dfrac{2}{3} + \dfrac{3}{5}$ **b** $\dfrac{2}{3} \times \dfrac{4}{5}$ **c** $\dfrac{2}{3} \div \dfrac{3}{5}$

4 Write the following to the number of decimal places (d.p.) or significant figures (s.f.) stated:

   a 3.141 5926   **i** 3 d.p.   **ii** 3 s.f.

   b $\sqrt{10}$ **i** 2 d.p.   **ii** 3 d.p.

   c 1.6449   **i** 2 d.p.   **ii** 3 d.p.

   d 16^5 **i** 1 s.f.   **ii** 3 s.f.

   e $2^{2^4} + 1$ **i** 3 s.f.   **ii** 2 s.f.

5 Compute the following correct to 2 decimal places:

   a $\dfrac{\pi}{4} + 1$

   b $\dfrac{-(-7) \pm \sqrt{(-7)^2 - (4 \times 1 \times 12)}}{2}$

   c $\dfrac{2}{3}\sqrt{(30 \times 5) + 1}$

   d $\left(\dfrac{5}{2.718}\right)^5 \sqrt{10\pi}$

6 Evaluate the following:

   a $\dfrac{1}{2\pi \times 50 \times 3 \times 10^{-6}}$

   b $\dfrac{(5 \times 10^6) \pm \sqrt{(5 \times 10^6)^2 - (16 \times 10^{12})}}{2}$

7 [*electrical principles*] Write the following electrical quantities in symbol notation of Tables 2 and 3:

   a 378 000 V   **b** 0.000 01 A   **c** 1300 Ω

8 [*electrical principles*] Compute the range of values of the following resistors:

   a 100 Ω ± 5%   **b** 5 kΩ ± 2.5%

   c 13 MΩ ± 0.1%

9 [*electrical principles*] The input to output current ratio of an amplifier is given by $1\dfrac{1}{2} : 10\dfrac{3}{4}$. Simplify this ratio.

10 Evaluate 15% of $0.5 \times 10^{-3} \times 50 \times 10^3$.

11 [*thermodynamics*] Thermal efficiency of a plant is given by

$$\text{thermal efficiency} = \frac{3.5 \times 10^6}{24 \times 10^6}$$

Write this as a percentage.

12 [*electrical principles*] The power produced by an electrical generator is 20 kW. If it loses 6% of its power, find the power loss.

13 Calculate the following, correct to 2 d.p., by using your calculator:

   a $\dfrac{20 \times 10^{11}}{1.5 \times 10^6}$ **b** $\dfrac{115 \times 10^3}{(15 + 1.8)^2}$

   c $500 + \dfrac{100^2 - 157^2}{3 \times 10^3} - 160$

   d $\dfrac{(2 \times 10^6 \times 0.015) - (2 \times 10^5 \times 0.075)}{1.44 - 1}$

   e $\dfrac{(20 \times 10^9)(1 \times 10^3)\left[(1.5 \times 10^3)^2 - (0.5 \times 10^3)^2\right]^{\frac{3}{2}}}{25 \times 10^{21}}$

CHAPTER 1
Engineering Formulae

In this chapter we look at the applications of basic algebra to engineering problems.

The word *algebra* comes from the Arabic *al-jabr* which occurs in Al-Khwarizmi's book *Hisab al-jabr w'al-muqabala* written in the early ninth century. *Al-jabr* means restoration (or transpose to remove the negative quantities of an equation, e.g. $3x + 1 = 8 - 4x$ goes to $7x + 1 = 8$). Al-Khwarizmi (780–850 AD) was born in Khwarizm, now called Khiva, a town located in Uzbekistan (a former Soviet republic which became independent in 1991).

SECTION A **Substitution and transposition**

By the end of this section you will be able to:
▶ evaluate formulae using BROIDMAS
▶ solve equations
▶ transpose formulae

A1 **Evaluating formulae**

A **formula** is a general rule or law of mathematics. The plural of formula is **formulae**.

In evaluating formulae, the mnemonic BROIDMAS gives the order of operation (see the **Introduction chapter**):

<u>B</u>rackets	} First
<u>RO</u>ots	
<u>I</u>ndices	} Second
<u>D</u>ivision	
<u>M</u>ultiplication	} Third
<u>A</u>ddition	
<u>S</u>ubtraction	} Fourth

It's imperative that you understand BROIDMAS because it tells us the rules of algebra and is used for evaluating and simplifying algebraic expressions. Moreover it can be useful for typing in an expression into a computer algebra package or a calculator.

In algebra, letters or symbols are used to represent numbers. These letters or symbols may be **constants**, that is fixed, or **variables**, which means they can take up various values.

No space between letters represents multiplication, for example

$$ab = a \times b = a \cdot b$$

So if $a = 3$ and $b = 7$ then $ab = 3 \times 7 = 21$. To evaluate a formula we substitute the given numbers in place of letters and then apply BROIDMAS to evaluate the arithmetical expression as in the **Introduction Chapter.**

For example, evaluate $a(b + c) + \dfrac{c(a + b)^2}{b}$ where $a = 2$, $b = 3$ and $c = 5$:

$$2(3 + 5) + \frac{5(2 + 3)^2}{3} = (2 \times 8) + \left(\frac{5 \times 5^2}{3}\right)$$

$$= 16 + \left(\frac{5 \times 25}{3}\right)$$

$$= 16 + \frac{125}{3}$$

$$= 57\frac{2}{3}$$

Example 1

Pythagoras theorem gives the length of the longest side, c, in terms of the other two sides of a right-angled triangle, a and b, as

$$c = \sqrt{a^2 + b^2}$$

Evaluate c for $a = 5$ and $b = 12$.

Solution

Substituting $a = 5$ and $b = 12$ into $c = \sqrt{a^2 + b^2}$ gives

$$c = \sqrt{5^2 + 12^2}$$

$$= \sqrt{25 + 144}$$

$$= \sqrt{169} = 13$$

Hence $c = 13$.

A2 Transposition of formulae

In the formula $v = u + at$, we say v is the **subject** of the formula. If we want to make t the subject of the formula, then we need to change the form to

$$t = \boxed{}$$

This process of changing the subject is called **transposition** of formulae.

When transposing we can

► add, subtract, multiply or divide by the same quantities on both sides of the formula (though we cannot divide by zero)

A3 Transposition applied to equations

An **equation** is a mathematical statement that says two expressions are equal. For example,

$$x - 3 = 7$$

is an equation where x is an unknown variable or place-holder. To solve this equation means we need to find the value (or values) of x so that

$$\text{Left-Hand Side (LHS)} = \text{Right-Hand Side (RHS)}$$

? Hence we need to make x the subject of $x - 3 = 7$. **That is we need to remove the 3 on the Left-Hand Side. How?**

Add 3. We need to add 3 to both sides because we have to maintain the **balance** of the equation:

$$x - 3 + 3 = 7 + 3$$

$$x - 0 = 10$$

$$x = 10$$

In this case $x = 10$ is a **solution**, or a **root**, of the above equation. The process of finding the value of x is the same as transposition of formulae. Let's try some examples in the field of electrical principles and mechanics.

Example 2 *electrical principles*

If the voltage, V, across a resistor $R = 100\ \Omega$ is 10 volts, find the current I through the resistor, given that $V = IR$.

Solution

Substituting $V = 10$ and $R = 100$ into $V = IR$ gives

$$10 = 100I$$

? **What are we trying to find?**

? The value of I. **How do we find I?**

Divide both sides by 100:

$$\frac{100I}{100} = \frac{10}{100}$$

Cancelling the 100's on the Left-Hand Side:

$$I = \frac{10}{100} = 0.1\ \text{amp (A)}$$

The unit for current is amp and will generally be denoted by A.

As a check you can substitute $I = 0.1$ into $10 = 100I$, thus $10 = 100 \times 0.1$.

Example 3 *mechanics*

A vehicle's speed, v, is given by

$$v = 14 + 5t$$

where t is time. Find the time taken in seconds to reach a speed of 23 m/s.

Solution

Substituting $v = 23$ gives

$$14 + 5t = 23$$

We need to find t. How?

Subtract 14 from both sides:

$$5t = 23 - 14 = 9$$

How do we remove the 5 from the Left-Hand Side?

Divide both sides by 5:

$$\frac{5t}{5} = \frac{9}{5}$$

Cancelling 5's on the Left-Hand Side gives:

$$t = \frac{9}{5} = 1.8 \text{ s}$$

We use SI units throughout the book – see the **Introduction Chapter.** For example, velocity is given in m/s, acceleration in m/s^2, time in s, etc.

The above equations, $14 + 5t = 23$ and $100I = 10$, are examples of **linear equations**.

A linear equation is an equation where the unknown ($x, y, t, \ldots$) is of degree 0 or 1.

A4 **Transposition applied to engineering formula**

Algebraic expressions can be simplified by adding, subtracting or cancelling like terms, for example

$$x + x = 2x, \quad x - x = 0, \quad x + 5x = 6x \quad \text{and} \quad \frac{x}{x} = 1 \quad [\text{provided } x \neq 0]$$

We can **only** add and subtract like terms. We cannot simplify the following:

$$x + y = x + y, \quad x - y = x - y, \quad x + 5y = x + 5y \quad \text{and} \quad \frac{x}{y} = \frac{x}{y}$$

The procedure for applying transposition to formulae is very similar to that used in solving equations. Let's try some engineering examples.

Example 4 *electronics*

The power, P, dissipated in a circuit is given by

$$P = IV$$

where I is current and V is voltage. Transpose the formula for V.

Solution

We want to get $V = \underline{\hspace{1cm}}$. **How can we achieve this from $P = IV$?**

We need to remove I from the Right-Hand Side. How?

Divide both sides by I. Thus

$$\frac{P}{I} = \frac{IV}{I} \qquad \text{[the } I\text{'s on the Right-Hand Side cancel out because } \frac{I}{I} = 1\text{]}$$

$$\frac{P}{I} = V \text{ or } V = \frac{P}{I}$$

In Example 4, how do we know that we need to divide both sides by I?

The subject that we want to obtain is V.

What does the formula $P = IV$ do to V?

It is multiplied by I.

We want to remove the I and find V on its own. **How can we remove I?**

We can divide by I.

One way of obtaining the subject in many cases is to see what the formula does to the subject and then do the opposite on the other side. In **Example 4** the subject (V) is multiplied by I so we need to divide the other side by I, thus obtaining $V = \dfrac{P}{I}$.

Example 5 *mechanics*

The velocity, v, of an object with an initial velocity u and constant acceleration a after time t is given by

$$v = u + at$$

Transpose to make t the subject of the formula.

Solution

We need to get from $v = u + at$ to $t = \underline{\hspace{1cm}}$

Example 5 *continued*

We need to remove the u first. How?

Subtract u from both sides:

$$v - u = (u + at) - u$$
$$= \underbrace{u - u}_{=0} + at$$
$$v - u = at$$

How can we obtain t from $v - u = at$?

Divide both sides by a:

$$\frac{v - u}{a} = \frac{at}{a}$$

The a's on the Right-Hand Side cancel out to give

$$t = \frac{v - u}{a}$$

We will assume that the variable we are dividing by is **not** zero because we cannot divide by zero. So in **Example 5**, a is not zero.

SUMMARY

In evaluating the formula we use the mnemonic BROIDMAS: <u>B</u>rackets, <u>RO</u>ots, <u>I</u>ndices, <u>D</u>ivision, <u>M</u>ultiplication, <u>A</u>ddition, <u>S</u>ubtraction.

We can transpose a formula to make a certain variable the subject of the formula. Transposing involves arithmetical operations carried out on both sides of the formula. We use transposition to solve equations.

Exercise 1(a)

Solutions at end of book. Complete solutions available at www.palgrave.com/engineering/singh

1 Given that

$$c = \sqrt{a^2 + b^2}$$

evaluate c for $a = 24$ and $b = 7$.

2 [*electrical principles*]   If the voltage, V, across a resistor $R = 1000\ \Omega$ is 15 V, then find the current I given that
$$V = IR.$$

3 [*thermodynamics*]   A gas is expanded from an initial pressure P_1 and volume V_1 of 5×10^6 N/m^2 and 2×10^{-4} m^3 respectively, to a final pressure $P_2 = 2 \times 10^7$ N/m^2. Find the new volume V_2 given that $P_1 V_1 = P_2 V_2$.

4 [*thermodynamics*]   A gas has pressure $P = 5.6 \times 10^5$ N/m^2, volume $V = 0.015$ m^3 and is at a temperature $T = 312$ K. If there are $n = 34.6$ mole of gas, determine the mass, m, given that

$$PV = nmRT$$

where $R = 8.31$ J/(K mole) (R is called the universal or molar gas constant).

5 🚗 [*mechanics*]   The distance, s, travelled in time t is related by

$$s = ut + \frac{1}{2}at^2$$

where u is the initial velocity and a is constant acceleration. Determine a, given that $s = 30$ m, $u = 2$ m/s and $t = 5$ s.

6 [*electrical principles*]   The resistance, R, of a wire at $t\,°C$ is given by

$$R = R_0(1 + \alpha t)$$

where R_0 is the resistance at $0\,°C$ and α is the temperature coefficient of resistance. Determine α, given that $R_0 = 33\ \Omega$, $R = 35\ \Omega$ and $t = 89\,°C$. (The units of α are $/\,°C$.)

7 [*electrical principles*]   A battery with e.m.f. $E = 12$ V and an internal resistance $r = 1\ \Omega$ is connected across a resistor $R = 20\ \Omega$. Find the voltage V across R, given that

$$E = \frac{V(R + r)}{R}$$

8 🚗 [*mechanics*]   The velocity, v, of an object is given by

$$v = u + at$$

where u is the initial velocity, a is constant acceleration and t is time. Transpose to make
i u the subject
ii a the subject.

9 [*electrical principles*]   Ohm's law states that

$$I = \frac{V}{R}$$

where I is current, V is voltage and R is resistance. Transpose to make R the subject of the formula.

10 [*thermodynamics*]   The characteristic equation of a perfect gas is given by

$$PV = mRT$$

(P = pressure, V = volume, R = gas constant, m = mass and T = temperature). Make T the subject.

11 🚗 [*mechanics*]   The slip, S (%), of a vehicle is given by

$$S = \left(1 - \frac{r\omega}{v}\right) \times 100$$

where r = radius of tyre, ω = angular velocity and v = velocity. Make ω the subject of the formula.

SECTION B **Transposing engineering formulae**

By the end of this section you will be able to:
▶ transpose complicated formulae
▶ evaluate formulae by substitution

Transposition and substitution are two very important concepts of algebra, and it is critical that you fully understand the processes involved because the remaining chapters in this book rely upon these techniques.

This section is more difficult than **Section A** in the sense that the formulae involve roots and inverse quantities. Make sure you fully understand **Section A** before embarking on your journey into **Section B**.

B1 Formulae involving roots

As discussed in the **Introduction Chapter,** the square root and the nth root are denoted by $\sqrt{}$ and $\sqrt[n]{}$ respectively. We can also write these as

| 1.1 | | $\sqrt{a} = a^{1/2}$ | [square root] |
| 1.2 | | $\sqrt[n]{a} = a^{1/n}$ | [nth root] |

For example,

$\sqrt{49} = 7$ [positive square root]

$\sqrt[3]{8} = 2$ [because $2 \times 2 \times 2 = 8$ or $2^3 = 8$; $\sqrt[3]{}$ denotes the cube root]

? What is $8^{1/3}$ and $256^{1/4}$ equal to?

$8^{1/3} = \sqrt[3]{8} = 2$

$256^{1/4} = 4$ [because $4 \times 4 \times 4 \times 4 = 256$]

Now let's take a look at roots where letters represent variables.

We also have:

$$\sqrt{a^2} = \left(\sqrt{a}\right)^2 = a$$
$$\sqrt[n]{a^n} = \left(\sqrt[n]{a}\right)^n = a$$

(These can be demonstrated by using the rules of indices which are explored in the next section.)

Example 6 *aerodynamics*

The lift force, L, on an aircraft is given by

$$L = \frac{1}{2}\rho v^2 AC$$

where ρ is density, v is speed, A is area and C is lift coefficient. Make v the subject of the formula.

Solution

? **How can we get $v = $ ——?**

We can first isolate v^2 and then take the square root of both sides. ($\sqrt{v^2} = v$)

? **How do we get $v^2 = $ —— ?**

? **First we need to remove the $\frac{1}{2}$. How?**

Multiply both sides by 2: $2L = \rho v^2 AC$

? **Next we need to remove ρAC from the Right-Hand Side. How?**

Divide through by ρAC:

$$\frac{2L}{\rho AC} = v^2$$

> ### Example 6 *continued*
>
> **How can we find v?**
>
> Take the square root of both sides (because $\sqrt{v^2} = v$):
>
> $$\sqrt{v^2} = \sqrt{\frac{2L}{\rho AC}}$$
>
> $$v = \sqrt{\frac{2L}{\rho AC}} \text{ which we may write as } \underbrace{\left(\frac{2L}{\rho AC}\right)^{1/2}}$$
>
> by 1.1 †

† We adopt this approach of quoting the reference number of a previously established rule throughout the book, and the formula itself will either be in the main text or at the bottom of the page below a horizontal line so that you do not need to flick over pages to find the reference.

> ### Example 7 *materials*
>
> The second moment of area, I, of a rectangle of height h and breadth b is given by
>
> $$I = \frac{1}{12}bh^3$$
>
> Make h the subject of the formula.
>
> Solution
>
> **First we need to remove $\dfrac{1}{12}$ from the Right-Hand Side. How?**
>
> Multiply both sides by 12: $12I = bh^3$
>
> **How can we isolate h?**
>
> We can initially obtain h^3 and then find h. So divide both sides by b:
>
> $$\frac{12I}{b} = h^3$$
>
> and now take the cube root, $\left(\sqrt[3]{\ }\right)$, of both sides:
>
> $$\sqrt[3]{\frac{12I}{b}} = h \ \left(\text{because } \sqrt[3]{h^3} = h\right)$$
>
> or
>
> $$h = \left(\frac{12I}{b}\right)^{1/3}$$
>
> by 1.2

1.1 $\sqrt{a} = a^{1/2}$ 1.2 $\sqrt[3]{a} = a^{1/3}$

As discussed in the Introduction Chapter, in many engineering examples it is appropriate to give your final answer to the smallest number of significant figures consistent with the data. The intermediate working has to be one more decimal point (d.p.) or significant figure (s.f.) than is needed. Thus in order to give your final answer to 1 d.p. (or 1 s.f.) you need to work to 2 d.p. (or 2 s.f.).

In the next example we use substitution and transposition of formulae to evaluate the capacitance, C. It is more difficult than the above examples. Go through it carefully.

Example 8 *electronics*

The impedance, Z, of a circuit containing a resistor of resistance R, a capacitor of capacitance C and an inductor of inductance L is given by

$$Z = \sqrt{R^2 + (X_L - X_C)^2}$$

where $X_L = 2\pi f L$ and $X_C = \dfrac{1}{2\pi f C}$ (f represents frequency).

Determine C if $R = 100\,\Omega$, $Z = 104\,\Omega$, $L = 0.1$ henry and $f = 50$ Hz.

Solution

Substituting $f = 50$ and $L = 0.1$ into $X_L = 2\pi f L$ gives

$$X_L = 2\pi \times 50 \times 0.1 = 10\pi$$

Substituting $X_L = 10\pi$, $Z = 104$ and $R = 100$ into the given formula, $Z = \sqrt{R^2 + (X_L - X_C)^2}$, results in

$$104 = \sqrt{100^2 + (10\pi - X_C)^2}$$

What are we trying to find?

We need to determine C but first we find X_C and then obtain C.

Squaring both sides gives

$$104^2 = 100^2 + (10\pi - X_C)^2$$

Transposing

$$104^2 - 100^2 = 816 = (10\pi - X_C)^2$$

We have

$$(10\pi - X_C)^2 = 816$$

Taking square roots of both sides:

$$10\pi - X_C = \sqrt{816} = 28.566$$

Hence $X_C = 10\pi - 28.566 = 2.850$

Since $X_C = \dfrac{1}{2\pi f C}$ we have

$$\frac{1}{2\pi f C} = 2.850$$

Example 8 *continued*

Transposing

$$C = \frac{1}{2\pi f \times 2.85}$$

$$\underset{\substack{\text{substituting} \\ f = 50}}{=} \frac{1}{2\pi \times 50 \times 2.85} = 0.0011$$

Hence $C = 0.0011$ farad or 1.1×10^{-3} farad $= 1.1$ millifarad (mF). Remember that the prefix milli, m, represents 10^{-3}.

B2 **Formulae involving the inverse**

The multiplicative inverse of x ($\neq 0$) is denoted by x^{-1} and defined as

1.3 $\qquad x^{-1} = \dfrac{1}{x} \ (= 1 \div x)$

Example 9

Show that

$$\left(\frac{a}{b}\right)^{-1} = \frac{b}{a} \quad (a \neq 0, b \neq 0)$$

Solution

We have

$$\left(\frac{a}{b}\right)^{-1} = \frac{1}{\left(\dfrac{a}{b}\right)} \quad \left[\text{by } \boxed{1.3} \text{ with } x = \frac{a}{b}\right]$$

$$= 1 \div \frac{a}{b}$$

$$= 1 \times \frac{b}{a} = \frac{b}{a}$$

Remember $1 \div \dfrac{a}{b} = 1 \times \dfrac{b}{a}$ because when we divide fractions we turn the second fraction upside down and multiply.

We give this important result a reference number:

1.4 $$\left(\frac{a}{b}\right)^{-1} = \frac{b}{a}$$

We also have $\left(\dfrac{a}{b}\right)^{2} = \dfrac{a^2}{b^2}$, but more of this in the next section on indices.

Example 10 *thermodynamics*

A gas in a cylinder in state 1 with pressure P_1, temperature T_1 and volume V_1 expands to state 2 with pressure P_2, temperature T_2 and volume V_2. A formula relating these variables is given by $\dfrac{P_1V_1}{T_1} = \dfrac{P_2V_2}{T_2}$. Make T_1 the subject of the formula.

Solution

From $\dfrac{P_1V_1}{T_1} = \dfrac{P_2V_2}{T_2}$ we need $T_1 = $ ——. Taking the inverse, $(\ )^{-1}$, of both sides gives

$$\left(\frac{P_1V_1}{T_1}\right)^{-1} = \left(\frac{P_2V_2}{T_2}\right)^{-1}$$

$$\frac{T_1}{P_1V_1} = \frac{T_2}{P_2V_2} \quad [\text{by } \boxed{1.4}\ ]$$

How can we find $T_1 = $ ——?

Multiply both sides by P_1V_1:

$$T_1 = \frac{P_1V_1T_2}{P_2V_2}$$

SUMMARY

The square root, $\sqrt{\ }$, and the nth root, $\sqrt[n]{\ }$, are defined as

1.1 $\sqrt{a} = a^{1/2}$

1.2 $\sqrt[n]{a} = a^{1/n}$

The inverse of x ($\neq 0$) is defined as

1.3 $x^{-1} = \dfrac{1}{x}$

1.4 $\left(\dfrac{a}{b}\right)^{-1} = \dfrac{b}{a}$ ($a \neq 0,\ b \neq 0$)

It is well worth spending some time learning these definitions, 1.1 to 1.4, because they are used throughout the book.

Solutions at end of book. Complete solutions available at www.palgrave.com/engineering/singh

1 [electrical principles]   The power P dissipated in a resistor of resistance R is given by

$$P = \frac{V^2}{R} \quad [V \text{ is voltage}]$$

Make V the subject of the formula.

2 [acoustics]   The speed, c, of sound in air is given by

$$c = \sqrt{\frac{\gamma P}{\rho}}$$

where γ is the specific heat ratio, P is the pressure and ρ is the density. Make P the subject of the formula.

3 [mechanics]   The airflow over a vehicle causes drag D, which is given by

$$D = \frac{1}{2}\rho C v^2 A$$

where ρ is density, C is drag coefficient, v is velocity and A is the frontal area of the vehicle. Make A the subject of the formula.

4 [electrical principles]   Evaluate the total resistance, R, in a circuit containing two resistors in parallel of resistances $R_1 = 100\,\Omega$ and $R_2 = 270\,\Omega$, where

$$\frac{1}{R} = \frac{1}{R_1} + \frac{1}{R_2}$$

(Ω is the SI unit ohm used to measure electrical resistance).

5 [mechanics]   The time, T, taken for a pendulum to make a complete swing is given by

$$T = 2\pi\sqrt{\frac{l}{g}}$$

where l = length of pendulum and $g = 9.81$ m/s^2. Determine l, if $T = 0.5$ s.

6 [electronics]   The impedance, Z, of a circuit containing a resistor of resistance R, a capacitor of capacitance C and an inductor of inductance L is given by

$$Z = \sqrt{R^2 + (X_L - X_C)^2}$$

where $X_L = 2\pi f L$ and $X_C = \dfrac{1}{2\pi f C}$

(f represents frequency).

Determine C if $R = 50\,\Omega$, $Z = 100\,\Omega$, $L = 1$ henry and $f = 50$ Hz.

7 [aerodynamics]   The power, P, required to drive an air screw of diameter D is given by

$$P = 2\pi k\rho n^3 D^5$$

(k = torque coefficient, ρ = density, n = number of revolutions per second). Make D the subject of the formula.

8 [electronics]   A system with feedback β and gain A has an input voltage v_{in} given by

$$v_{in} = \left(\frac{1}{A} - \beta\right)v_{out}$$

(v_{out} = output voltage). Show that

$$\frac{v_{out}}{v_{in}} = \frac{A}{1 - A\beta}$$

9 [materials]   A cylinder of radius r is subject to a torque T at each end, which causes it to twist. The shear stress τ is given by

$$\tau = \frac{T}{\frac{1}{2}\pi r^3}$$

Make r the subject of the formula.

10 The following formulae occur in various engineering fields. Make the letter in the square brackets the subject of the formula.

a $V = \dfrac{ER}{R + r}$ $\quad\quad [r]$

b $v^2 = u^2 + 2as$ $\quad\quad [u]$

c $v = \left(\dfrac{K + 4a/3}{\rho}\right)^{1/2}$ $\quad [K]$

Solutions at end of book. Complete solutions available at www.palgrave.com/engineering/singh

d $\eta = \dfrac{\pi P r^4 t}{8 v L}$ $\qquad$ [r]

e $T = 2\pi \sqrt{\dfrac{l}{g}}$ $\qquad$ [l]

f $RT = \left(P + \dfrac{a}{V^2}\right)(V - b)$ $\qquad$ [P]

g $W = \dfrac{P_1 V_1 - P_2 V_2}{n - 1}$ $\qquad$ [V_1]

h $PV^n = C$ $\qquad$ [V]

i $f = C \dfrac{W}{D} \sqrt{\dfrac{h}{u}}$ $\qquad$ [u]

SECTION C **Indices**

By the end of this section you will be able to:
► use the laws of indices to simplify expressions
► use the laws of indices in applications of thermodynamics

Do you remember what 3^5 represents?

It is $\underbrace{3 \times 3 \times 3 \times 3 \times 3}_{\text{5 copies}}$ which is equal to 243.

The 5 in 3^5 is called the **index** or the **power**. The plural of index is **indices**. In this section we will predominantly apply the rules of indices to letters rather than numbers.

The topic of indices is very important for engineers but many students do find this a difficult topic – invariably because they don't know the rules well enough. It is worth making the effort to remember the rules as they are introduced.

C1 **Some rules of indices**

We have already stated some rules of indices in the previous section, 1.1 to 1.4 . Other important rules of indices are

1.5 $\qquad a^m\, a^n = a^{m+n}$

1.6 $\qquad a^m \div a^n = \dfrac{a^m}{a^n} = a^{m-n} \quad (a \neq 0)$

1.7 $\qquad (a^m)^n = a^{m \times n}$

1.8 $\qquad a^0 = 1 \quad (a \neq 0)$

1.9 $\qquad a^1 = a$

1.10 $\qquad a^{-n} = \dfrac{1}{a^n} \quad (a \neq 0)$

1.11 $\qquad (ab)^n = a^n\, b^n$

1.12 $\qquad \left(\dfrac{a}{b}\right)^n = \dfrac{a^n}{b^n} \quad (b \neq 0)$

Example 11

Simplify the following:

a $x^3 x^2$ **b** $\dfrac{x^3}{x^2}$ **c** $\dfrac{x}{\sqrt{x}}$ **d** $\left(\sqrt[3]{x}\right)^2 \sqrt[3]{x}$

Solution

Using the above rules we have

a $x^3 x^2 \underset{\text{by } \boxed{1.5}}{=} x^{3+2} = x^5$

b $\dfrac{x^3}{x^2} \underset{\text{by } \boxed{1.6}}{=} x^{3-2} = x^1 \underset{\text{by } \boxed{1.9}}{=} x$

c $\dfrac{x}{\sqrt{x}} = \dfrac{x^1}{x^{1/2}} \underset{\text{by } \boxed{1.6}}{=} x^{1-1/2}$

$\qquad = x^{1/2} = \sqrt{x} \qquad$ [by $\boxed{1.1}$]

d $\left(\sqrt[3]{x}\right)^2 \sqrt[3]{x} \underset{\text{by } \boxed{1.2}}{=} \left(x^{1/3}\right)^2 \left(x^{1/3}\right)$

$\qquad \underset{\text{by } \boxed{1.7}}{=} x^{2/3} x^{1/3} \underset{\text{by } \boxed{1.5}}{=} x^{(2/3)+(1/3)} = x^1 = x$

As **Example 11** shows, the rules of indices, $\boxed{1.1}$ to $\boxed{1.12}$, can be used to simplify algebraic expressions. We can also apply these to show results that we have already used, such as $\sqrt{a^2} = a$:

$$\sqrt{a^2} = (a^2)^{1/2} = a^{2 \times 1/2} = a^1 = a$$

Similarly we have

$$\sqrt[n]{a^n} = (a^n)^{1/n} = a^{n \times 1/n} = a^1 = a$$

Note that if $x^n = a$ then taking the nth root of both sides gives

$$(x^n)^{1/n} = a^{1/n}$$

Thus we have

$\boxed{\dagger} \qquad x = a^{1/n}$

We call this $\boxed{\dagger}$ because we will refer to it later on.

Let's try an engineering example.

$\boxed{1.1}$ $\sqrt{x} = x^{1/2}$ $\boxed{1.2}$ $\sqrt[n]{a} = a^{1/n}$ $\boxed{1.5}$ $a^m a^n = a^{m+n}$

$\boxed{1.6}$ $\dfrac{a^m}{a^n} = a^{m-n}$ $\boxed{1.7}$ $(a^m)^n = a^{m \times n}$ $\boxed{1.9}$ $a^1 = a$

Example 12 *thermodynamics*

A gas in a cylinder is compressed according to the law

$$P_1 V_1^{1.5} = P_2 V_2^{1.5}$$

where P is pressure and V is volume. If the gas has an initial volume of $V_1 = 0.16$ m^3 and pressure of $P_1 = 140 \times 10^3$ N/m^2 and is then compressed to a pressure of $P_2 = 750 \times 10^3$ N/m^2, find the new volume, V_2.

Solution

Substituting $P_1 = 140 \times 10^3$, $P_2 = 750 \times 10^3$ and $V_1 = 0.16$ into

$$P_1 V_1^{1.5} = P_2 V_2^{1.5}$$

gives

$$(140 \times 10^3) \times (0.16)^{1.5} = (750 \times 10^3) \times V_2^{1.5}$$

$$V_2^{1.5} = \frac{(140 \times 10^3) \times (0.16)^{1.5}}{750 \times 10^3} \quad \text{[Dividing by } 750 \times 10^3\text{]}$$

$$= \frac{140 \times (0.16)^{1.5}}{750} \quad \text{[Cancelling } 10^3\text{'s]}$$

$$V_2^{1.5} = 0.0119$$

Applying the index 1/1.5 to both sides and using [†] from the previous page yields:

$$V_2 = (0.0119)^{1/1.5} = 0.052 \text{ m}^3 \text{ (2 s.f.)}$$

SUMMARY

We can use the rules of indices, [1.1] to [1.12], to simplify algebraic expressions and in engineering applications from the field of thermodynamics.

Exercise **1(c)**

Solutions at end of book. Complete solutions available at www.palgrave.com/engineering/singh

1 Simplify the following:

a $x^5 x^2$ **b** $x^{1/5} x^{1/2}$ **c** $\dfrac{x^3}{x^3}$ **d** $\dfrac{x^7}{x^9}$

e $\left(\sqrt[5]{x}\right)^2 \sqrt[3]{x}$ **f** $\sqrt{x^2}\,\sqrt[3]{x^3}$ **g** $\sqrt[7]{x^3 x^4}$

2 Simplify

a $(1+y)^2(1+y)$ **b** $\dfrac{(1+x^2)^5}{(1+x^2)^3}$

c $\left(\sqrt[3]{x^2+x+1}\right)^5 \sqrt[3]{x^2+x}$

d $\left(\sqrt[3]{x^2+x+1}\right)^5 \sqrt[3]{x^2+x+1}$

e $\sqrt{x^2+2x}\,\sqrt{x^2+x}$

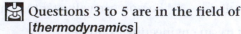 Questions 3 to 5 are in the field of [*thermodynamics*]

3 A gas in an engine obeys the law

$$P_1 V_1^{1.45} = P_2 V_2^{1.45}$$

where P represents pressure and V represents volume.

Solutions at end of book. Complete solutions available at www.palgrave.com/engineering/singh

If $P_1 = 2 \times 10^6$ N/m^2, $V_1 = 0.15$ m^3 and $P_2 = 2 \times 10^5$ N/m^2, find V_2.

4 The work done, W, on the face of a piston by a gas is given by

$$W = \frac{CV_2^{-0.35} - CV_1^{-0.35}}{-0.35}$$

where $C = P_1 V_1^{1.35} = P_2 V_2^{1.35}$ (P and V are pressure and volume respectively and C is a constant). Show that

$$-0.35W = P_2 V_2 - P_1 V_1$$

5 The state of a gas changes from P_1, V_1 and T_1 to P_2, V_2 and T_2 (pressure, volume and temperature respectively). The characteristic equation is given by

$$\frac{P_1 V_1}{T_1} = \frac{P_2 V_2}{T_2}$$

By a thermodynamics law we have

$$P_1 V_1^{\,n} = P_2 V_2^{\,n}$$

By using these formulae, show that

$$\frac{T_1}{T_2} = \left(\frac{P_1}{P_2}\right)^{1 - \frac{1}{n}}$$

6 ✈ [aerodynamics] In aerodynamics the following equation holds:

$$\frac{\rho_2}{\rho_1} = \left(\frac{T_2}{T_1}\right)^{g/LC} \left(\frac{T_1}{T_2}\right)$$

where ρ_1, ρ_2, T_1 and T_2 represent the densities and temperatures at altitude 1 and 2 respectively. L is the rate of decrease of temperature with altitude. C is a constant and g is acceleration due to gravity. Show that

$$\frac{\rho_2}{\rho_1} = \left(\frac{T_2}{T_1}\right)^{\frac{g - LC}{LC}}$$

SECTION D Dimensional analysis

By the end of this section you will be able to:
► apply the rules of indices to check equations which involve physical quantities

D1 Dimensional analysis

There are **three** fundamental dimensions: Mass, Length and Time (M, L and T respectively). All mechanical quantities can be expressed in terms of powers of M, L and T. (Non-mechanical quantities such as electrical current can also be expressed in terms of M, L and T, but it is easier to introduce a fourth dimension – charge Q.)

We use the following notation:

[force] represents the dimension of force.

Example 13

Obtain the fundamental dimensions of velocity (units m/s), acceleration (units m/s²) and force (= mass × acceleration).

Solution

We know the units of velocity are m/s so the dimensions are

$$\frac{\text{Length}}{\text{Time}} = \frac{L}{T} = L\left(\frac{1}{T}\right) \underset{\text{by } 1.3}{=} LT^{-1}$$

Similarly acceleration has units m/s², so the dimensions are

$$\frac{\text{Length}}{(\text{Time})^2} = \frac{L}{T^2} = L\left(\frac{1}{T^2}\right) \underset{\text{by } 1.10}{=} LT^{-2}$$

? **What about force?**

$$\text{force} = \text{mass} \times \text{acceleration}$$
$$[\text{force}] = M \times (LT^{-2})$$
$$= MLT^{-2}$$

Similarly we can evaluate the dimensions of the other quantities as shown in Table 1. Try verifying all of these in your own time.

	Quantity	Units	Dimensions
TABLE 1	Area	m²	L^2
	Volume	m³	L^3
	Velocity	m/s	LT^{-1}
	Acceleration	m/s²	LT^{-2}
	Force	newton (N)	MLT^{-2}
	Work (or energy)	joule (J)	$ML^2 T^{-2}$
	Power	watt (W)	$ML^2 T^{-3}$
	Pressure	N/m²	$ML^{-1} T^{-2}$
	Density	kg/m³	ML^{-3}
	Frequency	hertz (Hz)	T^{-1}

1.3 $\dfrac{1}{x} = x^{-1}$	**1.10** $\dfrac{1}{a^n} = a^{-n}$

Dimensional analysis is a method used to check the validity of an equation by establishing the same dimension formula on each side of the equation, that is

[Left-Hand Side] = [Right-Hand Side]

Numbers with no units attached to them are dimensionless.

Example 14 *fluid mechanics*

Bernoulli's equation is given by

$$P + \frac{1}{2}\rho v^2 + \rho g z = \text{constant}$$

where P = pressure, ρ = density, v = velocity, z = height and g = acceleration due to gravity.

Find the dimensions of the constant.

Solution

Using Table 1 we have $\left(\text{remember } \dfrac{1}{2} \text{ is dimensionless}\right)$

$$\underbrace{ML^{-1}\,T^{-2}}_{P} + \underbrace{ML^{-3}}_{\rho}\underbrace{(LT^{-1})^2}_{v^2} + \underbrace{ML^{-3}}_{\rho}\underbrace{LT^{-2}}_{g}\underbrace{L}_{z}$$

$$= ML^{-1}\,T^{-2} + ML^{-3}\underbrace{L^2\,T^{-2}}_{\text{by }1.7} + ML^{\underbrace{-3+2}_{\text{by }1.5}}\,T^{-2}$$

$$= ML^{-1}\,T^{-2} + \underbrace{ML^{-1}T^{-2}}_{=\,L^{-3+2}} + ML^{-1}T^{-2}$$

Hence the constant has the dimensions $ML^{-1}T^{-2}$.

A physical requirement is that **dimensional homogeneity** holds, that is both sides of an equation must have the same dimensions.

Example 15 *mechanics*

The period T of a pendulum of length l is given by

$$T = 2\pi\sqrt{\frac{l}{g}}$$

where g is acceleration due to gravity. Show that the formula has dimensional homogeneity.

1.5 $a^m\,a^n = a^{m+n}$ 1.7 $(a^m)^n = a^{m\times n}$

Example 15 *continued*

Solution

Remember 2π is dimensionless. By Table 1, g has the dimensions LT^{-2}. So we have

$$[T] = \sqrt{\dfrac{L}{LT^{-2}}} = \left(\dfrac{\cancel{L}}{\cancel{L}T^{-2}}\right)^{\frac{1}{2}} \underset{\text{by } 1.12}{=} \dfrac{1^{\frac{1}{2}}}{(T^{-2})^{\frac{1}{2}}} = \dfrac{1}{\underset{\text{by } 1.7}{\underbrace{T^{-2\times\frac{1}{2}}}}} = \dfrac{1}{T^{-1}} = T$$

The last step is justified by

$$\dfrac{1}{T^{-1}} \underset{\text{by } 1.3}{=} (T^{-1})^{-1} \underset{\text{by } 1.7}{=} T^{(-1)\times(-1)} = T^{1} \underset{\text{by } 1.9}{=} T$$

Clearly period T has dimensions T.

SUMMARY

There are three fundamental dimensions – mass M, length L and time T. We can apply the rules of indices to check dimensional homogeneity.

Exercise 1(d)

Solutions at end of book. Complete solutions available at www.palgrave.com/engineering/singh

 All questions in this exercise belong to [*dimensional analysis*].

1 Show that the dimensions of

a pressure $\left(= \dfrac{\text{force}}{\text{area}}\right)$ are $ML^{-1}T^{-2}$

b density $\left(= \dfrac{\text{mass}}{\text{volume}}\right)$ are ML^{-3}

c momentum (= mass × velocity) are MLT^{-1}

d power (= force × velocity) are ML^2T^{-3}

e impulse (= force × time) are MLT^{-1}

f kinetic energy
$\left[= \dfrac{1}{2} \times \text{mass} \times (\text{velocity})^2\right]$ are ML^2T^{-2}

g potential energy (= mass × acceleration × height) are ML^2T^{-2}

2 The pressure, P, at a depth d of a fluid of density ρ is given by
$$P = \rho g d \quad (g = \text{acceleration})$$

Show that the formula has dimensional homogeneity.

3 Which of the following are dimensionally correct (have dimensional homogeneity)?

a $F = mgl$ **b** $s = ut + \dfrac{1}{2}gt^2$

c $v^2 = u^2 + 2gs$ **d** $W = F \times v$

e $P = F \times l$

(m = mass, g = acceleration, l = length, t = time, s = distance, u and v = velocities, F = force, W = work and P = power).

1.3 $\dfrac{1}{x} = x^{-1}$ 1.7 $(a^m)^n = a^{m\times n}$ 1.9 $a^1 = a$ 1.12 $\left(\dfrac{a}{b}\right)^n = \dfrac{a^n}{b^n}$

Exercise **1(d) continued**

4 The dynamic coefficient of viscosity μ (viscosity of a fluid) is found from

$$F = \frac{\mu Av}{d}$$

where v = velocity, d = distance, F = force and A = area. Find the dimensions of μ.

5 Show that the following are dimensionless parameters by checking that the dimensions of each are equal to 1:

a Reynolds Number = $\dfrac{\rho vl}{\mu}$

b Mach Number = $\dfrac{v}{c}$

c Euler Number = $\dfrac{p}{\rho v^2}$

d Froude Number = $\dfrac{v}{\sqrt{gl}}$

e Weber Number = $\dfrac{v^2 l \rho}{\sigma}$

(ρ is density, v is velocity, g is acceleration due to gravity, l is length, μ is viscosity, p is pressure, c is speed of sound and σ is surface tension whose units are N/m.)

SECTION E **Expansion of brackets**

By the end of this section you will be able to:
▶ expand brackets
▶ use expansion of brackets in engineering applications
▶ expand brackets of the type $(a + b)(c + d)$ using FOIL

E1 **Revision of brackets**

? **What does $5(x + 3)$ mean?**

All the terms inside the bracket are multiplied by 5:

$$5(x + 3) = (5 \times x) + (5 \times 3)$$
$$= 5x + 15$$

Let's do a few examples.

Example 16

Multiply out the brackets of the following:

a $5(2x + 1)$ **b** $3(3x - 2)$ **c** $-(x - 1)$ **d** $-2(-x - 4)$

Solution

a $5(2x + 1) = (5 \times 2x) + (5 \times 1) = 10x + 5$
b $3(3x - 2) = (3 \times 3x) - (3 \times 2) = 9x - 6$
c Remember that 'minus times minus equals plus':

> ## Example 16 continued
>
> $$-(x - 1) = -1(x - 1) = (-1 \times x) - \underbrace{[1 \times (-1)]}_{=-1} = -x + \underset{\substack{\text{because}\\-(-1)=1}}{1} = 1 - x$$
>
> The result of taking a negative sign inside a bracket is to change **all** the signs inside the bracket.
>
> **d** $-2(-x - 4) = [-2 \times (-x)] - \underbrace{[2 \times (-4)]}_{=-8} = 2x + 8$

> ## Example 17
>
> Simplify the following:
>
> **a** $3(x + 2) + 5(2x + 3)$ **b** $(x + 5) - 2(x - 1)$ **c** $-(2x + 3) + (2x + 3)$
>
> Solution
>
> We add all the like terms:
>
> **a** $3(x + 2) + 5(2x + 3) = (3x + 6) + (10x + 15)$
> $$= \underbrace{3x + 10x}_{\text{collecting all the } x \text{ terms}} + (6 + 15)$$
> $$= 13x + 21$$
> **b** Multiplying out the brackets gives
> $$(x + 5) - 2(x - 1) = (x + 5) - (2 \times x) - (2 \times (-1))$$
> $$= (x + 5) - 2x + 2$$
> $$= x - 2x + (5 + 2)$$
> $$= 7 - x$$
> **c** $-(2x + 3) + (2x + 3) = 0$

>
>
> ## Example 18 structures
>
> The deflection, y, at a distance x from one end of a beam of length l is given by
> $$y = \frac{wx^2}{6EI}(3l - x)$$
> where w is the load per unit length and EI is the flexural rigidity of the beam. Remove the brackets of this expression.
>
> Solution
>
> We have
> $$y = \frac{wx^2}{6EI}(3l - x) = \frac{wx^2}{6EI}3l - \frac{wx^2}{6EI}x$$
> $$= \frac{3wx^2 \, l}{6EI} - \frac{wx^2 \, x}{6EI}$$
> $$= \frac{wx^2 \, l}{2EI} - \frac{wx^3}{6EI} \qquad \left[\text{Because } \frac{3}{6} = \frac{1}{2} \text{ and } x^2 x = x^3 \right]$$

E2 Using FOIL

? **How do we remove the brackets from an expression like $(x + 3)(x + 2)$?**

Each term in the first bracket (x and 3) multiplies the second bracket ($x + 2$):

$$(x + 3)(x + 2) = x(x + 2) + 3(x + 2)$$
$$= (x \times x) + (x \times 2) + (3 \times x) + (3 \times 2)$$
$$= x^2 + \underbrace{2x + 3x}_{= 5x} + 6$$
$$= x^2 + 5x + 6$$

Another way is to use FOIL, which is a mnemonic for <u>F</u>irst, <u>O</u>utside, <u>I</u>nside and <u>L</u>ast:

$$(x + 3)(x + 2) = \underbrace{(x \times x)}_{F} + \underbrace{(x \times 2)}_{O} + \underbrace{(3 \times x)}_{I} + \underbrace{(3 \times 2)}_{L}$$

$$= x^2 + 5x + 6$$

Multiply

The <u>F</u>irst terms in each bracket

The <u>O</u>utside terms

The <u>I</u>nside terms

The <u>L</u>ast terms

The process of multiplying brackets is also known as **expanding brackets**.

Example 19

Expand the following:

a $(x + 4)(x + 5)$ **b** $(x + 5)(x - 1)$ **c** $(2x + 3)(3x + 5)$ **d** $(3x - 1)(4x - 2)$

Solution

Using FOIL in each case gives

a $(x + 4)(x + 5) = \underbrace{(x \times x)}_{F} + \underbrace{(x \times 5)}_{O} + \underbrace{(4 \times x)}_{I} + \underbrace{(4 \times 5)}_{L}$

$$= x^2 + \underbrace{5x + 4x}_{= 9x} + 20$$

$$= x^2 + 9x + 20$$

b $(x + 5)(x - 1) = \underbrace{(x \times x)}_{F} + \underbrace{(x \times (-1))}_{O} + \underbrace{(5 \times x)}_{I} + \underbrace{(5 \times (-1))}_{L}$

$$= x^2 - x + 5x - 5$$

$$= x^2 + \underbrace{4x}_{= 5x - x} - 5$$

c $(2x + 3)(3x + 5) = \underbrace{(2x \times 3x)}_{F} + \underbrace{(2x \times 5)}_{O} + \underbrace{(3 \times 3x)}_{I} + \underbrace{(3 \times 5)}_{L}$

$$= 6x^2 + \underbrace{10x + 9x}_{= 19x} + 15$$

$$= 6x^2 + 19x + 15$$

Example 19 *continued*

d $(3x - 1)(4x - 2) = \underbrace{(3x \times 4x)}_{F} + \underbrace{(3x \times (-2))}_{O} + \underbrace{((-1) \times 4x)}_{I} + \underbrace{((-1) \times (-2))}_{L}$

$$= 12x^2 \underbrace{- 6x - 4x}_{= -10x} + 2$$

$$= 12x^2 - 10x + 2$$

E3 **Important expansions**

Important expansions are $(a + b)^2$ and $(a - b)^2$. Let's use FOIL to expand these.

We have

$$(a + b)^2 = (a + b)(a + b) = \underbrace{(a \times a)}_{F} + \underbrace{(a \times b)}_{O} + \underbrace{(b \times a)}_{I} + \underbrace{(b \times b)}_{L}$$

$$= a^2 + \underbrace{ab + ba}_{= 2ab} + b^2$$

$$= a^2 + 2ab + b^2$$

Similarly we have

$$(a - b)^2 = (a - b)(a - b) = \underbrace{(a \times a)}_{F} - \underbrace{(a \times b)}_{O} - \underbrace{(b \times a)}_{I} + \underbrace{(b \times b)}_{L}$$

$$= a^2 - ab - ba + b^2$$

$$= a^2 - 2ab + b^2$$

Note the following:

$$(a + b)^2 \neq a^2 + b^2 \quad \text{[Not equal]}$$

$$(a - b)^2 \neq a^2 - b^2 \quad \text{[Not equal]}$$

The symbol '$\neq$' means 'does not equal'. These are common misconceptions held by many students.

It is important to remember these results because they are used throughout mathematics:

| 1.13 | $(a + b)^2 = a^2 + 2ab + b^2$ |
| 1.14 | $(a - b)^2 = a^2 - 2ab + b^2$ |

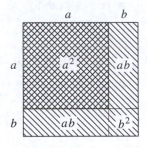

Fig.1 Shows why
$(a+b)^2 = a^2 + 2ab + b^2$

For example, by using 1.13 with $a = 2x$ and $b = 3$ we have

$$(2x + 3)^2 = (2x)^2 + (2 \times 2x \times 3) + 3^2$$

$$= 2^2 x^2 + (4x \times 3) + 9$$

$$= 4x^2 + 12x + 9$$

Similarly, using 1.14 with $a = 5x$ and $b = 2$ we have

$$(5x - 2)^2 = (5x)^2 - (2 \times 5x \times 2) + 2^2$$

$$= 5^2 x^2 - (10x \times 2) + 4$$

$$= 25x^2 - 20x + 4$$

Another important result which will be discussed in **Exercise 1(e)** is

`1.15` $(a - b)(a + b) = a^2 - b^2$

This is illustrated in Fig. 2. Expansions of the type `1.13` to `1.15` are prevalent in many fields of engineering and are worth learning until they become second nature to you. Make sure you know these results `1.13` to `1.15` from left to right and right to left.

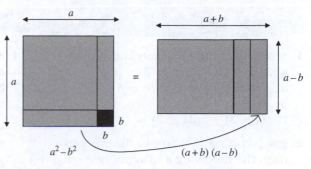

Fig. 2 Explains why $(a^2 - b^2) = (a - b)(a + b)$

Example 20 *electrical principles*

The impedance, Z, of a circuit is given by

$$Z^2 = R^2 + \left(\omega L - \frac{1}{\omega C}\right)^2$$

where R is resistance, L is inductance, C is capacitance and ω is angular frequency. Expand the brackets and simplify.

Solution

Substituting $a = \omega L$ and $b = \dfrac{1}{\omega C}$ into $(a - b)^2 = a^2 - 2ab + b^2$ produces

$$\left(\omega L - \frac{1}{\omega C}\right)^2 = (\omega L)^2 - \left(2 \times \cancel{\omega}L \times \frac{1}{\cancel{\omega}C}\right) + \left(\frac{1}{\omega C}\right)^2$$

$$= \omega^2 L^2 - \left(\frac{2 \times L \times 1}{C}\right) + \frac{1^2}{(\omega C)^2}$$

$$= \omega^2 L^2 - \frac{2L}{C} + \frac{1}{\omega^2 C^2} \qquad \text{[Simplifying]}$$

Substituting this into the original formula gives

$$Z^2 = R^2 + \omega^2 L^2 - \frac{2L}{C} + \frac{1}{\omega^2 C^2}$$

SUMMARY

Expand brackets of the form $(a + b)(c + d)$ by using FOIL (**F**irst, **O**utside, **I**nside, **L**ast). Important expansions which you should learn are

`1.13` $(a + b)^2 = a^2 + 2ab + b^2$

`1.14` $(a - b)^2 = a^2 - 2ab + b^2$

`1.15` $(a + b)(a - b) = a^2 - b^2$

Exercise 1(e)

1 Multiply out the brackets and simplify:
  a $2(3x + 1)$ **b** $-(2x + 1)$
  c $-3(5y + 1)$ **d** $x(3x + 5)$
  e $3(y - 1) - (2y + 1)$
  f $x(x - 3) + x(3x + 2)$

2 [structures]   Remove the brackets from the following and simplify:

  a $y = \dfrac{w}{2EI}(Lx^3 - x^4)$

  b $y = \dfrac{wx^3}{8EI}(2L - 3x)$

  c $y = \dfrac{wx^2}{48EI}(3L^2 - 2x^2)$

  d $y = -\dfrac{w}{12EI}\left(Lx^3 - \dfrac{x^4}{2} - \dfrac{L^3x}{2}\right)$

(L is length of beam, x is distance along the beam, EI is the flexural rigidity, y is deflection of the beam and w is the load per unit length).

3 Expand the following brackets and simplify:
  a $(x + 1)(x + 2)$
  b $(2x + 3)(3x + 5)$
  c $(2x - 1)^2$
  d $(a + b)^2 - (a - b)^2$
  e $(xy + 1)^2 - x(y + 1)$

4 By expanding the brackets show that
  a $(x - 5)(x + 5) = x^2 - 25$
  b $(2x - 3)(2x + 3) = 4x^2 - 9$
  c $(9x - 7)(9x + 7) = 81x^2 - 49$

What do you notice about the above results?

Show that

  1.15 $(a + b)(a - b) = a^2 - b^2$

This is known as the **difference between two squares** and is an important result.

5 [electrical principles]   Expand and simplify the following:

  a $(R + \omega L)(R - \omega L)$

  b $\dfrac{1}{R^2} + \left(\omega C - \dfrac{1}{\omega L}\right)^2$

(R is resistance, ω is angular frequency, L is inductance and C is capacitance).

6 Find

  $(x - a)(x - b)(x - c) \cdots (x - z)$

where $\cdots$ means (x − number represented by next letter of the alphabet).

SECTION F **Factorization**

By the end of this section you will be able to:
▶ factorize simple expressions
▶ factorize quadratic expressions

This section is a lot more demanding than the previous section because it expects you to find common factors which is **no** easy task.

F1 **Factorizing expressions**

We investigated factors in the **Introduction Chapter. What are the factors of 10?**

        5 and 2 because $5 \times 2 = 10$

Of course there are other factors of 10, namely 1 and 10.

Similarly $2 \times 5 \times 7 = 70$, and we say that 2, 5 and 7 are factors of 70.

? **What are the factors of 30 031?**

This is a difficult question and one of the reasons why this section is challenging. In fact $59 \times 509 = 30\ 031$, therefore 59 and 509 are the factors of 30 031.

In this section we look at factors of algebraic expressions.

Example 21

Factorize $5x + 5y + 5z$.

Solution

? **What do you notice about $5x + 5y + 5z$?**

The number 5 is common to all the terms in $5x + 5y + 5z$. We write

$$5x + 5y + 5z = 5(x + y + z)$$

and say that 5 and $x + y + z$ are **factors** of $5x + 5y + 5z$.

Factorization is the reverse process of expansion discussed in **Section E**.

? **How do we factorize an expression like**

$$5x - 4x^2?$$

We know x is common in both terms because $x^2 = xx$, thus

$$5x - 4x^2 = 5x - 4xx = x(5 - 4x)$$

? **How do we factorize an engineering expression such as**

$$y = \frac{wx^2}{EI} - \frac{wx^3}{EI}?$$

We know from the rules of indices that x^3 can be written as x^2x, so we have:

$$y = \frac{wx^2}{EI} - \frac{wx^2\,x}{EI}$$

? **What is common between the two terms on the Right-Hand Side?**

Clearly it is $\dfrac{wx^2}{EI}$. So we can take out this common factor and write y as

$$y = \frac{wx^2}{EI}1 - \frac{wx^2}{EI}x = \frac{wx^2}{EI}(1 - x)$$

Let's do another example.

Example 22 *structures*

The deflection y of a beam of length L at distance x is given by

$$y = \frac{wx^2 L^2}{16EI} + \frac{wx^4}{16EI}$$

where w is the load per unit length and EI is the flexural rigidity. Factorize this expression.

Example 22 *continued*

Solution

From the rules of indices we have $x^4 = x^2 x^2$, so we can write y as

$$y = \frac{wx^2 L^2}{16EI} + \frac{wx^2 x^2}{16EI}$$

$\frac{wx^2}{16EI}$ is common to **both** terms on the Right-Hand Side, so we can take this factor out:

$$y = \frac{wx^2}{16EI}(L^2 + x^2)$$

The following example is a lot more difficult because it involves an algebraic fraction with different denominators.

Example 23 *structures*

The deflection, y, of a beam of length L at distance x is given by

$$y = \frac{wx^2 L^2}{8EI} - \frac{wx^4}{24EI}$$

where w is the load per unit length and EI is the flexural rigidity. Factorize this expression.

Solution

From the rules of indices we have $x^4 = x^2 x^2$, so we can write y as

$$y = \frac{wx^2 L^2}{8EI} - \frac{wx^2 x^2}{24EI}$$

$\frac{wx^2}{EI}$ is common to both terms on the Right-Hand Side, so we can take this factor out:

★
$$y = \frac{wx^2}{EI}\left(\frac{L^2}{8} - \frac{x^2}{24}\right)$$

? **Can we factorize this further?**

Yes. The bracket term $\frac{L^2}{8} - \frac{x^2}{24}$ is an example of an algebraic fraction. It is dealt with in the same way as an arithmetic fraction.

? **How do you evaluate $\frac{1}{8} - \frac{1}{24}$?**

We need a common denominator, 24. Hence

$$\frac{1}{8} - \frac{1}{24} = \underbrace{\frac{3}{24}}_{=1/8} - \frac{1}{24}$$

Example 23 *continued*

Similarly we have

$$\frac{L^2}{8} - \frac{x^2}{24} = \frac{3L^2}{24} - \frac{x^2}{24} = \frac{3L^2 - x^2}{24}$$

(Of course we cannot simplify $3L^2 - x^2$ any further because they are **not** like terms.)

Substituting $\dfrac{L^2}{8} - \dfrac{x^2}{24} = \dfrac{3L^2 - x^2}{24}$ into ▨* gives

$$y = \frac{wx^2}{EI}\left(\frac{3L^2 - x^2}{24}\right)$$

$$= \frac{wx^2}{24EI}(3L^2 - x^2)$$

In **Example 23** the examination of the fraction, $\dfrac{1}{8} - \dfrac{1}{24}$, might seem like a diversion, but to deal with the algebraic fraction, $\dfrac{L^2}{8} - \dfrac{x^2}{24}$, we need to consider the arithmetic fraction.

F2 **Factorizing quadratics** ($ax^2 + bx + c$)

This subsection is more difficult than what has gone before. Take your time.

An expression of the form $ax^2 + bx + c$ (where a is not zero) is called a **quadratic.**

If we use FOIL to expand $(x + 2)(x + 5)$ we get:

$$(x + 2)(x + 5) = \underbrace{(x \times x)}_{F} + \underbrace{(x \times 5)}_{O} + \underbrace{(2 \times x)}_{I} + \underbrace{(2 \times 5)}_{L}$$

$$= x^2 + 5x + 2x + 10$$

$$= x^2 + 7x + 10$$

Therefore the factors of $x^2 + 7x + 10$ are $(x + 2)$ and $(x + 5)$.

Remember in this section we are factorizing, so we go in the opposite direction and this is challenging.

How can we obtain $(x + 2)(x + 5)$ given the quadratic $x^2 + 7x + 10$? (Or how do we factorize $x^2 + 7x + 10$?)

Let's assume we don't know the factors of $x^2 + 7x + 10$. We know
$x^2 + 7x \underset{\uparrow}{+} 10 = (x \pm \ast)(x \pm \bullet)$ because $x \times x$ gives x^2.

1 If the sign in front of 10 is

  +, then $\pm$ and $\pm$ in the resultant brackets are the **same** sign
  −, then $\pm$ and $\pm$ in the resultant brackets are **different** signs

  In this example, $\pm$ and $\pm$ are the same sign but we have to establish which sign.

2 If the signs are the same then

$$x^2 \pm 7x + 10 = (x \pm *)(x \pm \bullet)$$

    this first sign tells you what the sign is, hence

$$x^2 + 7x + 10 = (x + *)(x + \bullet)$$

3 Now we look at the factors of 10 (because 10 is the only term in the quadratic which does not contain an x). **What are the factors of 10?**

    10 and 1 or 5 and 2

We have a $7x$ on the Left-Hand Side, therefore

$$7x = x\bullet + *x \qquad \text{[Expanding]}$$

$$= (\bullet + *)x \qquad \text{[Factorizing]}$$

Since we want 7 in the middle, $*$ must be 5 and $\bullet$ must be 2 (or vice versa). So we have

$$x^2 + 7x + 10 = (x + 5)(x + 2)$$

These three rules are very complicated in comparison to previous concepts but try the following examples.

Example 24

Factorize $x^2 - 2x - 3$.

Solution

Using the above procedure we have

$$x^2 - 2x - 3 = (x + \ )(x - \ )$$

    Because of this, the signs are **different**. Next we look at the factors of 3.

What are the factors of 3?

    1 and 3

Hence we have

$$(x + 1)(x - 3) \text{ or } (x + 3)(x - 1)$$

Since we want -2 in the middle it is -3 and $+1$. Thus

$$x^2 - 2x - 3 = (x + 1)(x - 3)$$

How do we factorize $x^2 + 5x - 3$?

Since the only factors of 3 are 1 and 3 we can only have

$$(x + 1)(x - 3) \text{ or } (x - 1)(x + 3)$$

Multiplying out either of these does not give

$$x^2 + 5x - 3$$

? **Where have we made a mistake?**

There is **no** mistake. Simply, **not** all quadratics, $ax^2 + bx + c$, can be factorized into whole numbers. The actual factorization is

$$x^2 + 5x - 3 = \left(x + \frac{5 - \sqrt{37}}{2}\right)\left(x + \frac{5 + \sqrt{37}}{2}\right)$$

Of course this looks horrendous, and you are not expected to attempt this factorization in this chapter. The quadratic $x^2 + 5x - 3$ **cannot** be factorized into simple whole numbers.

Example 25

Factorize $2x^2 + 7x - 15$.

Solution

We know:

$$2x^2 + 7x \underset{\uparrow}{-} 15 = (2x \pm *)(x \pm \bullet) \qquad \text{[Because we want } 2x^2\text{]}$$

This sign tells us that the signs in the middle of the resultant brackets ($\pm$ and $\pm$) are different.

So we have

$$(2x - *)(x + \bullet) \text{ or } (2x + *)(x - \bullet)$$

Let's consider the case $(2x - *)(x + \bullet)$. The factors of 15 are 15 and 1 or 5 and 3. We need a 7 in the middle (the x term). In this example we need to be careful because the middle term is obtained by

$$(2x - *)(x + \bullet) = \dots \underbrace{2x \times \bullet}_{\text{outside}} + \underbrace{x \times (-*)}_{\text{inside}} \dots \overset{\text{only the middle term}}{}$$

Clearly the factors 15 and 1 are useless because we will never get 7. They need to be 5 and 3 because $(2 \times 5) - 3 = 7$. So $\bullet$ is 5 and $*$ is 3, that is the x term is made from $(2 \times 5) - 3 = 7$. We have

$$2x^2 + 7x - 15 = (2x - 3)(x + 5)$$

You can always check your result; expanding $(2x - 3)(x + 5)$ gives $2x^2 + 7x - 15$. Also note that if you change the signs such that we have $(2x + 3)(x - 5)$ then you get $-7x$ in the middle and not $+7x$ as required. You can only judge the placement of signs by practising a number of factorizations. It's good practice to expand your final factorization to check your result.

F3 Important factorization

? **How do we factorize $x^2 - 25$?**

It is a quadratic because the highest power term is x^2 and it doesn't matter if there is no x. Remember a quadratic is $ax^2 + bx + c$ where a is **not** zero but b or c may be zero.

? **How do we factorize this, $x^2 - 25$?**

We can use

1.15 $\qquad a^2 - b^2 = (a + b)(a - b)$

It's easier than the above, so we have

$$x^2 - 25 = x^2 - 5^2$$

$$\underset{\text{by } 1.15}{\equiv} (x + 5)(x - 5)$$

Similarly we have

$$x^2 - 9 = (x + 3)(x - 3)$$
$$x^2 - 16 = (x + 4)(x - 4)$$
$$x^2 - 5 = x^2 - \left(\sqrt{5}\right)^2 \qquad \left[\text{Rewriting } 5 = \left(\sqrt{5}\right)^2\right]$$

$$= \left(x + \sqrt{5}\right)\left(x - \sqrt{5}\right)$$

Let's investigate a more challenging example.

🚗 Example 26 *mechanics*

The equation of an object falling in air is given by

$$ma = mg - kv^2$$

where m ($\neq 0$) is the mass of the object, a is acceleration, v is velocity, g is acceleration due to gravity and k is a constant. Show that

$$a = \left(\sqrt{g} - cv\right)\left(\sqrt{g} + cv\right) \text{ where } c = \sqrt{\frac{k}{m}}$$

Solution

Dividing the given equation, $ma = mg - kv^2$, by m gives

$$a = \frac{mg}{m} - \frac{k}{m}v^2 = g - \frac{k}{m}v^2$$

From earlier work we know that $g = \left(\sqrt{g}\right)^2$ and $\dfrac{k}{m} = \left(\sqrt{\dfrac{k}{m}}\right)^2$ so we have

$$a = \left(\sqrt{g}\right)^2 - \left(\sqrt{\frac{k}{m}}\right)^2 v^2$$

$$= \left(\sqrt{g}\right)^2 - c^2 v^2 \qquad \text{where } c = \left(\sqrt{\frac{k}{m}}\right)$$

$$= \left(\sqrt{g}\right)^2 - \underset{\text{by } 1.11}{\underline{(cv)^2}}$$

1.11 $\quad a^n b^n = (ab)^n$

Example 26 *continued*

? How can we place $\left(\sqrt{g}\right)^2 - (cv)^2$ inside two brackets?

Use 1.15 , hence

$$a = \left(\sqrt{g}\right)^2 - (cv)^2 = \left(\sqrt{g} - cv\right)\left(\sqrt{g} + cv\right)$$

SUMMARY

To factorize $ax^2 + bx + c$ we need to look at factors of a and c and signs inside the expression. A common factorization is

1.15 $a^2 - b^2 = (a - b)(a + b)$

You need to know this result in both directions, that is from left to right and right to left.

Exercise **1(f)**

Solutions at end of book. Complete solutions available at www.palgrave.com/engineering/singh

1 Factorize the following:
  a $4x + 4y + 4z$
  b $8x + 8xy$
  c $2x - 4y$
  d $3x - 2x^2$
  e $x^2 - xy$
  f $16x + 4x^2$
  g $9x^2 - 27x^3$

2 🚗 [*mechanics*]   The following formulae occur in mechanics. Factorize each of them.

  a $s = ut + \dfrac{1}{2}at^2$

  b $F = \dfrac{mv_2}{t} - \dfrac{mv_1}{t}$

  c $F = \rho Av_2 v_1 - \rho Av_1^2$

3 The surface area, S, of a cone of radius r and height h is given by

$$S = \pi r^2 + \pi r(r^2 + h^2)^{1/2}$$

Factorize this formula.

4 Factorize the following:
  a $x^2 + 7x + 10$
  b $x^2 + 5x + 4$
  c $x^2 - 5x + 4$
  d $x^2 - 4x - 12$

  e $2x^2 + x - 1$
  f $x^2 - 3x - 4$
  g $21x^2 + 29x - 10$
  h $6x^2 + x - 12$

5 Factorize the following:
  a $x^2 - 2x + 1$
  b $x^2 + 2x + 1$
  c $x^2 - 36$
  d $x^2 - 7$
  e $4x^2 + 12x + 9$

6 🔲 [*electrical principles*]   Factorize the following:

  a $Z^2 - R^2$
  b $\omega^2 L^2 - \dfrac{1}{\omega^2 C^2}$

(Z is impedance, R is resistance, L is inductance, C is capacitance and ω is angular frequency).

7 ✈ [*aerodynamics*]   The Froude efficiency, F, of a propulsive system is given by

$$F = \frac{2(VV_s - V^2)}{V_s^2 - V^2}$$

[V_s and V are velocities].

1.15 $a^2 - b^2 = (a - b)(a + b)$

Exercise **1(f) continued**

Solutions at end of book. Complete solutions available at www.palgrave.com/engineering/singh

Show that

$$F = \frac{2V}{V_s + V}$$

8 🏞 [*structures*]   Factorize the following:

a $y = \dfrac{3wLx^2}{6EI} - \dfrac{wx^3}{6EI}$

b $y = \dfrac{wLx^3}{4EI} - \dfrac{3wx^4}{8EI}$

c $y = \dfrac{wx^4}{24EI} - \dfrac{wLx^3}{12EI} + \dfrac{wL^2 x^2}{24EI}$

where y is the deflection at a distance x along a beam of length L, w is load per unit length and EI is flexural rigidity.

9 🚗 [*mechanics*]   The acceleration, a, of an object in vibration is given by

$$a = g - k^2\omega^2$$

where g is acceleration due to gravity, ω is angular frequency and k is a constant.

Show that $a = \left(\sqrt{g} - k\omega\right)\left(\sqrt{g} + k\omega\right)$.

10 🏗 [*structures*]   The deflection, y, of a beam of length l at a distance x from one end is given by

$$y = \frac{wx^3}{12EI} - \frac{lx^2 w}{8EI} + \frac{l^2 wx}{24EI}$$

where EI is flexural rigidity and w is load per unit length on the beam. Show that

$$y = \frac{wx}{24EI}(2x - l)(x - l)$$

11 📟 [*electrical principles*]   Show that

a If $N = \dfrac{Z_0 + \frac{1}{2}Z_1}{Z_0 - \frac{1}{2}Z_1}$ then

$$Z_1 = 2Z_0\left(\frac{N - 1}{N + 1}\right)$$

b If $Z_1(N - 1)^2 + 2Z_0(N^2 - 1)$

$$= Z_1(N + 1)^2$$

then $Z_1 = Z_0\left(\dfrac{N^2 - 1}{2N}\right)$

(Z_1, Z_0 are impedances and N is a number).

SECTION G **Quadratic equations**

By the end of this section you will be able to:
► solve some quadratic equations of the form $ax^2 = b$
► solve some quadratic equations by factorization
► solve all quadratic equations by formula

In **Section A3** we considered linear equations. In this section we consider the different methods involved in solving quadratic equations.

A **quadratic equation** is an equation with the unknown variable to the second power. It has the form

$$ax^2 + bx + c = 0 \qquad [a \neq 0]$$

where x is the unknown variable. In **Example 27** below both equations are quadratics.

G1 **Solving quadratics using factorization**

We use the process of factorization described in the previous section to solve quadratic equations.

We know from the **Introduction Chapter** that if the result of multiplying two numbers is zero then one of the numbers must be zero. This can be stated as:

If A and B are numbers and $A \times B = 0$ then $A = 0$ or $B = 0$.

We use this to solve various equations, for example to solve $x^2 - 2x = 0$.

Since x is common in both terms we can factorize, thus

$$x^2 - 2x = xx - 2x = 0$$
$$x(x - 2) = 0 \quad \text{[Factorizing]}$$
$$x = 0 \ \text{ or } \ x - 2 = 0$$
$$x = 0 \ \text{ or } \ x = 2$$

Example 27

Solve the following equations:

a $x^2 - x - 6 = 0$ **b** $27x^2 - 6x - 5 = 0$

Solution

a What are we trying to find?

The value(s) of x satisfying $x^2 - x - 6 = 0$. **Can we factorize $x^2 - x - 6$?**
$$x^2 - x - 6 = (x + 2)(x - 3)$$

So we have
$$(x + 2)(x - 3) = 0$$

What can we say about $(x + 2)(x - 3) = 0$?
$$(x + 2) = 0 \quad \text{or} \quad (x - 3) = 0$$

Hence we have
$$x + 2 = 0 \qquad \text{or} \qquad x - 3 = 0$$
$$x = -2 \qquad \text{or} \qquad x = 3$$

b Factorizing $27x^2 - 6x - 5$ is more difficult and may be intimidating but it can be factorized into whole numbers:
$$27x^2 - 6x - 5 = (3x + 1)(9x - 5)$$
$$(3x + 1)(9x - 5) = 0$$

which gives
$$3x + 1 = 0 \qquad \text{or} \qquad 9x - 5 = 0$$
$$3x = -1 \qquad \text{or} \qquad 9x = 5$$
$$x = \frac{-1}{3} = -\frac{1}{3} \qquad \text{or} \qquad x = \frac{5}{9}$$

Remember that **not all** quadratics **can** be factorized into simple whole numbers.

G2 **Solving quadratics using formula**

The universal formula for solving any quadratic equation

$$ax^2 + bx + c = 0$$

where x is a unknown variable is given by

1.16
$$x = \frac{-b \pm \sqrt{b^2 - 4ac}}{2a}$$

Subsequently we will show this result in **Chapter 2.** (See question **8** in **Exercise 2(d)**.)

If the factorization is difficult or impossible then we use 1.16 . Generally students prefer to use this formula rather than factorization even when they shouldn't, for example to solve $x^2 - 2x = 0$.

Example 28 *structures*

The bending moment, M, of a beam is given by

$$M = 0.3x^2 + 0.35x - 2.6$$

where x is the distance (in m) along a beam from one end. Find the value of x for which $M = 0$.

Solution

We have

$$0.3x^2 + 0.35x - 2.6 = 0$$

It is **not** easy to factorize this, so we use formula 1.16 to determine x. For the formula, a is the number next to x^2, b is the number next to x and c is the number without any x attached to it. Hence

$$a = 0.3, \quad b = 0.35 \quad \text{and} \quad c = -2.6$$

Substituting this, $a = 0.3$, $b = 0.35$ and $c = -2.6$, into

1.16
$$x = \frac{-b \pm \sqrt{b^2 - 4ac}}{2a}$$

gives

$$x = \frac{-0.35 \pm \sqrt{0.35^2 - (4 \times 0.3 \times (-2.6))}}{2 \times 0.3}$$

$$= \frac{-0.35 \pm \sqrt{3.243}}{0.6}$$

$$= \frac{-0.35 \pm 1.801}{0.6}$$

$$x = \frac{-0.35 + 1.801}{0.6} \quad \text{or} \quad \frac{-0.35 - 1.801}{0.6}$$

$$x = 2.42 \text{ (2 d.p.)} \quad \text{or} \quad x = -3.59 \text{ (2 d.p.)}$$

Since we cannot have a distance of -3.59 m on the beam, the bending moment $M = 0$ is at $x = 2.42$ m.

SUMMARY

For a quadratic equation, $ax^2 + bx + c = 0$, first seek factorization. If this fails then try the formula

1.16 $$x = \frac{-b \pm \sqrt{b^2 - 4ac}}{2a}$$

Exercise 1(g)

Solutions at end of book. Complete solutions available at www.palgrave.com/engineering/singh

1 Solve the following equations:

a $2x - 1 = 0$ **b** $x^2 + 5x + 6 = 0$
c $x^2 - 10x + 21 = 0$ **d** $6x^2 - 13x - 5 = 0$
e $5x^2 + 14x - 3 = 0$ **f** $x^2 - 1 = 0$
g $x^2 - 2x + 1 = 0$ **h** $4x^2 + 8x + 4 = 0$
i $3x^2 + 9x + 6 = 0$ **j** $-2x^2 + 6x - 4 = 0$

2 [*mechanics*] A vehicle with velocity v and constant acceleration a is related by

$$v^2 = u^2 + 2as$$

where s is the distance and u is the initial velocity. If $a = 3.2$ m/s², $s = 187$ m and $v = 35$ m/s, find u.

3 [*structures*] The maximum deflection of a beam occurs at x satisfying

$$2x^2 - 3xL + L^2 = 0$$

where L is the length of the beam and x is the distance along the beam from one end. Find x at maximum deflection.

4 [*mechanics*] The displacement, s, of a particle is given by

$$s = 1.9t + 4.3t^2 \quad (t \geq 0)$$

where t is time. Find the time taken for a displacement of 50 m.

5 A rectangular conservatory has length l and its width is 5 m shorter than its length. Given that the area of the floor is 84 m², find the dimensions of the floor.

6 [*structures*] The bending moment, M, of a beam is given by

$$M = 3000 - 500x - 20x^2$$

where x is the distance along the beam from one end. At what distance is the bending moment $M = 0$.

7 [*structures*] A simply supported beam has the bending moment, M, given by

$$M = \frac{15}{8}x - \frac{29}{4}\left(x - \frac{1}{2}\right)^2$$

where x is the distance along the beam from one support. Find the value(s) of x for $M = 0$.

8 [*mechanics*] The height h (above the ground level) of a ball thrown vertically upwards is given by

$$h = -4.9t^2 + 55t + 12$$

where t is time. Find the time taken to reach the ground.

9 [*mechanics*] A ball is thrown vertically upwards from a height h_0. The height h above ground level is given by

$$h = h_0 + ut - \frac{1}{2}gt^2$$

where t is time and u is initial velocity. Find an expression of t for the ball to reach the ground.

***10** [*aerodynamics*] The following equation occurs in aerodynamics:

$$\frac{-T}{2wL^{3/2}} = \frac{4kL^{5/2} - 3(DL^{1/2} + kL^{5/2})}{2L^3}$$
$$(L \neq 0, w \neq 0)$$

where T is thrust, w is weight, L is lift coefficient, D is drag coefficient and k ($\neq 0$) is a constant. Show that

$$L = \frac{-T \pm \sqrt{T^2 + 12kDw^2}}{2kw}$$

*This starred problem is a little more difficult than are the others in this exercise. Try to do it without looking at the complete solutions online.

SECTION H **Simultaneous equations**

By the end of this section you will be able to:
▶ solve a pair of linear simultaneous equations

H1 **Solving simultaneous linear equations**

Simultaneous means occurring together. **Simultaneous equations** are a set of equations such that the unknown variables $x, y, z, \ldots$ have values that satisfy every equation in the set. In this section we solve two simultaneous linear equations.

Example 29

Solve the simultaneous equations:

$$150x + 140y = 10.4$$

$$150x + 100y = 10$$

Solution

What are we trying to find?

The values of x and y satisfying the above equations:

† $150x + 140y = 10.4$

†† $150x + 100y = 10$

If we have one equation with one unknown then it is easy, we can apply a simple transposition of formulae. **Can we possibly get one equation with one unknown from** † **and** †† **?**

If we subtract these equations, † − †† :

† $150x + 140y = 10.4$

†† $-(150x + 100y = 10)$

we get $0 + 40y = 0.4$

Can we solve $40y = 0.4$?

This is just a **linear equation** with one unknown, y:

$$40y = 0.4$$

$$y = \frac{0.4}{40} = 0.01$$

Example 29 *continued*

Have we completed this problem?

No, we need to find x. **How do we find x?**

Substitute $y = 0.01$ into †† (or †):

$$150x + (100 \times 0.01) = 10$$
$$150x + 1 = 10$$
$$150x = 9$$
$$x = \frac{9}{150} = 0.06$$

Hence

$$x = 0.06 \text{ and } y = 0.01$$

We can check our solution by substituting $x = 0.06$ and $y = 0.01$ into the original equations:

$$150x + 140y = 10.4$$
$$150x + 100y = 10$$

We get

$$(150 \times 0.06) + (140 \times 0.01) = 10.4$$
$$(150 \times 0.06) + (100 \times 0.01) = 10$$

The procedure outlined in the above example is a process of **elimination**. We eliminate one of the unknown variables and then solve for the remaining unknown variable.

Example 30 *mechanics*

The distance, s, travelled by an object is given by

$$s = ut + \frac{1}{2}at^2$$

where a is constant acceleration, u is initial velocity and t is time.

An experiment produces the following results: After times of 3 s and 5 s the distances travelled by the object are 66 m and 160 m respectively.

Determine the values of u and a.

Solution

Substituting $t = 5$ and $s = 160$ into $ut + \frac{1}{2}at^2 = s$ gives

$$5u + \underbrace{\frac{1}{2}5^2a}_{=12.5} = 160$$

Example 30 *continued*

Substituting $t = 3$ and $s = 66$ into $ut + \dfrac{1}{2}at^2 = s$ gives

$$3u + \underbrace{\dfrac{1}{2}3^2 a}_{= 4.5} = 66$$

Rewriting these as

> † $5u + 12.5a = 160$

> †† $3u + 4.5a = 66$

How can we get one equation with one unknown from † **and** †† **?**

In the previous example we had the same number of x's so, when we subtracted, the x's vanished. **Can we remove the u's from** † **and** †† **?**

Yes, we need to make the numbers in front of u (the coefficients of u) to be the same. In † we have $5u$, and in †† we have $3u$.

If we multiply $5u$ by 3 we get $15u$, and if we multiply $3u$ by 5 we also get $15u$. Thus multiplying † through by 3 gives

$$(3 \times 5u) + (3 \times 12.5a) = 3 \times 160$$

> * $15u + 37.5a = 480$

and multiplying †† by 5 yields

> ** $15u + 22.5a = 330$

Why?

Because when we subtract, * − ** , the u's are eliminated:

> * $15u + 37.5a = 480$

> ** $-(15u + 22.5a = 330)$

$$\overline{0 + 15a = 150}$$

$$a = \dfrac{150}{15}$$

$$a = 10$$

Substituting $a = 10$ into $3u + 4.5a = 66$ gives the linear equation

$$3u + (4.5 \times 10) = 66$$

$$3u + 45 = 66$$

$$3u = 66 - 45 = 21$$

$$u = \dfrac{21}{3} = 7$$

We have $a = 10$ m/s^2 and $u = 7$ m/s. You can check your result by plugging these numbers into the original equations.

? **Why do you think we remove the u's in the above example?**

It is straightforward to find a common multiple of 3 and 5, that is 15, rather than find a common multiple of 4.5 and 12.5. We say that the 5 of $5u$ is the **coefficient** of u.

SUMMARY

For two simultaneous linear equations, eliminate one of the unknown variables and the result is a linear equation with the other unknown. Solve for this unknown, substitute this value into one of the original equations and solve for the remaining unknown.

Exercise 1(h)

Solutions at end of book. Complete solutions available at www.palgrave.com/engineering/singh

1 Solve the simultaneous equations:

a $8x + 5y = 13$
$x + 5y = 6$

b $x + y = 2$
$x - y = 0$

c $2x - 3y = 5$
$x - y = 2$

d $2x - 3y = 35$
$x - y = 2$

e $5x - 7y = 2$
$9x - 3y = 6$

f $\pi x - 5y = 2$
$\pi x - y = 1$

2 [mechanics] A lifting machine obeys the law

$$E = aW + b$$

where E is effort force, W is load and a, b are constants. An experiment produces the following results: Effort forces of 45.5 N and 53 N lift loads of 70 N and 120 N respectively. Find the values of the constants a and b.

3 [mechanics] The displacement, s, of a body is given by

$$s = ut + \frac{1}{2}at^2$$

where t is time, u is initial velocity and a is acceleration. If at $t = 2$ s then $s = 33$ m and at $t = 3$ s then $s = 64.5$ m, find the initial velocity (u) and acceleration (a).

4 [electrical principles] By applying Kirchhoff's law in a circuit we obtain

$$25(I_1 - I_2) + 56I_1 = 2.225$$
$$17I_2 - 3(I_1 - I_2) = 1.31$$

where I_1 and I_2 represent currents. Find I_1 and I_2.

5 [materials] The length, ℓ, of an alloy varies with temperature t according to the law

$$\ell = \ell_0(1 + \alpha t)$$

where ℓ_0 is the original length of the alloy and α is the coefficient of linear expansion. An experiment produces the following results:

At $t = 55\,°C$ $\ell = 20.11$ m

At $t = 120\,°C$ $\ell = 20.24$ m

Determine ℓ_0 and α. (The units of α are /°C.)

[Hint: Eliminate ℓ_0 by division]

6 [electrical principles] Resistors R_1 and R_2 are parallel in a circuit and satisfy

$$\frac{1}{R_1} + \frac{1}{R_2} = 1.2 \times 10^{-3}$$
$$\frac{5}{R_1} + \frac{8}{R_2} = 6.6 \times 10^{-3}$$

Determine R_1 and R_2.

[Hint: Let $x = 1/R_1$ and $y = 1/R_2$]

7 [dimensional analysis] The force, F, of a jet is a function of density ρ, area A and velocity v. By assuming

$$F = K\rho^a A^b v^c$$

Exercise **1(h) continued**

Solutions at end of book. Complete solutions available at www.palgrave.com/engineering/singh

and dimensional homogeneity, find a, b and c and express F in terms of ρ A and v. (K, a, b and c are real numbers. *Hint*: Use the fact that the equations must be

dimensionally homogeneous to write three simultaneous equations by referring back to Table 1 on page 70.)

Examination questions **1**

Solutions at end of book. Complete solutions available at www.palgrave.com/engineering/singh

The questions in this exercise are not generally ranked in order of difficulty but grouped together by universities and year of examination.

1 Multiply out and simplify:
(2x − 5)(x − 2)

2 Simplify $\dfrac{a^3\sqrt{b^5}}{a^{-1/3}b^{3/5}}$

3 Find the value of s, given that

$v^2 = u^2 + 2as$, when $v = 3.5$, $u = 2.3$ and $a = -9.8$.

4 Solve the simultaneous equations:
$$2x + 3y = 14$$
$$5x - 2y = 25.5$$

5 Solve the following equation:
$$\frac{2}{x} + \frac{3}{x+1} = \frac{4}{x}$$

6 Rearrange the following equation to form a quadratic, and hence solve
$$\frac{2}{3x} - \frac{3x}{4} = 5$$

Questions 1 to 6 are from University of Portsmouth, UK, 2007

7 Write as a single power of y: $\dfrac{y^3\sqrt{y}}{y^4}$

8 Solve the following equation:
2(x + 7) = 5(x − 1) + 8

9 Solve, to 3 d.p., the quadratic equation
$$3x^2 - 4x - 1 = 0$$

10 Rearrange to make m the subject of the formula and hence find m when $M = 0.5$ and $t = 0.004$:
$$t = \frac{M - 2m}{3m - 5}$$

Questions 7 to 10 are from University of Portsmouth, UK, 2009

11 Solve, by substitution or otherwise, the following simultaneous equations, leaving your answer as fractions:
$$2x = 3 + \frac{4}{y}$$
$$5y - \frac{2}{x} = 4$$

University of Portsmouth, UK, 2006

12 Use the laws of indices to simplify each of the following, giving answers with positive indices only:

i $\sqrt[3]{\dfrac{9^{x-1}2^{-2x-1}}{5^{x+2}32^{x+4}20^{7-x}}}$ ii $\dfrac{3^{(3x+2)}2^{(4x+1)}}{4^{(2x-5)}27^{(x-1)}}$

13 Make x the subject of the following formula:

$$v = \frac{1}{k}\sqrt{\left[\frac{1 + \left(\frac{1}{x} + 1\right)^2}{Lg}\right]}$$

Hence calculate x given the following values:

$k = 1.26$, $L = 48.55$, $g = 32.21$, $v = 0.0483$

Questions 12 and 13 are from Cork Institute of Technology, Ireland, 2007

Miscellaneous exercise **1**

Solutions at end of book. Complete solutions available at www.palgrave.com/engineering/singh

The starred problems in this exercise (and subsequent exercises in this book) are 'tough nuts to crack'. Try to do them without looking at the complete solutions online.

1 [aerodynamics]   The pressure coefficient C is defined by

$$C = \frac{\frac{1}{2}\rho(v^2 - u^2)}{\frac{1}{2}\rho v^2}$$

where u, v are velocities and ρ is density. Simplify this formula.

2 [fluid mechanics]   The pressures P_1 and P_2 at depths d_1 and d_2 respectively are given by

$$P_1 = \rho g(d - d_1)$$
$$P_2 = \rho g(d - d_2)$$

where d is depth of the fluid, ρ is the density of fluid and g is acceleration due to gravity. Show that

$$P_2 - P_1 = -\rho g(d_2 - d_1)$$

3 [fluid mechanics]   The head loss, h, of a fluid in a pipe is given by

$$h = \frac{v_2}{g}(v_2 - v_1) - \frac{v_2^2 - v_1^2}{2g}$$

(g is acceleration due to gravity and v_1, v_2 are velocities of fluid). Show that

$$h = \frac{(v_1 - v_2)^2}{2g}$$

4 Evaluate $x^2 + x + 41$ for $x = 0, 1, 2, 3, 4$ and 5. What do your results have in common?

5 [electronics]   The resonant frequency, f_0, of a tuned circuit is given by

$$f_0 = \frac{1}{2\pi\sqrt{LC}}$$

a Evaluate f_0 for $L = 5 \times 10^{-3}$ henry and $C = 1 \times 10^{-6}$ farad.

b If $f_0 = 1000$ Hz and $L = 1 \times 10^{-3}$ henry then determine C.

6 [electrical principles]   The total resistance, R, of two resistors of resistances R_1 and R_2, in parallel, is given by

$$\frac{1}{R} = \frac{1}{R_1} + \frac{1}{R_2}$$

Show that

$$R = \frac{R_1 R_2}{R_1 + R_2}$$

7 [electrical principles]   Find the total resistance, R, of a circuit consisting of three resistors of resistances $R_1 = 10$ kΩ, $R_2 = 15$ kΩ and $R_3 = 1.2$ kΩ, connected in parallel. (The total resistance, R, is given by

$$\frac{1}{R} = \frac{1}{R_1} + \frac{1}{R_2} + \frac{1}{R_3}.)$$

8 [aerodynamics]   An aircraft's drag, D, at speed v in a medium of density ρ is given by

$$D = \frac{1}{2}\rho v^2 A C_D + \frac{1}{2}\rho v^2 A k C_L^2$$

where C_D, C_L are drag coefficients, A is area and k is a constant. Transpose to make v the subject of the formula.

9 [mechanics]   The excess energy, E, of an engine between the points of maximum speed, v_1, and minimum speed, v_2, is given by

$$E = \frac{1}{2}Iv_1^2 - \frac{1}{2}Iv_2^2$$

where I is the moment of inertia. Make I the subject of the formula.

10 [fluid mechanics]   The flow of liquid from location 1 to location 2 can be described by Bernoulli's equation:

$$\frac{p_1}{\rho g} + \frac{v_1^2}{2g} + h_1 = \frac{p_2}{\rho g} + \frac{v_2^2}{2g} + h_2$$

where v is flow velocity, p is pressure, h is height, g is acceleration due to gravity and ρ is density. Make v_1 the subject of the formula.

11 Solve the following equations:

  a $5x - 1 = 0$ **b** $3x + 2 = 8$

  c $(x - 1)(x + 2) = 0$

  d $(3x - 1)(2x + 3) = 0$

12 [mechanics]   The vertical displacement, y, of a projectile in motion is given by

$$y = ut - \frac{1}{2}gt^2$$

where u is the initial velocity of the projectile. Find t for $y = 10$ m and $u = 14$ m/s. (Take $g = 9.8$ m/s^2.)

13 [thermodynamics]   The exit velocity, u, of a fluid from a nozzle is given by

$$u = \left\{ \frac{2\gamma P_1 V_1}{\gamma - 1} \left[1 - \frac{P_2 V_2}{P_1 V_1} \right] \right\}^{\frac{1}{2}}$$

where P_1, V_1 represent the entrance pressure and specific volume respectively and P_2, V_2 represent the exit pressure and specific volume respectively. γ is the ratio of specific heat capacities. Given that

$$P_1 V_1^\gamma = P_2 V_2^\gamma$$

show that

$$u^2 = \frac{2\gamma P_1 V_1}{\gamma - 1} \left[1 - \left(\frac{P_2}{P_1} \right)^{1 - 1/\gamma} \right]$$

Find u (correct to 1 d.p.) given that $\gamma = 1.39$, $P_1 = 5.2 \times 10^6$ N/m^2, $V_1 = 3.1 \times 10^{-3}$ m^3/kg and $V_2 = 5 \times 10^{-3}$ m^3/kg.

*14 [electrical principles]   In an electrical circuit a resistor of resistance R satisfies

$$R = 1 + \frac{3(9 + R)}{12 + R}$$

Determine R.

15 Solve the following equations:

  a $x^2 - 7x + 10 = 0$

  b $x^2 - 1 = 0$

  c $2x^2 - 3x + 1 = 0$

  d $15x^2 - x - 2 = 0$

  e $-100x^2 + 400x - 300 = 0$

16 [dimensional analysis]   The following coefficients occur in aerodynamics:

  Lift coefficient $C_\mathrm{L} = \dfrac{L}{PA}$

  Drag coefficient $C_\mathrm{D} = \dfrac{D}{PA}$

  Moment coefficient $C_\mathrm{M} = \dfrac{M}{PA\ell}$

where L is the lift (in N), D is the drag (in N), P is the pressure, A is the area, M is the moment (in Nm) and l is the length. Show that C_L, C_D and C_M are dimensionless.

17 [structures]   The deflection, y, of a beam of length L is given by

$$y = \frac{wx^4}{36EI} - \frac{wLx^3}{9EI} + \frac{wL^4}{36EI}$$

Solutions at end of book. Complete solutions available at www.palgrave.com/engineering/singh

Miscellaneous exercise 1 continued

where w is the load per unit length, EI is the flexural rigidity and x is the distance along the beam from one end. Factorize this expression.

18 [structures]   The critical load, P, of a steel column can be obtained from

$$L\sqrt{\frac{P}{EI}} = n\pi$$

where L is the length, EI is flexural rigidity and n is a positive whole number.

i Transpose to make P the subject of the formula.
ii Determine P (correct to 2 d.p.) for $n = 1$, $E = 0.2 \times 10^{12}$ N/m^2, $I = 6.95 \times 10^{-6}$ m^4 and $L = 1.07$ m.

19 Solve the following equations:

a $x^2 + 3x + 1 = 0$
b $x^2 + 4x + 2 = 0$
c $5x^2 + 2x - 1 = 0$
d $1 - 3x - 2x^2 = 0$

*20 [vibrations]   A constant, C, in a vibrational problem is defined as

$$C = \frac{F_0}{k - m\alpha^2} \qquad \left(\alpha \neq \sqrt{k/m}\right)$$

where F_0 is the magnitude of the forcing function, k is the spring stiffness, m is the mass and α is the angular frequency.

If $\omega = \sqrt{\dfrac{k}{m}}$ and $r = \dfrac{\alpha}{\omega}$ then show that

$$C = \frac{F_0/k}{1 - r^2}$$

21 [electrical principles]   Applying Kirchhoff's law to a circuit gives

$$12(I_1 + I_2) + 67I_2 = 5.794$$
$$3I_1 - 5(I_1 - I_2) = 0.306$$

where I_1 and I_2 represent currents. Determine I_1 and I_2.

22 [fluid mechanics]   The acoustic velocity, v, is given by

$$v = (\gamma k \rho^{\gamma - 1})^{1/2}$$

Using $k\rho^\gamma = P$ and $\dfrac{P}{\rho} = RT$, show that

$$v = \sqrt{\gamma RT}$$

(k, R are constants, T is temperature, P is pressure, ρ is density and γ is the specific heat capacity ratio).

23 [fluid mechanics]   The ratio of depths $\dfrac{d_2}{d_1}\left(\dfrac{\text{upstream depth}}{\text{downstream depth}}\right)$ of water flowing through a channel can be derived from

$$d_1^2 - d_2^2 = \frac{2Q}{g}\left(\frac{Q}{d_2} - \frac{Q}{d_1}\right)$$
$$(d_1 \neq 0, \; d_2 \neq 0, \; d_1 \neq d_2)$$

where Q is the flow rate and g is acceleration due to gravity. Given that the Froude number, F, is defined as

$$F = \frac{Q}{\sqrt{gd_1^3}}$$

show that

$$\frac{d_2}{d_1} = \frac{1}{2}\left[-1 \pm \sqrt{1 + 8F^2}\right]$$

24 Solve the following simultaneous equations:

a $2x + 3y = 5$ **b** $3x + 8y = -18$
   $x + 2y = 3$ $5x + 5y = 25$
c $3x + 2y = 7$
   $x + 5y = 6$

*25 [vibrations]   The natural frequency, ω, of a flywheel is given by

$$\omega^2 = \frac{JG}{IL}$$

where I and J are moments of inertia, G is shear modulus of elasticity and L is length.

If mass m is placed at a distance r from the centre, then the natural frequency, α, of the flywheel becomes

$$\alpha^2 = \frac{JG}{(I + 2mr^2)L}$$

From these two formulae, show that

$$I = \frac{2mr^2\alpha^2}{\omega^2 - \alpha^2}$$

*26 [vibrations]   When looking at vibrational problems, we often need to solve the quadratic equation

$$\qquad \star \qquad mx^2 + \zeta x + k = 0$$

where m is the mass, ζ is the damping coefficient and k is the spring constant. Show that

$$x = \frac{\zeta}{2m}\left(-1 \pm \sqrt{1 - \frac{4mk}{\zeta^2}}\right)$$

For critical damping, we need $b^2 - 4ac = 0$ of the quadratic $ax^2 + bx + c = 0$. For what value of $\zeta \,(>0)$ in (*) does critical damping occur?

*27 [thermodynamics]   The specific heat at constant volume c_v and the specific heat at constant pressure c_p are related by

$$c_p - c_v = R$$

where R is the gas constant. If $k = \dfrac{c_p}{c_v}$, show that

a $c_v = \dfrac{R}{k-1}$ **b** $c_p = \dfrac{Rk}{k-1}$

*28 [vibrations]   The four natural frequencies, ω_1, ω_2, ω_3 and ω_4, of a system are given by the roots of the equation

$$\omega^4 - 402\omega^2 + 800 = 0$$

Determine ω_1, ω_2, ω_3 and ω_4.

29 [vibrations]   The natural frequencies, ω_1, ω_2, ω_3 and ω_4, of two pendulums connected by a spring can be obtained from the equation

$$\omega^4 - 2\left(\frac{g}{\ell} + \frac{kx^2}{m\ell^2}\right)\omega^2 + \frac{g^2}{\ell^2} + \frac{2kx^2 g}{m\ell^3} = 0$$

where ℓ, x are lengths, m is mass, k is spring stiffness and g is acceleration due to gravity. Show that

$$\omega_1 = \sqrt{\frac{g}{\ell}}, \quad \omega_2 = -\omega_1,$$

$$\omega_3 = \sqrt{\frac{g}{\ell} + \frac{2kx^2}{m\ell^2}} \text{ and } \omega_4 = -\omega_3$$

30 [aerodynamics]   Minimum drag occurs when the lift coefficient, L, satisfies

$$kL^2 = Z$$

where Z is the zero lift coefficient and k is a constant. The velocity, v, of an aircraft satisfies

$$w = \frac{1}{2}\rho v^2\, LA$$

where w is weight, ρ is density and A is area. Show that

$$v^4 = \left(\frac{2w}{\rho A}\right)^2 \cdot \frac{k}{Z}$$

Solutions at end of book. Complete solutions available at www.palgrave.com/engineering/singh

Miscellaneous exercise **1 continued**

*31 [*structures*]   The deflection, y, of a beam of length, L, is given by

$$EIy = \frac{w(L - x)L}{6} - \frac{w}{6L}(L - x)^3$$

$[EI \neq 0, w \neq 0]$

where w is the load per unit length, EI is the flexural rigidity and x is the distance along the beam. Determine the value(s) of x for which the deflection is zero.

CHAPTER 2
Visualizing Engineering Formulae

SECTION A **Graphs**

By the end of this section you will be able to:
▶ draw graphs of the form $y = mx + c$
▶ define the gradient and intercept of a straight line graph
▶ observe what effect the change in m and c has on the graph of $y = mx + c$

A1 **Linear graphs**

? **What does the term *linear equation* mean?**

It is an equation which gives a straight line when plotted on a graph.

? **What does the word *graph* mean?**

It is a set of points connected together which shows a relationship between two variables.

Here is an example to remind you of the terminology of graphs.

Example 1

Plot the graph of $y = 2x + 1$ for $x = -2, -1, 0, 1$ and 2.

Solution

We first find the y values by substituting the given x values into $y = 2x + 1$ as shown in Table 1.

TABLE 1	x	$y = 2x + 1$
	-2	$[2 \times (-2)] + 1 = -3$
	-1	$[2 \times (-1)] + 1 = -1$
	0	$[2 \times 0] + 1 = 1$
	1	$[2 \times 1] + 1 = 3$
	2	$[2 \times 2] + 1 = 5$

Next we plot the points on a graph. For example the first point $(-2, -3)$ means -2 horizontal (i.e. 2 to the left) and -3 vertical (i.e. 3 down). Similarly the last point $(2, 5)$ means 2 horizontal (i.e. 2 to the right) and 5 vertical (i.e. 5 up).

Example 1 *continued*

The *x* values are plotted horizontally and the *y* values vertically. To obtain a graph we connect all the points in the table as the graph in Fig. 1 shows.

Fig. 1 Portion of the graph of $y = 2x + 1$

Generally the horizontal axis is called the *x* axis and the vertical axis is called the *y* axis. The point (0, 0) is called the *origin*. This is known as the rectangular Cartesian co-ordinate system, $x - y$ axes. We say *x* is the **independent variable** and *y* is the **dependent variable**

The name *Cartesian* is derived from the French philosopher **René Descartes** born in 1596. Descartes attended a Jesuit college and because of his poor health he was allowed to remain in bed until 11 o'clock in the morning, a habit he continued until his death in 1650.

(depends on *x*). Generally the independent variable is plotted along the horizontal axis and the dependent variable along the vertical axis.

? **What does the term *gradient* mean?**

It gives a figure to the steepness of a hill but in terms of graphs it's a value which describes the slope of a line on a graph.

? **How do we measure this slope?**

It is the ratio between the vertical distance between two points to the horizontal distance between them (Fig. 2).

Fig. 2

$$\text{Gradient} = \frac{\text{Vertical}}{\text{Horizontal}} \left(= \frac{\text{rise}}{\text{run}} \right)$$

? **What is the gradient of the straight line $y = 2x + 1$ shown in Fig. 1?**

The first step is to select any two points. Let's choose the two end points $(-2, -3)$ and $(2, 5)$. Then we draw horizontal and vertical lines as shown in Fig. 3.

Fig. 3 Portion of the graph of $y = 2x + 1$

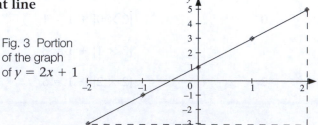

$$\text{Gradient} = \frac{\text{Vertical}}{\text{Horizontal}} = \frac{5 + 3}{2 + 2} = 2$$

So the gradient of the line $y = 2x + 1$ is 2.

? **Is there any relationship between $y = 2x + 1$ and the gradient, 2?**

Consider other examples:

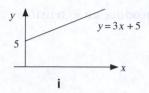

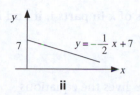

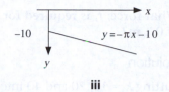

i   ii   iii

i The gradient of $y = 3x + 5$ is 3.

ii The gradient of $y = -\frac{1}{2}x + 7$ is $-\frac{1}{2}$.

iii The gradient of $y = -\pi x - 10$ is $-\pi$.

In fact every relationship of the form $y = mx + c$ is a straight line and the gradient of the general straight line

$$y = mx + c$$

is m, the number next to the x variable. This number is called the **x coefficient.**

? **Consider another aspect of the graph. At what value of y does the graph $y = 2x + 1$ cross the y axis?**

At 1. This is known as the y-intercept. Similarly

i The y-intercept of $y = 3x + 5$ is 5, see graph **i.**

ii The y-intercept of $y = -\frac{1}{2}x + 7$ is 7, see graph **ii.**

iii The y-intercept of $y = -\pi x - 10$ is -10, see graph **iii.**

? **What rule can you gather from these examples?**

The y-intercept is the value of the constant, c, of the graph $y = mx + c$.

So summarizing the above, we have the graph of the straight line

2.1 $$y = mx + c$$

where m is the gradient and c is the y-intercept.

Next we look at how a change in m and c changes the graph. **Example 2** illustrates the effect of changing the gradient on a straight line graph. It also shows that we can deal with other variables besides x and y. In this example the horizontal and vertical axes are x and f respectively.

Example 2 *mechanics*

The force, f, required to produce an extension, x, of a spring is given by

$$f = kx$$

where k is the spring stiffness. On the same axes, sketch the graphs of f against x for $0 \le x \le 0.2$ and

Example 2 *continued*

i $k = 10$ N/m   **ii** $k = 20$ N/m   **iii** $k = 40$ N/m

What force, f, is required for values of k in parts i, ii and iii to produce an extension of 0.1 m?

Solution

Putting $k = 10$, 20 and 40 into $f = kx$ gives the equations

$$f = 10x, f = 20x \text{ and } f = 40x$$

In each case the y-intercept is 0 and the gradients of $f = 10x$, $f = 20x$ and $f = 40x$ are 10, 20 and 40 respectively. We have the graphs shown in Fig. 4.

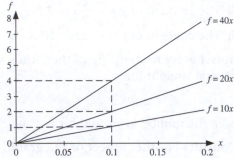

Notice that as the gradient increases, the line gets steeper. For an extension of 0.1 m we read the values from the graph where $x = 0.1$, which gives

Fig. 4
Graphs of
force against
extension

$$k = 10 \text{ N/m}, f = 1 \text{ N}$$

$$k = 20 \text{ N/m}, f = 2 \text{ N}$$

$$k = 40 \text{ N/m}, f = 4 \text{ N}$$

This example also shows that the scales on each axis do not need to be the same. The scale on the x axis is 0.05 whilst that on the f axis is 1.

Example 3 *fluid mechanics*

The streamlines of a fluid flow are given by the equation

$$y - x = C$$

where C is a constant. On the same axes, sketch the graphs for $C = 0$, 1, -1, 2, -2, 3 and -3.

Do you notice any relationship between the graph and the value for C?

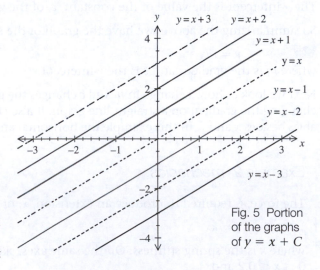

Solution

By rearrangement, we have $y = x + C$, which is a straight line with gradient equal to 1 since the x coefficient is 1. (Remember that $x = 1x$.) For the given C values, we have the straight line graphs shown in Fig. 5. The y-intercept is the value of C.

Fig. 5 Portion of the graphs of $y = x + C$

From **Examples 2** and **3**, observe how a change in m and c alters the straight line graph $y = mx + c$. As m gets larger, the line gets steeper and c is the point where the line cuts the y axis.

 What effect does a negative gradient (negative m value) have on the straight line $y = mx + c$?

A negative gradient means that the line slopes downwards as you move to the right, that is, in the direction of increasing x.

Example 4 illustrates the impact of a negative gradient. In this example, the horizontal and vertical axes are t and a respectively. This is a more difficult example to comprehend so go over it carefully.

Example 4 *mechanics*

The acceleration, a, of a particle is given by

$$a = 1 - \frac{t}{3} \qquad 0 \le t \le 3$$

Sketch the graph of a against t.

Solution

The vertical a-intercept is 1 and the gradient is $-\frac{1}{3}$ because this is the number next to the t variable. Since the gradient is negative the line slopes downwards to the right, that is, as t increases so a decreases.

Substituting $a = 0$ gives $0 = 1 - \frac{t}{3}$. We have $\frac{t}{3} = 1$ which gives $t = 3$.

Hence when $t = 3$, $a = 0$, that is, the graph touches the t axis at 3. Joining the points on the axes gives the straight line graph shown in Fig. 6.

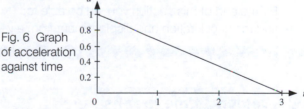

Fig. 6 Graph of acceleration against time

If $m = 0$ in $y = mx + c$, then the graph of $y = c$ is a horizontal line with value c.
If $m = \pm\infty$ (plus or minus infinity), then we have the vertical line $x = c$.

SUMMARY

A straight line graph is defined by

2.1 $y = mx + c$

where m is the gradient and c is the y-intercept. Negative m means the line slopes downwards as you move to the right.

Exercise 2(a)

1 Find the gradient, m, and y-intercept, c, of the following graphs, and sketch them:

   a $y = x + 1$ **b** $y = x - 1$

   c $y = 3x + 5$ **d** $y = \dfrac{1}{2}x - 7$

   e $y - \pi = 0$ **f** $y + \pi - x = 0$

2 Sketch the following graphs:

   a $y = 3$
   b $x = 3$
   c $2x + 2y = 6$
   d $x + 2y = -3$
   e $x - y - 3 = 0$
   f $2x + 7y - 5 = 0$

3 [mechanics] Hooke's law states that the force F (N) exerted on a spring for an extension x (m) is given by

$$F = kx$$

where k is the spring stiffness (constant) (N/m). Plot the graphs of F against x for $k = 10$, 50 and 100 N/m. What effect does the spring stiffness k have? (*Hint*: Consider an extension of 0.5 m.)

4 [fluid mechanics] The streamlines of a fluid flow are given by

$$y = 3x + c$$

where c is a constant. On the same axes, sketch the graphs for $c = -10, -5, -1, 0, 1, 5$ and 10.

5 [electrical principles] The voltage, V, across a capacitor is given by

$$V = \begin{cases} 2t & 0 \le t < 1 \\ 4 - 2t & t \ge 1 \end{cases}$$

Sketch the graph of V against t for $0 \le t \le 2$.

[*Hint*: The graph of V is given in two parts. You sketch $V = 2t$ for $0 \le t < 1$ and then sketch $V = 4 - 2t$ for $t \ge 1$.]

SECTION B Applications of graphs

By the end of this section you will be able to:
▶ find the acceleration and displacement from a v–t graph

B1 Velocity–time graphs

There are many engineering applications of graphs but in this section we only examine the area of mechanics. We state how you can obtain the acceleration and displacement from a velocity–time graph. You may have covered the material on a mechanics module.

Fig. 7 ─────┼──────┼─
                    O   P

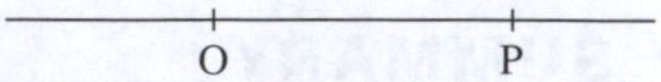

Consider a particle which starts from O and moves in a straight line as shown in Fig. 7.

The distance O–P travelled by the particle is called the **displacement** of the particle. If we plot the velocity–time graph of the particle, then the area under the graph gives the total displacement of the particle and the gradient of the graph gives the acceleration of the particle.

For example, let the graph in Fig. 8 be the velocity–time, v–t, graph of a particle moving along the horizontal line shown in Fig. 7. From this graph we can obtain the acceleration and total displacement. These quantities are given by

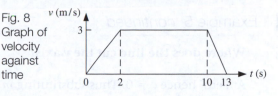

Fig. 8
Graph of velocity against time

2.2   acceleration = gradient of the v–t graph

2.3   total displacement = area under the graph

For the next example we need to know the area of a trapezium. Figure 9 shows a trapezium.

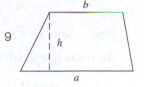

Fig. 9

*   Area of the trapezium = $\frac{1}{2}(a + b)h$

Example 5 *mechanics*

Consider the v–t graph of Fig. 8.
a Determine the acceleration for
  i $0 \le t \le 2$
  ii $2 < t \le 10$
  iii $10 < t \le 13$
b Determine the total displacement.
c Obtain an equation for v in terms of t.

Solution

By examining the graph of Fig. 8, we have

a i For $0 \le t \le 2$, the gradient = $\frac{3}{2}$, hence acceleration = $\frac{3}{2}$ m/s².

  ii For $2 < t \le 10$, there is no slope, therefore the gradient = 0, hence acceleration = 0 m/s².

  iii For $10 < t \le 13$, the line slopes downwards to the right so we have a negative gradient. The gradient = $-\frac{3}{3} = -1$, thus acceleration = −1 m/s².

b Total displacement = Area of trapezium

$$= \frac{1}{2}(13 + 8) \times 3 \quad [\text{By } \boxed{*}]$$

$$= 31.5 \text{ m}$$

c Remember that m of **2.1** is the gradient = acceleration in this example and c is the v-intercept. Since the graph consists of three straight lines, we can write each line in the form v = mt + c where m is the gradient and c is the v-intercept. Let's find the equation of each line separately.

The first line is when t is between 0 and 2 seconds, $0 \le t \le 2$.

What is the value of the gradient, m?

$$m = \frac{3}{2} \quad [\text{From } \boxed{\textbf{a i}}]$$

Example 5 *continued*

Where does the line cut the *v* axis?

At zero, hence $c = 0$. Thus substituting $m = \dfrac{3}{2}$ and $c = 0$ into $v = mt + c$ gives

$$v = \frac{3}{2}t \quad \text{for} \quad 0 \le t \le 2$$

The second line is when t is between 2 and 10 seconds, $2 < t \le 10$.

What is the gradient, *m*, equal to in this case?

$$m = 0 \quad [\text{From } \boxed{\text{a ii}}\,]$$

The v-intercept is at $c = 3$ because if the horizontal line is extended to touch the v axis, then it will cross the axis at 3. Substituting $m = 0$, $c = 3$ into $v = mt + c$ gives

$$v = 3 \quad \text{for} \quad 2 < t \le 10$$

Consider the last line for t between 10 and 13 seconds, $10 < t \le 13$.

What is the gradient, *m*, equal to?

$$m = -1 \quad [\text{From } \boxed{\text{a iii}}\,]$$

What is the *v*-intercept, *c*, equal to?

We cannot read off the v-intercept from the graph so at the moment we only have

$$\boxed{\dagger} \qquad v = -t + c$$

How can we find *c*?

We know that when $t = 13$, $v = 0$ because this is where the line cuts the t axis. Substituting $t = 13$, $v = 0$ into $\boxed{\dagger}$ gives

$$0 = -13 + c$$

Adding 13 to make c the subject, establishes

$$c = 13$$

Hence the equation of the third line is

$$v = -t + 13 \text{ or } v = 13 - t \qquad 10 < t \le 13$$

Assembling these three equations together gives

$$v = \begin{cases} \dfrac{3}{2}t & 0 \le t \le 2 \\ 3 & 2 < t \le 10 \\ 13 - t & 10 < t \le 13 \end{cases}$$

This is the equation for v in terms of t.

SUMMARY

From a v–t graph, the acceleration and total displacement are given by

| 2.2 | acceleration = gradient of the v–t graph |
| 2.3 | total displacement = area under the graph |

Exercise 2b

Solutions at end of book. Complete solutions available at www.palgrave.com/engineering/singh

🚗 All questions in this exercise are in the field of [*mechanics*].

1 The velocity, v, of a vehicle is given by

$$v = \begin{cases} \dfrac{3}{2}t & 0 \le t \le 2 \\ 3 & 2 < t \le 100 \\ 103 - t & 100 < t \le 103 \end{cases}$$

Sketch the v–t graph.

a Determine the acceleration for $2 < t \le 100$.

b Determine the total displacement of the vehicle.

2 Figure 10 shows a velocity–time graph for a vehicle.

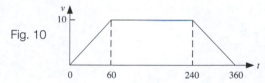

Fig. 10

Write the equation of the graph.

3

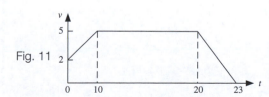

Fig. 11

Figure 11 shows a v–t graph for a train travelling along a track. Determine an equation for v in terms of t.

4 The velocity, v, of an object is given by

$$v = \begin{cases} 2t & 0 \le t \le 2 \\ 4 & 2 < t \le 5 \\ 9 - t & 5 < t \le 9 \end{cases}$$

a Sketch the graph of v against t.

b The acceleration, a, can be obtained by evaluating the gradient of the velocity, v. Sketch the graph of a against t.

5 An object has an initial velocity of 20 m/s at $t = 0$. For the first 10 seconds it has **no** acceleration and then it has a constant acceleration of $a = -5$ m/s².

i Sketch the velocity–time graph for $0 \le t \le 15$.

ii At what time t is the velocity equal to zero?

6 Figure 12 shows a velocity–time graph for an object.

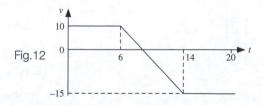

Fig. 12

a State the durations for which the acceleration $a = 0$.

b What is the acceleration of the object for $6 < t < 14$?

Solutions at end of book. Complete solutions available at www.palgrave.com/engineering/singh

Exercise **2b continued**

7

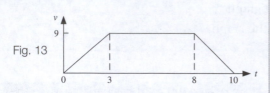

Fig. 13

Figure 13 shows a v–t graph of a particle moving in a straight line. Sketch the corresponding a–t graph.

SECTION C **Quadratic graphs**

By the end of this section you will be able to:
- ► sketch graphs of the form $y = kx^2 + c$
- ► see what effect changes in k and c have on the graph of $y = kx^2 + c$

C1 **Quadratic graph $y = x^2$**

We looked at quadratic equations in **Chapter 1.** In this section we examine quadratic curves. First we plot the simplest quadratic curve, $y = x^2$.

Example 6

Plot the graph of $y = x^2$ for x between -3 and $+3$.

Solution

Considering whole numbers between -3 and $+3$ gives the values shown in Table 2.

TABLE 2

x	$y = x^2$
-3	$(-3)^2 = 9$
-2	$(-2)^2 = 4$
-1	$(-1)^2 = 1$
0	$0^2 = 0$
1	$1^2 = 1$
2	$2^2 = 4$
3	$3^2 = 9$

Example 6 *continued*

Connecting the points $(-3, 9)$, $(-2, 4)$,... and $(3, 9)$ gives the graph of $y = x^2$ (Fig. 14).

Note that the graph of $y = x^2$ is not a straight line.

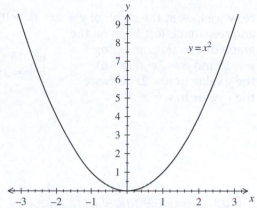

Fig. 14 Graph of $y = x^2$ for x between -3 and 3

What else do you notice about the graph $y = x^2$?

 i It is symmetrical about the y axis; for negative and positive values of x it gives the same y value. For example, $x = 2$ and $x = -2$ gives $y = 4$.

ii It goes through the origin $(0, 0)$.

Property **i** gives the definition of an **even** function, but we will discuss this in **Chapter 3** (**Miscellaneous exercise 3**).

C2 **Graph of $y = x^2 + c$**

Next we look at the graph of $y = x^2 + c$ where c is a constant. This graph is similar to $y = x^2$ but shifted up or down according to the value of c.

Example 7 *fluid mechanics*

The streamlines of fluid flow are given by:

$$y = x^2 + c$$

where c is a constant. Sketch the streamlines for $c = 0, -1, 1, -2, 2, -3$ and 3.

Solution

The graphs of $y = x^2 + c$ for $c = 0, -1, 1, -2, 2, -3$ and 3 are $y = x^2$, $y = x^2 - 1$, $y = x^2 + 1$, $y = x^2 - 2$, $y = x^2 + 2$, $y = x^2 - 3$ and $y = x^2 + 3$ respectively and are shown in Fig. 15.

Fig. 15 Portion of the graphs of $y = x^2 + c$

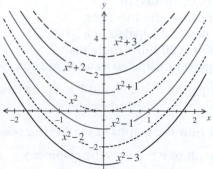

Notice how the graph of $y = x^2 + c$ varies as c changes. The c is where the curve cuts the y axis.

C3 **Graph of $y = kx^2$**

Now we look at the graph of $y = kx^2$ $(k > 0)$ and see what effect k has on the graph of $y = x^2$. Comparing $y = x^2$ and $y = 2x^2$ (Fig. 16), the y value in $y = 2x^2$ is twice the y value in $y = x^2$.

Fig. 16 Graphs of $y = x^2$ and $y = 2x^2$

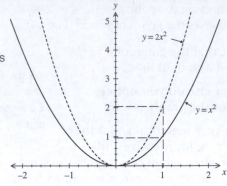

Example 8 *electrical principles*

The power dissipated, P, in a resistor of resistance R, is given by

$$P = i^2R$$

where i is the current flowing through the resistor. Sketch on the same axes the graphs of P against i for $0 \le i \le 1$ and

a $R = 1\,\Omega$ **b** $R = 5\,\Omega$ **c** $R = 10\,\Omega$ **d** $R = 20\,\Omega$

What happens to the power dissipated for an increase in resistance R?

Solution

The formulae of the graphs for the given values of R are obtained by substituting $R = 1$, $R = 5$, $R = 10$ and $R = 20$ into $P = i^2R$:

$$P = i^2, \quad P = 5i^2, \quad P = 10i^2 \text{ and } P = 20i^2$$

These graphs are the quadratic curves, with each curve getting steeper as the coefficient of i^2 gets larger (Fig. 17).

Fig. 17 Graphs of $P = i^2R$ for i between 0 and 1

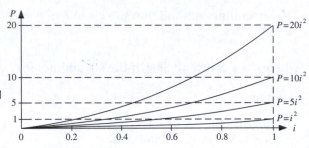

As R increases, the power dissipated in the resistor also increases. The power dissipated in a $20\,\Omega$ resistor is 20 times more than the power dissipated in a $1\,\Omega$ resistor.

Comparing $P = 20i^2$ and $P = i^2$, the P value in $P = 20i^2$ is 20 times the P value for $P = i^2$.

Notice how the graphs of $y = kx^2$ (for positive k) are stretched by a factor of k.

The graph of $y = -x^2$ is the graph of $y = x^2$ inverted (Fig. 18).

Notice that with $y = -x^2$, the curve has a peak, $\cap$, whilst $y = x^2$ gives a trough, $\cup$. We say the graph $y = -x^2$ has a maximum value at the peak, $\cap$, and the graph $y = x^2$ has a

minimum value at the trough, ∪.
If the number next to x^2 is positive
then we have a trough, ∪,
and if the x^2 number is
negative then we have a
peak, ∩. The number next
to x^2 is called the **coefficient**
of x^2.

Fig. 18 Portion
of the graphs of
$y = x^2$ and
$y = -x^2$

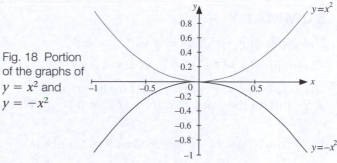

C4 **Graph of $y = kx^2 + c$**

We can sketch the graph of $y = kx^2 + c$ by locating the points where it cuts the axes and
then drawing a smooth curve through these points.

Where does $y = kx^2 + c$ cut the y axis?

At c

Where does $y = kx^2 + c$ cut the x axis?

This is where $y = 0$, that is, $kx^2 + c = 0$

We try to find real values of x which satisfy $kx^2 + c = 0$.

For example, $y = x^2 - 16$ cuts the y axis at -16, and the x axis when

$$x^2 - 16 = 0$$

Solving for x gives

$$x^2 = 16$$

$$x = \pm\sqrt{16}$$

$$x = 4, -4$$

Hence $y = x^2 - 16$ cuts the x axis at -4 and $+4$.
The graph of $y = x^2 - 16$ is essentially the graph
of $y = x^2$ but shifted down by 16 (Fig. 19).

Fig. 19

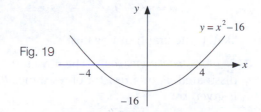

**What does the graph of $y = x^2 + 16$
look like?**

It is basically the graph of $y = x^2$
but shifted up by 16 (Fig. 20).

Fig. 20

Note that this graph, $y = x^2 + 16$,
does **not** cut the x axis.

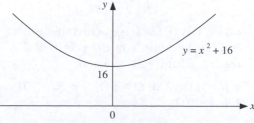

SUMMARY

The graph of $y = x^2$ is a curve with a trough ∪.

The graph of $y = -x^2$ is a curve with a peak ∩.

The graph of $y = x^2 + c$ is similar to the graph of $y = x^2$ but shifted up or down according to the value of c. If $c > 0$ then it is shifted up and if $c < 0$ then it is shifted down.

The graph of $y = kx^2$ is stretched by a factor of k ($k > 0$).

The graph of $y = kx^2 + c$ cuts the y axis at c.

Not all graphs of the form $y = kx^2 + c$ cut the x axis, as seen in Fig. 20.

However if there are real values of x which satisfy $kx^2 + c = 0$, then the graph $y = kx^2 + c$ cuts the x axis at these real values.

Exercise 2(c)

Solutions at end of book. Complete solutions available at www.palgrave.com/engineering/singh

1 [*fluid mechanics*]   The following table shows the velocities of fluid between two parallel plates 100 mm apart:

Distance x (mm)	0	10	20	30	40	50
Velocity v (m/s)	0	15	18	21	25	29

Distance x (mm)	60	70	80	90	100
Velocity v (m/s)	26	20	16	12	0

Plot the graph of v against x.

2 [*fluid mechanics*]   The velocity, v, of a fluid through a pipe is given by

$$v = x^2 - 9 \qquad -3 \le x \le 3$$

Sketch the graph of v against x.

3 [*electrical principles*]   The power dissipated, P, in a resistor of resistance R, is given by

$$P = \frac{V^2}{R}$$

where V is the voltage across the resistor. Plot on the same axes the graphs of P against V for $0 \le V \le 5$ and

a $R = 1\,\Omega$ **b** $R = 5\,\Omega$ **c** $R = 10\,\Omega$
d $R = 20\,\Omega$

4 [*fluid mechanics*]   The velocity, v, of a fluid through a pipe is given by

$$v = \frac{1}{40}(400 - x^2) \qquad -20 \le x \le 20$$

where x is the distance from the central axis. Sketch the graph of v against x.

5 [*mechanics*]   The velocity, v, of a particle is given by

$$v = 48 - 3t^2 \qquad \text{where } 0 \le t \le 4$$

Sketch the graph of v against t. What can you conclude about the acceleration of the particle?

6 [*electrical principles*]   The voltage, V, of a circuit is given by

$$V = \begin{cases} 0.1t^2 & 0 \le t \le 5 \\ 2.5 & t > 5 \end{cases}$$

Sketch the graph of V against t.

7 [*mechanics*]   The position, s, of a particle along a line is given by

$$s = \begin{cases} t^2 + 1 & 0 \le t < 2 \\ 2t + 1 & 2 \le t < 10 \end{cases}$$

Sketch the graph of s against t for $0 \le t < 10$.

SECTION D **Quadratics revisited**

By the end of this section you will be able to:
▶ complete the square on the expression $ax^2 + bx + c$
▶ sketch the graph of $y = Ax^2 + Bx + C$, locating the minimum, maximum and the points where the graph cuts the axes

D1 **Completing the square**

Completing the square is a challenging topic and you may have to spend some time familiarizing yourself with the algebra involved. It might seem like a huge leap from the earlier sections but completing the square can be used to solve quadratic equations and to sketch quadratic curves, and is an important technique in more advanced engineering mathematics.

Example 9

Expand $(x + 1)^2 + 5$.

Solution

Opening up the brackets and simplifying gives

$$(x + 1)^2 + 5 = \underbrace{(x^2 + 2x + 1)}_{\text{by } \boxed{1.13}} + 5$$

$$= x^2 + 2x + 6$$

By looking at **Example 9**, note that $x^2 + 2x + 6$ can be written as $(x + 1)^2 + 5$. In this section, we are given the expanded version $x^2 + 2x + 6$ and we try to rewrite this as $(x + 1)^2 + 5$. In other words we travel from the solution to the question of **Example 9**:

$$x^2 + 2x + 6 = (x + 1)^2 + 5$$

This is an example of an **identity.** The term *identity* means that two mathematical expressions are equal for **all** values of the variables. Other examples of identities are

$$x^2 + 4x + 1 = (x + 2)^2 - 3$$

$$x^2 - 5x + 2 = \left(x - \frac{5}{2}\right)^2 - \frac{17}{4}$$

$$x^2 + 7x + 13 = \left(x + \frac{7}{2}\right)^2 + \frac{3}{4}$$

? **Do you notice any relationship between the Left-Hand Side and the Right-Hand Side of these examples?**

$\boxed{1.13}$ $(a + b)^2 = a^2 + 2ab + b^2$

Look at the x coefficient (the number next to the x) on the left and the number in the bracket on the right.

The number within the brackets is half the x coefficient on the left (Table 3).

TABLE 3	x coefficient	Number within the bracket
	4	2
	-5	$-5/2$
	7	$7/2$

Let's consider $x^2 + 4x + 1 = (x + 2)^2 - 3$.

The 2 in the brackets is half of 4 but where does -3 on the Right-Hand Side materialize from?

By applying $(a + b)^2 = a^2 + 2ab + b^2$ with $a = x$ and $b = 2$, we have

$$(x + 2)^2 = x^2 + 2x(2) + 2^2 = x^2 + 4x + 4$$

So we have

$$x^2 + 4x = (x + 2)^2 - 4$$

We need to subtract 4 because we have produced an extra 4 within the $(x + 2)^2$ term. Hence

$$x^2 + 4x + 1 = (x + 2)^2 - 4 + 1$$
$$= (x + 2)^2 - 3$$

In a similar manner, we have

$$x^2 - 5x + 2 = \left(x - \frac{5}{2}\right)^2 - \left(\frac{5}{2}\right)^2 + 2$$

$$= \left(x - \frac{5}{2}\right)^2 - \underbrace{\frac{17}{4}}_{= -\left(\frac{5}{2}\right)^2 + 2}$$

and

$$x^2 + 7x + 13 = \left(x + \frac{7}{2}\right)^2 - \left(\frac{7}{2}\right)^2 + 13$$

$$= \left(x + \frac{7}{2}\right)^2 + \underbrace{\frac{3}{4}}_{= -\left(\frac{7}{2}\right)^2 + 13}$$

This method is called **completing the square** and results in an algebraic identity. To understand completing the square, you have to ensure that the following result is second nature to you:

1.13 $\qquad (a + b)^2 = a^2 + 2ab + b^2$

The following illustration will help you visualize completing-the-square method:

Consider the quadratic $x^2 + bx = c$.

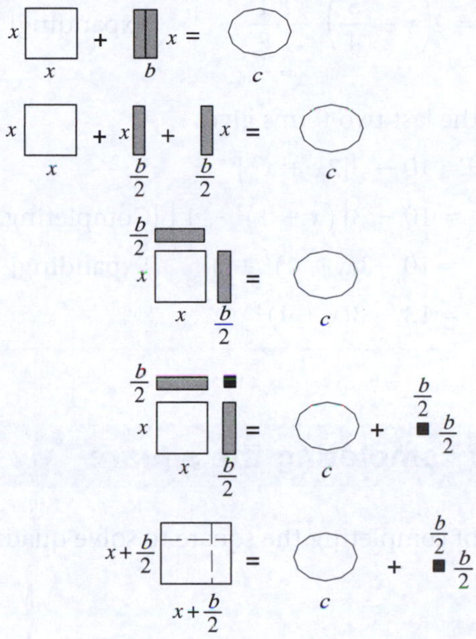

Fig. 21 This diagram shows that $x^2 + bx = c$ is equivalent to

$$\left(x + \frac{b}{2}\right)^2 = c + \left(\frac{b}{2}\right)^2$$

Let's try some more examples.

Example 10

Complete the square on

a $x^2 - 8x + 3$ **b** $2x^2 - 5x + 3$ **c** $10 - 6x - 3x^2$

Solution

a $x^2 - 8x + 3 = (\underbrace{x - 4}_{\text{half of } -8})^2 - 16 + 3$

$\qquad\qquad\quad = (x - 4)^2 - 13$

For **b** and **c** we need to take out the x^2 coefficient.

b We first take a factor of 2 out:

$$2x^2 - 5x + 3 = 2\left[x^2 - \frac{5}{2}x + \frac{3}{2}\right]$$

$$= 2\left[\left(x - \underbrace{\frac{5}{4}}_{\text{half of } 5/2}\right)^2 - \left(\frac{5}{4}\right)^2 + \frac{3}{2}\right]$$

$$= 2\left[\left(x - \frac{5}{4}\right)^2 - \underbrace{\frac{1}{16}}_{= -(^5/_4)^2 + ^3/_2}\right]$$

Example 10 continued

$$= 2\left(x - \frac{5}{4}\right)^2 - \underbrace{\frac{1}{8}}_{= \,^2/_{16}} \qquad \text{[Expanding]}$$

c Factorizing the -3 in the last two terms gives

$$10 - 6x - 3x^2 = 10 - 3[2x + x^2]$$

$$= 10 - 3\big[(x + 1)^2 - 1\big] \quad \text{[Completing the square]}$$

$$= 10 - 3(x + 1)^2 + 3 \qquad \text{[Expanding]}$$

$$= 13 - 3(x + 1)^2$$

D2 Applications of completing the square

We can use the procedure of completing the square to solve quadratic equations and to sketch quadratic curves.

Example 11

Solve the quadratic equation

$$10 - 6x - 3x^2 = 0$$

Solution

We have already completed the square in **Example 10(c)** so we have the identity

$$10 - 6x - 3x^2 = 13 - 3(x + 1)^2$$

Putting the Right-Hand Side to zero and solving for x we have

$$13 - 3(x + 1)^2 = 0$$

$$3(x + 1)^2 = 13 \qquad \text{[Transposing]}$$

$$(x + 1)^2 = \frac{13}{3} \qquad \text{[Dividing by 3]}$$

$$(x + 1) = \pm\sqrt{\frac{13}{3}} \quad \text{[Taking square root]}$$

$$x = -1 \pm \sqrt{\frac{13}{3}}$$

The roots of $10 - 6x - 3x^2 = 0$ are $x = -1 + \sqrt{\dfrac{13}{3}}$ or $x = -1 - \sqrt{\dfrac{13}{3}}$

Generally if the quadratic expression can be factorized easily, then it is **not** worth investing your effort into solving the quadratic equation by completing the square because the solution of the factorized equation is straightforward. However most quadratic equations, $ax^2 + bx + c = 0$, cannot be easily factorized, so we are compelled to use the completing-the-square procedure or the quadratic equation formula

1.16
$$x = \frac{-b \pm \sqrt{b^2 - 4ac}}{2a}$$

as discussed in Chapter 1. You are asked to show result 1.16 in **Exercise 2(d).**

D3 Sketching quadratics

One advantage of the completing-the-square procedure over the quadratic formula is in sketching quadratic curves. Completing the square tells us where the minimum or the maximum point of the quadratic curve occurs. For example, say we want to sketch the quadratic $y = (x - a)^2 + b$. It is basically a x^2 graph but shifted horizontally by a units and vertically by b units (part **i** of Fig. 22). Why it is shifted horizontally by a units and vertically by b units will be explained in **Chapter 3.**

Similarly the graph of $y = b - (x - a)^2$ is shown in part **ii** of Fig. 22.

In both parts **i** and **ii** of Fig. 22 we have assumed a and b to be positive. This transformation of graphs is more thoroughly analyzed in Chapter 3.

Fig. 22 **i**
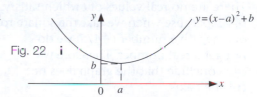

Thus we have

2.4   The quadratic $y = (x - a)^2 + b$
        has a minimum value of b at $x = a$

Fig. 22 **ii**
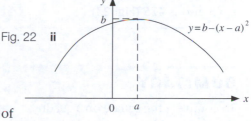

2.5   The quadratic $y = b - (x - a)^2$ has
        a maximum value of b at $x = a$

We use 2.4 or 2.5 to help us sketch the graph of $y = Ax^2 + Bx + C$.

Note that if the x^2 coefficient is positive then we have a minimum, but if the x^2 coefficient is negative then we have a maximum.

We execute the following to sketch the general quadratic curve $y = Ax^2 + Bx + C$:

 i Mark C on the y axis; this is the point where the graph cuts the y axis.

 ii If we can find real values of x which satisfy $Ax^2 + Bx + C = 0$, then the graph $y = Ax^2 + Bx + C$ cuts the x axis at these real values. The graph may not cut the x axis.

 iii Complete the square to ascertain the points of minimum or maximum by using 2.4 or 2.5 . Mark these on the $x-y$ plane.

Combining **i**, **ii** and **iii** gives us a good sketch of $y = Ax^2 + Bx + C$.

Example 12 *aerodynamics*

The height, h, of an aircraft during a manoeuvre is given by

$$h = t^2 - 10t + 500$$

where t is time. Find the minimum height reached by the aircraft and sketch the graph of h against t.

Solution

By completing the square we have

$$h = t^2 - 10t + 500 = (t - 5)^2 - 25 + 500$$

$$= (t - 5)^2 + 475$$

Hence by 2.4 the minimum height reached by the aircraft is 475 m.

The graph, $h = t^2 - 10t + 500$, cuts the h axis at 500 but where does it cut the t axis?

Since the minimum value is 475, the graph cannot cut the t axis. We can show this algebraically. We need to solve

$$(t - 5)^2 + 475 = 0$$

$$(t - 5)^2 = -475$$

There are no real values of t which satisfy
★ because when we take the square root
of a negative number (-475) we do
not get a real number. Therefore we
can conclude that the graph does not
cut the t axis.

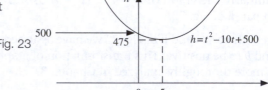
Fig. 23

By 2.4 the minimum value of h occurs at
$t = 5$ and is 475 (Fig. 23).

SUMMARY

To complete the square on a quadratic expression we consider half the x coefficient and then subtract the extra square term produced by completing the square. If the x^2 coefficient is not 1 then we need to factorize this out first. This method gives an algebraic identity.

To sketch $y = Ax^2 + Bx + C$ we first mark C on the y axis then we solve the quadratic equation for x:

$$Ax^2 + Bx + C = 0$$

These x values are where the graph cuts the x axis. Not all quadratics cut the horizontal axis as in Fig. 23. Next we complete the square on $Ax^2 + Bx + C$ and then use 2.4 or 2.5 to locate the maximum or minimum.

2.4 The quadratic $y = (x - a)^2 + b$ has a minimum value of b at $x = a$

1 Complete the square on the following:

a $x^2 - 4x + 3$ **b** $x^2 + 8x + 9$

c $x^2 - 6x + 8$ **d** $x^2 - 10x + 2$

*e $9 + 8x - x^2$ **f** $x^2 + 7x + 1$

g $2x^2 + 7x + 1$

2 Solve the following equations:

a $x^2 - 4x + 3 = 0$ **b** $x^2 + 8x + 9 = 0$

c $x^2 - 6x + 8 = 0$ **d** $x^2 - 10x + 2 = 0$

e $9 + 8x - x^2 = 0$ **f** $x^2 + 7x + 1 = 0$

g $2x^2 + 7x + 1 = 0$

3 [electrical principles]   The voltage, V, of a circuit is defined by:

$$V = t^2 - 5t + 6 \quad (t \geq 0)$$

Sketch the graph of V against t, indicating the minimum value of V.

4 [electrical principles]   The voltage, V, of a circuit is given by

$$V = t^2 - t \quad (t \geq 0)$$

Sketch the graph of V against t.

5 [mechanics]   The height h of a projectile fired from the ground with respect to time t is given by

$$h = 12t - t^2$$

Sketch this curve. Find the maximum height reached by the projectile.

6 [mechanics]   The height h (m) of a projectile fired vertically upwards is given by

$$h = 200t - 4.9t^2$$

where t is time in seconds. Evaluate the maximum height reached by the projectile and sketch the curve.

7 [materials]   Sketch the graph of M against x, where

$$M = \frac{W}{24EI}(Lx - x^2)$$

indicating the maximum or minimum value of M.

(M is the bending moment of the beam, L is the length of the beam, EI (> 0) is the flexural rigidity, x is the distance along the beam and W is the uniform load per unit length.)

*8 By completing the square, show that the solution of the general quadratic equation $ax^2 + bx + c = 0$ is given by

1.16 $$x = \frac{-b \pm \sqrt{b^2 - 4ac}}{2a}$$

(*Hint*: First, divide through by a)

SECTION E **Further graphs**

By the end of this section you will be able to:

▶ understand the term *asymptote*

▶ sketch the non-linear graph $\dfrac{1}{x}$

▶ sketch graphs involving variations of $\dfrac{1}{x}, x^3, x^4, \dots$

▶ plot more complicated graphs using a symbolic manipulator or a graphical calculator

E1 **Non-linear graphs**

Non-linear graphs are graphs which do not have the shape of a straight line, thus the quadratic graphs $y = ax^2 + bx + c$ just considered are examples of non-linear graphs. First we look at the graph of $y = \dfrac{1}{x}$.

> ### Example 13 *electrical principles*

The relationship between frequency, f, and periodic time, T, is defined as

$$f = \frac{1}{T} \qquad (T > 0)$$

Sketch the graph of f against T.

Solution

We have $f = \frac{1}{T}$. Note that as T increases, f decreases. For example if $T = 2$, $f = \frac{1}{2}$ and if $T = 10$, $f = \frac{1}{10}$. Consequently f gets smaller and smaller as T gets larger and larger.

How is this reflected in the graph of $f = \frac{1}{T}$?

It means that f gets closer and closer to the T axis for large values of T without ever touching it (Fig. 24). It never touches the T axis because $f = \frac{1}{T}$ cannot equal zero. We say that the T axis is an **asymptote** to the graph $f = \frac{1}{T}$.

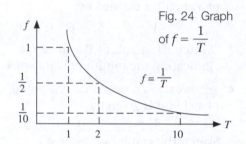

Fig. 24 Graph of $f = \frac{1}{T}$

Similarly for small positive values of T, f is large and the graph gets steeper and steeper.

For example if $T = \frac{1}{2}$, $f = \frac{1}{1/2} = 2$

and if $T = \frac{1}{10}$, $f = 10$. Moreover

the graph gets closer to the f axis.

Hence the f axis is also an asymptote to the graph $f = \frac{1}{T}$. The resulting graph, $f = \frac{1}{T}$, is shown in Fig. 25.

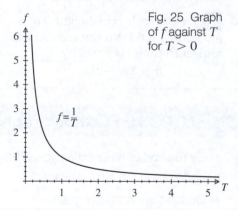

Fig. 25 Graph of f against T for $T > 0$

> ### Example 14

Sketch the graph of $y = \frac{1}{x}$.

Solution

For positive values of x, $x > 0$, we have the graph of Fig. 25 with the T, f axes labelled as x, y respectively. For negative values of x, $x < 0$, we obtain the graph shown in Fig. 26:

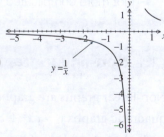

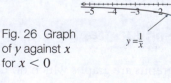

Fig. 26 Graph of y against x for $x < 0$

Example 14 *continued*

Note that the graph of $y = \dfrac{1}{x}$ is not defined at $x = 0$.

Why not?

Because $\dfrac{1}{0}$ does not equal a real number.
Combining the graphs of Figs 25 and 26 gives that shown in Fig. 27.

Fig. 27
Graph of y
against x

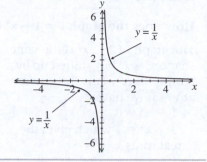

Other non-linear graphs which occur in engineering, such as x^3, $\sqrt{x}$ and x^4, are sketched in Figs 28, 29 and 30 respectively.

Fig. 28

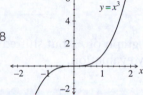

Fig. 29

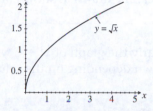

Fig. 30

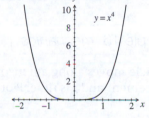

Next we sketch graphs involving variations of these terms.

Example 15 *fluid mechanics*

The velocity, v, of a fluid through a pipe is given by

$$v = 1 - x^4 \quad (-1 \le x \le 1)$$

where x is the distance from the central axis. Sketch this graph. (In fluid mechanics, this graph is called the *velocity profile*.)

Solution

By the above Fig. 30 we know the graph of x^4.

What then does the graph of $-x^4$ look like?

Fig. 31 Graph of $v = -x^4$

The graph of $-x^4$ is the inverted graph of x^4 (Fig. 31).

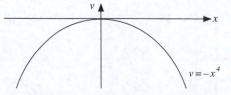

Example 15 *continued*

How does the graph $v = 1 - x^4$ differ from $v = -x^4$?

The graph of $1 - x^4$ is the same shape as $-x^4$ but shifted up by 1 unit. Since v cuts the x axis at $v = 0$ we have

$$1 - x^4 = 0$$
$$x^4 = 1 \text{ which gives the}$$

real roots $x = 1, -1$

Fig. 32 Graph of $v = 1 - x^4$

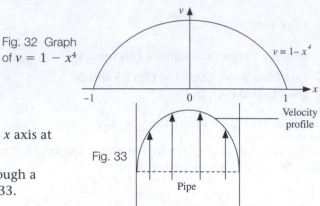

Fig. 33

Hence the graph $v = 1 - x^4$ cuts the x axis at -1 and 1, as shown in Fig. 32.

The velocity profile, $v = 1 - x^4$, through a pipe can be drawn as shown in Fig. 33.

Similarly the graph of $x^3 + C$ will have the same shape as the graph of x^3, but shifted up or down depending on the value of C. Next we plot graphs by evaluation.

E2 Plotting graphs

Example 16 *mechanics*

A particle moves along a horizontal line according to $x = t^3 - 6t^2 + 7t - 1$ where x is the displacement and t is time. Plot the graph of x against t for $0 \le t \le 5$.

Solution

We evaluate x by substituting whole numbers for t between 0 and 5 into $x = t^3 - 6t^2 + 7t - 1$ (Table 4).

t	$x = t^3 - 6t^2 + 7t - 1$
0	$0^3 - (6 \times 0^2) + (7 \times 0) - 1 = -1$
1	$1^3 - (6 \times 1^2) + (7 \times 1) - 1 = 1$
2	$2^3 - (6 \times 2^2) + (7 \times 2) - 1 = -3$
3	$3^3 - (6 \times 3^2) + (7 \times 3) - 1 = -7$
4	$4^3 - (6 \times 4^2) + (7 \times 4) - 1 = -5$
5	$5^3 - (6 \times 5^2) + (7 \times 5) - 1 = 9$

TABLE 4

Example 16 *continued*

Using this table we plot the points and connect them to give us the graph of
$x = t^3 - 6t^2 + 7t - 1$
(Fig. 34).

Fig. 34 Graph of
$x = t^3 - 6t^2 + 7t - 1$
for *t* between 0 and 5

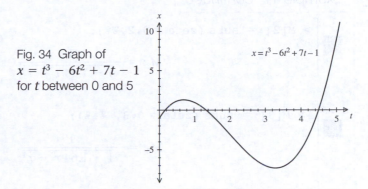

Graphs are a particular fruitful topic for computer algebra packages and graphical calculators. By changing the coefficients of variables we can visualize how the graph changes.

The remaining examples show the plotting of graphs on a symbolic manipulator (MAPLE) and a graphical calculator. If you have not used MAPLE then you might find MAPLE syntax difficult to follow, so try these examples on a package you have available (Derive, Matlab, Mathcad, Mathematica, Maxima, etc.). Matlab is a numerical manipulator and Maxima is at present FREE to download onto your machine.

Example 17 *vibrations*

In the theory of vibrations, the magnification factor *M* is defined as

$$M = \frac{1}{\left[\left(1 - r^2\right)^2 + \left(2\zeta r\right)^2\right]^{\frac{1}{2}}}$$

where *r* is the frequency ratio and ζ (zeta) is the damping ratio.

Plot, on the same axes, the graphs of *M* versus *r* for the following values of ζ: $\zeta = 0.1$, 0.2 and 0.3 and for *r* between 0 and 2.

Solution

The MAPLE output for this example is shown below. The % symbol in MAPLE refers to the previous output and %% refers to the penultimate output.

```
> M:= 1/sqrt((1 − r^2)^2 + (2*zeta*r)^2);
```
$$M := \frac{1}{\sqrt{1 - 2\zeta^2 + r^4 + 4\zeta^2 r^2}}$$

```
> M[1]:= subs (zeta = 0.1,%);
```
$$M_1 := \frac{1}{\sqrt{1 - 1.96\, r^2 + r^4}}$$

Example 17 *continued*

```
> M[2]:= subs (zeta = 0.2,%%);
```
$$M_2: = \frac{1}{\sqrt{1 - 1.84\, r^2 + r^4}}$$

```
> M[3]:= subs (zeta = 0.3, %%%);
```
$$M_3: = \frac{1}{\sqrt{1 - 1.64\, r^2 + r^4}}$$

```
> plot ({M[1], M[2], M[3]}, r = 0..2, color = [black, gold, gray]);
```

Fig. 35

For commands in MATLAB see the book's website (www.palgrave.com/engineering/singh).

Example 18 *reliability engineering*

The failure density function, f, is given by

$$f = \frac{0.5}{(1 + kt)^5} \qquad t \geq 0$$

where t is in years and $k \geq 0$. Plot, on the same axes, the graphs of f against t for $t \geq 0$ with

i $k = 0.01$ **ii** $k = 0.1$ **iii** $k = 1$

What happens as k gets large?

Solution

The MAPLE commands are similar to **Example 17**. The portability of the graphical calculator is a great advantage over a computer algebra system. Try this example on your graphical calculator. Your screen should show something similar to Fig. 36.

If the screen fails to show graphs of Fig. 36 then adjust the range. Also the calculator screen may not show the labels attached to the graph.

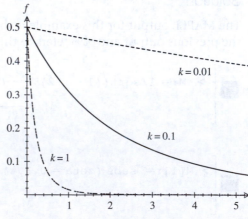

Fig. 36 Graphs of $f = \dfrac{0.5}{(1 + kt)^5}$ for different values of k

Notice that the graph decays more rapidly for large k.

SUMMARY

Use the assistance of a computer algebra system or a graphical calculator to plot complicated graphs.

Graphs can also be plotted by evaluating in a table.

Graphs of $y = x^3$, $y = \sqrt{x}$ and $y = x^4$ are shown in Figs 28–30.

Exercise 2(e)

Solutions at end of book. Complete solutions available at www.palgrave.com/engineering/singh

1 [mechanics]   The displacement, s, of an object is given by $s = t^3 - 2t^2 + 5$.
Plot the graph of s against t for t between 0 and 3.

2 [mechanics]   The velocity, v, of a particle is given by

$$v = t^3 - t \qquad (t \geq 0)$$

Plot the graph of v against t.

3 [electrical principles]   The current, I, through a variable load resistor of resistance R, is given by $I = \dfrac{10}{5 + R}$. Plot the graph of I versus R for $0 \leq R \leq 10$.

4 [electrical principles]   The voltage, V, across a variable load resistor of resistance R, is given by $V = \dfrac{10R}{5 + R}$. Plot the graph of V versus R for $0 \leq R \leq 10$.

5 [fluid mechanics]   The velocity profile, v, of a fluid is given by

$$v = 3x^2 - x^3 \qquad (x \text{ is between 0 and 3})$$

where x represents distance. Plot the graph of v against x, considering $x = 0, 0.5, 1, 1.5, 2.0, 2.5$ and 3.0.

6 [fluid mechanics]   The following equations are the velocity distributions v for a fluid:

a $v = 2x^2 - x^3$

b $v = 4x^2 - 2x^3$

where x is the distance and is between 0 and 2. Sketch the graphs of v against x on the same axes.

7 [fluid mechanics]   The flow rate, Q, of a liquid through a discharge valve is given by

$$Q = kP^{\frac{1}{2}}$$

where P is the pressure across the valve and k is a constant. Sketch the graphs, on the same axes, of Q versus P for the following values of k:

  i $k = 0.5$

 ii $k = 0.75$

iii $k = 1$

iv $k = 2$

 v $k = 10$

8 [fluid mechanics]   The velocity, v, of a fluid flow is given by $v = \dfrac{k}{x}$. Sketch, on the same axes, the graphs of v against x for x between 0 and 10 and

Exercise **2(e) continued**

Solutions at end of book. Complete solutions available at www.palgrave.com/engineering/singh

a $k = 1$

b $k = 10$

c $k = 100$

What happens as k increases?

9 [heat transfer]   The thermal resistance, R, of a material is defined as

$$R = \frac{t}{kA}$$

where t is the thickness, A is the cross-sectional area and k is the thermal conductivity of the material. For $t = 0.01$ m, $A = 5$ m^2, sketch the graph of R against k for $k > 0$. What happens to the thermal resistance, R, as k increases? What value does R tend to as k goes to infinity?

10 [heat transfer]   For $t = 0.05$ m, $k = 5 \times 10^{-3}$ (W/m K), sketch the graph of the thermal resistance R against area A for $A > 0$ and $R = \frac{t}{kA}$. What effect does large area, A, have on the thermal resistance, R?

11 [fluid mechanics]   The velocity, v, of a fluid through a pipe is given by

$$v = 64 - x^4 \quad \left(-64^{\frac{1}{4}} \le x \le 64^{\frac{1}{4}} \right)$$

where x is the distance from the central axis. Sketch the graph of v versus x.

12 [fluid mechanics]   The velocity, v, of a fluid through a pipe is given by

$$v = \sqrt{25 - x^2} \,, \quad -5 \le x \le 5$$

where x is the distance from the central axis. Sketch the profile of v.

For the remaining questions use a symbolic manipulator or a graphical calculator.

13 [electrical principles]   For the following, P represents power and R

represents variable loads in an electrical circuit. Plot the graphs (on different axes) of P versus R for the corresponding R values.

a $P = \dfrac{400R}{(10 + R)^2}$, $0 \le R \le 20$

b $P = \dfrac{500\,R}{[(10 \times 10^3) + R]^2}$, $0 \le R \le 20 \times 10^3$

c $P = \dfrac{2500R}{(1500 + R)^2}$, $0 \le R \le 3000$

By using the various facilities on your graphical calculator or symbolic manipulator, determine the R values for maximum power P in each of the above cases. What do you notice about your results?

14 [electrical principles]   The voltage, V, across a variable load resistor of resistance R_L, is given by

$$V = \frac{ER_L}{R + R_L}$$

where E is the source e.m.f. and R is the source resistance. Plot the graphs (on different axes) of V versus R_L for the corresponding values:

a $E = 60$ volts, $R = 10\ \Omega$ for $0 \le R_L \le 20$

b $E = 15$ volts, $R = 3 \times 10^3\ \Omega$ for $0 \le R_L \le 6 \times 10^3$

c $E = 10$ volts, $R = 15 \times 10^3\ \Omega$ for $0 \le R_L \le 30 \times 10^3$

From each of your graphs, determine the value of V at $R_L = R$. Do you notice any relationship between V and E at $R_L = R$? Show algebraically that if

$$R_L = R \text{ then } V = \frac{E}{2}.$$

SECTION F Binomial expansion

By the end of this section you will be able to:

- expand $(1 + x)^n$ where n is a small positive whole number
- use Pascal's triangle to find coefficients
- expand $(a + b)^n$ where n is a small positive whole number

F1 Expansion of $(1 + x)^n$

The prefix *bi* in binomial refers to two or twice, and *nomial* is from the Greek word *nomen* meaning name or term. Thus the word *binomial* means two terms.

? What does binomial expansion refer to?

It's expanding an expression with two terms; expanding expressions such as $(1 + x)^2$, $(5 + 2x)^7$, $(3 - 7x)^{10}$, etc. Let's examine $(1+x)^n$ where n is a positive whole number.

? How do we expand $(1 + x)^2$?

By using $(a + b)^2 = a^2 + 2ab + b^2$ with $a = 1$ and $b = x$ as in **Chapter 1.**

$$(1 + x)^2 = (1 + x)(1 + x)$$
$$= 1 + 2x + x^2$$

By expanding further we can also obtain the following:

$$(1 + x)^3 = 1 + 3x + 3x^2 + x^3$$

$$(1 + x)^4 = 1 + 4x + 6x^2 + 4x^3 + x^4$$

$$(1 + x)^5 = 1 + 5x + 10x^2 + 10x^3 + 5x^4 + x^5$$

Remember $(1 + x)^5 = (1 + x)(1 + x)(1 + x)(1 + x)(1 + x)$

? Can you see a pattern in the index powers of x in the final expansion above?

The x index increases by 1 each time you move across the terms. First term is $x^0 = 1$, 2nd term involves $x^1 = x$, 3rd term involves x^2 etc, up to x^5:

$$(1 + x)^5 = 1 + 5x + 10x^2 + 10x^3 + 5x^4 + x^5$$

The numbers in front of x, the **coefficients** of x, can be determined by Pascal's triangle, which is shown in Fig. 37.

Each row begins and ends with 1. Every other number in a row is obtained by adding the two numbers in the row above which are immediately to the left and right of it.

Fig. 37

$n = 0$ 1
$n = 1$ 1 1
$n = 2$ 1 2 1
$1+2=3$
$n = 3$ 1 3 3 1
$3+3=6$
$n = 4$ 1 4 6 4 1
$n = 5$ 1 <u>5</u> 10 10 5 1
$n = 6$ 1 6 15 <u>20</u> 15 6 1
$n = 7$ 1 7 21 35 35 21 7 1

For example, the 5 underlined in Fig. 37 is determined by adding 1 and 4. Similarly the 20 is established by summing 10 and 10. Furthermore each of the rows gives the coefficients for the binomial expansion of the corresponding values of n. For example, $n = 5$ gives 1, 5, 10, 10, 5 and 1. Hence

$$(1 + x)^5 = 1 + (5 \times x) + (10 \times x^2) + (10 \times x^3) + (5 \times x^4) + (1 \times x^5)$$
$$= 1 + 5x + 10x^2 + 10x^3 + 5x^4 + x^5$$

This method is painless compared to multiplying out:

$$(1 + x)(1 + x)(1 + x)(1 + x)(1 + x)$$

? **By using Pascal's triangle can you find the expansion of $(1 + x)^7$?**

? **The x index increases by 1 but how do we find the coefficients of x?**

Looking at the row $n = 7$ in Pascal's triangle (Fig. 37) gives 1, 7, 21, 35, 35, 21, 7 and 1:

> **Blaise Pascal** was a French mathematician born in 1623. Pascal developed the first mechanical calculator in 1645. He also worked on hydrostatics, projective geometry and probability theory. Pascal suffered from poor health and died at the young age of 39 in 1662. He is often referred to as the great 'might have been' mathematician.

$$(1 + x)^7 = 1 + 7x + 21x^2 + 35x^3 + 35x^4 + 21x^5 + 7x^6 + x^7$$

Note the symmetrical nature of Pascal's triangle.

F2 Expanding $(a + b)^n$

We now investigate the expansion of the general binomial expression $(a + b)^n$. By using Pascal's triangle for $n = 5$ we have

$$(a + b)^5 = a^5 + 5a^4 b + 10a^3 b^2 + 10a^2 b^3 + 5ab^4 + b^5$$

The coefficients in front of the a, b are obtained from Pascal's triangle in Fig. 37 above for $n = 5$.

The a, b powers are determined by reducing a by an index of 1, giving a^5, a^4, a^3, . . . , a^0; and increasing b by an index of 1, giving b^0, b, b^2, . . . , b^5.

Generally

2.6 $$(a + b)^n = C_n a^n + C_{n-1} a^{n-1} b + C_{n-2} a^{n-2} b^2 + \cdots + C_0 b^n$$

where C_n, C_{n-1}, . . . are coefficients found from Pascal's triangle. Note that the two end coefficients $C_n = C_0 = 1$. Also note that the coefficients next to the ends, C_{n-1} and C_1, are both equal to n. Let's consider some examples.

Example 19

Expand **a** $(2 + x)^3$ **b** $(5 - x)^4$ **c** $(3 + 2x)^5$

Solution

a From Fig. 37 with $n = 3$ we have the coefficients 1, 3, 3 and 1. Using $\boxed{2.6}$ with $a = 2$, $b = x$ and $n = 3$ gives

$$(2 + x)^3 = (1 \times 2^3) + (3 \times 2^2 \times x) + (3 \times 2 \times x^2) + (1 \times x^3)$$

$$= 8 + 12x + 6x^2 + x^3$$

b The coefficients in the expansion of $(5 - x)^4$ are 1, 4, 6, 4 and 1. Putting $a = 5$, $b = -x$ and $n = 4$ into $\boxed{2.6}$ gives

$$(5 - x)^4 = [5 + (-x)]^4$$

$$= (1 \times 5^4) + [4 \times 5^3 \times (-x)] + [6 \times 5^2 \times (-x)^2]$$
$$+ [4 \times 5 \times (-x)^3] + [1 \times (-x)^4]$$

$$= 625 + [500 \times (-x)] + [150 \times x^2] + [20 \times (-x^3)] + [1 \times x^4]$$

$$= 625 - 500x + 150x^2 - 20x^3 + x^4$$

c The coefficients in the expansion of $(3 + 2x)^5$ are 1, 5, 10, 10, 5 and 1. (From row $n = 5$ in Fig. 37.) Putting $a = 3$, $b = 2x$ and $n = 5$ into $\boxed{2.6}$ gives

$$(3 + 2x)^5 = [1 \times 3^5] + [5 \times 3^4 \times 2x] + \left[10 \times 3^3 \times \underbrace{(2x)^2}_{= 4x^2} \right]$$

$$+ \left[10 \times 3^2 \times \underbrace{(2x)^3}_{= 8x^3} \right] + \left[5 \times 3 \times \underbrace{(2x)^4}_{= 16x^4} \right] + [1 \times (2x)^5]$$

$$= 243 + (5 \times 3^4 \times 2)x + (10 \times 3^3 \times 4)x^2$$

$$+ (10 \times 3^2 \times 8)x^3 + (5 \times 3 \times 16)x^4 + 2^5 x^5$$

$$= 243 + 810x + 1080x^2 + 720x^3 + 240x^4 + 32x^5$$

$$= 32x^5 + 240x^4 + 720x^3 + 1080x^2 + 810x + 243$$

We use Pascal's triangle to find the binomial expansion if n is a small positive whole number.

? **Why can't we use Pascal's triangle to expand terms such as $(a + b)^{20}$?**

It's a laborious effort to evaluate the row $n = 20$ of Pascal's triangle. The binomial expansion for other values of n will be covered in **Chapter 7.**

SUMMARY

Binomial expansion is the expansion of two terms to a given index.

For a small positive whole number, n, we have

2.6 $(a + b)^n = C_n \, a^n + C_{n-1} \, a^{n-1} \, b + C_{n-2} \, a^{n-2} \, b^2 + \cdots + C_0 \, b^n$

where C_n, C_{n-1}, . . . represent the coefficients determined by Pascal's triangle. Also $C_n = C_0 = 1$; $C_{n-1} = C_1 = n$.

Exercise 2(f)

Solutions at end of book. Complete solutions available at www.palgrave.com/engineering/singh

1 Expand **a** $(x + 1)^6$ **b** $(a + b)^6$

   c $(1 - x)^6$

2 Expand **a** $(5 + x)^4$ **b** $(2 + 3x)^5$

   c $(4 - 3x)^6$ **d** $(2x - y)^4$ **e** $\left(x + \dfrac{1}{x}\right)^5$

   f $\left(x - \dfrac{1}{x}\right)^4$ **g** $(1 + x^2)^7$

3 📱 [*communications*]   The amplitudes of a n-element array are determined by the coefficients of the binomial expansion of

$$(1 + x)^{n-1}$$

Find the amplitudes for a 10-element array.

4 Determine the expansion of $\left(\dfrac{w}{4} - \dfrac{x}{3}\right)^7$.

Examination questions 2

Solutions at end of book. Complete solutions available at www.palgrave.com/engineering/singh

1 For the function $y = x^2 - 2x - 4$:
   a Use the quadratic formula to find the roots.
   b Plot the graph from $x = -4$ to $x = 4$.

   University of Portsmouth, UK, 2007

2 For the function $y = 3x^2 - 2x - 9$:
   a Use the quadratic formula to find the roots to 3 significant figures.
   b Plot the graph from $x = -3$ to $x = 4$.

   University of Portsmouth, UK, 2009

3 Use the binomial theorem to express

$$(1 + x)^6$$

as a sum of powers of x. Use your result to obtain an approximation to $(11/10)^6$ which is correct to three decimal places.

   University of Manchester, UK, 2009

4 Write down the binomial expansion for

$$(a + b)^5$$

Hence, or otherwise, find the expansion for

$$\left(x^2 + \frac{1}{x}\right)^5$$

expressing your answer as simply as you can.

   University of Manchester, UK, 2008

5 By completing the square solve the equation

$$3x^2 - 6x + 1 = 0$$

Use your calculator to give the two roots to three decimal places.

   University of Manchester, UK, 2008

Examination questions **2** *continued*

Solutions at end of book. Complete solutions available at www.palgrave.com/engineering/singh

6 Consider the function

$$f(x) = 9x^2 - 12x - 5.$$

a Factor $f(x)$, and use this factored form to determine its x-intercepts.

b Write $f(x)$ in standard form by completing the square, and use this form to identify the vertex of the graph.

[The vertex of the graph is its maximum or minimum point.]

Memorial University of Newfoundland, Canada, 2009

7 Given the graphs of the quadratic functions, determine their definition (formula) from their graph.

a

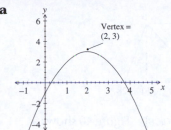

b

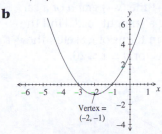

Arizona State University, USA (Review)

Miscellaneous exercise **2**

Solutions at end of book. Complete solutions available at www.palgrave.com/engineering/singh

1 🚗 [*mechanics*]   The displacement, s, of a vehicle is given by $s = t^2$. Sketch the s–t graph.

2 〰 [*fluid mechanics*]   The velocity, v, of a fluid is given by

$$v = 36 - x^2 \qquad -6 \le x \le 6$$

where x is the distance. Sketch the graph of v against x.

3 🔥 [*thermodynamics*]   The pressure, P, is related to the specific volume, V, by

$$P = \frac{C}{V}$$

where C is a constant. Sketch the graphs on the same axes for $V > 0$ and

a $C = 90$ **b** $C = 120$
c $C = 150$ **d** $C = 180$

4 〰 [*fluid mechanics*]   The velocity distribution, v, of a fluid varies between $v = \sqrt{x}$ and $v = \sqrt{2(x - 3)}$. Sketch the velocity profile.

5 〰 [*fluid mechanics*]   The velocity profile, v, of a fluid through a pipe is given by

$$v = 5\sqrt{1 - \frac{x^2}{9}}$$

where $-3 \le x \le 3$

(x is the distance from the central axis). Sketch the velocity profile.

6 🚗 [*mechanics*]

Fig. 38

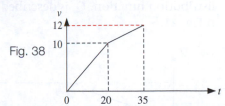

Figure 38 shows a velocity–time (v–t) graph for a vehicle moving along a straight track. Find the equation giving v in terms of t.

7 🚗 [*mechanics*]   Figure 39 shows a v–t graph of a car travelling along a straight road. Sketch the a–t graph.

Miscellaneous exercise **2** *continued*   Solutions at end of book. Complete solutions available at www.palgrave.com/engineering/singh

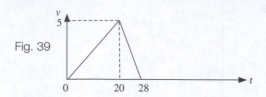

Fig. 39

8 [mechanics]   Figure 40 shows a displacement–time (*s*–*t*) graph of a car moving along a straight road. Find the equation of *s* in terms of *t*. Sketch the *v*–*t* and *a*–*t* graphs for $0 \le t \le 20$.

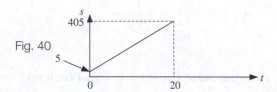

Fig. 40

9 By comparing coefficients of $\sqrt{-2}$ in

$$z + \sqrt{-2} = \left(x + y\sqrt{-2}\right)^3$$

show that $1 = y(3x^2 - 2y^2)$.

10 Sketch the graphs of $y = \dfrac{x^2}{a^2}$, on the same axes, for $a = 1, 5$ and 10. (Take your *x* values to be between -10 and 10.) What effect does *a* have on *y*?

11 [reliability engineering]   The failure distribution function, *Q*, is described as in Fig. 41.

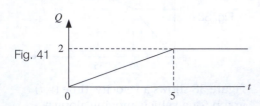

Fig. 41

Write an expression for *Q*.

12 [reliability engineering]   The hazard function, *h*, of a component is described as in Fig. 42.

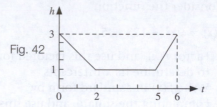

Fig. 42

Write an expression for *h*.

13 [control engineering]   The amplitude ratio, *r*, of the output and the input of a system is given by

$$r = \frac{1}{\sqrt{1 + \dfrac{\omega^2}{9}}}$$

where ω is the angular frequency. Plot the graph of *r* against ω for $0 \le \omega \le 10$.

For the remaining questions use a computer algebra system or a graphical calculator.

14 [vibrations]   The magnification factor, *M*, in the theory of vibrations is defined as

$$M = \begin{cases} \dfrac{1}{1 - r^2} & r < 1 \\[2mm] \dfrac{1}{r^2 - 1} & r > 1 \end{cases}$$

where *r* is the frequency ratio. Plot the graph of *M* versus *r* for $r \ge 0$.

15 [thermodynamics]   The pressure, *P*, and volume, *V*, are related by

$$P = \frac{100}{V^n}$$

where *n* is a constant. Plot the graphs of *P* against *V* on the same axes for $V = 0.1$ to 0.9 and
a $n = 1$ **b** $n = 1.3$ **c** $n = 1.6$.

16 🖳 [*thermodynamics*]   The pressure, P, and volume, V, of a gas are related by

$$PV^n = 1000, \qquad 0.01 \leq V \leq 0.1$$

Taking V to be the horizontal axis, plot the graphs, on the same axes, of P against V for $n = 1.2$, 1.3, 1.4 and 1.5.

17 〰 [*fluid mechanics*]   For hydrogen gas, the viscosity, η, is determined by

$$\eta = (0.7 \times 10^{-6}) \left[\frac{T^{1.5}}{T + 70} \right]$$

where T is the temperature in Kelvin. Plot the graph of η against T for $273 \leq T \leq 373$.

18 〰 [*fluid mechanics*]   The viscosity, η, of a fluid at temperature, θ, is given by

$$\eta = \frac{1.8 \times 10^{-3}}{1 + (34 \times 10^{-3})\theta + (22 \times 10^{-6})\theta^2}$$

Plot the graph of η against θ for $\theta = 0$ to 1000.

19 〰 [*fluid mechanics*]   The viscosity, η, of water at temperatures between $0\,°C$ and $100\,°C$ is given by

$$\eta = \frac{1.8 \times 10^{-3}}{1 + (22 \times 10^{-3})\theta + (22 \times 10^{-6})\theta^2}$$

Plot the graph of η against θ for this range.

CHAPTER 3

Functions in Engineering

SECTION A **Concepts of functions**

By the end of this section you will be able to:
▶ understand the terms *function, domain, range, one–one* and *many–one*
▶ evaluate functions
▶ find simplified expressions for functions

A1 **Definition of a function**

? **In general, what does the word *function* mean in everyday life?**

It's an aim of a person or an object, for example switching on a television is a function. You can think of a function as an assigned purpose or activity.

We are interested in mathematical examples.

? **Can you think of any mathematical examples of functions?**

Adding 5, subtracting 2, dividing by 10, etc.

Adding 5 can be represented by

$$2 \overset{+5}{\rightarrow} 7$$

$$3 \overset{+5}{\rightarrow} 8$$

$$4 \overset{+5}{\rightarrow} 9$$

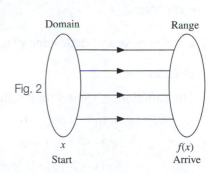

Fig. 1

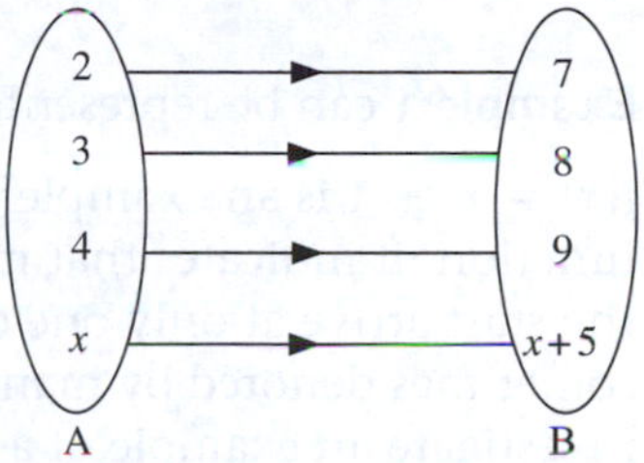

? **Where does adding 5 take the variable x?**

$$x \overset{+5}{\rightarrow} ?$$

$x + 5$. We say that the function is to add 5 for any given number. This can be drawn as shown in Fig. 1.

The function, adding 5 to any given number, is normally denoted by f as follows:

$$f(x) = x + 5$$

? **What does $f(x) = x + 5$ mean?**

Fig. 2

It means that the function, f, takes x to $x + 5$. The aim is to add 5 to any given value. The function $f(x)$ is pronounced 'f of x'. The set A of Fig. 1 where we start from is called the **domain** of the function. The set B of Fig. 1 where we arrive at is called its **range** (Fig. 2). *Set* means a collection of objects.

The domain and range do not just consist of a few selected points shown in Fig. 1 but could be all the real numbers.

? The variable x is called the **independent** variable because $f(x)$ depends on x. **What does the term *function* mean in mathematics?**

A **function** is a rule that gives every x in the domain **only** one value in the range, $f(x)$.

A2 Evaluating functions

Example 1

Let $f(x) = x^2 + 1$, and evaluate the following:

$$f(-3), f(3), f(-2), f(2), f(-1) \text{ and } f(1)$$

Solution

The function squares each value and then adds 1, thus

$$f(-3) = (-3)^2 + 1 = 10$$
$$f(3) = 3^2 + 1 = 10$$
$$f(-2) = (-2)^2 + 1 = 5$$
$$f(2) = 2^2 + 1 = 5$$
$$f(-1) = (-1)^2 + 1 = 2$$
$$f(1) = 1^2 + 1 = 2$$

Example 1 can be represented as shown in Fig. 3.

$f(x) = x^2 + 1$ is an example of a many-to-one function. It indicates that many (i.e. at least 2) from the start arrive at only one destination. This is sometimes denoted by many → one. Let's investigate an example of a one-to-one function, one → one.

Fig. 3

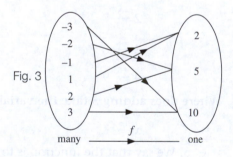

Example 2

Let $g(x) = 3x$ and determine $g(0)$, $g(1)$, $g(-1)$, $g(2)$ and $g(-2)$.

Solution

The function, g, trebles (multiplies by 3) each given value:

$$g(0) = 3 \times 0 = 0$$
$$g(1) = 3 \times 1 = 3$$
$$g(-1) = 3 \times (-1) = -3$$
$$g(2) = 3 \times 2 = 6$$
$$g(-2) = 3 \times (-2) = -6$$

Note that $g(x)$ depends on the given values of x.

This can be drawn as shown in Fig. 4.

One element from the start arrives at only one element (or one input has only one output). Remember that functions can only have one destination (or one output).

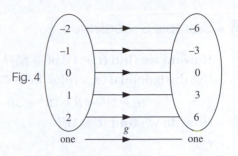

Fig. 4

So functions can be many → one or one → one but **not** one → many or many → many. The essential property is that you must arrive at only one station (or one output) for any given input. An input always has only **one** output if it is a function.

A3 **Engineering applications of functions**

? **Why do engineering and mathematics books use *f*(*x*) to denote functions?**

Because '*f*' is the first letter of the word *function* and *x* normally represents an independent variable. In the next example, we use the Greek letter '*ϕ*', pronounced 'fi', and *t* as the independent variable representing time.

Example 3 *mechanics*

The displacement, $\phi(t)$, of a particle at time t is given by

> * $\qquad \phi(t) = 2t^3 + t^2 - 10t + 10$

a Evaluate $\phi(2)$, $\phi(3)$ and $\phi(5)$.

b Find simplified expressions for
  i $\phi(t^2)$ **ii** $\phi(t + 1)$

Solution

a We have

$$\phi(2) = (2 \times 2^3) + 2^2 - (10 \times 2) + 10 = 10$$

$$\phi(3) = (2 \times 3^3) + 3^2 - (10 \times 3) + 10 = 43$$

$$\phi(5) = (2 \times 5^3) + 5^2 - (10 \times 5) + 10 = 235$$

b i To find $\phi(t^2)$ we replace t with t^2 in $\phi(t) = 2t^3 + t^2 - 10t + 10$:

$$\phi(t^2) = 2 \times (t^2)^3 + (t^2)^2 - 10t^2 + 10$$
$$= \underline{2t^6} + \underline{t^4} - 10t^2 + 10$$
$$\text{by } \boxed{1.7} \text{ by } \boxed{1.7}$$

ii To find $\phi(t + 1)$ we replace t with $t + 1$ in * :

> † $\qquad \phi(t + 1) = 2(t + 1)^3 + (t + 1)^2 - 10(t + 1) + 10$

$\boxed{1.7}$ $(a^m)^n = a^{m \times n}$

Example 3 *continued*

How do we find $(t + 1)^3$ of ⬚† ?
Use the binomial expansion from **Chapter 2**:

$$(t + 1)^3 = t^3 + 3t^2 + 3t + 1$$

How do we find $(t + 1)^2$?

Use:

1.13
$$(a + b)^2 = a^2 + 2ab + b^2$$
$$(t + 1)^2 = t^2 + 2t + 1$$

Substituting these into ⬚† gives

$$\phi(t + 1) = 2(t^3 + 3t^2 + 3t + 1) + (t^2 + 2t + 1) - 10t - 10 + 10$$
$$= 2t^3 + 6t^2 + 6t + 2 + t^2 + 2t + 1 - 10t$$
$$= 2t^3 + \underbrace{6t^2 + t^2}_{= 7t^2} + \underbrace{6t + 2t - 10t}_{= -2t} + 2 + 1$$
$$= 2t^3 + 7t^2 - 2t + 3$$

In **Example 3**, the displacement, $\phi(t)$, depends on the time t, hence t is the independent variable.

SUMMARY

A *function* is a rule which gives only **one** output for every given input. The set where we start from is called the *domain* and the set where we finish is called the *range*. There are one → one and many → one functions. There is **no** such thing as one → many or many → many functions.

Exercise 3(a)

Solutions at end of book. Complete solutions available at www.palgrave.com/engineering/singh

1 Let $f(x) = x^2$ and find $f(1)$, $f(2)$, $f(3)$, $f(-1)$, $f(-2)$ and $f(-3)$. What type, one → one or many → one, of function is $f(x)$?

2 Let $f(t) = \dfrac{9}{5}t + 32$. This function converts temperature from °C to °F. Evaluate $f(0)$, $f(100)$ and $f(24)$.

3 Let $g(x) = x^3$ and find $g(1)$, $g(-1)$, $g(2)$, $g(-2)$, $g(3)$ and $g(-3)$. What do you notice about your results?

4 Let $g(t) = -4.9t^2$, and evaluate $g(1)$, $g(\pi)$, $g\left(\dfrac{1}{\sqrt{4.9}}\right)$.

5 Is $f(h) = \pm\sqrt{20h}$ a function? If it is a function then is it a one → one or a many → one function?

6 [mechanics]   The velocity, $v(t)$, of an object is given by
$$v(t) = 9.8t$$
  a Evaluate $v(0)$, $v(1)$, $v(2)$ and $v(5)$.
  b Find simplified expressions for $v(t^2)$ and $v(t + 1)$.

7 [mechanics]   The displacement, $s(t)$, of a particle is given by
$$s(t) = t^3 - t^2 + 5$$

Solutions at end of book. Complete solutions available at www.palgrave.com/engineering/singh

Exercise **3(a) continued**

a Evaluate $s(0)$, $s(1.5)$ and $s(2.5)$.
b Simplify $s(t + 1)$.

8 [*electrical principles*]   The power dissipated, $P(V)$, by a resistor is given by

$$P(V) = \frac{V^2}{R}$$

Find $P(IR)$.

9 [*electrical principles*]   The voltage, $V(R)$, of a battery is given by

$$V(R) = \frac{ER}{R + r}$$

where r is the internal resistance and E is the electromotive force. Find $V(r)$ and $V(2r)$.

10 [*mechanics*]   The height, $h(t)$, of an object is given by

$$h(t) = 200t - 4.9t^2$$

a Determine t for $h(t) = 0$.
b Simplify $h(t + 2)$.

SECTION B **Inverse functions**

By the end of this section you will be able to:
▶ understand what is meant by 'inverse functions'
▶ find inverse functions

B1 **Inverse of a function**

? What does the word *inverse* mean in everyday life?

Reverse or opposite.

? What is the inverse of:

a turning a light on?
b united?
c multiplying by 3?
d adding 2 and then multiplying by 3?

The inverses are

a turning a light off.
b divided.
c dividing by 3.
d dividing by 3 and then subtracting 2.

Inverse functions simply unlock what you have done. Note the inverse of **d**. You undo the last operation first.

In the next example, the notation {2, 3, 4} means that the set contains only the numbers 2, 3 and 4.

Example 4

Consider $f(x) = 2x$ with domain {2,3,4}. Determine the inverse function.

Solution

f is the function 'multiply by 2'. **Where does f take 2, 3 and 4?**

f takes $2 \to 4$, $3 \to 6$ and $4 \to 8$ (Fig. 5). **What should the inverse function do?**

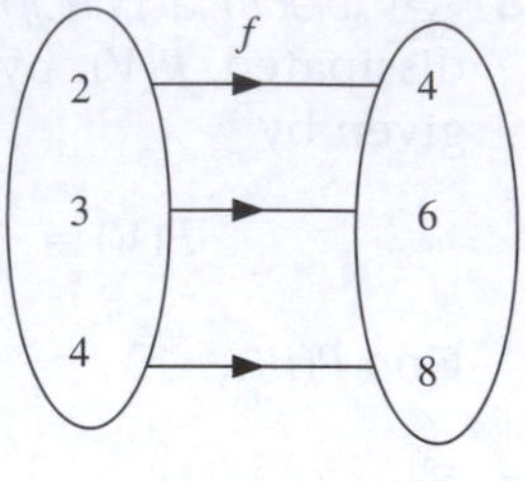

Since the inverse function undoes f, it should take $4 \to 2$ (Fig. 6), $6 \to 3$ and $8 \to 4$.

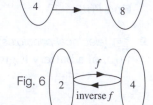

Fig. 5

Fig. 6

The opposite of multiplying by 2 is dividing by 2, thus

$$\text{inverse} \quad f(x) = \frac{x}{2}$$

Instead of writing inverse $f(x)$, we can write this as $f^{-1}(x)$ so $f^{-1}(x) = \frac{x}{2}$. f^{-1} does **not** mean f to the power of -1; it is the conventional notation for inverse functions.

There are various ways of finding the inverse function – one method is to transpose formulae. To find the inverse function, $f^{-1}(x)$, we introduce a new variable, y, and let $f(x) = y$. Then we transpose to make x the subject, $x = \dots$. Let's do an example.

Example 5

If $f(x) = 2x + 3$, find $f^{-1}(x)$.

Solution

Let $f(x) = y$, then we have

$$y = 2x + 3$$

We need to make x the subject, $x = \dots$. How?

We need to remove 3 and 2 from the Right-Hand Side. How?

Subtract 3:

$$y - 3 = 2x$$

Divide by 2:

$$\frac{y - 3}{2} = x$$

We replace the y with x to obtain $f^{-1}(x)$ (a function of x, not y):

$$f^{-1}(x) = \frac{x - 3}{2}$$

Example 6

Consider the domain of f to be real numbers. If $f(x) = \dfrac{x^3}{12} + 5$, find $f^{-1}(x)$ and the real number $f^{-1}(23)$.

Solution

Let $y = f(x) = \dfrac{x^3}{12} + 5$. Therefore

$$\frac{x^3}{12} + 5 = y$$

We need to transpose this formula to obtain $x = \dots$. **How?**

First we remove the $+5$ from the Left-Hand Side, by subtracting 5 from both sides:

$$\frac{x^3}{12} = y - 5$$

Next how do we remove the divide by 12?

Multiply both sides by 12:

$$x^3 = 12(y - 5)$$

By what means do we remove the power 3 on the x term?

Take the cube root, $\sqrt[3]{\ }$, of both sides:

$$x = \sqrt[3]{12(y - 5)}$$

We replace the y with x to obtain the inverse function, $f^{-1}(x)$:

$$f^{-1}(x) = \sqrt[3]{12(x - 5)}$$

The domain of f^{-1} is also the real numbers.

How do we evaluate $f^{-1}(23)$?

We replace x with 23 in $f^{-1}(x) = \sqrt[3]{12(x - 5)}$:

$$f^{-1}(23) = \sqrt[3]{12(23 - 5)}$$
$$= \sqrt[3]{12 \times 18}$$
$$= \sqrt[3]{216}$$
$$f^{-1}(23) = 6$$

This can be represented by Fig. 7.

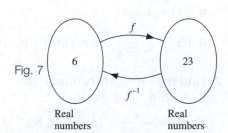

Fig. 7

We need to be careful with statements such as

$$f(x) = \frac{1}{x - 2}$$

? This f is not a function because it is not defined at $x = 2$. **Why not?**

Because the denominator is zero at $x = 2$ and of course we cannot divide by zero. So we write

$$f(x) = \frac{1}{x - 2} \quad \text{(provided } x \neq 2)$$

Now f is a function with domain of all the real numbers apart from 2.

Similarly, for inverse functions we cannot divide by zero. The inverse of f is

$$f^{-1}(x) = 2 + \frac{1}{x} \quad \text{(provided } x \neq 0)$$

More functions of this nature occur in **Exercise 3(b)**.

Not **all** functions have an inverse function. It is only **one** → **one** functions which have an inverse function.

SUMMARY

Inverse functions, denoted by $f^{-1}(x)$, undo $f(x)$ (reverse process). To find $f^{-1}(x)$ of $f(x)$, let $y = f(x)$ and then obtain $x = \ldots$. Finally, replace the y with x to give $f^{-1}(x)$.

Exercise 3(b)

Solutions at end of book. Complete solutions available at www.palgrave.com/engineering/singh

Let the domain of all the functions be real numbers with the restriction shown in brackets.

1 Obtain the inverse function, $f^{-1}(x)$, in each of the following cases:

a $f(x) = 7x - 2$

b $f(x) = \frac{1}{x}$ (provided $x \neq 0$)

c $f(x) = 1 - x$

d $f(x) = \frac{3}{5 - x}$ (provided $x \neq 5$)

2 Find the inverse function of the following:

a $g(t) = \frac{t}{t + 2}$ (provided $t \neq -2$)

b $h(x) = \frac{x - 1}{2 - x}$ (provided $x \neq 2$)

c $f(x) = \frac{2x - 1}{x + 1}$ (provided $x \neq -1$)

3 i If f is defined as $f(x) = \frac{x + 1}{3x + 1}$ then show that

$$f^{-1}(x) = \frac{1 - x}{3x - 1} \quad \text{(provided } 3x - 1 \neq 0)$$

ii If $g(t) = \frac{1 - t}{3t - 1}$ $\left(\text{provided } t \neq \frac{1}{3}\right)$ then find $g^{-1}(t)$.

4 i Let $f(x) = \frac{x^3 + 3}{2 - x^3}$ (provided $2 - x^3 \neq 0$). Show that the inverse function, $f^{-1}(x)$, is given by

$$f^{-1}(x) = \left(\frac{2x - 3}{1 + x}\right)^{\frac{1}{3}} \quad \text{(provided } x \neq -1)$$

ii Let $g(t) = \left(\frac{2t - 3}{1 + t}\right)^{\frac{1}{3}}$ (provided $t \neq -1$), then find $g^{-1}(t)$.

5 What do you notice about your results to $g^{-1}(t)$ in questions **3 ii** and **4 ii**?

SECTION C **Graphs of functions**

By the end of this section you will be able to:
▶ plot graphs of functions
▶ sketch the inverse function, f^{-1}, from the function f
▶ sketch graphs by transformation

C1 **Graphing functions**

Graphs of functions are drawn on the rectangular Cartesian co-ordinate system, x–y axes. Normally the domain is plotted on the horizontal axis and the range on the vertical axis, as shown in Fig. 8.

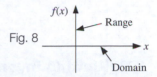
Fig. 8

Remember that x is the independent variable and $f(x)$ is the dependent variable (depends on x).

Much of this section is similar to the work on graphs in **Chapter 2**. However in this section we examine graphs from a function viewpoint.

Example 7

Plot the graph of the function $f(x) = x^2 - x + 5$ with the domain between -3 and 3.

Solution

We can substitute whole numbers for x between -3 and $+3$ into the function $x^2 - x + 5$ (see Table 1).

TABLE 1

x	$f(x) = x^2 - x + 5$
-3	$(-3)^2 - (-3) + 5 = 17$
-2	$(-2)^2 - (-2) + 5 = 11$
-1	$(-1)^2 - (-1) + 5 = 7$
0	$0^2 - 0 + 5 = 5$
1	$1^2 - 1 + 5 = 5$
2	$2^2 - 2 + 5 = 7$
3	$3^2 - 3 + 5 = 11$

Example 7 *continued*

Thus the points on the graph are $(-3, 17)$, $(-2, 11)$, $(-1, 7)$, $(0, 5)$, $(1, 5)$, $(2, 7)$ and $(3, 11)$. Connecting these points gives the graph of the function $y = f(x)$ (see Fig. 9.)

We plot like the normal x–y graph with $y = f(x) = x^2 - x + 5$.

Fig. 9 Graph of $f(x) = x^2 - x + 5$ for x between -3 and $+3$

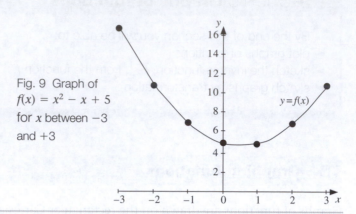

C2 Plotting inverse functions

Example 8

Plot the line $y = x$ and $f(x) = 2x$. Find the inverse function, $f^{-1}(x)$, and plot it on the same axes.

Solution

Since the function f multiplies by 2, the inverse function, f^{-1}, divides by 2. So we have

$$f^{-1}(x) = \frac{x}{2}$$

What sort of graphs are the functions $f(x) = 2x$, $y = x$ and $f^{-1}(x) = \dfrac{x}{2}$?

Fig. 10
Portion of the graphs of $f(x) = 2x$, $y = x$ and $f^{-1}(x) = \dfrac{x}{2}$

They are all straight lines going through the origin with gradient 2, 1 and $\dfrac{1}{2}$ respectively. (See Fig. 10.)

Example 9

Using a graphical calculator, plot the graph of the function $f(x) = x^3$ between $x = -1$ and $x = 1$. On the same diagram plot $y = x$. The inverse of $f(x)$ is

$$f^{-1}(x) = \sqrt[3]{x}$$

Plot $f^{-1}(x)$ on the same axes.

Example 9 *continued*

Solution

On a graphical calculator plot all three functions. It should show the plots given in Fig. 11.

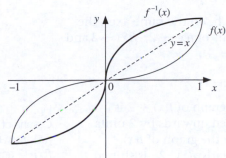

Fig. 11
Graphs of
$y = x$,
$f(x) = x^3$ and
$f^{-1}(x) = \sqrt[3]{x}$
for x between
-1 and 1

? By observing the graphs of Figs 10 and 11, can you see a relationship between f, f^{-1} and $y = x$?

3.1   The inverse function, f^{-1}, is the reflection of f in the line $y = x$

So if f is a function with inverse f^{-1} then f^{-1} is obtained by reflecting the graph of f in the line $y = x$ (Fig. 12).

Fig. 12
Graphs of $f(x)$,
$y = x$ and
$f^{-1}(x)$

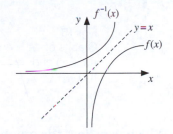

C3 Transformation of graphs

We sketched graphs of some important engineering functions in **Chapter 2** such as $y = x^2$, $y = x^3$, $y = \dfrac{1}{x}$, etc. These can also be written as $f(x) = x^2$, $f(x) = x^3$, $f(x) = \dfrac{1}{x}$, etc. respectively. From these basic functions we can sketch more complicated functions as was discussed in **Chapter 2**.

In the following cases take c to be a positive constant.

From the graph of $f(x)$ we can sketch the graph of $f(x) + c$ and $f(x) - c$.

Figure 13 shows how the graph of $f(x) + c$ is shifted upwards by c from the graph of $f(x)$. Similarly the graph of $f(x) - c$ is shifted downwards by c.

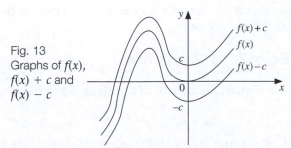

Fig. 13
Graphs of $f(x)$,
$f(x) + c$ and
$f(x) - c$

Example 10 *fluid mechanics*

The streamline, $f(x)$, of a fluid flow is sketched in
Fig. 14.

Sketch, on the same axes, the
graphs of $f(x) + 2$, $f(x) - 2$ and
$f(x) - 4$.

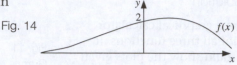

Fig. 14

Solution

The graph of $f(x) + 2$ is
shifted upwards by 2 units
from the graph of $f(x)$.
Similarly $f(x) - 2$ is shifted
down by 2 units and $f(x) - 4$
is shifted down by 4 units
(Fig. 15).

Fig. 15
Graphs of
$f(x)$, $f(x) + 2$,
$f(x) - 2$ and
$f(x) - 4$

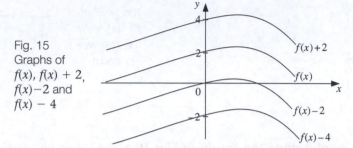

The graph of $f(x + c)$ is the same as the graph of $f(x)$ but translated to the **left** by c units
(Fig. 16).

Fig. 16
Graphs of
$f(x + c)$ and
$f(x)$

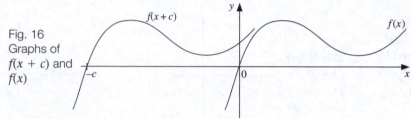

Which way does the graph $f(x - c)$ shift in relation to $f(x)$?

It is translated to the **right** by c units, as shown in Fig. 17.

Fig. 17
Graphs of
$f(x - c)$ and
$f(x)$

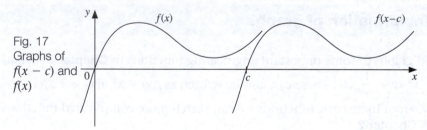

Note that if we have $f(x + c)$ we move to the left by c units and if we have $f(x - c)$ we move
to the right by c units.

Example 11 *mechanics*

The displacement, $x(t)$, of a particle is given by

$$x(t) = (t - 3)^2$$

i Sketch the graph of displacement versus time.
ii At what time(s) is $x(t) = 0$?

Example 11 *continued*

Solution

i How do we sketch the graph of
$x(t) = (t - 3)^2$**?**
It is the same shape as the quadratic
graph t^2 but shifted to the right by
3 units (Fig. 18).

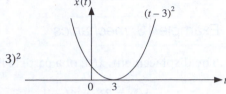

Fig. 18
Graph of
$x(t) = (t - 3)^2$

ii $x(t) = 0$ at $t = 3$.

For $c > 1$, the graph of $cf(x)$ is stretched vertically by a factor of c (Fig. 19).

For $0 < c < 1$, the graph of $cf(x)$ is squashed vertically by a factor of $1/c$ (Fig. 20).

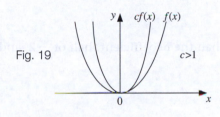

Fig. 19

Fig. 20

Example 12

Sketch, on the same axes, the graphs of $f(x) = x^3$ and $\frac{1}{8}f(x)$.

Solution

From **Chapter 2**, we have the graph of $f(x) = x^3$. **How do we sketch** $\frac{1}{8}x^3$**?**

It is the graph of x^3 but squashed vertically by a factor of 8 (Fig. 21).

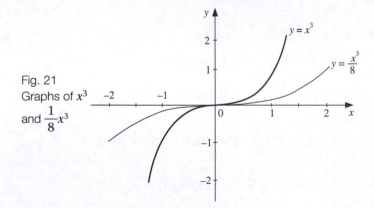

Fig. 21
Graphs of x^3
and $\frac{1}{8}x^3$

The function $\frac{1}{8}x^3$ has a value of one-eighth of $f(x) = x^3$.

The next example is more challenging than the previous examples because of the algebra involved in tackling the problem. You will need to recall your work on completing the square from Chapter 2.

> ### Example 13 *mechanics*
>
> The displacement, $x(t)$, of a particle is given by
>
> $$x(t) = 2t^2 - 4t - 1$$
>
> Sketch the graph of $x(t)$ indicating where it cuts the t axis.
>
> Solution
>
> **How do we sketch the graph of $x(t) = 2t^2 - 4t - 1$?**
>
> We rewrite $2t^2 - 4t - 1$ as
>
> $$2t^2 - 4t - 1 = 2[t^2 - 2t] - 1$$
>
> We first complete the square on $t^2 - 2t$. We take half the t coefficient, half of -2, and then adjust the arithmetic:
>
> $$2[t^2 - 2t] - 1 = 2[(t - 1)^2 - 1] - 1$$
> $$= 2(t - 1)^2 - 2 - 1$$
> $$= 2(t - 1)^2 - 3$$
>
> **How do we sketch the graph of $x(t) = 2(t - 1)^2 - 3$?**
>
> We know it has the same shape as the basic graph of t^2.
>
> **How do we adjust the basic graph to make it into $2(t - 1)^2 - 3$?**
>
> The $2(t - 1)^2$ shifts the graph of t^2 to the right by 1 unit and stretches it vertically by a factor of 2, and the -3 shifts it down by 3 units. Moreover the graph crosses the t axis at $x(t) = 0$. We have
>
> $$2(t - 1)^2 - 3 = 0$$
> $$2(t - 1)^2 = 3 \qquad \text{[Adding 3]}$$
> $$(t - 1)^2 = \frac{3}{2} \qquad \text{[Dividing by 2]}$$
> $$t - 1 = \pm\sqrt{\frac{3}{2}} \qquad \text{[Taking the square root]}$$
> $$t = 1 \pm \sqrt{\frac{3}{2}} \qquad \text{[Adding 1]}$$
> $$t = 1 + \sqrt{\frac{3}{2}} \quad \text{or} \quad t = 1 - \sqrt{\frac{3}{2}}$$

Example 13 *continued*

The graph crosses the t axis at $t = 1 + \sqrt{\dfrac{3}{2}}$ and $t = 1 - \sqrt{\dfrac{3}{2}}$ (Fig. 22).

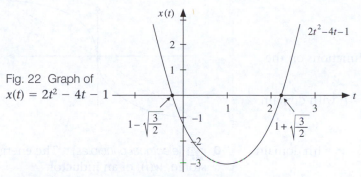

Fig. 22 Graph of
$x(t) = 2t^2 - 4t - 1$

SUMMARY

Graphs are plotted with domain on the horizontal axis and range on the vertical axis.
Inverse functions, f^{-1}, of f can be determined by reflecting f in the line $y = x$.
From the graph of $f(x)$ we can sketch the following (c is a positive constant):

▶ $f(x) + c$ shifts $f(x)$ up by c units
▶ $f(x) - c$ shifts $f(x)$ down by c units
▶ $f(x + c)$ translates $f(x)$ to the left by c units
▶ $f(x - c)$ translates $f(x)$ to the right by c units
▶ for $c > 1$, the graph of $cf(x)$ is stretched vertically by a factor of c
▶ for $0 < c < 1$, the graph of $cf(x)$ is squashed vertically by a factor of $1/c$
▶ for $c > 1$, the graph of $f(cx)$ is squashed horizontally by a factor of c
▶ for $0 < c < 1$, the graph of $f(cx)$ is stretched horizontally by a factor of $1/c$

Examples of the last two points will be given in the next chapter.

Exercise **3(c)**

Solutions at end of book. Complete solutions available at www.palgrave.com/engineering/singh

1 Plot, on different diagrams, each of the following functions:

  a $f(x) = 2x + 3$ with domain between -5 and 5

  b $f(I) = 2.54I$ with domain between -5 and 5

  c $f(t) = -9.8t$ with domain between 0 and 5

  d $g(f) = 2\pi f$ with domain between 0 and 50

  e $f(C) = \dfrac{9}{5}C + 32$ with domain between 0 and 100.

2 Sketch the following functions with domain between -1 and 1:

  i $f(x) = x^2$ **ii** $g(x) = -x^2$

Do you notice any relationship between **i** and **ii**?

Exercise **3(c) continued**

Solutions at end of book. Complete solutions available at www.palgrave.com/engineering/singh

3 Sketch the following functions:

  i $f(x) = x^2 + 2x + 1$

  ii $f(x) = (x + 1)^2$

4 Sketch the following functions on the same axes:

  a $f(r) = \pi r^2$ and $g(r) = r^2$ with domain $0 \le r \le 10$

  b $f(v) = \dfrac{v^2}{20}$ and $g(v) = v^2$ with domain $0 \le v \le 10$

  c $f(T) = \dfrac{10}{T}$ and $g(T) = \dfrac{1}{T}$ with domain $0 < T \le 10$.

Can you see any connection between the graphs of f and g?

5 The following diagrams show the graphs of $f(x)$. Sketch these and their inverse function, $f^{-1}(x)$, on the same axes.

  a

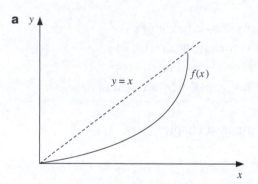

  b

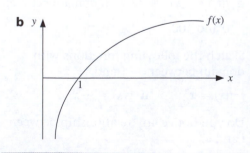

c

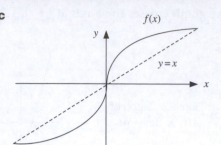

6 [*electrical principles*]   The energy stored, $w(i)$, of an inductor, L, is given by $w(i) = \dfrac{1}{2}Li^2$. Sketch the graph of $w(i)$ versus i for $L = 0.1$ H and $i \ge 0$.

7 [*electronics*]   The current, $i(v)$, through a semiconductor with voltage, v, is given by

$$i(v) = 2v^2 - 3$$

Sketch the graph of $i(v)$ against v, labelling the points where the graph cuts the axes.

8 [*electrical principles*]   The current, $i(R)$, through a resistor, R, is given by

$$i(R) = \dfrac{1}{R + 5}$$

Sketch the graph of $i(R)$ against R.

9 [*mechanics*]   The displacements, $x(t)$, of a particle are given by

  a $x(t) = (t - 1)^3$

  b $x(t) = 5(t - 1)^3$

  c $x(t) = \dfrac{1}{2}(t - 1)^3$

Sketch the displacement–time (t) function in each case on the same axes.

SECTION D **Combinations of functions**

By the end of this section you will be able to:
▶ understand composite functions
▶ simplify functions using combinations such as addition, multiplication, etc.
▶ apply combination rules to engineering examples such as reliability and control engineering

D1 **Composition of functions**

Let f and g be functions, then $f \circ g$ (f composite g) is called the **composite** function and is defined as

3.2 $f \circ g = f[g(x)]$

First apply the function g to x and then apply function f to $g(x)$.

Example 14

If $f(x) = 2x$ (double) and $g(x) = x^2$ (squared), then find

i $f \circ g$ **ii** $g \circ f$ **iii** $f \circ f$

Solution

i $f \circ g = f[g(x)]$

$= f(\underbrace{x^2}_{\text{squared}})$

$f \circ g \underset{\text{double}}{=} 2x^2$

Hence you first apply g which squares x, then you apply f which doubles your result of squaring.

ii $g \circ f = g[f(x)]$

$\underset{\text{double}}{=} g(2x)$

$\underset{\substack{\text{squares} \\ \text{the result}}}{=} (2x)^2$

$g \circ f = 4x^2$

iii $f \circ f = f[f(x)]$

$\underset{\text{double}}{=} f(2x)$

$\underset{\substack{\text{double} \\ \text{the result}}}{=} 4x$

In part **i** we square first and then double. In part **ii** we double first and then square.

?   **What can you conclude from your answers i and ii of Example 14?**

The result **i** $f \circ g$ is different from **ii** $g \circ f$:

$$f \circ g \neq g \circ f$$

In general, we have the property

3.3 $f \circ g \neq g \circ f$ [Not equal]

D2 Other combinations of functions

We have

3.4 $$(f \pm g)(x) = f(x) \pm g(x)$$

3.5 $$\frac{f}{g}(x) = \frac{f(x)}{g(x)} \text{ provided } g(x) \neq 0$$

3.6 $$f \cdot g(x) = f(x) \times g(x)$$

Note that there is a **difference** between $f \circ g$ and $f \cdot g$ as highlighted in   3.3   and   3.6
respectively.

Example 15

Let f and g be functions defined by

$$f(x) = x - 1$$
$$g(x) = 3x + 5$$

Determine

i $f + g$ **ii** $f - g$ **iii** $g - 5$ **iv** $f \cdot g$ **v** $\dfrac{f}{g}$ **vi** $f \circ g$

Solution

i Using 3.4 :

$$f + g = (f + g)(x)$$
$$= f(x) + g(x)$$
$$= (x - 1) + (3x + 5)$$
$$= (x + 3x) + (5 - 1)$$
$$f + g = 4x + 4$$

ii Using 3.4 :

$$f - g = (f - g)(x)$$
$$= f(x) - g(x)$$
$$= (x - 1) - (3x + 5)$$
$$= (x - 3x) + (-1 - 5)$$
$$f - g = -2x - 6$$

iii Using 3.4 with a constant function, 5, taken away from g:

$$g - 5 = g(x) - 5$$
$$= (3x + 5) - 5$$
$$= 3x$$

Example 15 *continued*

iv $f \cdot g$ means

$$f \cdot g(x) = f(x) \times g(x) \qquad [\text{By } \boxed{3.6}]$$
$$= (x - 1)(3x + 5)$$
$$\underline{=} (x \times 3x) + (x \times 5) + (-1 \times 3x) + (-1 \times 5)$$
$$\underset{\text{FOIL}}{}$$
$$= 3x^2 + \underbrace{5x - 3x}_{= 2x} - 5$$
$$f \cdot g(x) = 3x^2 + 2x - 5$$

v Using $\boxed{3.5}$:

$$\frac{f}{g} = \frac{f(x)}{g(x)}$$

$$= \frac{x - 1}{3x + 5} \qquad [\text{Provided } 3x + 5 \neq 0]$$

vi Applying $\boxed{3.3}$ we have:

$$f \circ g = f[g(x)]$$
$$= f(3x + 5) = 3x + 5 - 1 = 3x + 4$$

Note that $(f \circ g)(x) = 3x + 4$ but $f \cdot g(x) = 3x^2 + 2x - 5$ as shown above in part **iv**.

D3 Engineering applications

We now apply the above combinations of functions to reliability and control engineering. The reliability example is straightforward but the control example is more challenging. You need to use your algebraic skills to deal with the control problem.

Example 16 *reliability engineering*

The failure density function, $f(t)$, for a component is given by

$$f(t) = \frac{1}{8} \text{ where } 0 < t < 8 \text{ years}$$

Find $F(t)$, $R(t)$ and $h(t)$ where these are defined as

$$F(t) = tf(t) \qquad \text{(Failure distribution function)}$$

$$R(t) = 1 - F(t) \qquad \text{(Reliability function)}$$

$$h(t) = \frac{f(t)}{R(t)} \qquad \text{(Hazard rate function)}$$

and $0 < t < 8$ years.

Example 16 *continued*

Solution

We have

$$F(t) = t\left(\frac{1}{8}\right) = \frac{t}{8}$$

$$R(t) = 1 - F(t) = 1 - \frac{t}{8}$$

$$h(t) = \frac{f(t)}{R(t)} = \frac{1/8}{1 - \dfrac{t}{8}}$$

$$= \frac{1}{8 - t} \qquad \begin{bmatrix} \text{Multiplying numerator} \\ \text{and denominator by 8} \end{bmatrix}$$

Example 17 *control engineering*

The transfer function, $G(s)$, of a system is given by

$$G(s) = \frac{\dfrac{1}{(2s + 1)}}{\dfrac{7}{(s^2 + 5s + 2)}}$$

Simplify $\dfrac{G(s)}{1 + G(s)N(s)}$ where $N(s) = 10s$.

Solution

We need to first simplify $G(s)$. How?

$$G(s) = \frac{1}{(2s + 1)} \div \frac{7}{(s^2 + 5s + 2)}$$

$$= \frac{1}{(2s + 1)} \times \frac{(s^2 + 5s + 2)}{7} \qquad \begin{bmatrix} \text{Turning the second} \\ \text{fraction upside down} \\ \text{and multiplying} \end{bmatrix}$$

$$= \frac{1 \times (s^2 + 5s + 2)}{7(2s + 1)}$$

$$G(s) = \frac{s^2 + 5s + 2}{14s + 7}$$

Putting $G(s) = \dfrac{s^2 + 5s + 2}{14s + 7}$ and $N(s) = 10s$ into

$$\frac{G(s)}{1 + N(s)G(s)} = \frac{\left(\dfrac{s^2 + 5s + 2}{14s + 7}\right)}{1 + 10s\cdot\left(\dfrac{s^2 + 5s + 2}{14s + 7}\right)}$$

Example 17 *continued*

How do you simplify * ?

Multiply numerator and denominator by $14s + 7$:

$$\frac{G(s)}{1 + G(s)N(s)} = \frac{s^2 + 5s + 2}{(14s + 7) + 10s(s^2 + 5s + 2)}$$

$$= \frac{s^2 + 5s + 2}{14s + 7 + \underbrace{10s \cdot s^2 + (10s \times 5s) + (2 \times 10s)}_{\text{expanding } 10s(s^2 + 5s + 2)}}$$

$$= \frac{s^2 + 5s + 2}{14s + 7 + 10s^3 + 50s^2 + 20s}$$

$$\frac{G(s)}{1 + G(s)N(s)} = \frac{s^2 + 5s + 2}{10s^3 + 50s^2 + \underbrace{34s}_{= 20s + 14s} + 7}$$

Notice how we only need to use the algebra of Chapter 1 to simplify $\dfrac{G(s)}{1 + G(s)N(s)}$ of **Example 17**.

SUMMARY

The composition of functions $f(x)$ and $g(x)$ denoted by $f \circ g$ is defined as

3.2 $\qquad f \circ g = f[g(x)]$

First apply g, then f. Also in general

3.3 $\qquad f \circ g \neq g \circ f$

Other combinations are

3.4 $\qquad (f \pm g)(x) = f(x) \pm g(x)$

3.5 $\qquad \dfrac{f}{g}(x) = \dfrac{f(x)}{g(x)} \qquad$ [Provided $g(x) \neq 0$]

3.6 $\qquad f \cdot g(x) = f(x) \times g(x)$

Exercise **3(d)**

Solutions at end of book. Complete solutions available at www.palgrave.com/engineering/singh

Let the domain of all the functions be real numbers with any restriction in brackets.

1 Let f and g be functions defined by
$$f(x) = x + 1, \quad g(x) = 2x + 3$$
Find

i $f \circ g$ $\qquad$ **ii** $f \cdot g$

iii $g - 1$ $\qquad$ **iv** $f + g$

v $f - g$ $\qquad$ **vi** $\dfrac{f}{g} \ (g(x) \neq 0)$

2 Let f be the function $f(x) = 2x + 3$. Find

i $f \circ f$ $\qquad$ **ii** $f \circ f \circ f$ $\qquad$ **iii** $f\{f[f(-3)]\}$

Exercise **3(d) continued**

Solutions at end of book. Complete solutions available at www.palgrave.com/engineering/singh

3 Let $f(x) = ax^2 + bx + c$ where a, b and c are non-zero constants.

Find and simplify

  i $f(0)$

 ii $f(x + 1)$

iii $f(x + 1) - f(x)$

4 Let f and g be functions defined by

$$f(x) = x^2, \quad g(x) = x + 1$$

Find

  i $g \circ g$

 ii $f \circ g$

iii $g \circ f$

5 If $f(x) = x^2 - 1$ and $g(x) = x$ (identity function), find

  i $f \circ g$

 ii $g \circ f$

iii What do you notice about the results of **i** and **ii**?

Determine **iv** $g \circ g$ **v** $f \circ f$

6 The function f is defined by

$$f(x) = \frac{6}{3 - x} \quad (x \neq 3)$$

Find

  i f^{-1}

 ii $f \circ f^{-1}$

iii $f^{-1} \circ f$

7 If $f(x) = \dfrac{x^3 + 3}{2 - x^3}$ $(2 - x^3 \neq 0)$ then

$$f^{-1}(x) = \left(\frac{2x - 3}{1 + x} \right)^{\frac{1}{3}} \quad (1 + x \neq 0).$$

Find

  i $f \circ f^{-1}$

 ii $f^{-1} \circ f$

8 What do your results to questions **6 ii**, **iii** and **7 i**, **ii** have in common?

9 Let $f(x) = x - 1$ and $g(x) = \sqrt{x}$ $(x \geq 0)$ be functions. Express the following in terms of f and g:

  i $\sqrt{x} - 1$ **ii** $\sqrt{x - 1}$ **iii** $\sqrt{x - 1} + 7$

iv $\sqrt{x^2 - 2x + 1}$

10 [reliability engineering]   The failure density function, $f(t)$, for a system is given by

$$f(t) = \frac{1}{5} \text{ where } 0 < t < 5 \text{ years}$$

Find $F(t)$, $R(t)$ and $h(t)$ where these functions are defined as in **Example 16** on page 155.

11 [control engineering]   A system transfer function, $T(s)$, is defined by

$$T(s) = \frac{G(s)}{1 + N(s)G(s)}$$

Simplify $T(s)$ for the following (assume the denominator $\neq 0$):

a $G(s) = \dfrac{k}{s(s + 1)}$ where k is a constant and $N(s) = 0.01$

b $G(s) = \dfrac{1}{s + k_1}$, $N(s) = k_2$ where k_1 and k_2 are constants

c $G(s) = \dfrac{s + 1}{s^2 + 3s + 2}$ and $N(s) = 0.3$.

12 [control engineering]   A transfer function, $G(s)$, of a system is given by

$$G(s) = \frac{10}{(s - 2)} \times \frac{s}{(s^2 + 2s - 5)}$$

Simplify $\dfrac{G(s)}{1 + G(s)H(s)}$ where

$$H(s) = s + 3.$$

SECTION E **Limits of functions**

By the end of this section you will be able to:
▶ evaluate limits of functions
▶ apply limits of functions to engineering problems

Limits of functions is a complex and difficult topic. This section might seem like a colossal jump from the previous sections but functions and limits are powerful concepts in engineering mathematics, particularly calculus.

E1 **Finding limits of functions**

The phrase 'x tends to zero' is denoted in mathematical terms by
$$'x \to 0'$$
which means x is as close as possible to zero.

Example 18

By using your calculator, investigate to 5 d.p.
$$f(x) = 2^x \text{ as } x \to 0$$

Solution

How can we find an approximation to $f(x) = 2^x$ as x goes to zero?

We can substitute certain values of x.

What values of x do we substitute into f?

We can use x values close to zero such as $x = 1$, 0.1, 1×10^{-3}, 1×10^{-6} and 1×10^{-9}. Let's establish a table of values for $f(x) = 2^x$ (see Table 2).

TABLE 2	x	$f(x) = 2^x$
	1	$2^1 = 2$
	0.1	$2^{0.1} = 1.071773$
	1×10^{-3}	$2^{1 \times 10^{-3}} = 1.000693$
	1×10^{-6}	$2^{1 \times 10^{-6}} = 1.000000$
	1×10^{-9}	$2^{1 \times 10^{-9}} = 1.000000$

From the last two rows, notice that as x gets closer and closer to zero, $f(x)$ gets closer and closer to 1.00000. Trying even smaller values of x, you will find $f(x)$ is 1.00000 rounded to 5 d.p.

We say the function $f(x)$ has limit 1.00000 as $x \to 0$ and write this as
$$\lim_{x \to 0} f(x) = 1.00000 \quad \text{(5 d.p.)}$$

Similarly we can write 'x tends to infinity' as '$x \to \infty$', which means x is made as large as possible. Let's examine a function for large values of x, ($x \to \infty$).

Example 19

By using your calculator, investigate to 5 d.p.

$$g(x) = \left(1 + \frac{1}{x}\right)^x \quad \text{as } x \to \infty$$

Solution

How do we find an approximation for $g(x)$ as x goes to infinity?

We can substitute certain values of x.

What values of x do we substitute into g?

We can use $x = 1, 10, 1000, 1 \times 10^6$ and 1×10^{10}. To input 1×10^6 on a calculator we PRESS

| 1 | EXP | 6 | or | 1 | EE | 6 |

Let's establish a table of values for $g(x)$ – see Table 3.

TABLE 3	x	$g(x) = \left(1 + \dfrac{1}{x}\right)^x$
	1	$\left(1 + \dfrac{1}{1}\right)^1 = 2$
	10	$\left(1 + \dfrac{1}{10}\right)^{10} = 2.593742$
	1000	$\left(1 + \dfrac{1}{1000}\right)^{1000} = 2.716924$
	1×10^6	$\left(1 + \dfrac{1}{1 \times 10^6}\right)^{1 \times 10^6} = 2.718280$
	1×10^{10}	$\left(1 + \dfrac{1}{1 \times 10^{10}}\right)^{1 \times 10^{10}} = 2.718282$

From the last two rows, notice that as x gets larger and larger, $g(x)$ gets closer and closer to 2.71828. Trying even larger values of x, you will find $g(x)$ is 2.71828 rounded to 5 d.p.

We say the function $g(x)$ has limit 2.71828 as $x \to \infty$ and write this as

$$\lim_{x \to \infty} g(x) = 2.71828 \qquad \text{(5 d.p.)}$$

Actually the function $g(x)$ closes in on an irrational number called e ($= 2.718281828\ldots$). Thus

$$\lim_{x \to \infty} \left(1 + \frac{1}{x}\right)^x = e$$

Remember that an irrational number cannot be written as a fraction of two whole numbers. Moreover it has non-repeating decimals:

$$e = 2.71828182846...$$

So no technological gadget can evaluate the exact value of numbers like e. Computer algebra packages and calculators use a rounded value of e. We revisit this important number, e, in **Chapter 5**.

We can plot the graph of $\left(1 + \dfrac{1}{x}\right)^x$ for large x on a computer algebra system. In MAPLE we use the commands shown in Fig. 23. Notice how the graph gets closer and closer to 2.71828... .

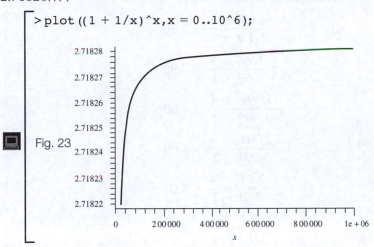

Fig. 23

In general, the notation

$$\lim_{x \to a} f(x) = L$$

means as x gets close to a, $f(x)$ gets close to the limit L.

Let's examine what happens to the function

$$f(x) = \frac{x^2 - 4}{x - 2}$$

as x gets close to 2.

Example 20

By using your calculator, investigate to 1 d.p.

$$f(x) = \frac{x^2 - 4}{x - 2} \text{ as } x \to 2$$

Solution

How do you evaluate $f(x)$ as x gets close to 2?

Substitute x values near to 2 into $f(x)$.

Example 20 *continued*

Which values should we substitute for x?

We could try 1.9, 1.99, 1.999, 2.001, 2.01, 2.1, etc. Substituting these x values into $f(x) = \dfrac{x^2 - 4}{x - 2}$ and rounding to 2 d.p. we have the results shown in Table 4.

x	$f(x) = \dfrac{x^2 - 4}{x - 2}$
1.9	$\dfrac{1.9^2 - 4}{1.9 - 2} = 3.90$
1.99	$\dfrac{1.99^2 - 4}{1.99 - 2} = 3.99$
1.999	$\dfrac{1.999^2 - 4}{1.999 - 2} = 4.00$
2.001	$\dfrac{2.001^2 - 4}{2.001 - 2} = 4.00$
2.01	$\dfrac{2.01^2 - 4}{2.01 - 2} = 4.01$
2.1	$\dfrac{2.1^2 - 4}{2.1 - 2} = 4.1$

TABLE 4

We can see from the table that the function $f(x)$ gets closer and closer to 4 as $x \rightarrow 2$. Trying even closer values to 2, you will find $f(x)$ goes to 4. We say the function $f(x)$ has limit 4 as $x \rightarrow 2$. This is generally written as

$$\lim_{x \to 2} f(x) = 4.0 \qquad \text{(1 d.p.)}$$

We can plot the function $f(x) = \dfrac{x^2 - 4}{x - 2}$ for x close to 2 in MAPLE. The output from MAPLE is shown in Fig. 24.

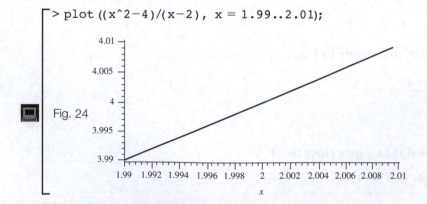

```
> plot ((x^2-4)/(x-2), x = 1.99..2.01);
```

Fig. 24

Actually the function does not exist at $x = 2$ because we have $\dfrac{0}{0}$ which is not defined.

? **Why does the MAPLE output show a value of 4 at $x = 2$ for $f(x) = \dfrac{x^2 - 4}{x - 2}$?**

MAPLE is a powerful computer algebra package but it can be misleading, and the value 4 shown for $f(x)$ is for x values close to 2, and not at $x = 2$.

E2 Employing algebra to find limits

You can also find limits of functions algebraically. We show the above result

$$\lim_{x \to 2} f(x) = 4$$

by using algebra in the next example.

Example 21

Evaluate $\lim\limits_{x \to 2} \dfrac{x^2 - 4}{x - 2}$ by using algebra.

Solution

? **How can we find the value of $\dfrac{x^2 - 4}{x - 2}$ close to 2?**

Remember that we cannot substitute $x = 2$ because we end up with $\dfrac{0}{0}$ which is not defined. We can rearrange some algebra on the function $\dfrac{x^2 - 4}{x - 2}$.

? **What can we use?**

Apply

1.15
$$a^2 - b^2 = (a - b)(a + b)$$

to factorize the numerator:

$$x^2 - 4 = x^2 - 2^2 = (x - 2)(x + 2)$$

Thus

$$\lim_{x \to 2} \frac{x^2 - 4}{x - 2} = \lim_{x \to 2} \frac{(x - 2)(x + 2)}{(x - 2)}$$

†
$$\underset{\substack{\text{cancelling} \\ x - 2}}{=} \lim_{x \to 2} (x + 2)$$

We can cancel $x - 2$ because in the limit x never equals 2.

Example 21 *continued*

As *x* gets close to 2, what does (*x* + 2) get close to?

As *x* approaches 2 (*x* → 2) so *x* + 2 approaches 4 (*x* + 2 → 4). Thus

$$\lim_{x \to 2}(x + 2) = 4$$

Substituting this into ▮ † ▮ gives

$$\lim_{x \to 2}\frac{x^2 - 4}{x - 2} = \lim_{x \to 2}(x + 2) = 4$$

Example 22 *control engineering*

The steady-state error, e_{ss}, of a system is given by

$$e_{ss} = \lim_{s \to 0}\frac{1}{s[1 + G(s)]}$$

Evaluate e_{ss} for $G(s) = \frac{5(s + 1)}{s(s + 2)}$.

Solution

Substituting $G(s) = \frac{5(s + 1)}{s(s + 2)}$ into e_{ss} gives

$$e_{ss} = \lim_{s \to 0}\frac{1}{s\left[1 + \frac{5(s + 1)}{s(s + 2)}\right]}$$

$$= \lim_{s \to 0}\frac{1}{\left[s + s\frac{5(s + 1)}{s(s + 2)}\right]} = \lim_{s \to 0}\frac{1}{\left[s + \frac{5(s + 1)}{(s + 2)}\right]}$$

As *s* gets close to 0, *s* goes to 0 and $\frac{5(s + 1)}{(s + 2)}$ goes to $\frac{5(1)}{(2)} = \frac{5}{2}$. Therefore

$$e_{ss} = \frac{1}{0 + {}^5/_2} = \frac{2}{5}$$

SUMMARY

The notation *x* → 0 means *x* can be made as close as possible to zero. The notation

$$\lim_{x \to a} f(x) = L$$

means *f(x)* tends to the value of *L* as *x* goes to *a*.

Exercise 3(e)

1 By using your calculator, estimate the following limits:

a $\lim\limits_{x \to 0} 5^x$

b $\lim\limits_{x \to \infty} 2.71828^{-x}$

2 🚗 [mechanics]   The terminal velocity, v, of an object falling in air is given by

$$v = \lim_{t \to \infty} 10\left(1 - 2.71828^{-0.5t}\right)$$

By using your calculator, estimate v.

3 Employ algebra to evaluate the following limits:

a $\lim\limits_{x \to 3} \dfrac{x^2 - 9}{x - 3}$

b $\lim\limits_{x \to -\frac{1}{2}} \dfrac{2x^2 - x - 1}{2x + 1}$

c $\lim\limits_{t \to 1} \dfrac{t - 1}{\sqrt{t} - 1}$

4 🛠 [control engineering]   The steady-state error, e_{ss}, of a closed loop system is given by

$$e_{ss} = \lim_{s \to 0} \frac{1}{s[1 + G(s)]}$$

For $G(s) = \dfrac{3(s + 1)}{s(s + 2)}$, evaluate e_{ss}.

5 🛠 [control engineering]   The steady-state error, e_{ss}, of a closed loop system is given by

$$e_{ss} = \lim_{s \to 0} \frac{s}{1 + G(s)}$$

Evaluate e_{ss} for $G(s) = \dfrac{5(s + 2)}{s(s + 5)}$.

6 🛠 [control engineering]   The steady-state error, e_{ss}, of a system is given by

$$e_{ss} = \lim_{s \to 0} \frac{1}{s\left(1 + \dfrac{k}{s}\right)}$$

where k is a positive constant. Evaluate e_{ss}.

7 🛠 [control engineering]   The steady-state error, e_{ss}, of a system is given by

$$e_{ss} = \lim_{s \to 0} \frac{M}{1 + G(s)}$$

where M is a constant. Given that

$$G(s) = \left(\frac{k_1 + k_2}{s}\right) \times \left(\frac{1}{1 + s\tau}\right)$$

where k_1, k_2 and τ are non-zero constants, evaluate e_{ss}.

Employ algebra to find the limits of questions **8** and **9**.

8 Let $f(x) = x^2$. Find $\lim\limits_{h \to 0} \dfrac{f(x + h) - f(x)}{h}$.

9 Let $g(x) = x^3$, determine

$$\lim_{h \to 0} \frac{g(x + h) - g(x)}{h}$$

SECTION F Modulus function

By the end of this section you will be able to:
▶ understand and sketch the modulus function

F1 Definition and sketching

The modulus function, denoted $|x|$, is defined as

3.7
$$|x| = \begin{cases} x & \text{if } x \geq 0 \\ -x & \text{if } x < 0 \end{cases} \qquad \begin{array}{l} \text{(if } x \text{ is positive or zero)} \\ \text{(if } x \text{ is negative)} \end{array}$$

The modulus function gives the size of a number, and it is always positive or zero. We say $|3|$, pronounced mod 3, is equal to 3.

? **What is the value of $|-3|$?**

We use the second line of 3.7 given above with $x = -3$:

$$|-3| = -(-3) = 3$$

Thus $|-3| = 3$.

? **What are the values of $|5|$ and $|-5|$?**

In the same way we have $|5| = 5$ and $|-5| = 5$. Let's plot the graph of $|x|$.

Example 23

Plot the graph of $|x|$ for x between -3 and 3.

Solution

Substituting whole numbers between -3 and 3 into 3.7 gives the results shown in Table 5.

TABLE 5	x	$|x|$
	-3	$|-3| = -(-3) = 3$
	-2	$|-2| = -(-2) = 2$
	-1	$|-1| = -(-1) = 1$
	0	$|0| = 0$
	1	$|1| = 1$
	2	$|2| = 2$
	3	$|3| = 3$

Example 23 *continued*

Connecting the points $(-3, 3)$, $(-2, 2)$,..., $(3, 3)$ gives the graph of Fig. 25.

Fig. 25 Graph
of $|x|$ for x
between -3
and 3

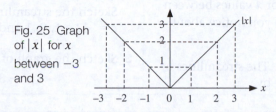

Example 24

Sketch, on different axes, the graphs of

$$g(x) = |x - 1| \quad \text{and} \quad h(x) = |x + 3| - 2$$

Solution

We know the graph of $g(x) = |x - 1|$ is the same shape as $|x|$ but it has been modified in some way.

?   **What adjustments do we need to make for the graph of $|x - 1|$?**

The graph of $|x|$ is shifted to the right by 1 unit to give $|x - 1|$ (see Fig. 26).

Fig. 26 Graph
of $|x - 1|$

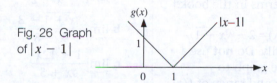

The $|x + 3|$ in $h(x)$ shifts the graph $|x|$ to the left by 3 units and the -2 drops it down by 2 units. Thus the graph of $h(x) = |x + 3| - 2$ is as shown in Fig. 27.

Fig. 27
Graph of
$|x + 3| - 2$

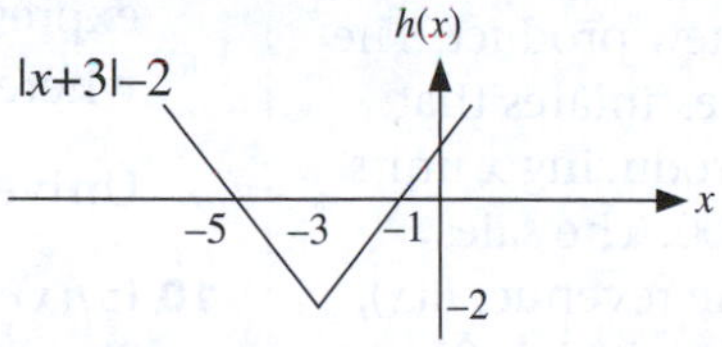

SUMMARY

The modulus function, $|x|$, is defined by

3.7 $|x| = \begin{cases} x & \text{if} \quad x \geq 0 \\ -x & \text{if} \quad x < 0 \end{cases}$

Exercise 3(f)

1 On the same axes plot the graphs of $|1 - x|$ and $|x - 1|$ for x values between -3 and $+3$. What do you notice about your results?

2 [*fluid mechanics*]   The streamlines of a fluid flow are given by

$$f(x) = |x| + c$$

Sketch the streamlines for $c = -2, -1, 0, 1$ and 2.

3 Sketch the graph of
$$|x - 3| + 1$$

Examination questions 3

1 Sketch the graph of $f(x) = -3 - 2x$. Label the graph carefully. Do **not** use a table of values. What are the domain and range of $f(x)$?

2 a Compared to the graph of $y = |x|$, the graph of $y = -3|x| + 2$ has undergone what graphical operation? Sketch the graph.
b Identify the equation of the axis of symmetry of the quadratic function $f(x) = (x + 3)^2 - 4$. Sketch the graph. [This question has been edited to make it compatible with the terms in the book.]

3 Sketch the graph of $f(x) = 2 - \sqrt{x + 1}$. Label the graph carefully. Do **not** use a table of values. Use the graph to determine the domain and range of $f(x)$.

Questions 1 to 3 are from Memorial University of Newfoundland, Canada, 2009

4 A firm is considering a new product. The accounting department estimates that the total cost, $C(x)$, of producing x units will be $C(x) = 65x + 3500$. The sales department estimates the revenue, $R(x)$, from selling x units will be $R(x) = 85x$, but no more than 600 units can be sold at that price.
Find and interpret $(R - C)(600)$.

5 Find and simplify the difference quotient, $\dfrac{f(x + h) - f(x)}{h}$, $h \neq 0$ for the function $f(x) = \dfrac{1}{3x}$.

6 Given $f(x) = \dfrac{1 - 2x}{3x}$ and $g(x) = 3x^2 - 9x$, find the following
a $(g \circ f)(-1)$ **b** $(f \circ g)(2)$

7 Find the inverse function, $f^{-1}(x)$, given the one-to-one function $f(x) = \dfrac{x - 5}{2x + 3}$.

8 Algebraically find the following limits:
a $\lim\limits_{x \to -2} \dfrac{x^2 - x - 6}{x + 2}$
b $\lim\limits_{x \to 8} \dfrac{\sqrt{x + 1} - 3}{x - 8}$
c $\lim\limits_{x \to \infty} \dfrac{1 - 12x}{4x + 5}$

Questions 4 to 8 are from Arizona State University, USA (Review)

9 If $f(x) = 4 - x^2$ and $g(x) = \dfrac{3}{x} + 2$, find expressions, which you should simplify where possible, for $fg(x)$, $ff(x)$ and $g^{-1}(x)$.

University of Sussex, UK, 2006

10 If $f(x) = 2 - x^2$, $x \leq 0$, find $f^{-1}(x)$. Sketch graphs of $f(x)$ and $f^{-1}(x)$ on the same axes, labelling points where the graphs cross the axes.

University of Sussex, UK, 2007

11 Jennifer's position at time t relative to Stone Cold Creamery is given by the function

$$p(t) = 5 + 2t - t^2$$

Solutions at end of book. Complete solutions available at www.palgrave.com/engineering/singh

Examination questions 3 **continued**

(positive indicates west and negative east). Explain why Jennifer must have visited the creamery between the hours of $t = 0$ and $t = 4$.

12 Compute the following limits; justify your answers.

a $\displaystyle\lim_{x \to (-1)} \frac{x^4 - 3x^2 - 8}{x^2 - \sqrt{4x + 8}}$

b $\displaystyle\lim_{x \to (-2)} \frac{x^4 - 3x - 10}{x^2 + x - 2}$

c $\displaystyle\lim_{h \to 0} \frac{h + \sqrt{4 + h} - 2}{h}$

Stanford University, USA, 2008

Miscellaneous exercise 3

Solutions at end of book. Complete solutions available at www.palgrave.com/engineering/singh

Let the domain of all the functions be real numbers with any restriction in brackets.

1 If $f(x) = \dfrac{2x - 1}{1 - x}$ $(x \neq 1)$, find

i $\left[f\left(\dfrac{1}{2}\right)\right]^2$ **ii** $\dfrac{6f(2) - 3f(5)}{f(5)}$ **iii** $f \circ f$

2 Let f be the function defined by

$$f(t) = 4.9t^2$$

Determine the following with $h \neq 0$:

i $f(t + h)$

ii $\dfrac{f(t + h) - f(t)}{h}$

iii $\dfrac{f(1 + h) - f(1)}{h}$

iv $\dfrac{f(t + h) - f(t)}{h}$ for $t = 1$ and $h = 10^{-20}$

v $\displaystyle\lim_{h \to 0} \dfrac{f(t + h) - f(t)}{h}$

3 Let $\phi(t) = t^3$. Find expressions for the following with $h \neq 0$:

i $\dfrac{\phi(t + h) - \phi(t)}{h}$

ii $\displaystyle\lim_{h \to 0} \dfrac{\phi(t + h) - \phi(t)}{h}$

(*Hint*: $(a + b)^3 = a^3 + 3a^2 b + 3ab^2 + b^3$)

4 Sketch: **i** $|x - 3|$ **ii** $|x + 1|$

5 Let $f(x) = 2x^4 - x^2 + 1001$. Show that $f(-a) = f(a)$. (A function with this property is called an **even** function.)

6 Let $g(t) = t^5 - t^3 - t$. Show that $g(-a) = -g(a)$. (A function with this property is called an **odd** function.)

7 Let $f(x) = x^2 + 5$ and $g(x) = 6x$. Find the roots of the equation $f(x) = g(x)$.

8 Let $f(x) = x^3 - 2x^2 - x - 0.625$. Evaluate **i** $f(2)$ **ii** $f(3)$

Without plotting the graph of f, what can you say about the graph of f between $x = 2$ and $x = 3$. Use a computer algebra package (symbolic manipulator) to plot $f(x)$ and find a root of the equation $f(x) = 0$.

9 Sketch the graph $f(x) = \dfrac{1}{x}$ $(x \neq 0)$.

Find its inverse f^{-1} and sketch it on the same axes.

10 Let ϕ be the function defined by

$$\phi(x) = \frac{ax - b}{cx - a} \quad (cx - a \neq 0)$$

Show that $\phi^{-1}(x) = \phi(x)$. If

$$f(x) = \frac{2x - 3}{5x - 2} \text{ then find } f^{-1}(x).$$

(Assume the denominator $5x - 2 \neq 0$.)

You may use a computer algebra system, or a graphical calculator, for questions 11 and 12.

11 Plot, on the same axes, the graphs of $f(x) = x^3$ and $g(x) = 4 + 4x - x^2$. Find the roots of $f(x) - g(x) = 0$.

12 Let $f(x) = 2.718281828^x$. For the following, h is not zero:

  i Plot $f(x)$ for x between -2 and 2.

  ii Estimate $\lim\limits_{h \to 0} \dfrac{f(h) - 1}{h}$.

  iii Find $\lim\limits_{h \to 0} \dfrac{f(x + h) - f(x)}{h}$.

13 [*mechanics*]   The displacement, $x(t)$, of a particle is given by

$$x(t) = t^2 - t - 2$$

Sketch the graph of $x(t)$ against t, marking the points where the graph crosses the axes.

14 [*control engineering*]   The transfer function, $G(s)$, of a system is given by

$$G(s) = \left(\frac{3}{s + 2} \right) \times \left(\frac{s + 1}{s^2 + 5s + 6} \right)$$

Show that

$$G(s) = \frac{3s + 3}{s^3 + 7s^2 + 16s + 12}$$

15 [*control engineering*]   The poles of a system occur where the **denominator** of the transfer function, $G(s)$, is zero. That is, if

$$G(s) = \frac{Y(s)}{X(s)}$$

then the system poles are the values of s which satisfy $X(s) = 0$.

Find the system poles for

a $G(s) = \dfrac{s^2 + 1}{s^2 + 4s + 3}$

b $G(s) = \dfrac{s + 1}{s^2 + 5s - 7}$

c $G(s) = \dfrac{s}{3}$

16 [*reliability engineering*]   The failure density function, $f(t)$, for a system is given by

$$f(t) = \frac{1}{10} \quad 0 < t < 10 \text{ years}$$

Find $F(t)$, $R(t)$ and $h(t)$ where

$F(t) = tf(t)$ (Failure density function)

$R(t) = 1 - F(t)$ (Reliability function)

$h(t) = \dfrac{f(t)}{R(t)}$ (Hazard rate function)

Plot, by using a symbolic manipulator or otherwise, on different axes, $F(t)$, $R(t)$ and $h(t)$ for $0 < t < 10$.

17 [*reliability engineering*]   In reliability engineering we have the following relationship:

$$R(t) = 1 - Q(t)$$

where $R(t)$ is called the reliability function, $Q(t)$ is called the failure distribution function and t is time in years. For a set of components, the failure distribution function, $Q(t)$, is given by

$$Q(t) = 0.3t - 0.015t^2$$

Sketch $Q(t)$ for $t \geq 0$. Also find $R(t)$.

Use a computer algebra system for question 18.

18 [*control engineering*]   The overall transfer function, $T(s)$, of a system is given by

$$T(s) = \frac{G(s)}{1 + G(s)H(s)}$$

Simplify $T(s)$ for

$$G(s) = \frac{2}{s + 1} \times \frac{s}{(s + 2)(s + 5)}$$

and $H(s) = 0.001$.

Trigonometry and Waveforms

The word *trigonometry* is a combination of the Greek words *trigono* meaning triangle and *metro* meaning to measure. It was the early Greeks who laid the foundation stones of trigonometry around 140 BC. However the sine of an angle was developed by Aryabhata, a Hindu.

Trigonometry is one of the most widely used mathematical techniques. It crops up in surveying, navigation, GPS in cars, radio waves, digital signal processing, etc.

SECTION A **Trigonometric functions**

By the end of this section you will be able to:
▶ find sin, cos and tan of angles
▶ solve right-angled triangles
▶ calculate inverse trigonometric functions
▶ find cosec, sec and cot of angles

A1 **Review of trigonometric ratios**

Consider the right-angled triangle ABC (Fig. 1) with an angle θ (theta) at A (it is convention to use θ to denote 'angle').

A right-angled triangle is a triangle that has one angle equal to 90° and this is represented by a square, □, as seen in Fig. 1.

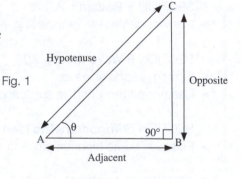

Fig. 1

We define

adjacent = side AB, **adjacent** to the angle θ

opposite = side BC, **opposite** to the angle θ

hypotenuse = side AC, the **longest** side of the right-angled triangle

We characterize sin (short for sine), cos (short for cosine) and tan (short for tangent) as follows:

4.1 $$\sin(\theta) = \frac{\text{opposite length}}{\text{hypotenuse length}}$$

4.2 $$\cos(\theta) = \frac{\text{adjacent length}}{\text{hypotenuse length}}$$

4.3 $$\tan(\theta) = \frac{\text{opposite length}}{\text{adjacent length}}$$

The well-known mnemonic 'SOHCAHTOA' can be used to remember these ratios.

4.1 to 4.3 are called the trigonometric ratios and as each of sin, cos and tan are functions of the angle θ, they are generally referred to as the **trigonometric functions**. Let *opp* = opposite, *adj* = adjacent and *hyp* = hypotenuse, then by using Fig. 1 we can rewrite these trigonometric functions in terms of *opp*, *adj* and *hyp* (Fig. 2):

4.1 $\sin(\theta) = \dfrac{opp}{hyp}$

4.2 $\cos(\theta) = \dfrac{adj}{hyp}$

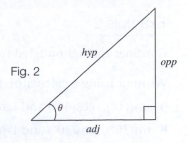
Fig. 2

4.3 $\tan(\theta) = \dfrac{opp}{adj}$

We can use these ratios to find an unknown angle or length of a side of any right-angled triangle. Note that we **must** have a right-angled triangle. Before we proceed to an example, we need to recall Pythagoras's theorem. **By considering Fig. 2 what does Pythagoras tell us about the relationship between the lengths *adj*, *opp* and *hyp*?**

P $hyp^2 = adj^2 + opp^2$

We give this formula reference (P) because it is central to this chapter and you should remember it.

In order to find *hyp* we take the square root of both sides. By transposing formula (P) we can find *adj* and *opp*. Often by using this theorem we can find the length of a side we need.

Example 1

Find the length AC of Fig. 3.

Without using a calculator, evaluate $\sin(45°)$, $\cos(45°)$ and $\tan(45°)$.

Fig. 3

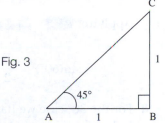

Solution

How can we find AC?

Since opposite = 1 and adjacent = 1 we can apply Pythagoras:

$$AC^2 = 1^2 + 1^2 = 2$$

Take the square roots of both sides: $AC = \sqrt{2}$

Thus hypotenuse = $\sqrt{2}$. With $adj = 1$, $opp = 1$ and $hyp = \sqrt{2}$ we have

$$\sin(45°) = \frac{opp}{hyp} = \frac{1}{\sqrt{2}}$$

$$\cos(45°) = \frac{adj}{hyp} = \frac{1}{\sqrt{2}}$$

$$\tan(45°) = \frac{opp}{adj} = \frac{1}{1} = 1$$

Example 2

Consider the right-angled triangle shown in Fig. 4.

Without using a calculator, find the values of

i sin(60°), cos(60°) and tan(60°)

ii sin(30°), cos(30°) and tan(30°).

Fig. 4

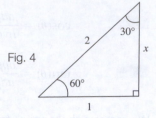

Solution

i Considering the 60° case we have the situation shown in Fig. 5.

What is sin(60°) equal to?

For 60°, opposite = x, hypotenuse = 2 and adjacent = 1.

Fig. 5

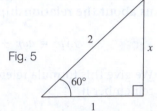

By 4.1 **, sin(60°) = $\dfrac{x}{2}$ but what is the value of x?**

Rearranging Pythagoras:

$$x^2 = 2^2 - 1^2 = 3 \text{ which gives } x = \sqrt{3}$$

Applying 4.1 , 4.2 and 4.3 with *opp* = $\sqrt{3}$, *adj* = 1 and *hyp* = 2 respectively gives

$$\sin(60°) = \frac{\sqrt{3}}{2}, \ \cos(60°) = \frac{1}{2} \text{ and } \tan(60°) = \frac{\sqrt{3}}{1} = \sqrt{3}$$

ii For the 30° case we have the situation shown in Fig. 6.

Now with reference to 30°, adjacent = $\sqrt{3}$, opposite = 1 and hypotenuse = 2:

Fig. 6

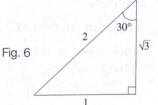

$$\sin(30°) = \underset{\text{by } 4.1}{\frac{1}{2}}, \ \cos(30°) = \underset{\text{by } 4.2}{\frac{\sqrt{3}}{2}} \text{ and } \tan(30°) = \underset{\text{by } 4.3}{\frac{1}{\sqrt{3}}}$$

The values obtained in **Examples 1** and **2** are the exact values of the corresponding trigonometric ratios. Thus combining these with trigonometric ratios for angles 0°, 90° we obtain Table 1.

4.1   sin(θ) = *opp/hyp*   4.2   cos(θ) = *adj/hyp*   4.3   tan(θ) = *opp/adj*

TABLE 1		$\theta = 0°$	$\theta = 30°$	$\theta = 45°$	$\theta = 60°$	$\theta = 90°$
	$\sin(\theta)$	0	$\dfrac{1}{2}$	$\dfrac{1}{\sqrt{2}}$	$\dfrac{\sqrt{3}}{2}$	1
	$\cos(\theta)$	1	$\dfrac{\sqrt{3}}{2}$	$\dfrac{1}{\sqrt{2}}$	$\dfrac{1}{2}$	0
	$\tan(\theta)$	0	$\dfrac{1}{\sqrt{3}}$	1	$\sqrt{3}$	undefined

This table shows the exact trigonometric ratios for some special angles.

The angles are read from the top row. Note that $\sin(0°) = 0$, $\cos(0°) = 1$ and $\tan(0°) = 0$. To evaluate any of the trigonometric functions, sin, cos or tan, we can use a calculator. For example, to evaluate $\sin(30°)$, place the calculator into degrees mode first and then PRESS [sin] [30]. The calculator should show 0.5. On some older calculators you enter the argument [30] first and then the function [sin]. Hence $\sin(30°) = 0.5 = \dfrac{1}{2}$. Similarly by using a calculator we have

$$\cos(22°) = 0.927, \quad \tan(15°) = 0.268,$$
$$\sin(53.1°) = 0.800, \text{ etc.}$$

Note that these are **not** exact values but rounded to 3 decimal places.

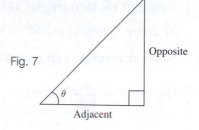

Fig. 7

? **Why is tan(90°) not defined?**

$$\tan(\theta) = \frac{\text{opposite}}{\text{adjacent}}$$

? **If $\theta = 90°$ in Fig. 7 then what is the size of the adjacent length?**

Adjacent must be zero and so

$$\tan(90°) = \frac{\text{opposite}}{0}$$

Of course we cannot divide by zero. This is why $\tan(90°)$ is **not** defined.

A2 Solving right-angled triangles

Transposing the standard formulae, 4.1 to 4.3, allows us to evaluate many other unknown values of the right-angled triangle. By transposing we have the following:

4.4 $opp = hyp \times \sin(\theta)$

4.5 $adj = hyp \times \cos(\theta)$

4.6 $opp = adj \times \tan(\theta)$

4.7 $hyp = \dfrac{opp}{\sin(\theta)}$

4.8	$hyp = \dfrac{adj}{\cos(\theta)}$

4.9	$adj = \dfrac{opp}{\tan(\theta)}$

Example 3

Figure 8 shows a **symmetrical** roof truss. Span BF is of length 8 m. Rafters AB and AC are inclined at angles 40° and 65° respectively. Find the lengths AF and DE.

In this example the critical word is **symmetrical**.

Fig. 8

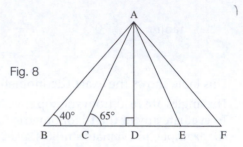

Solution

Since the roof truss is **symmetrical**, angle $F = 40°$ and $DF = 4$ m (Fig. 9).

? **How can we find length AF?**

AF is the hypotenuse and we know the adjacent = 4.

? **So what formula can we use?**

Use $hyp = \dfrac{adj}{\cos(\theta)}$:

Fig. 9

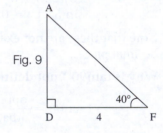

$$AF = \frac{4}{\cos(40°)} = \frac{4}{0.7660} = 5.222 \text{ m (3 d.p.)}$$

For length DE, consider triangle ADE. We know by **symmetry** that angle $E = 65°$ and length AD can be found from Fig. 9. In Fig. 9, AD is the opposite, so using $opp = adj \times \tan(\theta)$ we have

$$AD = 4 \times \tan(40°)$$

$$= 4 \times 0.8391 = 3.356$$

(see Fig. 10)

Fig. 10   3.356

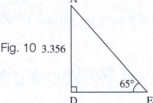

? **How can we find length DE?**

DE is adjacent to 65° and we have $opp = 3.356$, so using $adj = \dfrac{opp}{\tan(\theta)}$ gives

$$DE = \frac{3.356}{\tan(65°)} = \underbrace{\frac{3.356}{2.1445}}_{\text{by calculator}} = 1.565 \text{ m (3 d.p.)}$$

A3 Inverse trigonometric functions

Let the sine of angle A be x, that is

$$\sin(A) = x$$

then the angle A is given by

$$A = \sin^{-1}(x)$$

where $\sin^{-1}$ denotes **inverse sine**. Remember from **Chapter 3** that the inverse function is represented by the index -1. This is why the inverse of the sine function is denoted by

? $\sin^{-1}$. **What is** $\sin^{-1}\left(\dfrac{1}{\sqrt{2}}\right)$ **equal to?**

From Table 1 we know that $\sin(45°) = \left(\dfrac{1}{\sqrt{2}}\right)$, thus

$$\sin^{-1}\left(\dfrac{1}{\sqrt{2}}\right) = 45°$$

We can also use a calculator to determine inverse sin. For example, to find angle A from $\sin(A) = 0.791$ we PRESS $\boxed{\textbf{SHIFT}}$ $\boxed{\textbf{SIN}}$ $\boxed{\textbf{0.791}}$ $\boxed{\textbf{=}}$, which should show 52.27906149. (Make sure you put your calculator into degrees mode.)

On many basic scientific calculators, the $\sin^{-1}$ is a secondary function so we need to press the $\boxed{\textbf{SHIFT}}$ or $\boxed{\textbf{INV}}$ or $\boxed{\textbf{2nd FUN}}$ button prior to entering the sin button. (This is just like pressing the $\boxed{\textbf{SHIFT}}$ key on a computer keyboard to access capital letters.) Likewise, the inverse cos is denoted by $\cos^{-1}$ and inverse tan by $\tan^{-1}$ and defined similarly. On some calculators and computer algebra systems, the inverse sin, $\sin^{-1}$, is denoted by arcsin. Similarly

$$\cos^{-1} = \text{arccos and } \tan^{-1} = \text{arctan}$$

In MATLAB the command for inverse sine is 'asin'.

Example 4

Evaluate the following:

$$\sin^{-1}(0.23), \cos^{-1}(0.75) \text{ and } \tan^{-1}(1.73)$$

Solution

By using a calculator and rounding to 2 decimal places we have

$$\sin^{-1}(0.23) = 13.30°, \cos^{-1}(0.75) = 41.41° \text{ and } \tan^{-1}(1.73) = 59.97°$$

Example 5

Figure 11a shows the **symmetrical** part of a bridge. Point X is the midpoint of ABCD. Find the size of angle θ if CD = 10 m and AC = 18 m.

Solution

Consider triangle AXY where Y is the midpoint of AB.

? **What is the length of AY and XY?**

Since X is the midpoint, AY = 5 and XY = 9.

? **How can we find θ?**

Angle AXY is half of θ (Fig. 11b).

? **How can we find θ/2?**

We know with reference to angle θ/2 that opposite = 5 and adjacent = 9, so we can use $\boxed{4.3}$:

* $$\tan\left(\frac{\theta}{2}\right) = \frac{5}{9}$$

? **How do we extract θ/2 from $\boxed{*}$?**

Take inverse tan of both sides:

$$\frac{\theta}{2} = \tan^{-1}\left(\frac{5}{9}\right) = 29.055° \text{ [By calculator]}$$

So θ = 2 × 29.055° = 58.11°

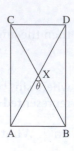

a

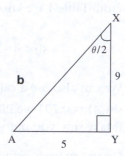

b

Fig. 11

A4 Additional trigonometric functions

These additional trigonometric functions may seem abstract compared with sin, cos, and tan, but these terms are used in complex trigonometric problems.

The other trigonometric functions are called cosecant (cosec), secant (sec) and cotangent (cot), and are determined by

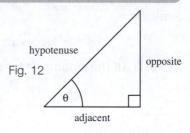

Fig. 12

| 4.10 | $$\operatorname{cosec}(\theta) = \frac{1}{\sin(\theta)} = \frac{\text{hypotenuse}}{\text{opposite}}$$ | (provided $\sin(\theta) \neq 0$) |

| 4.11 | $$\sec(\theta) = \frac{1}{\cos(\theta)} = \frac{\text{hypotenuse}}{\text{adjacent}}$$ | (provided $\cos(\theta) \neq 0$) |

| 4.12 | $$\cot(\theta) = \frac{1}{\tan(\theta)} = \frac{\text{adjacent}}{\text{opposite}}$$ | (provided $\tan(\theta) \neq 0$) |

4.3 $\tan(x) = opp/adj$

We can evaluate these new trigonometric functions, cosec, sec and cot, by using a calculator. Some calculators evaluate these directly, on others you have to write them in terms of the basic trigonometric functions and then evaluate them.

Example 6

Evaluate the following to 3 d.p.:

$$\text{cosec}(22°), \sec(19.61°) \text{ and } \cot(89°)$$

Solution

By using a calculator we have

$$\text{cosec}(22°) = \frac{1}{\sin(22°)} = \frac{1}{0.3746} = 2.669 \text{ (3 d.p.)}$$

$$\sec(19.61°) = \frac{1}{\cos(19.61°)} = \frac{1}{0.9420} = 1.062 \text{ (3 d.p.)}$$

$$\cot(89°) = \frac{1}{\tan(89°)} = \frac{1}{57.2900} = 0.018 \text{ (3 d.p.)}$$

SUMMARY

The basic trigonometric functions are defined as

4.1	$\sin(\theta) = opp/hyp$
4.2	$\cos(\theta) = adj/hyp$
4.3	$\tan(\theta) = opp/adj$

Inverse sin is denoted by $\sin^{-1}$.

Exercise 4(a)

Solutions at end of book. Complete solutions available at www.palgrave.com/engineering/singh

1 Evaluate the following to 2 d.p.:

$\sin(35°), \cos(47°) \text{ and } \tan(21°)$

2 By using Table 1, find the exact values of the following:

  i $\cos(45°)\sin(45°)$

  ii $\tan(60°) + \tan(30°)$

  iii $\sin(60°) + \tan(60°)$

  iv $\dfrac{\sin(45°)}{\tan(45°)\cos(45°)}$

  v $\cos(45°) + \sin(60°)$

  vi $\sin(60°)\cos(30°) + \cos(60°)\sin(30°)$

Exercise **4(a) continued**

Solutions at end of book. Complete solutions available at www.palgrave.com/engineering/singh

3 Figure 13 shows a symmetrical roof truss, with rise $r = 4$ m and rafter AC inclined at 25°.

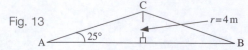

Fig. 13

Evaluate the length of the rafter AC and span AB.

4 By using $\sin(\theta) = \dfrac{opp}{hyp}$ and $\cos(\theta) = \dfrac{adj}{hyp}$

show that $\tan(\theta) = \dfrac{\sin(\theta)}{\cos(\theta)}$.

5 Figure 14 shows a garage roof with $r = 3$ m and $s = 6$ m. Find the size of the angle θ.

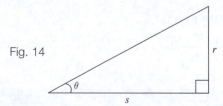

Fig. 14

6 Figure 15 shows a symmetrical part of a bridge lattice. Determine the length h, given that AD = 5 m, angle DAB = 55° and X is the midpoint of ABCD.

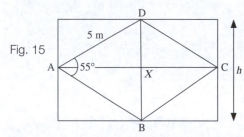

Fig. 15

7 Figure 16 shows a cross-section of a rain gutter with PQ = QR = RS = 180 mm and angle $\theta = 60°$. Find the area of the cross-section PQRS.

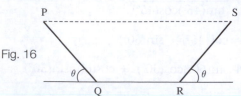

Fig. 16

8 Determine the value of x in each of the following cases:

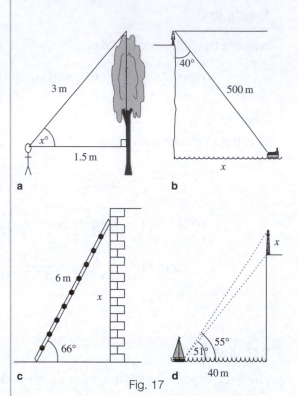

a   **b**

c   **d**   40 m

Fig. 17

*9 Without using a calculator, find the five other trigonometric ratios [including $\sec(\theta)$, $\csc(\theta)$ and $\cot(\theta)$] given:

i $\sin(\theta) = \dfrac{1}{\sqrt{3}}$ **ii** $\cos(\theta) = \dfrac{2}{3}$

iii $\tan(\theta) = \dfrac{3}{4}$

10 Evaluate the following to 3 d.p.:

$$\sec(36.4°), \cot(29.17°)$$

and $\csc(44.44°)$

11 By using any mathematical software plot the following graphs between 0° and 360°.

a $y = \csc(\theta)$ **b** $y = \sec(\theta)$
c $y = \cot(\theta)$

SECTION B **Angles and graphs**

By the end of this section you will be able to:
▶ find the trigonometric functions for angles of any size
▶ plot the graphs of these functions

B1 **Angles of any size**

Let P be the point (x, y) in the Cartesian co-ordinate system at a distance r from the origin O, at an angle θ. Consider the triangle in Fig. 18.

By the formula adjacent = hypotenuse × cos(θ) we have

$$x = r \times \cos(\theta)$$

By the formula opposite = hypotenuse × sin(θ) we have

$$y = r \times \sin(\theta)$$

If P rotates **anticlockwise** about the origin, then we define the trigonometric functions (sin, cos and tan) for any angle θ as follows:

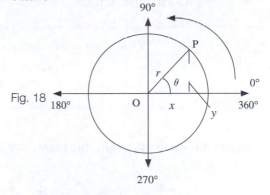

Fig. 18

4.13 $\sin(\theta) = \dfrac{y}{r}$

4.14 $\cos(\theta) = \dfrac{x}{r}$

4.15 $\tan(\theta) = \dfrac{y}{x}$

Here θ can be of any size (magnitude). One complete revolution is 360°, two complete revolutions are $2 \times 360° = 720°$. You can think of P in Fig. 18 as a point on the edge of a wheel, and θ as the angle made by P from the 0° position. Thus if the wheel makes 10 complete revolutions then it has completed 3600°.

By convention, the angle θ is normally measured from the positive x axis in an anticlockwise direction. The angle θ can be divided into four equal quadrants:

0°−90°, 90°−180°, 180°−270° and 270°−360° (Fig. 19).

Fig. 19

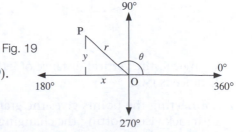

B2 Graphs of sine, cosine and tangent functions

Let us assume that we are interested in plotting the changing length of y (Fig. 18) as P completes one full revolution anticlockwise. For convenience, let r be 1 unit in length. Thus we know from formula 4.1 , that for any position of point P, the length of y is given by

$$\frac{y}{r} = \sin(\theta), \text{ so when } r = 1 \text{ we have } y = \sin(\theta)$$

By using a graphical calculator or a computer algebra package, the graphs of sin, cos and tan between 0° and 360° can be created very quickly. However, we can use a basic calculator to evaluate some points and connect them to make a graph, as the next example shows.

Example 7

Plot the graph of $y = \sin(\theta)$ for $\theta = 0°{-}360°$.

Solution

By taking intervals of 30° and evaluating the sine of each angle on a calculator, rounding to 3 d.p., we establish Table 2.

TABLE 2	θ	0°	30°	60°	90°	120°	150°	180°	210°	240°	270°	300°	330°	360°
	$y = \sin(\theta)$	0	0.5	0.866	1	0.866	0.5	0	−0.5	−0.866	−1	−0.866	−0.5	0

Figure 20a shows (at right) the graph of $y = \sin(\theta)$ plotted for θ between 0° and 360°.

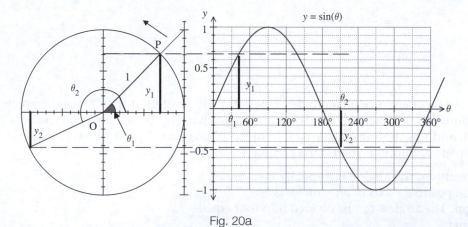

Fig. 20a

At angle θ_1 the vertical distance of P (Fig. 20a – at left) is y_1 and at angle θ_2 the vertical distance is y_2.

Connecting the points gives the graph of $y = \sin(\theta)$ (Fig. 20a). As the point P rotates anticlockwise, plotting the changing value of y traces out a sine curve. Take a

Example 7 *continued*

moment to ensure that you understand how the diagram on the left generates the graph on the right.

This distinctive curve is known as a **sine wave**.

We can plot the graphs of $\cos(\theta)$ and $\tan(\theta)$ in a similar way. The graphs are shown below, (Figs 20b and 20c) but you are asked to tabulate some values in **Exercise 4(b)**.

Here we plot the changing length of x as P rotates anticlockwise, using $x = \cos(\theta)$:

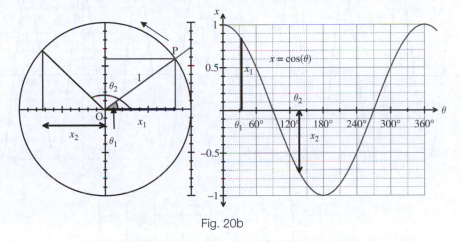

Fig. 20b

The resulting graph demonstrates a **cosine wave** because the graph is showing how x changes as θ is varied. At angle θ_1 the horizontal distance of P is x_1 and at angle θ_2 the horizontal distance is x_2.

Note that x lies on the horizontal axis within the circle, but is plotted vertically on the right-hand graph.

Finally, we can plot the changing ratio between lengths x and y as P rotates anticlockwise using $\tan(\theta) = \dfrac{y}{x}$:

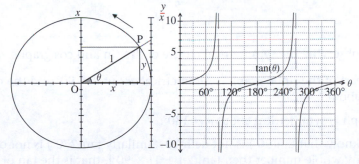

Fig. 20c

Note that when $\theta = 90°$ and $\theta = 270°$, $x = 0$, so $y/x = y/0$, and is therefore undefined. At these points the graph has a discontinuity or jump shown by the vertical broken lines on the graph.

Example 7 *continued*

Note that the graphs $y = \sin(\theta)$ and $y = \cos(\theta)$ perform smooth oscillations between 1 and -1.

If we extend the θ-axis domain, then the sine and cosine graphs are repeated at 360° intervals, and the tan graph is repeated at 180° intervals. You can observe this by looking at the following graphs (Fig. 21a–c) where each of these functions is plotted between $-720°$ and 720°:

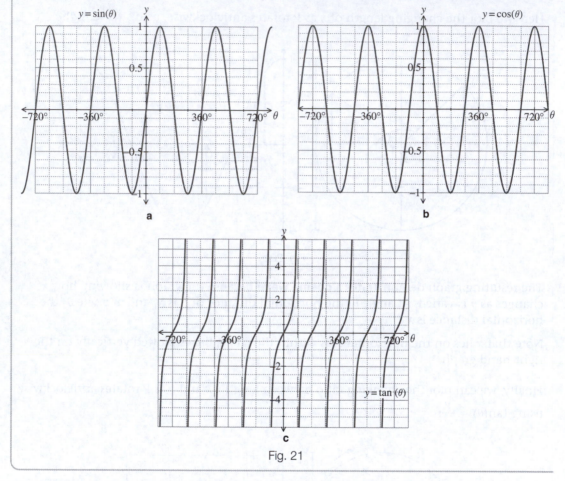

Fig. 21

Also it is worth observing the symmetrical nature of the sin and cos graphs:

i the positive part of the $\sin(\theta)$ graph is symmetrical about $\theta = 90°$ and the negative part is symmetrical about $\theta = 270°$

ii the $\cos(\theta)$ graph is symmetrical about $\theta = 180°$.

From above we know that $\tan(90°)$ is not defined. Similarly $\tan(270°)$ is not defined. In general, if n is any whole number then $\tan[(2n + 1) \times 90°]$, that is the tan of an odd multiple of 90°, is not defined.

If we extend the θ-axis domain, it can be seen by plotting the graphs that the sin and cos graphs are repeated every 360° interval and the tan graph is repeated every 180° interval (Fig. 22 **a–c**).

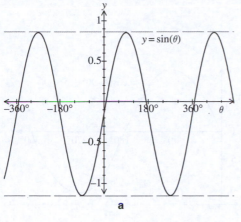

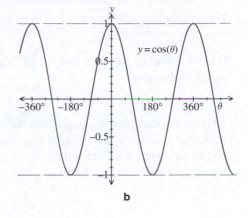

a

b

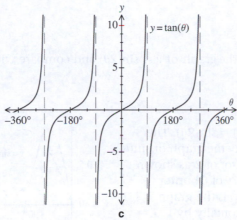

c

Fig. 22

The importance of the graphs we have created must **not** be underestimated. This type of oscillating motion is found repeatedly throughout fields as diverse as mathematics, engineering, science and medicine. It is well worth taking the time to ensure you are familiar with the nature of these graphs before moving on.

B3 Transformation of graphs

We can apply the transformation of graphs covered in Chapter 3 to the transformation of trigonometric graphs.

Example 8

Sketch the graph of $y = \cos(\theta - 90°)$ for $0° \le \theta \le 360°$.

What do you notice about your sketch?

Solution

Let $f(\theta) = \cos(\theta)$ then $f(\theta - 90°) = \cos(\theta - 90°)$, so we shift the cos graph in Fig. 22**b** to the right by 90° (Fig. 23). (Remember from **Chapter 3** that $f(x - c)$ is the graph of $f(x)$ shifted to the right by c units.)

Example 8 *continued*

It is identical to the sine graph (see Fig. 20a). Actually for all θ:

$$\cos(\theta - 90°) = \sin(\theta)$$

This is an example of a trigonometric identity. There are many other trigonometric identities which we will examine later in this chapter.

Fig. 23

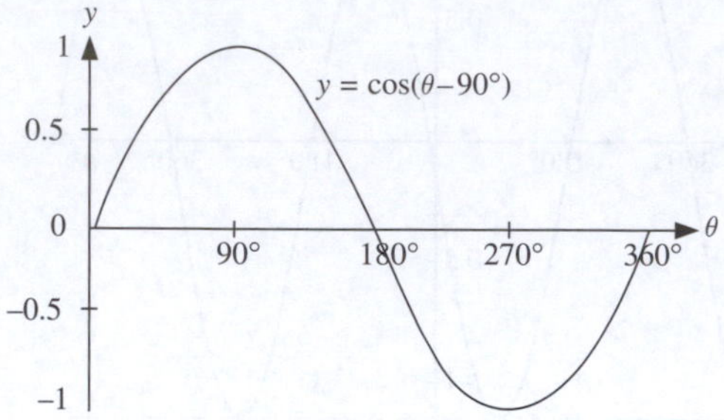

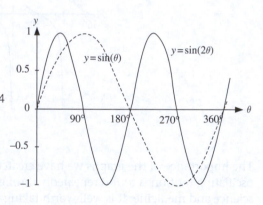

Example 9

For $0° \leq \theta \leq 360°$, sketch the graph of $y = \sin(2\theta)$ and compare with the graph of $y = \sin(\theta)$.

Solution

Let $f(\theta) = \sin(\theta)$ then $f(2\theta) = \sin(2\theta)$. The graph of $\sin(2\theta)$ is basically the graph of $\sin(\theta)$, but it is squeezed by a factor of 2 as shown in Fig. 24. (In the summary of **Chapter 3** we pointed out that $f(cx)$ is the graph of $f(x)$ but squashed horizontally by a factor of c if $c > 1$.)

Fig. 24

The graph $\sin(2\theta)$ has two cycles between 0° and 360° compared to the one cycle for $\sin(\theta)$. Similarly, a graph of $\sin(3\theta)$ will have three cycles between 0° and 360°.

SUMMARY

We can define the trigonometric functions for angles of any size. The graphs of $\sin(\theta)$ and $\cos(\theta)$ lie between -1 and $+1$; note the symmetrical nature of these graphs. Also $\tan(\theta)$ is not defined for odd whole number multiples of 90°.

Exercise **4(b)**

Solutions at end of book. Complete solutions available at www.palgrave.com/engineering/singh

1 Plot the graph of $y = \cos(\theta)$ for $\theta = 0°$ to 360°.

2 Plot the graph of $y = \tan(\theta)$ for $\theta = 0°$ to 360°.

3 Sketch, on different axes, the graphs of
  a $\cos(2\theta)$ **b** $\sin(\theta + 90°)$
  c $\tan(\theta - 180°)$ **d** $\sin(\theta + 360°)$
  e $\cos(3\theta)$ **f** $\tan(2\theta)$

SECTION C **Trigonometric equations**

By the end of this section you will be able to:
► find particular solutions of trigonometric equations
► find general solutions of trigonometric equations

Solving trigonometric equations is a highly sophisticated and difficult topic. Finding the general solution of these equations requires special proficiency in algebra. This section is more challenging than the previous two sections of this chapter.

C1 **Particular solutions of trigonometric equations**

Example 10

Solve:

a $\sin(\theta) = 0.5$ **b** $\cos(\theta) = -\dfrac{3}{4}$

where θ is between $0°$ and $360°$.

Solution

a What are we trying to find?

We need to find the angle θ which satisfies $\sin(\theta) = 0.5$. **How?**

Take inverse sin, $\sin^{-1}$, of both sides:

$$\theta = \sin^{-1}(0.5) = 30°$$

Ideally you should know $\sin(30°) = 0.5$ (from Table 1) but you can use a calculator. We PRESS $\boxed{\text{SHIFT}}$ $\boxed{\text{SIN}}$ $\boxed{0.5}$ and the calculator should show 30. (On some calculators we PRESS $\boxed{\text{INV}}$ or $\boxed{\text{2nd FUN}}$ instead of $\boxed{\text{SHIFT}}$.)

By looking at the graph of $\sin(\theta)$ there is another θ value between $0°$ and $180°$ which satisfies $\sin(\theta) = 0.5$ (Fig. 25).

The other angle is obtained by subtracting $30°$ from $180°$ because the sine graph is symmetrical about $\theta = 90°$ (Fig. 25). So

$$\theta = 30° \text{ or } \theta = 180° - 30° = 150°$$

Checking our result, we have $\sin(30°) = \sin(150°) = 0.5$.

Example 10 *continued*

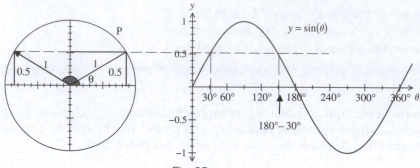

Fig. 25

b We need to determine angle θ which satisfies $\cos(\theta) = -\dfrac{3}{4}$. Initially we can use our

calculator to find the angle, that is $\theta = \cos^{-1}\left(-\dfrac{3}{4}\right) = 138.6°$.

By examining the cos graph we notice that there is another angle θ between 180° and

360° which satisfies $\cos(\theta) = -\dfrac{3}{4} = -0.75$:

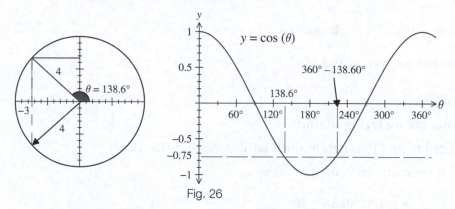

Fig. 26

The other angle is determined by subtracting 136.6° from 360° because as stated earlier the cos graph is **symmetrical** about the $\theta = 180°$ as you can see from Fig. 26. Hence

$$\theta = 138.6° \text{ or } \theta = 360° - 138.6° = 221.4°$$

? **Do we need to plot the graph of $\sin(\theta)$ to solve $\sin(\theta) = 0.5$ between 0° and 360°, and find both solutions?**

No.

Next we examine another technique for solving trigonometric equations like

$\sin(\theta) = 0.5$ or $\cos(\theta) = -\dfrac{3}{4}$.

In the previous section we divided the interval 0° to 360° into four equal quadrants: 0°−90°, 90°−180°, 180°−270° and 270°−360°. By the graphs of sin, cos and tan we have Fig. 27, part **a**:

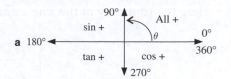

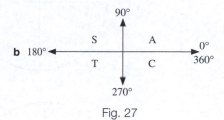

All [$\sin(\theta)$, $\cos(\theta)$ and $\tan(\theta)$] are positive, if 0° < θ < 90°.

Only $\sin(\theta)$ is positive, if 90° < θ < 180°.

Only $\tan(\theta)$ is positive, if 180° < θ < 270°.

Only $\cos(\theta)$ is positive, if 270° < θ < 360°.

Fig. 27

The mnemonic CAST can be used to remember the signs of the trigonometric functions as shown in Fig. 27, part **b**. By convention, θ is negative if we measure clockwise from the positive x axis (Fig. 28).

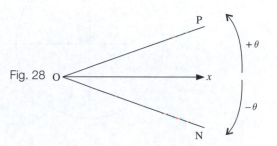

Fig. 28

The equation $\sin(\theta) = 0.5$ from **Example 10a** can also be solved by looking at CAST (Fig. 29, part **a**).

We know θ lies in quadrants S and A because those are the only quadrants where $\sin(\theta)$ is positive, $\sin(\theta) = +0.5$. Hence, $\theta = 30°$ or $150°$. One of the angles can be found by using your calculator and the other angle(s) can be established using CAST.

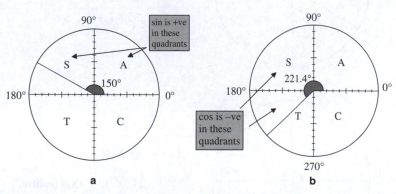

Fig. 29

So for **Example 10b**, $\cos(\theta) = -\dfrac{3}{4}$, therefore θ lies in the quadrants S and T because these are the only quadrants where $\cos(\theta)$ is negative (see Fig. 29, part **b**). Hence we have $\theta = 138.6°$ or $\theta = 221.4°$.

Here is a summary of the sine, cosine and tangent values for angles in different quadrants:

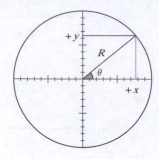

$$\sin(\theta) = \frac{y}{R} = POS$$

$$\cos(\theta) = \frac{x}{R} = POS$$

$$\tan(\theta) = \frac{y}{x} = POS$$

ALL positive

a

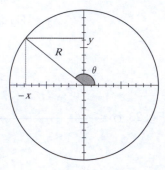

$$\sin(\theta) = \frac{y}{R} = POS$$

$$\cos(\theta) = -\frac{x}{R} = NEG$$

$$\tan(\theta) = \frac{y}{-x} = NEG$$

SIN positive

b

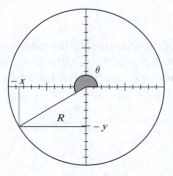

$$\sin(\theta) = -\frac{y}{R} = NEG$$

$$\cos(\theta) = -\frac{x}{R} = NEG$$

$$\tan(\theta) = \frac{-y}{-x} = POS$$

TAN positive

c

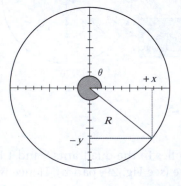

$$\sin(\theta) = -\frac{y}{R} = NEG$$

$$\cos(\theta) = \frac{x}{R} = POS$$

$$\tan(\theta) = -\frac{y}{x} = NEG$$

COS positive

d

Fig. 30

It's worth remembering the mnemonic CAST.

Example 11

Solve:

$$2\cos^2(\theta) = 2 - 3\cos(\theta)$$

where $-180° \leq \theta \leq 180°$.

Solution

? **What are we trying to find?**

The value of θ which satisfies the above equation. Rearranging gives

$$\boxed{*} \qquad 2\cos^2(\theta) + 3\cos(\theta) - 2 = 0$$

? The term $\cos^2(\theta) = [\cos(\theta)]^2$. **So how do we solve this equation?**

Let $x = \cos(\theta)$, so we have

$$2x^2 + 3x - 2 = 0 \quad \text{[Quadratic equation]}$$

? We need to find x. **How?**

Factorize:

$$(2x - 1)(x + 2) = 0$$

$$x = \frac{1}{2} \text{ or } x = -2 \qquad \text{[Solving]}$$

$$\cos(\theta) = \frac{1}{2} \text{ or } \cos(\theta) = -2 \quad \text{[Replacing } x = \cos(\theta)\text{]}$$

The result $\cos(\theta) = -2$ is impossible because $\cos(\theta)$ lies between -1 and $+1$, as can be seen by the graph of $\cos(\theta)$ in Fig. 31a.

So $\cos(\theta) = \frac{1}{2} = 0.5$.

Fig. 31a

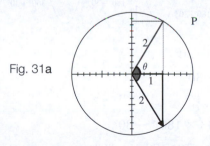
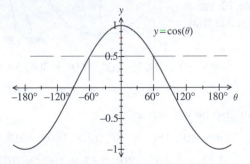

? **How do we find θ?**

Take inverse cos, $\cos^{-1}$, of both sides: $\theta = \cos^{-1}(1/2) = 60°$.

Example 11 *continued*

The other angle lies in the 4th quadrant because that is the only other quadrant where $\cos(\theta)$ is positive, $\cos(\theta) = +0.5$ (Fig. 31b).

Since θ needs to be between $-180°$ and $180°$, according to the question, we measure the other angle clockwise from $0°$. The other angle is $-60°$.

Fig. 31b

Therefore $\theta = 60°$ or $-60°$. Check your result by substituting $\theta = 60°$ and $\theta = -60°$ into given equation ▮ * ▮ .

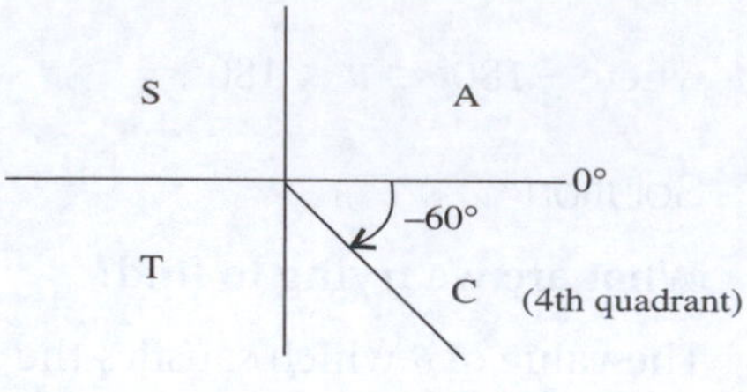

C2 General solutions of trigonometric equations

Example 12

Find the general solution of

$$\sin(\theta) = 0.5$$

Solution

Here, there is no restriction on the size of θ. By looking at the graph of $\sin(\theta)$, and using our results from **Example 10**, we find (Fig. 32) that

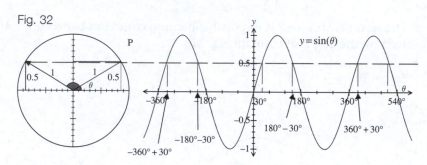

Fig. 32

$$\theta = 30°, \ 180° - 30°, \ 360° + 30°, \ 540° - 30°, \ \ldots$$
$$-180° - 30°, \ -360° + 30°, \ -540° - 30°, \ \ldots$$

This can also be written as

▮ * ▮　$\theta = 30°, \ 180° - 30°, \ (2 \times 180)° + 30°, \ (3 \times 180)° - 30°, \ \ldots$
$$(-1 \times 180)° - 30°, \ (-2 \times 180)° + 30°, \ (-3 \times 180)° - 30° \ \ldots$$

By looking at ▮ * ▮ , can you see a pattern?

The general solution of $\sin(\theta) = 0.5$ is

$$\theta = (n \times 180)° + [(-1)^n \times 30]°$$

where n is any whole number.

? The $(-1)^n$ simply changes the sign alternately. **What is $[(-1)^n \times 30]°$ if $n=1$?**

$$[(-1)^1 \times 30]° = -30°$$

Similarly if $n = 2$ then $[(-1)^2 \times 30]° = +30°$. For $n = 3$, we have $-30°$, and for $n = 4$ we have $+30°$, etc.

In general, if $\sin(\theta) = R$ (where R is a real number between -1 and 1 inclusive) then

4.16 $\theta = (180 \times n)° + (-1)^n\alpha$

where α is the angle shown in Fig. 33.

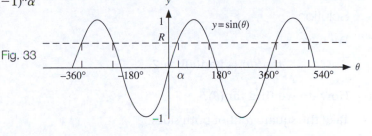

Fig. 33

4.16 is called the general solution of $\sin(\theta) = R$.

To find angle α, use your calculator.

Similarly the general solution of $\cos(\theta) = R$ is

4.17 $\theta = (360 \times n)° \pm \alpha$

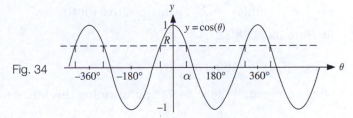

Fig. 34

where α is the angle shown in Fig. 34.

The general solution of $\tan(\theta) = R$ (in this case R is any real number) is

4.18 $\theta = (180 \times n)° + \alpha$

where α is the angle shown in Fig. 35.

Fig. 35

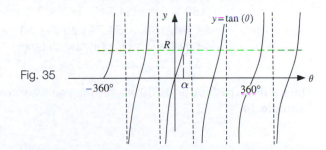

Example 13

Find the general solution of

$$\cos(\theta) = 0.927$$

Solution

By using the above formula **4.17** :

$$\theta = (360 \times n)° \pm \alpha$$

where $\alpha = \cos^{-1}(0.927) = 22.03°$. (This angle is obtained by using a calculator.) For any whole number n

$$\theta = (360 \times n)° \pm 22.03°$$

Example 14

Solve

$$\tan^2(\theta) = 2$$

where $0° \le \theta \le 720°$ and $\tan(\theta)$ is positive.

Solution

We have

$$\tan^2(\theta) = [\tan(\theta)]^2 = 2$$

How do we find $\tan(\theta)$?

Take the square root of both sides:

$$\tan(\theta) = \sqrt{2} \quad \text{[Only positive root]}$$

By formula 4.18 :

$$\theta = (180 \times n)° + \alpha$$

where $\alpha = \tan^{-1}(\sqrt{2}) = 54.74°$. Substituting this into θ for $n = 0, 1, 2$ and 3:

$$\theta = (180 \times 0)° + 54.74°, (180 \times 1)° + 54.74°,$$
$$(180 \times 2)° + 54.74°, (180 \times 3)° + 54.74°$$
$$= 54.74°, 180° + 54.74°, 360° + 54.74°, 540° + 54.74°$$
$$= 54.74°, 234.74°, 414.74°, 594.74°$$

The result is that you add $180°$ each time you want to find a new positive angle in
Example 14.

SUMMARY

We can solve trigonometric equations by examining the appropriate graph or using
CAST.
For general solutions of $\sin(\theta) = R$, $\cos(\theta) = R$ and $\tan(\theta) = R$ we apply 4.16 , 4.17
and 4.18 respectively.

Exercise 4(c)

Solutions at end of book. Complete solutions available
at www.palgrave.com/engineering/singh

1 Solve the following for θ, where
$0° \le \theta \le 360°$:

  a $\sin(\theta) = 0.707$ **b** $\cos(\theta) = 0.5$

  c $\tan(\theta) = \sqrt{3}$ **d** $\tan^2(\theta) - 3 = 0$

  e $\cos^2(\theta) - \cos(\theta) + \dfrac{1}{4} = 0$

  f $2\sin^2(\theta) - 3\sin(\theta) = 0$

  g $7\sin(\theta) - 3 = 0$

Exercise **4(c) continued**

Solutions at end of book. Complete solutions available at www.palgrave.com/engineering/singh

2 Find the general solution (if possible) of

a $\tan(\theta) = 0.8$ **b** $\cos(\theta) = -0.125$

c $\sin(\theta) = 0.22$ **d** $\sin(3\theta) = 0.72$

e $\cos(2\theta + 60°) = 2$

f $\cos(2\theta + 60°) = \dfrac{1}{2}$

3 Solve (if possible) the following for θ in the given domain:

a $\tan^2(\theta) - 2\tan(\theta) = -1$

$$0° \le \theta \le 720°$$

b $\cos(\theta)\sin(\theta) + \cos(\theta) = 0$

$$0° \le \theta \le 360°$$

c $2\cos^2(\theta) - 4\cos(\theta) - 5 = 0$

$$0° \le \theta \le 180°$$

d $\tan^2(\theta) - \tan(\theta) - 1 = 0$

$$0° \le \theta \le 360°$$

e $\sin^2(\theta) - 6\sin(\theta) = -4$

$$0° \le \theta \le 720°$$

f $\cos^2(\theta) - 10\cos(\theta) + 23 = 0$

$$-360° \le \theta \le 360°$$

4 Find the exact values of θ, where $0° \le \theta \le 360°$, which satisfies the following equations.

a $2\cos^3(\theta) - \cos(\theta) = 0$

b $4\sin^2(\theta) + \sqrt{2}\sin(\theta) = \sqrt{6}\sin(\theta)$

c $16\cos^2(\theta)\sin^2(\theta) - 12\cos^2(\theta) - 12\sin^2(\theta) + 9 = 0$

SECTION D **Trigonometric rules**

By the end of this section you will be able to:

► apply the sine rule to solve a non-right-angled triangle

► apply the cosine rule to solve a non-right-angled triangle

? **What property do the triangles of Section A have in common?**

Each of these triangles contains a 90° angle. The formulae of that section **only** apply to a right-angled triangle. You cannot use the formulae for any triangle. In this section we examine some new formulae, called the sine and cosine rules, that can be applied to **any** triangle. These rules are used to find unknown sides and angles.

In this section we use the property:

The sum of the angles in a triangle is equal to 180°

The following are good handy rules to use.

D1 Sine rule

Consider any triangle ABC with **side** a opposite **angle** A, **side** b opposite **angle** B and **side** c opposite **angle** C (Fig. 36). Then the following is true:

4.19
$$\frac{a}{\sin(A)} = \frac{b}{\sin(B)} = \frac{c}{\sin(C)}$$

4.19 is called the **sine rule**. The sine rule can be used provided we are given either:

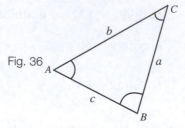

Fig. 36

i two angles and one side or
ii two sides and one angle but the angle is **not** between the two given sides.

Note that we can apply the sine rule for any triangle; it doesn't need to be a right-angled triangle.

Example 15

Figure 37 shows part of a crane. Find the length of the inclined jib BC.

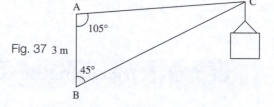

Fig. 37 3 m

Solution

Since we are given two angles and one side, we can use the sine rule 4.19 :

We need to find length a (Fig. 38).
From 4.19 :

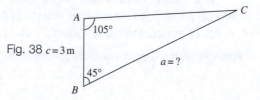

Fig. 38 $c = 3$ m

★
$$\frac{a}{\sin(A)} = \frac{c}{\sin(C)}$$

We don't need to use the central part of 4.19 because length b is not required. **We know length $c = 3$, but what is the value of angle C?**

Since the sum of the angles of a triangle is 180° we have

$$\text{angle } C = 180° - (105° + 45°) = 30°$$

Substituting $A = 105°$, $C = 30°$ and $c = 3$ into ★ gives

$$\frac{a}{\sin(105°)} = \frac{3}{\sin(30°)} = 6$$
$$a = 6 \times \sin(105°)$$
$$= 5.80 \text{ m}$$

Length BC = 5.8 m (1 d.p.).

D2 Ambiguous case of the sine rule

Example 16

Find angle B in the case shown in Fig. 39.

Fig. 39

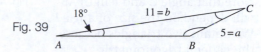

Solution

? **Can we apply the sine rule?**

Since we have two sides and an angle, we can use the sine rule. Labelling the sides and angles according to the sine rule we have $a = 5$, $b = 11$ and angle $A = 18°$. Using 4.19 :

$$\frac{a}{\sin(A)} = \frac{b}{\sin(B)}$$

with $a = 5$, $b = 11$ and $A = 18°$ we have

$$\frac{5}{\sin(18°)} = \frac{11}{\sin(B)}$$

$$16.18 = \frac{11}{\sin(B)}$$

$$\sin(B) = \frac{11}{16.18} = 0.68$$

? **How do we find angle B?**

Take the inverse sin, $\sin^{-1}$, of both sides:

$$B = \sin^{-1}(0.68) = 42.84° \quad \text{[By calculator]}$$

By looking at Fig. 39 we can see angle B is much larger than $42.84°$.

? **What mistake(s) have we made in our calculation?**

There are no mistakes but this is **not** the **only** solution for angle B. We know from the previous section that the trigonometric equation

Fig. 40

$$\sin(B) = 0.68$$

has many solutions. Figure 40 shows the sine graph.

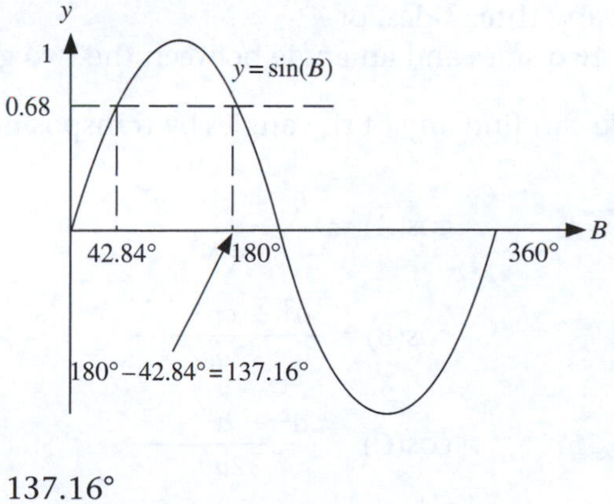

Examining Fig. 40, angle B can have two values:

$$B = 42.84° \text{ or } B = 180° - 42.84° = 137.16°$$

The triangle of Fig. 39 has angle $B = 137.16°$ (2 d.p.).

Remember that the sum of the angles in a triangle is 180°. Thus we only examine the sine graph between 0° and 180°.

In **Example 16** if you are **not** shown the triangle but just given the data, angle A is 18°, a is 5 and b is 11, then angle B could be either 42.84° or 137.16°. Thus we obtain two triangles (Fig. 41).

Note that angle C and length AB will depend on the size of angle B. We obtain two triangles because the general trigonometric equation

$\sin(\theta)$ = positive real number

has two solutions between 0° and 180°.

This situation where two triangles are obtained is called the **'ambiguous case'** of the sine rule.

Fig. 41

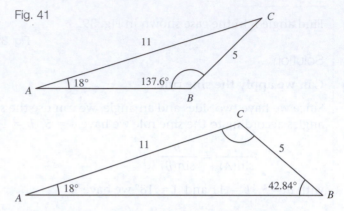

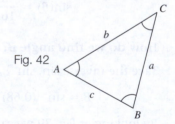

D3 Cosine rule

With reference to Fig. 42, we also have

| 4.20 | $a^2 = b^2 + c^2 - 2bc\cos(A)$ |

Fig. 42

| 4.21 | $b^2 = a^2 + c^2 - 2ac\cos(B)$ |

| 4.22 | $c^2 = a^2 + b^2 - 2ab\cos(C)$ |

4.20 to 4.22 are called the **cosine rule.** The cosine rule can be used provided we are given either

i any three sides, or
ii two sides and an angle between the two given sides.

We can find any of the angles by transposing formulae 4.20 to 4.22 :

| 4.23 | $\cos(A) = \dfrac{b^2 + c^2 - a^2}{2bc}$ |

| 4.24 | $\cos(B) = \dfrac{a^2 + c^2 - b^2}{2ac}$ |

| 4.25 | $\cos(C) = \dfrac{a^2 + b^2 - c^2}{2ab}$ |

Again, the cosine rule can be used for any triangle.

Example 17

Figure 43 shows a crank mechanism. Crank arm OA, of length 150 mm, rotates clockwise about O. Connecting rod AB is of length 350 mm. If angle OAB = 75°, find length OB.

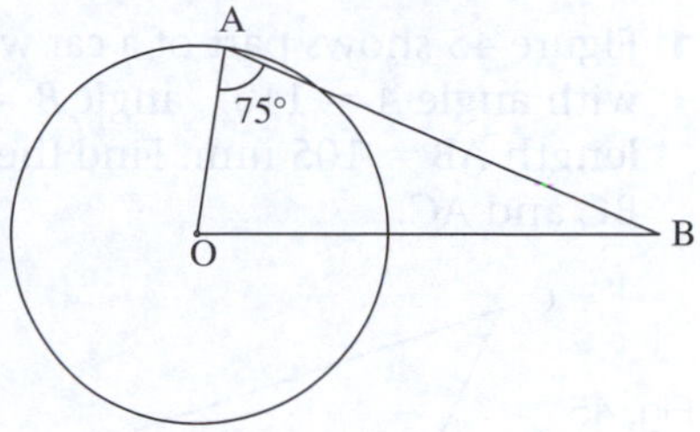

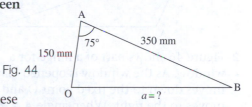

Fig. 43

Solution

We have the situation shown in Fig. 44.

Since we are given two sides and the angle between them, which rule do we employ?

The cosine rule 4.20

Let $c = 350$, $b = 150$ and angle $A = 75°$. We label side OB as a because OB is opposite angle A. We need to find length a. Substituting these values into $a^2 = b^2 + c^2 - 2bc\cos(A)$ gives

Fig. 44

$$a^2 = 150^2 + 350^2 - \left[2 \times 150 \times 350 \times \cos(75°)\right]$$
$$= 117\,824$$

Taking the square root of both sides:

$$a = \sqrt{117\,824} = 343 \text{ mm} \quad (3 \text{ s.f.})$$

SUMMARY

If we define sides a, b and c to be the lengths opposite angles A, B and C respectively, then we have

4.19 $$\frac{a}{\sin(A)} = \frac{b}{\sin(B)} = \frac{c}{\sin(C)} \qquad \text{(sine rule)}$$

The sine rule is used when given two angles and one side, or two sides and one angle, but the angle is **not** between the two given sides.

4.20 $$a^2 = b^2 + c^2 - 2bc\cos(A)$$

4.21 $$b^2 = a^2 + c^2 - 2ac\cos(B) \left.\right\} \text{(cosine rule)}$$

4.22 $$c^2 = a^2 + b^2 - 2ab\cos(C)$$

The cosine rule is used when given any three sides, or, two sides and an angle between the given sides.

We can transpose 4.20 – 4.22 to find any of the angles.

Exercise **4(d)**

Solutions at end of book. Complete solutions available at www.palgrave.com/engineering/singh

1 Figure 45 shows part of a car window with angle $A = 110°$, angle $B = 40°$ and length AB = 105 mm. Find the lengths BC and AC.

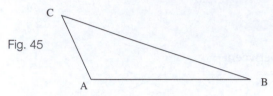

Fig. 45

2 Figure 46 shows part of a hinge for a window. As the window is opened, part B moves down to the fixed point O and A moves to the right. When angle $A = 22°$, find length OB.

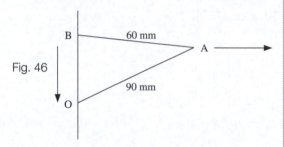

Fig. 46

3 A crane consists of an inclined jib BC = 6.0 m, a vertical part AB = 2.0 m and a tie AC = 5.0 m. Find angle A.

4 Figure 47 shows a robot at C moving an object from A to B. Lengths AC = 1.40 m, BC = 1.90 m and angle $A = 50.0°$. Find the distance the robot moves the object.

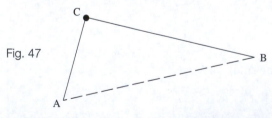

Fig. 47

5 A roof (of a triangular shape) has a span (or width) of 10.0 m and the two sides make an angle of 40° and 45° at each end. Find the lengths of these sides.

6 Figure 48 shows a slider crank mechanism. Arm OA = 60 mm,

connecting rod AB = 180 mm and length OB = 200 mm. Find the size of angle A.

Fig. 48

7 Let ABC be a triangle with AC = 1 m, AB = 2 m and angle $C = 45°$. Find the length BC.

8 A crank mechanism is shown in Fig. 49. It consists of an arm OA = 30 mm rotating anticlockwise about O and a connecting rod AB = 60 mm. B moves along the horizontal line OB. What is the distance OB when OA has rotated by $\frac{1}{8}$ of a revolution? (A complete revolution is 360°.)

Fig. 49

9 A surveyor wants to know the distance from location A to location B. She knows AC = 291 m, BC = 405 m and angle $C = 79°$, where C is another location. From these measurements, find the distance AB.

10 Figure 50 shows a **symmetric** framework about OC. Using the measurements shown, find the lengths CD and DE.

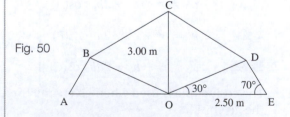

Fig. 50

11 Find all the unknown sides and angles of a triangle ABC for the following:
   a AC = 10, BC = 6 and angle $A = 30°$
   b AB = 7, BC = 9 and angle $C = 35°$.

SECTION E **Radians**

By the end of this section you will be able to:
► convert angles from degrees to radians
► convert angles from radians to degrees

E1 **Radian measure**

Up until now we have been measuring angles in degrees, but the SI unit of angular measure is called the **radian** and is derived from the words radius and angle, denoted by rad.

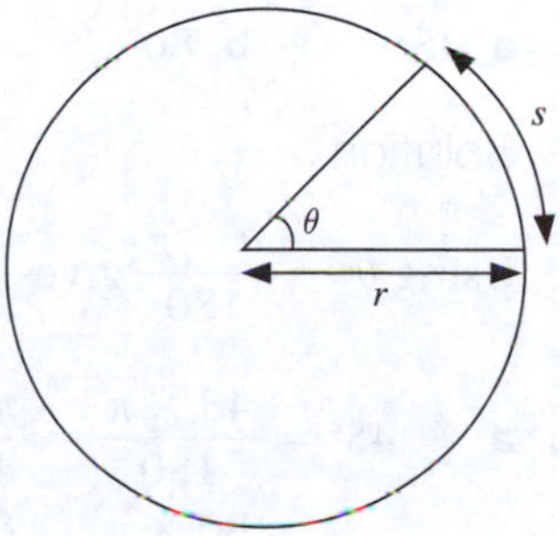

Let's recall some of the definitions connected with a circle.

The **radius**, r, is the distance from the centre of the circle to the edge.

Fig. 51

The **circumference** is the length of the perimeter of the circle and is equal to $2\pi r$.

The **arc** (s) is part of the circle (or curve).

The angle θ of Fig. 51 in radians is defined as

$$\theta = \frac{\text{arc length}}{\text{radius}} = \frac{s}{r} \text{ radians}$$

If the line of the radius does one complete revolution, $\theta = 360°$, then

$$\theta = \frac{\text{circumference of circle}}{\text{radius}}$$

$$= \frac{2\pi r}{r} = 2\pi \text{ radians} \qquad \text{[Cancelling } r\text{'s]}$$

Thus we have

$$360° = 2\pi \text{ radians}$$

? **What is 180° in radians?**

For 180°, the radius line only completes half a revolution, so $180° = \pi$ radians.

? **What is 1° in radians?**

$$1° = \frac{\pi}{180} \text{ radians}$$

? **What is $\theta°$ in radians?**

$$\theta° = \theta \times \frac{\pi}{180} \text{ radians}$$

To convert from $\theta°$ to radians we use

4.26 $$\theta° = \frac{\theta \times \pi}{180} \text{ radians}$$

Example 18

Express the following in radians (leaving π in your answer, wherever appropriate):

a 45°   **b** 80°   **c** 135°   **d** 153.2°

Solution

Using $\theta° = \dfrac{\theta \times \pi}{180}$ gives

a $45° = \dfrac{45 \times \pi}{180} = \dfrac{\pi}{4} \text{ rad} \left(\text{because } \dfrac{45}{180} = \dfrac{1}{4}\right)$

b $80° = \dfrac{80 \times \pi}{180} = \dfrac{8\pi}{18} = \dfrac{4\pi}{9} \text{ rad}$ [Cancelling 0's]

c $135° = \dfrac{135 \times \pi}{180} = \dfrac{3\pi}{4} \text{ rad} \left(\text{because } \dfrac{135}{180} = \dfrac{3}{4}\right)$

d $153.2° = \dfrac{153.2 \times \pi}{180} = 2.67 \text{ rad}$ (2 d.p.)

? **How do we convert θ radians to degrees?**

By transposing formula 4.26 :

4.27 $$\theta \text{ radians} = \left(\frac{\theta \times 180}{\pi}\right)°$$

We use 4.27 to convert from radians to degrees.

Example 19

Express the following radians in degrees:

a $\dfrac{\pi}{2}$ **b** 2.5   **c** 1   **d** $\dfrac{3\pi}{2}$

Example 19 *continued*

Solution

Using $\theta = \left(\dfrac{\theta \times 180}{\pi}\right)^{\circ}$, by replacing θ with the appropriate value above, gives

a $\dfrac{\pi}{2} = \left(\dfrac{\pi \times 180}{2\pi}\right)^{\circ} = 90°$ [Cancelling π's]

b $2.5 = \left(\dfrac{2.5 \times 180}{\pi}\right)^{\circ} = 143.24°$ (2 d.p.)

c $1 = \left(\dfrac{1 \times 180}{\pi}\right)^{\circ} = 57.30°$ (2 d.p.)

What does 1 radian mean?

This is the angle (57.3°) where the radius = arc length.

d $\dfrac{3\pi}{2} = \left(\dfrac{3\pi \times 180}{2\pi}\right)^{\circ} = 270°$ [Cancelling π's]

You may use your calculator to convert from degrees to radians or vice versa.

SUMMARY

To convert from degrees to radians we use

4.26 $x° = \dfrac{x \times \pi}{180}$ radians

To convert from radians to degrees we use

4.27 x radians $= \left(\dfrac{x \times 180}{\pi}\right)^{\circ}$

Exercise 4(e)

Solutions at end of book. Complete solutions available at www.palgrave.com/engineering/singh

(You may use your calculator to convert.)

1 Express the following in radians:

  i 123°   **ii** 13°
  iii 131.67°   **iv** 333.3°

2 Convert the following into radians (leave π in your answers):

  i 90°   **ii** 30°   **iii** 330°   **iv** 22.5°
  v 10°   **vi** 27°   **vii** 3°   **viii** 144°

3 Express the following radians in degrees:

  i $\dfrac{\pi}{6}$ **ii** $\dfrac{3\pi}{10}$ **iii** 1.45   **iv** $\dfrac{\pi}{126}$

  v 200   **vi** $\dfrac{\pi}{9}$ **vii** $\dfrac{7\pi}{81}$ **viii** $\dfrac{13\pi}{120}$

4 Evaluate the following:

  i $\sin\left(\dfrac{\pi}{10}\right)$ **ii** $\cos\left(\dfrac{7\pi}{2}\right)$ **iii** $\dfrac{\cos(7\pi)}{2}$

Solutions at end of book. Complete solutions available at www.palgrave.com/engineering/singh

Exercise **4(e) continued**

5 Find the exact values of

$$\textbf{i}\quad \frac{\pi}{2}\sin\!\left(\frac{\pi}{4}\right)\qquad \textbf{ii}\quad \frac{1}{\sqrt{3}}\tan\!\left(\frac{\pi}{3}\right)$$

$$\textbf{iii}\quad \sin\!\left(\frac{\pi}{4}\right)\cos\!\left(\frac{\pi}{4}\right)$$

6 The following are car engine speeds in revolutions per minute. Convert these into radians per second.

    i 2500   **ii** 500

    iii 6300

SECTION F **Wave theory**

By the end of this section you will be able to:
▶ find the amplitude, period, frequency and phase of a wave
▶ sketch graphs of waves
▶ determine the time displacement between two waves

F1 **Waves**

In many engineering applications, it is appropriate to graph the sine of an angle vertically versus time horizontally. The graph of $y = \sin(t)$ is called the **sine wave** (Fig. 52).

Fig. 52

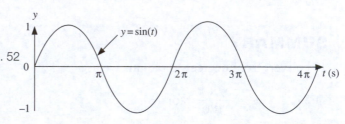

Notice that the graph of $y = \sin(t)$ looks like a wave, and it does one complete cycle from 0 to 2π seconds.

Similarly the graph of $y = \cos(t)$ is called the **cosine wave** (Fig. 53).

Fig. 53

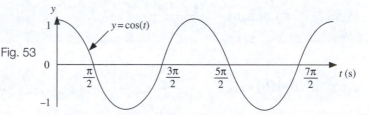

F2 **Amplitude, period and frequency of a wave**

Here we look in more detail at the sine and cosine waves introduced in **Section B2**, and define the terms that allow us to describe various properties of waves.

Consider a point P on a wheel of radius R rotating anticlockwise about O as shown in Fig 54. The **angular velocity** of point P as it travels around the circumference is given by the distance travelled (given by the magnitude of θ), divided by time taken (in seconds). The SI unit of angular velocity is therefore radians/second, and is denoted by the symbol ω (omega).

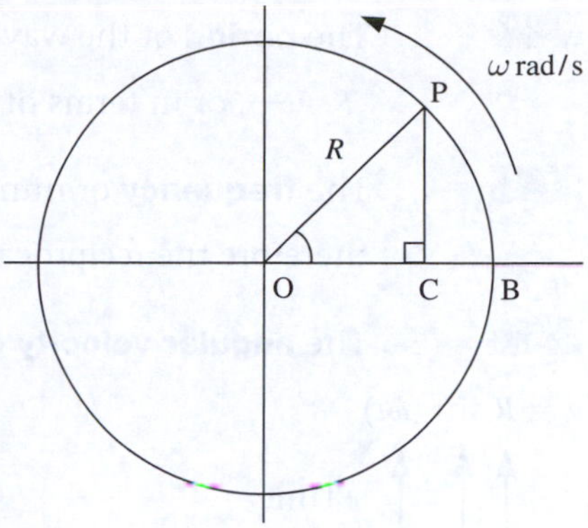

$$\text{Angular velocity} = \frac{\text{radians}}{\text{second}} \quad \text{or in symbolic}$$

Fig. 54

notation $\omega = \dfrac{\theta}{t}$

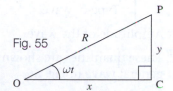

We can rearrange this equation to make θ the subject:

$$\theta = \omega t$$

This means that in t seconds the point P would have rotated through an angle of ωt radians.

Let x and y be the horizontal and vertical components of OP respectively (Fig. 55). Recollect that with reference to the angle θ, we have x = adjacent, y = opposite and R = hypotenuse.

Fig. 55

By 4.5 :   adjacent = hypotenuse $\times \cos(\theta)$, $x = R\cos(\theta)$. Since the angle $\theta = \omega t$ we can redefine the changing length of x as

$$x = R\cos(\omega t)$$

By 4.4 :   opposite = hypotenuse $\times \sin(\theta)$, so by the same substitution $\theta = \omega t$:

$$y = R\sin(\omega t)$$

Thus as point P completes a revolution, the changing length of y is plotted on a graph and produces a sine wave, while the graph describing the changing length of x describes a cosine wave.

Consider the sine wave, $y = R\sin(\omega t)$ and the cosine wave, $x = R\cos(\omega t)$.

The time taken for point P to complete one revolution is known as the **period** and is denoted by T. This is also the time taken to complete one full cycle of the wave.

The **frequency** f of the wheel is given by the number of times point P rotates during an interval of 1 second. This is also the number of cycles the wave makes during an interval of 1 second.

$$\text{frequency} = \frac{1 \text{ (full cycle)}}{T \text{ (time taken for 1 cycle, or the period)}} \quad \text{that is } f = \frac{1}{T}$$

Since the angular velocity of point P is ω rad/s and there are 2π radians in each revolution, frequency is often written $f = \dfrac{\omega}{2\pi}$.

The SI unit of frequency is hertz, denoted by Hz.

For the waves $x = R\sin(\omega t)$ and $y = R\cos(\omega t)$, we define the following:

4.28   The **amplitude** of the wave is governed by the magnitude of R (the displacement of point P from the origin O)

4.29   The **period** of the wave (the time taken to complete one full cycle) is given by $T = \dfrac{1}{f}$, or in terms of angular velocity over distance, $T = \dfrac{2\pi}{\omega}$.

4.30   The **frequency** or number of cycles the wave completes during a period is therefore the reciprocal of the period, given by $f = \dfrac{1}{T}$, or $f = \dfrac{\omega}{2\pi}$.

4.31   The **angular velocity** of point P (or any point on the wave) is given by $\omega = \dfrac{\theta}{t}$.

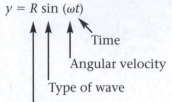

$y = R \sin(\omega t)$

Time
Angular velocity
Type of wave
Amplitude of the wave

Fig. 56

These quantities are shown for a general wave in Fig. 56.

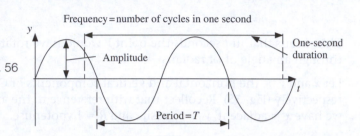

Frequency = number of cycles in one second

Amplitude

Period = T

One-second duration

Example 20

Find the exact values for the amplitude, frequency and period of

a $y = \dfrac{1}{5}\sin(3t)$ **b** $y = A\cos(100\pi t)$

Solution

a What is the amplitude of $y = \dfrac{1}{5}\sin(3t)$?

By 4.28 :   amplitude = 1/5

By 4.29 with $\omega = 3$:   period $T = \dfrac{2\pi}{\omega} = \dfrac{2\pi}{3}$ seconds

By 4.30 :   frequency $f = \dfrac{\omega}{2\pi} = \dfrac{3}{2\pi}$ Hz

b The amplitude of $y = A\cos(100\pi t)$ is A. Comparing $A\cos(100\pi t)$ with $R\cos(\omega t)$ gives $\omega = 100\pi$, and applying 4.29 and 4.30 we have

period $T = \dfrac{2\pi}{100\pi} = \dfrac{1}{50} = 0.02\,$s or 20 milliseconds

frequency $f = \dfrac{100\pi}{2\pi} = 50$ Hz   [Cancelling π's]

Example 21

State the equation of the waves shown in Fig. 57.

Solution

a By looking at Fig. 57, part **a**, we know this is a sine wave. **What is the value of the amplitude?**

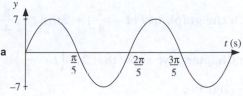

Fig. 57

Amplitude = 7

thus $y = 7\sin(\omega t)$. We need to find ω; the sine wave covers one cycle in $\dfrac{2\pi}{5}$ seconds.

Period, T, is the time taken to complete one cycle;

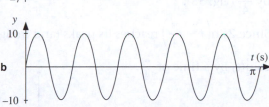

$$\text{period } T = \frac{2\pi}{\omega} = \frac{2\pi}{5}, \text{ giving } \omega = 5$$

Substituting $\omega = 5$ into $y = 7\sin(\omega t)$ gives the equation of the wave as $y = 7\sin(5t)$.

b Examining Fig. 57, part **b**, we have another sine wave. The amplitude of this graph is 10. The equation of the wave is of the form $y = 10\sin(\omega t)$, and we need to find the value of ω. By observing the graph, there are five cycles in π seconds so

$$T = \frac{\pi}{5} \qquad \text{[Time taken to complete one cycle]}$$

By 4.29 $T = \dfrac{2\pi}{\omega}$. Equating these implies

$$\frac{2\pi}{\omega} = \frac{\pi}{5}, \text{ and transposing gives } \omega = 5 \times 2 = 10$$

Substituting $\omega = 10$ into $y = 10\sin(\omega t)$ gives the equation $y = 10\sin(10t)$.

F3 Phase and time displacement of waves

Example 22

Sketch on the same axes the graphs of:

$$y = 2\sin(t), \quad y = 2\sin\left(t - \frac{\pi}{4}\right)$$

208   4 ► Trigonometry and Waveforms

Example 22 *continued*

Solution

Let $f(t) = 2\sin(t)$, then $f\left(t - \dfrac{\pi}{4}\right) = 2\sin\left(t - \dfrac{\pi}{4}\right)$, replacing t with $t - \dfrac{\pi}{4}$. **How do we**

sketch the graph of $f\left(t - \dfrac{\pi}{4}\right) = 2\sin\left(t - \dfrac{\pi}{4}\right)$?

From **Chapter 3** we know that $2\sin\left(t - \dfrac{\pi}{4}\right)$ is the graph of $2\sin(t)$ but shifted to the right

by $\dfrac{\pi}{4}$ (Fig. 58).

Since $2\sin\left(t - \dfrac{\pi}{4}\right)$ reaches its peaks (and troughs) $\dfrac{\pi}{4}$ **after** $2\sin(t)$, we say $2\sin\left(t - \dfrac{\pi}{4}\right)$ **lags**

$2\sin(t)$ by a time of $\dfrac{\pi}{4}$.

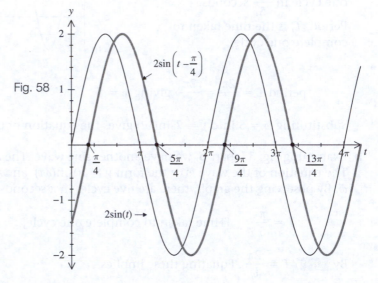

Fig. 58

If you had difficulty in following **Example 22** it might help to revise the transformation of graphs section of **Chapter 3**. Remember that the graph of $f(t - c)$ is the graph of $f(t)$ but shifted to the right by c units.

What is the relationship between the graphs $f(t)$ and $f(t + c)$?

The graph of $f(t + c)$ is the graph of $f(t)$ but shifted to the left by c units.

We use this concept in the next example.

Example 23

Sketch on the same axes:

i $y = 5\sin(2t)$ **ii** $y = 5\sin\left(2t + \dfrac{\pi}{3}\right)$

Example 23 *continued*

Solution

We can rewrite **ii** $y = 5\sin\left(2t + \dfrac{\pi}{3}\right)$ as

$$y = 5\sin\left(2t + \frac{\pi}{3}\right)$$

$$= 5\sin\left(2t + \frac{2\pi}{6}\right) \quad \left[\text{Remember } \frac{1}{3} = \frac{2}{6}\right]$$

$$= 5\sin\left[2\left(t + \frac{\pi}{6}\right)\right] \quad \text{[Taking out a factor of 2]}$$

Let $f(t) = 5\sin(2t)$ then $f\left(t + \dfrac{\pi}{6}\right) = 5\sin\left[2\left(t + \dfrac{\pi}{6}\right)\right]$, replacing t with $t + \dfrac{\pi}{6}$.

By the above statement, the graph

of $f\left(t + \dfrac{\pi}{6}\right) = 5\sin\left[2\left(t + \dfrac{\pi}{6}\right)\right]$ is

the graph of $5\sin(2t)$ but shifted to

the left by $\dfrac{\pi}{6}$ (Fig. 59).

In this example $5\sin\left(2t + \dfrac{\pi}{3}\right)$

reaches its peak $\dfrac{\pi}{6}$ **before**

$5\sin(2t)$, so we say that

$5\sin\left(2t + \dfrac{\pi}{3}\right)$ **leads** $5\sin(2t)$

by a time of $\dfrac{\pi}{6}$

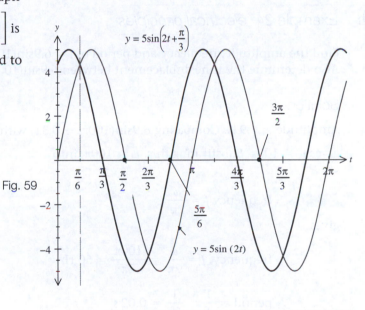

Fig. 59

By using a graphical calculator, or otherwise, you can find the following results:

a $\sin\left(5t + \dfrac{\pi}{6}\right)$ **leads** $\sin(5t)$ by $\dfrac{\pi/6}{5} = \dfrac{\pi}{5 \times 6} = \dfrac{\pi}{30}$

b $100\sin(7t + 2)$ **leads** $100\sin(7t)$ by $\dfrac{2}{7}$

This lead or lag time is called the **time displacement**.

?   **By examining the above examples, can you predict a general rule for the time displacement of $R\sin(\omega t)$ and $R\sin(\omega t + \alpha)$?**

4.32 $R\sin(\omega t + \alpha)$ **leads** $R\sin(\omega t)$ by $\dfrac{\alpha}{\omega}$

?   **What is the time displacement between $R\sin(\omega t)$ and $R\sin(\omega t - \alpha)$?**

4.33 $R\sin(\omega t - \alpha)$ **lags** $R\sin(\omega t)$ by $\dfrac{\alpha}{\omega}$

4.34   The **angle** α in $R\sin(\omega t \pm \alpha)$ is generally known as the **phase angle** or just the **phase**

These wave definitions also apply to the cosine wave and are commonly used in electrical principles, as the following example shows.

Example 24 *electrical principles*

Find the amplitude, frequency and period of $i = 6.9\sin(100\pi t - 0.31)$ amps (t in s). Also determine the time displacement between $6.9\sin(100\pi t - 0.31)$ and $6.9\sin(100\pi t)$.

Solution

Amplitude = 6.9 A. Comparing $6.9\sin(100\pi t - 0.31)$ with $R\sin(\omega t \pm \alpha)$

gives $\omega = 100\pi$. Substituting this, $\omega = 100\pi$, into

4.30   frequency $f = \dfrac{\omega}{2\pi}$

gives

$$\text{frequency, } f = \frac{\omega}{2\pi} = \frac{100\pi}{2\pi} = 50 \text{ Hz}$$

$$\text{period} = \frac{1}{f} = \frac{1}{50} = 0.02 \text{ s}$$

By applying

4.33 $R\sin(\omega t - \alpha)$ lags $R\sin(\omega t)$ by $\dfrac{\alpha}{\omega}$

we find that $6.9\sin(100\pi t - 0.31)$ lags $6.9\sin(100\pi t)$ by:

$$\frac{0.31}{100\pi} = 9.87 \times 10^{-4} \text{ or } 0.987 \text{ ms (milliseconds)}$$

SUMMARY

If a wave has the equation $y = R\sin(\omega t \pm \alpha)$ then

R = amplitude

ω = angular velocity (rad/s)

4.30   frequency $f = \dfrac{\omega}{2\pi} = \dfrac{1}{T}$ (Hz)

α = phase angle of lag or lead compared with $R\sin(\omega t)$ in radians or degrees

$\dfrac{\alpha}{\omega}$ = time displacement

Exercise 4(f)

Solutions at end of book. Complete solutions available at www.palgrave.com/engineering/singh

1 State the amplitude, period and frequency of the following waves:

a $10\sin(2t)$

b $\dfrac{1}{10}\cos\left(\dfrac{t}{10}\right)$

c $H\sin(\pi t)$

d $\dfrac{\cos(\pi^2 t)}{\pi}$

e $10\,000\sin\left(\dfrac{t}{\pi^2}\right)$

f $\dfrac{\cos(t/\pi e)}{\pi e}$

g $220\cos(1000\pi t)$

h $\sin\left(\dfrac{t}{\pi/6}\right)$

f $\sin\left(\dfrac{t}{10} - \pi\right)$

g $\sin\left(\dfrac{t - \pi}{10}\right)$

h $\cos\left[\dfrac{1}{e\pi}\left(\dfrac{t+1}{e\pi}\right)\right]$

4 What is the time displacement between the following waves?

a $A\sin\left(5t + \dfrac{\pi}{5}\right)$ and $A\sin\left(5t - \dfrac{\pi}{7}\right)$

b $\cos(7t - 0.26)$ and $5\cos(7t + \pi)$

c $\cos\left(2t + \dfrac{\pi}{2}\right)$ and $\cos\left(2t + \dfrac{\pi}{3}\right)$

2 Determine the amplitude, period and phase (state lead or lag) of the following:

a $4\sin\left(t + \dfrac{\pi}{3}\right)$ **b** $\sin(314.159t - 0.25)$

c $20\sin\left[\left(100t + \dfrac{1}{5}\right)\pi\right]$

d $315\cos(7t + 30°)$

e $5 + 2\sin\left(2\pi t - \dfrac{\pi}{2}\right)$

5 State the equations of the waveforms shown in Fig. 60.

3 Find the time displacement (stating lead or lag) of the following waves:

a $\sin(t)$ **b** $7\sin\left(t + \dfrac{\pi}{7}\right)$

c $3\cos\left(2t + \dfrac{\pi}{2}\right)$

d $\cos[(100t - 1)\pi]$ **e** $31\cos\left(t - \dfrac{\pi}{2}\right)$

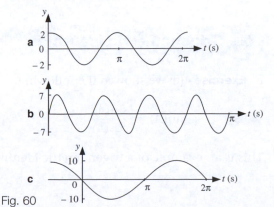

Fig. 60

Exercise **4(f) continued**

Solutions at end of book. Complete solutions available at www.palgrave.com/engineering/singh

6 [electrical principles]   Sketch the voltage waveform, $v = 5\cos(\omega t)$, as t varies over two complete cycles, 0 to $4\pi/\omega$, showing important points.

7 Sketch one cycle for the following graphs starting with $t = 0$. Show important points.

  a $7\sin(3t)$ **b** $2\cos\left(\dfrac{t}{2}\right)$

  c $3\sin\left(t + \dfrac{\pi}{3}\right)$

8 [electrical principles]   An alternating current $i = 5\sin(120\pi t + 0.52)$, ($i$ in mA, t in s). Find

  a The amplitude, period, frequency and phase (in degrees) of i.

  b The current i at: **i** $t = 0$ **ii** $t = \dfrac{1}{60}$
  iii $t = \dfrac{1}{30}$ **iv** $t = \dfrac{1}{15}$

What do you notice about your results? Why is this?

  c Find the time, t, when current, i, is first a maximum.

  d Find the time displacement between $i = 5\sin(120\pi t + 0.52)$ and $i = 5\sin(120\pi t)$.

  e Sketch one complete cycle of the waveform $i = 5\sin(120\pi t + 0.52)$.

SECTION G **Trigonometric identities**

By the end of this section you will be able to:
▶ establish some trigonometric identities
▶ use these identities to evaluate the sine, cosine and tangent of particular angles

In this section we state a lot of trigonometric identities because they are used in subsequent chapters. Some of these identities are shown in this section but you are asked to verify many of them in **Exercise 4(g)**.

The applications for the trignometric identities you learn here are discussed in **Section H**.

G1 **Trigonometric identities**

In **Exercise 4(a)** we showed the following:

4.35 $\tan(A) = \dfrac{\sin(A)}{\cos(A)}$

This is an example of a trigonometric identity.

? **What do we mean by an identity?**

It's two expressions that are equal for all values of the variables. Another example is

4.36 $\quad\cot(A) = \dfrac{\cos(A)}{\sin(A)}$

Example 25

Show the identity $\cot(A) = \dfrac{\cos(A)}{\sin(A)}$.

Solution

From 4.12 we have $\cot(A) = \dfrac{1}{\tan(A)}$, thus

$$\cot(A) = \dfrac{1}{\tan(A)} = \underbrace{\dfrac{1}{\sin(A)/\cos(A)}}_{\text{by } 4.35}$$

$$= 1 \div \dfrac{\sin(A)}{\cos(A)}$$

$$= 1 \times \dfrac{\cos(A)}{\sin(A)} \qquad \text{[Inverting the second fraction]}$$

$$= \dfrac{\cos(A)}{\sin(A)}$$

The following are more examples of trigonometric identities:

4.37 $\quad \sin(A + B) = \sin(A)\cos(B) + \cos(A)\sin(B)$

4.38 $\quad \sin(A - B) = \sin(A)\cos(B) - \cos(A)\sin(B)$

4.39 $\quad \cos(A + B) = \cos(A)\cos(B) - \sin(A)\sin(B)$

4.40 $\quad \cos(A - B) = \cos(A)\cos(B) + \sin(A)\sin(B)$

4.41 $\quad \tan(A + B) = \dfrac{\tan(A) + \tan(B)}{1 - \tan(A)\tan(B)}$

4.42 $\quad \tan(A - B) = \dfrac{\tan(A) - \tan(B)}{1 + \tan(A)\tan(B)}$

We can use these identities to establish other identities or to evaluate sin, cos and tan of an angle, as the following example shows.

Example 26

By using the above identities and Table 1 on page 175, find the exact value of $\sin(15°)$.

Solution

By examining Table 1 we only have the sin of angles 0°, 30°, 45°, etc. So we need to write 15° in terms of these angles.

Example 26 *continued*

[?] **How can we rewrite 15°?**

$$15° = 45° - 30°$$

Substituting this we have

$$\sin(15°) = \sin(45° - 30°) \underset{\text{using } 4.38}{=} [\sin(45°)\cos(30°)] - [\cos(45°)\sin(30°)]$$

$$\underset{\text{by Table 1}}{=} \left(\frac{1}{\sqrt{2}} \cdot \frac{\sqrt{3}}{2}\right) - \left(\frac{1}{\sqrt{2}} \cdot \frac{1}{2}\right)$$

$$\sin(15°) = \frac{1}{2\sqrt{2}}\left(\sqrt{3} - 1\right) \qquad \left[\text{Factorizing } \frac{1}{2\sqrt{2}}\right]$$

Example 27

Show that $\cos(90° - A) = \sin(A)$.

Solution

[?] We can expand $\cos(90° - A)$. **How?**

Use

4.40 $\cos(A - B) = \cos(A)\cos(B) + \sin(A)\sin(B)$

$$\cos(90° - A) = \left[\cos(90°)\cos(A)\right] + \left[\sin(90°)\sin(A)\right]$$
$$= \left[0 \times \cos(A)\right] + \left[1 \times \sin(A)\right] \text{ [Remember } \cos(90°) = 0 \text{ and } \sin(90°) = 1]$$
$$= \sin(A)$$

Similarly we can establish the following identities. Try verifying them.

4.43	$\sin(A) = \cos(90° - A)$
4.44	$\cos(A) = \sin(90° - A)$
4.45	$\sin(A) = \sin(180° - A)$
4.46	$\cos(A) = -\cos(180° - A)$
4.47	$\tan(A) = -\tan(180° - A)$
4.48	$\sin(A) = -\sin(360° - A)$
4.49	$\cos(A) = \cos(360° - A)$
4.50	$\sin(-A) = -\sin(A)$
4.51	$\cos(-A) = \cos(A)$
4.52	$\tan(-A) = -\tan(A)$

4.38 $\sin(A - B) = \sin(A)\cos(B) - \cos(A)\sin(B)$

Example 28

Verify the trigonometric identity:

$$\sin(2A) = 2\sin(A)\cos(A)$$

Solution

We can rewrite $2A$ as $A + A$.

$$\sin(2A) = \sin(A + A)$$
$$\underline{= \sin(A)\cos(A) + \cos(A)\sin(A)}$$
$$\text{using } 4.37$$
$$= 2\sin(A)\cos(A)$$

Example 28 is an example of the **double-angle formula**. The following are further examples of double-angle formulae:

4.53	$\sin(2A) = 2\sin(A)\cos(A)$
4.54	$\cos(2A) = \cos^2(A) - \sin^2(A) = 2\cos^2(A) - 1 = 1 - 2\sin^2(A)$
4.55	$\tan(2A) = \dfrac{2\tan(A)}{1 - \tan^2(A)}$

You are asked to verify the last two, 4.54 and 4.55 , in **Exercise 4(g)**.

Example 29

Show that

$$2\sin(A)\cos(B) = \sin(A + B) + \sin(A - B)$$

Solution

Expanding the Right-Hand Side by applying 4.37 and 4.38 gives

$$\sin(A + B) + \sin(A - B) = \underbrace{\sin(A)\cos(B) + \cos(A)\sin(B)}_{\text{using } 4.37} + \underbrace{\sin(A)\cos(B) - \cos(A)\sin(B)}_{\text{using } 4.38}$$
$$= \sin(A)\cos(B) + \sin(A)\cos(B) = 2\sin(A)\cos(B)$$

Similarly we can establish the following:

4.56	$2\sin(A)\cos(B) = \sin(A + B) + \sin(A - B)$
4.57	$2\cos(A)\sin(B) = \sin(A + B) - \sin(A - B)$
4.58	$2\cos(A)\cos(B) = \cos(A + B) + \cos(A - B)$
4.59	$2\sin(A)\sin(B) = \cos(A - B) - \cos(A + B)$

4.37 $\sin(A + B) = \sin(A)\cos(B) + \cos(A)\sin(B)$ 4.38 $\sin(A - B) = \sin(A)\cos(B) - \cos(A)\sin(B)$

| 4.60 | $\sin(A) + \sin(B) = 2\sin\left(\dfrac{A + B}{2}\right)\cos\left(\dfrac{A - B}{2}\right)$ |

| 4.61 | $\sin(A) - \sin(B) = 2\cos\left(\dfrac{A + B}{2}\right)\sin\left(\dfrac{A - B}{2}\right)$ |

| 4.62 | $\cos(A) + \cos(B) = 2\cos\left(\dfrac{A + B}{2}\right)\cos\left(\dfrac{A - B}{2}\right)$ |

| 4.63 | $\cos(A) - \cos(B) = 2\sin\left(\dfrac{A + B}{2}\right)\sin\left(\dfrac{B - A}{2}\right)$ |

Remember

$$\sin(A)\sin(A) = [\sin(A)]^2 = \sin^2(A)$$

Other trigonometric functions can be represented in a similar manner.

Example 30

Show that

| * | $\cos^2(A) + \sin^2(A) = 1$ |

Solution

This is a **very** important identity which is used throughout the book. **How can we verify this identity?**

Remember that $\cos(0) = 1$, so we rewrite the 1 in ▨ * as $\cos(0)$. **But how does that serve us?**

Well, $\cos(0) = \cos(A - A)$, so by using

| 4.40 | $\cos(A - B) = \cos(A)\cos(B) + \sin(A)\sin(B)$ |

with $B = A$ we have

$$1 = \cos(A - A)$$
$$= \cos(A)\cos(A) + \sin(A)\sin(A)$$
$$= [\cos(A)]^2 + [\sin(A)]^2$$
$$1 = \cos^2(A) + \sin^2(A)$$

This identity can also be proved directly from Pythagoras's theorem. Consider a right-angled triangle and try proving it.

Example 31

Show that

$$1 + \tan^2(A) = \sec^2(A)$$

Solution

Dividing both sides of

$$\cos^2(A) + \sin^2(A) = 1$$

by $\cos^2(A)$ gives

$$\frac{\cos^2(A)}{\cos^2(A)} + \frac{\sin^2(A)}{\cos^2(A)} = \frac{1}{\cos^2(A)}$$

$$1 + \frac{\sin^2(A)}{\cos^2(A)} = \frac{1}{\cos^2(A)}$$

We can rewrite tan as $\dfrac{\sin}{\cos}$ and so $\tan^2(A) = \dfrac{\sin^2(A)}{\cos^2(A)}$. Also $\dfrac{1}{\cos(A)} = \sec(A)$

which implies $\dfrac{1}{\cos^2(A)} = \sec^2(A)$. Using these, we have

$$1 + \frac{\sin^2(A)}{\cos^2(A)} = \frac{1}{\cos^2(A)}$$

$$1 + \tan^2(A) = \sec^2(A)$$

We can establish the following trigonometric identities:

4.64	$\cos^2(A) + \sin^2(A) = 1$
4.65	$1 + \tan^2(A) = \sec^2(A)$
4.66	$1 + \cot^2(A) = \operatorname{cosec}^2(A)$
4.67	$\cos^2(A) = \dfrac{1}{2}\left[1 + \cos(2A)\right]$
4.68	$\sin^2(A) = \dfrac{1}{2}\left[1 - \cos(2A)\right]$

The next three identities are examples of **multiple-angle formulae**:

4.69	$\sin(3A) = 3\sin(A) - 4\sin^3(A)$
4.70	$\cos(3A) = 4\cos^3(A) - 3\cos(A)$
4.71	$\tan(3A) = \dfrac{3\tan(A) - \tan^3(A)}{1 - 3\tan^2(A)}$

Below, 4.72 to 4.74 are examples of **half-angle formulae**.

If $t = \tan(A/2)$ then

4.72 $\sin(A) = \dfrac{2t}{1 + t^2}$

4.73 $\cos(A) = \dfrac{1 - t^2}{1 + t^2}$

4.74 $\tan(A) = \dfrac{2t}{1 - t^2}$

Many of these results will be shown in **Exercise 4(g)**.

You don't have to remember all these trigonometric identities, but some of these are more important than others and worth remembering. For example,

$$\cos^2(A) + \sin^2(A) = 1$$

$$\tan(A) = \frac{\sin(A)}{\cos(A)}$$

$$\sin(2A) = 2\sin(A)\cos(A)$$

$$\cos(2A) = \cos^2(A) - \sin^2(A)$$

Sometimes these are referred to as **fundamental trigonometric identities**.

SUMMARY

We have established some trigonometric identities. Remember that an identity is an expression which is equal for all values of the variables.

Exercise **4(g)**

Solutions at end of book. Complete solutions available at www.palgrave.com/engineering/singh

1 By using the identities 4.37 to 4.42 and the values in Table 1, find the exact values of

   i $\sin(75°)$ **ii** $\sin(120°)$ **iii** $\cos(105°)$

   iv $\cos(150°)$ **v** $\tan(15°)$

2 Show that

   i $\sin(90° - A) = \cos(A)$

   ii $\sin(-A) = -\sin(A)$

   iii $\cos(-A) = \cos(A)$

   iv $\tan(-A) = -\tan(A)$

3 Show that

   i $\cos(2\theta) = \cos^2(\theta) - \sin^2(\theta)$

   ii $\tan(2\theta) = \dfrac{2\tan(\theta)}{1 - \tan^2(\theta)}$

4 Show that $1 + \cot^2(\theta) = \csc^2(\theta)$.

5 Show that

   i $\dfrac{1}{2}\left[1 + \cos(2\theta)\right] = \cos^2(\theta)$

   ii $\dfrac{1}{2}\left[1 - \cos(2\theta)\right] = \sin^2(\theta)$

   [Can also be written as

   $\cos(2\theta) = 2\cos^2(\theta) - 1 = 1 - 2\sin^2(\theta)$.]

Solutions at end of book. Complete solutions available at www.palgrave.com/engineering/singh

Exercise **4(g) continued**

6 Show that

i $\sin(3\theta) = 3\sin(\theta) - 4\sin^3(\theta)$

ii $\cos(3\theta) = 4\cos^3(\theta) - 3\cos(\theta)$

iii $\tan(3\theta) = \dfrac{3\tan(\theta) - \tan^3(\theta)}{1 - 3\tan^2(\theta)}$

7 Show that

i $\sin(A + B) - \sin(A - B)$
$= 2\cos(A)\sin(B)$

ii $\cos(A + B) + \cos(A - B)$
$= 2\cos(A)\cos(B)$

iii $\cos(A - B) - \cos(A + B)$
$= 2\sin(A)\sin(B)$

8 Show that

i $2\sin\left(\dfrac{A + B}{2}\right)\cos\left(\dfrac{A - B}{2}\right)$
$= \sin(A) + \sin(B)$

ii $2\cos\left(\dfrac{A + B}{2}\right)\sin\left(\dfrac{A - B}{2}\right)$
$= \sin(A) - \sin(B)$

iii $2\cos\left(\dfrac{A + B}{2}\right)\cos\left(\dfrac{A - B}{2}\right)$
$= \cos(A) + \cos(B)$

iv $2\sin\left(\dfrac{A + B}{2}\right)\sin\left(\dfrac{B - A}{2}\right)$
$= \cos(A) - \cos(B)$

SECTION H **Applications of identities**

By the end of this section you will be able to:
▶ use identities of **Section G** to show various engineering results

This section might seem like a colossal jump from previous sections, but most of it involves applying algebra to trigonometric formulae.

H1 **Engineering applications of identities**

We can simplify or rewrite engineering trigonometric formulae by using the identities of the previous section, as the following two examples show.

Example 32 *electrical principles*

The power, *p*, in a circuit is given by

$$p = iv$$

where *i* and *v* are the a.c. current and voltage respectively. Show that, if

> ### Example 32 *continued*
>
> $i = I\sin(\omega t - \phi)$ and $v = V\sin(\omega t)$ then
>
> $$p = \frac{IV}{2}\left[\cos(\phi) - \cos(2\omega t - \phi)\right]$$
>
> (I = current amplitude, V = voltage amplitude, ω = angular frequency and ϕ = phase).
> Don't let the Greek letters put you off. The symbol ϕ is the Greek letter phi pronounced
> 'fi' (to rhyme with tie).
>
> ### Solution
>
> Substituting $i = I\sin(\omega t - \phi)$ and $v = V\sin(\omega t)$ into $p = iv$ gives
>
> † $p = IV\sin(\omega t - \phi)\sin(\omega t)$
>
> By putting $A = \omega t - \phi$ and $B = \omega t$ into
>
> 4.59 $2\sin(A)\sin(B) = \cos(A - B) - \cos(A + B)$
>
> we have
>
> $$2\sin(\omega t - \phi)\sin(\omega t) = \cos(\omega t - \phi - \omega t) - \cos(\omega t - \phi + \omega t)$$
>
> $$= \underbrace{\cos(-\phi)}_{=\cos(\phi) \text{ by } 4.51} - \cos(2\omega t - \phi)$$
>
> $$= \cos(\phi) - \cos(2\omega t - \phi)$$
>
> Dividing both sides by 2 gives
>
> $$\sin(\omega t - \phi)\sin(\omega t) = \frac{1}{2}\left[\cos(\phi) - \cos(2\omega t - \phi)\right]$$
>
> Substituting this into the Right-Hand Side of † gives the required result:
>
> $$p = \frac{IV}{2}\left[\cos(\phi) - \cos(2\omega t - \phi)\right]$$

The next example is a bit tricky in comparison with earlier examples. Just take your time
and follow the process.

> ### Example 33 *mechanics*
>
> A projectile is fired with a velocity u from an inclined angle of β. The projectile makes
> an angle of α with the horizontal.
>
> **i** If $\left(u\sin(\alpha) - \frac{1}{2}gt\right)\cos(\beta) = u\sin(\beta)\cos(\alpha)$, show that
>
> $$t = \frac{2u\sin(\alpha - \beta)\sec(\beta)}{g}$$
>
> where g is acceleration due to gravity and t is time of flight.

4.51 $\cos(-A) = \cos(A)$

Example 33 *continued*

ii If $gR = u^2 \sec^2(\beta)\left[1 - \sin(\beta)\right]$, show that $R = \dfrac{u^2}{g\left[1 + \sin(\beta)\right]}$ where R is the range of the projectile.

Solution

i Transposing the given equation:

$$u \sin(\alpha)\cos(\beta) - \frac{1}{2}gt \cos(\beta) = u \sin(\beta)\cos(\alpha)$$

yields

$$u \sin(\alpha)\cos(\beta) - u \sin(\beta)\cos(\alpha) = \frac{1}{2}gt \cos(\beta) \quad \text{[Transposing]}$$

$$u\left[\sin(\alpha)\cos(\beta) - \cos(\alpha)\sin(\beta)\right] = \frac{1}{2}gt \cos(\beta) \quad \text{[Factorizing]}$$

$$u \underbrace{\sin(\alpha - \beta)}_{\text{by } \boxed{4.38}} = \frac{1}{2}gt \cos(\beta)$$

$$\frac{2u \sin(\alpha - \beta)}{g \cos(\beta)} = t \qquad \text{[Transposing]}$$

Thus

$$t = \frac{2u \sin(\alpha - \beta)}{g \cos(\beta)} = \frac{2u \sin(\alpha - \beta)}{g} \cdot \underbrace{\left(\frac{1}{\cos(\beta)}\right)}_{= \sec(\beta)}$$

$$t = \frac{2u \sin(\alpha - \beta)}{g}\sec(\beta) = \frac{2u \sin(\alpha - \beta)\sec(\beta)}{g}$$

ii Examining the Right-Hand Side of $gR = u^2 \sec^2(\beta)\left[1 - \sin(\beta)\right]$:

$$u^2 \sec^2(\beta)\left[1 - \sin(\beta)\right] = \frac{u^2\left[1 - \sin(\beta)\right]}{\cos^2(\beta)} \quad \left[\text{Because } \sec^2(\beta) = \frac{1}{\cos^2(\beta)}\right]$$

$$= \frac{u^2\left[1 - \sin(\beta)\right]}{1 - \sin^2(\beta)} \qquad \left[\text{By } \boxed{4.64} \text{ on denominator}\right]$$

$$= \frac{u^2\left[1 - \sin(\beta)\right]}{\underbrace{\left[1 - \sin(\beta)\right]\left[1 + \sin(\beta)\right]}_{\text{by } \boxed{1.15}}} \qquad \text{[Factorizing denominator]}$$

$$u^2 \sec^2(\beta)\left[1 - \sin(\beta)\right] = \frac{u^2}{1 + \sin(\beta)} \qquad \left[\text{Cancelling } 1 - \sin(\beta)\right]$$

$\boxed{1.15}$ $a^2 - b^2 = (a - b)(a + b)$ $\boxed{4.38}$ $\sin(A)\cos(B) - \cos(A)\sin(B) = \sin(A - B)$

$\boxed{4.64}$ $\sin^2(A) + \cos^2(A) = 1$

🚗 Example 33 *continued*

Substituting this into the original equation gives

$$gR = \frac{u^2}{[1 + \sin(\beta)]}$$

$$R = \frac{u^2}{g[1 + \sin(\beta)]} \qquad \text{[Dividing through by } g]$$

SUMMARY

We can use the identities of **Section G** to simplify or rewrite engineering trigonometric formulae.

Exercise **4(h)**

Solutions at end of book. Complete solutions available at www.palgrave.com/engineering/singh

1 🚗 [*mechanics*]   A vehicle is tested on a circular track with a bank at an angle θ. By doing the dynamic analysis we obtain the equations

$$N\cos(\theta) - mg = 0$$

$$N\sin(\theta) - \frac{mv^2}{r} = 0$$

where N = normal reaction of road surface, m = mass of vehicle, r = track radius of curvature and v = neutral steer speed (the speed at which the car goes round the track without steering). Show that

$$v = \sqrt{rg \tan(\theta)}$$

2 📱 [*communications*]   An amplitude modulated (a.m.) signal, v, is given by

$$v = V_c\cos(\omega_c t)$$
$$+ V_m\cos(\omega_m t)\cos(\omega_c t)$$

where V_c = carrier amplitude, V_m = modulating signal amplitude, ω_c = angular frequency of carrier and

ω_m = angular frequency of signal. Show that

$$v = V_c\left\{\cos(\omega_c t) + \frac{d}{2}\Big(\cos\big[(\omega_m + \omega_c)t\big]\right.$$
$$\left. + \cos\big[(\omega_m - \omega_c)t\big]\Big)\right\}$$

where $d = \dfrac{V_m}{V_c}$ (sometimes called the depth of modulation if given as a percentage).

3 📱 [*communications*]   An amplitude modulated signal, v, is given by

$$v = \big[V_c + V_m\sin(\omega_m t)\big]\sin(\omega_c t)$$

where V_c, V_m, ω_m, ω_c and d are defined as in the above question. Show that

$$v = \frac{V_c}{2}\Big(2\sin(\omega_c t) + d\cos\big[(\omega_c - \omega_m)t\big]$$
$$- d\cos\big[(\omega_c + \omega_m)t\big]\Big)$$

Exercise **4(h) continued**

Solutions at end of book. Complete solutions available at www.palgrave.com/engineering/singh

4 ⊞ [*electrical principles*]   The power, p, in a circuit is given by $p = iv$ where i = current and v = voltage. Show the following:

a If $i = I\sin(\omega t)$ and $v = V\sin(\omega t)$ then

$$p = \frac{IV}{2}\left[1 - \cos(2\omega t)\right]$$

b If $i = I\sin(\omega t)$ and $v = V\cos(\omega t)$ then

$$p = \frac{IV}{2}\sin(2\omega t)$$

c If $i = I\sin(\omega t)$ and $v = V\sin\left(\omega t - \frac{\pi}{2}\right)$

then $p = -\dfrac{IV}{2}\sin(2\omega t)$

d If $i = I\cos(\omega t)$ and $v = V\cos\left(\omega t + \dfrac{\pi}{2}\right)$

then $p = \dfrac{IV}{2}\cos\left(2\omega t + \dfrac{\pi}{2}\right)$

e If $i = I\sin\left(\omega t - \dfrac{\pi}{2}\right)$ and $v = V\sin(\omega t)$

then $p = -\dfrac{IV}{2}\cos\left(2\omega t - \dfrac{\pi}{2}\right)$

f If $i = 12\sin\left(100\pi t - \dfrac{\pi}{3}\right)$ and

$v = 240\sin(100\pi t)$ then

$$p = 720 - 1440\cos\left(200\pi t - \frac{\pi}{3}\right)$$

SECTION I **Conversion**

By the end of this section you will be able to:
▶ rewrite the function $a\cos(\theta) + b\sin(\theta)$ as $R\cos(\theta - \beta)$
▶ sketch the graph of $a\cos(\theta) + b\sin(\theta)$
▶ solve engineering applications with $a\cos(\theta) + b\sin(\theta)$

Another challenging section. This section relies on **one** particular formula, 4.75 , to be described below. The most difficult part is verifying this formula which is carried out in **Example 35**. As an engineering student, you should at least be able to apply this formula.

I1 **The function** $a\cos(\theta) + b\sin(\theta)$

We can write $a\cos(\theta) + b\sin(\theta)$ just in terms of the cosine function, as the following examples show.

Example 34

Show that

$$2\cos(\theta) + 3\sin(\theta) = \sqrt{13}\cos(\theta - 56.31°)$$

Solution

Expanding the Right-Hand Side by using:

4.40 $\cos(A - B) = \cos(A)\cos(B) + \sin(A)\sin(B)$

with $A = \theta$ and $B = 56.31°$ we have

$$\sqrt{13}\cos(\theta - 56.31°) = \sqrt{13}\left[\cos(\theta)\cos(56.31°) + \sin(\theta)\sin(56.31°)\right]$$

$$= \underbrace{\sqrt{13}\cos(56.31°)}_{=2}\cos(\theta) + \underbrace{\sqrt{13}\sin(56.31°)}_{=3}\sin(\theta)$$

$$= 2\cos(\theta) + 3\sin(\theta)$$

In this section we are **not** given $\sqrt{13}\cos(\theta - 56.31°)$ but we need to find it from $2\cos(\theta) + 3\sin(\theta)$. Let's show the general case which is an important result and then try some examples.

Example 35

Show that

* $a\cos(\theta) + b\sin(\theta) = R\cos(\theta - \beta)$ where $R = \sqrt{a^2 + b^2}$ and $\beta = \tan^{-1}\left(\dfrac{b}{a}\right)$

Solution

We can expand the Right-Hand Side of * by using

4.40 $\cos(A - B) = \cos(A)\cos(B) + \sin(A)\sin(B)$

$$R\cos(\theta - \beta) = R\left[\cos(\theta)\cos(\beta) + \sin(\theta)\sin(\beta)\right]$$

$$= R\cos(\beta)\cos(\theta) + R\sin(\beta)\sin(\theta) \qquad \text{[Multiplying through by } R\text{]}$$

So we have

$$a\cos(\theta) + b\sin(\theta) = \left[R\cos(\beta)\right]\cos(\theta) + \left[R\sin(\beta)\right]\sin(\theta)$$

Equating coefficients of $\cos(\theta)$ gives

$$a = R\cos(\beta)$$

and equating coefficients of $\sin(\theta)$ gives

$$b = R\sin(\beta)$$

How can we show $R = \sqrt{a^2 + b^2}$?

We can find $a^2 + b^2$:

$$a^2 + b^2 = \left[R\cos(\beta)\right]^2 + \left[R\sin(\beta)\right]^2$$

$$= R^2\cos^2(\beta) + R^2\sin^2(\beta)$$

$$= \underset{\substack{\text{factorizing} \\ R^2}}{R^2}\underbrace{\left[\cos^2(\beta) + \sin^2(\beta)\right]}_{=1 \text{ by } \boxed{4.64}} = R^2$$

4.64 $\cos^2(A) + \sin^2(A) = 1$

Example 35 *continued*

Taking square root gives $R = \sqrt{a^2 + b^2}$. Next we show $\beta = \tan^{-1}\left(\dfrac{b}{a}\right)$.

$$\frac{b}{a} = \frac{R\sin(\beta)}{R\cos(\beta)} = \frac{\sin(\beta)}{\cos(\beta)} \underset{\text{by } \boxed{4.35}}{=} \tan(\beta)$$

Taking the inverse tan gives

$$\beta = \tan^{-1}\left(\frac{b}{a}\right)$$

Fig. 61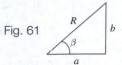

where β is the angle shown in Fig. 61.

The result of **Example 35** is useful in many engineering applications:

$\boxed{4.75}$ $\qquad a\cos(\theta) + b\sin(\theta) = R\cos(\theta - \beta)$ where $R = \sqrt{a^2 + b^2}$ and $\beta = \tan^{-1}\left(\dfrac{b}{a}\right)$

It is used in engineering fields because it converts $a\cos(\theta) + b\sin(\theta)$ into **amplitude–phase form**, $R\cos(\theta - \beta)$, where R is the amplitude and β is the phase.

We need to be careful when calculating the phase β using $\tan^{-1}\left(\dfrac{b}{a}\right)$. The signs of a and b also affect the sine and cosine waves they are associated with. This may result in your calculator displaying an answer which is 180° out. The angle β can be found as follows:

Examine the signs of a and b and then plot β in the correct quadrant (Fig. 62).

Fig. 62

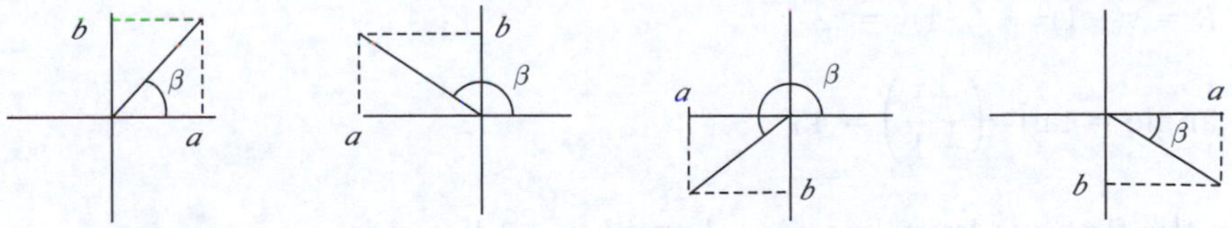

This method is further explored in **Example 36**.

Example 36

Write the following in the form $R\cos(\theta \pm \beta)$:

i $2\cos(\theta) + 3\sin(\theta)$ $\qquad$ **ii** $-5\cos(\theta) + 8\sin(\theta)$ $\qquad$ **iii** $-\cos(\theta) - \sin(\theta)$

iv $-\sqrt{3}\sin(\theta) + \cos(\theta)$

$\boxed{4.35}$ $\sin(A)/\cos(A) = \tan(A)$

Example 36 *continued*

Solution

We use 4.75 with a as the coefficient of $\cos(\theta)$ and b as the coefficient of $\sin(\theta)$:

i Substituting $a = 2$ and $b = 3$ into 4.75 gives:

$$R = \sqrt{2^2 + 3^2} = \sqrt{13}$$

$$\beta = \tan^{-1}\left(\frac{3}{2}\right) = 56.31° \quad \text{[By calculator]}$$

Thus $2\cos(\theta) + 3\sin(\theta) = \sqrt{13}\cos(\theta - 56.31°)$. This is the result in **Example 34**.

ii Putting $a = -5$ and $b = 8$ into 4.75 :

$$R = \sqrt{(-5)^2 + 8^2} = \sqrt{89}$$

$$\text{angle} = \tan^{-1}\left(\frac{8}{-5}\right) = -57.99° \quad \text{[By calculator]}$$

$-57.99°$ is the angle shown in Fig. 63 but β is in the second quadrant because $a = -5$ and $b = 8$. Therefore

$$\beta = -57.99° + 180° = 122.01°$$

Using $R = \sqrt{89}$ and $\beta = 122.01°$ we have

$$-5\cos(\theta) + 8\sin(\theta) = \sqrt{89}\cos(\theta - 122.01°)$$

Fig. 63

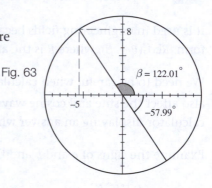

iii We have $a = -1$ and $b = -1$. Applying 4.75 :

$$R = \sqrt{(-1)^2 + (-1)^2} = \sqrt{2}$$

$$\text{angle} = \tan^{-1}\left(\frac{-1}{-1}\right) = 45°$$

$45°$ is in the first quadrant, but $a = -1$ and $b = -1$ lies in the third quadrant as shown in Fig. 64. Therefore

$$\beta = 45° + 180° = 225°$$

Using 4.75 :

$$-\cos(\theta) - \sin(\theta) = \sqrt{2}\cos(\theta - 225°)$$

Note that $225° = -135°$ (Fig. 64).

Fig. 64

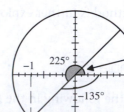

 4.75 $a\cos(\theta) + b\sin(\theta) = R\cos(\theta - \beta)$ with $R = \sqrt{a^2 + b^2}$ and $\beta = \tan^{-1}(b/a)$

Example 36 *continued*

Thus we can also write $-\cos(\theta) - \sin(\theta)$ as

$$-\cos(\theta) - \sin(\theta) = \sqrt{2}\cos(\theta + 135°)$$

iv In this case $b = -\sqrt{3}$ and $a = 1$. (Remember that a and b are the cosine and sine coefficients respectively.)

$$R = \sqrt{1^2 + (-\sqrt{3})^2} = 2$$

$$\beta = \tan^{-1}(-\sqrt{3}) = -60°$$

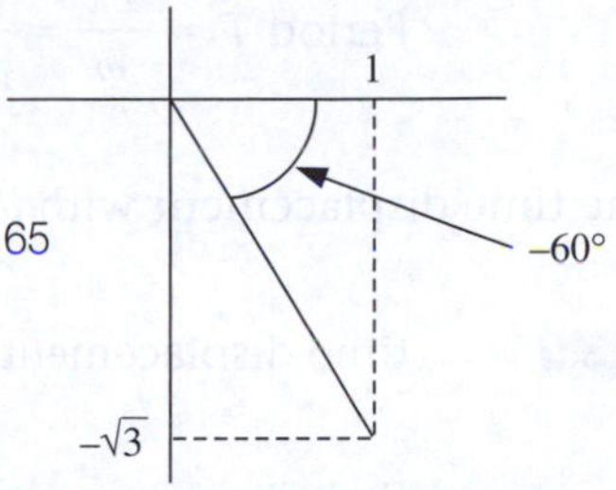

Fig. 65

The plot of $-\sqrt{3}\sin(\theta) + \cos(\theta)$ is shown in Fig. 65.

By 4.75 we have

$$-\sqrt{3}\sin(\theta) + \cos(\theta) = 2\cos[\theta - (-60°)] = 2\cos(\theta + 60°)$$

| 2 **Engineering applications**

Example 37 *electrical principles*

An a.c. current, i, is given by

$$i = 3\cos(\omega t) + 4\sin(\omega t)$$

Find the amplitude, and sketch one complete cycle for $\omega = 100\pi$ rad/s.

Solution

Using 4.75 with $a = 3$ and $b = 4$, we have

$$R = \sqrt{3^2 + 4^2} = 5$$
$$\beta = \tan^{-1}\left(\frac{4}{3}\right) = 0.9273$$

(Better to use radians since ω is in radians.) Hence

$$3\cos(\omega t) + 4\sin(\omega t) = 5\cos(\omega t - 0.9273)$$

What is the amplitude of the waveform?

Amplitude = 5 A

4.75 $a\cos(\theta) + b\sin(\theta) = R\cos(\theta - \beta)$ with $R = \sqrt{a^2 + b^2}$ and $\beta = \tan^{-1}(b/a)$

Example 37 *continued*

For sketching, we need to find period T and the time displacement from $5\cos(\omega t)$. **How do we find the period?**

Use 4.29 with $\omega = 100\pi$:

$$\text{Period } T = \frac{2\pi}{\omega} = \frac{2\pi}{100\pi} = 20 \times 10^{-3} = 20 \text{ ms}$$

The time displacement with $\omega = 100\pi$ and the phase $\alpha = 0.9273$ is found by using

4.33 $$\text{time displacement} = \alpha/\omega$$

$$\text{time displacement} = \frac{0.9273}{100\pi} = 2.95 \times 10^{-3} = 2.95 \text{ ms}$$

So $5\cos(100\pi t - 0.9273)$ lags $5\cos(100\pi t)$ by 2.95 ms (Fig. 66).

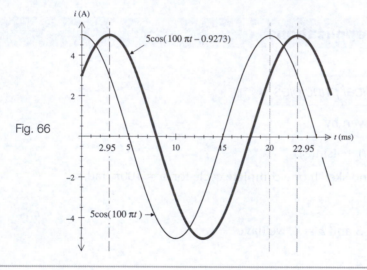

Fig. 66

Example 38 *vibrations*

Consider a spring of length L which is fixed at one end and has a mass m attached at the other end. If the spring is extended by a length x from its natural position, then the motion of the mass can be described by

$$x = A\cos\left(\sqrt{\frac{\lambda}{Lm}}t\right) + B\sin\left(\sqrt{\frac{\lambda}{Lm}}t\right)$$

where t is the time after the mass, m, has been released, and λ, A, B are constants.

Find the amplitude and frequency of the vibrations, if $A = 3$ and $B = \sqrt{3}$.

Example 38 *continued*

Solution

Substituting $A = 3$ and $B = \sqrt{3}$ gives

$$x = 3\cos\left(\sqrt{\frac{\lambda}{Lm}}t\right) + \sqrt{3}\sin\left(\sqrt{\frac{\lambda}{Lm}}t\right)$$

Using 4.75 :

$$R = \sqrt{3^2 + (\sqrt{3})^2} = \sqrt{12} = \sqrt{4 \times 3} = 2\sqrt{3}$$

and

$$\beta = \tan^{-1}\left(\frac{\sqrt{3}}{3}\right) = \tan^{-1}\left(\frac{1}{\sqrt{3}}\right) = \frac{\pi}{6} \text{ (or 30°)}$$

Thus

$$x = 2\sqrt{3}\cos\left(\sqrt{\frac{\lambda}{Lm}}t - \frac{\pi}{6}\right)$$

What is the amplitude and frequency?

$$\text{Amplitude} = 2\sqrt{3}$$

We can find the frequency by using

4.30 $\qquad f = \dfrac{\omega}{2\pi}$

with $\omega = \sqrt{\dfrac{\lambda}{Lm}}$:

$$\text{frequency } f = \frac{\omega}{2\pi} = \frac{\sqrt{\lambda/Lm}}{2\pi} = \sqrt{\frac{\lambda}{4\pi^2 Lm}} \qquad \left[\text{Because } (2\pi)^2 = 4\pi^2\right]$$

SUMMARY

We can convert $a\cos(\theta) + b\sin(\theta)$ into amplitude–phase form by

4.75 $\qquad a\cos(\theta) + b\sin(\theta) = R\cos(\theta - \beta)$ where $R = \sqrt{a^2 + b^2}$ and $\beta = \tan^{-1}\left(\dfrac{b}{a}\right)$

Remember, to find β we have to check the signs of a and b.

We can use 4.75 to convert $a\cos(\theta) + b\sin(\theta)$ into amplitude–phase form and to sketch graphs.

Exercise **4(i)**

Solutions at end of book. Complete solutions available at www.palgrave.com/engineering/singh

1 Write the following in the form $R\cos(\omega t \pm \beta)$:

   i $10\cos(\omega t) + 10\sin(\omega t)$

   ii $\sqrt{2}\cos(\omega t) + 2\sin(\omega t)$

   iii $-2\sin(\omega t) + 5\cos(\omega t)$

   iv $\cos(\omega t) - \sin(\omega t)$

   v $-\cos(\omega t) - \sqrt{3}\sin(\omega t)$

   vi $-5\cos(\omega t) + 5\sin(\omega t)$

2 Sketch one complete cycle of the following waveforms:

   i $y = \sin(\theta) + \cos(\theta)$

   ii $y = \sqrt{3}\cos(\theta) - \sin(\theta)$

   iii $y = \sin(\theta) - 2\cos(\theta)$

3 [electrical principles]   An a.c. current, i, is represented by

$$i = 3\cos(3t) - \sqrt{3}\sin(3t)$$

Find the amplitude and period.

4 [vibrations]   The motion of a bob of a pendulum can be described by

$$\theta = A\cos(\omega t) + B\sin(\omega t)$$

where $\omega = \sqrt{\dfrac{g}{l}}$, where g = acceleration due to gravity, l = length, θ the angle made by the bob and A, B are constants.

For $l = 1$ m, $g = 10$ m/s², $A = 1$, and $B = 0.5$, write θ in the form

$$R\cos(\omega t \pm \beta)$$

State amplitude, period of oscillation and phase.

5 [vibrations]   In a spring–mass system, the motion of the mass is described by

$$x = A\cos(\omega t) + B\sin(\omega t)$$

where x is the distance of the mass from its natural position, ω is the natural frequency of vibration and A, B are constants. For $A = \sqrt{3}$, $B = 1$ and $\omega = 10$:

   i Write x in the form $R\cos(\omega t \pm \beta)$ and state the amplitude of x.

   ii Sketch one complete cycle of x.

6 [thermodynamics]   Figure 67 shows a piston moving in a cylinder with rod AB. Point B is rotating clockwise with an angular velocity of ω.

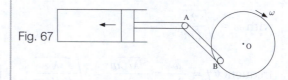

Fig. 67

The motion of B can be described by

$$r = C\cos(\omega t) + D\sin(\omega t)$$

where r is the (horizontal + vertical) distance of B from O, and C, D are constants. For $C = D = 4$ cm and $\omega = 2$ rad/s, write r in the form $R\cos(\omega t \pm \beta)$ and state the amplitude and period of r. Sketch one complete cycle of r.

Solutions at end of book. Complete solutions available at www.palgrave.com/engineering/singh

Examination questions 4

1 For the function $y = 3\cos\left(\dfrac{x}{2}\right)$:

  a Sketch the graph between $x = -\pi$ and $x = 2\pi$.

  b State the period, frequency and amplitude of the function.

University of Portsmouth, UK, 2009

2 Find the general solution to $\cos(3\theta) = -\dfrac{1}{2}$.

3 Write $\sin x - \sqrt{3}\cos x$ in the form $r\cos(x - \alpha)$.

University of Manchester, UK, 2009

4 Show that, for any angles A and B, the following identity holds

$$\frac{\sin A}{\sin B} + \frac{\cos A}{\cos B} = 2\frac{\sin(A + B)}{\sin 2B}$$

[you may assume any of the standard trigonometric identities].
Use the identity to evaluate

$$\frac{\sin 75°}{\sin 60°}$$

without using a calculator [merely giving the correct answer will attract no credit].

University of Manchester, UK, 2007

5 Find an exact expression for

$$\cos\left(\arctan\left(\frac{24}{7}\right)\right).$$

Memorial University of Newfoundland, Canada, 2010

6 Give all the solutions t in the interval $[0, 2\pi]$ of the equation

$$\cos(2t) - 2\sin^2(t) = 0$$

7 Express $\dfrac{13\pi}{12}$ as the sum or difference of two special angles. Use this expression to find the exact value of $\cos\left(\dfrac{13\pi}{12}\right)$.

8 Given that $\tan(\theta) = -2$ and θ is the angle in the fourth quadrant, find $\sin\left(\dfrac{\theta}{2}\right)$.

9 If θ is the angle in the second quadrant for which $\sin(\theta) = \dfrac{3}{5}$, determine the other five trigonometric ratios of θ.

10 Verify the identity

$$\frac{1}{\cos(x)\operatorname{cosec}^2(x)} = \sec(x) - \cos(x).$$

Questions 6 to 10 are from Memorial University of Newfoundland, Canada, 2009

11 By writing the Left-Hand Side of the equation in the form $R\sin(x + \alpha)$, or otherwise, solve for the values of x between 0° and 360°:

$$\sin x + 3\cos x = 2.4$$

University of Portsmouth, UK, 2007

12 **a** Show that

$$\sin\left(x - \frac{3\pi}{2}\right) = \cos(x)$$

  b Show that $\sec x - \cos x = \sin x \tan x$.

13 Solve for b in the triangle below:

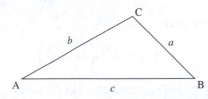

when $a = 20$, $A = 30°$ and $B = 45°$.

14 Write $\cos(\arctan 1 + \arccos x)$ as an algebraic expression of x.

Questions 12 to 14 are from University of California, Davis, USA

Examination questions **4 continued**

Solutions at end of book. Complete solutions available at www.palgrave.com/engineering/singh

15 The instantaneous value of voltage in an a.c. circuit at any time t seconds is given by $v(t) = 57 \sin(67\pi t - 0.26)$ volts. Determine the

  i amplitude, period time, frequency, phase angle in degrees and phase time

  ii value of the voltage when $t = 0$

  iii value of the voltage when $t = 8$ ms

  iv time when the voltage is first a maximum

  v time when the voltage first reaches 40 volts

Draw a sketch of $v(t)$ over one cycle, showing relevant points.

Cork Institute of Technology, Ireland, 2007

Miscellaneous exercise **4**

Solutions at end of book. Complete solutions available at www.palgrave.com/engineering/singh

1 Figure 68 shows a sheet of metal which is **symmetrical** about the broken line. Length AB = x , length BC = y and angle BAE = α.

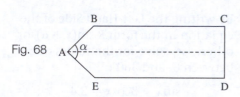

Fig. 68

Show that the area of the sheet ABCDE is

$$\frac{x}{2}\left[x\sin(\alpha) + 4y\sin\left(\frac{\alpha}{2}\right) \right]$$

2 [electrical principles]   Let

$$i_1 = \sin\left(\omega t + \frac{\pi}{3} \right) \text{ and } i_2 = \cos\left(\omega t - \frac{\pi}{6} \right).$$

Show that

$$i_1 + i_2 = 2\cos\left(\omega t - \frac{\pi}{6} \right)$$

3 [signal processing]   Let $f_1(t)$ and $f_2(t)$ be two signals defined by

$$f_1(t) = A_1\sin(\omega t) \text{ and }$$

$$f_2(t) = A_2\sin[\omega(t + \tau) + \phi]$$

where A_1 = amplitude of $f_1(t)$, A_2 = amplitude of $f_2(t)$, τ = time shift, ϕ = phase shift, ω = frequency and t = time. Show that

$$f_1(t)f_2(t) = \frac{A_1 A_2}{2}\left[\cos(\omega\tau + \phi) - \cos(2\omega t + \omega\tau + \phi) \right]$$

4 [electrical principles]   The power in a circuit is given by $P = \dfrac{v^2}{R}$. If

$$v = V\cos\left(\omega t + \frac{\pi}{4} \right) \text{ show that}$$

$$P = \frac{V^2}{2R}\left[1 - \sin(2\omega t) \right]$$

5 [electrical principles]   Let

$$v = 4\sin\left(\omega t + \frac{\pi}{4} \right) \text{ and } i = \sin(\omega t) \text{ in a}$$

circuit. Show that power, $P = iv$, is given by

$$P = \sqrt{2}\left[2\sin^2(\omega t) + \sin(2\omega t) \right]$$

6 [mechanics]   A projectile fired at an angle α has its horizontal (x) and vertical (y) distances given by

$$x = ut \cos(\alpha) \text{ and }$$

$$y = ut \sin(\alpha) - \frac{gt^2}{2} \text{ respectively}$$

where u is the initial velocity of the projectile, g is the acceleration due to gravity and t is time. Show that

$$y = \frac{1}{2u^2}\left[2u^2 x \tan(\alpha) - gx^2\sec^2(\alpha) \right]$$

Miscellaneous exercise **4 continued**   Solutions at end of book. Complete solutions available at www.palgrave.com/engineering/singh

7 Find the lengths of the two unknown sides in Fig. 69.

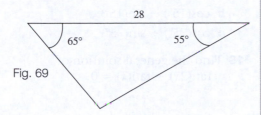

Fig. 69

8 Three sides of a triangle are 14, 16 and 18. Find all the angles of the triangle.

9 By using the identity $\sin^2(\theta) + \cos^2(\theta) = 1$, deduce

$$1 + \cot^2(\theta) = \operatorname{cosec}^2(\theta)$$

10 Let x be an angle between $0°$ and $90°$, with

$$\tan(x) = \sqrt{2} - 1$$

Without using a calculator, find the value of x.

$$\left[\textit{Hint}: \text{Use } \tan(2x) = \frac{2\tan(x)}{1 - \tan^2(x)} \right]$$

11 Solve, for x between 0 and 2π, the equations

a $\cos(2x) = 0$

b $\cos\left(x + \dfrac{\pi}{6}\right) = \sin(x)$

c $\tan(2x) \cdot \tan(x) = 1$

d $4\sin(x)\cos(x) - 4\cos^2(x) = -2$

12 [*electronics*]   An a.c. voltage, v, across a circuit is given by

$$v = 20\sin(50\pi t + 0.6)$$

i Find the amplitude, period and frequency of v.

ii Calculate the time displacement of v with respect to $20\sin(50\pi t)$.

iii Sketch one complete cycle of v.

13 [*electronics*]   An a.c. current through a circuit is given by

$$i = 200\sin\left(\omega t + \frac{\pi}{3}\right)$$

Sketch one complete cycle of i against t.

14 [*vibrations*]   Let x_1 and x_2 represent the horizontal displacement of two simple pendulums from their central position. Let

$$x_1 = 0.5\sin\left(\frac{\pi t}{5}\right) \text{ and } x_2 = 0.1\sin\left(\frac{\pi t}{4}\right)$$

where t is in seconds. If they both start together at $t = 0$, find the time taken for them to meet at $x_1 = x_2 = 0$ again.

*15 [*vibrations*]   As in the above question, let

$$x_1 = 0.3\sin\left(\frac{\pi t}{3}\right) \text{ and } x_2 = 0.3\cos\left(\frac{\pi t}{3}\right)$$

At $t = 0$ they are not together. Find the time taken for them to meet for the first time.

*16 [*structures*]   The displacement curve of a uniform pin-ended strut is given by

$$y = A\cos(kx) + B\sin(kx)$$
where k is a constant

By using the conditions $x = 0$, $y = e$ and at $x = L$, $y = e$, find A and B. Also show that

$$y = e\left(\sin(kx) \tan\left(\frac{kL}{2}\right) + \cos(kx) \right)$$

Show that the central displacement y_c at $x = \dfrac{L}{2}$ is $y_c = e\sec\left(\dfrac{kL}{2}\right)$.

$$\left[\textit{Hint}: \frac{1 - \cos(x)}{\sin(x)} = \tan\left(\frac{x}{2}\right) \right]$$

Miscellaneous exercise **4 continued**   Solutions at end of book. Complete solutions available at www.palgrave.com/engineering/singh

17 Solve, if possible, the following equations for x where $0° \leq x \leq 360°$:

  i $\cos(x) - \sec(x) = 1$

  ii $\cos(x) + \sec(x) = 1$

  iii $\cos^2(x) = 2\cos(x) + \tan^2(x)$

18 Find, by using Table 1 on page 175, the exact values of

  a $\dfrac{\tan(40°) + \tan(20°)}{1 - \tan(40°)\tan(20°)}$

  b $\cos(15°) - \sin(15°)$

  c $\cos(75°) + \sin(75°)$

*19 Find the general solution of
$$\tan(2x) + \tan(x) = 0$$

CHAPTER 5
Logarithmic, Exponential and Hyperbolic Functions

SECTION A **Indices revisited**

By the end of this section you will be able to:
▶ use the laws of indices
▶ apply these rules to thermodynamics

A1 **Review of indices**

We first visited indices in **Chapter 1**. The rules of indices, stated below, are **critical** in most engineering disciplines. We will use these rules throughout this chapter.

5.1	$a^m \, a^n = a^{m+n}$
5.2	$\dfrac{a^m}{a^n} = a^{m-n} \quad (a \neq 0)$
5.3	$(a^m)^n = a^{m \times n}$
5.4	$\dfrac{1}{a^n} = a^{-n} \quad (a \neq 0)$
5.5	$a^{\frac{1}{n}} = \sqrt[n]{a}$
5.6	$a^{\frac{m}{n}} = \sqrt[n]{a^m}$
5.7	$a^0 = 1 \quad (a \neq 0)$
5.8	$a^1 = a$

Let's look at some applications of indices in the field of thermodynamics.

Example 1 *thermodynamics*

A gas in a cylinder expands according to the law

$$P_1(V_1)^{1.35} = P_2(V_2)^{1.35}$$

where P_1, P_2 represent pressure and V_1, V_2 represent volume. Find an expression for V_2.

Solution

We have

$$P_2(V_2)^{1.35} = P_1(V_1)^{1.35}$$

$$(V_2)^{1.35} = \frac{P_1(V_1)^{1.35}}{P_2} \quad \text{[Dividing by } P_2\text{]}$$

†

$$(V_2)^{1.35} = \frac{P_1}{P_2} \cdot (V_1)^{1.35}$$

Example 1 *continued*

 How do we remove the power of V_2, 1.35?

By taking both sides to the power of $\dfrac{1}{1.35}$ because

$$\left(V_2^{1.35}\right)^{\frac{1}{1.35}} \underset{\text{by } 5.3}{=} V_2^{1.35 \times \frac{1}{1.35}} = V_2^{1} \underset{\text{by } 5.8}{=} V_2$$

Applying the power of $\dfrac{1}{1.35}$ to both sides of ⟨†⟩ gives

$$V_2 = \left[\frac{P_1}{P_2} \cdot V_1^{1.35}\right]^{\frac{1}{1.35}}$$

$$= \left[\frac{P_1}{P_2}\right]^{\frac{1}{1.35}} V_1$$

 Example 2 *thermodynamics*

A gas in a piston has an initial volume of 0.01 m³ at a pressure of 250 kN/m². The gas is then compressed to a pressure of 800 kN/m² according to the law $P(V)^{1.2} = C$, where C is a constant. Find the final volume.

Solution

We have $P_1(V_1)^{1.2} = P_2(V_2)^{1.2}$ and we need to find V_2.

Substituting $V_1 = 0.01$, $P_1 = 250 \times 10^3$ and $P_2 = 800 \times 10^3$ (remember k = kilo = 10^3) gives

$$250 \times 10^3 \times (0.01)^{1.2} = 800 \times 10^3 \times (V_2)^{1.2}$$

 How do we find V_2?

First we obtain $V_2^{1.2}$:

$$(V_2)^{1.2} = \frac{250 \times 10^3}{800 \times 10^3} \times (0.01)^{1.2} \quad \text{[Dividing by } 800 \times 10^3\text{]}$$

$$= \frac{25}{80} \times (0.01)^{1.2} \quad \text{[Cancelling]}$$

As in **Example 1**, taking both sides to the power of $\dfrac{1}{1.2}$ gives

$$V_2 = \left[\frac{25}{80}\right]^{\frac{1}{1.2}} \times 0.01 = 3.79 \times 10^{-3}$$

Hence the final volume is 0.0038 m³ (2 s.f.).

SUMMARY

We can apply the rules of indices, 5.1 to 5.8, to engineering applications such as those shown in **Examples 1** and **2**.

Exercise 5(a)

Solutions at end of book. Complete solutions available at www.palgrave.com/engineering/singh

1 [thermodynamics] A gas with an initial volume $V_1 = 0.3$ m^3 is compressed from $P_1 = 100$ kPa to $P_2 = 350$ kPa.
From the formula

$$P_1(V_1)^{5/3} = P_2(V_2)^{5/3}$$

evaluate the final volume V_2.
(Pascal, Pa = N/m^2, is the named SI unit of pressure.)

2 [thermodynamics] A gas in a cylinder with an initial volume $V_1 = 0.25$ m^3 and pressure $P_1 = 100$ kPa is compressed to a final pressure $P_2 = 450$ kPa. Given that

$$P_1(V_1)^{1.33} = P_2(V_2)^{1.33}$$

determine the final volume V_2.

3 [thermodynamics] A volume, 0.1 m^3, of gas at a pressure of 200 kN/m^2 is compressed to a pressure of 632 kN/m^2. If the gas obeys the law

$$P_1 V_1^{1.5} = P_2 V_2^{1.5}$$

then find the new volume.

*4 [thermodynamics] For an ideal gas, the PVT equations are given by

$$P_1(V_1)^k = P_2(V_2)^k \text{ and } \frac{P_1 V_1}{T_1} = \frac{P_2 V_2}{T_2}$$

where P, V, T are pressure, volume, temperature respectively and k is a constant. Show that

i $\dfrac{T_2}{T_1} = \left(\dfrac{V_1}{V_2}\right)^{k-1}$

ii $\dfrac{T_1}{T_2} = \left(\dfrac{V_2}{V_1}\right)^{k-1}$

*5 Show that $x^{\frac{1}{\Psi}}\left(y^{\frac{\Psi-1}{\Psi}} - x^{\frac{\Psi-1}{\Psi}}\right)$ can be

simplified to $x\left[\left(\dfrac{y}{x}\right)^{\frac{\Psi-1}{\Psi}} - 1\right]$.

6 [fluid mechanics] The velocity, v, of a fluid in a channel of slope s and radius r is defined as $v = \dfrac{r^{1/6}}{\eta}\sqrt{rs}$ where η is a constant. Show that

$$v = \frac{r^{2/3}s^{1/2}}{\eta}$$

SECTION B **The exponential function**

By the end of this section you will be able to:
▶ use a calculator to evaluate exponential expressions
▶ plot the graphs of e^x and e^{-x}
▶ apply exponential functions to engineering problems

B1 **Exponential function**

We sometimes hear the statement – the population is growing exponentially.

? **What does this mean?**

It means that the population is rapidly increasing. For example, if the population starts from 100 and doubles every year, we have the situation shown in Table 1.

Year (x)	Population (y)
0	100
1	200
2	400
3	800
4	1600

TABLE 1

We can predict the future population, y, by using the equation, $y = 100 \times (2^x)$. The initial population is 100 and it doubles every year, (2^x).

? **However if the population trebles every year then what equation can we use?**

$$y = 100 \times (3^x)$$

Year (x)	Population (y)
0	100
1	300
2	900
3	2700
4	8100

TABLE 2

This doubling and trebling of population is known as **exponential growth.** A particular type of exponential growth is defined as $y = e^x$ where

$$e = 2.71828182846\ldots$$

This number e is an irrational number (like π) and so we cannot write down the exact value of e. The numbers e and π are the most important irrational numbers in mathematics because they crop up all over the mathematical landscape.

? **Can you recall where you have seen this number before?**

In the **Introduction Chapter** and in **Chapter 3** under **Section E** (Limits of functions) where it was defined as

$$\lim_{x \to \infty} \left(1 + \frac{1}{x}\right)^x = e$$

? **Why is e^x important?**

The function e^x occurs in many engineering disciplines such as aerodynamics, mechanics, electrical principles, etc. It also crops up in natural sciences, for example population growth.

The population growth for the exponential function $y = 100 \times (e^x)$ is given in Table 3.

Year (x)	Population (y) (nearest whole number)
0	100
1	272
2	739
3	2009
4	5460

TABLE 3

The exponential function, e^x, more than doubles but does not quite treble over a period. Compare Tables 1, 2 and 3.

Plots of the general exponential functions $y = 2^x$, 3^x and e^x are shown in Fig. 1.

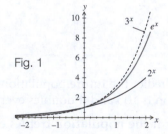

Fig. 1

The exponential function, e^x, is sometimes denoted by $\exp(x)$. It was the Swiss mathematician **Euler** who introduced the notation e^x in 1727. This function, e^x, is extensively used throughout engineering. Leonhard Euler (pronounced 'Oiler') is often credited as being the most prolific mathematician of all time. It is said that eighteenth century mathematics belonged to Euler. He was responsible for developing Graph Theory, a branch of mathematics resulting from a puzzle he solved called 'Bridges of Königsberg', as well as giving us the symbols $f(x)$, $\sin(x)$, Σ and e^x.

Euler had 13 children, although sadly only five survived. Incredibly, despite being blind for the last 17 years of his life, he continued working in the field of mathematics until his death in 1783.

Example 3

Complete Table 4.

	x	−2	−1	0	1	2	3	4
TABLE 4	$y = e^x$							

Hence plot the function $y = e^x$.

Solution

To find the y values we use our calculator. Most calculators have the exponential button, e^x; it might be a secondary function under the 'ln' button. For example to find e^2, PRESS:

$\boxed{\text{SHIFT}}$ $\boxed{\text{ln}}$ $\boxed{2}$ $\boxed{=}$ and the calculator should show 7.389...

On some calculators we need to replace the $\boxed{\text{SHIFT}}$ button with the $\boxed{\text{INV}}$ or $\boxed{\text{2nd FUN}}$ button (see the handbook of your calculator).

We obtain Table 5.

	x	−2	−1	0	1	2	3	4
TABLE 5	$y = e^x$	0.135	0.368	1.000	2.718	7.389	20.086	54.598

Plotting these points gives the graph shown in Fig. 2.

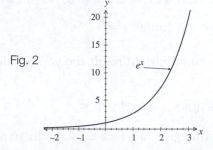

Fig. 2

Example 4 *aerodynamics*

The pressure, P, of a gas at an altitude, Z, and temperature, T, is given by

$$P = Ae^{-\left(\frac{g}{RT}\right)Z}$$

Find P given that $A = 20 \times 10^3$ N/m^2, $R = 287$ J/(kg K), $Z = 5 \times 10^3$ m, $T = 200$ K and $g = 9.81$ m/s^2.

Example 4 continued

Solution

Substituting these values into the above formula with e = exp and using a calculator gives

$$P = (20 \times 10^3)\exp\left[-\frac{9.81}{(287 \times 200)} \times (5 \times 10^3)\right]$$

$$= (20 \times 10^3)\exp(-0.8545)$$

The screen of the calculator should show 8509.918. Remember that k = kilo = 10^3 and the unit of pressure, P, is N/m². Thus

$$P = 8510 \text{ N/m}^2 = 8.51 \text{ kN/m}^2 \text{ (3 s.f.)}.$$

B2 Properties of the exponential function

We have the following results for the exponential function:

5.9 $e^0 = 1$ [This is 5.7 with $a = e$]

5.10 $\dfrac{1}{e^x} = e^{-x}$ [This is 5.4 with $a = e$]

The graph of $y = e^{-x}$ is a decaying graph (Fig. 3).

 Can you think of a decaying population?

Possibly black rhinos or giant pandas.

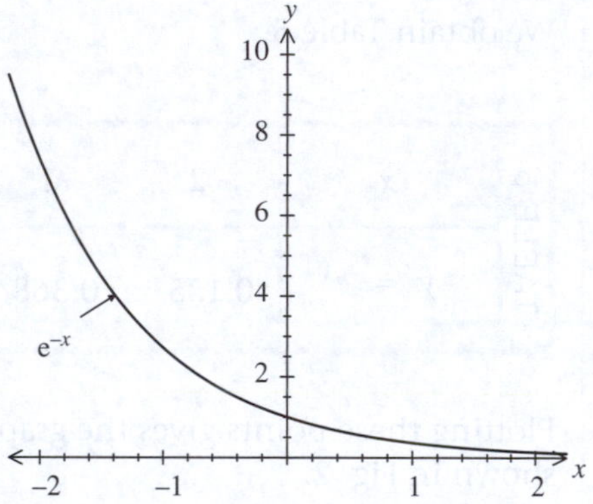

As x becomes positively large then e^{-x} tends towards zero. That is, as $x \to \infty$, $e^{-x} \to 0$. We can write this as

5.11 $\lim\limits_{x \to \infty}(e^{-x}) = 0$

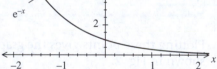
Fig. 3

This is an important result and you should try to remember it.

Example 5 electronics

The voltage, v, across a capacitor in an RC circuit is given by

$$v = 100e^{-t/\tau}$$

where τ (pronounced 'taw') is the time constant. Find the percentages of voltage, v, at $t = \tau, 2\tau, 3\tau, 4\tau$ and 5τ of the initial voltage at $t = 0$. Sketch the graph of v versus t for $0 \le t \le 5\tau$.

Solution

Some engineering students find difficulty in following this example because of the Greek letter τ. Generally in electronics and electrical principles τ is used to represent the time constant.

Example 5 *continued*

Substituting $t = 0$ into the given equation, $v = 100e^{-t/\tau}$, yields

$$v = 100e^0 = 100$$

The following results are correct to 1 d.p.

At $t = \tau$: $\quad v = 100e^{-\frac{\tau}{\tau}} = 100e^{-1} = 36.8$

At $t = 2\tau$: $\quad v = 100e^{-\frac{2\tau}{\tau}} = 100e^{-2} = 13.5$

At $t = 3\tau$: $\quad v = 100e^{-\frac{3\tau}{\tau}} = 100e^{-3} = 5.0$

At $t = 4\tau$: $\quad v = 100e^{-\frac{4\tau}{\tau}} = 100e^{-4} = 1.8$

At $t = 5\tau$: $\quad v = 100e^{-\frac{5\tau}{\tau}} = 100e^{-5} = 0.7$

The percentages of voltage after $t = \tau, 2\tau, 3\tau, 4\tau$ and 5τ are 36.8%, 13.5%, 5.0%, 1.8% and 0.7% respectively (Fig. 4).

Fig. 4

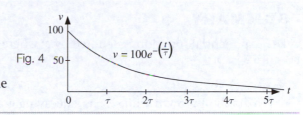

Note that after 5 time constants (5τ), the voltage v is approximately 0 volts.

How can we simplify expressions like $e^{-2t}e^{-4t}$?

We can apply the rule of indices, $a^m a^n = a^{m+n}$ with $a = e$:

$$e^{-2t}e^{-4t} = e^{-2t-4t} = e^{-6t}$$

Simplify $(e^{-200t})^2$.

Applying $(a^m)^n = a^{m \times n}$ with $a = e$ gives

$$(e^{-200t})^2 = e^{-200t \times 2} = e^{-400t}$$

The following example is more demanding because it involves a number of algebraic topics such as the rules of indices, expansion of brackets, substitution, etc. To follow this example, you must know your work from **Chapter 1** thoroughly.

Example 6 *electrical principles*

The energy, w, of an inductor, L, is defined as

$$w = \frac{1}{2}Li^2$$

For $L = 2 \times 10^{-3}$ H and $i = 5(e^{-100t} - e^{-200t})$ A, show that

$$w = 0.025\,(e^{-200t} - 2e^{-300t} + e^{-400t})$$

Example 6 *continued*

Solution

Substituting $L = 2 \times 10^{-3}$ and $i = 5(e^{-100t} - e^{-200t})$ into $w = \dfrac{1}{2}Li^2$ gives

$$w = \frac{1}{2}(2 \times 10^{-3}) \times \left[5(e^{-100t} - e^{-200t})\right]^2$$

$$= (1 \times 10^{-3}) \times 25 \times (e^{-100t} - e^{-200t})^2 \qquad \text{[Taking out } 5^2 = 25\text{]}$$

$$= (25 \times 10^{-3}) \times \underbrace{\left[(e^{-100t})^2 - 2 \cdot (e^{-100t}e^{-200t}) + (e^{-200t})^2\right]}_{\text{by } \boxed{1.4}}$$

$$= (25 \times 10^{-3}) \times \left[\underbrace{e^{-200t}}_{\text{by } \boxed{5.3}} - \underbrace{2e^{-300t}}_{\text{by } \boxed{5.1}} + \underbrace{e^{-400t}}_{\text{by } \boxed{5.3}}\right] \qquad \text{[Simplifying]}$$

$$w = 0.025(e^{-200t} - 2e^{-300t} + e^{-400t})$$

SUMMARY

The exponential function is denoted by e^x where e is $2.71828\ldots$. Also

$\boxed{5.9}$ $e^0 = 1$

$\boxed{5.10}$ $e^{-x} = 1/e^x$

e^x is a graph of growth while e^{-x} is a decaying graph.

Exercise 5(b)

Solutions at end of book. Complete solutions available at www.palgrave.com/engineering/singh

1 [*mechanics*]   A pulley is driven by a flat belt with an angle of lap $\theta = \dfrac{2\pi}{3}$. Given that T_1, T_2 are tensions on each end of the belt and

$$T_2 = T_1 e^{\mu\theta}$$

Determine T_2 for $T_1 = 1000$ N and $\mu = 0.2$ (coefficient of friction).

2 Complete the following table. Plot the graph of $y = e^{-x}$.

x	-2	-1	0	1	2
$y = e^{-x}$			1.00		

3 [*electronics*]   The current, i, through a diode is given by

$$i = (0.1 \times 10^{-6})(e^{20v} - 1)$$

where v is the voltage. Complete the following table:

v	0	0.1	0.2	0.3	0.4	0.5	0.6	0.7
i								

Plot the graph of i versus v for the values in the table.

4 [*aerodynamics*]   The pressure, p, of air at an altitude, z, is given by

$$p = (20 \times 10^3)e^{-(1.25 \times 10^{-4})z}$$

$\boxed{1.4}$ $(a - b)^2 = a^2 - 2ab + b^2$ $\boxed{5.1}$ $a^m a^n = a^{m+n}$ $\boxed{5.3}$ $(a^m)^n = a^{m \times n}$

Exercise **5(b) continued**

Solutions at end of book. Complete solutions available at www.palgrave.com/engineering/singh

Find p to the nearest whole number for
a $z = 5 \times 10^3$ **b** $z = 7 \times 10^3$
c $z = 10 \times 10^3$

5 [*heat transfer*] The temperature, θ (°C), at time t (min) of a body is given by

$$\theta = 300 + 100e^{-0.1t}$$

Evaluate θ for $t = 0$, 1, 2 and 5.

6 [*electrical principles*] The energy, W, of a capacitor of capacitance C, is defined by

$$W = \frac{1}{2}CV^2$$

For $C = 1\ \mu F$ and $V = e^{-(1 \times 10^3)t}$, find W.

*7 [*electrical principles*] The instantaneous power, p, of a capacitor is defined as

$$p = vi$$

a For $v = 1 - e^{-t/2 \times 10^{-6}}$ and $i = 0.4e^{-t/2 \times 10^{-6}}$, find p.

b For $v = e^{-(5 \times 10^3)(t-1)}$ and $i = -0.05e^{-(5 \times 10^3)(t-1)}$, find p.

8 [*electrical principles*] The voltage, v, and current, i, of an inductor is given by

$$i = 5(e^{-200t} - e^{-800t}) \text{ and }$$
$$v = e^{-200t} + 400e^{-800t}$$

Find an expression for the power, $p = vi$, of the inductor.
[*Hint*: Use FOIL]

9 [*electrical principles*] The current, i (mA), through a capacitor at time t (s) is given by $i = 100e^{-(5 \times 10^3)t}$. Sketch the graph of i against t for $t \geq 0$.

10 [*mechanics*] The velocity, v, of a train along a straight track is given by

$$v = 20(1 - e^{-t})$$

Sketch the graph of v against t for $t \geq 0$. What is the value of v for large t?

SECTION C **The logarithmic function**

By the end of this section you will be able to:
▶ use a calculator to find the logarithm of a number
▶ apply the rules of logarithms to engineering problems
▶ change the base of a logarithm

C1 **Natural logarithms**

The most important logarithm in engineering is the **natural** logarithm and is denoted by 'ln' which stands for logarithm natural. The logarithmic function, $\ln(x)$, is the inverse of the exponential function, e^x; that is, $y = e^x$ and $\ln(y) = x$ are equivalent. For example, $e^2 = 7.389$ therefore $\ln(7.389) = 2$. Similarly $e^{3.142} = 23.150$ and so $\ln(23.150) = 3.142$.

The word *logarithm* is made up of two Greek words – *logos* (expression) and *arithmos* (number). It was **John Napier** (1550–1617) of Edinburgh who developed the logarithm while trying to simplify complex multiplication and division problems to the level of addition and subtraction. The logarithmic unit of ratio, the neper, is named after him. He was also the first person to use a dot for the decimal point.

Figure 5 shows the graphs of $y = e^x$ and $y = \ln(x)$. Note that $\ln(x)$ is not defined for negative x and $x = 0$.

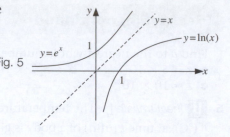

Fig. 5

At one time you had to find the logarithm of a number by using tables, but since the introduction of the hand-held calculator in the 1970s these logarithmic tables, along with the slide rule, have been exiled to history.

You can evaluate the ln function on your calculator. To evaluate $\ln(3)$

PRESS: [In][3][=], which should show 1.09861.

Example 7 *electronics*

The current gain, G, in neper of a system is defined as

$$G = \ln\left(\frac{I_{out}}{I_{in}}\right)$$

where I_{out} and I_{in} are the output and input current respectively. Evaluate G for $I_{out} = 0.5$ mA and $I_{in} = 0.1$ mA. (Remember that m = milli = 10^{-3}.)

Solution

Substituting the given values into G:

$$G = \ln\left(\frac{0.5 \times 10^{-3}}{0.1 \times 10^{-3}}\right) = \ln(5) = 1.609 \text{ nep}$$

Example 8

Evaluate the following:

a $\ln(2) + \ln(3)$ and $\ln(6)$ **b** $\ln(7) + \ln(5)$ and $\ln(35)$

 What do you notice about your results?

Solution

By using a calculator and rounding to 3 d.p.:

a $\ln(2) + \ln(3) = 1.792$ and $\ln(6) = 1.792$

b $\ln(7) + \ln(5) = 3.555$ and $\ln(35) = 3.555$

We have

$$\ln(2) + \ln(3) = \ln(2 \times 3) = \ln(6)$$
$$\ln(7) + \ln(5) = \ln(7 \times 5) = \ln(35)$$

In general we have the following laws of logarithms:

> 5.12 $\ln(A) + \ln(B) = \ln(A \times B) = \ln(AB)$

> 5.13 $\ln(A) - \ln(B) = \ln\left(\dfrac{A}{B}\right)$

> 5.14 $\ln(A^n) = n \ln(A)$

In 5.12 to 5.14 , A and B are positive because the ln function is not defined for negative numbers. We can employ these to simplify our results, as the next example shows.

These laws of logs, 5.12 to 5.14 , are used to simplify arithmetic calculations.

5.12 says that by considering logs we can convert a **multiplication** problem into **addition**.

5.13 says that by considering logs we can convert a **division** problem into **subtraction**.

5.14 says that by considering logs we can convert an **index** problem into **multiplication**.

From the seventeenth century to the mid-twentieth century virtually all scientific calculations, especially astronomical ones, employed logs.

From the 1960s and 1970s, computers and calculators made logs for calculation obsolete. However logs found applications in calculus, complex numbers, physical and biological processes.

> ### Example 9 *thermodynamics*
>
> The pressure, P, and volume, V, of a gas are related by
>
> $$\ln(P) = -n \ln(V) + \ln(C)$$
>
> where n and C are constants. Show that $\ln(PV^n) = \ln(C)$.
>
> Solution
>
> Adding $n \ln(V)$ to both sides gives
>
> $$\ln(P) + n \ln(V) = \ln(C)$$
>
> By 5.14 , $n \ln(V) = \ln(V^n)$ so we have
>
> $$\ln(P) + \ln(V^n) = \ln(C)$$
>
> $$\underbrace{\ln(PV^n)}_{\text{by } 5.12} = \ln(C)$$

? **What is $\ln(e)$ equal to?**

Since ln and e are inverses of each other:

$$\ln(e) = 1$$

? **What is $\ln(e^x)$ equal to?**

$$\ln(e^x) = x \ln(e) \qquad [\text{By } 5.14]$$

$$= x \qquad [\text{Because } \ln(e) = 1]$$

Thus by noting that $\ln(x)$ is the inverse function of e^x we have the following results:

5.15 $\ln(e) = 1$

5.16 $\ln(e^x) = x$

5.17 $e^{\ln(x)} = x$ (provided x is positive)

We can use these results to solve equations involving the exponential function or show other results such as

$$e^{x \ln(a)} = a^x$$

[See **Exercise 5(c)**, question **15**.]

> ## Example 10 *mechanics*
>
> The velocity, v, of a vehicle during the application of brakes according to time t is given by
>
> $$v = 10e^{-kt}$$
>
> where k is a friction constant of the brakes. At $t = 30$ s, the velocity is reduced to 2.5 m/s.
> Determine k.
>
> Solution
>
> Substituting $t = 30$ and $v = 2.5$ into $v = 10e^{-kt}$ gives
>
> $$2.5 = 10e^{-30k}$$
>
> $$\frac{2.5}{10} = e^{-30k} \quad \text{[Dividing by 10]}$$
>
> $$0.25 = e^{-30k}$$
>
> **How can we find k?**
>
> By taking natural logs, ln, of both sides because ln is the inverse of the exponential function. Applying $\ln(A^n) = n \ln(A)$ to the Right-Hand Side gives
>
> $$\ln(e^{-30k}) = -30k \underbrace{\ln(e)}_{=1} = -30k$$
>
> Therefore we have
>
> $$\ln(0.25) = -30k$$
>
> $$k = \frac{\ln(0.25)}{-30} \quad \text{[Dividing by } -30\text{]}$$
>
> $$= 0.046 \quad \text{(2 s.f.)}$$

C2 Common logarithms

Apart from natural logarithms, ln, there are other logarithms which are also used in engineering. The other important logarithm is called the **common** logarithm and is denoted by log.

The logarithmic function, $\log(x)$, is the inverse function of 10^x. In other words the log of a number is the power of 10 needed to make that number. For example, $1000 = 10^3$ thus $\log(1000) = \log(10^3) = 3$. Similarly $10 = 10^1$ so $\log(10) = \log(10^1) = 1$.

? **What is log(100) equal to?**

$$\log(100) = \log(10^2) = 2$$

? **What is log(199.53) given that $10^{2.3} = 199.53$?**

$$\log(199.53) = 2.3$$

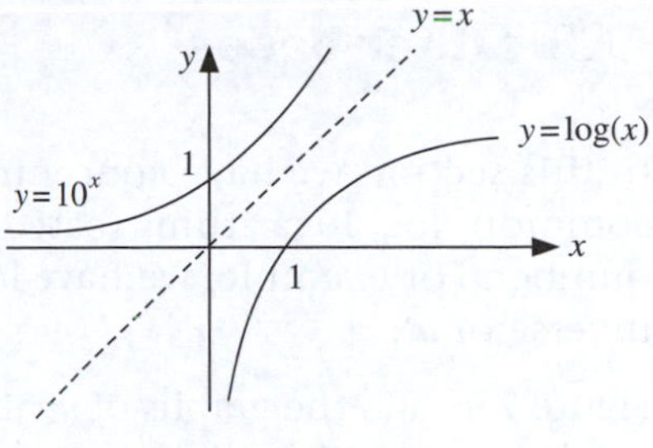

Fig. 6

Figure 6 shows the graphs of $y = 10^x$ and $y = \log(x)$. Note that $\log(x)$ is also not defined for negative x and $x = 0$.

Example 11 *electronics*

The power gain, G, in decibels (dB) of a system is defined as

$$G = 10 \log\left(\frac{P_{\text{out}}}{P_{\text{in}}}\right)$$

where P_{out} is the output power and P_{in} is the input power. Evaluate G for the following:

a $P_{\text{out}} = 5$ W, $P_{\text{in}} = 0.5$ W   **b** $P_{\text{out}} = 1$ W, $P_{\text{in}} = 1 \times 10^{-3}$ W

Solution

Remember that W = watt is the SI unit of power.

a $G = 10 \times \log\left(\dfrac{5}{0.5}\right) = 10 \times \underbrace{\log(10)}_{=1} = 10$ dB

b $G = 10 \times \log\left(\dfrac{1}{1 \times 10^{-3}}\right) = 10 \times \log(1000) = 10 \times 3 = 30$ dB

The **same laws** apply to the log function.

5.18	$\log(A) + \log(B) = \log(A \times B) = \log(AB)$
5.19	$\log(A) - \log(B) = \log\left(\dfrac{A}{B}\right)$
5.20	$\log(A^n) = n \log(A)$
5.21	$\log(10) = 1$
5.22	$\log(10^x) = x$
5.23	$10^{\log(x)} = x$

We can use these laws to evaluate G of **Example 11**. For example, if we had $P_{\text{out}} = 1$ mW, $P_{\text{in}} = 1$ W then

$$G = 10 \log\left(\frac{1 \times 10^{-3}}{1}\right) \underset{\text{by } 5.20}{=} 10 \log(10^{-3}) = -3 \times 10 \log(10) = -30 \underset{=1 \text{ by } 5.21}{\log(10)} = -30 \text{ dB}$$

Of course the easiest way to evaluate log is to use a calculator. For example, to evaluate $\log(3)$ on a calculator PRESS

$\boxed{\log}$ $\boxed{3}$ $\boxed{=}$ and the screen should show 0.4771213. Try **Example 11** on your calculator.

C3 **Other bases**

In this section we have concentrated on natural, ln, logarithms (based on e) and common, log, logarithms (based on 10). The logarithm can be **based** on any positive number. For example, we have logarithm to the base a written $\log_a(x)$ which is the inverse of a^x.

Figure 7 shows the graphs of a^x and $\log_a(x)$ where a can be any positive number. Consider the case $a = 2$.

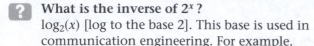

Fig. 7

What is the inverse of 2^x?
$\log_2(x)$ [log to the base 2]. This base is used in communication engineering. For example,

$$2^3 = 8$$
$$\log_2(8) = 3$$

Similarly the inverse of 3^x is $\log_3(x)$ [log to the base 3].

Given that $3^5 = 243$, what is $\log_3(243)$ equal to?

$$\log_3(243) = 5$$

In general $a^x = N$ is equivalent to $\log_a(N) = x$.

We can change the base of the logarithm by using the following rule:

5.24
$$\log_a(N) = \frac{\log_b(N)}{\log_b(a)}$$

5.24 changes the logarithm from base a to base b. We can use this change-of-base rule to evaluate logarithm to any base, as the following example shows.

Find the values of

$$\log_2(5), \log_3(7) \text{ and } \log_{8.7}(10.2)$$

Solution

Which base should we convert these logs to?

Since the calculator evaluates the natural and common logarithms, we have to convert to one of these bases. Converting to the natural logarithm by applying 5.24 and rounding to 2 d.p. we have

$$\log_2(5) = \frac{\ln(5)}{\ln(2)} = 2.32$$

$$\log_3(7) = \frac{\ln(7)}{\ln(3)} = 1.77$$

$$\log_{8.7}(10.2) = \frac{\ln(10.2)}{\ln(8.7)} = 1.07$$

Of course you can convert the logarithms of **Example 12** into the common logarithms with exactly the same results.

Example 13

Solve the following equations:

a $\log_x(x^4 - 3x + 3) = 4$ **b** $\log_2(x^2 + 2x) = 3$ **c** $3^{x^2-1} = 27$

Solution

a How do we solve $\log_x(x^4 - 3x + 3) = 4$?

By using the definition of log we have $x^4 - 3x + 3 = x^4$. Subtracting x^4 from both sides gives

$$-3x + 3 = 0 \text{ so that } 3x - 3 = 0$$

From $3x - 3 = 0$ we have $x = 1$.

b Similarly using the definition of log we have

$$x^2 + 2x = 2^3 = 8$$

Subtracting 8 from both sides gives

$$x^2 + 2x - 8 = 0$$

How do we solve this quadratic equation?

$$x^2 + 2x - 8 = (x + 4)(x - 2) = 0 \text{ so that } x = -4, x = 2$$

The solution to $\log_2(x^2 + 2x) = 3$ is $x = -4$, $x = 2$.

c How do we solve $3^{x^2-1} = 27$?

Note that $27 = 3^3$ therefore we have

$$3^{x^2-1} = 3^3$$

Taking $\log_3$ of both sides gives

$$\log_3(3^{x^2-1}) = \log_3 3^3$$
$$(x^2 - 1)\log_3(3) = 3\log_3(3) \qquad [\text{Applying } \log(A^n) = n\log(A)]$$

Remember $\log_3(3) = 1$ therefore

$$x^2 - 1 = 3$$
$$x^2 = 4 \text{ so that } x = \pm 2$$

The solution to $3^{x^2-1} = 27$ is $x = \pm 2$.

Logarithms to any base a, $\log_a$, also obey **all** the rules of logarithms stated for the common and natural logarithms.

SUMMARY

The natural logarithm, $\ln(x)$, is the inverse of the exponential function, e^x. The common logarithm, $\log(x)$, is the inverse of 10^x. All logarithms obey the rules:

| 5.12 | $\ln(A) + \ln(B) = \ln(A \times B) = \ln(AB)$ |

| 5.13 | $\ln(A) - \ln(B) = \ln\left(\dfrac{A}{B}\right)$ |

| 5.14 | $\ln(A^n) = n \ln(A)$ |

For any positive number a, the inverse of a^x is $\log_a(x)$. We can change the base of a logarithm by using

| 5.24 | $\log_a(N) = \dfrac{\log_b(N)}{\log_b(a)}$ |

Exercise 5(c)

Solutions at end of book. Complete solutions available at www.palgrave.com/engineering/singh

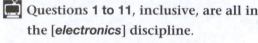 **Questions 1 to 11, inclusive, are all in the [electronics] discipline.**

1 By using the formula of **Example 7,**
$G = \ln\left(\dfrac{I_{out}}{I_{in}}\right)$, determine the current gain, G, in neper for

a $I_{out} = 0.1$ A, $I_{in} = 0.01$ A
b $I_{out} = 10$ mA, $I_{in} = 0.1$ mA
c $I_{out} = 5$ mA, $I_{in} = 15$ mA

2 The power gain, G, in neper is defined as

$$G = \frac{1}{2}\ln\left(\frac{p_o}{p_i}\right)$$

where $p_o =$ output power and $p_i =$ input power. Find G for

a $p_o = 1$ W, $p_i = 0.2$ W
b $p_o = 20$ mW, $p_i = 1$ mW

3 If $V_{out} = 1$ V and $V_{in} = 1$ mV then without using your calculator evaluate G, where

$$G = 20 \log\left(\frac{V_{out}}{V_{in}}\right)$$

4 If $G = 5$ dB and $P_{in} = 0.1$ W, evaluate P_{out} by using the formula of **Example 11:**
$G = 10 \log\left(\dfrac{P_{out}}{P_{in}}\right).$

5 In a cascaded electronic circuit the power gains at various stages are p_1, p_2 and p_3. Given that the overall power gain, G, is defined as

$$G = 10 \log(p_1) + 10 \log(p_2) + 10 \log(p_3)$$

show that $G = 10 \log(p_1 \times p_2 \times p_3)$.

6 If $p_{out} = I_{out}^2 R$ and $p_{in} = I_{in}^2 R$, show that

$$G = 20 \log\left(\frac{I_{out}}{I_{in}}\right)$$

($p_{out} =$ output power, $p_{in} =$ input power, $I_{out} =$ output current, $I_{in} =$ input current, $R =$ resistance and $G =$ gain.)

7 Given that $G = 20$ dB and $I_{in} = 1$ mA, determine I_{out} by using the formula of question **6**.

8 Given that $p_{out} = \dfrac{V_{out}^2}{R}$ and $p_{in} = \dfrac{V_{in}^2}{R}$, show that

$$G = 20 \log\left(\frac{V_{out}}{V_{in}}\right)$$

($V_{out} =$ output voltage, $V_{in} =$ input voltage, $p_{out} =$ output power,

Solutions at end of book. Complete solutions available at www.palgrave.com/engineering/singh

Exercise **5(c) continued**

p_{in} = input power, R = resistance and G = gain.)

*9 For the following values of G find the power ratio $\left(\dfrac{p_o}{p_i}\right)$ and the voltage ratio $\left(\dfrac{V_o}{V_i}\right)$ where

$$G = 10 \log\left(\frac{p_o}{p_i}\right) = 10 \log\left[\left(\frac{V_o}{V_i}\right)^2\right]:$$

a $G = 3$ dB **b** $G = 10$ dB
c $G = 20$ dB

10 The voltage, V, across a capacitor is given by

$$V = 9 - 10e^{-t}$$

Find t for $V = 0$.

11 The voltage, V, across a capacitor is given by

$$V = 9(1 - e^{-0.1t})$$

Evaluate the time taken, t, for $V = 3$.

12 Evaluate the following:

a $\log_2(7.6)$ **b** $\log_{7.6}(2)$
c $\log_7(6.3)$ **d** $\log_{6.3}(7)$

13 Evaluate $\ln\big[\ln(N)\big]$ for the following values of N:

a $N = 1000$ **b** $N = 10^{10}$ **c** $N = 10^{50}$

What do you notice about your results?

14 [thermodynamics]   In a thermodynamic system we have the relationship

$$\log(P) + n \log(V) = \log(C)$$

where P represents pressure, V represents volume, C is a constant and n is an index. Show that

$$PV^n = C$$

15 Show that $e^{x \ln(a)} = a^x$.

16 Given that $\ln(1 + y) = x + \dfrac{x^2}{2}$, show that

$$y = e^{x^2/2}e^x - 1$$

17 Solve the following equations:

a $27^x = 3$ **b** $5 \ln(x^2 - 9) = 3$
c $\log_9(3^{x-1}) = x$ **d** $3^x = 18$

*18 [heat transfer]   The temperatures θ_1 and θ_2 of a pipe with inner radius r_1 and outer radius r_2 are given by

$$\theta_1 = -\frac{Q}{2\pi kL} \ln(r_1) \text{ and}$$

$$\theta_2 = -\frac{Q}{2\pi kL} \ln(r_2)$$

where Q is the heat transfer rate, L is the length of the pipe and k is the thermal conductivity. Show that

$$Q = \frac{2\pi kL(\theta_1 - \theta_2)}{\ln\left(\dfrac{r_2}{r_1}\right)}$$

*19 [electronics]   The voltage, v, across a diode and the current, i, through it are related by

$$i = I_s(e^{11600v/\eta T} - 1)$$

where I_s is the reverse saturation current, η is the emission coefficient and T is the temperature. Make v the subject of the formula. Evaluate v for $\eta = 2$, $T = 330$ K, $I_s = 0.1$ µA and $i = 5$ mA.

SECTION D Applications of logarithms

By the end of this section you will be able to:
▶ choose appropriate variables for data to obtain a straight line
▶ apply laws of logs to experimental data
▶ plot the graph of $\log(y)$ against $\log(x)$ for data which follows the law $y = kx^n$
▶ find the values of k and n in $y = kx^n$

D1 Experimental law

Suppose x and y are related by $y = kx^2 + b$ where k and b are constants. If we plot the graph of y against x the result is a curve. For a straight line graph, $y = mx + c$, we can find the values of m (= gradient) and c (= intercept). **Can we find the values of k and b in $y = kx^2 + b$?**

If we plot y against x^2 instead of y against x then we should obtain a straight line. Compare

$$y = kx^2 + b \qquad (y \text{ against } x^2)$$
and
$$y = mx + c \qquad (y \text{ against } x)$$

The gradient of the resulting line gives the value of k and the vertical intercept gives b. Let's do an example.

Example 14

TABLE 6					
x	1	2	3	4	5
y	7.1	10.4	15.9	23.6	33.5

Suppose the experimental data in Table 6 is related by $y = kx^2 + b$. By plotting an appropriate graph find the approximate values of k and b.

Solution

We plot y vertically against x^2 horizontally.

TABLE 7					
x^2	1	4	9	16	25
y	7.1	10.4	15.9	23.6	33.5

We draw the most fitting straight line through the points (Fig. 8).

The gradient of this line gives the value of k:

$$k = 1.08$$

Example 14 *continued*

The vertical intercept gives the value of b:

$$b = 6$$

By substituting $k = 1.08$ and $b = 6$ into $y = kx^2 + b$ we obtain the law relating x and y:

$$y = 1.08x^2 + 6$$

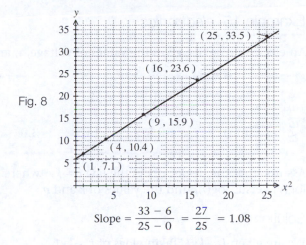

Fig. 8

$$\text{Slope} = \frac{33 - 6}{25 - 0} = \frac{27}{25} = 1.08$$

For other curves such as $y = \dfrac{k}{x} + b$ and $y = k\sqrt{x} + b$ we can obtain straight lines by plotting y against $\dfrac{1}{x}$ and y against $\sqrt{x}$ respectively. From the resulting straight lines the gradient gives the value of k and the vertical intercept gives b. Similarly other straight lines can also be obtained by plotting the appropriate variables.

D2 Determination of law by using logarithms

Consider an equation of the form $y = kx^n$ (where k and n are constants and $n \neq 1$).

? **Is it possible to find values of k and n given certain values of x and y from the graph of $y = kx^n$?**

? This is **not** a straight line of the form $y = mx + c$ where m is the gradient and c is the intercept. **How can we find k and n in $y = kx^n$?**

? We need to get $y = kx^n$ into linear form. **How?**

By taking common logs of both sides:

$$\log(y) = \log\big(kx^n\big) \underset{\text{by } 5.18}{=} \log\big(x^n\big) + \log(k)$$

$$\log(y) = \underbrace{n \log(x)}_{\text{by } 5.20} + \log(k)$$

This is similar to the equation of a straight line, $y = mx + c$, except we have $\log(y)$ and $\log(x)$ instead of y and x respectively. If we plot $\log(y)$ against $\log(x)$, we obtain a straight line with the gradient equal to n and the intercept equal to $\log(k)$. This plot is called a log–log plot.

Natural logarithms, ln, can also be taken and we would still obtain the same results. However in this section we restrict ourselves to common logarithms.

5.18 $\log(AB) = \log(A) + \log(B)$ **5.20** $\log(A^n) = n \log(A)$

Example 15 *electronics*

An electronic device gives the values of voltage, v, and current, i, shown in Table 8.

TABLE 8						
v (volts)	1	2	4	6	8	10
i (mA)	0.5	1.0	3.0	4.5	5.9	7.0

Assume that the device obeys the law $i = kv^n$ where k and n are constants. By plotting $\log(v)$ versus $\log(i)$, find the values of k and n.

Solution

We are given $i = kv^n$. Taking logs of $i = kv^n$:

$$\log(i) = \log(kv^n) \underset{\text{by } 5.18}{=} \log(v^n) + \log(k) = \underbrace{n\log(v)}_{\text{by } 5.20} + \log(k)$$

Evaluating $\log(v)$ for each of the v values in Table 8 and $\log(i)$ for the i values gives the results shown in Table 9.

TABLE 9						
$\log(v)$	0	0.30	0.60	0.78	0.90	1.00
$\log(i)$	−0.30	0.00	0.48	0.65	0.77	0.85

Plotting the graph of $\log(v)$ against $\log(i)$ yields Fig. 9.

$$\text{gradient} = \frac{0.9 + 0.3}{1.0 - 0} = \frac{1.2}{1} = 1.2$$

The gradient of the straight line in Fig. 9 gives the value of n. Hence $n = 1.2$. The intercept, where the line cuts the vertical axis, gives $\log(k)$. So $\log(k) = -0.3$. **How can we find k?**

We need to remove the log from the Left-Hand Side of $\log(k) = -0.3$. **How?**

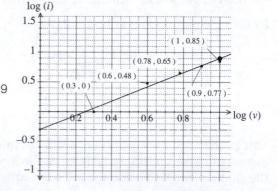

Fig. 9

Taking exponentials to the base 10 because 10^x is the inverse of $\log(x)$:

$$10^{\log(k)} = 10^{-0.3}$$

$$k = \underset{\text{by } 5.23}{10^{-0.3}} = 0.5$$

Substituting $k = 0.5$ and $n = 1.2$ into $i = kv^n$ gives the law $i = 0.5v^{1.2}$.

5.18 $\log(AB) = \log(A) + \log(B)$ **5.20** $\log(A^n) = n\log(A)$ **5.23** $10^{\log(x)} = x$

Example 16 *thermodynamics*

A gas in a cylinder expands according to the law $PV^n = C$ where P is pressure, V is volume and C, n are constants. An experiment produces the results shown in Table 10.

TABLE 10	P (kN/m²)	100	200	300	400	500	600
	V (m³)	0.160	0.100	0.076	0.063	0.054	0.048

Find the relationship between P and V.

Solution

As in **Example 15**, taking common logs of $PV^n = C$ gives

$$\log(PV^n) = \log(C)$$

$$\underbrace{\log(P) + \log(V^n)}_{\text{by } 5.18} = \log(C)$$

$$\underbrace{\log(P) + n\log(V)}_{\text{by } 5.20} = \log(C)$$

Transposing gives

$$\log(P) = -n\log(V) + \log(C)$$

Table 11 shows values of $\log(P)$ against $\log(V)$.

TABLE 11	$\log (P)$	2.000	2.301	2.477	2.602	2.699	2.778
	$\log (V)$	-0.796	-1.000	-1.119	-1.201	-1.268	-1.319

What do we plot on the horizontal axis?

Comparing

$$\log(P) = -n\log(V) + \log(C)$$
and $$y = mx + c$$

Fig. 10

shows that we plot $\log(V)$ on the horizontal axis and $\log(P)$ on the vertical axis (Fig. 10).

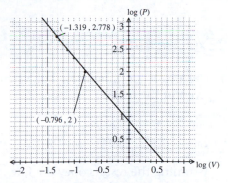

$$\text{Slope} = -\frac{3-0}{1.5+0.6} = -1.43$$

5.18 $\log(AB) = \log(A) + \log(B)$ 5.20 $\log(A^n) = n\log(A)$

Example 16 *continued*

The gradient (slope) of the straight line is $-n$, so we have

$$-n = -1.43$$
$$n = 1.43$$

How do we find C?

The value of $\log(C)$ is where the line cuts the vertical axis, therefore

$$\log(C) = 0.9$$

Taking exponentials to the base 10:

$$C = 10^{0.9} = 7.94$$

Substituting $n = 1.43$, $C = 7.94$ into $PV^n = C$ gives the relationship

$$PV^{1.43} = 7.94$$

SUMMARY

We need to select suitable variables to obtain a straight line, so that the new equation is of the form $y = mx + c$. If the experimental data follows the law $y = kx^n$ then we can obtain the values of k and n by:
1 plotting the graph of $\log(y)$ against $\log(x)$ which results in a straight line
2 the gradient of the straight line $= n$
3 the vertical intercept $= \log(k)$.
To obtain k, we take exponentials of both sides of **3** to the base 10.

Exercise 5(d)

Solutions at end of book. Complete solutions available at www.palgrave.com/engineering/singh

1 The area A for dimension d of a figure is given in the table:

d (cm)	2	3	4	5	6
A (cm²)	12.57	28.27	50.27	78.54	113.10

Assuming the results obey the law $A = kd^2 + b$, find k and b by plotting an appropriate graph. What do you notice about your results? What shape is the figure?

2 [*mechanics*] The force, F, between two objects at a distance d apart is given in the table:

d (m)	2	3	4	5	6	7
F ($\times 10^{-9}$) (N)	2.510	1.110	0.630	0.400	0.280	0.200

Assuming the results obey the rule $F = \dfrac{k}{d^2}$, find k by plotting a suitable graph.

Solutions at end of book. Complete solutions available at www.palgrave.com/engineering/singh

Exercise **5(d) continued**

3 [*thermodynamics*]   The pressure, P, and volume, V, of a gas are given in the table:

V (m³)	0.1	0.15	0.2	0.25	0.3	0.35	0.4
P (kPa)	416	277	208	166	139	119	104

   i Plot P vertically against V

   ii Plot P vertically against $\dfrac{1}{V}$

   iii Show that $P = \dfrac{k}{V}$ and find the value of k.

4 [*aerodynamics*]   The drag coefficient, C, is related to Reynolds number, R, by

$$C = kR^n$$

where k and n are constants. For the data in the table, find k and n.

R	3	4	5	6	7	8	9	10
C	0.060	0.056	0.054	0.052	0.050	0.049	0.048	0.047

5 [*structures*]   The table shows the deflection, x, of a beam under the corresponding loads.

x (mm)	5.00	7.00	10.00	15.00	20.00
L (kN)	4.39	5.49	6.95	9.12	11.05

Deduce that the results obey $L = kx^n$ and find k and n.

SECTION E **Hyperbolic functions**

By the end of this section you will be able to:
▶ plot graphs of these functions
▶ investigate properties of hyperbolic functions
▶ examine engineering applications of these functions

This is a particularly demanding section requiring a good grasp of the exponential function and algebra. If you come up against some difficulties and get 'stuck', do not give up too quickly. Try following the steps more slowly and go over them several times if necessary.

E1 **Hyperbolic functions**

Hyperbolic functions are characterized in terms of the exponential function. We define the hyperbolic sine, written sinh and pronounced 'shine', as

5.25

$$\sinh(x) = \frac{e^x - e^{-x}}{2}$$

We define the hyperbolic cosine, written cosh and pronounced 'cosh', as

> The German mathematician **Lambert** (1728–1777) was the first person who seriously developed hyperbolic functions.

5.26 $$\cosh(x) = \frac{e^x + e^{-x}}{2}$$

These are related by the fundamental identity $\cosh^2(x) - \sinh^2(x) = 1$ which will be shown in **Example 21**.

The hyperbolic tangent, written tanh and pronounced 'than' is defined as

5.27 $$\tanh(x) = \frac{\sinh(x)}{\cosh(x)} = \frac{e^x - e^{-x}}{e^x + e^{-x}}$$

You can use your calculator to find values of these hyperbolic functions.

Example 17

Evaluate, correct to three d.p., $\sinh(2.1)$, $\tanh(1)$, $\cosh(0.3)$, $\cosh(15)$, $\tanh(5000)$, $\tanh(10^6)$, $\sinh(0.01)$ and $\sinh(10^{-6})$.

Solution

To evaluate $\sinh(2.1)$ on a calculator, PRESS [**hyp**] [**sin**] [**2.1**] [**=**] which should show 4.021856742.

Correct to 3 d.p., $\sinh(2.1) = 4.022$. Also note that $\sinh(2.1) = 4.022$ means

$$\frac{e^{2.1} - e^{-2.1}}{2} = 4.022$$

(You can check this result on your calculator.)

Evaluating the others on a calculator gives

$$\tanh(1) = 0.762, \cosh(0.3) = 1.045, \cosh(15) = 1634508.686,$$

$$\tanh(5000) = 1, \tanh(10^6) = 1, \sinh(0.01) = 0.01 \text{ and } \sinh(10^{-6}) = 10^{-6}$$

? **What do you notice about the last four results of Example 17?**

It seems that tanh of a large number is approximately 1 and sinh of a small number gives the same small number. These statements are approximately correct, as can be seen by the graphs of these functions.

E2 Graphs of hyperbolic functions

> **Example 18**
>
> Plot the graph of $y = \cosh(x)$ for $x = -4$ to $+4$.
>
> **Solution**
>
> We select whole numbers of x between -4 and $+4$ and evaluate $\cosh(x)$ on a calculator for each of these values (Table 12).

	x	-4	-3	-2	-1	0	1	2	3	4
TABLE 12	$y = \cosh(x)$	27.31	10.07	3.76	1.54	1.00	1.54	3.76	10.07	27.31

> Connecting the points in the table gives the graph shown in Fig. 11.
>
> **What do you notice about your graph?**
>
> **1** $\cosh(-x) = \cosh(x)$, symmetrical about the y axis
> **2** $\cosh(0) = 1$
> **3** $\cosh(x)$ is always bigger than or equal to 1.
>
> You can confirm the first two statements by using the definition of $\cosh(x)$.

Fig. 11

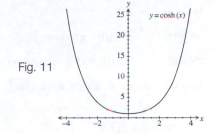

The sinh and tanh graphs can be plotted similarly to the cosh graph (see Figs 12 and 13)

You are asked to plot these in **Exercise 5(e)**.

Fig. 12

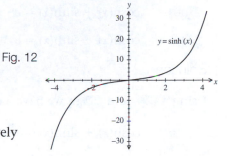

Observe from Fig. 13 that as x gets positively or negatively large then $\tanh(x)$ gets close to 1 or -1 respectively.

(More difficult to observe is the property stated earlier about $\sinh(x) \approx x$ for small x, but if you plot $\sinh(x)$ on a computer algebra system and zoom in close to zero you should be able to read off the x values on the horizontal axis and the corresponding $\sinh(x)$ values.)

Fig. 13

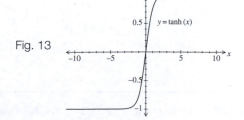

E3 Properties of hyperbolic functions

Example 19

Show that

$$\cosh(-x) = \cosh(x)$$

Solution

This is statement **1** from the previous page. **How can we show** $\cosh(-x) = \cosh(x)$**?**

By substituting $-x$ for x into 5.26 :

$$\cosh(-x) = \frac{e^{-x} + e^{-(-x)}}{2}$$

$$= \frac{e^{-x} + e^{+x}}{2}$$

$$= \frac{e^{x} + e^{-x}}{2} = \cosh(x) \qquad \left[\text{By } 5.26\right]$$

We have

5.28 $\cosh(-x) = \cosh(x)$

5.29 $\sinh(-x) = -\sinh(x)$

You can also use 5.28 and 5.29 to deduce $\tanh(-x) = -\tanh(x)$.

Example 20

Show that

5.30 $\cosh(x) + \sinh(x) = e^{x}$

5.31 $\cosh(x) - \sinh(x) = e^{-x}$

Solution

Using 5.25 and 5.26 we have a common denominator:

$$\cosh(x) + \sinh(x) = \frac{e^{x} + e^{-x}}{2} + \frac{e^{x} - e^{-x}}{2}$$

$$= \frac{e^{x} + e^{-x} + e^{x} - e^{-x}}{2}$$

$$= \frac{2e^{x}}{2} \qquad \text{[Adding numerator]}$$

$$= e^{x} \qquad \text{[Cancelling 2's]}$$

5.25 $\sinh(x) = (e^{x} - e^{-x})/2$ 5.26 $\cosh(x) = (e^{x} + e^{-x})/2$

Example 20 *continued*

Similarly we can show the other result 5.31 :

$$\cosh(x) - \sinh(x) = \frac{e^x + e^{-x}}{2} - \frac{e^x - e^{-x}}{2}$$

$$= \frac{e^x + e^{-x} - (e^x - e^{-x})}{2}$$

$$= \frac{e^x + e^{-x} - e^x + e^{-x}}{2}$$

$$= \frac{2e^{-x}}{2} = e^{-x}$$

Example 21

Show the fundamental identity

5.32 $\quad \cosh^2(x) - \sinh^2(x) = 1$

where $\cosh^2(x) = \left[\cosh(x)\right]^2$ and $\sinh^2(x) = \left[\sinh(x)\right]^2$.

Solution

How can we show $\cosh^2(x) - \sinh^2(x) = 1$?

We can use the difference of two squares:

1.15 $\quad a^2 - b^2 = (a + b)(a - b)$

You really need to **know** this identity, otherwise it would not occur to us to use this identity in this context:

$$\cosh^2(x) - \sinh^2(x) = \left[\cosh(x)\right]^2 - \left[\sinh(x)\right]^2$$

$$= \left[\cosh(x) + \sinh(x)\right]\left[\cosh(x) - \sinh(x)\right]$$

by 1.15

$$= e^x \times e^{-x} = e^{x-x} = e^0 = 1$$

by 5.30   by 5.31   by 5.1

Similar to trignometric functions we have other hyperbolic functions such as:

5.33 $\quad \operatorname{sech}(x) = \dfrac{1}{\cosh(x)}$

5.34 $\quad \operatorname{cosech}(x) = \dfrac{1}{\sinh(x)}$

5.35 $\quad \coth(x) = \dfrac{1}{\tanh(x)}$

5.1 $\quad a^m a^n = a^{m+n}$ 5.30 $\quad \cosh(x) + \sinh(x) = e^x$ 5.31 $\quad \cosh(x) - \sinh(x) = e^{-x}$

Use your calculator to find cosech(2).

$$\text{cosech}(2) = \frac{1}{\sinh(2)} = \frac{1}{3.627} = 0.276$$

E4 Engineering applications of hyperbolic functions

The following example is an increase in the level of sophistication compared to the previous examples.

Example 22 *electrical principles*

A transmission line of length L has voltage V_x at a distance x from the sending end and is given by

$$V_x = \left(\frac{V + IZ_0}{2}\right)e^{-\gamma x} + \left(\frac{V - IZ_0}{2}\right)e^{\gamma x}$$

where V is voltage, I is current, Z_0 is characteristic impedance and γ is propagation coefficient. Show that at $x = L$:

$$V_L = V\cosh(\gamma L) - IZ_0 \sinh(\gamma L)$$

Solution

Don't be put off by all the different symbols used, it is really quite straightforward.
Putting $x = L$ into ⬛ * ⬛ gives

$$V_L = \left(\frac{V + IZ_0}{2}\right)e^{-\gamma L} + \left(\frac{V - IZ_0}{2}\right)e^{\gamma L}$$

$$= V\underbrace{\left(\frac{e^{-\gamma L} + e^{\gamma L}}{2}\right)}_{\text{collecting the } V \text{ terms}} + IZ_0\underbrace{\left(\frac{e^{-\gamma L} - e^{\gamma L}}{2}\right)}_{\text{collecting the } IZ_0 \text{ terms}}$$

$$= V\underbrace{\cosh(\gamma L)}_{\text{by } 5.26} - IZ_0\left(\frac{e^{\gamma L} - e^{-\gamma L}}{2}\right)$$

$$V_L = V\cosh(\gamma L) - IZ_0\underbrace{\sinh(\gamma L)}_{\text{by } 5.25}$$

The next example is particularly difficult. See if you can follow it through. It uses a noteworthy 'trick' which is to write the number 1 as

$$1 = \frac{2}{2} = e^0 = e^{kx - kx}\ldots$$

There are an infinite number of ways you can write any number.

The next example is a little tricky but do-able.

5.25 $\sinh(x) = (e^x - e^{-x})/2$ 5.26 $\cosh(x) = (e^x + e^{-x})/2$

Example 23 *electronics*

In a semiconductor, a force, F, exerted on an electron is given by

$$\boxed{\;\ast\;} \qquad F = \frac{Qcke^{-kx}}{(1 + e^{-kx})^2}$$

where c and k are constants, x is the distance from the pn junction and Q is charge. Show that

$$F = \frac{Qck}{2[1 + \cosh(kx)]}$$

Solution

Expanding the denominator of $\boxed{\;\ast\;}$ by using

$\boxed{1.13}$
$$(a + b)^2 = a^2 + 2ab + b^2$$
$$(1 + e^{-kx})^2 = 1 + 2e^{-kx} + (e^{-kx})^2$$
$$= 1 + 2e^{-kx} + e^{-kx}e^{-kx}$$

Also we can write 1 as $1 = e^0 = e^{kx - kx} \underset{\text{by } \boxed{5.1}}{=} e^{kx}e^{-kx}$, so in the denominator we have

$$(1 + e^{-kx})^2 = \underbrace{e^{kx}e^{-kx}}_{\text{replacing 1}} + 2e^{-kx} + e^{-kx}e^{-kx}$$
$$= e^{-kx}(e^{kx} + 2 + e^{-kx}) \qquad \text{[Taking out } e^{-kx}\text{]}$$

Substituting this, $(1 + e^{-kx})^2 = e^{-kx}(2 + e^{kx} + e^{-kx})$, into $\boxed{\;\ast\;}$ gives

$$F = \frac{Qcke^{-kx}}{e^{-kx}(2 + e^{kx} + e^{-kx})} = \frac{Qck}{2 + e^{kx} + e^{-kx}} \qquad \text{[Cancelling } e^{-kx}\text{]}$$

Examining the denominator we have

$$2 + e^{kx} + e^{-kx} = 2 + (e^{kx} + e^{-kx})$$
$$= 2 + 2\left(\frac{e^{kx} + e^{-kx}}{2}\right) \qquad \left[\text{Writing } 1 = \frac{2}{2}\right]$$
$$= 2 + 2\underbrace{\cosh(kx)}_{\text{by } \boxed{5.26}} = 2[1 + \cosh(kx)]$$

Substituting $2 + e^{kx} + e^{-kx} = 2[1 + \cosh(kx)]$ into the denominator of F gives the required result:

$$F = \frac{Qck}{2[1 + \cosh(kx)]}$$

$\boxed{5.1}$ $\quad a^m a^n = a^{m+n}$

SUMMARY

We define the hyperbolic functions in terms of the exponential functions:

5.25 $\sinh(x) = (e^x - e^{-x})/2$

5.26 $\cosh(x) = (e^x + e^{-x})/2$

5.27 $\tanh(x) = \dfrac{\sinh(x)}{\cosh(x)}$

Exercise 5(e)

Solutions at end of book. Complete solutions available at www.palgrave.com/engineering/singh

1 Evaluate the following (use your calculator): $\sinh(10)$, $\sinh(-3)$, $\cosh(10)$, $\cosh(1 \times 10^{-3})$, $\tanh(5 \times 10^7)$ and $\tanh(0.5)$.

2 a Plot the graph of $y = \sinh(x)$ for $x = -4$ to $x = 4$.
   b Plot the graph of $y = \tanh(x)$ for $x = -4$ to $x = 4$.

3 Show that $\sinh(-x) = -\sinh(x)$.

4 Show that $\tanh(-x) = -\tanh(x)$.

5 Show that $\cosh^2(x) - 2\cosh(x)\sinh(x) + \sinh^2(x) = e^{-2x}$.

6 Show that $\cosh^2(x) + 2\cosh(x)\sinh(x) + \sinh^2(x) = e^{2x}$.

7 If $x = a\cosh(t)$ and $y = b\sinh(t)$, show that

$$\frac{x^2}{a^2} - \frac{y^2}{b^2} = 1$$

(Equation of hyperbola, hence the name hyperbolic functions.)

8 Determine $\operatorname{sech}(1.5)$, $\operatorname{cosech}(1.5)$ and $\coth(1.5)$.

9 [wave theory] The speed, v (m/s), of water waves is given by

$$v^2 = 1.8\lambda \tanh\left(\frac{6.3d}{\lambda}\right)$$

where d is depth and λ is wavelength. If $d = 35$ m and $\lambda = 300$ m, evaluate v.

10 [heat transfer] The heat transfer rate, Q, of a rod of length L is given by

$$Q = 15\left[\frac{\sinh(2.56L) + (6 \times 10^{-3})\cosh(2.56L)}{\cosh(2.56L) + (6 \times 10^{-3})\sinh(2.56L)}\right]$$

If $L = 30 \times 10^{-3}$ m, find Q.

11 [electrical principles] A transmission line of length L has current I_x at a distance x from the sending end given by

$$I_x = \left(\frac{v/z + I}{2}\right)e^{\gamma x} - \left(\frac{v/z - I}{2}\right)e^{-\gamma x}$$

where v is the receiving end voltage, I is the receiving end current, z is the characteristic impedance and γ is the propagation coefficient. Show that at $x = L$

$$I_L = \frac{v}{z}\left[\sinh(\gamma L)\right] + I\left[\cosh(\gamma L)\right]$$

12 [mechanics] The tension, T, in a cable is given by

$$T^2 = H^2\left[1 + \sinh^2\left(\frac{wx}{H}\right)\right]$$

where H is horizontal tension, w is weight per unit length and x is horizontal base distance. Show that

$$T = H\cosh\left(\frac{wx}{H}\right).$$

Exercise **5(e) continued**

Solutions at end of book. Complete solutions available at www.palgrave.com/engineering/singh

*13 ▦ [*electrical principles*] The current, I_x, at a distance x along a transmission line of length L is given by

$$I_x = Ie^{-\gamma L}\left[e^{\gamma x} + \left(\frac{Z_0 - Z}{Z_0 + Z}\right)e^{-\gamma x}\right]$$

where I is current, Z_0 is characteristic impedance, Z is load impedance and γ is propagation coefficient. Show that

$$I_x = \frac{2Ie^{-\gamma L}}{Z_0 + Z}\left[Z_0\cosh(\gamma x) + Z\sinh(\gamma x)\right]$$

*14 ▦ [*electrical principles*] The impedance, Z_x, at a distance x from the sending end of a transmission line is given by

$$Z_x = Z_0\left[\frac{e^{\gamma x} + \left(\frac{Z - Z_0}{Z_0 + Z}\right)e^{-\gamma x}}{e^{\gamma x} + \left(\frac{Z_0 - Z}{Z_0 + Z}\right)e^{-\gamma x}}\right]$$

where Z_0 is the characteristic impedance, Z is the load impedance and γ is the propagation coefficient. Show that

$$Z_x = Z_0\left[\frac{Z_0\sinh(\gamma x) + Z\cosh(\gamma x)}{Z_0\cosh(\gamma x) + Z\sinh(\gamma x)}\right]$$

Examination questions **5**

Solutions at end of book. Complete solutions available at www.palgrave.com/engineering/singh

1 Showing intermediate steps, find $\sinh(1.5)$, $\cosh(0.3)$ and $\operatorname{sech}(1.2)$.

University of Manchester, UK, 2008

2 A stereo amplifier's power output P (in watts) is related to its decibel voltage gain d by the formula: $P(d) = 25e^{0.1d}$.

 a What is the power output when the decibel voltage gain is 4 decibels?

 b What is the power output when the decibel voltage gain is 12 decibels?

Arizona State University, USA (Review)

3 Solve for x in each of the following equations:

 i $\log_2 3 + \log_2(x + 1) = \log_2(x + 11)$

 ii $2[\ln(x^2 + 4)] = 3.9$

 iii $5(2^{6x+1}) = 42$

4 The following table gives values of p and V which are believed to be related by the law of the form $p = aV^2 + bV$, where a and b are constants.

p	4.5	38.5	121.4	231.8	318.2	469.7	565.2
V	0.5	2.6	5.3	7.7	9.2	11.4	12.7

Verify the law exists. Find the approximate values of a and b. Hence state the law.

Questions 3 and 4 are from Cork Institute of Technology, Ireland, 2007

5 Solve for x:

 i $8^x = \dfrac{1}{2}$

 ii $\log_4 x = 2$

University of Manchester, UK, 2009

6 Solve the following equations for x:

 i $8^x = 2$ **ii** $\log_3\left(\dfrac{27}{x + 1}\right) = 2$

 iii $\log_2(4^{x+1}) = x$

 iv $\log_x(x^3 - 2x + 1) = 3$

University of Manchester, UK, 2010

Examination questions 5 continued

Solutions at end of book. Complete solutions available at www.palgrave.com/engineering/singh

7 Solve the equation $9^{3-2x} = 3.27^{x-3}$.

Memorial University of Newfoundland, Canada, 2009

8 a Write in terms of $\log A$, $\log B$, $\log C$

$$\log \left(\frac{A^2 \sqrt{C}}{B^5} \right)$$

 b Solve the equation $3^{2x-1} = 18$.

 c For the function $y = \ln(x^2 - 3)$, find

  i The value of y when $x = 2$

  ii The value of x when $y = 1$

University of Portsmouth, UK, 2009

9 A semiconductor diode can be modelled by the equation $I = I_S(e^{40v} - 1)$, where v is voltage and I is current.

 i Given that $I = 2$ at $v = 0.003$, find the original current, I_S.

 ii Find the value of v when $I = 3$.

University of Portsmouth, UK, 2006

10 What is the limit $\lim_{y \to 0} (e^{2y} - 1)/(e^y - 1)$ equal to?

11 What is the ratio $(1 - e^{2/x})/(1 + e^{2/x})$ equal to?

[These questions have been slightly edited.]

Questions 10 and 11 are from King's College London, UK, 2008

12 The proportion of US households that own a DVD player can be represented by the logistic model function $P(t) = \dfrac{0.9}{1 + 6e^{-0.32t}}$. Let $t = 0$ represent the year 2000.

 a When will the proportion of households reach 0.6 (60%)?

 b When will the proportion of households reach 0.8 (80%)?

Arizona State University, USA (Review)

13 Using the definition of $\sinh x$ and $\cosh x$ in terms of e^x, prove that

 a $1 + \sinh^2 x = \cosh^2 x$

 b $\cosh 2x = 1 + 2\sinh^2 x$

University of Sussex, UK, 2006

14 Given that $\sinh x = \dfrac{12}{5}$, find

 a $\cosh x$ **b** $\tanh x$ **c** $\sinh 2x$

University of Surrey, UK, 2006

15 Solve for x, given $\sinh(x) + 3\cosh(x) = 4$.

University of Portsmouth, UK, 2006

Miscellaneous exercise 5

Solutions at end of book. Complete solutions available at www.palgrave.com/engineering/singh

1 [electronics] Evaluate V for $I = 10$ mA, $I_0 = 1$ μA:

$$V = (20 \times 10^{-3}) \ln \left(\frac{I}{I_0} + 1 \right)$$

2 [electronics] The ideal diode equation is given by

$$I = I_0 \left[\exp \left(\frac{qV}{kT} \right) - 1 \right]$$

where I, I_0 represent currents, V represents voltage, T is temperature, q is charge and k is a constant. Show that

$$V = \frac{kT}{q} \ln \left(\frac{I}{I_0} + 1 \right)$$

(Remember that $\exp(x) = e^x$.)

3 [vibrations] In a vibrational system, the logarithmic decrement, δ (delta), is defined as

$$\delta = \ln\left(\frac{ke^{-\zeta\omega t_1}}{ke^{-\zeta\omega t_2}}\right)$$

where ζ (zeta) is the damping ratio, ω is the angular frequency, t_1, t_2 represent time and k is a constant. Show that

$$\delta = \zeta\omega(t_2 - t_1)$$

4 [electrical principles]   During a charging phase of a RC circuit, the voltage, V, across a capacitor of capacitance C, is given by

$$V = V_0(1 - e^{-t/\tau})$$

where $V_0 = 100$ volts, t is time (in s) and τ is the time constant (s) [$\tau = RC$].

Plot the graph of V versus t for $0 \le t \le 5\tau$ and state the percentage of voltage V of the initial voltage V_0 at $t = \tau, 2\tau, 3\tau, 4\tau$ and 5τ.

5 [control engineering]   The output, y, of a system is given by

$$y = 1 - e^{-t/\tau}$$

where $\tau > 0$ is the time constant and t is time. The final value, Y, of the output is defined by

$$Y = \lim_{t \to \infty}(1 - e^{-t/\tau})$$

 i Determine Y.

 ii Evaluate the values of y for $t = \tau, 2\tau, 3\tau, 4\tau$ and 5τ.

 iii What percentage of the final value, Y, is reached at $t = \tau, 2\tau, 3\tau, 4\tau$ and 5τ.

 iv Sketch the graph of y versus t.

6 [mechanics]   The velocity, v, of an object falling in air is given by

$$v = \frac{10}{k}(1 - e^{-kt})$$

where k is a positive constant (air resistance).

 i How does v change as k increases if t is large?

 ii What happens to v as $t \to \infty$? (This is called the terminal velocity.)

 iii Plot on the same axes the graphs of v against t for $k = 1, 2$ and 5 and for t between 0 and 10.

 iv Do the graphs satisfy your answers in parts **i** and **ii**?

 v Determine the terminal velocity for $k = 1, 2$ and 5.

7 [aerodynamics]   The friction coefficient, C, is related to Reynolds number, R, by

$$C = kR^n$$

where k and n are constants. Find k and n for the data in the following table.

R	3	4	5	6	7	8
$C\ (\times 10^{-3})$	48	45	43	42	40	39

Use a symbolic manipulator or a graphical calculator for questions 8 to 12.

8 [structures]   A cable is suspended between two level points, $L = 270$ m apart, and has a sag $y = 55$ m. If the load per unit length $w = 0.16$ kN/m, find the horizontal tension H, given that

$$y = \frac{H}{w}\left[\cosh\left(\frac{wL}{2H}\right) - 1\right]$$

9 [heat transfer]   Given that

$\theta = 300 + 100e^{kt}$ where θ is the temperature of a body at time t, plot on the same axes the graphs of θ against t for $t \ge 0$ and $k = -0.01, -0.1$ and -1.

How does the temperature θ change with different values of k?

What is the temperature θ in each case as $t \to \infty$? (This is called the steady-state temperature.)

Miscellaneous exercise 5 continued   Solutions at end of book. Complete solutions available at www.palgrave.com/engineering/singh

10 🗲 [*control engineering*]   The output response, y, of a feedback control system is given by

$$y = \frac{k}{1 + k}\left[1 - e^{-(1 + k)t}\right]$$

Plot on the same axes the graphs of y against t (> 0) for $k = 1, 5, 7$ and 10.

11 The normal distribution with zero mean is described by

$$f(x) = \frac{1}{\sigma\sqrt{2\pi}} \exp\left(\frac{-x^2}{2\sigma}\right)$$

Plot the graphs of $f(x)$ against x for $\sigma = 0.1, 0.2$ and 0.3 on the same axes for x between -1 and 1. As σ increases, how does the graph change?

12 🗇 [*reliability engineering*]   The reliability function, $R(t)$, for a particular type of component is given by

$$R(t) = e^{-\sqrt{t}} \qquad t \geq 0$$

i Plot $R(t)$ and $F(t) = 1 - R(t)$ on the same axes for $t \geq 0$.

ii What are the values of $R(t)$ and $F(t)$ as $t \to \infty$?

13 By using the exponential definition of sinh and cosh, show that

$$\cosh(A)\cosh(B) + \sinh(A)\sinh(B)$$
$$= \cosh(A + B)$$

For questions 14 and 15 use a symbolic manipulator (or a graphical calculator).

14 🚗 [*mechanics*]   The shape of a hanging cable between two fixed points is given by

$$y = k \cosh\left(\frac{x}{k}\right)$$

where $k = T/W$, T is the tension in the cable, W is the load per unit length and x is the horizontal distance. On the same axes, plot the graphs of y against x for x values between -10 and $+10$ and $k = 10, 20, 30, 40$ and 50. What do you notice about the graph as k increases, that is the tension, T, in the cable increases?

15 🚗 [*mechanics*]   A design engineer wants to design a bridge whose arch is a shape of the form

$$h = c - c \cosh\left(\frac{x}{c}\right)$$

(h gives the height (m) of the arch at a distance x (m) along the base.)

Plot graphs of h, on the same axes, for $-20 \leq x \leq 20$ and for the following values of c:

i 50   **ii** 100   **iii** 150   **iv** 200
v 250   **vi** 300

16 Determine where the following damped oscillations y are equal to zero.

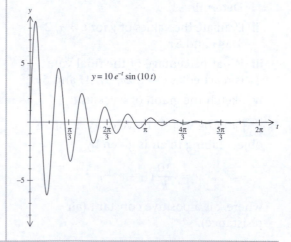

$y = 10\,e^{-t}\sin(10t)$

CHAPTER 6
Differentiation

SECTION A **The derivative**

By the end of this section you will be able to:
▶ evaluate the gradient of a curve at a point
▶ understand the notation $\frac{dy}{dx}$, $f'(x)$ and $\dot{x}$
▶ sketch the derivative graph

In the next four chapters we introduce the concepts of calculus which are differentiation and integration. Calculus is perhaps the most powerful mathematical tool ever developed because it brings the whole physical world within the scope of mathematics. Many physical systems involve rates of change which can be described with calculus. Every aircraft, car, suspension bridge and earthquake-proof building owe their design in part to calculus.

A1 **Gradient**

❓ **What does the term *gradient* mean?**

Approaching a hill we sometimes see the warning sign

❓ **What does this mean?**

The slope of the approaching hill is 4 units across for every
1 unit up (Fig. 1). The fraction $\frac{1}{4}$ (or 25%) is called the gradient

Fig. 1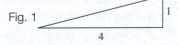

of the hill. Gradient is a measure of the steepness of a hill. Below we examine the gradients of straight lines on graphs. We covered these in **Chapter 2** where we defined gradient as

$$\text{Gradient} = \frac{\text{vertical}}{\text{horizontal}} \text{ or } \frac{\text{rise}}{\text{run}}$$

Example 1

Find the gradient of $y = 2x$

Solution

$$\text{Gradient} = \frac{\text{change in } y}{\text{change in } x} = \frac{4 - 2}{2 - 1} = 2$$

Fig. 2

Of course the gradient in **Example 1** could easily have been evaluated by using 2.1, $y = mx + c$, where m is the gradient and c is the y-intercept. We have $m = 2$ so the gradient is 2.

Now let's examine the gradient of the curve $y = x^2$. Suppose we draw the graph of $y = x^2$ (Fig. 3).

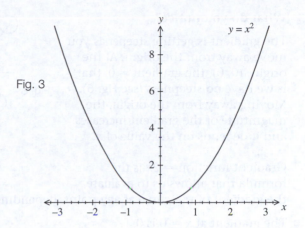
Fig. 3

? **How do we find the gradient of this curve? More precisely the question should be – _where_ do we want to find the gradient of this curve?**

As we move away from the origin, (0, 0), the gradient gets steeper. The gradient is changing as you move along the x axis.

? **Can we find the gradient at an arbitrary point x?**

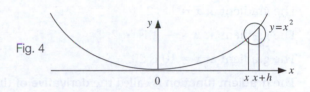
Fig. 4

Consider the graph of $y = x^2$ close to a point x. With graphical calculators or computer algebra packages we can zoom in very close to a point x and see what the graph looks like around x and $x + h$, where h is small (Fig. 4). If this small part is magnified, then it will look almost like a straight line (Fig. 5).

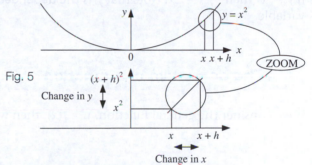
Fig. 5

Remember the gradient of a straight line:

$$\text{Gradient} = \frac{\text{change in } y}{\text{change in } x}$$

$$= \frac{(x + h)^2 - x^2}{(x + h) - x}$$

$$= \frac{(x^2 + 2hx + h^2) - x^2}{h} = \frac{2hx + h^2}{h} = \frac{h(2x + h)}{h}$$

$$\text{Gradient} = 2x + h \qquad [\text{Cancelling } h\text{'s}]$$

In order to treat this section of the curve as a straight line, h needs to be very small. This can be written as $\lim\limits_{h \to 0}$.

Remember $\lim\limits_{h \to 0}$ means we can make h as close as possible to zero. We have

$$\text{Gradient} = \lim_{h \to 0}(2x + h) = 2x$$

? **When we found the gradient of a straight line it was a real number. But why is the gradient of the curve, $y = x^2$, a function, $2x$?**

Observe the graph in Fig. 6.

? **What do you notice?**

The gradient is getting steeper as you move away from the origin. At the origin, (0, 0), the gradient = 0, that is we have **no** steepness (see Fig. 6). Moving away from the origin, the magnitude of the gradient increases and it depends on the value of x.

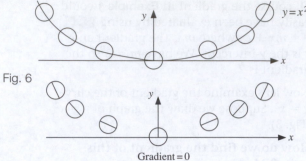

Fig. 6

Gradient = 0

Gradient function = $2x$ is the formula that allows us to evaluate the gradient of the curve at **any** point, depending on the value of x.

The gradient at $x = 0$ is 0.

The gradient at $x = 1$ is 2.

The gradient at $x = -1$ is -2.

The gradient at $x = 10$ is 20.

The **gradient function** is called the **derivative** of the original function and is denoted by $\dfrac{dy}{dx}$. If $y = x^2$ then $\dfrac{dy}{dx} = 2x$. Note that y is the **dependent** variable and x is the **independent** variable.

A2 **The derivative of a function**

If we consider the general function, $y = f(x)$, then we have (Fig. 7):

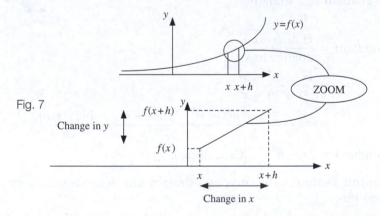

Fig. 7

$$\frac{dy}{dx} = \lim_{h \to 0} \frac{\text{change in } y}{\text{change in } x}$$

$$= \lim_{h \to 0} \frac{f(x + h) - f(x)}{(x + h) - x}$$

$$\frac{dy}{dx} = \lim_{h \to 0} \frac{f(x + h) - f(x)}{h}$$

Isaac Newton 1642–1727

Gottfried Leibniz 1647–1716

Two men, the Englishman Isaac Newton (1642–1727) and the German Gottfried Leibniz (1647–1716), can claim to have developed calculus as we recognize it today. Evidence suggests Newton discovered his form of calculus first, but Leibniz published his own work on calculus earlier.

Leibniz's notation $\frac{dy}{dx}$, $\int$ was far superior,

and his algebraic method led to swift advances in this field, while Newton's geometric style of calculus was difficult to use.

Newton claimed that his work had been plagiarised while Leibniz strongly refuted the accusation, and both men and their supporters argued fiercely over who should be credited. This argument became a slanging match between Continental Europe and Britain. Incredibly, by the middle of the nineteenth century, Britain had become a mathematical backwater as a result of this dispute.

By the age of 24, Newton had become the greatest mathematician of Europe. His first great development was the Binomial Theorem. While he is less celebrated, Leibniz had an equally illustrious career. As well as significant contributions to many areas of science, he invented the binary system, which is still used internally by today's computers.

So for the general function, $y = f(x)$, the derivative is defined to be

6.1 $$\frac{dy}{dx} = \lim_{h \to 0} \frac{f(x + h) - f(x)}{h}$$

The notation, $\frac{dy}{dx}$, pronounced 'dee y by dee x'

was developed by Leibniz in 1684. This is more popular than Newton's notation, which is $\dot{x}$, pronounced 'x dot' and represents $\frac{dx}{dt}$. The $\dot{x}$ symbol is often found in applied mathematics:

$$\frac{dx}{dt} = \dot{x}$$

(Sometimes called the time derivative because t represents time.)

There are also other notations for the derivative. If $y = f(x)$, then the derivative can also be denoted as $f'(x)$, pronounced 'f dashed x':

$$\frac{dy}{dx} = f'(x)$$

Most students find differentiation pretty straightforward, but it is the associated algebra that they have difficulty with. However algebra is the bread and butter of calculus, so make sure you are comfortable with the concepts covered in **Chapter 1**.

Example 2

If $y = 2x$, determine $\dfrac{dy}{dx}$ by using 6.1 .

Solution

Let $f(x) = 2x$ then $f(x + h) = 2(x + h)$. Putting these into 6.1 gives

$$\frac{dy}{dx} = \lim_{h \to 0} \frac{2(x + h) - 2x}{h}$$

$$= \lim_{h \to 0} \frac{2x + 2h - 2x}{h} \quad \text{[Expanding numerator]}$$

$$= \lim_{h \to 0} \frac{2\cancel{h}}{\cancel{h}} = 2 \quad \text{[Cancelling } h\text{'s]}$$

The derivative of $y = 2x$ is 2. Sometimes this is written as

$$\frac{d}{dx}(2x) = 2$$

Note that this is the gradient of the straight line, $y = 2x$, as found in **Example 1**.

Example 3

If $y = x^3$, find $\dfrac{dy}{dx}$ by using 6.1 .

Solution

This is a more challenging example because it uses complicated algebra such as the binomial expansion.

With $f(x) = x^3$ then $f(x + h) = (x + h)^3$. Substituting these into 6.1 gives

$$\boxed{\dagger} \qquad \frac{dy}{dx} = \lim_{h \to 0} \frac{(x + h)^3 - x^3}{h}$$

By using the binomial expansion on $(x + h)^3$ we have

$$(x + h)^3 = x^3 + 3x^2 h + 3xh^2 + h^3$$

Substituting this, $(x + h)^3 = x^3 + 3x^2h + 3xh^2 + h^3$, into † gives

$$\frac{dy}{dx} = \lim_{h \to 0} \frac{(x^3 + 3x^2h + 3xh^2 + h^3) - x^3}{h}$$

Example 3 *continued*

$$= \lim_{h \to 0} \frac{3x^2h + 3xh^2 + h^3}{h} \quad \text{[Simplifying numerator]}$$

$$= \lim_{h \to 0} \frac{h(3x^2 + 3xh + h^2)}{h} \quad \text{[Factorizing numerator]}$$

$$= \lim_{h \to 0} (3x^2 + 3xh + h^2) = 3x^2$$

$$\frac{dy}{dx} = 3x^2$$

The gradient of the curve $y = x^3$ can be evaluated at any point from the formula $3x^2$.

So the derivative of $y = x^3$ is $3x^2$. It might also be written as $\dfrac{d}{dx}(x^3) = 3x^2$.

$\dfrac{dy}{dx}$ means differentiating the **dependent** variable, y, with respect to the **independent** variable, x. Other variables can also be used:

$$\frac{ds}{dt}, \; \frac{dx}{da}, \; \frac{d\theta}{d\alpha}, \; \frac{dM}{dx}, \text{ etc.}$$

Example 4

Using a computer algebra system:

i Plot the function $y = x^3$ for $x = 0$ to 6.

ii Plot, on different axes, the function $y = x^3$ for $x = 2.5$ to 3.5, $x = 2.9$ to 3.1 and $x = 2.99$ to 3.01.
Notice how the graph gets closer and closer to a straight line as x gets close to 3.

iii Evaluate the gradient, $\lim\limits_{h \to 0} \dfrac{(3 + h)^3 - 3^3}{h}$. This value is the derivative, $\dfrac{dy}{dx}$, of $y = x^3$ at $x = 3$.

Solution

The following are the MAPLE commands and output for this example. Even if you don't have access to MAPLE or any other computer algebra system you should be able to follow this example through. It demonstrates how a curve becomes a straight line as two points become closer and closer.

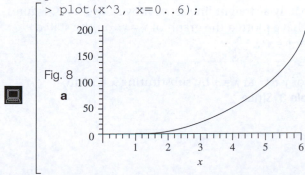

Fig. 8a

Example 4 *continued*

```
> plot(x^3, x=2.5..3.5);
```

Fig. 8
b

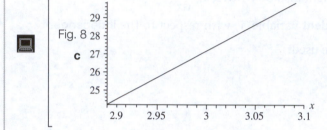

```
> plot(x^3, x=2.9..3.1);
```

Fig. 8
c

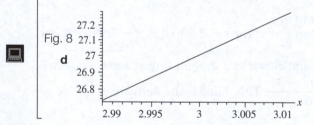

```
> plot(x^3, x=2.99..3.01);
```

Fig. 8
d

```
> simplify(((3+h)^3 − 3^3)/h);
```
$27 + 9h + h^2$

```
> grad:=limit(%,h=0);
```
grad: = 27

The graph of $y = x^3$ becomes approximately a straight line as we move closer to 3. Figures 8a–8d show the MAPLE output as we have plotted the graph of $y = x^3$ on the stated domains. The gradient is 27 at $x = 3$ for $y = x^3$.

We could also have found the gradient of $y = x^3$ at $x = 3$ by substituting $x = 3$ into the derivative function found in **Example 3**. Since

$$\frac{dy}{dx} = 3x^2$$

at $x = 3$, $\dfrac{dy}{dx} = 3 \times 3^2 = 27$

Example 5 *structures*

The bending moment, M, of a beam of length 10 m is given by

$$M = \begin{cases} x & 0 \leq x < 6 \\ 15 - 1.5x & 6 \leq x \leq 10 \end{cases}$$

where x is the distance along the beam.
Sketch M for $0 \leq x \leq 10$.

The shear force, F, is given by $F = \dfrac{dM}{dx}$.

Sketch F for $0 \leq x \leq 10$.

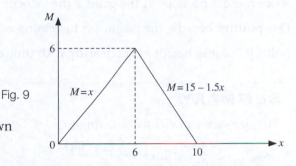

Fig. 9

Solution

First, we must create the graph of M shown in Fig. 9.

Since $F = \dfrac{dM}{dx}$, we need to evaluate the gradient of M to find F.

For $0 \leq x < 6$, the gradient of $M = x$ is 1. Hence

$$F = 1$$

For $6 \leq x \leq 10$, the gradient of $M = 15 - 1.5x$ is -1.5. Hence

$$F = -1.5$$

Note that the gradient does not exist at $x = 6$ (Fig. 10). We say M is **not** differentiable at $x = 6$.

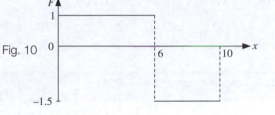

Fig. 10

Notice that not **all** functions are differentiable.

A3 **Rate of change**

The derivative describes the **rate of change** of one quantity with respect to another. The laws of physical sciences and engineering can be described in the language of calculus because a lot of these laws are based on instances of a rate of change. For example, mechanics is the mathematical study of moving bodies, therefore rates of change are central to this field. Suppose a ball is thrown up and the height s is given by $s = 15t - 4.9t^2$, the graph of which is shown below in Fig. 11.

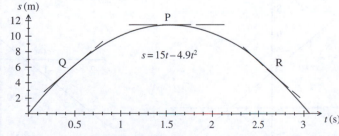

Fig. 11 This graph shows how the height s changes with time t

? The value of the gradient function $\dfrac{ds}{dt}$ gives us the **rate of change** of height s with respect to time t. This rate of change $\dfrac{ds}{dt}$ is the velocity of the ball at time t. **What is velocity of the ball at the point P?**

Since there is **no** slope at the point P the velocity $\left(= \dfrac{ds}{dt}\right)$ is **zero**. The velocity at the point Q is positive because the height s is increasing with time t, and the velocity is **negative** at point R because height s is decreasing with time t.

SUMMARY

The derivative of $y = f(x)$ is defined as

6.1
$$\frac{dy}{dx} = \lim_{h \to 0} \frac{f(x + h) - f(x)}{h}$$

Also derivatives can be denoted by

$$\frac{dy}{dx} = f'(x) \qquad \dot{x} = \frac{dx}{dt} \quad \text{(time derivative)}$$

Exercise **6(a)**

Solutions at end of book. Complete solutions available at www.palgrave.com/engineering/singh

1 Find $\dfrac{dy}{dx}$ of the graphs shown below in Fig. 12:

Fig. 12

a

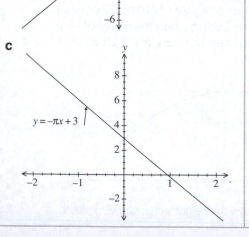

$y = 3x$

b

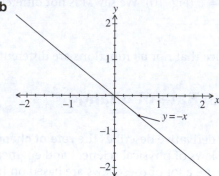

$y = -x$

c

$y = -\pi x + 3$

d

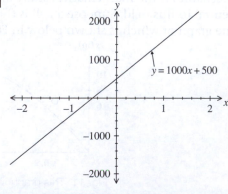

$y = 1000x + 500$

Exercise **6(a) continued**

Solutions at end of book. Complete solutions available at www.palgrave.com/engineering/singh

2 [*structures*]   The bending moment, M (kN m), of a beam of length 8 m is given by

$$M = \begin{cases} 2.16x & 0 \le x < 5 \\ 28.8 - 3.6x & 5 \le x \le 8 \end{cases}$$

where x is the distance along the beam.

The shear force, V, is given by $V = \dfrac{dM}{dx}$.

Sketch V for $0 \le x \le 8$.

3 [*electrical principles*]   The current, i, through an inductor is given by

$$i = \begin{cases} t & 0 \le t \le 2 \\ 4 - t & 2 < t \le 4 \end{cases}$$

Sketch the current i.

The voltage, V, applied across the inductor is given by $V = 0.25\dfrac{di}{dt}$.

Sketch the voltage, V, on different axes.

4 [*electrical principles*]   The voltage, V, applied across an inductor of inductance L is

$$V = L\frac{di}{dt}$$

where i represents the current through the inductor. If the current i is as shown in Fig. 13, find V.

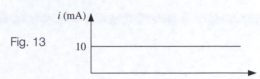
Fig. 13

5 By using the definition   6.1 , find $\dfrac{dy}{dx}$ for the following:

a $y = x$ **b** $y = 5x$

*c $y = x^2 + x$ **d** $y = \dfrac{1}{x}$

Use a computer algebra system or a graphical calculator for question 6.

6 **i** Plot the function $y = 10x^6$ for $x = 0$ to 3.

  ii Plot on different axes the function
  $y = 10x^6$ for $x = 1.4$ to 1.6,
  $x = 1.49$ to 1.51 and
  $x = 1.499$ to 1.501.

  iii Evaluate $\lim\limits_{h \to 0} \dfrac{10(1.5 + h)^6 - 10(1.5)^6}{h}$.

  This is $\dfrac{dy}{dx}$ of $y = 10x^6$ at $x = 1.5$.

SECTION B **Derivatives of functions**

By the end of this section you will be able to:
▶ find the derivatives of common engineering functions
▶ apply the chain rule

B1 **General formulae for derivatives**

To find a derivative of a function using $\boxed{6.1}$ can be very tedious. For example, if $y = x^{10}$ and you want to obtain an expression for $\dfrac{dy}{dx}$ then you would have to expand $(x + h)^{10}$ and then find $\dfrac{dy}{dx}$:

$$\frac{dy}{dx} = \lim_{h \to 0} \frac{(x + h)^{10} - x^{10}}{h}$$

However by repeating the earlier derivation for the power function, $y = x^n$, we find that there is a general formula for $\dfrac{dy}{dx}$. If $y = x^n$ where n is any real number then

$$\frac{dy}{dx} = nx^{n-1}$$

We will not prove this result but it can be derived by using $\boxed{6.1}$.

If $y = x^{10}$ then $\dfrac{dy}{dx} = 10x^{10-1} = 10x^9$.

There are general formulae for finding the derivative of many functions such as exponential, trigonometric, logarithmic, etc. Some of these common formulae are set out in Table 1. Generally it is easier to use this table rather than $\boxed{6.1}$.

TABLE 1

Reference number	$y = f(x)$	$\dfrac{dy}{dx} = f'(x)$
$\boxed{6.2}$	x^n	nx^{n-1}
$\boxed{6.3}$	kx^n	nkx^{n-1}
$\boxed{6.4}$	Constant	0
$\boxed{6.5}$	e^x	e^x

TABLE 1 CONTINUED		
6.6	$\sin(x)$	$\cos(x)$
6.7	$\cos(x)$	$-\sin(x)$
6.8	$\tan(x)$	$\sec^2(x)$
6.9	$\ln(x)$	$\dfrac{1}{x}$

For 6.6 to 6.8 , x is in radians.

For $y = x^3$ we have

$$\frac{dy}{dx} = 3x^{3-1} = 3x^2 \qquad [\text{By } 6.2 \text{ with } n = 3]$$

The process in differentiating x^3 is to multiply x by the given index, 3, and reduce the index by 1; $3 - 1 = 2$. Thus differentiating x^3 gives $3x^2$.

For $y = 5x^2$ we have

$$\frac{dy}{dx} = (2 \times 5)x^{2-1} \qquad [\text{By } 6.3 \text{ with } n = 2 \text{ and } k = 5]$$

$$= 10x^1 = 10x$$

In Table 1 the variables x and y can be replaced by other variables such as s, t, u, v, α, β, etc., and the same formulae apply. For example, if $y = u^n$ then by 6.2 :

$$\frac{dy}{du} = nu^{n-1}$$

B2 Derivatives of sums and differences

In addition to the formulae given in Table 1, we also define:

$$\frac{d}{dx}(u \pm v) = \frac{du}{dx} \pm \frac{dv}{dx}$$

where u and v are functions of x. This formula means that you can differentiate each part separately and then add or subtract the resulting functions.

For example, if $y = x^3 + \sin(x)$ then

$$\frac{dy}{dx} = 3x^2 + \cos(x)$$

Also

$$\frac{d}{dx}(ku) = k\frac{du}{dx} \qquad (\text{where } k \text{ is a constant})$$

which means you can take out the constant term, k, and differentiate the remaining expression. For example, if $y = 10\cos(x)$ then

$$\frac{dy}{dx} = 10[-\sin(x)] = -10\sin(x)$$

Example 6

Find $\dfrac{dy}{dx}$ in each of the following cases:

a $y = x^6$ **b** $y = 3x^7 + x^6$ **c** $y = e^x + \dfrac{3}{\sqrt{x}} + \tan(x)$

Solution

a Applying $\boxed{6.2}$ on $y = x^6$ we have

$$\frac{dy}{dx} = 6x^{6-1} = 6x^5$$

b $y = 3x^7 + x^6$

$$\frac{dy}{dx} = \underbrace{(7 \times 3)x^{7-1}}_{\text{by } 6.3} + \underbrace{6x^5}_{\text{by part a}} = 21x^6 + 6x^5$$

c $y = e^x + \dfrac{3}{\sqrt{x}} + \tan(x)$

We can find the derivatives of e^x and $\tan(x)$ from Table 1.

? **How do we find the derivative of $\dfrac{3}{\sqrt{x}}$?**

? **Can we write this term so that we can use Table 1?**

We need to convert $3/\sqrt{x}$ to a power:

$$\frac{3}{\sqrt{x}} = \frac{3}{x^{\frac{1}{2}}} = 3x^{-\frac{1}{2}} \quad [\text{Remember } \sqrt{x} = x^{1/2}]$$

We have

$$y = e^x + 3x^{-\frac{1}{2}} + \tan(x)$$

$$\frac{dy}{dx} = \underbrace{e^x}_{\text{by } 6.5} + \underbrace{\left(-\frac{1}{2} \times 3\right)x^{-\frac{1}{2}-1}}_{\text{by } 6.3} + \underbrace{\sec^2(x)}_{\text{by } 6.8}$$

$$= e^x - \frac{3}{2}x^{-\frac{3}{2}} + \sec^2(x)$$

$\boxed{6.2}$ $\dfrac{d}{dx}(x^n) = nx^{n-1}$ $\boxed{6.3}$ $\dfrac{d}{dx}(kx^n) = nkx^{n-1}$ $\boxed{6.5}$ $\dfrac{d}{dx}(e^x) = e^x$

$\boxed{6.8}$ $\dfrac{d}{dx}[\tan(x)] = \sec^2(x)$

Example 7 *mechanics*

The displacement, s, of a particle is given by

a $s = 80t - t^3$ **b** $s = t^3 - 12t^2 + 18t$

In each case find the velocity, v (rate of change of displacement with respect to time) where $v = \dfrac{\mathrm{d}s}{\mathrm{d}t}$.

Solution

a $v = \dfrac{\mathrm{d}s}{\mathrm{d}t} = 80 - 3t^2$

b We have

$$\dfrac{\mathrm{d}s}{\mathrm{d}t} = 3t^2 - 24t + 18 = \underset{\text{factorizing}}{3(t^2 - 8t + 6)} \quad \text{[3 is a common factor]}$$

Notice that: $\dfrac{\mathrm{d}}{\mathrm{d}t}(18t) = \dfrac{\mathrm{d}}{\mathrm{d}t}(18t^1) \underset{\text{by } \boxed{6.3}}{=} (18 \times 1)t^{1-1} = 18t^0 = 18$. Similarly

$$\dfrac{\mathrm{d}}{\mathrm{d}t}(80t) = 80$$

In general, if k is a constant then

$$\dfrac{\mathrm{d}}{\mathrm{d}t}(kt) = k$$

Differentiation itself is **not** difficult but you might find the algebra more demanding. If the examples have been difficult to follow, then invest some time in the revision of **indices (Sections 1C and 5A)**.

Example 8

If $y = x^{3/2} + 5\cos(x)$, find $\dfrac{\mathrm{d}y}{\mathrm{d}x}$.

Solution

Using Table 1 we have:

$$\dfrac{\mathrm{d}y}{\mathrm{d}x} = \underset{\text{by } \boxed{6.2}}{\dfrac{3}{2}x^{\frac{3}{2}-1}} + \underset{\text{by } \boxed{6.7}}{5[-\sin(x)]}$$

$$= \dfrac{3}{2}x^{1/2} - 5\sin(x)$$

$\boxed{6.2}$ $\dfrac{\mathrm{d}}{\mathrm{d}x}(x^n) = nx^{n-1}$ $\boxed{6.3}$ $\dfrac{\mathrm{d}}{\mathrm{d}x}(kx^n) = nkx^{n-1}$ $\boxed{6.7}$ $\dfrac{\mathrm{d}}{\mathrm{d}x}[\cos(x)] = -\sin(x)$

B3 Chain rule

The **chain rule** says that if y is a function of u and u is a function of x then

6.10
$$\frac{dy}{dx} = \frac{dy}{du} \cdot \frac{du}{dx}$$

The 'chain rule' is also called a 'function of a function rule'. Let's apply the chain rule to an example.

Example 9

If $y = (x^2 + 2)^3$, find $\frac{dy}{dx}$.

Solution

At present we **cannot** use any of the derivative formulae of Table 1. Let $u = x^2 + 2$ then we have $y = u^3$.

If we differentiate this function, $y = u^3$, we obtain the derivative $\frac{dy}{du}$:

$$\frac{dy}{du} = 3u^2 \qquad \left[\text{By } \frac{d}{du}(u^n) = nu^{n-1} \right]$$

We need to find $\frac{dy}{dx}$, not $\frac{dy}{du}$. **How can we obtain $\frac{dy}{dx}$?**

From the chain rule, 6.10 , we have $\frac{dy}{dx} = \frac{dy}{du} \cdot \frac{du}{dx} = 3u^2 \cdot \frac{du}{dx}$, thus we need $\frac{du}{dx}$.

Differentiating $u = x^2 + 2$ gives

$$\frac{du}{dx} = 2x$$

Substituting into the chain rule:

$$\frac{dy}{dx} = \frac{dy}{du} \cdot \frac{du}{dx}$$

$$= 3u^2 \cdot 2x = 6xu^2 \quad [\text{Because } 3 \times 2x = 6x]$$

Is this our final answer?

No, because $\frac{dy}{dx} = 6xu^2$, but u was not part of the original function. We need to replace u.

What is u?

$u = x^2 + 2$. Hence $\frac{dy}{dx} = 6x(x^2 + 2)^2$.

Example 10

If $y = \dfrac{(x^2 + 2)^{3/2}}{3}$, find $\dfrac{dy}{dx}$.

Solution

We **cannot** use any of the derivative formulae of Table 1. Let $u = x^2 + 2$ then $y = \dfrac{u^{3/2}}{3}$.

If we differentiate this function we obtain the derivative $\dfrac{dy}{du}$:

$$\frac{dy}{du} = \frac{3}{2}\frac{u^{\frac{3}{2}-1}}{3} \qquad \left[\text{By } \frac{d}{du}\left(ku^n\right) = nku^{n-1}\right]$$

$$= \frac{u^{1/2}}{2} \qquad \text{[Cancelling 3's]}$$

? Again we need to find $\dfrac{dy}{dx}$, not $\dfrac{dy}{du}$. **How can we obtain $\dfrac{dy}{dx}$?**

From the chain rule, $\boxed{6.10}$, we have $\dfrac{dy}{dx} = \dfrac{dy}{du}\cdot\dfrac{du}{dx} = \dfrac{u^{1/2}}{2}\cdot\dfrac{du}{dx}$, thus we need $\dfrac{du}{dx}$.

Differentiating $u = x^2 + 2$ gives

$$\frac{du}{dx} = 2x$$

Substituting into the chain rule:

$$\frac{dy}{dx} = \frac{dy}{du}\cdot\frac{du}{dx}$$

$$= \frac{u^{1/2}}{2}\cdot 2x = xu^{1/2} \quad \text{[Cancelling 2's]}$$

Since $u = x^2 + 2$ we have $\dfrac{dy}{dx} = x(x^2 + 2)^{1/2}$.

Example 11

If $y = e^{x^2 - 2x}$, find $\dfrac{dy}{dx}$.

Solution

Again in Table 1 we have the derivative of e^x but **not** e^{x^2-2x}. Therefore let $u = x^2 - 2x$ and we have

$$y = e^u \qquad\qquad \frac{dy}{du} = e^u$$

> **Example 11** *continued*
>
> **How can we obtain $\dfrac{dy}{dx}$?**
>
> Use the chain rule. We need to find $\dfrac{du}{dx}$:
>
> $$u = x^2 - 2x$$
> $$\frac{du}{dx} = 2x - 2$$
>
> Applying
>
> | 6.10 | $$\dfrac{dy}{dx} = \dfrac{dy}{du} \cdot \dfrac{du}{dx}$$ |
>
> gives
>
> $$\frac{dy}{dx} = e^u \cdot (2x - 2) = (2x - 2)e^{x^2 - 2x} \qquad \text{[Replacing } u = x^2 - 2x\text{]}$$

> **Example 12**
>
> If $y = \sin(kx)$ where k is a constant, show that
>
> $$\frac{dy}{dx} = k \cos(kx)$$
>
> **Solution**
>
> Let $u = kx$ then $y = \sin(u)$ and $\dfrac{dy}{du} = \cos(u)$.
>
> It is worth remembering that the derivative of sin is cos. (Derivative of cos is $-$ sin.)
>
> $$u = kx$$
> $$\frac{du}{dx} = k$$
>
> Applying the chain rule, 6.10 , gives
>
> $$\frac{dy}{dx} = k \cos(u) = k \cos(kx) \quad \text{[Replacing } u = kx\text{]}$$

Thus by using the chain rule we can easily verify the results shown in Table 2.

	Reference number	$y = f(x)$	$\dfrac{dy}{dx} = f'(x)$
TABLE 2	6.11	e^{kx}	$k\,e^{kx}$
	6.12	$\sin(kx)$	$k\cos(kx)$
	6.13	$\cos(kx)$	$-k\sin(kx)$

Rather than applying the chain rule each time you can use Table 2 to differentiate functions of the type shown there. For example, e^{10x} is like e^{kx} and we have the formula

6.11
$$\frac{d}{dx}(e^{kx}) = k\,e^{kx}$$

Put $k = 10$ and we get

$$\frac{d}{dx}(e^{10x}) = 10e^{10x}$$

Similarly we have

$$\frac{d}{dx}[\sin(5x)] \equiv 5\cos(5x)$$
by 6.12

$$\frac{d}{dx}[\cos(3x)] \equiv -3\sin(3x)$$
by 6.13

It is worth investing time in learning the formulae from Table 2 because these functions crop up in so many examples.

SUMMARY

It is easier to use Tables 1 and 2 to find the derivatives rather than to use formula 6.1 .
The **chain rule** says that if y is a function of u and u is a function of x then

6.10
$$\frac{dy}{dx} = \frac{dy}{du} \cdot \frac{du}{dx}$$

Exercise 6(b)

Solutions at end of book. Complete solutions available at www.palgrave.com/engineering/singh

1 Find $\dfrac{dy}{dx}$ in each of the following cases:

a $y = x^3$

b $y = 3x^5 + x^3$

c $y = \sqrt{x} + \sin(x)$

d $y = \dfrac{1}{\sqrt[3]{x}} + e^x + \cos(x)$

e $y = \dfrac{3}{\sqrt{x}} + \dfrac{5}{\sqrt[3]{x}}$

2 [mechanics]   The displacement, s, of a particle is given by

$$s = 75t - t^3 \quad (t \geq 0)$$

   i Find the velocity v where $v = \dfrac{ds}{dt}$.

   ii At what value of t is $v = 0$?

3 [fluid mechanics]   The density, ρ, is defined as $\rho = \dfrac{1}{V}$ where V represents volume of unit mass (specific volume).

Find an expression for $\dfrac{dV}{d\rho}$.

4 Find $\dfrac{dy}{dx}$ in each of the following cases:

   a $y = \cos(3x)$ **b** $y = \sin(\pi x)$

   c $y = e^{11x}$ **d** $y = \sin\left(\dfrac{\pi}{2}x\right)$

   e $y = \cos\left(\dfrac{\pi}{2}x\right)$

5 [electromagnetism]   Faraday's law states that the electromotive force, E, induced by N turns of a coil with a flux, ϕ, passing through it, is given by

$$E = -N\dfrac{d\phi}{dt}$$

If $\phi = K\sin(2\pi ft)$ where K and f are constants, determine E.

6 Find $\dfrac{dy}{dx}$ in each of the following cases:

   a $y = \sin(x^2)$ **b** $y = e^{\sin(x)}$

   c $y = \ln(x^2)$ **d** $y = \cos(2x^3 - 3x)$

   e $y = (x^2 + 1)^{10}$ **f** $y = \tan(5x^7 + 3x^4)$

SECTION C **Chain rule revisited**

By the end of this section you will be able to:
▶ apply the chain rule to differentiate complicated functions

C1 **Applications of the chain rule**

Applying the chain rule to differentiate functions like

$$\sin(x^3 + 2x), \quad e^{\cos(x) + x^2}, \quad \ln[\tan(x) + \sec(x)]$$

can be laborious. Using the rule on the following sine functions we obtain:

$$\frac{d}{dx}\left[\sin(x^2)\right] = 2x\cos(x^2)$$

$$\frac{d}{dx}\left[\sin(x^5)\right] = 5x^4\cos(x^5)$$

$$\frac{d}{dx}\left[\sin(x^3 + 2x)\right] = (3x^2 + 2)\cos(x^3 + 2x)$$

? **What do you notice?**

$$\frac{\mathrm{d}}{\mathrm{d}x}\big[\sin(u)\big] = (\text{derivative of } u) \times \cos(u)$$

$$\frac{\mathrm{d}}{\mathrm{d}x}\big[\sin(u)\big] = \frac{\mathrm{d}u}{\mathrm{d}x}\cos(u)$$

where u is a function of x and $\dfrac{\mathrm{d}u}{\mathrm{d}x}$ is its derivative. Note the similarity with the derivative given in Table 1:

$$\frac{\mathrm{d}}{\mathrm{d}x}\big[\sin(x)\big] = \cos(x)$$

$$\frac{\mathrm{d}}{\mathrm{d}x}\big[\sin(u)\big] = \cos(u)\frac{\mathrm{d}u}{\mathrm{d}x}$$

We find this similarity with each derivative of Table 1, for example:

$$\frac{\mathrm{d}}{\mathrm{d}x}\big[\cos(u)\big] = -\sin(u)\frac{\mathrm{d}u}{\mathrm{d}x} \qquad \frac{\mathrm{d}}{\mathrm{d}x}\big[\cos(x)\big] = -\sin(x)$$

$$\frac{\mathrm{d}}{\mathrm{d}x}(e^u) = e^u\frac{\mathrm{d}u}{\mathrm{d}x} \qquad \frac{\mathrm{d}}{\mathrm{d}x}(e^x) = e^x$$

where u is a function of x and $\dfrac{\mathrm{d}u}{\mathrm{d}x}$ is the derivative of u.

Thus we can use Table 1 in a slightly modified form to differentiate complicated functions. The modifications are

1 replace the x in Table 1 with u where u is a function of x

2 multiply your result of **1** by $\dfrac{\mathrm{d}u}{\mathrm{d}x}$.

By implementing these two steps and adding more functions we obtain the results shown in Table 3.

TABLE 3			
	Reference number	y (u is a function of x)	$\dfrac{\mathrm{d}y}{\mathrm{d}x}$ (or y')
	6.14	u^n	$nu^{n-1}\dfrac{\mathrm{d}u}{\mathrm{d}x}$
	6.15	ku^n	$nku^{n-1}\dfrac{\mathrm{d}u}{\mathrm{d}x}$
	6.16	e^u	$e^u\dfrac{\mathrm{d}u}{\mathrm{d}x}$
	6.17	a^u	$a^u\ln(a)\dfrac{\mathrm{d}u}{\mathrm{d}x}$
	6.18	$\ln(u)$	$\dfrac{1}{u}\dfrac{\mathrm{d}u}{\mathrm{d}x}$

Reference number	y (u is a function of x)	$\dfrac{dy}{dx}$ (or y')
6.19	$\sin(u)$	$\cos(u)\dfrac{du}{dx}$
6.20	$\cos(u)$	$-\sin(u)\dfrac{du}{dx}$
6.21	$\tan(u)$	$\sec^2(u)\dfrac{du}{dx}$
6.22	$\sec(u)$	$\sec(u)\tan(u)\dfrac{du}{dx}$
6.23	$\operatorname{cosec}(u)$	$-\operatorname{cosec}(u)\cot(u)\dfrac{du}{dx}$
6.24	$\cot(u)$	$-\operatorname{cosec}^2(u)\dfrac{du}{dx}$
6.25	$\sinh(u)$	$\cosh(u)\dfrac{du}{dx}$
6.26	$\cosh(u)$	$\sinh(u)\dfrac{du}{dx}$
6.27	$\tanh(u)$	$\operatorname{sech}^2(u)\dfrac{du}{dx}$
6.28	$\operatorname{sech}(u)$	$-\operatorname{sech}(u)\tanh(u)\dfrac{du}{dx}$
6.29	$\operatorname{cosech}(u)$	$-\operatorname{cosech}(u)\coth(u)\dfrac{du}{dx}$
6.30	$\coth(u)$	$-\operatorname{cosech}^2(u)\dfrac{du}{dx}$

TABLE 3 CONTINUED

So in the last column we have the chain rule, $\dfrac{dy}{du} \times \dfrac{du}{dx}$. The first part gives $\dfrac{dy}{du}$ and then we multiply this by $\dfrac{du}{dx}$. For example, 6.14 says

$$\frac{d}{dx}\big[u^n\big] = nu^{n-1}\frac{du}{dx}$$

$$\frac{d}{dx}\big[u^n\big] = nu^{n-1} \times (\text{derivative of } u)$$

where u is any function of x.

This chapter on differentiation is more straightforward if we use Table 3, despite the horrendous appearance of formulae. In practice it is usual to use a table of standard derivatives. This book will use Table 3 as its table of derivatives.

Examples 13–17 use Table 3.

Example 13

Find the derivatives with respect to x of

a $(x^3 + 2x)^5$ **b** $\tan(x^7 - 5x)$ **c** $\sinh(x^2 + 3x)$

Solution

a By

6.14
$$\frac{d}{dx}(u^n) = nu^{n-1}\frac{du}{dx}$$

with $u = x^3 + 2x$ gives

$$\frac{d}{dx}\left[(x^3 + 2x)^5\right] = 5(x^3 + 2x)^4 \times \underbrace{(3x^2 + 2)}_{\text{derivative of } x^3 + 2x}$$

b Similarly by

6.21
$$\frac{d}{dx}\left[\tan(u)\right] = \sec^2(u)\frac{du}{dx}$$

with $u = x^7 - 5x$ gives

$$\frac{d}{dx}\left[\tan(x^7 - 5x)\right] = \sec^2(x^7 - 5x) \times \underbrace{(7x^6 - 5)}_{\text{derivative of } x^7 - 5x} = (7x^6 - 5)\sec^2(x^7 - 5x)$$

c Also using

6.25
$$\frac{d}{dx}\left[\sinh(u)\right] = \cosh(u)\frac{du}{dx}$$

with $u = x^2 + 3x$ gives

$$\frac{d}{dx}\left[\sinh(x^2 + 3x)\right] = \cosh(x^2 + 3x) \times \underbrace{(2x + 3)}_{\text{derivative of } x^2 + 3x} = (2x + 3)\cosh(x^2 + 3x)$$

Example 14

Determine $\dfrac{d}{dx}\left[\sin(x^3 + 2x)\right]$

Solution

By 6.19 , $\dfrac{d}{dx}\left[\sin(u)\right] = \cos(u)\dfrac{du}{dx}$ with $u = x^3 + 2x$ we have

Example 14 *continued*

$$\frac{du}{dx} = 3x^2 + 2$$

$$\frac{d}{dx}\left[\sin(x^3 + 2x)\right] = \cos(x^3 + 2x) \cdot (3x^2 + 2)$$

$$= (3x^2 + 2)\cos(x^3 + 2x)$$

Example 15

Obtain $\dfrac{d}{dx}[e^{x^2 + \cos(x)}]$

Solution

Using **6.16**, $\dfrac{d}{dx}(e^u) = e^u \dfrac{du}{dx}$, with

$$u = x^2 + \cos(x)$$

$$\frac{du}{dx} = 2x - \sin(x)$$

Thus

$$\frac{d}{dx}[e^{x^2 + \cos(x)}] = e^{x^2 + \cos(x)}[2x - \sin(x)]$$

$$= [2x - \sin(x)]e^{x^2 + \cos(x)} \quad \text{[Writing the square bracket term first]}$$

Example 16

Find $\dfrac{dy}{dx}$ given that $y = \ln[\cos(x)]$.

Solution

Applying **6.18**, $\dfrac{d}{dx}\left[\ln(u)\right] = \dfrac{1}{u}\dfrac{du}{dx}$, with $u = \cos(x)$ and therefore

$$\frac{du}{dx} = -\sin(x)$$

$$\frac{dy}{dx} = \frac{1}{\cos(x)}\left[-\sin(x)\right]$$

$$= -\frac{\sin(x)}{\cos(x)} \underset{\text{by } \textbf{4.35}}{=} -\tan(x)$$

4.35 $\sin(x)/\cos(x) = \tan(x)$

Example 17 *electrical principles*

The voltage, $v = 1 - e^{-t/(2 \times 10^{-6})}$, across a capacitor of capacitance $C = 0.2 \times 10^{-6}$ F, has a current i, given by

$$i = C\frac{dv}{dt}$$

Find i.

Solution

Substituting the given values yields

$$\boxed{\ *\ } \qquad i = (0.2 \times 10^{-6})\frac{d}{dt}\left[1 - e^{-t/(2 \times 10^{-6})}\right]$$

Differentiating the 1 inside the square brackets gives 0 so we are left with differentiating the exponential function. Use $\boxed{6.16}$, $\dfrac{d}{dt}[e^u] = e^u\dfrac{du}{dt}$, with

$$u = -\frac{t}{2 \times 10^{-6}}$$

$$\frac{du}{dt} = -\frac{1}{2 \times 10^{-6}} \qquad \left[\text{Using } \frac{d}{dt}(kt) = k\right]$$

$$\frac{d}{dt}\left[1 - e^{-t/(2 \times 10^{-6})}\right] = 0 - e^{-t/(2 \times 10^{-6})}\left(-\frac{1}{2 \times 10^{-6}}\right)$$

$$= \frac{1}{2 \times 10^{-6}}e^{-t/(2 \times 10^{-6})}$$

Substituting this into $\boxed{\ *\ }$ gives

$$i = (0.2 \times 10^{-6})\underbrace{\frac{1}{2 \times 10^{-6}}}_{=\,0.1}e^{-t/(2 \times 10^{-6})} = 0.1e^{-t/(2 \times 10^{-6})}$$

SUMMARY

Rather than applying the chain rule to differentiate complicated functions it is easier to use Table 3.

Exercise **6(c)**

Solutions at end of book. Complete solutions available at www.palgrave.com/engineering/singh

1 Find $\dfrac{dy}{dx}$ in each of the following cases:

a $y = (x^2 + 2)^3$ **b** $y = \sin(x^2)$

c $y = \cos(x^2 + 2x)$ **d** $y = \ln(x^3 + x)$

e $y = \sinh(3x^5 + x^3)$

2 Differentiate the following with respect to x and simplify your answer:

a $\ln[\sin(x)]$ **b** $\ln[\cosh(x)]$

c $\tan[\cos(x)]$ **d** 10^{x^2}

e $\sin[\ln(x^2)]$

3 [electrical principles]   The current, i, in a circuit is given by

$$i = (5 \times 10^{-3})\left[1 - e^{-(1 \times 10^3)t}\right]$$

Find the voltage, v, where

$$v = (3 \times 10^{-3})\dfrac{di}{dt}$$

4 [electrical principles]   The voltage, v, across a capacitor of capacitance C, in series with a resistor of resistance R, is given by

$$v = Ee^{-t/RC}$$

where $E > 0$ is a constant. Determine i where $i = C\dfrac{dv}{dt}$.

*5 [electronics]   The current, i, through a diode is given by

$$i = I_s(e^{11600V/\eta T} - 1)$$

where V is the voltage, η is the emission coefficient, T is the temperature and I_s is the saturation current. The a.c. resistance, r, of a diode is defined as

$$r = \dfrac{1}{di/dV}$$

Find r by considering η, T and I_s as constants.

*6 [electronics]   The flow of charge, q, is given by

$$q = \tau I_s\, e^{\alpha v/\eta}$$

where τ is the mean lifetime, I_s is the saturation current, v is the voltage, η is the emission coefficient and α is a constant. The diffusion capacitance $C_D = \dfrac{dq}{dv}$. Find C_D by considering τ, I_s, α and n as constants.

*7 [reliability engineering]   The reliability function, $R(t)$, of a component is given by

$$R(t) = e^{-\sqrt{t}} \quad (t \geq 0)$$

Find $\dfrac{d}{dt}[1 - R(t)]$.

*8 [reliability engineering]   The distribution function, F, for a set of components is given by

$$F = 4 - \dfrac{20}{5 + t/3}$$

where t is in years. Determine the density function $\dfrac{dF}{dt}$.

* These are 'tough nuts' to crack because of the algebra involved.

SECTION D **Product and quotient rules**

By the end of this section you will be able to:
▶ apply the product rule
▶ apply the quotient rule

D1 Product rule

How can we differentiate functions like $x^2 \ln(x)$?

Note that $x^2 \ln(x)$ is a product of two functions, x^2 and $\ln(x)$. We can find the derivative of x^2 and $\ln(x)$ on their own, but

$$\frac{d}{dx}\left[x^2 \ln(x)\right] \neq \frac{d}{dx}(x^2) \times \frac{d}{dx}\left[\ln(x)\right] \qquad \text{[Not equal]}$$

That is, the derivative of $x^2 \ln(x)$ **does not equal** the derivative of x^2 times the derivative of $\ln(x)$. We use the **product rule** to find $\frac{d}{dx}\left[x^2 \ln(x)\right]$.

The product rule is applied to functions of the type

$$y = u \times v = uv$$

where u and v are functions of x. Note that y is a product of u and v. The product rule is

6.31
$$\frac{d}{dx}(uv) = \left(\frac{du}{dx} \cdot v\right) + \left(u \cdot \frac{dv}{dx}\right) \text{ or } (uv)' = u'v + uv'$$

Remember the symbol ' means differentiate and we could just as well use $(uv)' = u'v + uv'$. It is the same formula but using different notation.

Pictorial explanation of the product rule

Figure 14 shows how the area increases when you heat a rectangular plate of width u and height v.

The symbol Δ is used in calculus to represent a very small change in a value.

Fig. 14

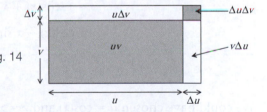

The expansion in the width u is given by Δu where Δ represents a small change and the expansion in the height v is given by Δv.

How does the area change with expansion?

Since the initial area is uv, the change (increase) in area is

$$u\Delta v + v\Delta u + \Delta u \Delta v$$

Remember Δ represents a very small change therefore the multiplication $\Delta u \Delta v$ is extremely small and can be neglected. Hence the change in area due to expansion is approximately equal to

$$u\Delta v + v\Delta u$$

To obtain the exact increase in area we need Δu and Δv to be very close to zero, which is denoted by $\Delta u \to 0$ and $\Delta v \to 0$. Hence increase in area is

$$u dv + v du \qquad \text{[Product rule]}$$

where dv has replaced Δv as $\Delta v \to 0$ and similarly du has replaced Δu as $\Delta u \to 0$. This of course is the product rule.

Example 18

If $y = x \cos(x)$, find $\dfrac{dy}{dx}$.

Solution

Since y is the product of x and $\cos(x)$ we apply the product rule:

$$6.31 \qquad \frac{d}{dx}(uv) = \left(\frac{du}{dx} \cdot v\right) + \left(u \cdot \frac{dv}{dx}\right)$$

Let

$$u = x \qquad\qquad v = \cos(x)$$

In order to use the product rule we have to differentiate these:

$$\frac{du}{dx} = 1 \qquad\qquad \frac{dv}{dx} = -\sin(x) \qquad \left[\text{Because } \frac{d}{dx}[\cos(x)] = -\sin(x)\right]$$

Substituting these into 6.31 we have

$$\frac{d}{dx}\left[x\cos(x)\right] = \left(\frac{du}{dx} \cdot v\right) + \left(u \cdot \frac{dv}{dx}\right)$$

$$= 1 \cdot \cos(x) + x \cdot \left[-\sin(x)\right]$$

$$= \cos(x) - x\sin(x)$$

We could have chosen $u = \cos(x)$ and $v = x$.

You can remember the verbal form of the product rule as 'differentiate the first and leave the second plus differentiate the second and leave the first'.

Example 19

If $y = x^2 \ln(x)$, find $\dfrac{dy}{dx}$.

Solution

Since y is a product of two functions x^2 and $\ln(x)$, we use the product rule, 6.31 . Let

$$u = x^2 \qquad\qquad v = \ln(x)$$

6 ▶ Differentiation **299**

> ### Example 19 *continued*
>
> **?** **In order to use 6.31 , what else do we need to find?**
>
> $$\frac{du}{dx} = 2x \qquad \frac{dv}{dx} \underset{\text{by } 6.9}{=} \frac{1}{x}$$
>
> Substituting these into $\dfrac{d}{dx}(uv) = \dfrac{du}{dx} \cdot v + u \cdot \dfrac{dv}{dx}$ gives
>
> $$\frac{d}{dx}\left[x^2 \ln(x)\right] = 2x \cdot \ln(x) + x^2 \cdot \frac{1}{x}$$
>
> $$= 2x \ln(x) + x \qquad \text{[Cancelling } x\text{'s]}$$
>
> $$= x[2 \ln(x) + 1] \qquad \text{[Factorizing]}$$
>
> We could select u to be $\ln(x)$ and v to be x^2, it doesn't make any difference. For the product rule u can be the second function and v the first, or vice versa.

> ### Example 20 *electrical principles*
>
> The voltage, E, applied across an inductor with inductance L is given by
>
> $$E = L\frac{di}{dt}$$
>
> where i is the current through the inductor. For $i = te^{-t}$ and $L = 5 \times 10^{-3}$, find E.
>
> Solution
>
> Substituting $L = 5 \times 10^{-3}$ and $i = te^{-t}$ into $E = L\dfrac{di}{dt}$ gives
>
> $$E = (5 \times 10^{-3}) \frac{d}{dt}(te^{-t})$$
>
> **?** **How do we differentiate te^{-t}?**
>
> Since te^{-t} is a product of t times e^{-t}, we need to use the product rule:
>
> 6.31 $\qquad (uv)' = u'v + uv'$
>
> You may find it easier to multiply diagonals as shown below for the product rule:
>
> $$\begin{array}{ll} u = t & v = e^{-t} \\ u' = 1 & v' = -e^{-t} \end{array}$$
> $$\text{by } 6.11$$

6.9 $\dfrac{d}{dx}[\ln(x)] = \dfrac{1}{x}$ 　　 6.11 $(e^{kt})' = ke^{kt}$

Example 20 *continued*

Hence $E = (5 \times 10^{-3})\Big[\underbrace{1 \cdot e^{-t} + t \cdot (-e^{-t})}_{\text{by } u' \cdot v + u \cdot v'}\Big]$

$ = (5 \times 10^{-3})(e^{-t} - te^{-t})$

$ = (5 \times 10^{-3})(1 - t)e^{-t}$ [Factorizing out e^{-t}]

D2 **Quotient rule**

How can we differentiate functions like $\dfrac{\sin(x)}{x^2}$?

There is a formula called the **quotient rule** which gives the derivative of quotients. The quotient rule is applied to functions of the type

$$y = \frac{u}{v} \quad (v \neq 0)$$

where u and v are functions of x. Note that y is the quotient of u by v. The quotient rule is

6.32 $\dfrac{d}{dx}\left(\dfrac{u}{v}\right) = \dfrac{\left(\dfrac{du}{dx} \cdot v\right) - \left(u \cdot \dfrac{dv}{dx}\right)}{v^2}$ or $\left(\dfrac{u}{v}\right)' = \dfrac{u'v - uv'}{v^2}$

Example 21

Find $\dfrac{d}{dx}\left(\dfrac{\sin(x)}{x^2}\right)$.

Solution

Since we have a quotient, we apply   6.32 :

$$u = \sin(x) \qquad v = x^2$$

$$\frac{du}{dx} = \cos(x) \qquad \frac{dv}{dx} = 2x$$

Putting these into

$$\frac{d}{dx}\left(\frac{u}{v}\right) = \frac{\dfrac{du}{dx} \cdot v - u \cdot \dfrac{dv}{dx}}{v^2}$$

yields

$$\frac{d}{dx}\left(\frac{\sin(x)}{x^2}\right) = \frac{\cos(x) \cdot x^2 - \sin(x) \cdot 2x}{(x^2)^2}$$

$$= \frac{x^2\cos(x) - 2x\sin(x)}{x^4}$$

$$= \frac{x\cos(x) - 2\sin(x)}{x^3} \qquad \text{[Cancelling } x's\text{]}$$

Example 22

By using 6.32 , show that $\dfrac{d}{dx}[\tan(x)] = \sec^2(x)$.

Solution

From

4.35 $\tan(x) = \dfrac{\sin(x)}{\cos(x)}$

we have $\dfrac{d}{dx}[\tan(x)] = \dfrac{d}{dx}\left(\dfrac{\sin(x)}{\cos(x)}\right)$.

How do we derive the given result?

We use 6.32 with $u = \sin(x)$ $v = \cos(x)$

$\dfrac{du}{dx} = \cos(x)$ $\dfrac{dv}{dx} = -\sin(x)$

[The numerator is found by multiplying one diagonal and subtracting the other.]

Substituting these into

6.32 $\dfrac{d}{dx}\left(\dfrac{u}{v}\right) = \dfrac{\dfrac{du}{dx}\cdot v - u \cdot \dfrac{dv}{dx}}{v^2}$

gives

$\dfrac{d}{dx}\left(\dfrac{\sin(x)}{\cos(x)}\right) = \dfrac{\cos(x)\cdot\cos(x) - \sin(x)\cdot[-\sin(x)]}{[\cos(x)]^2}$

$= \dfrac{\cos^2(x) + \sin^2(x)}{\cos^2(x)} = \dfrac{1}{\cos^2(x)}$ [By 4.64]

$= \sec^2(x)$ [By 4.11]

You often need to use algebra, particularly indices, or trigonometric identities, to simplify your result after differentiating. It might be worth revising **Sections 1C**, **4G** and **5A** if you are finding it difficult to follow the above examples.

SUMMARY

Product rule: If $y = uv$ where u and v are functions of x then

6.31 $\dfrac{d}{dx}(uv) = \left(\dfrac{du}{dx}\cdot v\right) + \left(u\cdot\dfrac{dv}{dx}\right)$ or $(uv)' = u'v + uv'$

4.11 $1/\cos(x) = \sec(x)$ 4.64 $\cos^2(x) + \sin^2 = 1$

SUMMARY continued

Quotient rule: If $y = \dfrac{u}{v}$ where u and v are functions of x then

6.32 $\dfrac{\mathrm{d}}{\mathrm{d}x}\left(\dfrac{u}{v}\right) = \dfrac{\left(\dfrac{\mathrm{d}u}{\mathrm{d}x}\cdot v\right) - \left(u\cdot\dfrac{\mathrm{d}v}{\mathrm{d}x}\right)}{v^2}$ or $\left(\dfrac{u}{v}\right)' = \dfrac{u'v - uv'}{v^2}$

Exercise 6(d)

Solutions at end of book. Complete solutions available at www.palgrave.com/engineering/singh

1 [electrical principles]   The current, i, through an inductor is given by

$i = 5te^{-5t}$. Find $\dfrac{\mathrm{d}i}{\mathrm{d}t}$.

2 [electrical principles]   The current, i, through a capacitor of capacitance C is given by

$$i = C\dfrac{\mathrm{d}v}{\mathrm{d}t}$$

where v is the voltage across the capacitor. For $v = (t+1)e^{-(1\times10^3)t}$ and $C = 1\ \mu\mathrm{F}$, find i.

In question 3 don't be put off by the different symbols used for variables; the same differentiation rules apply.

3 Differentiate each of the following expressions with respect to the variables used, simplify your result where possible:

a $t\sin(t)$ **b** $x\ln(x)$ **c** θe^{θ}

d $e^u \sin(u)$ **e** $\sin(\alpha)\cos(\alpha)$

f $\dfrac{q}{1+q}$ **g** $\dfrac{\cos(a)}{1+\sin(a)}$

h $\dfrac{e^z}{\cos(z)}$ **i** $M^3\ln(M)$

j $\dfrac{\beta^2}{\sin(\beta)}$ ***k** $\dfrac{h}{\sqrt{1+h^2}}$

l $J\ln[\sin(J)]$ **m** $e^{\Sigma}\sin(2\Sigma)$

n $\sqrt{\dfrac{Z+1}{Z-1}}$ ***o** $\Sigma e^{\Sigma}\sin(5\Sigma)$

*p $\ln\left(\dfrac{\delta^2+1}{\delta^2-1}\right)$

4 [mechanics]   A damped oscillator has displacement, s, given by

$$s = e^{-kt}\cos(\omega t)$$

where ω is the angular velocity and k is a constant. Find the velocity $v = \dfrac{\mathrm{d}s}{\mathrm{d}t}$.

SECTION E Higher derivatives

By the end of this section you will be able to:

► understand second-derivative notation $\dfrac{\mathrm{d}^2y}{\mathrm{d}x^2}$, $f''(x)$ and $\ddot{x}$

► apply these to engineering problems

► understand other higher derivatives

Remember the derivative describes the rate of change.

In mechanics the rate of change of displacement s, gives us the velocity, v, at time t, that is
$v = \dfrac{ds}{dt}$.

Similarly the rate of change of velocity v gives us the acceleration a at time t, that is
$a = \dfrac{dv}{dt}$.

E1 Second derivatives

 Example 23 *mechanics*

A particle moves along a horizontal line according to the equation

$$s = 2t^3 + t^2 - 10t$$

where s is the displacement at time t. Find the velocity, $v = \dfrac{ds}{dt}$. Also determine the
acceleration, $a = \dfrac{dv}{dt}$ (differentiate v with respect to t).

Solution

$$v = \frac{d}{dt}\left(2t^3 + t^2 - 10t\right) = 6t^2 + 2t - 10$$

To find the acceleration, $a = \dfrac{dv}{dt}$, we need to differentiate $6t^2 + 2t - 10$ with respect to t:

$$\frac{dv}{dt} = \frac{d}{dt}\left(6t^2 + 2t - 10\right) = 12t + 2$$

$\dfrac{dv}{dt}$ in the above example is called the **second derivative** of s with respect to t and is
denoted by $\dfrac{d^2s}{dt^2}$ because we differentiate $s = 2t^3 + t^2 - 10t$ twice.

In general, if $y = f(x)$, a function of x, then $\dfrac{d^2y}{dx^2}$, pronounced 'dee two y by dee x squared',
denotes the second derivative of y with respect to x. $f''(x)$, pronounced 'f double dashed x',
also denotes the second derivative. Additionally there is Newton's notation, often found in
applied mathematics: $\ddot{x} = \dfrac{d^2x}{dt^2}$ where $\ddot{x}$ is pronounced 'x double dot'. In **Example 23**,
$\ddot{s} = 12t + 2$ and represents the acceleration of the particle. Differentiating displacement once
gives the velocity and twice the acceleration. We will explore differentiation applied to
mechanics in **Chapter 7**.

Example 24 *vibrations*

The displacement, x, of a particle in simple harmonic motion is given by

$$x = A\cos(\omega t) + B\sin(\omega t)$$

where ω represents the angular frequency and A, B are constants. Show that

$$\frac{d^2x}{dt^2} + \omega^2 x = 0$$

Solution

We have to differentiate x twice with respect to t:

$$x = A\cos(\omega t) + B\sin(\omega t)$$

$$\frac{dx}{dt} = \underbrace{-\omega A \sin(\omega t)}_{\text{by } 6.13} + \underbrace{\omega B \cos(\omega t)}_{\text{by } 6.12} \qquad \text{[Differentiating]}$$

$$\frac{d^2x}{dt^2} = -\omega \cdot \underbrace{\omega A \cos(\omega t)}_{\text{by } 6.12} - \omega \cdot \underbrace{\omega B \sin(\omega t)}_{\text{by } 6.13} \quad \text{[Differentiating again]}$$

$$= -\omega^2 A \cos(\omega t) - \omega^2 B \sin(\omega t) \qquad [\text{Because } \omega \cdot \omega = \omega^2]$$

$$= -\omega^2 \underbrace{\left[A\cos(\omega t) + B\sin(\omega t) \right]}_{=\,x} \qquad \text{[Factorizing]}$$

$$\frac{d^2x}{dt^2} = -\omega^2 x$$

We have

$$\frac{d^2x}{dt^2} + \omega^2 x = -\omega^2 x + \omega^2 x = 0$$

Example 25 *mechanics*

The displacement, s, of a particle is given by

$$s = 2te^{-0.5t}$$

Find the value of t at which the acceleration $\ddot{s} = 0$.

6.12 $[\sin(kt)]' = k \cos(kt)$ 6.13 $[\cos(kt)]' = -k \sin(kt)$

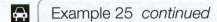

Example 25 *continued*

Solution

First we find $\dot{s}$. **How?**

By using the product rule 6.31 on $s = 2te^{-0.5t}$. Let

$$u = 2t \qquad v = e^{-0.5t}$$
$$u' = 2 \qquad v' = -0.5e^{-0.5t} \qquad \text{[Differentiating]}$$

Substituting these into $\dot{s} = u'v + uv'$ gives

$$\dot{s} = 2e^{-0.5t} + 2t(-0.5e^{-0.5t})$$

$$= 2e^{-0.5t} - te^{-0.5t}$$

$$= (2 - t)e^{-0.5t} \qquad \text{[Factorizing]}$$

To find the acceleration, $\ddot{s}$, we need to use the product rule again because $\dot{s} = (2 - t)e^{-0.5t}$ is the product of $2 - t$ times $e^{-0.5t}$:

$$u = 2 - t \qquad v = e^{-0.5t}$$
$$u' = -1 \qquad v' = -0.5e^{-0.5t} \qquad \text{[Differentiating]}$$

Putting these into $\ddot{s} = u'v + uv'$ yields

$$\ddot{s} = (-1)e^{-0.5t} + (2 - t)(-0.5)e^{-0.5t}$$

$$= -e^{-0.5t} + (-1 + 0.5t)e^{-0.5t}$$

$$= (-1 - 1 + 0.5t)e^{-0.5t} \qquad \text{[Taking out common factor } e^{-0.5t}\text{]}$$

$$= (-2 + 0.5t)e^{-0.5t}$$

For $\ddot{s} = 0$ we have

$$(-2 + 0.5t)e^{-0.5t} = 0$$

From the properties of the exponential function we know

$$e^{-0.5t} \neq 0 \quad \text{[Cannot equal zero]}$$

Thus the bracket term, $(-2 + 0.5t)$, must be zero:

$$-2 + 0.5t = 0$$

$$0.5t = 2 \text{ which gives } t = \frac{2}{0.5} = 4$$

The acceleration is zero, $\ddot{s} = 0$, at $t = 4$.

E2 Other higher derivatives

If $y = f(x)$ then differentiating $\dfrac{dy}{dx}$ gives

| 6.33 | $$\frac{d}{dx}\left(\frac{dy}{dx}\right) = \frac{d^2y}{dx^2} \text{ or } f''(x)$$ |

Other higher derivatives are defined as

| 6.34 | $$\frac{d}{dx}\left(\frac{d^2y}{dx^2}\right) = \frac{d^3y}{dx^3} \text{ or } f'''(x)$$ |

| 6.35 | $$\frac{d}{dx}\left(\frac{d^3y}{dx^3}\right) = \frac{d^4y}{dx^4} \text{ or } f^{iv}(x)$$ |

and so on.

Of course these definitions apply to other variables such as i, t, u, v, z, etc., and not just to x and y.

Example 26

If $f(x) = \sin(x)$, determine $f'''(0)$.

Solution

$f'''(0)$ means differentiate $\sin(x)$ three times and substitute zero into your result:

$$f(x) = \sin(x)$$
$$f'(x) = \cos(x) \qquad \text{[Differentiating]}$$
$$f''(x) = -\sin(x) \qquad \text{[Differentiating]}$$
$$f'''(x) = -\cos(x) \qquad \text{[Differentiating]}$$

Substituting $x = 0$ into $f'''(x) = -\cos(x)$ gives $f'''(0) = -\cos(0) = -1$

SUMMARY

The second derivative is defined as

| 6.33 | $$\frac{d^2y}{dx^2} = \frac{d}{dx}\left(\frac{dy}{dx}\right) = f''(x)$$ |

$\ddot{x}$ also denotes the second derivative. Other higher derivatives are defined in a similar manner.

Solutions at end of book. Complete solutions available at www.palgrave.com/engineering/singh

Exercise **6(e)**

🚗 **Questions 1 to 5, inclusive, are in the field of [*mechanics*], with all distances in metres and all times in seconds.**

1 In the following, x represents the displacement of a particle in a horizontal line at time t. In each case find expressions for the velocity, $v = \dfrac{dx}{dt}$, and acceleration, $a = \dfrac{d^2x}{dt^2}$ at time t:

a $x = t^3 - 10t^2 + 40t - 30$

b $x = 7t^3 - 50t^2 + 50t$

c $x = 5\cos(2t)$

d $x = 3\cos(10\pi t) + 5\sin(10\pi t)$

2 The displacement, x, of a particle moving in a straight line at time t is given by

$$x = 2t^3 - 15t^2 - 36t \quad (t \geq 0)$$

i Find expressions for the velocity $v = \dot{x}$ and acceleration $a = \ddot{x}$.

ii Find the value of t when the particle comes to rest ($v = 0$ m/s).

3 A projectile is fired vertically upwards and the height, h, is given by

$$h = 10t - 4.9t^2$$

Find the acceleration $a = \ddot{h}$. Comment upon your result.

4 A stone is dropped from a bridge and the distance fallen, h, at time t is given by

$$h = 4.9t^2 \quad 0 \leq t \leq 3$$

Determine the velocity $v = \dot{h}$ and acceleration $a = \ddot{h}$ at $t = 1.5$ s.

5 Find the velocity, $v = \dot{x}$, and acceleration, $a = \ddot{x}$, for the following displacements, x:

a $x = 3t - \sin(3t)$ **b** $x = \dfrac{1 - e^{-t/5}}{5}$

c $x = te^{-t}$ **d** $x = \ln(1 + 3t^2)$

*e $x = 0.5\ln\left[\cosh\left(\sqrt{20}t\right)\right]$

6 〝Y〞[*vibrations*] The displacement, x, of a particle in simple harmonic motion is given by

$$x = \cos(3t) - \sin(3t)$$

Show that

$$\frac{d^2x}{dt^2} + 9x = 0$$

7 Determine $f^v(0)$ for $f(x) = \cos(x)$.

*8 🏭 [*thermodynamics*] The pressure, P, and volume, V, of a gas in a cylinder are related by

$$PV^n = C \quad \text{(where } C \text{ is a constant)}$$

Show that

$$\frac{d^2P}{dV^2} = \frac{n(n+1)P}{V^2}$$

*9 〝Y〞[*vibrations*] The displacement, x, of a mass in a vibrating system is given by

$$x = (1 + t)e^{-\omega t}$$

where ω is the natural frequency of vibration. Show that

$$\ddot{x} + 2\omega\dot{x} + \omega^2 x = 0$$

*10 〝Y〞[*vibrations*] The amplitude, u, at displacement x along a bar satisfies

$$u = A\sin\left(\frac{\omega x}{k}\right) + B\cos\left(\frac{\omega x}{k}\right)$$

where ω = natural frequency of vibration and k, A, B are constants. Show that

$$\frac{d^2u}{dx^2} + \frac{\omega^2}{k^2}u = 0$$

SECTION F **Parametric differentiation**

By the end of this section you will be able to:
▶ plot graphs of parametric equations
▶ find $\dfrac{dy}{dx}$ and $\dfrac{d^2y}{dx^2}$ of parametric equations
▶ investigate applications of parametric equations

Parametric differentiation might seem like a huge leap in sophistication from previous sections because of the notation used. Also many of the examples rely on trigonometry, particularly trigonometric identities. Follow each example carefully and try not to skip anything you don't understand.

F1 **Graphs of parametric equations**

Parametric equations arise when the co-ordinate variables x and y are expressed in terms of a common third variable, t, θ, ϕ, etc. The third variable is called a **parameter**.

Figure 15 shows a circle of radius r and centre origin. The point (x, y) is defined by

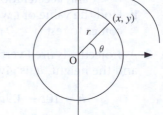

$$x = r\cos(\theta), \ \ y = r\sin(\theta)$$ Fig. 15

where r is of fixed length and θ varies as shown in Fig. 15. (We derived these in **Chapter 4**.) Hence x and y are expressed in terms of the **parameter** θ.

Parametric equations are commonly used in dynamics, when a particle moves in a plane and the x, y co-ordinates are usually defined in terms of a common third variable, say time t. That is, the x, y co-ordinates give the position of the particle at time t.

Example 27 *mechanics*

A particle moving in the $x-y$ plane has its displacement components, x and y, satisfying

$$x = t - \sin(t), \ \ y = 1 - \cos(t)$$

By using a computer algebra system or a graphical calculator, plot the path for $0 \le t \le 4\pi$.

Solution

By using MAPLE we have Fig. 16a.

Example 27 *continued*

```
> plot([t − sin(t), 1 − cos(t), t = 0..4*Pi]);
```

Fig. 16a

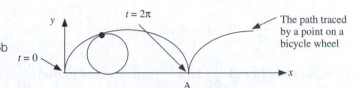

You can think of the graph of Fig. 16a as the path traced out by a point on the edge of a bicycle wheel (Fig. 16b).

Fig. 16b

After *t* seconds, the position of the point is

$$x = t - \sin(t), \quad y = 1 - \cos(t)$$

After 2π seconds, the point on the edge of the wheel reaches A whose co-ordinates are given by substituting $t = 2\pi$ into $x = t - \sin(t)$ and $y = 1 - \cos(t)$:

$$x = 2\pi - \underbrace{\sin(2\pi)}_{=0} = 2\pi$$

$$y = 1 - \underbrace{\cos(2\pi)}_{=1} = 1 - 1 = 0$$

As can be seen from the graph, the point A has the co-ordinates $(2\pi, 0)$.

This curve, $x = t - \sin(t)$, $y = 1 - \cos(t)$, is called the cycloid and is referred to as the 'Helen of Geometry'– the most beautiful curve in the world.

F2 **Parametric differentiation**

For the parametric equations (cycloid) of Example 27:

$$x = t - \sin(t), \quad y = 1 - \cos(t)$$

what does $\dfrac{dy}{dx}$ represent?

The gradient at time *t*. For example, these functions have the gradient graph shown in Fig. 17.

How can we find $\dfrac{dy}{dx}$?

Fig. 17

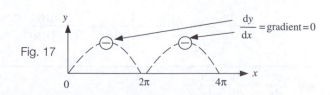

?   First we have $x = t - \sin(t)$. **What derivative can we find from this equation?**

$$\frac{dx}{dt}$$

$$\frac{dx}{dt} = 1 - \cos(t)$$

Secondly we have $y = 1 - \cos(t)$ [Given equation]

?   **What derivative can we find from this equation?**

$$\frac{dy}{dt}$$

$$\frac{dy}{dt} = 0 - [-\sin(t)] = \sin(t)$$

So far we have found $\dfrac{dx}{dt}$ and $\dfrac{dy}{dt}$ but we need to find $\dfrac{dy}{dx}$.

?   **How do we determine $\dfrac{dy}{dx}$?**

By the chain rule 6.10 , we have $\dfrac{dy}{dx} = \dfrac{dy}{dt} \cdot \dfrac{dt}{dx}$

We know $\dfrac{dy}{dt}$ but we do not have an expression for $\dfrac{dt}{dx}$. However we do have an expression

?   for $\dfrac{dx}{dt}$. **Can we find $\dfrac{dt}{dx}$ from $\dfrac{dx}{dt}$?**

$$\frac{dt}{dx} = 1 \div \frac{dx}{dt} = \frac{1}{dx/dt}$$

Remember that $\dfrac{dt}{dx}$ is a derivative and **not** a fraction.

Substituting $\dfrac{dx}{dt} = 1 - \cos(t)$ into $\dfrac{dt}{dx} = \dfrac{1}{dx/dt}$ yields

$$\frac{dt}{dx} = \frac{1}{1 - \cos(t)} \quad \text{(provided } 1 - \cos(t) \neq 0)$$

Putting $\dfrac{dy}{dt} = \sin(t)$ and $\dfrac{dt}{dx} = \dfrac{1}{1 - \cos(t)}$ into $\dfrac{dy}{dx} = \dfrac{dy}{dt} \cdot \dfrac{dt}{dx}$ gives

$$\frac{dy}{dx} = \sin(t) \cdot \frac{1}{1 - \cos(t)} = \frac{\sin(t)}{1 - \cos(t)}$$

6.10 $\dfrac{dy}{dx} = \dfrac{dy}{du} \cdot \dfrac{du}{dx}$

In general, if x and y are expressed in terms of a common parameter t then

$$\frac{dy}{dx} = \frac{dy}{dt} \cdot \frac{dt}{dx} = \frac{dy}{dt} \cdot \frac{1}{dx/dt} = \frac{dy/dt}{dx/dt}$$

Thus

6.36
$$\frac{dy}{dx} = \frac{dy/dt}{dx/dt}$$

Example 28

Find $\dfrac{dy}{dx}$ if

$$x = \sin^2(\theta), \quad y = \cos^2(\theta)$$

Comment upon your result. Plot the path traced out by x and y.

Solution

We have

$$x = \sin^2(\theta) = [\sin(\theta)]^2$$

$$\frac{dx}{d\theta} = 2\sin(\theta)\cos(\theta) \qquad \left[\text{By } (u^n)' = nu^{n-1}u'\right]$$

$$y = \cos^2(\theta) = [\cos(\theta)]^2$$

$$\frac{dy}{d\theta} = 2\cos(\theta)[-\sin(\theta)] \qquad \left[\text{By } (u^n)' = nu^{n-1}u'\right]$$

$$= -2\cos(\theta)\sin(\theta)$$

Substituting these into 6.36 gives

$$\frac{dy}{dx} = \frac{dy/d\theta}{dx/d\theta} = -\frac{2\cos(\theta)\sin(\theta)}{2\sin(\theta)\cos(\theta)} = -1 \quad \text{[Cancelling]}$$

The gradient at any value of the parameter θ is always -1. The graph of $x = \sin^2(\theta), \ y = \cos^2(\theta)$ is shown in Fig. 18.

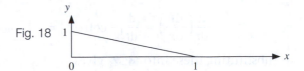

Fig. 18

Notice from the graph that the gradient of the line is -1.

The graph in Fig. 18 shows the line $y = 1 - x$ which in terms of θ is written as

$$\cos^2(\theta) = 1 - \sin^2(\theta)$$

[Substituting $y = \cos^2(\theta)$ and $x = \sin^2(\theta)$.]

This is a rearrangement of the well-known trigonometric identity

4.64 $$\sin^2(\theta) + \cos^2(\theta) = 1$$

The second derivative, $\dfrac{d^2y}{dx^2}$, of parametric equations x, y written in terms of a parameter t is

6.37 $$\frac{d^2y}{dx^2} = \frac{\dfrac{d}{dt}\left(\dfrac{dy}{dx}\right)}{\dfrac{dx}{dt}}$$

Example 29

Find $\dfrac{dy}{dx}$ and $\dfrac{d^2y}{dx^2}$ for $x = r\cos(\omega t)$, $y = r\sin(\omega t)$

Solution

This is a very challenging example, follow each step very carefully.

Differentiating these:

$$\underset{\text{by }6.13}{\frac{dx}{dt} = -\omega r\sin(\omega t)}, \quad \underset{\text{by }6.12}{\frac{dy}{dt} = \omega r\cos(\omega t)}$$

Putting these into 6.36 :

$$\frac{dy}{dx} = \frac{dy/dt}{dx/dt} = \frac{\omega r\cos(\omega t)}{-\omega r\sin(\omega t)}$$

$$= -\frac{\cos(\omega t)}{\sin(\omega t)} \qquad [\text{Cancelling } \omega r]$$

$$= -\cot(\omega t) \qquad [\text{Because } \cos/\sin = \cot]$$

To find $\dfrac{d^2y}{dx^2}$ we first determine the numerator of 6.37 :

$$\frac{d}{dt}\left(\frac{dy}{dx}\right) = \frac{d}{dt}\left[-\cot(\omega t)\right] = -\underset{\text{by }6.24}{\left[-\omega\,\text{cosec}^2(\omega t)\right]} = \omega\,\text{cosec}^2(\omega t)$$

Substituting these into 6.37 gives

$$\frac{d^2y}{dx^2} = \frac{\omega\,\text{cosec}^2(\omega t)}{-\omega r\sin(\omega t)} = -\frac{1}{r\sin(\omega t)}\cdot\text{cosec}^2(\omega t) \quad [\text{Cancelling } \omega]$$

6.12 $[\sin(kt)]' = k\cos(kt)$ 6.13 $[\cos(kt)]' = -k\sin(kt)$ 6.24 $[\cot(u)]' = -\text{cosec}^2(u)\,\dfrac{du}{dt}$

6.36 $\dfrac{dy}{dx} = \dfrac{dy/dt}{dx/dt}$

Example 29 *continued*

$$= -\frac{1}{r} \underbrace{cosec(\omega t)}_{by\ 4.10}.cosec^2(\omega t)$$

$$= -\frac{1}{r} cosec^3(\omega t)$$

A common misconception among students is $\dfrac{d^2y}{dx^2} = \dfrac{d}{dt}\left(\dfrac{dy}{dx}\right).$ ✗

This is wrong!

Remember you have to **divide** the Right-Hand Side by dx/dt to obtain $\dfrac{d^2y}{dx^2}$.

F3 **Applications**

Next we examine parametric equations in mechanics. We find the velocity and acceleration components by determining $\dot{x}$, $\ddot{x}$, $\dot{y}$ and $\ddot{y}$.

 Example 30 *mechanics*

The motion of a projectile is given by

$$x = 3t^2 - 6t, \ y = 18t^2 - 2t^3$$

The velocity components, v_x and v_y, and acceleration components, a_x and a_y, are defined by

$$v_x = \dot{x}, \quad v_y = \dot{y}, \quad a_x = \ddot{x}, \quad a_y = \ddot{y}$$

Find expressions for v_x, v_y, a_x and a_y.

Solution

Remember that $\dot{x}$ denotes the time derivative, $\dfrac{dx}{dt}$.

$$x = 3t^2 - 6t$$
$$\dot{x} = 6t - 6 \qquad [\text{Differentiating}]$$
$$\ddot{x} = 6 \qquad [\text{Differentiating again}]$$

Hence $v_x = 6t - 6$, $a_x = 6$. Similarly

$$y = 18t^2 - 2t^3$$
$$\dot{y} = 36t - 6t^2 \qquad [\text{Differentiating}]$$
$$\ddot{y} = 36 - 12t \qquad [\text{Differentiating again}]$$

Hence $v_y = 36t - 6t^2$, $a_y = 36 - 12t$.

4.10 $\dfrac{1}{sin(x)} = cosec(x)$

SUMMARY

If the parametric equations x, y are expressed in terms of a parameter t then

6.36 $$\frac{dy}{dx} = \frac{dy/dt}{dx/dt}$$

6.37 $$\frac{d^2y}{dx^2} = \frac{\frac{d}{dt}\left(\frac{dy}{dx}\right)}{\frac{dx}{dt}}$$

Velocity and acceleration components are defined by $v_x = \dot{x}$, $a_x = \ddot{x}$, $v_y = \dot{y}$ and $a_y = \ddot{y}$

Exercise 6(f)

Solutions at end of book. Complete solutions available at www.palgrave.com/engineering/singh

1 Find $\dfrac{dy}{dx}$ in each of the following cases:

  a $x = \sin(t)$, $y = \cos(t)$

  b $x = 2t^3 - t^2$, $y = 10t^2 - t^3$

  c $x = (t - 3)^2$, $y = t^3 - 1$

  d $x = e^t - 1$, $y = e^{t/2}$

🚗 **Questions 2 to 7 are in the area of [mechanics]. For questions 2 to 4 use a graphical calculator or a computer algebra system to plot the parametric equations. Note that you can still find the derivatives without plotting the parametric equations.**

2 A particle moves in the plane according to

$$x = t\sin(t), \quad y = t\cos(t)$$

  a Plot the path of the particle for $0 \le t \le 50$.

  b Find $\dfrac{dy}{dx}$.

3 A particle moving in the plane has its displacement components, x and y, given by

$$x = (t - 2)^2, \quad y = t^3 - 5$$

  i Plot the path for $0 \le t \le 3.5$.

  ii Determine

$$v_x = \dot{x}, \; v_y = \dot{y} \text{ and } v = \sqrt{v_x^2 + v_y^2}.$$

4 A particle moving in the x–y plane has its displacement components given by

$$x = \frac{(t^2 + 2)^{\frac{3}{2}}}{3}, \quad y = t$$

  i Plot the path for $0 \le t \le 5$.

  ii Determine the velocity components, $v_x = \dot{x}$ and $v_y = \dot{y}$.

  iii Determine the magnitude of the velocity (speed), v, defined by

$$v = \sqrt{v_x^2 + v_y^2}$$

5 A particle moving in the x–y plane has its displacement components, x and y, satisfying

$$x = 8t(2 - t^2), \quad y = t^3$$

  i Determine the velocity components, $v_x = \dot{x}$ and $v_y = \dot{y}$.

  ii Determine the magnitude of the velocity (speed), v, defined by

$$v = \sqrt{v_x^2 + v_y^2}$$

Solutions at end of book. Complete solutions available at www.palgrave.com/engineering/singh

Exercise **6(f) continued**

6 The x, y components of a particle moving in the x–y plane are given by
$$x = r\cos(t), \quad y = r\sin(t)$$
where r is a constant. Determine
$$v_x = \dot{x}, \ v_y = \dot{y}, \ a_x = \ddot{x} \ \text{and} \ a_y = \ddot{y}.$$

Show that $v = \sqrt{v_x^2 + v_y^2} = r$

and $a = \sqrt{a_x^2 + a_y^2} = r$.

7 A particle moves with curvilinear motion given by
$$x = t^3 - t^2, \quad y = 5t^2 - t^3$$

i Find
$$v_x = \dot{x}, \ v_y = \dot{y}, \ a_x = \ddot{x} \ \text{and} \ a_y = \ddot{y}.$$

ii Find the magnitudes of the velocity, v (m/s) and the acceleration, a (m/s²) where

$$v = \sqrt{v_x^2 + v_y^2}$$
$$a = \sqrt{a_x^2 + a_y^2}$$

at $t = 2$ s.

8 If $x = \cosh(t)$ and $y = \sinh(t)$, show that

$$\frac{d^2 y}{dx^2} = -\operatorname{cosech}^3(t)$$

9 The general parametric equations of a cycloid (Helen of Geometry) are given by
$$x = r[\theta - \sin(\theta)], \quad y = r[1 - \cos(\theta)]$$
where r is a constant. Show that

$$\frac{dy}{dx} = \frac{\sin(\theta)}{1 - \cos(\theta)}$$

and $\dfrac{d^2 y}{dx^2} = \dfrac{-1}{r[1 - \cos(\theta)]^2}$

SECTION G **Implicit and logarithmic differentiation**

By the end of this section you will be able to:

▶ find $\dfrac{dy}{dx}$ of an implicit function

▶ apply logarithmic differentiation

G1 **Implicit differentiation**

The following are examples of **explicit functions**:
$$y = 1 - x^2$$
$$y = x^3 + e^x$$
$$y = \cos^2(x) + e^x + 3x^2$$
The following are examples of **implicit functions**:
$$x^3 + y^3 + e^{xy} - 2 = 0$$
$$y + \tan(xy) + x^3 = 0$$
$$x^2 + y^2 - 4 = 0$$

? **What is the difference between implicit and explicit functions?**

If y can be explicitly expressed in terms of x then we say that y is an explicit function, as shown by the first three examples; that is, $y = f(x)$. However if the relationship $y = f(x)$ is hidden, then we say that the function is an implicit function, as shown by the last three examples. It is sometimes impossible to extract y in terms of x, $y = f(x)$, from an implicit function. An implicit function is denoted by $f(x, y) = 0$; that is, f is a function of x and y.

In this section we are interested in finding the derivative $\dfrac{dy}{dx}$ of implicit functions. We can use the formulae of Table 3.

Example 31

If
$$x^3 + y^2 + x + y = 2$$

then find $\dfrac{dy}{dx}$.

Solution

?
? Differentiating x^3 and x in the above expression is straightforward. **But how do we differentiate y and y^2 with respect to x? If you differentiate y with respect to x, what derivative do you obtain?**

$$\frac{dy}{dx}$$

This derivative is very important. We have done this numerous times before. For example, if $y = x^3$ then $\dfrac{dy}{dx} = 3x^2$.

Differentiating y gives $\dfrac{dy}{dx}$.

? **How do we differentiate y^2 with respect to x?**

Using $\dfrac{d}{dx}(u^n) = nu^{n-1}\dfrac{du}{dx}$ with $u = y$ and $n = 2$, we obtain

$$\frac{d}{dx}\left(y^2\right) = 2y\frac{dy}{dx}$$

So

$$\frac{d}{dx}\left(x^3 + y^2 + x + y = 2\right) = \frac{d}{dx}(x^3) + \frac{d}{dx}(y^2) + \frac{d}{dx}(x) + \frac{d}{dx}(y) = \frac{d}{dx}(2)$$

$$\ast \quad 3x^2 + 2y\frac{dy}{dx} + 1 + \frac{dy}{dx} = 0$$

Example 31 *continued*

? **How do we find $\dfrac{dy}{dx}$?**

By collecting the $\dfrac{dy}{dx}$ terms on one side of ▮ * ▮ and the remaining terms on the other side; that is, by transposing to make $\dfrac{dy}{dx}$ the subject of the formula.

$$2y\frac{dy}{dx} + \frac{dy}{dx} = -(3x^2 + 1)$$

$$(2y + 1)\frac{dy}{dx} = -(3x^2 + 1) \qquad \text{[Factorizing]}$$

$$\frac{dy}{dx} = -\frac{(3x^2 + 1)}{2y + 1} \qquad \text{[Dividing by } 2y + 1\text{]}$$

Example 32

If

$$x^2 y^2 + 2xy = 3$$

then find $\dfrac{dy}{dx}$.

Solution

? **How do we differentiate the first term, $x^2 y^2$?**

By using the product rule, ▮ 6.31 ▮, because $x^2 y^2 = x^2 \times y^2$:

$$u = x^2 \qquad\qquad\qquad v = y^2$$
$$u' = 2x \qquad\qquad\qquad v' = 2y\frac{dy}{dx}$$

Substituting into $\dfrac{d}{dx}\left(x^2 y^2\right) = u'v + uv'$ gives

▮ * ▮ $$\frac{d}{dx}\left(x^2 y^2\right) = 2xy^2 + x^2 \cdot 2y\frac{dy}{dx}$$

Similarly the other term, $2xy$, can be differentiated by using the product rule, ▮ 6.31 ▮, again:

$$u = x \qquad\qquad\qquad v = y$$
$$u' = 1 \qquad\qquad\qquad v' = \frac{dy}{dx}$$

Example 32 *continued*

$$\frac{d}{dx}(2xy) = 2\frac{d}{dx}(xy)$$

$$= 2[u'v + uv'] = 2\left[(1)y + x\frac{dy}{dx}\right]$$

**　$$\frac{d}{dx}(2xy) = 2\left(y + x\frac{dy}{dx}\right)$$

Hence $\frac{d}{dx}(x^2y^2 + 2xy = 3)$ gives

$$\underbrace{2xy^2 + 2x^2y\frac{dy}{dx}}_{\text{by } *} + \underbrace{2\left(y + x\frac{dy}{dx}\right)}_{\text{by } **} = 0$$

$$xy^2 + x^2y\frac{dy}{dx} + y + x\frac{dy}{dx} = 0 \qquad \text{[Dividing by 2]}$$

$$(x^2y + x)\frac{dy}{dx} + (xy^2 + y) = 0 \qquad \text{[Collecting like terms]}$$

$$(x^2y + x)\frac{dy}{dx} = -(xy^2 + y) \qquad \text{[Subtracting } xy^2 + y]$$

$$\frac{dy}{dx} = -\left[\frac{xy^2 + y}{x^2y + x}\right] \qquad \text{[Dividing by } x^2y + x]$$

$$= -\left[\frac{y(xy + 1)}{x(xy + 1)}\right] = -\frac{y}{x} \qquad \text{[Cancelling } xy + 1]$$

Example 33

Find $\frac{d}{dx}\left[\ln(y)\right]$.

Solution

Using

6.18　$$\frac{d}{dx}\left[\ln(u)\right] = \frac{1}{u}\frac{du}{dx}$$

with $u = y$ we have

$$\frac{d}{dx}\left[\ln(y)\right] = \frac{1}{y}\frac{dy}{dx}$$

G2 **Logarithmic differentiation**

Example 34

Find $\dfrac{dy}{dx}$ if $y = x^2(x^3 - 1)^5\, e^x$.

Solution

How do we differentiate functions like $y = x^2(x^3 - 1)^5 e^x$?

Although it is possible to use the product rule (try it), we wish to show a simpler method to differentiate the given function. We use logs because they convert products, quotients and indices into sums and differences. It is easier to differentiate a sum of functions rather than a product. We can apply logs to both sides of the given function:

$$\ln(y) = \ln\left[x^2(x^3 - 1)^5 e^x\right]$$

$$\underset{\text{by }5.12}{=} \ln(x^2) + \ln\left[(x^3 - 1)^5\right] + \ln(e^x) = \underset{\text{by }5.14}{2\ln(x)} + \underset{\text{by }5.14}{5\ln(x^3 - 1)} + \underset{\text{by }5.16}{x}$$

Differentiating this, $\ln(y) = 2\ln(x) + 5\ln(x^3 - 1) + x$, is much easier than applying the product rule:

$$\frac{d}{dx}\left[\ln(y)\right] = 2\frac{d}{dx}\left[\ln(x)\right] + 5\frac{d}{dx}\left[\ln(x^3 - 1)\right] + \frac{d}{dx}(x)$$

$$\frac{1}{y}\frac{dy}{dx} = 2\frac{1}{x} + 5\frac{1}{x^3 - 1}(3x^2) + 1 \qquad \left[\text{By } \frac{d}{dx}\left[\ln(u)\right] = \frac{1}{u}\frac{du}{dx}\right]$$

$$\frac{dy}{dx} = y\left(\frac{2}{x} + \frac{15x^2}{x^3 - 1} + 1\right) \qquad \text{[Multiplying by } y\,]$$

We can rewrite the algebraic expression in the brackets on the Right-Hand Side with a common denominator as follows:

$$\frac{2}{x} + \frac{15x^2}{x^3 - 1} + 1 = \frac{2(x^3 - 1) + 15x^2 x + x(x^3 - 1)}{x(x^3 - 1)}$$

$$= \frac{2x^3 - 2 + 15x^3 + x^4 - x}{x(x^3 - 1)} \qquad \text{[Expanding numerator]}$$

$$= \frac{x^4 + 17x^3 - x - 2}{x(x^3 - 1)} \qquad \text{[Simplifying numerator]}$$

Thus we have

$$\frac{dy}{dx} = y\left(\frac{x^4 + 17x^3 - x - 2}{x(x^3 - 1)}\right)$$

5.12 $\ln(AB) = \ln(A) + \ln(B)$ **5.14** $\ln(A^n) = n\ln(A)$ **5.16** $\ln(e^x) = x$

Example 34 *continued*

Substituting our given y, that is $y = x^2(x^3 - 1)^5 e^x$:

$$\frac{dy}{dx} = x^2(x^3 - 1)^5 e^x \left(\frac{x^4 + 17x^3 - x - 2}{x(x^3 - 1)} \right)$$

$$= x(x^3 - 1)^4 e^x \left(x^4 + 17x^3 - x - 2 \right) \qquad \text{[Cancelling like terms } x \text{ and } x^3 - 1]$$

$$= x(x^3 - 1)^4 \left(x^4 + 17x^3 - x - 2 \right) e^x \qquad \text{[Rearranging]}$$

This method is called **logarithmic differentiation** and turns difficult products and quotients into simple sums and differences. However the algebra is still quite demanding.

?   **How would you differentiate x^x?**

This is an arduous and tedious task. It can be made more straightforward by using logs, as the following example shows.

Example 35

Let $y = x^x$, find $\dfrac{dy}{dx}$.

Solution

Taking natural logs, ln, of both sides:

    ✳ $\ln(y) = \ln(x^x)$

?   **Can we rewrite the Right-Hand Side of ✳ ?**

Using

  5.14 $\ln(A^n) = n \ln(A)$

gives

        $\ln(x^x) = x \ln(x)$

?   **How do we differentiate $x \ln(x)$?**

By using the product rule, **6.31** , because we have a product of x and $\ln(x)$:

    $u = x$ $v = \ln(x)$

    $u' = 1$ $v' = \dfrac{1}{x}$

Substituting these into $\dfrac{d}{dx}\left[x \ln(x) \right] = u'v + uv'$ gives

> ### Example 35 *continued*
>
> $$\frac{d}{dx}\left[x\ln(x)\right] = 1\cdot\ln(x) + x\cdot\frac{1}{x} = \ln(x) + 1$$
>
> Therefore the derivative of the Right-Hand Side of ▮ * ▮ is $\ln(x) + 1$.
>
> **How do we differentiate $\ln(y)$?**
>
> From **Example 33** above we have $\dfrac{d}{dx}\left[\ln(y)\right] = \dfrac{1}{y}\dfrac{dy}{dx}$.
>
> Hence differentiating $\ln(y) = \ln(x^x)$ gives
>
> $$\frac{1}{y}\frac{dy}{dx} = \ln(x) + 1$$
>
> $$\frac{dy}{dx} = y[\ln(x) + 1] \qquad \text{[Multiplying by } y\text{]}$$
>
> $$= x^x[\ln(x) + 1] \qquad \text{[Substituting our given } y = x^x\text{]}$$

Logarithmic differentiation is applied to functions of the type:

$y = u^v$ [An example is x^x]

$y = \dfrac{u^m v^n}{w^k}$ $\left[\text{An example is } \dfrac{x^2(1-x)^3}{\ln(x)}\right]$

$y = u^m v^n k^w$ [An example is $x^2(x^3-1)^5 e^x$]

where u, v and w are functions of x and k, m and n are constants. Generally logarithmic differentiation is applied to cases where logs make differentiation simpler.

> ## SUMMARY
>
> An implicit function is of the form
>
> $$f(x, y) = 0$$
>
> When differentiating y with respect to x we obtain $\dfrac{dy}{dx}$.
>
> Use **Table 3** to differentiate the y terms.
> Logarithmic differentiation applies to cases where
>
> $$y = u^v, \quad y = \frac{u^m v^n}{w^k}, \quad y = u^m v^n k^w$$
>
> (u, v and w are functions of x).

Exercise 6(g)

Solutions at end of book. Complete solutions available at www.palgrave.com/engineering/singh

1 Find $\dfrac{dy}{dx}$ for the following:

a $x^2 + y^2 = 4$ **b** $x^3 + y^3 - 2x = 3$

c $\dfrac{x^2}{4} + \dfrac{y^2}{16} = 1$

d $x^2 + y^2 - 4x - 6y = -12$

2 Find $\dfrac{dy}{dx}$ for each of the following cases:

a $x^2 + y^2 - xy^2 = 5$

*b $\cos(x^2 + y) - 2xy = 0$

c $\ln(y) = \ln(x - 1) - \ln(x^2)$

d $\ln(xy) + 2y = 0$

e $e^{x+y} + \sin(x) = 0$

3 The folium of Descartes is given by $x^3 + y^3 - 3axy = 0$ whose graph for $a = 4$ is shown in Fig. 19.

Determine $\dfrac{dy}{dx}$ for **i** $a = 4$ and **ii** for general a.

What is the value of $\dfrac{dy}{dx}$ at the origin $(0,0)$?

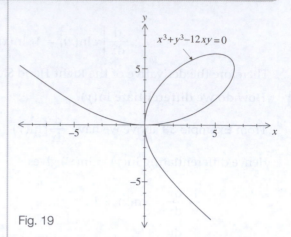

Fig. 19

4 Show that $\dfrac{d}{dx}(10^x) = 10^x \ln(10)$.

*5 Find $\dfrac{dy}{dx}$ for each of the following:

a $y = x^{\sin(x)}$

b $y = (x^2 + 1)^{1/2}(3x + 1)^{1/3}$

c $y = \dfrac{(x^2 + 1)^{1/2}(3x + 1)^{1/3}}{(x^4 - 2)^{1/5}}$

d $y = e^{(x^x)}$ **e** $y = x^{(x^x)}$

Examination questions 6

Solutions at end of book. Complete solutions available at www.palgrave.com/engineering/singh

1 Differentiate each of the following with respect to the variable in each case:

i $y = 5x^{-3} + \dfrac{7}{x} + \dfrac{2x^6}{3} - 9x + 26$

ii $y = 2x\ln\left(\dfrac{2}{x}\right)$

iii $y = \dfrac{4e^{-2x}}{\sin 5x}$

iv $y = \sqrt{1 + 5x^2 - 4x^3}$

Cork Institute of Technology, Ireland, 2007

2 For the function $v = 2\sin(t) + 3t^2$, find the gradient at the point where $t = 2$.

University of Portsmouth, UK, 2007

3 Use logarithmic differentiation to find dy/dx, given

$$y = \dfrac{e^x(x^2 + 1)}{\sin(x)}$$

University of Portsmouth, UK, 2006

4 Differentiate:

  i $f(x) = x^6 - 5x^3 - x$

  ii $w(\theta) = e^{2\theta}\cos(3\theta)$

  iii $s(x) = \dfrac{\ln(2x)}{x^2 + 4}$

  iv $g(t) = e^{\sin 2t}$

  v $y(x) = \sqrt{x^3 - x}$

  vi $L(z) = \dfrac{2}{1 + e^z}$

5 i Find $\dfrac{dy}{dx}$ at $x = \dfrac{\pi}{4}$ for $y = \dfrac{\sin\left(x + \dfrac{\pi}{4}\right)}{(1 - \cos x)}$

ii Find $\dfrac{dy}{dx}$ for $y = \dfrac{(x^2 + 1)^2(x + 7)^3}{(2x - 3)}$

iii Find $\dfrac{dy}{dx}$ given that $x^2 + y^2 + \sin(xy) = 0$.

Questions 4 and 5 are from University of Manchester, UK, 2009

The following two questions from King's College London have been slightly edited.

6 If $f(x) = \sin[\ln(2x)]$ then determine $f'(x)$.

King's College London, UK, 2007

7 If $f(x) = \ln[1/(e^{3x} + 1)]$ then determine $f'(x)$.

King's College London, UK, 2008

8 If $y = a\cosh(x/a) + c$, where a and c are constants, show that

$$\frac{d^2y}{dx^2} = \frac{1}{a}\sqrt{1 + \left(\frac{dy}{dx}\right)^2}$$

University of Surrey, UK, 2009

9 a If $f(x) = e^{\sin x}$, $g(x) = x^3 - \cos 2x$, find

$$\frac{df}{dx}, \frac{dg}{dx} \text{ and } \frac{d}{dx}[f(x).g(x)]$$

b If $x = 2t - \cos t$, $y = \sin t$, find $\dfrac{dy}{dx}$ and $\dfrac{d^2y}{dx^2}$ at $t = \dfrac{\pi}{2}$.

Loughborough University, UK, 2006

10 i Given that

$$xe^{x^2} - yx^{3/2} = y^2 \cos x - 5$$

find an expression for $\dfrac{dy}{dx}$ in terms of x and y.

ii A curve is given, parametrically, by

$$x = t - \sin\left(\frac{\pi t}{2}\right), \ y = \cos\left(\frac{\pi t}{2}\right) - t^2$$

Find the x and y co-ordinates of the point P which corresponds to the parameter $t = 1$ and find the value of $\dfrac{dy}{dx}$ at P.

University of Aberdeen, UK, 2003

11 Find the following:

$\dfrac{dy}{dx}$ at the point $(x, y) = (0,0)$, if $\ln(e + y) = e^{\sin(x+y)}$.

University of Toronto, Canada, 2005

12 Compute the following derivatives. You do not have to use the definition of the derivative. If you can 'do them in your head' instead of showing every step that is up to you (though if you get it wrong we cannot give you partial credit).

a Let $f(x) = 3^{x \ln(x)}$. Find $f'(x)$.

b Find $\dfrac{d}{dx}\left[\sqrt{x}\ln(x^4)\right]$.

c Find $\dfrac{d}{dx}\left[(\ln(x))^{\cos(x)}\right]$.

d Find $\dfrac{d}{dx}\left[\dfrac{(4x-1)^3}{(2x^2 - 1)^{3/2}(x + 1)^2}\right]$

e Find $\dfrac{d}{dx}\left[x^2\cos(x)e^{3x}\sin(\pi/2)\right]$

Stanford University, USA, 2008

1 🖳 [*electrical principles*]   The voltage, v, across an inductor with inductance L is defined as

$$v = L\frac{di}{dt}$$

where i represents current. Given that the current i is constant, find v.

2 🚗 [*mechanics*]   The velocity, v, of a particle is given by $v = 3e^{2t}$. Show that

$$\frac{dv}{dt} - 2v = 0$$

3 🚗 [*mechanics*]   If the velocity of a particle is given by $v = ae^{bt}$, show that

$$\frac{dv}{dt} - bv = 0$$

4 [heat transfer]   The heat transfer rate, Q, is given by

$$Q = -kA\frac{dT}{dx}$$

where k is the thermal conductivity of the material, A is the cross-sectional area and $T = T(x)$ is the temperature distribution.

For $T(x) = -10x^3 - 500x + 600$, $k = 0.7$ and $A = 10$, find Q at $x = t$ (t represents thickness of material).

5 [heat transfer]   The heat transfer rate, Q, is determined by Fourier's law:

$$Q = -kA\frac{dT}{dx}$$

where k, A and T are defined in question **4**. For $k = 370$, $A = 0.01\pi$ and $T = 100e^{-3x}$, find Q for $x = t$.

*6 [vibrations]   A system consists of a rod of length l and mass m attached to a spring of stiffness k. The potential energy, V, is given by

$$V = \frac{1}{2}kl^2\cos^2(\theta) + mgl\sin(\theta)$$

where θ is an angle such that $0 \le \theta \le \pi/2$ and g is acceleration due to gravity. A system is in equilibrium if

$$\frac{dV}{d\theta} = 0$$

Find the value(s) of θ for the system to be in equilibrium.

7 [vibrations]   A mass, m, is held by a spring with a stiffness constant k. The potential energy, P, of the system is given by

$$P = \frac{1}{2}kx^2 - mgx$$

where x is displacement and g is acceleration due to gravity. A system is said to be in equilibrium if

$$\frac{dP}{dx} = 0$$

Determine x for which the above system is in equilibrium.

8 [electrical principles]   The voltage, v, in a circuit is given by

$$v = -6e^{-(2\times10^3)t} + 10e^{-(8\times10^3)t}$$

Find the current, i, given that

$$i = C\frac{dv}{dt}$$

where $C = 3\ \mu F$ $(3 \times 10^{-6}\ F)$.

*9 [reliability engineering]   The reliability function, $R(t)$, is given by

$$R(t) = e^{e^t - 1}$$

Show that $\dfrac{d}{dt}[1 - R(t)] = -e^{e^t + t - 1}$

10 [electrical principles]   The current, i, through a capacitor of capacitance C is given by

$$i = C\frac{dv}{dt}$$

Find, i, given that $C = 10\ \mu F$ $(10 \times 10^{-6}\ F)$ and $v = 50\sin(75t)$.

*11 [structures]   A cantilever beam of length L has a bending moment M given by

$$M = -\frac{\omega_0 L^3}{6} - \frac{kL^6}{24} + \omega_0 Lx + \frac{kL^4 x}{4} - \frac{\omega_0 x^2}{2} - \frac{kx^5}{20}$$

where k, ω_0 are constants and x is the distance along the beam. The shear force, V, is given by $V = \dfrac{dM}{dx}$. Show that

$$V = (L - x)\left[\omega_0 + \frac{k}{4}(L^3 + L^2 x + Lx^2 + x^3)\right]$$

[Hint: $a^4 - b^4 = (a - b)(a^3 + a^2 b + ab^2 + b^3)$]

12 [structures]   The bending moment, M, of a beam of length L is given by

$$M = \frac{\omega_0 L}{\pi}\left[L + x + \frac{L}{\pi} \sin\left(\frac{\pi x}{L}\right)\right]$$

The shear force, V, is given by $V = \dfrac{dM}{dx}$.

Show that

$$V = \frac{\omega_0 L}{\pi}\left[1 + \cos\left(\frac{\pi x}{L}\right)\right]$$

*13 [vibrations]   The displacement, x, of a mass from its equilibrium position in a vibrating system is given by

$$x = e^{(-2+\sqrt{3})\omega t} + e^{-(2+\sqrt{3})\omega t}$$

where ω is the natural frequency of vibration. Show that

$$\ddot{x} + 4\omega\dot{x} + \omega^2 x = 0$$

14 [electronics]   At a pn-junction in a semiconductor the barrier potential, v, is given by

$$v = \frac{A}{1 + e^{-kx}}$$

where x is the distance into the n-type material and A, k are constants. Show that the electric field E where

$$E = -\frac{dv}{dx}$$

is given by $E = \dfrac{-kA}{2[1 + \cosh(kx)]}$

15 [mechanics]   The x co-ordinate of a particle in motion is given by

$$x = t^2 \sin(t)$$

Find $v_x = \dot{x}$ and $a_x = \ddot{x}$.

16 [mechanics]   A particle moves in a curvilinear motion described by

$$x = t \cos(t), \quad y = t \sin(t)$$

i Determine v_x, v_y, a_x and a_y.

ii The magnitude of the velocity, v, and acceleration, a, are defined by

$$v = \sqrt{v_x^2 + v_y^2} \quad \text{and} \quad a = \sqrt{a_x^2 + a_y^2}$$

Show that

$$v = \sqrt{1 + t^2} \quad \text{and} \quad a = \sqrt{4 + t^2}.$$

iii Find $\dfrac{dy}{dx}$.

*17 If $y = 10^{(x^x)}$, find $\dfrac{dy}{dx}$.

CHAPTER 7

Engineering Applications of Differentiation

A1 **Stationary points**

Figure 1 shows a graph of a function with stationary points, A, B and C.

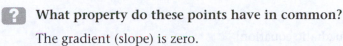

[?] **What property do these points have in common?**

The gradient (slope) is zero.

[?] **What do you think is the definition of a *stationary point*?**

A point on the curve $y = f(x)$ where the gradient is zero. Remember that $\dfrac{dy}{dx}$ is the gradient function, so we can say that a stationary point is a point on the curve such that

7.1 $$\frac{dy}{dx} = 0 \ (\text{or } y' = 0)$$

In Fig. 1, A is called a local maximum of $y = f(x)$.

[?] **Why do we use the term *local* maximum rather than just maximum?**

Local maximum is a stationary point where the value of y is greater than all the other values of y in the neighbourhood of that point. There may be other greater values of y elsewhere. For example, in Fig. 1 A is a local maximum, but as you can observe there are other values of y greater than A.

Figure 2 shows a local maximum at K.

Similarly B of Fig. 1 is called a **local minimum**. A stationary point where the value of y is smaller than all the other values of y in the neighbourhood of that point is called a local minimum.

Figure 3 shows a local minimum at K.

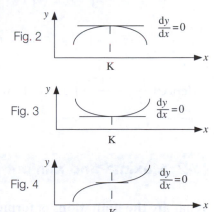

Generally we will call local maximum and local minimum just maximum and minimum respectively.

The point C of Fig. 1 is called a (horizontal) point of inflexion. A horizontal point of inflexion is a stationary point which is neither a local maximum nor a local minimum, as shown in Fig. 4.

We will return to points of inflexion later in this section.

Example 1

Find the stationary points of $y = x^3 - x^2 - 5x$.

Solution

? **How do we find the stationary points?**

Use 7.1 because at a stationary point the gradient is 0. Differentiating an expression gives us the gradient, so equating that to 0 will give the stationary points.

$$\frac{dy}{dx} = 3x^2 - 2x - 5$$

For stationary points

$$3x^2 - 2x - 5 = 0 \qquad \text{[Quadratic equation]}$$

We need to solve this quadratic equation:

$$(3x - 5)(x + 1) = 0 \quad \text{[Factorizing]}$$

$$x = \frac{5}{3}, x = -1 \qquad \text{[Solving]}$$

? **What else do we need to find?**

? The corresponding y values. **How?**

By substituting our x values $\frac{5}{3}$ and -1 into $y = x^3 - x^2 - 5x$.

$$x = \frac{5}{3}, \ y = \left(\frac{5}{3}\right)^3 - \left(\frac{5}{3}\right)^2 - 5\left(\frac{5}{3}\right) = -\frac{175}{27}$$

$$x = -1, \ y = (-1)^3 - (-1)^2 - [5 \times (-1)] = 3$$

Hence $\left(\frac{5}{3}, -\frac{175}{27}\right)$ and $(-1, 3)$ are the stationary points of $y = x^3 - x^2 - 5x$.

A2 Maxima and minima

? **What are the limitations of formula 7.1 ?**

 7.1 gives the stationary point(s) but does not reveal whether the stationary point is a local maximum, a local minimum or a point of inflexion.

Example 2

By using a graphical calculator or a computer algebra system, plot the graphs of $y = x^3 - x^2 - 5x$, $\dfrac{dy}{dx}$ and $\dfrac{d^2y}{dx^2}$ against x. Observe what happens at stationary points of y.

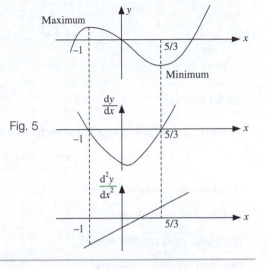

Fig. 5

Solution

Using a technical gadget produces Fig. 5.

Notice that at stationary points, $\dfrac{dy}{dx} = 0$. Also

at a maximum, $\dfrac{d^2y}{dx^2} < 0$, and at a

minimum, $\dfrac{d^2y}{dx^2} > 0$.

Let's examine the general case. Consider a local maximum at a point $x = C$. Let us assume we have plotted the graphs of y against x, $\dfrac{dy}{dx}$ against x and $\dfrac{d^2y}{dx^2}$ against x close to $x = C$, and obtained Fig. 6.

? **What do you notice about the graph of $\dfrac{d^2y}{dx^2}$ close to $x = C$?**

$$\dfrac{d^2y}{dx^2} < 0 \quad \text{[Negative]}$$

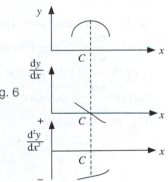

Fig. 6

Note from Fig. 6 that $\dfrac{dy}{dx}$ is decreasing, therefore $\dfrac{d^2y}{dx^2}$ is negative.

Generally, if at $x = C$ a function, $y = f(x)$, has the property

7.2 $\dfrac{dy}{dx} = 0$ and $\dfrac{d^2y}{dx^2} < 0$ [Negative]

then we say that y has a local maximum at $x = C$.

Similarly, if at $x = C$ a function, $y = f(x)$, has the property

7.3 $\dfrac{dy}{dx} = 0$ and $\dfrac{d^2y}{dx^2} > 0$ [Positive]

then we say that y has a local minimum at $x = C$.

The notation $\dfrac{dy}{dx}$ may be replaced by y' and $\dfrac{d^2y}{dx^2}$ by y''.

Summary of testing for maxima and minima

The following flow chart summarizes the above test for determining maxima and minima of a given function $y = f(x)$.

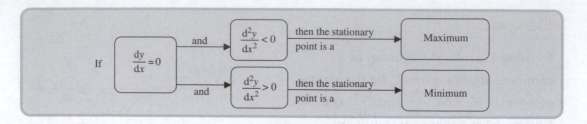

Example 3

By using a computer algebra system or a graphical calculator, plot the graph of

$$y = x^4 - 10x^3 + 35x^2 - 50x + 24$$

On the same axes plot the graph of $\dfrac{d^2y}{dx^2}$ against x.

What do you notice about the sign of $\dfrac{d^2y}{dx^2}$ close to the local maxima and minima points of y?

Fig. 7

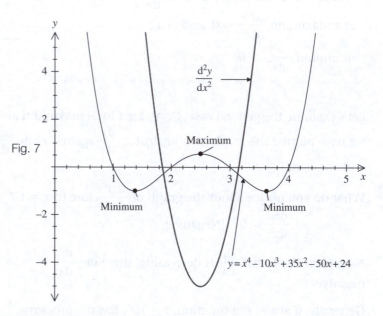

Solution

At the local maximum the sign of $\dfrac{d^2y}{dx^2}$ is negative, and at both local minima the sign of $\dfrac{d^2y}{dx^2}$ is positive (Fig. 7).

A3 Curve sketching

7.2 and 7.3 are used as a test for local maxima and local minima. Notice that we first establish a stationary point, $\dfrac{dy}{dx} = 0$, and then use the sign of the second derivative, $\dfrac{d^2y}{dx^2}$, to tell us whether the stationary point is a local maximum or a local minimum.

In this section the purpose of finding local maxima and minima is to sketch graphs of the form $y = f(x)$. In sketching graphs we can also establish where the graph cuts the x and y axes.

How can we find where $y = f(x)$ cuts the x axis?

The graph cuts the x axis at $y = 0$. This means that the x values satisfy $f(x) = 0$.

? **How can we find where $y = f(x)$ cuts the y axis?**

The graph cuts the y axis at $x = 0$ (Fig. 8). This means that the y value satisfies $y = f(0)$. Thus the graph

7.4   cuts the x axis where $f(x) = 0$, cuts the y axis where $y = f(0)$

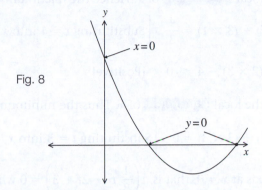

Fig. 8

🚗 Example 4 *mechanics*

A particle moves along a horizontal line according to

$$x = \frac{1}{3}t^3 - 2t^2 + 3t$$

where x is the displacement at time t.

  i Sketch the graph of x against t for $t \geq 0$.
  ii At what times do the stationary points of x occur?
 iii At what times is the particle at $x = 0$ (at rest)?

Solution

i

$$x = \frac{1}{3}t^3 - 2t^2 + 3t$$

$$\frac{dx}{dt} = t^2 - 4t + 3 \quad \text{[Differentiating]}$$

For stationary points $\dfrac{dx}{dt} = 0$, thus:

$$t^2 - 4t + 3 = 0 \quad \text{[Quadratic equation]}$$

$$(t - 3)(t - 1) = 0 \quad \text{[Factorizing]}$$

So stationary points occur at: $t = 3, \ t = 1$ [Solving]

To establish if these points represent a maximum or minimum we differentiate again and substitute our values $t = 1$ and $t = 3$.

$$\frac{d^2x}{dt^2} = 2t - 4$$

Example 4 *continued*

Substituting $t = 1$, $\dfrac{d^2x}{dt^2} = (2 \times 1) - 4 < 0$ [Negative]

By 7.2 , $t = 1$ gives local maximum of x. Hence the maximum value is

$x = \dfrac{1}{3}(1)^3 - (2 \times 1^2) + (3 \times 1) = \dfrac{4}{3}$ $\left[\text{Substituting } t = 1 \text{ into } x = \dfrac{1}{3}t^3 - 2t^2 + 3t\right]$

Putting $t = 3$, $\dfrac{d^2x}{dt^2} = (2 \times 3) - 4 > 0$ [Positive]

By 7.3 , $t = 3$ gives the local minimum of x. Thus the minimum value is

$x = \dfrac{1}{3}(3)^3 - (2 \times 3^2) + (3 \times 3) = 0$ $\left[\text{Substituting } t = 3 \text{ into } x = \dfrac{1}{3}t^3 - 2t^2 + 3t\right]$

The graph cuts the t axis at $x = 0$, that is, $t\left(\dfrac{1}{3}t^2 - 2t + 3\right) = 0$ which gives $t = 0$. Also the quadratic in the bracket is equal to zero.

$$\dfrac{1}{3}t^2 - 2t + 3 = 0$$

$$t^2 - 6t + 9 = 0   \left[\text{Multiplying by 3, because this removes } \dfrac{1}{3}\right]$$

$$(t - 3)^2 = 0   \text{ which gives } t = 3 \text{ (twice)}$$

Hence we have Fig. 9.

ii Stationary points of x occur when $\dfrac{dx}{dt} = 0$, which is at $t = 1$ and $t = 3$.

iii The particle is at rest, $x = 0$, at $t = 0$ and $t = 3$.

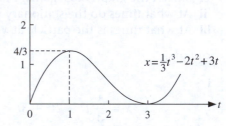

Fig. 9

You might ask the question, 'why are we devoting a section to the detailed technique of curve sketching when we have access to computer algebra systems and graphical calculators?'

Well generally these technical gadgets **only** give approximations to maxima and minima. Also you cannot visualize asymptotes (described in **Chapter 2**).

Example 5 *electrical principles*

The current, i, flowing through a circuit is

$$i = te^{-t} \ (t \geq 0)$$

i What is the value of the current i as t gets large, $t \to \infty$?
ii Determine the maximum value of the current.
iii Sketch the graph of i against t for $t \geq 0$.

Example 5 *continued*

Solution

i As t gets large, e^{-t} tends to zero. Hence as $t \to \infty$, $e^{-t} \to 0$ and therefore $i \to 0$. This is because e^{-t} decays to zero much faster than t increases. (Try large values of t on your calculator and note how close te^{-t} is to zero.)

ii First we find $\dfrac{di}{dt}$ and then equate this to zero to find the stationary points:

$$i = te^{-t}$$

We need to use the product rule, $\boxed{6.31}$, because i is t times e^{-t}:

$$u = t \qquad v = e^{-t}$$
$$u' = 1 \qquad v' = -e^{-t}$$

$$\frac{di}{dt} = u'v + uv' = (1)e^{-t} + t(-e^{-t}) = e^{-t}(1-t) \quad \text{[Factorizing]}$$

Equating this to zero for stationary points gives $e^{-t}(1 - t) = 0$.

By the properties of the exponential function, $e^{-t} \neq 0$ for any real values of t. Therefore $1 - t = 0$ which yields $t = 1$.

We use the 2nd derivative test to check for the nature of the stationary point. How?

Differentiating, $\dfrac{di}{dt} = e^{-t}(1-t)$. By using the product rule again:

$$u = 1 - t \qquad v = e^{-t}$$
$$u' = -1 \qquad v' = -e^{-t}$$

$$\frac{d^2i}{dt^2} = u'v + uv' = (-1)e^{-t} + (1-t)(-e^{-t}) \underset{\text{factorizing}}{=} (-1 - 1 + t)e^{-t} = (t-2)e^{-t}$$

Substituting $t = 1$ yields

$$\frac{d^2i}{dt^2} = (1-2)e^{-1} = -(1)e^{-1} < 0 \quad \text{[Negative]}$$

Hence i has a local maximum at $t = 1$. Moreover the maximum value is found by putting $t = 1$ into $i = te^{-t}$:

$$i = 1(e^{-1}) = e^{-1}$$

Fig. 10

iii At $t = 0$, the current $i = 0$, so the graph goes through the origin. Thus combining **i** and **ii** gives Fig. 10.

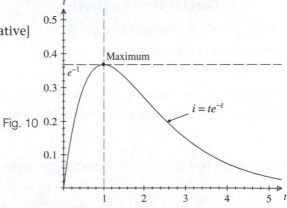

$\boxed{6.31}$ $(uv)' = u'v + uv'$

A4   Points of inflexion

So far we have described a test for maximum and minimum but we have not established a test for a point of inflexion.

Notice the change in direction of bending at the points of inflexion in Fig. 11.

Fig. 11

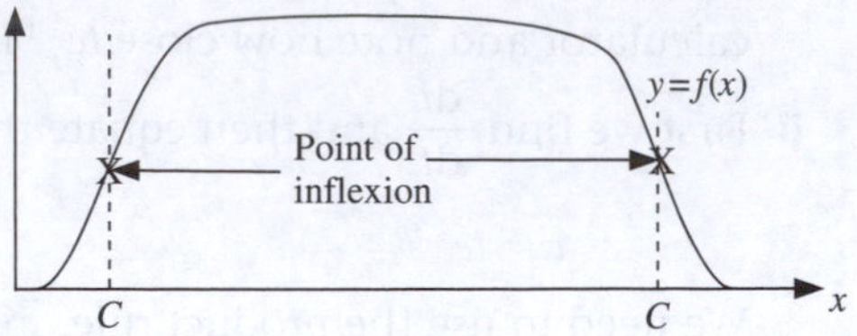

A **point of inflexion** at $x = C$ of $y = f(x)$ is where

7.5
$$\frac{d^2y}{dx^2} = 0 \text{ and } \frac{d^2y}{dx^2} \text{ changes sign close to } x = C$$

Example 6

Sketch on different axes the graphs of $y = x^3$ and $\dfrac{d^2y}{dx^2}$.

Solution

To plot the graph of $\dfrac{d^2y}{dx^2}$ we need to differentiate y twice:

$$y = x^3, \quad \frac{dy}{dx} = 3x^2 \text{ and}$$

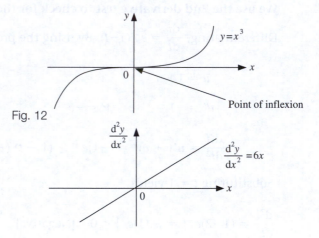

Fig. 12

$$\frac{d^2y}{dx^2} = 6x$$

We have sketched the graph of $y = x^3$ in **Chapter 2** and the graph of $\dfrac{d^2y}{dx^2} = 6x$ is a straight line going through the origin (Fig. 12).

Notice that $\dfrac{d^2y}{dx^2}$ changes sign close to the point of inflexion $(x = 0)$ and this point, shown in Fig. 12, is an example of a **horizontal point of inflexion**.

If at $x = C$ a function, $y = f(x)$, has the property

7.6
$$\frac{dy}{dx} = 0, \frac{d^2y}{dx^2} = 0 \text{ and also } \frac{d^2y}{dx^2} \text{ changes sign close to } x = C$$

then the function y has a horizontal point of inflexion at $x = C$.

If $\dfrac{d^2y}{dx^2}$ goes from negative to zero
to positive as x increases through C,
then the y graph has a horizontal
point of inflexion of the form shown
in Fig. 13a.

Fig. 13a

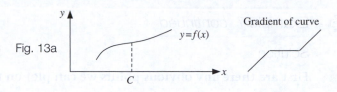

Gradient of curve

If $\dfrac{d^2y}{dx^2}$ goes from positive to zero
to negative as x increases through C,
then the y graph has a horizontal
point of inflexion of the form shown
in Fig. 13b.

Fig. 13b

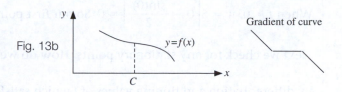

Gradient of curve

We can confirm that $y = x^3$ has a horizontal point of inflexion at $x = 0$ by checking the above. Thus differentiating and equating to zero we have

$$\frac{dy}{dx} = 3x^2 = 0 \text{ which gives } x = 0$$

At $x = 0$ we have a stationary point. Differentiating again:

$$\frac{d^2y}{dx^2} = 6x$$

At $x = 0$, $\dfrac{d^2y}{dx^2} = 6 \times 0 = 0$. It's **not** enough to say that since both $\dfrac{dy}{dx} = 0$ and $\dfrac{d^2y}{dx^2} = 0$ we have a horizontal point of inflexion at $x = 0$. We need to check the change of signs of $\dfrac{d^2y}{dx^2}$ for $x < 0$ and $x > 0$.

If $x < 0$ then $\dfrac{d^2y}{dx^2} = 6x < 0$ and if $x > 0$ then $\dfrac{d^2y}{dx^2} = 6x > 0$. Thus $\dfrac{d^2y}{dx^2}$ changes sign at $x = 0$ and therefore we have a horizontal point of inflexion.

The next example is particularly challenging because it involves a number of different concepts such as differentiation, solving trigonometric equations, substituting and sketching graphs.

Example 7 *electrical principles*

The energy, w, absorbed by a particular resistor at time t is given by

$$w = 5\left[t - \frac{\sin(2t)}{2}\right]$$

Sketch the graph of w against t for $0 \le t \le 2\pi$.

Example 7 *continued*

Solution

First are there any obvious points we can plot on the graph?

When $t = 0$, $w = 5\left[0 - \dfrac{\sin(0)}{2}\right] = 0$. So our first point on the graph is (0, 0).

Next we check for any stationary points. **How do we do that?**

By differentiating and finding values of t which satisfy $\dfrac{dw}{dt} = 0$:

$$\frac{dw}{dt} = 5\left[1 - \frac{2[\cos(2t)]}{2}\right] = 5[1 - \cos(2t)] \quad \text{[Cancelling 2's]}$$

For stationary points:

$$5[1 - \cos(2t)] = 0 \text{ gives } 1 - \cos(2t) = 0$$

Therefore $\cos(2t) = 1$ which gives $2t = 0$, 2π, 4π (by examining the cos graph, Fig. 53 of **Chapter 4** on page 204) and it follows that

$$t = 0, \pi, 2\pi \quad \text{[Dividing by 2]}$$

Differentiating again yields

$$\frac{d^2w}{dt^2} = \frac{d}{dt}[5(1 - \cos(2t))] = 5\{0 - [-2\sin(2t)]\} = 10\sin(2t)$$

Substituting $t = 0$, $\dfrac{d^2w}{dt^2} = 10\sin(2 \times 0) = 0$, unlike previous examples which have given us a positive or a negative answer.

Checking the graph at $t = 0$

We therefore need to check for a point of inflexion:

$$\text{For } t \text{ close to zero and } t < 0, \frac{d^2w}{dt^2} = 10\sin(2t) < 0 \quad \text{[Negative]}$$

$$\text{For } t \text{ close to zero and } t > 0, \frac{d^2w}{dt^2} = 10\sin(2t) > 0 \quad \text{[Positive]}$$

$$\left[\text{In each case you can try values of } t \text{ close to zero to check the sign of } \frac{d^2w}{dt^2}.\right]$$

By ▮7.6▮ the graph has a horizontal point of inflexion at $t = 0$ of the form ⌢—⌢ .

Checking the graph at $t = \pi$

Example 7 *continued*

Substituting $t = \pi$:

$$\frac{\text{d}^2w}{\text{d}t^2} = 10\sin(2\pi) = 0$$

For t close to π and $t < \pi$: $\dfrac{\text{d}^2w}{\text{d}t^2} = 10\sin(2t) < 0$ [Negative]

For t close to π and $t > \pi$: $\dfrac{\text{d}^2w}{\text{d}t^2} = 10\sin(2t) > 0$ [Positive]

By 7.6 , the graph has a horizontal point of inflexion at $t = \pi$ of the form $\diagup\!\!-\!\!\diagup$.

Checking the graph at $t = 2\pi$

Substituting $t = 2\pi$:

$$\frac{\text{d}^2w}{\text{d}t^2} = 10\sin(4\pi) = 0$$

For t close to 2π and $t < 2\pi$: $\dfrac{\text{d}^2w}{\text{d}t^2} < 0$ [Negative]

For t close to 2π and $t > 2\pi$: $\dfrac{\text{d}^2w}{\text{d}t^2} > 0$ [Positive]

We have a horizontal point of inflexion at $t = 2\pi$ of the form $\diagup\!\!-\!\!\diagup$.

We can find the w values at the three points of inflexion, at $t = 0$, π and 2π, by substituting these into

$$w = 5\left[t - \frac{\sin(2t)}{2}\right].$$

When $t = 0$, $w = 0$

$t = \pi$, $w = 5\left[\pi - \dfrac{\sin(2\pi)}{2}\right] = 5\pi$

$t = 2\pi$, $w = 5\left[2\pi - \dfrac{\sin(4\pi)}{2}\right] = 10\pi$

$(0, 0)$, $(\pi, 5\pi)$ and $(2\pi, 10\pi)$ are all horizontal points of inflexion of the form $\diagup\!\!-\!\!\diagup$, giving Fig. 14.

Fig. 14

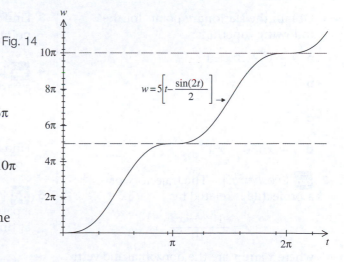

Remember that as well as horizontal points of inflexion we have other (vertical) points of inflexion, see Fig. 11.

SUMMARY

A stationary point is a point on the curve $y = f(x)$ such that

7.1 $\dfrac{dy}{dx} = 0$

If at $x = C$

7.2 $\dfrac{dy}{dx} = 0$ and $\dfrac{d^2y}{dx^2} < 0$ [Negative]

then y has a local maximum at $x = C$ and if

7.3 $\dfrac{dy}{dx} = 0$ and $\dfrac{d^2y}{dx^2} > 0$ [Positive]

then y has a local minimum at $x = C$.
However if at $x = C$

7.6 $\dfrac{dy}{dx} = 0$, $\dfrac{d^2y}{dx^2} = 0$ and also $\dfrac{d^2y}{dx^2}$ changes sign close to C then y has a

horizontal point of inflexion at $x = C$.

Exercise 7(a)

Solutions at end of book. Complete solutions available at www.palgrave.com/engineering/singh

1 Obtain the stationary points for the following functions:

   a $y = x^2 + 3x + 1$

   b $y = x^3 - 3x$

   c $y = \dfrac{x^3}{3} - x$

   d $y = x^2(50 - x^2)$

2 🚗 [mechanics]   The trajectory of a projectile is related by

   $$y = \frac{10x - x^2}{10}$$

   where x and y are the horizontal and vertical displacement components respectively. Sketch the graph of y against x.

3 Find the maximum and minimum points of $y = x^3 + x^2 - 5x$.

4 🚗 [mechanics]   The displacement, s, of a particle is given by

   $$s = t^3 - 6t^2 + 12t$$

   Find the maximum, minimum and points of inflexion of s.

5 🚗 [mechanics]   The displacement, s, of a particle moving along a horizontal line at time t is given by

   $$s = t^2(8 - t^2)$$

   Sketch the graph of s against t.

Exercise **7(a) continued**

▦ **The remaining questions in this exercise are in the field of [*electrical principles*].**

6 The energy, w, of an inductor at time t is given by

$$w = 1 - \cos(50t)$$

Sketch the graph of w against t for

$$0 \le t \le \frac{\pi}{25}.$$

7 The power, p, in a resistor is given by

$$p = 10\sin^2(t) \text{ where } 0 \le t \le 2\pi$$

Sketch the graph of p against t, marking all the points of maximum, minimum and inflexion.

*8 The energy, w, absorbed by a resistor at time t is given by

$$w = 5\sin(2t) + 10t$$

Sketch the graph of w against t for $0 \le t \le 2\pi$.

9 The power, p, of an inductor is given by

$$p = (t^2 - 2t)e^{-2t} \quad (t \ge 0)$$

 i What is the value of p as t gets large, $t \to \infty$?

 ii At what values of t is $p = 0$?

 iii Find the minimum value of p.

 iv Find the maximum value of p.

 v Sketch the graph of p against t for $t \ge 0$.

10 The current, i, through a circuit is given by

$$i = 5te^{-5t} \quad t \ge 0$$

 i What is the value of i for large t, $t \to \infty$?

 ii Find the maximum current through the circuit.

 iii Sketch the graph of i against t for $t > 0$.

SECTION B **Optimization problems**

By the end of this section you will be able to:
▶ formulate equations from given data
▶ use the derivative test technique for solving engineering problems which require maximizing or minimizing

B1 **General optimization**

In the previous section we used derivative tests to find the local maximum or local minimum in order to sketch graphs of various functions. In this section we use the same techniques to maximize or minimize a given engineering problem.

Example 8

Figure 15 shows a rectangular sheet of metal. A square of edge x (m) is cut from each corner and the remaining piece is folded along the broken line to make an open box of volume V. Show that

$$V = 4x^3 - 14x^2 + 10x$$

Example 8 *continued*

Find the value of x to 2 d.p. for which the volume V is a maximum.

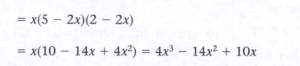

Fig. 15

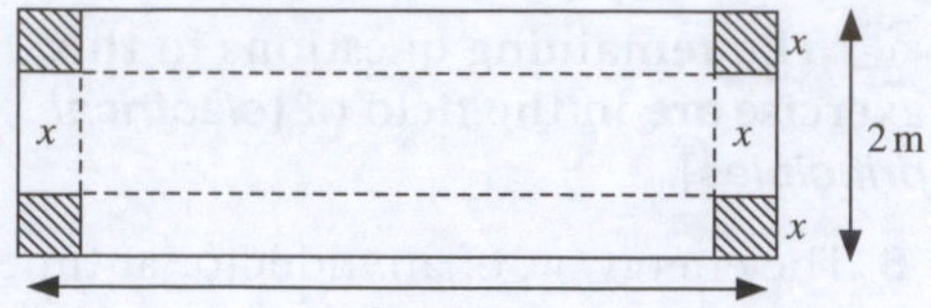

Solution

The base of the box after cutting away the corners is shown in Fig. 16.

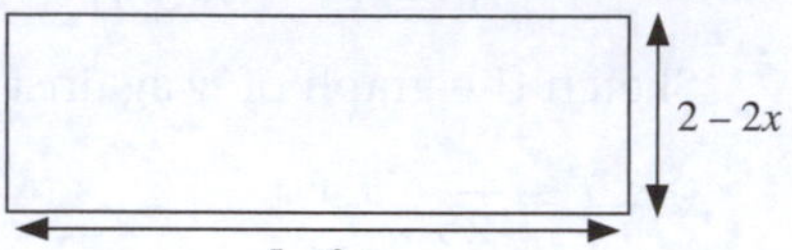

The height of the box is x, so the volume, V, is given by

Fig. 16

$$V = x \times [\text{base area}]$$

$$= x(5 - 2x)(2 - 2x)$$

$$= x(10 - 14x + 4x^2) = 4x^3 - 14x^2 + 10x$$

We have seen expressions similar to this, $V = 4x^3 - 14x^2 + 10x$, in **Section A**.

So how do we find when the volume is a maximum?

We need to find derivative $\dfrac{dV}{dx}$ and to solve the equation $\dfrac{dV}{dx} = 0$ for x.

$$V = 4x^3 - 14x^2 + 10x$$

$$\frac{dV}{dx} = 12x^2 - 28x + 10 = 0 \qquad [\text{Differentiating}]$$

Dividing by 2 gives

$$6x^2 - 14x + 5 = 0$$

How do we solve this quadratic equation?

By using the quadratic formula

1.16 $\qquad x = \dfrac{-b \pm \sqrt{b^2 - 4ac}}{2a}$

with $a = 6$, $b = -14$ and $c = 5$:

$$x = \frac{14 \pm \sqrt{(-14)^2 - (4 \times 6 \times 5)}}{(2 \times 6)} = 0.44, \ 1.89 \ (2 \text{ d.p.})$$

x cannot be 1.89 m. Why not?

Because the length of the base is $2 - 2x$. If $x = 1.89$ m then $2 - 2x$ will be negative. So we only need to check if $x = 0.44$ m gives a maximum volume.

Example 8 *continued*

?

How?

Use the second derivative test:

$$\frac{dV}{dx} = 12x^2 - 28x + 10$$

$$\frac{d^2V}{dx^2} = 24x - 28$$

Putting $x = 0.44$, $\dfrac{d^2V}{dx^2} = (24 \times 0.44) - 28 < 0$ [Negative]

By 7.2 , a square piece of edge 0.44 m gives the maximum volume of the box.

Example 9

Figure 17 shows a cylindrical container which holds a volume of 1 litre
(1000 cm^3). Show that the surface area, A, is given by $A = 2\pi r^2 + 2000r^{-1}$.

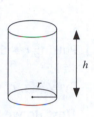

Fig. 17

?

Find the dimensions so that the surface area is a minimum (minimizing
the material used). **What do you notice about your results?**

(Volume $= \pi r^2 h$ and surface area $= 2\pi r^2 + 2\pi rh$)

Solution

Since the volume is 1000 cm^3 we have

$$\pi r^2 h = 1000$$

$$h = \frac{1000}{\pi r^2} \qquad \text{[Dividing by } \pi r^2\text{]}$$

Substituting $h = \dfrac{1000}{\pi r^2}$ into the surface area formula gives

$$A = 2\pi r^2 + 2\pi rh$$

$$= 2\pi r^2 + 2\pi r\left(\frac{1000}{\pi r^2}\right)$$

$$A = 2\pi r^2 + 2000r^{-1} \quad \text{[Cancelling]}$$

For stationary points we differentiate A with respect to r and equate the result to zero:

$$\frac{dA}{dr} = 4\pi r - 2000r^{-2} = 0$$

$$4\pi r = 2000r^{-2} = \frac{2000}{r^2} \qquad \left[\text{Remember } 2000r^{-2} = \frac{2000}{r^2}\right]$$

$$r^3 = \frac{2000}{4\pi\pi} = \frac{500}{\pi} \qquad \text{[Transposing]}$$

7.2 $V' = 0$, $V'' < 0$ max

Example 9 *continued*

How do we find *r*?

Take the cube root of both sides:

$$r = \sqrt[3]{\frac{500}{\pi}} = 5.42 \text{ cm} \quad (2 \text{ d.p.})$$

Using the second derivative test to ascertain that this value of *r*, *r* = 5.42, does indeed give minimum surface area:

$$\frac{\mathrm{d}A}{\mathrm{d}r} = 4\pi r - 2000r^{-2}$$

$$\frac{\mathrm{d}^2A}{\mathrm{d}r^2} = 4\pi + 4000r^{-3}$$

Putting *r* = 5.42, $\dfrac{\mathrm{d}^2A}{\mathrm{d}r^2} = 4\pi + \dfrac{4000}{(5.42)^3} > 0$ [Positive]

By 7.3 , *r* = 5.42 cm gives minimum *A*.

How do we find the other dimension, *h*?

Substituting *r* = 5.42 into $h = \dfrac{1000}{\pi r^2}$ gives

$$h = \frac{1000}{\pi(5.42)^2} = 10.84 \text{ cm} \quad (2 \text{ d.p.})$$

Note that the height is twice the radius.

The next example is a lot more complex than the above. Don't be put off by the different *i*'s.

B2 **Engineering problems**

Example 10 *electrical principles*

Consider an electrical circuit with resistors of resistance R_1, R_2 and R_3. In the following, treat R_1, R_2, R_3 and *i* as positive constants.

i The power, *p*, dissipated by R_1, R_2 and R_3 is given by

$$p = i_1^2 R_1 + i_1^2 R_2 + (i - i_1)^2 R_3$$

where i_1 and *i* represent current. Show that *p* is a minimum at

$$i_1 = \frac{R_3}{R_1 + R_2 + R_3}\, i$$

7.3 $A' = 0,\ A'' > 0$ min

Example 10 *continued*

ii The voltage, V, is given by $V = iR = i_1(R_1 + R_2)$ where R is the equivalent resistance of R_1, R_2 and R_3.

For minimum power dissipation, minimum p, show that

$$R = \frac{(R_1 + R_2)R_3}{R_1 + R_2 + R_3}$$

Solution

Don't be confused by the notation; we apply the same rules.

i Copying the equation for p we have

$$p = i_1{}^2 R_1 + i_1{}^2 R_2 + (i - i_1)^2 R_3$$

For stationary points we differentiate p with respect to i_1 and equate the result to zero:

$$\frac{dP}{di_1} = 2i_1R_1 + 2i_1R_2 + \underbrace{2(i - i_1)R_3(-1)}_{\text{chain rule}} = 0$$

$$2i_1R_1 + 2i_1R_2 - 2iR_3 + 2i_1R_3 = 0$$

$$i_1R_1 + i_1R_2 - iR_3 + i_1R_3 = 0 \qquad \text{[Dividing by 2]}$$

$$i_1(R_1 + R_2 + R_3) = iR_3 \qquad \text{[Adding } iR_3 \text{ and factorizing]}$$

$$i_1 = \frac{iR_3}{R_1 + R_2 + R_3} = \frac{R_3}{R_1 + R_2 + R_3}i$$

For minimum we need to differentiate again:

$$\frac{dp}{di_1} = 2i_1R_1 + 2i_1R_2 - 2iR_3 + 2i_1R_3$$

$$\frac{d^2p}{di_1{}^2} = 2R_1 + 2R_2 - 0 + 2R_3$$

(Remember that $2iR_3$ is a constant, so differentiating this gives 0.)

Since resistance values are positive, therefore $\dfrac{d^2p}{di_1{}^2} = 2R_1 + 2R_2 + 2R_3 > 0$.

By ⌈7.3⌉, $i_1 = \dfrac{iR_3}{R_1 + R_2 + R_3}$ gives minimum p.

ii We are given $iR = i_1(R_1 + R_2)$

Substituting $i_1 = \dfrac{iR_3}{R_1 + R_2 + R_3}$ into the given equation, $iR = i_1(R_1 + R_2)$, yields

$$iR = \frac{iR_3}{R_1 + R_2 + R_3}(R_1 + R_2)$$

Dividing both sides by i gives the required result:

$$R = \frac{(R_1 + R_2)R_3}{R_1 + R_2 + R_3}$$

7.3 $p' = 0$, $p'' > 0$ min

SUMMARY

We use the derivative tests from **Section A** to find which values maximize or minimize a given problem.

Exercise **7(b)**

Solutions at end of book. Complete solutions available at www.palgrave.com/engineering/singh

1

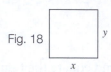

Fig. 18

y

x

Figure 18 shows a fence of dimensions x and y. The perimeter of the fence is 100 m. Show that the area, A, constrained by the fence is

$$A = 50x - x^2$$

Find the values of x and y for which the area A is a maximum. Is there a relationship between x and y which gives maximum area?

2 Figure 19 shows a field surrounded by 240 m of fencing on three sides.

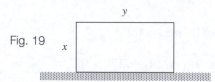

Fig. 19

y

x

The shaded part is a hedge where no fencing is required. Find the dimensions, x and y, that command maximum area of the field. What do you notice about your results?

3 [*aerodynamics*] Minimum drag occurs at the minimum value of

$$\frac{C_D + kC_L^2}{C_L}$$

where C_L = lift coefficient, C_D = zero lift drag coefficient and k = non-zero constant. For what value of C_L do we

have minimum drag? (Treat C_D and k as constants.)

4 Consider a cylindrical container without a top. If the volume of the container is 8 m³ then find the dimensions, radius (r in m) and height (h in m), so that the surface area required is a minimum. What do you notice about your results?

*5 Show that for any given volume, V, the minimum surface area required for a closed cylindrical can is when the height, h, is twice the radius, r.

6 [*materials*] The bending moment, M, at a distance x of a beam of length L is given by

$$M = \frac{WLx}{2} - \frac{Wx^2}{2}$$

where W is the weight per unit length. Find the value of x which gives the maximum bending moment and evaluate this maximum.

[*Hint*: Factorize first]

7 [*materials*] The downward deflection, y, of a cantilever of length L with a load W is given by

$$y = \frac{W}{6EI}(3L^2x - x^3) \quad \left(\frac{W}{EI} > 0\right)$$

where x is the horizontal distance from the fixed end. Find the maximum deflection and the value of x to achieve this. (See Fig. 20 in the next question.)

Solutions at end of book. Complete solutions available at www.palgrave.com/engineering/singh

Exercise **7(b) continued**

8  [*materials*]

Fig. 20

L

x

Figure 20 shows a cantilever of length L with uniform distributed loading of W per unit length. The downward deflection of the cantilever at a distance x is denoted by y and satisfies

$$y = \frac{W}{24EI}[x^4 - 4Lx^3 + 4L^2x^2]$$

where $EI > 0$ is the constant flexural rigidity of the cantilever. Show that the maximum deflection occurs at $x = L$ and determine this maximum deflection.

9 [*mechanics*]   A belt of mass m per unit length is wound partly around a pulley. The power, p, transmitted is given by

$$p = Tv - mv^3$$

where T is the tension and v is the velocity of the belt. Show that the maximum power transmitted occurs when $v = \sqrt{\dfrac{T}{3m}}$.

*10 [*electrical principles*]   The torque, T, of an electric machine is given by

$$T = 3\sin(2\theta) + 6\sin(\theta)$$

where θ is the torque position and $0 \le \theta < \pi$.

i Find the value of θ for which maximum torque occurs.
ii Determine the maximum torque. [You will need to use trigonometric identity for this question.]

*11 [*electrical principles*]   Consider an electrical circuit with resistors of resistance R_1 and R_2.

i The power, P, dissipated by R_1 and R_2 is given by

$$P = i_1^2 R_1 + (i - i_1)^2 R_2$$

where i_1 and i represent current.

Show that P is a minimum when

$$i_1 = \frac{R_2}{R_1 + R_2} i$$

ii The voltage, V, is given by

$$V = i_1 R_1 = iR$$

where R is the equivalent resistance of R_1 and R_2. For minimum power dissipation, show that

$$R = \frac{R_1 R_2}{R_1 + R_2}$$

SECTION C **First derivative test**

By the end of this section you will be able to:
▶ apply the first derivative test for maxima and minima
▶ investigate examples using the first derivative test

C1 **Maxima and minima problems**

Generally we use the second derivative test to establish whether a stationary point is a maximum or minimum. For complicated functions this can often be a laborious process. On these occasions we can extract the information from the first derivative, using the following method.

If we plot the graph of y against x for a **local maximum** at $x = C$, we obtain Fig. 21.

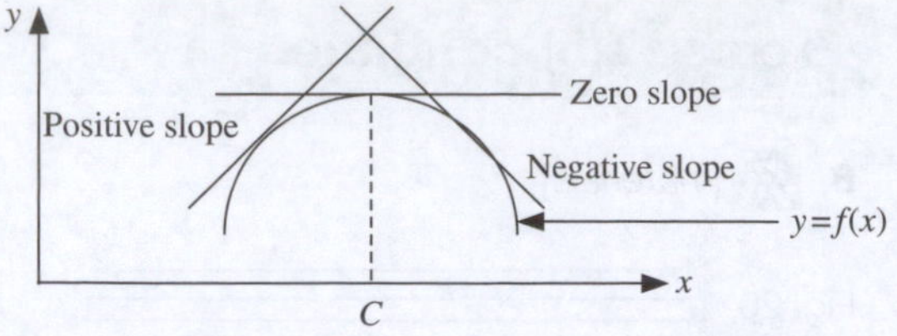

Fig. 21

? **What do you notice about the gradient, $\dfrac{dy}{dx}$, of y for x close to C?**

It changes sign from left to right – positive to zero to negative. In general a function, $y = f(x)$, with local maximum at $x = C$ means that

7.7

$$
\begin{cases}
\textbf{1} \quad \text{at } x = C, & \dfrac{dy}{dx} = 0 \quad \text{[Stationary point]} \\[2mm]
\textbf{2} \quad \text{for } x < C, & \dfrac{dy}{dx} > 0 \quad \text{[Positive]} \\[2mm]
\textbf{3} \quad \text{for } x > C, & \dfrac{dy}{dx} < 0 \quad \text{[Negative]}
\end{cases}
$$

$\dfrac{dy}{dx}$ changes sign as it passes through the stationary point, $x = C$. Also note that when we say $x < C$ or $x > C$ we mean x values close to C.

If we carry out the same procedure for a **local minimum** at $x = C$ we obtain Fig. 22.

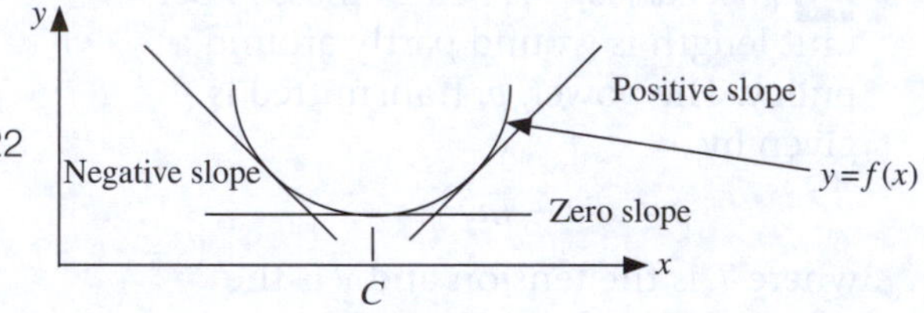

Fig. 22

Again $\dfrac{dy}{dx}$ changes sign as it passes through the stationary point, $x = C$.

It changes sign from left to right – negative to zero to positive.

In general a function, $y = f(x)$, with local minimum at $x = C$ means that

7.8

$$
\begin{cases}
\textbf{1} \quad \text{at } x = C, & \dfrac{dy}{dx} = 0 \quad \text{[Stationary point]} \\[2mm]
\textbf{2} \quad \text{for } x < C, & \dfrac{dy}{dx} < 0 \quad \text{[Negative]} \\[2mm]
\textbf{3} \quad \text{for } x > C, & \dfrac{dy}{dx} > 0 \quad \text{[Positive]}
\end{cases}
$$

Let's do an example.

Example 11

Find the points of maxima and minima of $y = \dfrac{x}{1 + x^2}$.

Solution

? For stationary points we need to differentiate y. **How?**

Example 11 *continued*

We need to apply the quotient rule, 6.32 , with

$$u = x \qquad v = 1 + x^2$$

$$u' = 1 \qquad v' = 2x$$

Substituting these gives

$$\frac{dy}{dx} = \frac{u'v - uv'}{v^2}$$

$$= \frac{1(1 + x^2) - 2x(x)}{(1 + x^2)^2}$$

$$= \frac{1 + x^2 - 2x^2}{(1 + x^2)^2} = \frac{1 - x^2}{(1 + x^2)^2} \qquad \text{[Simplifying numerator]}$$

Equating this to zero:

$$\frac{dy}{dx} = \frac{1 - x^2}{(1 + x^2)^2} = 0$$

The derivative $\dfrac{dy}{dx} = 0$ only when the numerator $1 - x^2 = 0$. Solving gives

$$x^2 = 1$$
$$x = \pm 1 \qquad \text{[Square root]}$$

Clearly we have two stationary points, $x = 1$ and $x = -1$. The denominator $(1 + x^2)^2$ is positive so we only need to check the sign of the numerator, $1 - x^2$:

If $x < 1$, try $x = 0.9$, then $1 - x^2 = 1 - 0.9^2 > 0$ so $\dfrac{dy}{dx} > 0$ [Positive gradient]

If $x > 1$, try $x = 1.1$, then $1 - x^2 = 1 - 1.1^2 < 0$ so $\dfrac{dy}{dx} < 0$ [Negative gradient]

Thus we have

$$\frac{dy}{dx} > 0 \qquad \diagup\diagdown \qquad \frac{dy}{dx} < 0$$
$$x = 1$$

By 7.7 , $x = 1$ gives a maximum.

Now we test the other point, $x = -1$:

If $x < -1$, try $x = -1.1$, then $1 - x^2 = 1 - (-1.1)^2 < 0$ so $\dfrac{dy}{dx} < 0$ [Negative gradient]

6.32 $(u/v)' = (u'\, v - uv')/v^2$

Example 11 *continued*

If $x > -1$, try $x = -0.9$, then $1 - x^2 = 1 - (-0.9)^2 > 0$ so $\dfrac{dy}{dx} > 0$ [Positive gradient]

Hence we have

$$\dfrac{dy}{dx} < 0 \qquad \diagdown\diagup \qquad \dfrac{dy}{dx} > 0$$
$$x = -1$$

By   7.8 , $x = -1$ gives a minimum.

? **Why would you *not* use the second derivative test for the above example?**

We have

$$\dfrac{dy}{dx} = \dfrac{1 - x^2}{(1 + x^2)^2}$$

Differentiating this to find $\dfrac{d^2y}{dx^2}$ by using the quotient rule is a horrendous task. The only other way to differentiate the original function, $y = \dfrac{x}{1 + x^2}$, twice is to use implicit differentiation which would involve significantly more work.

You may find it easier to use the first derivative test and **not** bother with the second derivative test at all.

SUMMARY

We can use a first derivative test for maxima and minima by finding the stationary points of a given function and then checking the sign of $\dfrac{dy}{dx}$ on both sides of the stationary point.

Remember that care needs to taken as we test x values close to the stationary point.

Exercise **7(c)**

Solutions at end of book. Complete solutions available at www.palgrave.com/engineering/singh

1 Find the stationary points of

$$y = x^2(8 - x^2)$$

and distinguish between them using the first derivative test.

2 🚗 [*mechanics*]   The acceleration, a, of a machine is given by

$$a = \dfrac{10r + 1}{5r^2 + 3150}$$

where r is the gear ratio. Determine the value of r for maximum acceleration to occur.

3 🔌 [*electrical principles*]

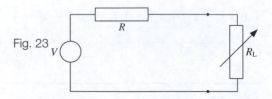

Fig. 23

Exercise **7(c) continued**

Figure 23 shows a circuit consisting of a source voltage V and resistor of resistance R applied across a variable load R_L. The power, P, delivered to the load R_L is given by

$$P = \frac{V^2 R_L}{(R + R_L)^2}$$

Show that maximum power occurs when $R_L = R$.

4 [*electrical principles*]   A capacitor is formed by two concentric spheres of radii a and b respectively. The electric stress, E, is given by

$$E = \frac{Vb}{ba - a^2} \quad (ba - a^2 \neq 0)$$

where V (> 0) is the potential difference between the spheres. Show that the minimum electric stress occurs at $a = \dfrac{b}{2}$.

[Differentiate E with respect to a.]

5 [*electromagnetism*]   The induced e.m.f., E, in a coil is given by

$$E = 2\pi f k \cos(2\pi f t)$$

Find the maximum value of E.

SECTION D **Applications to kinematics**

By the end of this section you will be able to:
▶ determine the velocity and acceleration of particles in rectangular (straight line) motion
▶ determine the angular velocity and angular acceleration of particles in circular motion

---

D1 **Rectangular motion**

Figure 24 shows a particle moving along a straight line. The displacement, s, from the origin is a function of time, t. The units of displacement are metres (m).

Fig. 24

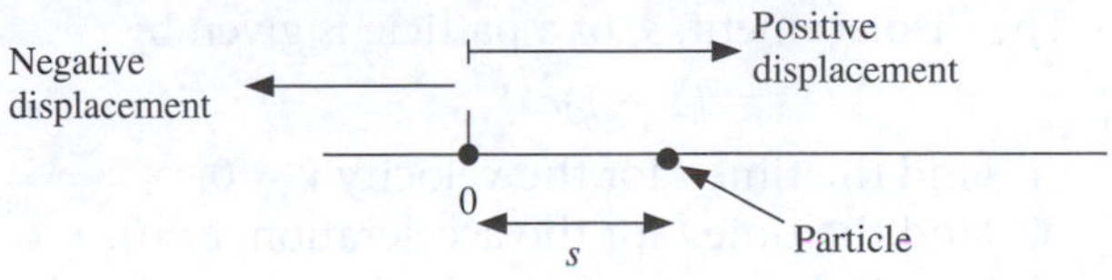

The velocity, v, of a particle at time t is defined as

7.9 $v = \dfrac{ds}{dt}$ [1st derivative]

Velocity is the **rate of change** of displacement with respect to time. The units of velocity, v, are metres per second (m/s). A particle with $v = 10$ m/s means that the particle covers 10 m for every second.

The acceleration, a, of a particle at time t is defined as

7.10 $a = \dfrac{dv}{dt} = \dfrac{d}{dt}\left(\dfrac{ds}{dt}\right) = \dfrac{d^2 s}{dt^2}$ [2nd derivative]

Acceleration is the **rate of change** of velocity with respect to time t. The units of acceleration, a, are metres per second^2 (m/s^2). A particle with an acceleration of 9.81 m/s^2 means that the velocity of the particle increases by 9.81 m/s for every second.

We have done many examples of finding the velocity and acceleration of particles in **Chapter 6**.

Example 12 *mechanics*

A particle moves along a horizontal line according to the equation

$$x = 2t^3 + t^2 - 10t + 10$$

where x is the displacement at time t. Find the displacement, velocity and acceleration at $t = 1$.

Solution

Differentiating x once gives velocity, and differentiating x twice gives acceleration:

$$x = 2t^3 + t^2 - 10t + 10$$
$$v = \frac{dx}{dt} = 6t^2 + 2t - 10 \qquad \text{[Differentiating]}$$
$$a = \frac{d^2x}{dt^2} = 12t + 2 \qquad \text{[Differentiating again]}$$

To find displacement, velocity and acceleration at $t = 1$, we substitute $t = 1$ into the above equations:

$$x = 2 + 1 - 10 + 10 = 3 \text{ m}$$
$$v = 6 + 2 - 10 = -2 \text{ m/s [The minus signifies the direction of the velocity]}$$
$$a = 12 + 2 = 14 \text{ m/s}^2$$

Example 13 *mechanics*

The displacement, s, of a particle is given by

$$s = (1 - t)e^{-t}$$

  i Find the time t for the velocity $v = 0$.
 ii Find the time t for the acceleration $a = 0$.
iii Sketch the graph of v against t showing all the points of maxima and minima.

Solution

  i We need to differentiate $s = (1 - t)e^{-t}$ to find the velocity. **How?**

Apply the product rule, 6.31 , because s is $1 - t$ times e^{-t}. We use the letters p and q because v is used for velocity. Let

$$p = 1 - t, \qquad q = e^{-t}$$
$$p' = -1, \qquad q' = -e^{-t}$$

6.31 $(pq)' = p'q + pq'$

Example 13 *continued*

Thus

$$v = \frac{ds}{dt} = p'q + pq'$$
$$= (-1)e^{-t} + (1-t)(-e^{-t})$$
$$= e^{-t}(-1-1+t) = e^{-t}(t-2)$$
$$v = e^{-t}(t-2)$$

For $v = 0$ we have

$$e^{-t}(t-2) = 0$$

Since the exponential function is **never zero** we have

$$t - 2 = 0 \text{ which gives } t = 2 \text{ s}$$

ii To find acceleration, a, we have to differentiate v. By applying the product rule, 6.31 , on $v = (t-2)e^{-t}$ we have

$$p = t - 2 \qquad q = e^{-t}$$
$$p' = 1 \qquad q' = -e^{-t}$$

Thus

$$a = \frac{d^2s}{dt^2} = p'q + pq'$$
$$= (1)e^{-t} + (t-2)(-e^{-t}) = \underbrace{e^{-t}(1-t+2)}_{\text{Taking out } e^{-t}} = e^{-t}(3-t)$$
$$a = e^{-t}(3-t)$$

We want to find t which satisfies $a = 0$:

$$(3-t)e^{-t} = 0 \text{ gives } 3 - t = 0, \text{ therefore } t = 3 \text{ s}$$

iii From part **i** we know that $v = (t-2)e^{-t}$.

Where does the v graph cut the t axis?

This is when $v = 0$ therefore $t = 2$. Also when $t = 0$, $v = (0-2)e^0 = -2$. Hence the graph goes through the points $(2, 0)$ and $(0, -2)$.

For sketching the graph we must find stationary points of v and their nature. **How can we find the stationary points of v?**

We covered this in part **ii**:

$$v' = \frac{dv}{dt} = (3-t)e^{-t} \left(= \frac{d^2s}{dt^2} \right)$$

For stationary points, v' needs to be zero, that is when $a = 0$:

$$(3-t)e^{-t} = 0 \text{ gives } t = 3$$

The exponential function e^{-t} is positive so we just need to check the sign of $(3-t)$:

If $t < 3$ then $3 - t > 0$, therefore $v' > 0$ [Positive]
If $t > 3$ then $3 - t < 0$, therefore $v' < 0$ [Negative]

6.31 $(pq)' = p'q + pq'$

Example 13 *continued*

Thus we have

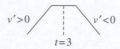

Hence v has a maximum at $t = 3$. The maximum value is found by substituting $t = 3$ into $v = (t - 2)e^{-t}$:

$$v = (3 - 2)e^{-3} = e^{-3}$$

Also what can you say about v as $t \to \infty$?

Since $v = (t - 2)e^{-t}$ and the e^{-t} term goes to zero as t approaches infinity, we can say as $t \to \infty$, $v \to 0$. Collecting all the above gives the graph shown in Fig. 25.

Fig. 25

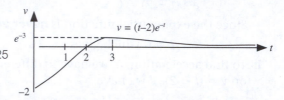

D2 **Angular motion**

So far we have dealt with a particle moving in a straight line. Now we discuss a particle moving with circular motion.

Consider a particle moving around a circle so that the angular displacement, θ, is a function of time t, $\theta = f(t)$ (Fig. 26.).

θ is the angle swept in time t and the units of θ are radians.

The angular velocity, ω, of a particle at time t is defined as

Fig. 26

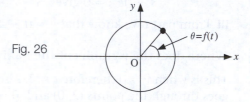

7.11 $\omega = \dfrac{d\theta}{dt}$ [1st derivative]

ω measures the rate at which the particle is going around the origin. The units of ω are radians per second, denoted by rad/s. A particle with angular velocity $\omega = \pi$ rad/s means that the particle goes through 180° (π radians) in every second.

The angular acceleration, α, of the particle at time t is defined as

7.12 $\alpha = \dfrac{d\omega}{dt} = \dfrac{d}{dt}\left(\dfrac{d\theta}{dt}\right) = \dfrac{d^2\theta}{dt^2}$ [2nd derivative]

The units of angular acceleration, α, are radians per second², denoted by rad/s². A particle with angular acceleration $\alpha = 3$ rad/s² means that the angular velocity increases by 3 rad/s for every second.

Example 14 *mechanics*

The angular displacement, θ, of a particle in circular motion is given as

$$\theta = \frac{1}{2}\cos\left(2t + \frac{\pi}{4}\right) \text{ where } 0 \le t \le \pi$$

i Obtain an expression for angular acceleration, α.
ii Find the first time t when $\alpha = 0$.

Solution

i By 7.12 we need to find $\dfrac{d^2\theta}{dt^2}$. First we obtain $\dfrac{d\theta}{dt}$:

$$\theta = \frac{1}{2}\cos\left(2t + \frac{\pi}{4}\right)$$

$$\frac{d\theta}{dt} = \frac{1}{2}(-2)\sin\left(2t + \frac{\pi}{4}\right) = -\sin\left(2t + \frac{\pi}{4}\right)$$

Differentiating again:

$$\frac{d^2\theta}{dt^2} = -2\cos\left(2t + \frac{\pi}{4}\right) = \alpha$$

ii For $\alpha = 0$ we have

$$-2\cos\left(2t + \frac{\pi}{4}\right) = 0$$

Dividing both sides by -2:

$$\cos\left(2t + \frac{\pi}{4}\right) = 0$$

How can we find t?
We need to remove the cos by taking inverse cos, $\cos^{-1}$, of both sides:

$$2t + \frac{\pi}{4} = \cos^{-1}(0) = \frac{\pi}{2}$$

$$2t = \frac{\pi}{2} - \frac{\pi}{4} = \frac{\pi}{4}$$

Divide both sides by 2: $t = \dfrac{\pi}{8}$ gives $\alpha = 0$.

SUMMARY

The velocity, v, and acceleration, a, of a particle in a straight line motion with displacement s at time t are defined as

 7.9 $$v = \frac{ds}{dt}$$

 7.10 $$a = \frac{d^2s}{dt^2}$$

The angular velocity, ω, and angular acceleration, α, of a particle in a circular motion with angle θ swept in time t are defined as

7.11 $\qquad \omega = \dfrac{d\theta}{dt}$

7.12 $\qquad \alpha = \dfrac{d^2\theta}{dt^2}$

Exercise 7(d)

Solutions at end of book. Complete solutions available at www.palgrave.com/engineering/singh

🚗 All questions in this exercise belong to the field of [*mechanics*] with units as defined in Sections D1 and D2.

1 A particle moves along a horizontal line with displacement, s, given by

$$s = t^2 - 4t + 3$$

  i Find the time(s) t for $s = 0$.
  ii Determine expressions for the velocity and acceleration.

2 A projectile is fired vertically upwards with height h given by

$$h = 25t - 4.9t^2$$
$$\text{where} \quad 0 \le t \le 5$$

Find expressions for the velocity and acceleration. Determine the maximum height reached by the projectile.

3 The displacement, s, of a particle is given by

$$s = 3t^3 - 9t^2 + 10$$

Find the time(s) t for the velocity $v = 0$ and acceleration $a = 0$.

4 The displacement, s, of an object is given by

$$s = 3t^3 - 10t^2 + t - 10$$

Find expressions for the velocity and acceleration.

5 The displacement, s, of a particle is given by

$$s = t^3 - 6t^2 + 12t$$

Find the minimum velocity and sketch the graph of v against t for $t \ge 0$.

6 The displacement, s, of a particle is given by

$$s = t^3 - 12t^2 + 18t$$

Find the velocity and sketch the graph of v against t for $t \ge 0$.

7 The displacement, s, of a particle is given by

$$s = \frac{1}{v} \quad (v > 0)$$

where v is the velocity of the particle. Find $\dfrac{ds}{dv}$ and sketch the graphs, on the same axes, of s and $\dfrac{ds}{dv}$ against v for $v > 0$.

8 The angular displacement, θ, of a flywheel at time t is given by

$$\theta = t^3 - 80\pi t$$

Determine the angular velocity, ω, and angular acceleration, α, at $t = 1$.

9 The angular displacement, θ, of a particle in circular motion is given by

$$\theta = \frac{1}{4}\sin\left(2t + \frac{\pi}{6}\right) \quad (0 \le t \le \pi)$$

  i Find an expression for angular acceleration, α.
  ii Determine the first time t for $\alpha = 0$.

10 The displacement, s, of a particle is given by

$$s = (1 + t)e^{-0.25t}$$

By using a computer algebra system or a graphical calculator, plot, on different axes, the displacement, velocity and acceleration graphs as functions of time t ($0 \le t \le 10$).

SECTION E **Tangents and normals**

By the end of this section you will be able to:
▶ obtain the equation of a tangent at a given point on a curve
▶ find the equation of the normal at a given point on a curve

E1 **Tangents and normals**

Figure 27 shows a curve $y = f(x)$ with a tangent and normal at the point P.

A **tangent** at point P is defined as a line which touches the curve at **only** P.

A **normal** at point P is defined as a line which is at right angles (90°) to the tangent at P.

Fig. 27

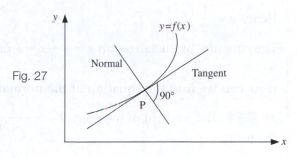

The equation of a tangent can be determined by using the general straight line formula $y = mx + c$, where m is the gradient of the tangent and c is the y-intercept.

? **How can we determine the gradient of a tangent?**

The derivative $\dfrac{dy}{dx}$ gives an expression for the gradient of a tangent.

For points where the gradient (m) of the tangent is not zero the gradient of the normal can be shown to be

7.13   Gradient of normal $= -\dfrac{1}{m}$

Example 15

Find the equation of the tangent and normal to the curve $y = \dfrac{1}{x}$ at $x = 2$.

Solution

Let $y = mx + c$ be the equation of the tangent.

? **What do we need to find?**

The gradient $m \left(= \dfrac{dy}{dx} \right)$ and the y-intercept c.

$$y = \frac{1}{x} = x^{-1}$$

Example 15 *continued*

$$\frac{dy}{dx} = -x^{-2} = -\frac{1}{x^2} \quad \text{[Differentiating]}$$

Substituting $x = 2$, $\dfrac{dy}{dx} = -\dfrac{1}{2^2} = -\dfrac{1}{4}$. Therefore the gradient $m = -\dfrac{1}{4}$.

? **How can we find c?**

We know the tangent goes through the point $\left(2, \dfrac{1}{2}\right)$ because when $x = 2$, $y = \dfrac{1}{x} = \dfrac{1}{2}$.

By substituting $x = 2$, $y = \dfrac{1}{2}$ and $m = -\dfrac{1}{4}$ into $y = mx + c$ we can find c:

$$\frac{1}{2} = \left(-\frac{1}{4} \times 2\right) + c = -\frac{1}{2} + c$$

Hence $c = \dfrac{1}{2} + \dfrac{1}{2} = 1$.

The equation of the tangent is $y = -\dfrac{1}{4}x + 1$ or $y = 1 - \dfrac{1}{4}x$.

? **How can we find the equation of the normal at the point $\left(2, \dfrac{1}{2}\right)$?**

By 7.13, the gradient of the normal $= -\dfrac{1}{-1/4} = 4$. The equation of the normal has the form

$$y = 4x + c_1$$

where c_1 is the y-intercept. Since the normal also goes through the **same** point, $\left(2, \dfrac{1}{2}\right)$,

we can substitute $x = 2$ and $y = \dfrac{1}{2}$ into $y = 4x + c_1$:

$$\frac{1}{2} = (4 \times 2) + c_1, \text{ which implies } \frac{1}{2} = 8 + c_1$$

$$c_1 = \frac{1}{2} - 8 = -7\frac{1}{2} = -\frac{15}{2}$$

Hence the equation of the normal is $y = 4x - \dfrac{15}{2}$.

Be careful not to be confused by the three different x and y relationships:

$$y = 1/x \quad \text{[Given equation]}$$

$$y = 1 - \frac{1}{4}x \quad \left[\text{Tangent equation at } \left(2, \frac{1}{2}\right)\right]$$

$$y = 4x - \frac{15}{2} \quad \left[\text{Normal equation at } \left(2, \frac{1}{2}\right)\right]$$

7.13 Gradient of normal $= -1/m$

Example 16 *electrical principles*

An electric circuit obeys the law

$$P = (10 \times 10^3)i^2$$

where P represents power in W and i represents current in A. Find the equation of the tangent relating P and i for a current $i = 5$ mA (5×10^{-3} A).

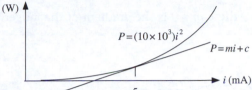

Fig. 28

Solution

Let $P = mi + c$ be the equation of the tangent at

$$i = 5 \text{ mA} = (5 \times 10^{-3}) \text{ A}$$

The gradient is obtained by differentiating $P = (10 \times 10^3)i^2$:

$$m = \frac{dP}{di} = 2 \times (10 \times 10^3)i = (20 \times 10^3)i$$

Substituting $i = 5 \times 10^{-3}$ into $m = (20 \times 10^3)i$ yields

$$m = (20 \times 10^3) \times (5 \times 10^{-3}) = 100 \quad \text{[Gradient]}$$

We have

  $P = 100i + c$

We need to find c. How?

We know the tangent goes through the value of $i = 5 \times 10^{-3}$ (Fig. 28). Substituting $i = 5 \times 10^{-3}$ into the given equation $P = (10 \times 10^3)i^2$:

$$P = (10 \times 10^3) \times (5 \times 10^{-3})^2 = 0.25$$

Hence the tangent goes through the point $(5 \times 10^{-3}, 0.25)$. Substituting $i = 5 \times 10^{-3}$ and $P = 0.25$ into ✱ yields

$$0.25 = (100 \times 5 \times 10^{-3}) + c \text{ giving } c = -0.25$$

Putting $c = -0.25$ into ✱ :

$$P = 100i - 0.25$$

This is the equation of the tangent relating P and i at $i = 5 \times 10^{-3}$ A.

Sometimes it is known as the linear (straight line) relationship between P and i at $i = 5 \times 10^{-3}$ A.

The concept of finding an equation of a tangent is **important** because it gives a linear relationship between the variables at a given point, that is, it simplifies functions in a given neighbourhood.

SUMMARY

The equation of the tangent line to a curve at a given point on that curve is found from the straight line equation

$$y = mx + c$$

SUMMARY continued

where m is the gradient and c is the vertical intercept.
The gradient of the normal is given by

7.13 $\text{gradient} = -\dfrac{1}{m}$

where $m \neq 0$ is the gradient of the tangent.

Exercise 7(e)

Solutions at end of book. Complete solutions available
at www.palgrave.com/engineering/singh

1 Obtain the equations of the tangent and normal of the following curves at the corresponding point:

a $y = x^2 - 3$, $x = 2$

b $y = \cos(x)$, $x = \dfrac{\pi}{2}$

c $y = e^x$, $x = 1$

d $y = \ln(x)$, $x = 1$

2 [electrical principles]   A non-linear resistor obeys the law
$$i = v^3$$

where v and i represent voltage and current respectively. Find the equation of the tangent relating v and i at $v = 1.6$ volts.

3 [electrical principles]   The current, i, through an inductor is given by

$$i = 2(1 - e^{-2000t})$$

Find the equation of the tangent relating i and t at $t = 1 \times 10^{-3}$ s.

SECTION F **Series expansion**

By the end of this section you will be able to:
▶ obtain the Maclaurin series for a given function
▶ understand the Maclaurin series expansion of some common engineering functions
▶ use these established series to evaluate limits of functions
▶ obtain the Taylor series of a given function

This section might seem like a huge jump from previous sections. One of the difficulties in this section is understanding the notation: we use $f'(x)$ to denote differentiation rather than $\dfrac{dy}{dx}$. Moreover, we differentiate a given function a number of times and then substitute $x = 0$ into the resulting derivatives.

F1 General Maclaurin series

> ### Example 17
>
> Find the equation of the tangent at $x = 0$ of $f(x) = e^x$.
>
> ### Solution
>
> Let the equation of the tangent be $y = mx + c$. We need to find the gradient, m, and y-intercept, c. Differentiating $f(x) = e^x$ gives
>
>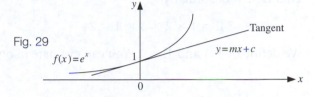
>
> Fig. 29
>
> $$m = f'(x) = e^x$$
>
> Since we want to find the gradient at $x = 0$, we substitute this value
>
> $$m = f'(0) = e^0 = 1$$
>
> By looking at Fig. 29 we have
>
> $$c = e^0 = 1$$
>
> Thus the equation of the tangent is given by substituting $m = 1$ and $c = 1$:
>
> $$y = mx + c = x + 1$$

? The tangent, $y = x + 1$, is **not** a good approximation to $f(x) = e^x$ away from x equals zero. **Why not?**

Because it is **only** accurate at $x = 0$, where it touches the curve, $f(x) = e^x$.

? **Can we adapt the tangent equation $y = x + 1$ to take account of the curved nature of e^x?**

We can add a quadratic term (x^2) to $x + 1$ (Fig. 30). Moreover we can achieve better and better approximations to a curve by adding more and more terms such as a cubic term (x^3), a quartic term (x^4), etc.

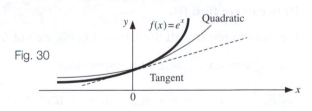

Fig. 30

We can write down the exact function if we consider an infinite number of terms and add these to make what is called an **infinite series.**

In this section we will express general engineering functions by using infinite series (power series). For example, the exponential function used above is given by

†
$$e^x = \underbrace{1 + x}_{\text{tangent}} + \underbrace{\frac{x^2}{2!}}_{\text{quadratic}} + \underbrace{\frac{x^3}{3!}}_{\text{cubic}} + \underbrace{\frac{x^4}{4!}}_{\text{quartic}} + \cdots$$

The sine function is

$$\sin(x) = x - \frac{x^3}{3!} + \frac{x^5}{5!} - \cdots$$

The '!' sign in the series is called factorial. For example, '3!', pronounced '3 factorial', is

$$1 \times 2 \times 3 = 6 \qquad \text{[Product of 1, 2 and 3]}$$

That is $3! = 6$. Similarly

$$4! = 1 \times 2 \times 3 \times 4 = 24$$

We define $0! = 1$ and $1! = 1$. You can evaluate $n!$ on a calculator.

? In this section we aim to determine the infinite series on the Right-Hand Side of equation █†█. **How can we determine the series for a specific function?**

Let's first try to identify the series for a general function $f(x)$ and then find the series for e^x and $\sin(x)$ in the examples.

We write down the general expression of the infinite series as

█*█ $$f(x) = A + Bx + Cx^2 + Dx^3 + Ex^4 + \cdots$$

? where $A,\ B,\ C,\ D,\ E, \ldots$ are constants and $f(x)$ is a general function such that $f(x)$ and its derivatives are defined at the origin and the series sums to a finite value. **What do we need to find?**

? The constants $A,\ B,\ C,\ D,\ E, \ldots$. **How can we find A?**

? By putting $x = 0$ into █*█. **Why?**

Because this will remove all the other terms on the Right-Hand Side of █*█ except A. We have

$$f(0) = A$$

? **How can we find B?**

? We have to remove all the terms of █*█ except B. **How?**

First differentiate █*█ with respect to x:

█**█ $$f'(x) = B + 2Cx + 3Dx^2 + 4Ex^3 + \cdots$$

Now substitute $x = 0$ into █**█:

$$f'(0) = B + 0$$

? Therefore $B = f'(0)$. **How can we find C?**

Differentiate █**█ with respect to x:

█***█ $$f''(x) = 0 + 2C + (3 \times 2)Dx + (4 \times 3)Ex^2 + \cdots$$

Substitute $x = 0$ into ***:

$$f''(0) = 2C + 0$$

$$C = \frac{f''(0)}{2} = \frac{f''(0)}{2!} \qquad \text{[Remember that } 2! = 2]$$

? How can we find D?

Differentiate *** with respect to x:

$$f'''(x) = 0 + (3 \times 2)D + (4 \times 3 \times 2)Ex + \cdots$$

Substitute $x = 0$:

$$f'''(0) = (3 \times 2)D + 0$$

$$D = \frac{f'''(0)}{3 \times 2} = \frac{f'''(0)}{3!} \qquad \text{[Remember that } 3! = 1 \times 2 \times 3]$$

Substituting $A = f(0)$, $B = f'(0)$, $C = \dfrac{f''(0)}{2!}$, $D = \dfrac{f'''(0)}{3!}$ into * gives

$$f(x) = f(0) + f'(0) + x\frac{f''(0)}{2!}x^2 + \frac{f'''(0)}{3!}x^3 + Ex^4 + \cdots$$

? By observing the pattern of the series, what do you think is the value of the constant E?

$$E = \frac{f^{(4)}(0)}{4!}$$

In the general case we have

7.14 $$f(x) = f(0) + f'(0)x + \frac{f''(0)}{2!}x^2 + \frac{f'''(0)}{3!}x^3 + \frac{f^{(4)}(0)}{4!}x^4 + \cdots \quad [f^{(4)} \text{ or } f^{(iv)} \text{ can be used}]$$

7.14 is called the Maclaurin series for the function $f(x)$. This is an infinite series with the $(n + 1)$th term given by

$$\frac{f^{(n)}(0)}{n!}x^n$$

where $f^{(n)}(0)$ represents the value of $\dfrac{d^n y}{dx^n}$ at $x = 0$.

We can only find the Maclaurin series if $f(x)$ and its derivatives are defined at $x = 0$ and the series approaches a finite value.

Colin Maclaurin was born in February 1698 in Argyllshire, Scotland and graduated from Glasgow University. In 1717 he became Professor of Mathematics at the University of Aberdeen and in 1725 he moved to Edinburgh University where he spent the remaining part of his career. The Maclaurin series is a special case, the case of approximating close to zero, of the general Taylor series which is described later in this section (F3).

F2 Maclaurin series of functions

Example 18

Obtain the Maclaurin series expansion of e^x.

Solution

Let $f(x) = e^x$. We need to differentiate this function a number of times to obtain the Maclaurin series and then substitute $x = 0$ into the resulting derivatives. Don't be confused by the f dashed notation.

$$f(x) = e^x \qquad\qquad\qquad\qquad\qquad f(0) = e^0 = 1$$

$$f'(x) = e^x \quad \text{[Differentiating } f] \qquad\qquad f'(0) = 1$$

$$f''(x) = e^x \quad \text{[Differentiating } f'] \qquad\qquad f''(0) = 1$$

$$f'''(x) = e^x \quad \text{[Differentiating } f''] \qquad\qquad f'''(0) = 1$$

$$f^{(4)}(x) = e^x \quad \text{[Differentiating } f'''] \qquad\qquad f^{(4)}(0) = 1$$

Substituting 1 for $f(0)$, $f'(0)$, $f''(0)$, $f'''(0)$, $f^{(4)}(0)$ into ▐ 7.14 ▐ gives

$$e^x = f(0) + f'(0)x + \frac{f''(0)}{2!}x^2 + \frac{f'''(0)}{3!}x^3 + \frac{f^{(4)}(0)}{4!}x^4 + \cdots$$

$$= 1 + (1)x + \frac{1}{2!}x^2 + \frac{1}{3!}x^3 + \frac{1}{4!}x^4 + \cdots \quad \text{[Substituting]}$$

$$= 1 + x + \frac{x^2}{2!} + \frac{x^3}{3!} + \frac{x^4}{4!} + \cdots \quad \text{[Simplifying]}$$

Use a computer algebra package, or a graphical calculator, to plot e^x, and $1 + x$, $1 + x + \dfrac{x^2}{2!}$, $1 + x + \dfrac{x^2}{2!} + \dfrac{x^3}{3!}$ and $1 + x + \dfrac{x^2}{2!} + \dfrac{x^3}{3!} + \dfrac{x^4}{4!}$ on the same axes, and observe how the graphs approach e^x (Fig. 31 on the next page).

By using MAPLE we have

```
> f: = 1 + x;
```
$$f: = 1 + x$$

```
> g:= 1 + x + (x^2)/2!;
```
$$g: = 1 + x + \frac{1}{2}x^2$$

```
> h:= 1 + x + (x^2)/2! + (x^3)/3!;
```
$$h: = 1 + x + \frac{1}{2}x^2 + \frac{1}{6}x^3$$

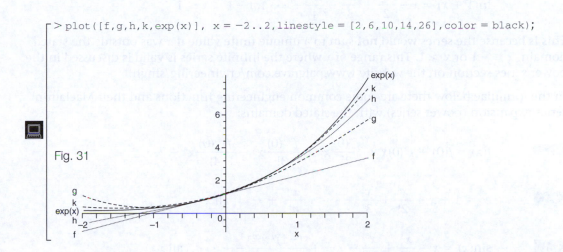

```
> k:=1 + x + (x^2)/2! + (x^3)/3! + (x^4)/4!;
```

$$k := 1 + x + \frac{1}{2}x^2 + \frac{1}{6}x^3 + \frac{1}{24}x^4$$

```
> plot([f,g,h,k,exp(x)], x = -2..2,linestyle = [2,6,10,14,26],color = black);
```

Fig. 31

Example 19

Obtain the Maclaurin series expansion of $\sin(x)$.

Solution

Let $f(x) = \sin(x)$, then differentiating and substituting $x = 0$ into the resulting derivatives gives

$$f(x) = \sin(x) \qquad\qquad f(0) = \sin(0) = 0$$
$$f'(x) = \cos(x) \qquad\qquad f'(0) = \cos(0) = 1$$
$$f''(x) = -\sin(x) \qquad\qquad f''(0) = -\sin(0) = 0$$
$$f'''(x) = -\cos(x) \qquad\qquad f'''(0) = -\cos(0) = -1$$
$$f^{(4)}(x) = \sin(x) \qquad\qquad f^{(4)}(0) = \sin(0) = 0$$
$$f^{(5)}(x) = \cos(x) \qquad\qquad f^{(5)}(0) = \cos(0) = 1$$

Substituting the above values into $\boxed{7.14}$ gives

$$\sin(x) = f(0) + f'(0)x + \frac{f''(0)}{2!}x^2 + \frac{f'''(0)}{3!}x^3 + \frac{f^{(4)}(0)}{4!}x^4 + \frac{f^{(5)}(0)}{5!}x^5 + \cdots$$

$$= 0 + (1)x + \frac{0}{2!}x^2 + \frac{(-1)}{3!}x^3 + \frac{0}{4!}x^4 + \frac{1}{5!}x^5 + \cdots \qquad [\text{Substituting}]$$

$$= x - \frac{x^3}{3!} + \frac{x^5}{5!} - \cdots \qquad [\text{Simplifying}]$$

The infinite series for e^x and $\sin(x)$ are evaluated in the above examples and are valid for all values of x. We have placed no restriction on x. However in some cases the infinite series may only be valid for a specific range of x values, for example

$$\ln(1 + x) = x - \frac{x^2}{2} + \frac{x^3}{3} - \frac{x^4}{4} + \cdots \text{ for } -1 < x \le 1$$

This is because the series would **not** sum to a unique finite value if x was outside the stated domain, $x \le -1$ or $x > 1$. This range of x where the infinite series is valid is discussed in the power series section on the website www.palgrave.com/engineering/singh.

In the formulae below there are some common engineering functions and their Maclaurin series expansion (power series) with the stated domains:

7.14
$$f(x) = f(0) + f'(0)x + \frac{f''(0)}{2!}x^2 + \frac{f'''(0)}{3!}x^3 + \frac{f^{(4)}(0)}{4!}x^4 + \cdots$$

7.15
$$e^x = 1 + x + \frac{x^2}{2!} + \frac{x^3}{3!} + \cdots + \frac{x^n}{n!} + \cdots \quad \text{all } x$$

7.16
$$\sin(x) = x - \frac{x^3}{3!} + \frac{x^5}{5!} - \cdots + \frac{(-1)^n x^{2n+1}}{(2n + 1)!} + \cdots \quad \text{all } x$$

7.17
$$\cos(x) = 1 - \frac{x^2}{2!} + \frac{x^4}{4!} - \cdots + \frac{(-1)^n x^{2n}}{(2n)!} + \cdots \quad \text{all } x$$

7.18
$$\tan(x) = x + \frac{x^3}{3} + \frac{2x^5}{15} + \frac{17x^7}{315} + \cdots \quad -\pi/2 < x < \pi/2$$

7.19
$$\sinh(x) = x + \frac{x^3}{3!} + \frac{x^5}{5!} + \cdots + \frac{x^{2n+1}}{(2n + 1)!} + \cdots \quad \text{all } x$$

7.20
$$\cosh(x) = 1 + \frac{x^2}{2!} + \frac{x^4}{4!} + \cdots + \frac{x^{2n}}{(2n)!} + \cdots \quad \text{all } x$$

7.21
$$\ln(1 + x) = x - \frac{x^2}{2} + \frac{x^3}{3} - \frac{x^4}{4} + \cdots + \frac{(-1)^{n+1} x^n}{n} \cdots \quad -1 < x \le 1$$

(The x for the trigonometric functions is in radians.)

? **Why do we need these infinite series for functions such as sin, cos, ln,…?**

Because electronic devices such as calculators and computers use these series to find such things as cos(1).

? **For example, how does a calculator or computer find cos(1), $\sqrt{2.2}$ and sin(2)?**

Electronic devices **cannot** store the infinite number of answers in their memory chip but tend to use infinite series such as Maclaurin or Taylor (to be discussed later). For example, to find cos(1) the calculator will use the Maclaurin series expansion of cosine   7.17   given above:

$$\cos(x) = 1 - \frac{x^2}{2!} + \frac{x^4}{4!} - \frac{x^6}{6!} + \frac{x^8}{8!} - \cdots$$

$$\cos(1) = 1 - \frac{1^2}{2!} + \frac{1^4}{4!} - \frac{1^6}{6!} + \frac{1^8}{8!} - \cdots \qquad \text{[Substituting } x = 1\text{]}$$

$$= 1 - \frac{1}{2} + \frac{1}{24} - \frac{1}{720} - \frac{1}{40320} \cdots = 0.540303 \text{ (6 d.p.)}$$

If you evaluate cos(1) on a calculator you get 0.540302 (6 d.p.).

Remember the Maclaurin series of cosine is an infinite series but calculators **cannot** use the infinite number of terms. **When does it stop?**

That depends on the accuracy of the calculator. If the calculator has 6 decimal place accuracy then it takes enough terms of the infinite series in order to give an answer correct to 6 d.p.

We can also use these series to evaluate limits of functions, as the following example shows.

Example 20

Determine $\lim\limits_{x\to 0} \dfrac{\sin(x)}{x}$.

Solution

Substituting $x = 0$, the function $\dfrac{\sin(x)}{x}$ gives $\dfrac{0}{0}$ which is not defined. **How can we find the value of** $\dfrac{\sin(x)}{x}$ close to 0?

Using 7.16 and substituting the series expansion for sin(x) we have

$$\frac{\sin(x)}{x} = \frac{x - \dfrac{x^3}{3!} + \dfrac{x^5}{5!} - \cdots}{x}$$

$$= \frac{x\left(1 - \dfrac{x^2}{3!} + \dfrac{x^4}{5!} - \cdots\right)}{x} \qquad \text{[Factorizing numerator]}$$

$$= 1 - \frac{x^2}{3!} + \frac{x^4}{5!} - \cdots \qquad \text{[Cancelling } x\text{'s]}$$

Hence

$$\lim_{x\to 0} \frac{\sin(x)}{x} = \lim_{x\to 0}\left(1 - \frac{x^2}{3!} + \frac{x^4}{5!} - \cdots\right) = 1$$

F3 Taylor series of functions

From the previous section we know we can obtain a Maclaurin series for basic engineering and science functions such as sin(x), cos(x) and e^x. **Why can't we find a Maclaurin series for the logarithmic function ln(x)?**

This is because ln(x) is **not** defined at $x = 0$. Recall the graph of ln(x), shown in Fig. 32:

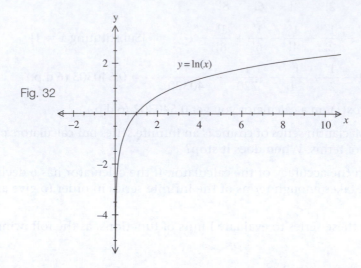

Fig. 32

ln(x) is **not** defined for any negative or zero x values. This means that we **cannot** find a Maclaurin series for ln(x) because the Maclaurin series is defined for functions that can be differentiated at $x = 0$. **Can we find a series representation for ln(x) at another point, say $x = a$?**

Yes and such a series is called a **Taylor series.**
The Taylor series has the same pattern as the Maclaurin series 7.14 because it is derived in the same manner. The Taylor series for a function $f(x)$ about a point $x = a$ is given by:

$$\boxed{7.22} \quad f(x) = f(a) + f'(a)(x - a) + \frac{f''(a)}{2!}(x - a)^2 + \frac{f'''(a)}{3!}(x - a)^3 + \cdots$$

We can only find the Taylor series if $f(x)$ and its derivatives are defined at $x = a$ and the infinite series approaches a finite value. **What happens at $a = 0$?**

Substituting $a = 0$ into the above Taylor series 7.22 gives the Maclaurin series.

Example 21

Obtain the Taylor series for ln(x) about the point $x = 2$.

Solution

Let $f(x) = \ln(x)$. **How do we find the Taylor series for $f(x) = \ln(x)$ about $x = 2$?**
We need to differentiate ln(x) and then substitute $x = 2$ into our result:

$$f(x) = \ln(x) \qquad\qquad\qquad f(2) = \ln(2)$$

$$f'(x) = \frac{1}{x} = x^{-1} \quad \text{[Differentiating]} \quad f'(2) = \frac{1}{2}$$

Example 21 *continued*

$$f''(x) = -x^{-2} = -\frac{1}{x^2} \qquad \text{[Differentiating]} \quad f''(2) = -\frac{1}{2^2} = -\frac{1}{4}$$

$$f'''(x) = 2x^{-3} = \frac{2}{x^3} \qquad \text{[Differentiating]} \quad f'''(2) = \frac{2}{2^3} = \frac{1}{4}$$

$$f^{(4)}(x) = -6x^{-4} = -\frac{6}{x^4} \qquad \text{[Differentiating]} \quad f^{(4)}(2) = -\frac{6}{2^4} = -\frac{3}{8}$$

Substituting these values into the general Taylor series

7.22 $$f(x) = f(a) + f'(a)(x - a) + \frac{f''(a)}{2!}(x - a)^2 + \frac{f'''(a)}{3!}(x - a)^3 + \cdots$$

with $a = 2$ and $f(x) = \ln(x)$ we have

$$\ln(x) = f(2) + f'(2)(x - 2) + \frac{f''(2)}{2!}(x - 2)^2 + \frac{f'''(2)}{3!}(x - 2)^3 + \frac{f^{(4)}(2)}{4!}(x - 2)^4 + \cdots$$

$$= \ln(2) + \frac{1}{2}(x - 2) + \frac{(-1/4)}{2!}(x - 2)^2 + \frac{1/4}{3!}(x - 2)^3 + \frac{(-3/8)}{4!}(x - 2)^4 + \cdots$$

$$= \ln(2) + \frac{(x - 2)}{2} - \frac{(x - 2)^2}{8} + \frac{(x - 2)^3}{24} - \frac{(x - 2)^4}{64} + \cdots \quad \text{[Simplifying]}$$

We can visualize some of the approximations to $\ln(x)$ near $x = 2$. Next we plot the first few terms of the series found in **Example 21**. By using a computer algebra package or graphical calculator, plot the following graphs on the same axes:

$\ln(x)$, $\ln(2) + \dfrac{(x - 2)}{2}$ (linear approximation), $\ln(2) + \dfrac{(x - 2)}{2} - \dfrac{(x - 2)^2}{8}$

(quadratic approximation) and

$\ln(2) + \dfrac{(x - 2)}{2} - \dfrac{(x - 2)^2}{8} + \dfrac{(x - 2)^3}{24}$ (cubic approximation).

We obtain Fig. 33:

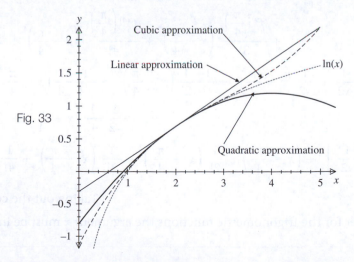

Fig. 33

Note that as we increase the number of terms we get better approximations to $\ln(x)$. Again if we consider the infinite series from **Example 21** then we can say that they are equal.

Example 22

Obtain the Taylor series for $\sin(x)$ about the point $x = \dfrac{\pi}{4}$.

Solution

Let $f(x) = \sin(x)$, then we have

$$f(x) = \sin(x) \qquad\qquad f\left(\frac{\pi}{4}\right) = \sin\left(\frac{\pi}{4}\right) = \frac{1}{\sqrt{2}}$$

$$f'(x) = \cos(x) \quad \text{[Differentiating]} \qquad f'\left(\frac{\pi}{4}\right) = \cos\left(\frac{\pi}{4}\right) = \frac{1}{\sqrt{2}}$$

$$f''(x) = -\sin(x) \quad \text{[Differentiating]} \qquad f''\left(\frac{\pi}{4}\right) = -\sin\left(\frac{\pi}{4}\right) = -\frac{1}{\sqrt{2}}$$

$$f'''(x) = -\cos(x) \quad \text{[Differentiating]} \qquad f'''\left(\frac{\pi}{4}\right) = -\cos\left(\frac{\pi}{4}\right) = -\frac{1}{\sqrt{2}}$$

$$f^{(4)}(x) = \sin(x) \quad \text{[Differentiating]} \qquad f^{(4)}\left(\frac{\pi}{4}\right) = \sin\left(\frac{\pi}{4}\right) = \frac{1}{\sqrt{2}}$$

Substituting these values into the Taylor series

$$7.22 \quad f(x) = f(a) + f'(a)(x - a) + \frac{f''(a)}{2!}(x - a)^2 + \frac{f'''(a)}{3!}(x - a)^3 + \cdots$$

with $f(x) = \sin(x)$ and $a = \dfrac{\pi}{4}$ we have

$$\sin(x) = f\left(\frac{\pi}{4}\right) + f'\left(\frac{\pi}{4}\right)\left(x - \frac{\pi}{4}\right) + \frac{f''\left(\frac{\pi}{4}\right)}{2!}\left(x - \frac{\pi}{4}\right)^2$$

$$+ \frac{f'''\left(\frac{\pi}{4}\right)}{3!}\left(x - \frac{\pi}{4}\right)^3 + \frac{f^{(4)}\left(\frac{\pi}{4}\right)}{4!}\left(x - \frac{\pi}{4}\right)^4 + \cdots$$

$$= \frac{1}{\sqrt{2}} + \frac{1}{\sqrt{2}}\left(x - \frac{\pi}{4}\right) - \frac{1}{\sqrt{2}}\frac{1}{2!}\left(x - \frac{\pi}{4}\right)^2 - \frac{1}{\sqrt{2}}\frac{1}{3!}\left(x - \frac{\pi}{4}\right)^3$$

$$+ \frac{1}{\sqrt{2}}\frac{1}{4!}\left(x - \frac{\pi}{4}\right)^4 + \cdots$$

$$= \frac{1}{\sqrt{2}}\left[1 + \left(x - \frac{\pi}{4}\right) - \frac{1}{2!}\left(x - \frac{\pi}{4}\right)^2 - \frac{1}{3!}\left(x - \frac{\pi}{4}\right)^3 + \frac{1}{4!}\left(x - \frac{\pi}{4}\right)^4 + \cdots\right]$$

[Taking out the common factor $1/\sqrt{2}$]

Remember for the trigonometric functions the argument x must be in **radians**.

SUMMARY

We can find the Maclaurin series of a function, $f(x)$, which has its derivatives defined at $x = 0$ and the series sums to a finite value by

$$\boxed{7.14} \qquad f(x) = f(0) + f'(0)x + \frac{f''(0)}{2!}x^2 + \frac{f'''(0)}{3!}x^3 + \frac{f^{(4)}(0)}{4!}x^4 + \cdots$$

We can use the established series to evaluate limits of functions.
The Taylor series for a function $f(x)$ about a point $x = a$ is given by

$$\boxed{7.22} \qquad f(x) = f(a) + f'(a)(x - a) + \frac{f''(a)}{2!}(x - a)^2 + \frac{f'''(a)}{3!}(x - a)^3 + \cdots$$

Exercise 7(f)

Solutions at end of book. Complete solutions available at www.palgrave.com/engineering/singh

1 Obtain the Maclaurin series expansion of $\cos(x)$ by using $\boxed{7.14}$.

2 Obtain the Maclaurin series expansion of $\ln(1 + x)$ by applying $\boxed{7.14}$ and where $-1 < x \le 1$.

3 By using the Maclaurin series, show that

$$(1 + x)^n = 1 + nx + \left[\frac{n(n - 1)}{2!} \right] x^2$$

$$+ \left[\frac{n(n - 1)(n - 2)}{3!} \right] x^3 + \cdots$$

$$(-1 < x < 1)$$

where n is any fraction or a negative number.

4 Find the Maclaurin series expansion of $\sin^2(x)$ and $\cos^2(x)$.
[*Hint*: Use a trigonometrical relation for $\cos^2(x)$ and substitute $2x$ into a known series.]

5 Determine

a $\displaystyle\lim_{x \to 0} \frac{\ln(1 + x)}{x}$ **b** $\displaystyle\lim_{x \to 0} \frac{1 - \cos(x)}{x^2}$

6 Determine the Taylor series for $f(x)$ at the given values of x.

a $f(x) = \ln(x)$ at $x = 1$

b $f(x) = \dfrac{1}{x}$ at $x = 1$

c $f(x) = e^x$ at $x = 3$

d $f(x) = \cos(x)$ at $x = \dfrac{\pi}{4}$

e $f(x) = \ln(x)$ at $x = e$

7 Find the Maclaurin series expansion of

$$e^{-\frac{1}{2}x^2}$$

*8 Determine the Maclaurin series of

$$\frac{1}{2} \ln \left(\frac{1 + x}{1 - x} \right)$$

SECTION G **Binomial revisited**

By the end of this section you will be able to:
▶ expand $(a + b)^n$ where n is a positive whole number
▶ expand $(1 + x)^n$ where n is a fraction or a negative whole number
▶ apply the binomial expansion to engineering examples

G1 **Expanding $(a + b)^n$**

We have covered binomial expansion (**Section 2F**) in **Chapter 2,** where we expanded terms like $(1 + x)^5$ by using Pascal's triangle.

? **However you would *not* expand $(1 + x)^{15}$ by using Pascal's triangle. Why not?**

Evaluating Pascal's triangle for $n = 15$ is a tedious task, so we need to examine some other technique.

Expansion of the general term $(a + b)^n$ is given by

7.23 $(a + b)^n = a^n + na^{n-1} b + \left[\dfrac{n(n - 1)}{2!} \right] a^{n-2} b^2$

$+ \left[\dfrac{n(n - 1)(n - 2)}{3!} \right] a^{n-3} b^3 + \cdots + b^n$

7.23 looks like an horrendous formula but once it has been applied a number of times you will become confident in using it. Also it might be easier to apply the formula if you notice the patterns that it contains.

The index of a starts with n and decreases by 1 in each term to the right:

$a^n, \ a^{n-1}, \ a^{n-2}, \ \cdots, \ a^1 = a, \ a^0 = 1$

Similarly the index of b increases by 1 with each term to the right:

$b^0 = 1, \ b^1 = b, \ b^2, b^3, \cdots, \ b^n$

Moreover notice the coefficients of a and b in the middle terms:

$n, \ \dfrac{n(n - 1)}{2!}, \ \dfrac{n(n - 1)(n - 2)}{3!}, \ \dfrac{n(n - 1)(n - 2)(n - 3)}{4!}, \ \cdots$

Example 23

Find the first four terms in the expansion of $(1 + x)^{15}$.

Solution

Substituting $a = 1$, $b = x$ and $n = 15$ into 7.23 gives

$(1 + x)^{15} = 1^{15} + 15(1)^{14} x + \left[\dfrac{15(15 - 1)}{2!} \right] (1)^{13} x^2$

$+ \left[\dfrac{15(15 - 1)(15 - 2)}{3!} \right] (1)^{12} x^3 + \cdots$

$= 1 + 15x + \left[\dfrac{15(14)}{2!} \right] x^2 + \left[\dfrac{15(14)(13)}{3!} \right] x^3 + \cdots$ [Remember that $1^{index} = 1$]

$= 1 + 15x + 105x^2 + 455x^3 + \cdots$ [Evaluating brackets]

Example 24

Determine up to and including the x^3 term in the expansion of $(5 - x)^{10}$.

Solution

Putting $a = 5$, $b = -x$ and $n = 10$ into 7.23 gives

$$(5 - x)^{10} = [5 + (-x)]^{10} = 5^{10} + [10 \times 5^9 \times (-x)] + \left[\frac{10(10-1)}{2!}\right]5^8(-x)^2$$

$$+ \left[\frac{10(10 - 1)(10 - 2)}{3!}\right]5^7(-x)^3 + \cdots$$

$$= 5^{10} + [10 \times 5^9](-x) + \left[\frac{10(9)}{2!} \times 5^8\right](-x)^2$$

$$+ \left[\frac{10(9)(8)}{3!} \times 5^7\right](-x)^3 + \cdots$$

$$\underset{\text{by calculator}}{=} 9765625 - 19531250x + 17578125x^2 - 9375000x^3 + \cdots$$

Notice how the numbers go down one step at a time on the numerator: 10, 9, 8 etc.

G2 **Binomial series**

In question **3** of **Exercise 7(f)** you were asked to show the following result by using the Maclaurin series:

7.24 $$(1 + x)^n = 1 + nx + \left[\frac{n(n - 1)}{2!}\right]x^2 + \left[\frac{n(n - 1)(n - 2)}{3!}\right]x^3 + \cdots \qquad -1 < x < 1$$

where n is not a positive whole number. 7.24 is called the binomial theorem and was Newton's first great discovery. If n is a positive whole number then the series terminates after $n + 1$ terms, as can be seen in 7.23 .

? **How many terms does $(1 + x)^{15}$ have?**

16

? **How many terms does $(5 - x)^{10}$ have?**

11

Example 25

Show that $\dfrac{1}{1+x} = 1 - x + x^2 - x^3 + x^4 - \cdots$ where $-1 < x < 1$.

Solution

$\dfrac{1}{1+x} = (1+x)^{-1}$. Substituting $n = -1$ into 7.24 gives

$$(1+x)^{-1} = 1 + (-1)x + \left[\frac{(-1)(-2)}{2!}\right]x^2 + \left[\frac{(-1)(-2)(-3)}{3!}\right]x^3$$
$$+ \left[\frac{(-1)(-2)(-3)(-4)}{4!}\right]x^4 \cdots$$
$$= 1 - x + x^2 - x^3 + x^4 \cdots \qquad \text{[Cancelling]}$$

Note that this is an infinite series.

We can visualize these approximations to $\dfrac{1}{1+x}$.

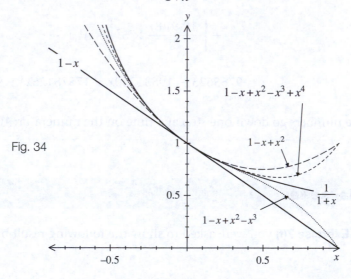

Fig. 34

Figure 34 shows $1-x$ (linear approximation), $1-x+x^2$ (quadratic approximation), $1-x+x^2-x^3$ (cubic approximation) and $1-x+x^2-x^3+x^4$ (quartic approximation) to $\dfrac{1}{1+x}$. Adding more terms will give better approximations to the given function $\dfrac{1}{1+x}$.

Example 26

Expand $(1+x)^{1/2}$.

Solution

Substituting $n = \dfrac{1}{2}$ into 7.24 yields

Example 26 *continued*

$$(1 + x)^{1/2} = 1 + \frac{1}{2}x + \left[\frac{\frac{1}{2}\left(\frac{1}{2} - 1\right)}{2!}\right]x^2 + \left[\frac{\frac{1}{2}\left(\frac{1}{2} - 1\right)\left(\frac{1}{2} - 2\right)}{3!}\right]x^3 + \cdots$$

$$= 1 + \frac{1}{2}x + \left[\frac{\frac{1}{2}\left(-\frac{1}{2}\right)}{2}\right]x^2 + \left[\frac{\frac{1}{2}\left(-\frac{1}{2}\right)\left(-\frac{3}{2}\right)}{6}\right]x^3 + \cdots$$

$$= 1 + \frac{1}{2}x - \frac{1}{8}x^2 + \frac{3}{48}x^3 + \cdots \qquad \text{[Evaluating brackets]}$$

$$= 1 + \frac{1}{2}x - \frac{1}{8}x^2 + \frac{1}{16}x^3 + \cdots \qquad \left[\text{Because } \frac{3}{48} = \frac{1}{16}\right]$$

This is also an infinite series.

G3 **Engineering applications**

Note that if n is a whole number in the expansion of $(1 + x)^n$ then we have a finite series, and if n is a fraction or a negative whole number then we have an infinite series.

The next example is very challenging. It involves using a number of algebraic techniques to simplify the result of applying 7.24 . As mentioned earlier, algebra is the bread and butter of calculus.

Example 27 *structures*

Figure 35a shows a cable with span L and sag h. The length of the cable can be determined by integrating the term

$$\left[1 + \left(\frac{8hx}{L^2}\right)^2\right]^{1/2} \qquad \text{Fig. 35a}$$

where x is the horizontal distance from the lowest point of the cable. Use the binomial expansion to expand the above expression up to and including the x^6 term, and state the range for which the expression is valid.

Solution

Squaring the bracket term:

$$\left(\frac{8hx}{L^2}\right)^2 = \frac{8^2h^2x^2}{(L^2)^2} = \frac{64h^2x^2}{L^4}$$

Substituting $n = \frac{1}{2}$ into 7.24 yields

Example 27 *continued*

$$\left[1 + \frac{64h^2x^2}{L^4}\right]^{1/2} = 1 + \frac{1}{2}\left(\frac{64h^2x^2}{L^4}\right) + \left[\frac{\frac{1}{2}\left(\frac{1}{2} - 1\right)}{2!}\right]\left(\frac{64h^2x^2}{L^4}\right)^2$$

$$+ \left[\frac{\frac{1}{2}\left(\frac{1}{2} - 1\right)\left(\frac{1}{2} - 2\right)}{3!}\right]\left(\frac{64h^2x^2}{L^4}\right)^3 \cdots$$

$$= 1 + \frac{1}{2}\left(\frac{64h^2x^2}{L^4}\right) + \left[\frac{\frac{1}{2}\left(-\frac{1}{2}\right)}{2}\right]\left(\frac{64^2(h^2)^2(x^2)^2}{(L^4)^2}\right)$$

$$+ \left[\frac{\frac{1}{2}\left(-\frac{1}{2}\right)\left(-\frac{3}{2}\right)}{6}\right]\left(\frac{64^3(h^2)^3(x^2)^3}{(L^4)^3}\right) \cdots$$

$$= 1 + \frac{64}{2}\frac{h^2x^2}{L^4} + \left[-\frac{1}{8}\right]64^2\frac{h^4x^4}{L^8} + \left[\frac{3}{48}\right]64^3\frac{h^6x^6}{L^{12}} \cdots$$
[Using indices]

$$= 1 + 32\frac{h^2x^2}{L^4} - 512\frac{h^4x^4}{L^8} + 16384\frac{h^6x^6}{L^{12}} \cdots \quad \text{[Simplifying]}$$

where $-1 < \dfrac{64h^2x^2}{L^4} < 1$

SUMMARY

For a positive whole number, n, we have a finite series:

7.23 $(a + b)^n = a^n + na^{n-1}b + \left[\dfrac{n(n-1)}{2!}\right]a^{n-2}b^2$

$$+ \left[\frac{n(n-1)(n-2)}{3!}\right]a^{n-3}b^3 + \cdots + b^n$$

If n is a fraction or a negative whole number we have an infinite series:

7.24 $(1 + x)^n = 1 + nx + \left[\dfrac{n(n-1)}{2!}\right]x^2 + \left[\dfrac{n(n-1)(n-2)}{3!}\right]x^3 + \cdots \quad -1 < x < 1$

Exercise 7(g)

Solutions at end of book. Complete solutions available at www.palgrave.com/engineering/singh

1 Find the first four terms in the binomial expansion of

a $(1 + x)^{20}$ **b** $(x + y)^{15}$

c $(2 - Z)^{10}$

2 Expand $\dfrac{1}{1 - Z}$ where $-1 < Z < 1$.

3 [structures] The length of a part of a cable of a suspension bridge can be found by integrating the term

$$\left[1 + \left(\frac{Wx}{T} \right)^2 \right]^{\frac{1}{2}}$$

where W is load per unit length, T is tension and x is distance along the bridge. Use the binomial expansion to expand the above expression up to and including the x^6 term, and state the range for which the expansion is valid.

4 i Show that

$$\sqrt{1 - x} = 1 - \frac{1}{2}x - \frac{1}{8}x^2$$
$$- \frac{1}{16}x^3 - \frac{5}{128}x^4 - \cdots$$

where $-1 < x < 1$

ii By writing $7 = 9\left(1 - \frac{2}{9} \right)$ determine an approximation to $\sqrt{7}$.

5 [structures] Figure 35b shows part of a suspension bridge cable where ℓ = total span and s = sag. The maximum tension, T, in the cable is given by

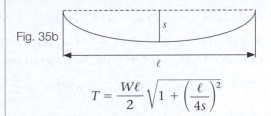

Fig. 35b

$$T = \frac{W\ell}{2} \sqrt{1 + \left(\frac{\ell}{4s} \right)^2}$$

where W is the load per unit length. Show that

$$T = \frac{W\ell}{2} + \frac{W\ell^3}{64s^2} - \frac{W\ell^5}{4096s^4} + \frac{W\ell^7}{131072s^6} \cdots$$

$$\left[\textit{Hint: Use the substitution } x = \frac{\ell}{4s} \right]$$

*6 [fluid mechanics] In fluid flow the pressure ratio P (between a stagnation point and free stream) is given by

$$P = \left(1 + \frac{1}{2}(\gamma - 1)M^2 \right)^{\frac{\gamma}{\gamma - 1}}$$

where γ is the specific heat ratio and M is the Mach number. Show that

$$P = 1 + \gamma \left(\frac{M^2}{2} + \frac{M^4}{8} + \frac{2 - \gamma}{48}M^6 + \cdots \right)$$

SECTION H **An introduction to infinite series**

By the end of this section you will be able to:
- ► understand what is meant by convergence and divergence of infinite series
- ► recognize a geometric series and test it for convergence
- ► test a series for convergence by applying the ratio test

You will need to ensure you are familiar with the concept of limits of functions (**Chapter 3**) to understand some of the examples in this section.

H1 Introduction

A series is of the form $a_1 + a_2 + a_3 + a_4 + \cdots$ where a_n is called the **nth term**.

For example, $1 + 2 + 3 + 4 + \cdots$ is a series with terms 1, 2, 3, etc. We use the Greek letter $\sum$, pronounced sigma, for writing the series in compact form. For example

$$1 + 2 + 3 + 4 + \cdots \text{ is written as } \sum_{n=1}^{\infty} n, \text{ that is, } 1 + 2 + 3 + 4 + \cdots = \sum_{n=1}^{\infty} n$$

In words, this $\sum_{n=1}^{\infty} n$ notation means 'the sum of all positive whole numbers between 1 and infinity'. **How would you write the following in $\sum$, sigma, notation?**

1 $1 + \dfrac{1}{2} + \dfrac{1}{3} + \dfrac{1}{4} + \dfrac{1}{5} + \cdots$

2 $1 + 2 + 4 + 8 + 16 + \cdots$

3 $\dfrac{1}{2} + \dfrac{2}{3} + \dfrac{3}{4} + \dfrac{4}{5} + \cdots$

1 can be written as $1 + \dfrac{1}{2} + \dfrac{1}{3} + \dfrac{1}{4} + \dfrac{1}{5} + \cdots = \sum_{n=1}^{\infty} \left[\dfrac{1}{n} \right]$

2 can be written as $1 + 2 + 4 + 8 + 16 + \cdots = \sum_{n=0}^{\infty} (2^n)$ [Start at $n = 0$ because $2^0 = 1$]

3 can be written as $\dfrac{1}{2} + \dfrac{2}{3} + \dfrac{3}{4} + \dfrac{4}{5} + \cdots = \sum_{n=1}^{\infty} \left(\dfrac{n}{n+1} \right)$

1, 2 and **3** are all examples of **infinite** series. **What does *infinite series* mean?**

A series which has an unlimited number of terms is called an infinite series. **What does *finite series* mean?**

A series which has a finite number of terms is called a **finite** series, for example $1 + 2 + 3$.

In this chapter we consider infinite series and ask **if we add an infinite number of terms, can we get a finite answer?**

Yes, provided the infinite series converges. **What does *convergent series* mean?**

It means that the sum of the infinite series $a_1 + a_2 + a_3 + \cdots + a_n + a_{n+1} + \cdots$ tends to a unique finite value. An example of an infinite series that can be added together to get a finite answer is

$$0.3 + 0.03 + 0.003 + 0.0003 + \cdots$$

What does this equal?

Clearly this is 0.3333... which is equivalent to 1/3. We can also write this series in fractions:

$$\frac{3}{10} + \frac{3}{100} + \frac{3}{1000} + \frac{3}{10000} + \cdots = \frac{1}{3}$$

In general, an infinite series is convergent if

$$\sum_{n=1}^{\infty} a_n = a_1 + a_2 + a_3 + \cdots + a_n + \cdots = S$$

where S is a real number. In this case we say that as n goes to infinity, then the series **converges** to S. S is also called the sum of the series.

For example

$$\frac{1}{5} + \frac{1}{5^2} + \frac{1}{5^3} + \frac{1}{5^4} + \cdots = \frac{1}{4}$$

This means adding infinitely many terms $\frac{1}{5}$, $\frac{1}{5^2}$, $\frac{1}{5^3}$, $\frac{1}{5^4}$, etc. gives the sum of $\frac{1}{4}$.

In Table 1 below we have evaluated the sum in the right-hand column for the corresponding number of terms in the left-hand column. We have evaluated our final summation to 10 d.p.:

	n	$\sum_{k=1}^{n} \left(\frac{1}{5^k} \right)$	$\left[\text{Sum of } \frac{1}{5^k} \text{ from } k = 1 \text{ to } n \right]$
TABLE 1	1		0.2000000000
	5		0.2499200000
	10		0.2499999744
	20		0.2500000000

Note that at $n = 20$, the sum is **approximately** 0.25, or $\frac{1}{4}$. We need to be careful because the sum of the first 20 terms is not 0.25, but something a tiny bit smaller. (Remember we have only given the summation to 10 d.p.)

We can sum a finite number of the terms by using a computer algebra system. If the sum does **not** approach a unique real number as n goes to infinity, then the series is said to **diverge**.

? For example, the harmonic series, $\sum_{k=1}^{\infty} \left(\frac{1}{k} \right) = 1 + \frac{1}{2} + \frac{1}{3} + \frac{1}{4} + \cdots$ diverges. **What does this mean?**

If we take enough terms of the series, then we can make the sum as large as we like.

	n	$\sum_{k=1}^{n} \left(\frac{1}{k} \right)$	[Sum of $\frac{1}{k}$ from $k=1$ to n]
TABLE 2	1		1
	10		2.928968254
	100		5.187377518
	1 000		7.485470861
	1 000 000		14.39272672

In Table 2 above the left-hand column is the number of terms (n) and the right-hand column gives the corresponding sum. That is, the sum of the first 100 terms of the series $\sum_{k=1}^{n} \left(\frac{1}{k} \right)$ is just over 5 and the sum of the first million terms (1 000 000) is about 14.4.

? **Is** $\sum_{n=1}^{\infty} \left(\frac{1}{2^n} \right)$ **a convergent series and if it is what does it converge to?**

? Actually we will show that $\sum_{n=1}^{\infty} \left(\frac{1}{2^n} \right) = 1$. **What does this** $\sum_{n=1}^{\infty} \left(\frac{1}{2^n} \right) = 1$ **mean?**

$\frac{1}{2} + \frac{1}{2^2} + \frac{1}{2^3} + \frac{1}{2^4} + \cdots = 1$. Adding infinitely many terms of the form $\frac{1}{2^n}$ gives the sum equal to 1. This is a **geometric series** which is discussed below.

We can demonstrate this as shown.

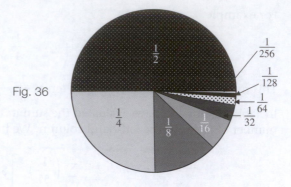

Fig. 36

Figure 36 illustrates the geometric series

$$\frac{1}{2} + \frac{1}{4} + \frac{1}{8} + \frac{1}{16} + \ldots = \frac{1}{2} + \frac{1}{2^2} + \frac{1}{2^3} + \frac{1}{2^4} + \ldots = 1.$$

The following is an important result for testing divergence of series:

7.25   If $\lim_{n \to \infty} (a_n) \neq 0$ [Not zero] then $\Sigma(a_n)$ diverges

? **What does this mean?**

If the terms do **not** tend to zero as $n \to \infty$ then the series **diverges**.

For example, $1 + 2 + 3 + 4 + \ldots$ diverges because the nth term, n, does **not** tend to zero.

However, the reverse is not true. We cannot assume a series **converges** because its nth term converges to zero. For example, the above harmonic series $\sum_{n=1}^{\infty} \left(\frac{1}{n}\right) = 1 + \frac{1}{2} + \frac{1}{3} + \frac{1}{4} + \ldots$ diverges while the nth term $\frac{1}{n}$ tends to zero as $n \to \infty$.

To get a feel for a series it is good practice to write out the first few terms of the series.

H2 Geometric series

? The general geometric series is given by $\sum_{n=1}^{\infty} ar^{n-1}$. **What are the first few terms of the series?**

$$\sum_{n=1}^{\infty} ar^{n-1} = ar^{1-1} + ar^{2-1} + ar^{3-1} + ar^{4-1} + ar^{5-1} + \ldots + ar^{n-1} + \ldots \qquad (a \neq 0)$$

$$= a + ar + ar^2 + ar^3 + \ldots + ar^{n-1} + \ldots \qquad [\text{Remember } r^0 = 1]$$

? **What do we mean by a *geometric* series?**

A geometric series is an infinite series in which each term is obtained from the preceding term by multiplying by r. For example the second term, ar, is obtained from the first term, a, by multiplying it by a constant r. The symbol r is called the **common ratio** and a is called the **first term**.

? What is r and what is a equal to for the above series $\dfrac{1}{2} + \dfrac{1}{2^2} + \dfrac{1}{2^3} + \dfrac{1}{2^4} + \cdots$?

Common ratio $r = \dfrac{1}{2}$ and first term $a = \dfrac{1}{2}$.

The geometric series is the infinite series (given above) defined as

7.26 $\displaystyle\sum_{n=1}^{\infty} ar^{n-1} = a + ar + ar^2 + ar^3 \cdots + ar^{n-1} + \cdots \qquad (a \neq 0)$

It can be shown that the geometric series, $\displaystyle\sum_{n=1}^{\infty} ar^{n-1}$, **converges** for $|r| < 1$ (less than 1) and the sum is

7.27 $\displaystyle\sum_{n=1}^{\infty} ar^{n-1} = \dfrac{a}{1-r} = \dfrac{[\text{First term}]}{1-[\text{Common ratio}]}$

If $|r| \geq 1$ then the geometric series **diverges**.

Since $r = \dfrac{1}{2} < 1$ and $a = \dfrac{1}{2}$ for the above series $\displaystyle\sum_{n=1}^{\infty}\left(\dfrac{1}{2^n}\right) = \dfrac{1}{2} + \dfrac{1}{2^2} + \dfrac{1}{2^3} + \dfrac{1}{2^4} + \cdots$ we have

$$\sum_{n=1}^{\infty}\left(\dfrac{1}{2^n}\right) = \dfrac{1/2}{1-1/2} = 1 \qquad \left[\text{Using formula } \boxed{7.27} \ \sum_{n=1}^{\infty} ar^{n-1} = \dfrac{a}{1-r}\right]$$

H3 **Examples of geometric series**

Example 28

Figure 37 illustrates the Cantor set which is the interval [0, 1] with the **middle third** removed each time:

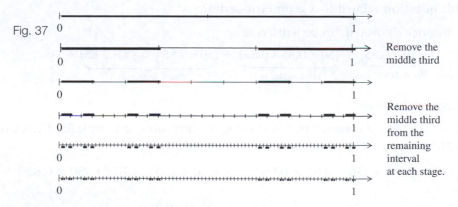

Fig. 37

Remove the middle third

Remove the middle third from the remaining interval at each stage.

The total length removed L is given by $L = \dfrac{1}{3} + \dfrac{2}{9} + \dfrac{4}{27} + \dfrac{8}{81} + \cdots = \displaystyle\sum_{n=1}^{\infty} \dfrac{1}{3}\left(\dfrac{2}{3}\right)^{n-1}$.

Determine L.

Example 28 *continued*

Solution

? **Is *L* a geometric series?**

Yes, it is a geometric series because each term is $\dfrac{2}{3}$ the previous term.

? The common ratio $r = \dfrac{2}{3}$ and the first term $a = \dfrac{1}{3}$. **Does this series converge?**

Yes, because the modulus (**Section 3F** of **Chapter 3**) of the common ratio

$$|r| = \left|\frac{2}{3}\right| = \frac{2}{3} < 1 \text{ (less than 1) so we conclude by}$$

7.27 $\displaystyle\sum_{n=1}^{\infty} ar^{n-1} = \frac{a}{1-r}$ if $|r| < 1$

that the infinite series *L* converges and the sum is

$$L = \frac{1/3}{1 - 2/3} = \frac{1/3}{1/3} = 1 \quad \left[\text{Substituting } a = \frac{1}{3} \text{ and } r = \frac{2}{3} \text{ into } \frac{a}{1-r}\right]$$

This means the infinite series *L* gives a total length (sum) equal to 1. That is, the whole interval $[0, 1]$ is eventually removed.

Example 29 *vibrations*

A spring vibrates 50 mm in the first oscillation and 85% of its previous value for each successive oscillation. Determine the total distance covered by the spring before stopping.

Solution

? **How is this question related to a geometric series?**

The total distance *d* covered can be written as

$$d = \underbrace{50}_{\text{First oscillation}} + \underbrace{(50 \times 0.85)}_{\text{Second oscillation}} + \underbrace{(50 \times 0.85 \times 0.85)}_{\text{Third oscillation}} + \underbrace{(50 \times 0.85 \times 0.85 \times 0.85)}_{\text{Fourth oscillation}} + \cdots$$

? **Is this a geometric series?**

Yes, because each term is obtained by multiplying the preceding term by 0.85. Therefore, it is a geometric series. This series can be written in compact form as

$$d = 50 + (50 \times 0.85) + (50 \times 0.85 \times 0.85) + \cdots = 50 + (50 \times 0.85) + (50 \times 0.85^2) + \cdots$$

$$= \sum_{n=1}^{\infty} 50\,(0.85)^{n-1}$$

? **What is the first term, *a*, and the common ratio, *r*, equal to?**

First term $a = 50$, common ratio $r = 0.85$ and because $|r| = |0.85| = 0.85 < 1$, therefore by

Example 29 *continued*

7.27 $$\sum_{n=1}^{\infty} ar^{n-1} = \frac{a}{1-r} \quad \text{if } |r| < 1$$

the series $\sum_{n=1}^{\infty} 50\,(0.85)^{n-1}$ converges. We are given $0.85 < 1$, therefore the distance d covered is

$$d = \sum_{n=1}^{\infty} 50\,(0.85)^{n-1} = \frac{50}{1-0.85} = 333.33 \text{ mm} \quad \left[= \frac{\text{First term}}{1-(\text{Common ratio})}\right]$$

The first element of a series may not start at $n = 1$ but at $n = 0$ or $n = 10$ or $n = 69$, and it is denoted by

$$\sum_{n=0}^{\infty} a_n, \quad \sum_{n=10}^{\infty} a_n \text{ or } \sum_{n=69}^{\infty} a_n, \text{ respectively}$$

Example 30

Determine whether the following series is convergent:

$$10 - \frac{10}{3} + \frac{10}{9} + \frac{10}{27} + \cdots$$

If it is convergent then find its sum.

Solution

Is the given series a geometric series?

If we divide two consecutive terms then we have:

$$\left[\text{Second term}\right] \div \left[\text{First term}\right] = \left(-\frac{10}{3}\right) \div 10 = -\frac{1}{3} \quad \text{or}$$

$$\left[\text{Third term}\right] \div \left[\text{Second term}\right] = \left(\frac{10}{9}\right) \div \left(-\frac{10}{3}\right) = -\frac{1}{3} \quad \text{or}$$

$$\left[\text{Fourth term}\right] \div \left[\text{Third term}\right] = \left(-\frac{10}{27}\right) \div \left(\frac{10}{9}\right) = -\frac{1}{3}$$

Hence there is a common ratio, $r = -\frac{1}{3}$, between two consecutive terms. So we have a geometric series with a common ratio $r = -\frac{1}{3}$ and the first term $a = 10$. Since

$$|r| = \left|-\frac{1}{3}\right| = \frac{1}{3} < 1 \qquad [\text{Less than 1}]$$

Example 30 *continued*

therefore by

7.27 $$\sum_{n=1}^{\infty} ar^{n-1} = \frac{a}{1-r} \quad \text{if } |r| < 1$$

the given series converges and the sum is equal to

$$10 - \frac{10}{3} + \frac{10}{9} - \frac{10}{27} + \cdots = \frac{10}{1-\left(-\dfrac{1}{3}\right)} \qquad \left[= \frac{\text{First term}}{1 - (\text{Common ratio})} \right]$$

$$= \frac{10}{4/3} = \frac{30}{4} = 7\frac{1}{2}$$

This means the sum of the infinite series is $7\frac{1}{2}$, that is, $10 - \frac{10}{3} + \frac{10}{9} - \frac{10}{27} + \cdots = 7\frac{1}{2}$.

Example 31

Determine whether the following series is convergent:

$$\sum_{n=1}^{\infty} \frac{(-3)^n}{2^n}$$

If it is convergent then find its sum.

Solution

Writing out the first few terms of the series we have

$$\sum_{n=1}^{\infty} \frac{(-3)^n}{2^n} = \sum_{n=1}^{\infty} \left(-\frac{3}{2}\right)^n \qquad \left[\text{Because } \frac{(-3)^n}{2^n} = \left(-\frac{3}{2}\right)^n \right]$$

$$= -\frac{3}{2} + \left(-\frac{3}{2}\right)^2 + \left(-\frac{3}{2}\right)^3 + \left(-\frac{3}{2}\right)^4 + \cdots$$

? **Is this a geometric series?**

Yes, because each term is obtained by multiplying the preceding term by the common ratio

? $r = -\dfrac{3}{2}$. **Does the series converge?**

No, because

$$|r| = \left|-\frac{3}{2}\right| = \frac{3}{2} \geq 1 \qquad [\text{Greater than or equal to } 1]$$

Example 31 *continued*

therefore the series **diverges**. What does this mean?

Adding infinitely many terms

$$-\frac{3}{2} + \left(-\frac{3}{2}\right)^2 + \left(-\frac{3}{2}\right)^3 + \left(-\frac{3}{2}\right)^4 + \cdots$$

does **not** approach a unique finite limit.

We can also use formula 7.25 to test the series shown in **Example 31**.

7.25 If $\lim\limits_{n\to\infty} (a_n) \neq 0$ [Not zero] then $\Sigma(a_n)$ diverges

That is, if the nth term does **not tend** to zero as n goes to infinity, then we conclude that the infinite series **diverges**. In the above example, because the nth term $a_n = \left(-\frac{3}{2}\right)^n$ does **not** tend to zero as $n \to \infty$, therefore the given series **diverges**. This is a simpler test to use than applying the geometric test of **Example 31**.

H4 Ratio test

Sometimes we will use the notation Σ to represent $\sum\limits_{n=1}^{\infty}$.

The ratio test can be used for testing convergence of a series. It depends on the limiting value of the ratio between successive terms in the series. In mathematical terms, we consider the limiting values of the ratio between the $(n + 1)$th term to the nth term as n goes to infinity.

Ratio test 7.28. Let $\sum\limits_{n=1}^{\infty} (a_n)$ be a series where a_n is real and positive. Let

$$\lim_{n\to\infty} \left(\frac{a_{n+1}}{a_n}\right) = L \quad \text{[Ratio of the } (n+1)\text{th term to the } n\text{th term]}$$

   i If $L < 1$ (less than 1), then the series $\Sigma(a_n)$ converges.

   ii If $L > 1$ (greater than 1), then the series $\Sigma(a_n)$ diverges.

   iii If $L = 1$ (equal to 1), then the test fails and we **cannot** conclude whether the series $\Sigma(a_n)$ converges or diverges.

Note that the ratio test does **not** give us the sum of the series.

Next we apply this ratio test to particular examples. The application of this method is generally straightforward but can involve a large number of algebraic steps.

H5 Application of the ratio test

Example 32

Determine whether the series $\sum \left(\dfrac{1}{n!} \right)$ converges or diverges.

Solution

? **Can we apply the ratio test?**

Yes, because we only need the terms of the series to be positive and, since $\dfrac{1}{n!} > 0$, we can

? use the ratio test. **What do $n!$ and $(n + 1)!$ mean?**

$$n! = 1 \times 2 \times 3 \times 4 \times \cdots \times n$$

$$(n+1)! = 1 \times 2 \times 3 \times 4 \times \cdots \times n \times (n + 1)$$

? Let $a_n = \dfrac{1}{n!}$. Then **what is a_{n+1} equal to?**

Replacing n with $n + 1$ in $a_n = \dfrac{1}{n!}$ gives

$$a_{n+1} = \dfrac{1}{(n + 1)!}$$

? **What do we need to find in order to use the ratio test?**

The limiting value of

$$\dfrac{a_{n+1}}{a_n} = a_{n=1} \div a_n = \dfrac{1}{(n+1)!} \div \left(\dfrac{1}{n!} \right) \qquad \text{[Substituting for } a_n \text{ and } a_{n+1}\text{]}$$

$$= \dfrac{1}{(n + 1)!} \times \left(\dfrac{n!}{1} \right) \qquad \left[\begin{array}{l} \text{Inverting the second} \\ \text{fraction and multiplying} \end{array} \right]$$

$$= \dfrac{n!}{(n + 1!)}$$

$$= \dfrac{1 \times 2 \times 3 \times 4 \times \cdots \times n}{1 \times 2 \times 3 \times 4 \times \cdots \times n \times (n + 1)} = \dfrac{1}{n + 1} \qquad \left[\begin{array}{l} \text{Cancelling all the common} \\ \text{factors } 1, 2, 3, \ldots \text{ and } n \end{array} \right]$$

Substituting $\dfrac{a_{n+1}}{a_n} = \dfrac{1}{n + 1}$ into $L = \lim\limits_{n \to \infty} \left(\dfrac{a_{n+1}}{a_n} \right)$ gives

$$L = \lim\limits_{n \to \infty} \left(\dfrac{1}{n + 1} \right) = 0 \qquad \left[\text{Because as } n \to \infty, \dfrac{1}{n + 1} \text{ tends to zero} \right]$$

Since $L = 0 < 1$ (less than 1) then by the ratio test:

7.28 ǀ **i** whereby if $L < 1$ then the series $\sum (a_n)$ converges

so the given series $\sum \left(\dfrac{1}{n!} \right)$ **converges.**

The ratio test states that if L is less than 1 then the series converges, but it does **not** give us the sum of the series.

Example 33

Discuss the convergence or divergence of $\sum\limits_{n=0}^{\infty} (n!)$.

Solution

Applying the ratio test with $a_n = n!$ then $a_{n+1} = (n+1)!$. Substituting these into $\dfrac{a_{n+1}}{a_n}$ gives

$$\frac{a_{n+1}}{a_n} = \frac{(n+1)!}{n!}$$

$$= \frac{1 \times 2 \times 3 \times 4 \times \cdots \times n \times (n+1)}{1 \times 2 \times 3 \times 4 \times \cdots \times n} = n+1 \qquad \left[\begin{array}{l}\text{Cancelling common} \\ \text{factors } 1, 2, 3, \ldots \text{ and } n\end{array}\right]$$

Substituting this $\dfrac{a_{n+1}}{a_n} = n+1$ into L gives

$$L = \lim_{n \to \infty} \left(\frac{a_{n+1}}{a_n}\right) = \lim_{n \to \infty} (n+1) = +\infty \qquad [\text{As } n \to \infty \text{ so does } n+1 \to \infty]$$

Since $L = +\infty > 1$ (greater than 1) therefore by the ratio test:

| 7.28 | **ii** whereby if $L > 1$ then the series $\sum (a_n)$ diverges

so the given series $\sum\limits_{n=0}^{\infty} (n!)$ **diverges**.

Since the nth term, $n!$, in **Example 33** does **not** tend to zero as $n \to \infty$, that is

$$\lim_{n \to} (n!) \neq 0 \qquad [\text{Not zero}]$$

therefore by | 7.25 | the given series $\sum n!$ diverges. This confirms the above ratio test result.

Example 34

Show that the ratio test fails for $\sum\limits_{n=0}^{\infty} \left(\dfrac{1}{n+1}\right)$.

Solution

Let $a_n = \dfrac{1}{n+1}$ then $a_{n+1} = \dfrac{1}{(n+1)+1} = \dfrac{1}{n+2}$. Substituting these into $\dfrac{a_{n+1}}{a_n}$ gives

Example 34 *continued*

$$\frac{a_{n+1}}{a_n} = \frac{1/(n+2)}{1/(n+1)}$$

$$= \frac{1}{(n+2)} \div \frac{1}{(n+1)} = \frac{1}{(n+2)} \times \frac{(n+1)}{1} = \frac{n+1}{n+2} \quad \left[\begin{array}{l}\text{Inverting the second} \\ \text{fraction and multiplying}\end{array}\right]$$

Substituting this into the limit of the ratio yields

$$L = \lim_{n \to \infty} \left(\frac{a_{n+1}}{a_n}\right) = \lim_{n \to \infty} \left(\frac{n+1}{n+2}\right)$$

$$= \lim_{n \to \infty} \left(\frac{1 + 1/n}{1 + 2/n}\right) \quad \text{[Dividing numerator and denominator by } n\text{]}$$

$$= \frac{1+0}{1+0} = 1 \quad \left[\text{Because as } n \to \infty \text{ then } \frac{1}{n} \to 0 \text{ and } \frac{2}{n} \to 0\right]$$

Since $L = 1$ the ratio test fails.

Example 35

Discuss the convergence or divergence of $\displaystyle\sum_{n=1}^{\infty} \left(\frac{5^n}{2^{n+1}\, n}\right)$.

Solution

Let $a_n = \dfrac{5^n}{2^{n+1}\, n}$ then $a_{n+1} = \dfrac{5^{n+1}}{2^{(n+1)+1}(n+1)} = \dfrac{5^{n+1}}{2^{n+2}(n+1)}$. **How do we find** $\dfrac{a_{n+1}}{a_n}$?

We multiply a_{n+1} with the inverted fraction of a_n:

$$\frac{a_{n+1}}{a_n} = \frac{5^{n+1}}{2^{n+2}(n+1)} \times \frac{2^{n+1}n}{5^n} \quad \left[\text{Substituting } \frac{1}{a_n} = \frac{2^{n+1}\, n}{5^n} \text{ and } a_{n+1} = \frac{5^{n+1}}{2^{n+2}(n+1)}\right]$$

$$= \frac{(5^n)\, 5 \times (2^n)\, 2n}{(2^n)\, 2^2\, (5^n)\, (n+1)} = \frac{5n}{2(n+1)} \quad \left[\begin{array}{l}\text{Cancelling common factors} \\ 5^n, \, 2^n \text{ and } 2\end{array}\right]$$

$$= \frac{5}{2}\left[\frac{n}{n+1}\right] \quad \left[\text{Taking out } \frac{5}{2}\right]$$

Substituting this into the evaluation of L gives:

$$L = \lim_{n \to \infty}\left(\frac{a_{n+1}}{a_n}\right) = \lim_{n \to \infty}\left[\frac{5}{2}\left(\frac{n}{n+1}\right)\right]$$

$$= \lim_{n \to \infty}\left[\frac{5}{2}\left(\frac{1}{1+1/n}\right)\right] \quad \left[\begin{array}{l}\text{Dividing numerator and} \\ \text{denominator by } n\end{array}\right]$$

$$= \frac{5}{2}\left(\frac{1}{1+0}\right) = \frac{5}{2} \quad \left[\text{Because as } n \to \infty \text{ then } \frac{1}{n} \to 0\right]$$

Example 35 *continued*

Since $L = 5/2 > 1$ (greater than 1) therefore by the ratio test

7.28 **ii** whereby if $L > 1$ then the series $\sum(a_n)$ diverges

so the given series $\sum\limits_{n=1}^{\infty} \left(\dfrac{5^n}{2^{n+1}n} \right)$ **diverges**.

SUMMARY

An infinite series is denoted in compact form as $\sum\limits_{n=1}^{\infty} a_n$ and is equal to

$a_1 + a_2 + a_3 + a_4 + \ldots$

A **geometric** series is defined as $\sum\limits_{n=1}^{\infty} ar^{n-1}$ where a is the first term and r is the common

ratio. This series converges if $|r| < 1$ and the sum is given by

7.27 $\sum\limits_{n=1}^{\infty} ar^{n-1} = \dfrac{a}{1-r} = \dfrac{[\text{First term}]}{1 - [\text{Common ratio}]}$

But the series diverges for $|r| \geq 1$.

Ratio test 7.28 . Let $\sum\limits_{n=1}^{\infty}(a_n)$ be a series where a_n is real and positive. Let $\lim\limits_{n \to \infty} \left(\dfrac{a_{n+1}}{a_n} \right) = L$

  i If $L < 1$, then the series $\sum(a_n)$ converges.

 ii If $L > 1$, then the series $\sum(a_n)$ diverges.

iii If $L = 1$, then the test fails and we **cannot** conclude whether the series $\sum(a_n)$ converges or diverges.

Exercise 7(h)

Solutions at end of book. Complete solutions available at www.palgrave.com/engineering/singh

1 Write the following series in $\sum$ notation:

  a $1 + \sqrt{2} + \sqrt{3} + 2 + \cdots$

  b $2 + 4 + 6 + 8 + \cdots$

  c $1 + 3 + 5 + 7 + \cdots$

 *d $1 - \dfrac{1}{2} + \dfrac{1}{3} - \dfrac{1}{4} + \dfrac{1}{5} - \cdots$

  e $1 + \dfrac{1}{3} + \dfrac{1}{9} + \dfrac{1}{27} + \dfrac{1}{81} + \cdots$

  f $\dfrac{2}{3} + \dfrac{4}{9} + \dfrac{8}{27} + \dfrac{16}{81} + \cdots$

2 Show that each of the following series is convergent and determine its sum:

  a $\sum\limits_{n=1}^{\infty} \left(\dfrac{1}{3} \right)^n$ **b** $\sum\limits_{n=1}^{\infty} \left(\dfrac{1}{4} \right)^n$

Exercise 7(h) continued

Solutions at end of book. Complete solutions available at www.palgrave.com/engineering/singh

c $\displaystyle\sum_{n=1}^{\infty}\left(\frac{1}{\pi}\right)^{n}$ **d** $\displaystyle\sum_{n=1}^{\infty}\left(\frac{1}{m}\right)^{n}$

where $m > 1$.

3 Determine whether the following series is convergent. If it is convergent, then find its sum.

a $\displaystyle\sum_{n=1}^{\infty}\left(\frac{1}{2^{2n-1}}\right)$ **b** $\displaystyle\sum_{n=1}^{\infty}\left(\frac{3}{2}\right)^{n}$

c $\displaystyle\sum_{n=1}^{\infty}(e)^{n}$ **d** $\displaystyle\sum_{n=1}^{\infty}10\left(\frac{1}{3}\right)^{n}$

4 🚗 [*Mechanics*]   A ball is dropped from a height of 10 m and bounces to 55% of its previous height. Determine the total distance travelled by the ball.

5 🚗 [*Mechanics*]   A hot air balloon rises 50 m in the first minute and then rises 65% of the distance travelled in the previous minute for subsequent minutes. Determine the maximum rise of the balloon.

6 The Sierpinski triangle shown in Fig. 38 below is created by continually removing the middle of a triangle. (The shaded part shows the triangle that is removed.)

Fig. 38

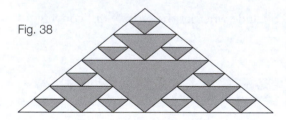

The area A removed is given by
$A = \displaystyle\sum_{n=0}^{\infty}\left(\frac{1}{4}\right)\left(\frac{3}{4}\right)^{n}$. Find A.

7 Determine the sum S of the following infinite series:

a $S = \dfrac{9}{10} + \dfrac{9}{100} + \dfrac{9}{1000} + \cdots$

b $S = \dfrac{3}{10} + \dfrac{3}{100} + \dfrac{3}{1000} + \cdots$

c $S = \dfrac{1}{10} + \dfrac{1}{100} + \dfrac{1}{1000} + \cdots$

What do you notice about your results?

8 A publishing company gives an academic a profit of £100 for the first year. For each subsequent year the profit falls by 9%. Determine the total possible profit.

9 Determine which of the following series converges. If it does converge then find its sum.

a $8 + 4 + 2 + 1 + \cdots$

b $3 + 6 + 12 + 24 + \cdots$

c $16 + 12 + 9 + \dfrac{27}{4} + \cdots$

10 Show that the following series converge and find the sum in each case:

a $\displaystyle\sum_{n=1}^{\infty}\frac{1}{x^{n}}$ where $|x| > 1$

b $\displaystyle\sum_{n=1}^{\infty}\left(\frac{x^{n}}{2^{n}}\right)$ where $|x| < 2$

c $\displaystyle\sum_{n=1}^{\infty}\frac{1}{(1+x)^{n}}$ where $x > 0$

d $\displaystyle\sum_{n=1}^{\infty}\frac{1}{(1+x^{2})^{n}}$ where $x \neq 0$

Throughout the remaining questions,
$\displaystyle\sum$ **represents** $\displaystyle\sum_{n=1}^{\infty}$.

11 Discuss convergence or divergence for each of the following series:

a $\displaystyle\sum\left(\frac{1}{(2n)!}\right)$ **b** $\displaystyle\sum_{n=0}^{\infty}\left(\frac{n!}{2^{n}}\right)$

c $\displaystyle\sum_{n=0}^{\infty}\left(\frac{n!}{3^{n}}\right)$ **d** $\displaystyle\sum\left(\frac{(n+1)^{2}}{2^{n}}\right)$

e $\displaystyle\sum(e^{-n})$ **f** $\displaystyle\sum\left(\frac{n^{2}}{3^{n}}\right)$

Exercise 7(h) continued

Solutions at end of book. Complete solutions available at www.palgrave.com/engineering/singh

g $\sum\left(\dfrac{10^n}{n!}\right)$ **h** $\sum\left(\dfrac{3^n n}{(n+1)^2}\right)$

i $\sum\left(\dfrac{n!}{(2n+1)!}\right)$ **j** $\sum\left(\dfrac{11^n}{2^{n+1}n}\right)$

12 Show that the ratio test fails for each of the following series:

a $\sum\left(\dfrac{1}{n^3}\right)$ **b** $\sum\left(\dfrac{1}{n+10}\right)$

c $\sum\left(\dfrac{1}{n^2+1}\right)$

*13 a Discuss convergence or divergence for each of the following series:

i $\sum\left(\dfrac{2^n n!}{n^n}\right)$ **ii** $\sum\left(\dfrac{3^n n!}{n^n}\right)$

b Determine the values of x for which the following series

$$\sum\left(\dfrac{x^n n!}{n^n}\right)$$

i converges

ii diverges $\left[\textit{Hint: } \lim\limits_{n\to\infty}\left(\dfrac{n}{1+n}\right)^n = e\right]$

SECTION I **Numerical solution of equations**

By the end of this section you will be able to:
▶ derive the Newton–Raphson formula
▶ apply the Newton–Raphson formula

Linear equations can be easily solved by using basic algebra. Non-linear equations such as quadratic equations can be solved by factorizing or by using the formula

$$x = \dfrac{-b \pm \sqrt{b^2 - 4ac}}{2a}$$

There is **no simple** formula for solving a cubic or equations of higher order.

Most of us have heard of Newton, but very little is known of **Joseph Raphson**. He was born in Middlesex, England, in 1648 and graduated from Cambridge. He wrote a book entitled *Analysis Aequationum Universals*, which contained Newton's method for approximating roots of an equation. Thus the name Newton–Raphson for solving equations came about.

We can solve these using a computer algebra system, a spreadsheet or a graphical calculator. Another approach is to solve these equations using numerical methods, which involves locating the roots of an equation by successive iterations. Computers and calculators use such methods to solve these equations. One such method is called the **Newton–Raphson**.

H1 **Newton–Raphson method**

We consider the non-linear equation $f(x) = 0$. The root(s) of this equation lie where the graph of $f(x)$ cuts the horizontal axis. Suppose the graph of $f(x)$ follows the path shown in Fig. 39.

Let r be the actual root of the equation $f(x) = 0$. Our aim is to find the root of the equation, r. If we first estimate the root of $f(x) = 0$ to be at r_1 (say) and we draw a tangent at point A as shown, then

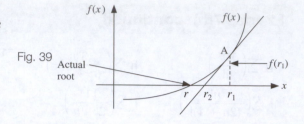

Fig. 39

$$f'(r_1) = \frac{f(r_1)}{r_1 - r_2} \quad \left[\text{Gradient} = \frac{\text{rise}}{\text{run}}\right]$$

Solving for r_2:

$$r_1 - r_2 = \frac{f(r_1)}{f'(r_1)} \text{ gives } r_2 = r_1 - \frac{f(r_1)}{f'(r_1)}$$

Fig. 40

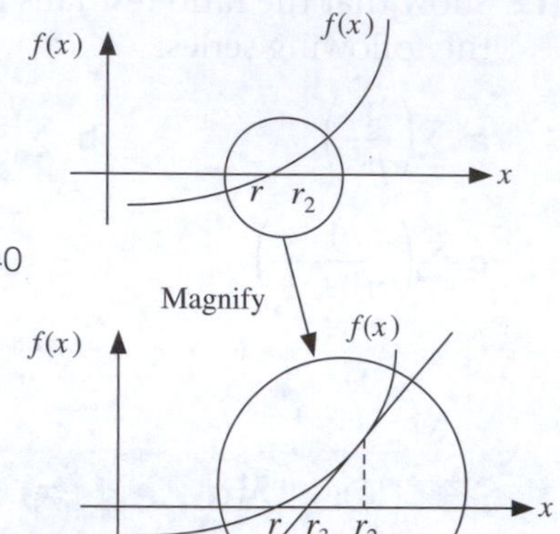

In Fig. 39 we can see that r_2 is closer to the actual root r than it is to r_1.

We can now magnify between r and r_2 and get a closer approximation to the root r (Fig. 40).

By the same procedure as above we get r_3. Now r_3 is closer to the actual root r than it is to r_2. So as above, we have

$$r_3 = r_2 - \frac{f(r_2)}{f'(r_2)}$$

By repeating we get

$$r_4 = r_3 - \frac{f(r_3)}{f'(r_3)}$$

$$r_5 = r_4 - \frac{f(r_4)}{f'(r_4)}$$

? **What do you think r_{n+1} is going to be?**

7.29
$$r_{n+1} = r_n - \frac{f(r_n)}{f'(r_n)}$$

The above method is known as the Newton–Raphson procedure. The number of iterations $n + 1$ is determined by the accuracy required.

Example 36

Find the root of the cubic equation

$$x^3 - 2x - 5 = 0$$

correct to 2 d.p. using the Newton–Raphson procedure with your first approximation, $r_1 = 2$.

Example 36 *continued*

Solution

Let

$$f(x) = x^3 - 2x - 5$$

$$f'(x) = 3x^2 - 2 \quad \text{[Differentiating]}$$

We are given $r_1 = 2$ and using $\boxed{7.29}$ $r_{n+1} = r_n - \dfrac{f(r_n)}{f'(r_n)}$ with $n = 1$ we have

$$r_2 = r_1 - \frac{f(r_1)}{f'(r_1)}$$

$\boxed{\dagger}$ $r_2 = 2 - \dfrac{f(2)}{f'(2)}$ $\Big[$Substituting the given $r_1 = 2\Big]$

$$f(2) = 2^3 - (2 \times 2) - 5 = -1 \quad \text{[Substituting } x = 2 \text{ into } f(x) = x^3 - 2x - 5\text{]}$$

$$f'(2) = 3(2)^2 - 2 = 10 \quad \text{[Substituting } x = 2 \text{ into } f'(x) = 3x^2 - 2\text{]}$$

Putting these values, $f(2) = -1$ and $f'(2) = 10$, into $\boxed{\dagger}$:

$$r_2 = 2 - \left(\frac{-1}{10}\right) = 2.10$$

To find r_3, use $\boxed{7.29}$ $r_{n+1} = r_n - \dfrac{f(r_n)}{f'(r_n)}$ with $n = 2$:

$$r_3 = r_2 - \frac{f(r_2)}{f'(r_2)}$$

and with $r_2 = 2.1$:

$\boxed{\dagger\dagger}$ $r_3 = 2.1 - \dfrac{f(2.1)}{f'(2.1)}$

$$f(2.1) = 2.1^3 - (2 \times 2.1) - 5 = 0.061 \quad \text{[Substituting } x = 2.1 \text{ into } f(x) = x^3 - 2x - 5\text{]}$$

$$f'(2.1) = 3 \times (2.1)^2 - 2 = 11.230 \quad \text{[Substituting } x = 2.1 \text{ into } f'(x) = 3x^2 - 2\text{]}$$

Putting these, $f(2.1) = 0.061$ and $f'(2.1) = 11.230$, into $\boxed{\dagger\dagger}$ gives

$$r_3 = 2.1 - \frac{0.061}{11.23} = 2.095 = 2.10 \quad \text{(2 d.p.)}$$

Since the r_2 and r_3 approximations agree to 2 d.p., the root of $x^3 - 2x - 5 = 0$ is approximately 2.10 (2 d.p.).

Sometimes we are not given the initial value, r_1, to start the Newton–Raphson procedure. To find r_1, we can either

 i plot a graph of the function and see approximately where the roots lie, or
ii evaluate the function at some obvious values.

Example 37 *mechanics*

The velocity, v, of an object is given by
$$v = t^3 - 3t^2 - 2 \quad \text{where} \quad 2 \le t \le 4$$
By using the Newton–Raphson procedure, find t correct to four significant figures for the velocity to be zero.

Solution

Here we are not given our first approximation r_1.

So what should we take r_1 to be?

Find v for obvious values of t, that is, $t = 2$, 3 and 4 because we are given $2 \le t \le 4$:

Substituting $t = 2$; $v = 2^3 - 3(2)^2 - 2 = -6$

Substituting $t = 3$; $v = 3^3 - 3(3)^2 - 2 = -2$

Substituting $t = 4$; $v = 4^3 - 3(4)^2 - 2 = 14$

Clearly by observing the last two values of v, there is a root of $t^3 - 3t^2 - 2 = 0$ between $t = 3$ and $t = 4$ because v goes from being negative to positive. Since -2 is closer to zero, take $r_1 = 3$. By 7.29 :

$$r_2 = r_1 - \frac{f(r_1)}{f'(r_1)}$$

With $r_1 = 3$, this gives

$$\boxed{*} \qquad r_2 = 3 - \frac{f(3)}{f'(3)}$$

Let $v = f(t)$:

$$f(t) = t^3 - 3t^2 - 2, \qquad\qquad f'(t) = 3t^2 - 6t$$

$$f(3) = 3^3 - 3(3)^2 - 2 = -2, \quad f'(3) = 3(3)^2 - (6 \times 3) = 9$$

Substituting $f(3) = -2$ and $f'(3) = 9$ into $\boxed{*}$:

$$r_2 = 3 - \left(-\frac{2}{9} \right) = 3.222$$

Using 7.29 with $r_2 = 3.222$:

$$\boxed{**} \qquad r_3 = 3.222 - \frac{f(3.222)}{f'(3.222)}$$

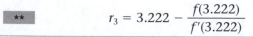

7.29 $\quad r_{n-1} = r_n - \dfrac{f(r_n)}{f'(r_n)}$

Example 37 *continued*

$f(3.222) = 3.222^3 - [3 \times (3.222)^2] - 2 = 0.3046$ [Substituting $t = 3.222$ into $f(t) = t^3 - 3t^2 - 2$]

$f'(3.222) = [3 \times (3.222)^2] - (6 \times 3.222) = 11.811$ [Substituting $t = 3.222$ into $f'(t) = 3t^2 - 6t$]

Putting $f(3.222) = 0.3046$ and $f'(3.222) = 11.811$ into ▮******▮ :

$$r_3 = 3.222 - \frac{0.3046}{11.811} = 3.196$$

Applying ▮7.29▮ with $r_3 = 3.196$:

▮*******▮ $$r_4 = 3.196 - \frac{f(3.196)}{f'(3.196)}$$

$f(3.196) = 3.196^3 - [3 \times (3.196)^2] - 2 = 0.002$ [Substituting $t = 3.196$]

$f'(3.196) = [3 \times (3.196)^2] - [6 \times 3.196] = 11.467$ [Substituting $t = 3.196$]

Putting these into ▮*******▮ :

$$r_4 = 3.196 - \frac{0.002}{11.467} = 3.1958 = 3.196 \ (4 \text{ s.f.})$$

Since r_3 and r_4 are identical, the velocity of the object is zero when $t = 3.196$ s.

The Newton–Raphson method normally requires a close initial estimate to the actual root, otherwise it may fail.

SUMMARY

We can use the Newton–Raphson procedure to solve non-linear equations.
With an initial guess, r_1, we obtain $r_2, r_3, \ldots$ by using

▮7.29▮ $$r_{n+1} = r_n - \frac{f(r_n)}{f'(r_n)}$$

The number of iterations depends on the accuracy required.

Exercise 7(i)

Solutions at end of book. Complete solutions available at www.palgrave.com/engineering/singh

1 Apply the Newton–Raphson procedure, with the first approximation $r_1 = -2$, to find a root of the following equations to 2 d.p.:

a $x^3 - 2x + 7 = 0$

b $x^4 - 3x^2 - 2 = 0$

c $e^x - 2x - 5 = 0$

▮7.29▮ $$r_{n-1} = r_n - \frac{f(r_n)}{f'(r_n)}$$

Exercise 7(i) continued

Solutions at end of book. Complete solutions available at www.palgrave.com/engineering/singh

2 🔲 [vibrations]   The natural frequencies of a three-storey building are related to λ where λ satisfies

$$\lambda^3 + 28\lambda^2 + 231\lambda + 541 = 0 \quad \text{(Fig. 41)}$$

By applying the Newton–Raphson procedure, determine all three roots of the above equation correct to 3 d.p.

$\lambda^3 + 28\lambda^2 + 231\lambda + 541$

Fig. 41

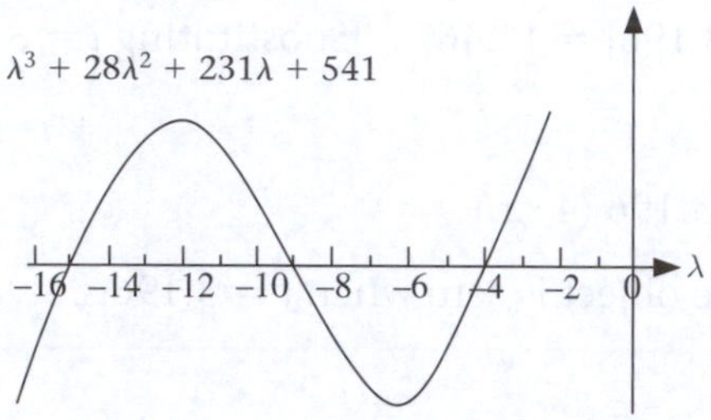

3 🔲 [fluid mechanics]   The velocity flow, v, of a liquid along a channel satisfies

$$v^3 - 6v^2 - 348v + 3112 = 0$$

Given that there is a root of this equation between $v = 10$ and $v = 11$, find this root correct to 3 d.p.

4 🔲 [materials]   The deflection, y, at a distance x along a beam of length L is given by

$$y = \frac{WL}{EI}(x^3 - 3x^2 + 2x - 1)$$

where W = weight, E = modulus of elasticity and I = second moment of area. These are all non-zero constants.

By applying the Newton–Raphson procedure, find the distance (correct to 3 d.p.), x, where the deflection y is zero.

Section J **Power series** is online (**www.palgrave.com/engineering/singh**).

Examination questions 7

Solutions at end of book. Complete solutions available at www.palgrave.com/engineering/singh

1 Given that the distance (s metres) of a particle from a point is given by the function $s = 30t - 6t^2$ at time t seconds, find (giving appropriate units)

  i the initial velocity

  ii the velocity after 3 seconds

  iii the (constant) acceleration.

University of Portsmouth, UK, 2009

2 a Find the local maximum and local minimum of the curve $y = x^3 - 3x + 1$ and distinguish between them.

  b The distance, x metres, of a particle from a fixed point Z after t seconds is given by $x = t^3 - 2t^2 + t$. Find

  i The particle's distance from Z after 2 seconds.

  ii The speed of the particle after 2 seconds.

  iii The acceleration of the particle after 2 seconds.

  iv The time when the particle is at rest.

  v The time when the acceleration is zero.

Cork Institute of Technology, Ireland, 2007

3 Determine the dimensions of the rectangle of largest area with a fixed perimeter of 20 feet.

4 You are to construct an open (no top) rectangular box with a square base. Material for the base costs 30 cents/in.^{2}

and material for the sides costs 10 cents/in.^{2}. If the box's volume is required to be 12 in.^{3}, what dimensions will result in the least expensive box?

Questions 4 and 5 are from University of California, Davis, USA

5 Find the first four non-zero terms of the

Maclaurin expansion of $\dfrac{e^{2x}}{1+x}$.

Loughborough University, UK, 2009

6 i Find the binomial expansion of $(1+3x)^{1/3}$ up to and including the term in x^3.

ii Use the expansion of e^x and $\cos x$ to find the Maclaurin expansion of $e^{-x^2}\cos 2x$ up to and including the term in x^4.

University of Surrey, UK, 2009

7 Sketch the graphs of the left and right hand sides of the equation $e^{2x} = 3 - x^2$, and show that the equation must have a root between 0 and $\sqrt{3}$. Show that the application of the Newton–Raphson method to this equation leads to the recursion formula

$$x_{n+1} = x_n - \left(\frac{e^{2x_n} + x_n^2 - 3}{2e^{2x_n} + 2x_n} \right)$$

Find an approximation to the root accurate to 4 d.p.

University of Surrey, UK, 2008

8 a Find the equation of the tangent line to the curve $y = x^3 - 6x$ at the point $(2, -4)$.

b Find the maximum value of the function $f(x) = 2x^2 - x^4$ and state the value(s) of x which give this maximum value.

9 Find the first four non-zero terms in the binomial series for

$$(1 - x^2)^{-1/2}$$

Using this series, find the value of $(0.99)^{-1/2}$ to 6 d.p.

Questions 8 and 9 are from University of Manchester, UK, 2009

The following question from King's College London has been slightly edited.

10 Determine the first three terms in the Taylor series expansion of $(1 + e^{x^2})(1 - x)$ around $x = 0$.

King's College London, UK, 2007

11 Find the equation of the line tangent to $y = (2 + x)e^{-x}$ at the point $(0, 2)$.

12 Find the equation of the tangent line to $x^2 + xy + y^2 = e^{y-2} + 6$ at $(1, 2)$.

Questions 11 and 12 are from Stanford University, USA, 2008

*13 The radial probability density function for the ground state of the hydrogen atom is

$$P(r) = \left(\frac{4r^2}{a^2} \right)e^{-2r/a}, \text{ for } r \geq 0,$$

where $a > 0$ is a constant. Sketch a graph of the function identifying its relative maximum and points of inflection.

University of Toronto, Canada, 2004

*14 A wire of length 100 cm is to be cut into two pieces. One piece will be bent into the shape of a circle, the other piece will be bent into the shape of a square. How should the wire be cut to

a maximize the combined area of the two shapes?

b minimize the combined area of the two shapes?

University of Toronto, Canada, 2006

Miscellaneous exercise **7**

Solutions at end of book. Complete solutions available at www.palgrave.com/engineering/singh

1 A rectangular tank with no top and a volume of 100 m³ is to be made from a sheet of metal. If one side has a length of 2 m, find the dimensions of the other sides so that minimum material is used (minimum surface area).

2 A parcel with a square cross-section is limited by the sum of its length and girth (perimeter of cross-section) to 2 m. Find the dimensions of the parcel which maximizes the volume.

3 Repeat the above question, showing that whatever may be the sum of length and girth of a square-based parcel, the maximum volume will always be when the length is twice the side of the square.

4 [materials]   The loading distribution, w, on a beam is given by

$$w = -36x^2 + 50x$$

where x is the distance in metres from one end of the beam. Find the value of x at which maximum loading occurs.

5 [materials]   A beam of length 8 m has a deflection y (m) at a distance x (m) from one end given by

$$y = \frac{1}{12 \times 10^3}(x^4 - 14x^3 + 36x^2)$$

Determine the maximum deflection, and the value of x to achieve this.

6 [mechanics]   A projectile is fired at an angle θ with the horizontal distance, x, given by

$$x = 2.5\sin(2\theta) \quad \text{where } 0 < \theta < \frac{\pi}{2}$$

Find the value of θ which gives maximum x.

7 [mechanics]   The horizontal distance, x, travelled by a projectile is given by

$$x = \frac{u^2}{25}\sin(\theta)\cos(\theta) \quad 0 < \theta < \frac{\pi}{2}$$

where u is the initial velocity and θ is the angle that the projectile makes with the horizontal. Determine the value of θ which maximizes x.

8 [mechanics]   A particle moves along a horizontal line with displacement, s, given by

$$s = 2 - te^{-t}$$

Find expressions for the velocity and acceleration of the particle. Sketch v against t for $t \geq 0$.

9 [mechanics]   The velocity, v, of a particle moving in a straight line at time t is given by

$$v = 4e^{-50t^2}$$

Sketch the graph of v against t, marking all points of maximum, minimum and inflexion.

(*Note*: The inflexion points are **not horizontal**, but are general points of inflexion as defined in the text: the point $x = c$ of $y = f(x)$ where $\frac{d^2y}{dx^2} = 0$ and $\frac{d^2y}{dx^2}$ changes sign close to $x = c$.)

*10 [heat transfer]   The critical insulation thickness t of a tube is defined to be the value of t which gives minimum resistance R where

$$R = \frac{\ln(t/t_1)}{2\pi k} + \frac{1}{2\pi th}$$

Miscellaneous exercise **7 continued**   Solutions at end of book. Complete solutions available at www.palgrave.com/engineering/singh

where $k > 0$ is thermal conductivity, h is heat transfer coefficient and t is inner thickness of insulation tube. Show that the critical thickness is

$$t = \frac{k}{h}.$$

(Treat t_1, k and h as constants.)

11 [mechanics]   The angular acceleration, α, of a machine is given by

$$\alpha = \frac{n^2 + 12}{3 - n}$$

where n is the gear ratio. Determine the value of n which gives the maximum acceleration α.

12 Evaluate $\lim\limits_{x \to 0} \dfrac{e^x - 1}{x}$.

13 Find the equations of the tangent and normal for $y = \sin^2(x)$ at $x = \pi/4$.

14 [signal processing]   The velocity, v, of a signal in a cable at a distance x is given by

$$v = kx \ln\left(\frac{1}{x}\right) \quad (0 < x < 1)$$

where k is a positive constant. Find the value of x which gives maximum velocity.

15 [electrical principles]   The voltage, v, across an inductor with inductance L, is given by

$$v = L\frac{di}{dt}$$

For $i = 5e^{-500t}$ and $L = 2\,\text{mH}\ (2 \times 10^{-3}\text{H})$, determine v. Sketch, on different axes, the graphs of v and i against t for $t \geq 0$, marking any maxima, minima and horizontal points of inflexion.

*16 [electromagnetism]   The magnetizing force, F, at a point halfway along the line joining the centres of two coils of equal radius is given by

$$F = \frac{Ir^2}{2(x^2 + r^2)^{3/2}}$$

where r is the radius of each coil, I is the current carried in each coil and x is the distance between the coils.

The distance x is varied and the graph of F against x has a general point of inflexion. Determine the point x in terms of r.

*17 [electromagnetism]   The efficiency, η, of a transformer with a voltage–ampere rating of s is given by

$$\eta = \frac{xs\cos(\phi)}{L_i + xs\cos(\phi) + x^2 L_c}$$

$$(s\cos(\phi) > 0)$$

where x is the fraction of output power of full load, $\cos(\phi)$ is the power factor, L_c is the copper loss and L_i is the iron loss. Show that the maximum efficiency occurs at

$$x = \sqrt{\frac{L_i}{L_c}}$$

(Treat s, ϕ, L_i and L_c as constants.)

18 i Obtain the first three non-zero terms of the Maclaurin series for $\sinh(x)$.

   ii Use a computer algebra system or a graphical calculator to plot $y = \sinh(x)$. On the same axes, plot the first three non-zero terms obtained in part **i**.

Miscellaneous exercise 7 continued   Solutions at end of book. Complete solutions available at www.palgrave.com/engineering/singh

19 i Show that the first three non-zero terms of the Maclaurin series for $\tan^{-1}(x)$ are given by

$$x - \frac{x^3}{3} + \frac{x^5}{5}$$

[*Hint*: $[\tan^{-1}(x)]' = \dfrac{1}{1 + x^2}$. Also the series is only valid for $-1 \le x \le 1$]

ii Show that

$$\frac{\pi}{4} = 1 - \frac{1}{3} + \frac{1}{5} \cdots$$

iii By using your calculator, evaluate the following:

$$\tan^{-1}\left(\frac{1}{2}\right) + \tan^{-1}\left(\frac{1}{3}\right),$$

$$\tan^{-1}\left(\frac{1}{2}\right) + \tan^{-1}\left(\frac{1}{5}\right) + \tan^{-1}\left(\frac{1}{8}\right),$$

$$4\tan^{-1}\left(\frac{1}{5}\right) - \tan^{-1}\left(\frac{1}{239}\right)$$

What do you notice about your results?

20 i Sketch the graph

$$y = \frac{1}{3}x^3 - 3x^2 + 8x - 3$$

ii By using the Newton–Raphson procedure, solve

$$\frac{1}{3}x^3 - 3x^2 + 8x - 3 = 0$$

correct to 3 d.p.

21 Figure 42 shows a cross-sectional area of a garage where WZ is 4 m high.

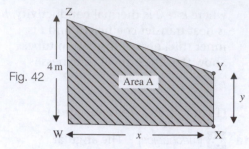

Fig. 42

You are informed that the sheeting for the lengths XY and YZ is 6 m, that is,

$$XY + YZ = 6$$

i Show that the cross-sectional area A is given by

$$A = \frac{1}{8}\,(24x - x^3)$$

ii Determine the dimensions x and y that give maximum cross-sectional area.

22 Expand the following:

$$\left(1 - \frac{v^2}{c^2}\right)^{1/2}$$

where v is velocity and c is speed of light.

23 The mass m of an object with velocity v is given by

$$m = \frac{m_0}{\sqrt{1 - \left(\dfrac{v}{c}\right)^2}}$$

where m_0 is the mass of the object at rest and c is the speed of light. The kinetic energy K is defined as

$$K = (m - m_0)c^2$$

Show that if the velocity v is very small in relation to c then

$$K = \frac{1}{2}m_0 v^2$$

CHAPTER 8

Integration

SECTION A **Introduction to integration**

By the end of this section you will be able to:
► understand an indefinite integral as the anti-derivative
► understand the physical significance of integration

A1 **Integration – inverse of differentiation**

Integration is the **reverse** process of **differentiation**. It is sometimes called *anti-differentiation*. Consider a function $y = f(x)$. The anti-derivative of y is called the **indefinite integral** and is denoted by $\int y\,\mathrm{d}x$. The integral notation, $\int y\,\mathrm{d}x$, means integrate, $\int$, y with respect to x, $\mathrm{d}x$. The function y is called the **integrand**.

For example, if we differentiate x^2 we obtain $2x$:

$$\frac{\mathrm{d}}{\mathrm{d}x}(x^2) = 2x$$

? **So does that mean the integral of $2x$ should be x^2?**

No.

$$\int 2x\,\mathrm{d}x = x^2 + C \text{ where } C \text{ is a constant}$$

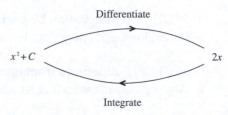

? **Why do we need to add a constant C?**

When we differentiate $x^2 + C$ we obtain $2x$ because differentiating a constant gives zero. It follows that

$$\int 2x\,\mathrm{d}x = x^2 + C \text{ holds for any constant } C$$

There are an infinite number of different functions which are the integral of $2x$, that is why $\int 2x\,\mathrm{d}x$ is called an **indefinite** integral.

Figure 1 shows some of the solutions to $\int 2x\,\mathrm{d}x$.

Fig. 1

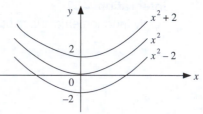

All these functions, $x^2 + C$, are similar in nature apart from the constant C. This C is called the *constant of integration*. Of course, other symbols can also represent the constant of integration. In general, if $y = f(x)$ and

$$\frac{\mathrm{d}y}{\mathrm{d}x} = f'(x) \text{ then integrating this gives } y = f(x) + C$$

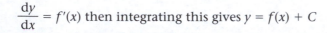

For example, if $y = x^4$ then

$$\frac{dy}{dx} = 4x^3 \text{ and } \int 4x^3 \, dx = x^4 + C$$

When we differentiate x^n we multiply by n and reduce the index of x by 1, thus

$$\frac{d}{dx}(x^n) = nx^{n-1}$$

? **What is the reverse process of this?**

We add 1 to the index of x and divide by the new index, $n + 1$, thus

$$\int x^n \, dx = \frac{x^{n+1}}{n+1} + C \text{ (provided } n \neq -1)$$

Example 1

Determine

a $\int x^3 \, dx$ **b** $\int x^7 \, dx$ **c** $\int x^{1/2} \, dx$

Solution

To integrate, we add 1 to the index of x and divide by 'index plus 1'.

a $\int x^3 \, dx = \dfrac{x^{3+1}}{3+1} + C = \dfrac{x^4}{4} + C$

b Similarly we have

$$\int x^7 \, dx = \frac{x^{7+1}}{7+1} + C = \frac{x^8}{8} + C$$

c Even if the index of x is not a whole number, we still apply the same rule:

$$\int x^{1/2} \, dx = \frac{x^{1/2+1}}{1/2+1} + C = \frac{x^{3/2}}{3/2} + C$$

$$= \frac{2x^{3/2}}{3} + C \quad \left[\text{Because } \frac{1}{3/2} = \frac{2}{3} \right]$$

A2 **Physical significance of integration**

In general we can calculate length, area and volume using integration.

Say we want to find the area A under the curve
? $y = f(x)$ between a and b of Fig. 2. **How can we obtain this area?**

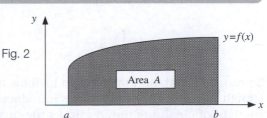

Fig. 2

It is not a regular shape like a circle or a rectangle so we cannot use any of the established formulae such as πr^2 or length $\times$ height.

We can cut the area A into smaller rectangular blocks as shown in Fig. 3.

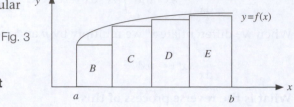

Fig. 3

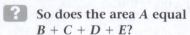 **So does the area A equal $B + C + D + E$?**

Not quite, because $B + C + D + E$ does **not** cover all the area A.

Some of the area under the curve is **not** covered by the rectangular blocks.

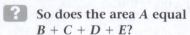 **How can we obtain a more accurate answer for the area A?**

The area A can be chopped into even smaller pieces of width Δx as shown in Fig. 4.

Fig. 4

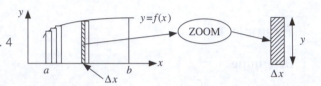

If the width Δx becomes small enough, then we can obtain an increasingly more accurate approximation for the entire area A.

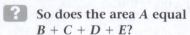 Consider the shaded area of Fig. 4. **How can we find this area?**

The shaded area is a rectangle of width Δx and height y:

$$\text{Rectangle area} = y\Delta x$$

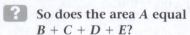 **How can we obtain an approximation to the original area A under the curve?**

By adding together all the rectangles of height y and width Δx.

Observe that the height y changes as you move along the curve. Hence the area A is approximately the sum from a to b of $y\Delta x$ and can be written as

$$\text{Area } A \simeq S_a^b \, y\Delta x$$

where S_a^b denotes the sum from a to b.

The approximation becomes closer and closer to the exact area A as Δx gets smaller and smaller. To obtain the exact area A we need Δx to be as close as possible to 0, without actually being 0, $\Delta x \to 0$. We have

$$\text{Area } A = \lim_{\Delta x \to 0} S_a^b \, y\Delta x = S_a^b \, y \, dx$$

where dx has replaced Δx as $\Delta x \to 0$. [In the notation, over the years, the letter S was replaced by an elongated S, $\int$.]

$$\text{Area } A = \int_a^b y \, dx$$

Some of this notation might seem baffling, but after a few examples it will become pretty straightforward. The notation was introduced by the German mathematician Liebniz who thought of obtaining the area under the curve by summing areas of rectangles. Let's define some of the terms.

$\int_a^b y\,dx$ is called the **definite integral** of y. The process of determining the area is called integration, and $x = a$, $x = b$ are called the **limits of integration.** Remember that y is a function of x, $y = f(x)$.

We can witness how the area under the curve can be determined by a definite integral by displaying graphs in a computer algebra system such as MAPLE.

If you don't have access to MAPLE or a similar package such as MATLAB, MATHEMATICA, DERIVE, etc., read through the example carefully and see if you understand the results.

Example 2

Plot the graph of $y = x^2$ from 0 to 3.

i Apply the leftbox command in MAPLE to draw 5 boxes on the graph of $y = x^2$ and find the total area of these boxes by using the command leftsum.

ii Repeat **i** for 10 boxes.

iii Repeat **i** for 50 boxes.

iv Repeat **i** for 100 boxes.

v Evaluate $\int_0^3 x^2\,dx$ by using the evalf (int(x^2, $x = 0..3$)) command

What do you notice about your graphs and their corresponding total area evaluation?

Solution

```
> with (student):
```
Warning, new definition for D

```
> leftbox (x^2,x=0..3,5);
```

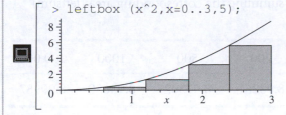

```
> evalf (leftsum(x^2,x=0..3,5));
```
6.480000000

```
> leftbox (x^2,x=0..3,10);
```

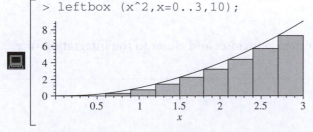

Example 2 *continued*

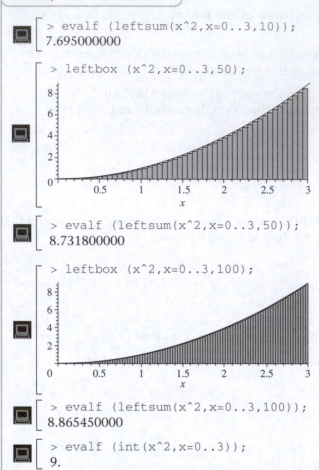

```
> evalf (leftsum(x^2,x=0..3,10));
7.695000000
```

```
> leftbox (x^2,x=0..3,50);
```

```
> evalf (leftsum(x^2,x=0..3,50));
8.731800000
```

```
> leftbox (x^2,x=0..3,100);
```

```
> evalf (leftsum(x^2,x=0..3,100));
8.865450000
```

```
> evalf (int(x^2,x=0..3));
9.
```

Notice as the number of boxes increases, more of the area under the curve becomes covered. Using more boxes in MAPLE and summing the corresponding area (leftsum) gives the results shown in Table 1.

	Number of boxes	100	200	500	1000	100 000
TABLE 1	Area (5 d.p.)	8.86545	8.93261	8.97302	8.98650	8.99987

If we consider more and more boxes, the limiting value of the sum will be 9. Thus the total area under the curve $y = x^2$ between 0 to 3 is 9. We have

$$\int_0^3 x^2 \, dx = 9$$

As the number of boxes increases, the area gets closer and closer to the integration of x^2 from 0 to 3.

We defined the area under the curve $y = f(x)$, between $x = a$ and $x = b$ (Fig. 5) as

$$\int_a^b f(x)\,dx \left[= \int_{x=a}^{x=b} f(x)\,dx \right]$$

This is called the definite integral.

If $\int f(x)\,dx = F(x)$ then

$$\int_a^b f(x)\,dx = F(b) - F(a)$$

In other words

Fig. 5   Shaded area $= \int_a^b f(x)\,dx$

$$\int_a^b f(x)\,dx = [\text{the value of the integral at } x = b] - [\text{the value of the integral at } x = a]$$

Example 3

?

Evaluate $\displaystyle\int_2^4 x^2\,dx$. **What does your result mean?**

Solution

We first integrate x^2 with respect to x. Then we evaluate the integral at $x = 4$ and $x = 2$ and subtract the two values. Notice how the 4 and 2 are transferred to the top right and bottom right of the square brackets after the integration is carried out.

$$\int_2^4 x^2\,dx = \left[\frac{x^3}{3} + C \right]_2^4 \qquad [\text{Integrating}]$$

$$= \underbrace{\left(\frac{4^3}{3} + C \right)}_{\text{substituting } x = 4} - \underbrace{\left(\frac{2^3}{3} + C \right)}_{\text{substituting } x = 2}$$

$$= \frac{64}{3} + C - \frac{8}{3} - C$$

$$\int_2^4 x^2\,dx = \frac{56}{3}$$

This is the area under the curve $y = x^2$ between $x = 2$ and $x = 4$ as shown in Fig. 6 over the page.

Example 3 *continued*

Note that the definite integral $\int_2^4 x^2 dx$ is a number, $\dfrac{56}{3}$ (area), and does not contain x. **Moreover there is *no* constant of integration (C). Why?**

Because the C's cancel each other out. When there is a definite integral you do not need to include the constant C.

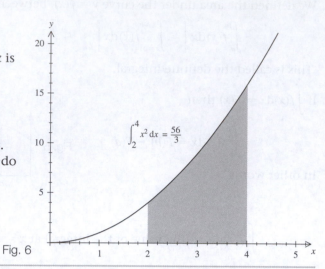

Fig. 6

SUMMARY

The **indefinite integral** of $y = f(x)$ is defined as the **anti-derivative** of y and is denoted by $\int y\, dx$.

The **definite integral** is defined as **the area under the curve**, $y = f(x)$, between $x = a$ and $x = b$. Area $= \int_a^b y\, dx$.

If $\int_a^b f(x) dx = F(x)$ then

$$\int_a^b f(x) dx = F(b) - F(a)$$

There is no constant (C) of integration in the evaluation of a definite integral.

Exercise 8(a)

Solutions at end of book. Complete solutions available at www.palgrave.com/engineering/singh

1 Determine the following integrals:

a $\displaystyle\int x\, dx$ **b** $\displaystyle\int u^2\, du$

c $\displaystyle\int z^3\, dz$ **d** $\displaystyle\int t^{1/2}\, dt$

e $\displaystyle\int (\omega t)\, d(\omega t)$ **f** $\displaystyle\int 3\, dt$

g $\displaystyle\int t^{-2}\, dt$ **h** $\displaystyle\int x^{-1/2}\, dx$

For question **2** you may use any appropriate mathematical software.

2 Plot the graph of $y = x^3$ from $x = 0$ to 1.

   i Use the leftbox command to draw four boxes on the graph of $y = x^3$.

Evaluate the total area of the four boxes by using the leftsum command.

   ii Repeat **i** for 8 boxes.

  iii Repeat **i** for 16 boxes.

  iv Repeat **i** for 32 boxes.

   v Repeat **i** for 64 boxes.

  vi Repeat **i** for 128 boxes.

 vii By applying the command evalf(int(x^3,x = 0..1)), evaluate $\int_0^1 x^3\, dx$. What do you notice about your results?

Solutions at end of book. Complete solutions available
at www.palgrave.com/engineering/singh

Exercise **8(a) continued**

3 Determine the shaded areas shown in Fig. 7 below:

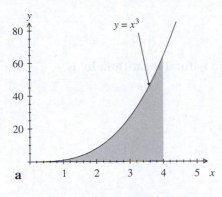

a

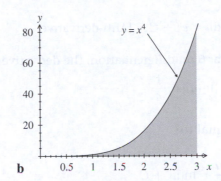

b

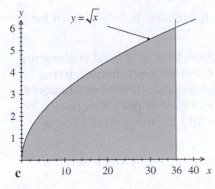

c

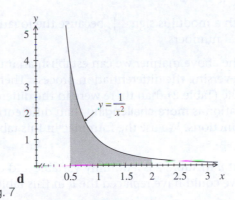

d

Fig. 7

SECTION B **Rules of integration**

By the end of this section you will be able to:

▶ apply the rules of integration
▶ recognize the anti-derivative of a given function

B1 **Creating a table of integrals**

? Let's consider other functions. **What is the derivative $\dfrac{\mathrm{d}}{\mathrm{d}u}[\sin(u)]$ equal to?**

$$\frac{\mathrm{d}}{\mathrm{d}u}[\sin(u)] = \cos(u)$$

? **What is $\displaystyle\int \cos(u)\,\mathrm{d}u$ equal to?**

Since integration is the **reverse of differentiation** we have

$$\int \cos(u)\,\mathrm{d}u = \sin(u) + C$$

❓ Hence the integral of cos is sin. **What is $\int e^u \, du$ equal to?**

Well $\dfrac{d}{du}(e^u) = e^u$, therefore

$$\int e^u du = e^u + C \quad \text{[Anti-derivative]}$$

Also from **Chapter 6** on differentiation, the derivative of a natural logarithm, ln, is

$$\frac{d}{du}[\ln(u)] = \frac{1}{u}$$

❓ **What is $\int \dfrac{du}{u}$ equal to?**

$$\int \frac{du}{u} = \ln|u| + C$$

We place u with a modulus sign, $|\ |$, because the logarithmic function is only defined for the positive real numbers.

Following in the above manner we can establish an integration table of general engineering functions by reversing the differentiation process. There are many more functions in the integration table (Table 2) than there were in the differentiation table (Table 3 of **Chapter 6**) because integration is **more challenging** than differentiation. Don't be put off by the many complicated functions. We use the formulae in this table to find the integrals of various functions.

Remember that u in Table 2 is a dummy variable and can be replaced by any letter. For example, we could have replaced the u in this table by x.

TABLE 2		Add a constant (C) to all of these
	8.1	$\int u^n du = \dfrac{u^{n+1}}{n+1} \ (n \neq -1)$
	8.2	$\int \dfrac{du}{u} = \ln|u|$
	8.3	$\int e^u du = e^u$
	8.4	$\int a^u du = a^u/\ln(a)$
	8.5	$\int \ln(u)du = u[\ln(u) - 1]$
	8.6	$\int \log(u)du = u[\log(u) - \log(e)]$
	8.7	$\int \sin(u)du = -\cos(u)$
	8.8	$\int \cos(u)du = \sin(u)$

TABLE 2 CONTINUED

8.9	$\int \tan(u)\mathrm{d}u = \ln|\sec(u)|$
8.10	$\int \sec(u)\mathrm{d}u = \ln|\sec(u) + \tan(u)|$
8.11	$\int \mathrm{cosec}(u)\mathrm{d}u = \ln|\mathrm{cosec}(u) - \cot(u)|$
8.12	$\int \cot(u)\mathrm{d}u = \ln|\sin(u)|$
8.13	$\int \sin^{-1}(u)\mathrm{d}u = u\sin^{-1}(u) + \sqrt{1 - u^2}$
8.14	$\int \cos^{-1}(u)\mathrm{d}u = u\cos^{-1}(u) - \sqrt{1 - u^2}$
8.15	$\int \tan^{-1}(u)\mathrm{d}u = u\tan^{-1}(u) - \frac{1}{2}\ln(1 + u^2)$
8.16	$\int \sinh(u)\mathrm{d}u = \cosh(u)$
8.17	$\int \cosh(u)\mathrm{d}u = \sinh(u)$
8.18	$\int \tanh(u)\mathrm{d}u = \ln[\cosh(u)]$
8.19	$\int \mathrm{sech}(u)\mathrm{d}u = \tan^{-1}[\sinh(u)]$
8.20	$\int \mathrm{cosech}(u)\mathrm{d}u = \ln|\tanh(u/2)|$
8.21	$\int \coth(u)\mathrm{d}u = \ln|\sinh(u)|$
8.22	$\int \sinh^{-1}(u)\mathrm{d}u = u\sinh^{-1}(u) - \sqrt{1 + u^2}$
8.23	$\int \cosh^{-1}(u)\mathrm{d}u = u\cosh^{-1}(u) - \sqrt{u^2 - 1}$ $\left(\cosh^{-1}(u) \geq 0\right)$
8.24	$\int \tanh^{-1}(u)\mathrm{d}u = u\tanh^{-1}(u) + \frac{1}{2}\ln(1 - u^2)$
8.25	$\int \dfrac{\mathrm{d}u}{\sqrt{a^2 - u^2}} = \sin^{-1}\left(\dfrac{u}{a}\right)$
8.26	$\int \dfrac{\mathrm{d}u}{u^2 + a^2} = \dfrac{1}{a}\tan^{-1}\left(\dfrac{u}{a}\right)$
8.27	$\int \dfrac{\mathrm{d}u}{u\sqrt{u^2 - a^2}} = \dfrac{1}{a}\sec^{-1}\left(\dfrac{u}{a}\right)$
8.28	$\int \dfrac{\mathrm{d}u}{\sqrt{a^2 + u^2}} = \sinh^{-1}\left(\dfrac{u}{a}\right) = \ln|u + \sqrt{u^2 + a^2}|$

(Table 2 continued on next page)

TABLE 2 CONTINUED

8.29	$\displaystyle\int \frac{du}{\sqrt{u^2 - a^2}} = \cosh^{-1}\left(\frac{u}{a}\right) = \ln\left|u + \sqrt{u^2 - a^2}\right|$	
8.30	$\displaystyle\int \frac{du}{a^2 - u^2} = \frac{1}{a}\tanh^{-1}\left(\frac{u}{a}\right) = \frac{1}{2a}\ln\left|\frac{a + u}{a - u}\right|$	
8.31	$\displaystyle\int \sqrt{u^2 \pm a^2}\,du = \frac{u}{2}\sqrt{u^2 \pm a^2} \pm \frac{a^2}{2}\ln\left|u + \sqrt{u^2 \pm a^2}\right|$	
8.32	$\displaystyle\int \sqrt{a^2 - u^2}\,du = \frac{u}{2}\sqrt{a^2 - u^2} + \frac{a^2}{2}\sin^{-1}\left(\frac{u}{a}\right)$	
8.33	$\displaystyle\int e^{au}\cos(bu)\,du = \frac{e^{au}}{a^2 + b^2}[a\cos(bu) + b\sin(bu)]$	
8.34	$\displaystyle\int e^{au}\sin(bu)\,du = \frac{e^{au}}{a^2 + b^2}[a\sin(bu) - b\cos(bu)]$	

We will show a number of these results later in this chapter.

Let p and q be functions of x, then we have

8.35
$$\int (p + q)\,dx = \int p\,dx + \int q\,dx$$

8.36
$$\int (p - q)\,dx = \int p\,dx - \int q\,dx$$

8.37
$$\int kp\,dx = k\int p\,dx \text{ where } k \text{ is a constant}$$

8.35 and 8.36 mean that we can take the integral of each function separately and then add or subtract the resulting integrals. 8.37 means that we can take a constant outside the integral and then multiply the result by the constant.

You can also use the differentiation table on pages 291 and 292 in reverse to find the integrals of functions not listed in Table 2 above.

Example 4

Determine

a $\displaystyle\int x^6\,dx$ **b** $\displaystyle\int (x^3 - 2x^2 + 5x)\,dx$ **c** $\displaystyle\int \frac{2}{\sqrt{x}}\,dx$

Solution

a $\displaystyle\int x^6\,dx = \frac{x^{6+1}}{6 + 1} + C = \frac{x^7}{7} + C$
by 8.1 with
$n = 6$

8.1 $\displaystyle\int u^n\,du = \frac{u^{n+1}}{n + 1} \quad (n \neq -1)$

Example 4 *continued*

b Splitting the integrand and taking out the constants we have

$$\int (x^3 - 2x^2 + 5x)\mathrm{d}x = \int x^3\mathrm{d}x - 2\int x^2\mathrm{d}x + 5\int x\,\mathrm{d}x$$

$$\underset{\text{by } 8.1}{\equiv} \frac{x^{3+1}}{3+1} - 2\left[\frac{x^{2+1}}{2+1}\right] + 5\left[\frac{x^{1+1}}{1+1}\right] + C$$

$$= \frac{x^4}{4} - \frac{2x^3}{3} + \frac{5x^2}{2} + C$$

c Which formula should we use to find $\int \dfrac{2}{\sqrt{x}}\mathrm{d}x$?

We need to rewrite $\dfrac{2}{\sqrt{x}}$: $\dfrac{2}{\sqrt{x}} = \dfrac{2}{x^{1/2}} = 2x^{-1/2}$ [Remember $\sqrt{x} = x^{1/2}$]

We have

$$\int \frac{2}{\sqrt{x}}\mathrm{d}x = \int 2x^{-1/2}\mathrm{d}x = 2\int x^{-1/2}\mathrm{d}x \qquad \text{[Taking out 2]}$$

$$= 2\left[\frac{x^{-\frac{1}{2}+1}}{-^1/_2 + 1}\right] + C \qquad \text{[Using } 8.1 \text{ with } n = 1/2]$$

$$= \frac{2x^{1/2}}{1/2} + C = 4x^{1/2} + C \qquad \left[\text{Because } \frac{2}{1/2} = 4\right]$$

Example 5 *thermodynamics*

Find **a** $\displaystyle\int PV^{1.3}\mathrm{d}V$ **b** $\displaystyle\int \frac{P}{V}\,\mathrm{d}V$

where P and V represent pressure and volume respectively. P is constant and V varies.

Solution

a $\displaystyle\int PV^{1.3}\mathrm{d}V = P\int V^{1.3}\mathrm{d}V$ [Taking out P]

$$= P\left[\underbrace{\frac{V^{1.3+1}}{1.3+1}}_{\substack{\text{by } 8.1 \text{ with}\\ n = 1.3}}\right] + C = \frac{PV^{2.3}}{2.3} + C$$

8.1 $\displaystyle\int x^n\mathrm{d}x = \frac{x^{n+1}}{n+1}$

Example 5 *continued*

b How do we determine $\int \dfrac{P}{V} \, dV$ **?**

$$\int \frac{P}{V} \, dV = P \int \frac{dV}{V} = P \underbrace{\ln(V)}_{\text{by } 8.2} + C$$

We can write $\ln|V| = \ln(V)$ because V is volume and therefore is positive.

Example 6

Determine

a $\displaystyle\int \sin(\omega t)\mathrm{d}(\omega t)$ **b** $\displaystyle\int [3e^t + 2\cos(t)]\mathrm{d}t$ **c** $\displaystyle\int \left(\frac{1}{x} - x^{-3} + 5x^{3/5}\right)\mathrm{d}x$

Solution

a From Table 2, using $\displaystyle\int \sin(u)\mathrm{d}u = -\cos(u) + C$ with $u = \omega t$ we have

$$\int \sin(\omega t)\mathrm{d}(\omega t) = -\cos(\omega t) + C$$

b We can integrate each part and take the constants out, thus

$$\int \left[3e^t + 2\cos(t)\right]\mathrm{d}t = 3\int e^t \mathrm{d}t + 2\int \cos(t)\mathrm{d}t$$

$$= \underbrace{3e^t}_{\text{by } 8.3} + \underbrace{2\sin(t)}_{\text{by } 8.8} + C$$

c Again splitting the integrand we have

$$\int \left(\frac{1}{x} - x^{-3} + 5x^{3/5}\right)\mathrm{d}x = \int \frac{\mathrm{d}x}{x} - \int x^{-3}\mathrm{d}x + 5\int x^{3/5}\mathrm{d}x$$

$$= \underbrace{\ln|x|}_{\text{by } 8.2} - \underbrace{\frac{x^{-3+1}}{-3+1}}_{\text{by } 8.1} + 5\underbrace{\frac{x^{3/5+1}}{3/5+1}}_{\text{by } 8.1} + C$$

$$= \ln|x| - \frac{x^{-2}}{-2} + 5\frac{x^{8/5}}{8/5} + C \quad \text{[Simplifying]}$$

$$= \ln|x| + \frac{x^{-2}}{2} + 25\frac{x^{8/5}}{8} + C \quad \left[\text{Because } \frac{5}{8/5} = \frac{25}{8}\right]$$

What is $\displaystyle\int K\mathrm{d}x$ **where K is a constant?**

Since there is **no** x in the integrand K we can write as $K = Kx^0$ because $x^0 = 1$. Using formula 8.1 we have

$$\int Kx^0 \mathrm{d}x = K\int x^0 \mathrm{d}x = K\left[\frac{x^{0+1}}{0+1}\right] + C = K\left[\frac{x^1}{1}\right] + C = Kx + C$$

8.1 $\displaystyle\int x^n \mathrm{d}x = \frac{x^{n+1}}{n+1}$ 8.2 $\displaystyle\int \mathrm{d}u/u = \ln|u|$ 8.3 $\displaystyle\int e^t \mathrm{d}t = e^t$ 8.8 $\displaystyle\int \cos(t)\mathrm{d}t = \sin(t)$

Hence $\int K\mathrm{d}x = Kx + C$ where K is a constant.

? What is $\int 6\mathrm{d}x$ equal to? and what is $\int 666\mathrm{d}x$ equal to?

$\int 6\mathrm{d}x = 6x+C$ and $\int 666\mathrm{d}x = 666x+C$

SUMMARY

Let p and q be functions of x, then

8.35	$\int (p + q)\mathrm{d}x = \int p\,\mathrm{d}x + \int q\,\mathrm{d}x$
8.36	$\int (p-q)\mathrm{d}x = \int p\,\mathrm{d}x - \int q\,\mathrm{d}x$
8.37	$\int kp\,\mathrm{d}x = k\int p\,\mathrm{d}x$ where k is a constant

---

Exercise 8(b)

Solutions at end of book. Complete solutions available at www.palgrave.com/engineering/singh

1 Determine the following:

a $\displaystyle\int \sin(t)\mathrm{d}t$ **b** $\displaystyle\int \cos(t)\mathrm{d}t$

c $\displaystyle\int \tan(t)\mathrm{d}t$ **d** $\displaystyle\int e^t\mathrm{d}t$

e $\displaystyle\int \cosh(t)\mathrm{d}t$ **f** $\displaystyle\int 9.81\mathrm{d}t$

g $\displaystyle\int 25\mathrm{d}x$ **h** $\displaystyle\int \frac{1}{2}\mathrm{d}x$

2 Find

*a $\displaystyle\int \cos(\omega t)\mathrm{d}(\omega t)$ **b** $\displaystyle\int 10^x\mathrm{d}x$

c $\displaystyle\int \sinh(t)\mathrm{d}t$ **d** $\displaystyle\int \sec(x)\mathrm{d}x$

3 🔬 [*thermodynamics*]   Find

a $\displaystyle\int PV^{1.35}\mathrm{d}V$ **b** $\displaystyle\int PV^{1.61}\mathrm{d}V$

(P is constant and V varies.)

4 Find

a $\displaystyle\int (x^2 + 2x)\mathrm{d}x$

b $\displaystyle\int \left(\frac{1}{\sqrt{x}} + \tan(x)\right)\mathrm{d}x$

c $\displaystyle\int (\sec^2(x) - 5e^x + 1)\mathrm{d}x$

[*Hint*: Use the differentiation table on page 292 in reverse to find the integral of $\sec^2(x)$]

d $\displaystyle\int (\sin(x) + 2\sqrt{x} - 1)\mathrm{d}x$

5 **i** Show that

$$\frac{\mathrm{d}}{\mathrm{d}x}\left(\frac{3 - 4x}{1 + x^2}\right) = \frac{4x^2 - 6x - 4}{(1 + x^2)^2}$$

ii Determine $\displaystyle\int \frac{4x^2 - 6x - 4}{(1 + x^2)^2}\mathrm{d}x$

SECTION C **Integration by substitution**

By the end of this section you will be able to:
▶ use substitution to find indefinite integrals
▶ apply integration to engineering problems

C1 **Using substitution**

Example 7

Find **a** $\displaystyle\int \cos(2x + 5)\mathrm{d}x$ **b** $\displaystyle\int \sin(2x)\mathrm{d}x$

Solution

? **a What formula of Table 2 can we use to determine**

$\displaystyle\int \cos(2x + 5)\mathbf{d}x$?

Using 8.8 , $\displaystyle\int \cos(u)\mathrm{d}u = \sin(u) + C$ with $u = 2x + 5$, we have

* $\displaystyle\int \cos(2x + 5)\mathrm{d}x = \int \cos(u)\mathrm{d}x = ?$

? We **cannot** use 8.8 to determine ***** because we have a $\mathrm{d}x$ in place of a $\mathrm{d}u$. **What can we substitute for $\mathrm{d}x$?**

We know $u = 2x + 5$ and we can differentiate this function to obtain

$$\frac{\mathrm{d}u}{\mathrm{d}x} = 2 \text{ which gives } \mathrm{d}u = 2\mathrm{d}x$$

because $\mathrm{d}u = \left(\dfrac{\mathrm{d}u}{\mathrm{d}x}\right)\mathrm{d}x$, where $\dfrac{\mathrm{d}u}{\mathrm{d}x}$ is the 'differential coefficient' and $\mathrm{d}u$ is the 'differential'.

Therefore we have $\mathrm{d}x = \dfrac{\mathrm{d}u}{2}$. Putting this into ***** :

$$\int \cos(u)\frac{\mathrm{d}u}{2} = \frac{1}{2}\int \cos(u)\mathrm{d}u = \frac{1}{2}\underset{\text{by } 8.8}{\sin(u)} + C$$

It follows that

$$\int \cos(2x + 5)\mathrm{d}x = \frac{1}{2}\sin(u) + C = \frac{1}{2}\underset{\substack{\text{substituting}\\ u = 2x+5}}{\sin(2x + 5)} + C$$

> **Example 7** *continued*
>
> **b** We can use 8.7 , $\int \sin(u)du = -\cos(u) + C$. Let $u = 2x$ then
>
> $$\frac{du}{dx} = 2 \qquad dx = \frac{du}{2}$$
>
> Substituting $u = 2x$ and $dx = \dfrac{du}{2}$ gives
>
> $$\int \sin(2x)dx = \int \sin(u)\frac{du}{2}$$
>
> $$= \frac{1}{2}\int \sin(u)\,du \quad \left[\text{Taking out } \frac{1}{2}\right]$$
>
> $$= \frac{1}{2}[-\cos(u)] + C = -\frac{\cos(2x)}{2} + C$$

In general, we have (k and m are constants):

8.38
$$\int \cos(kx + m)dx = \frac{\sin(kx + m)}{k} + C$$

8.39
$$\int \sin(kx + m)dx = -\frac{\cos(kx + m)}{k} + C$$

8.40
$$\int \sec^2(kx + m)dx = \frac{\tan(kx + m)}{k} + C$$

> **Example 8**
>
> Obtain $\displaystyle\int e^{7x+3}dx$
>
> Solution
>
> **Which formula can we use to find $\displaystyle\int e^{7x+3}dx$?**
>
> We use 8.3 , $\int e^u du = e^u + C$. Let $u = 7x + 3$ then we have
>
> $$\int e^{7x+3}dx = \int e^u dx$$
>
> **We need to replace the dx. How?**
>
> Differentiating $u = 7x + 3$ gives
>
> $$\frac{du}{dx} = 7 \qquad dx = \frac{du}{7}$$
>
> Substituting these: $\displaystyle\int e^{7x+3}dx = \int e^u \frac{du}{7} = \frac{1}{7}\int e^u du = \frac{1}{7}e^u + C = \frac{1}{7}e^{7x+3} + C$

In general

8.41 $\qquad \displaystyle\int e^{kx+m}\mathrm{d}x = \dfrac{e^{kx+m}}{k} + C$

Formulae 8.38, 8.39 and 8.41 are worth learning because they crop up frequently in engineering and science, especially in the case where $m = 0$:

$$\int \cos(kx)\mathrm{d}x = \frac{\sin(kx)}{k} + C, \int \sin(kx)\mathrm{d}x = \frac{-\cos(kx)}{k} + C \text{ and } \int e^{kx}\mathrm{d}x = \frac{e^{kx}}{k} + C.$$

You should learn these three so that when they reappear in a complex problem you will be comfortable using them.

C2 Engineering application

Let's investigate an engineering case. **Example 9** is particularly challenging. It is sophisticated compared to the above. Follow it through carefully.

Example 9 *structures*

The equation, y, of the catenary formed by a cable under its own weight is given by

$$y = \int \sinh\left(\frac{wx}{T}\right)\mathrm{d}x$$

where x is horizontal distance, T is horizontal tension and w is weight per unit length of cable. (T and w are constants.) Given that when $x = 0$, $y = 0$, show that

$$y = \frac{T}{w}\left[\cosh\left(\frac{wx}{T}\right) - 1\right]$$

Solution

We use 8.16, $\displaystyle\int \sinh(u)\mathrm{d}u = \cosh(u) + C$ with $u = \dfrac{wx}{T}$. Differentiating:

$$\frac{\mathrm{d}u}{\mathrm{d}x} = \frac{w}{T} \qquad \mathrm{d}x = \frac{T}{w}\mathrm{d}u$$

Putting these into the given integral:

$$y = \int \sinh\left(\frac{wx}{T}\right)\mathrm{d}x = \frac{T}{w}\int \sinh(u)\mathrm{d}u$$

$$= \frac{T}{w}\cosh(u) + C = \frac{T}{w}\cosh\left(\frac{wx}{T}\right) + C$$

Example 9 *continued*

Substituting the given $x = 0$, $y = 0$ into $y = \dfrac{T}{w}\cosh\left(\dfrac{wx}{T}\right) + C$ and using $\cosh(0) = 1$:

$$0 = \frac{T}{w} + C \text{ gives } C = -\frac{T}{w}$$

Hence

$$y = \frac{T}{w}\cosh\left(\frac{wx}{T}\right) - \frac{T}{w} = \frac{T}{w}\left[\cosh\left(\frac{wx}{T}\right) - 1\right] \quad \text{[Factorizing]}$$

C3 An important integral

Example 10

Obtain $\displaystyle\int \frac{3x^2}{x^3 - 5}\, dx$

Solution

? **What can we use to find** $\displaystyle\int \frac{3x^2}{x^3 - 5}\, dx$?

? We can try using $\displaystyle\int \frac{du}{u}$. Let $u = x^3 - 5$. We need to replace dx. **How can we find dx?**

Differentiating $u = x^3 - 5$ with respect to x gives

$$\frac{du}{dx} = 3x^2 \qquad\qquad dx = \frac{du}{3x^2}$$

$$\int \frac{3x^2}{x^3 - 5}\, dx = \int\left(\frac{3x^2}{u}\right)\frac{du}{3x^2} = \int \frac{du}{u} \quad \text{[Cancelling } 3x^2\text{]}$$

$$\underset{\text{by } \boxed{8.2}}{=} \ln|u| + C = \ln|x^3 - 5| + C$$

Take another look at $\displaystyle\int \frac{3x^2}{x^3 - 5}\, dx = \ln|x^3 - 5| + C$.

? **What do you notice about this integrand, $\dfrac{3x^2}{x^3 - 5}$?**

Similarly we have $\displaystyle\int \frac{2x}{x^2 - 1}\, dx = \ln|x^2 - 1| + C$

$\boxed{8.2}$ $\displaystyle\int du/u = \ln|u|$

$$\int \frac{2x - 2}{x^2 - 2x} \, dx = \ln|x^2 - 2x| + C$$

$$\int \frac{5x^4 - 6x}{x^5 - 3x^2} \, dx = \ln|x^5 - 3x^2| + C$$

? **What do you notice about all these integrands and associated solutions to the integral?**

In each case the numerator is the derivative of the denominator. That is, if you differentiate the denominator you get the numerator. In **Example 10** notice that the derivative, $3x^2$, of the denominator cancels out with the numerator and that is why we can use the $\int \dfrac{du}{u}$ formula.

Also observe that the solution of these integrals is the natural log, ln, of the denominator.

Example 11

Show that

$$\int \frac{f'(x)}{f(x)} \, dx = \ln|f(x)| + C$$

Solution

Let $u = f(x)$, then differentiating yields

$$\frac{du}{dx} = f'(x) \text{ therefore } dx = \frac{du}{f'(x)}$$

Substituting $u = f(x)$ and $dx = \dfrac{du}{f'(x)}$ into the given integral:

$$\int \frac{f'(x)}{f(x)} \, dx = \int \left(\frac{f'(x)}{u}\right) \frac{du}{f'(x)}$$

$$= \int \frac{du}{u} \quad [\text{Cancelling } f'(x)]$$

$$= \ln|u| + C = \ln|f(x)| + C \quad [\text{Replacing } u = f(x)]$$

In general, if $f(x)$ is a function of x and its derivative is $f'(x)$ then

8.42
$$\int \frac{f'(x)}{f(x)} \, dx = \ln|f(x)| + C$$

8.42 is a **very important** integral because it can be used to integrate a wide range of functions such as

$$\int \frac{2x}{x^2 + 1} \, dx = \ln|x^2 + 1| + C \qquad \left[\text{Because } \frac{d}{dx}[x^2 + 1] = 2x\right]$$

$$\int \frac{\cos(x)}{\sin(x)} \, dx = \ln|\sin(x)| + C \qquad \left[\text{Because } \frac{d}{dx}[\sin(x)] = \cos(x)\right]$$

$$\int \frac{2}{2x + 1} \, dx = \ln|2x + 1| + C \qquad \left[\text{Because } \frac{d}{dx}[2x + 1] = 2\right]$$

Say we had 3 in place of 2 in the above example, that is $\int \dfrac{3}{2x + 1} \, dx$.

? **Then how do we integrate this?**

Differentiating $2x + 1$ with respect to x gives 2, so we have to rewrite the integrand as follows:

$$\frac{3}{2x + 1} = \frac{3}{2} \frac{2}{2x + 1}$$

Thus

$$\int \frac{3}{2x + 1} \, dx = \frac{3}{2} \int \frac{2}{2x + 1} \, dx$$

$$= \frac{3}{2} \ln|2x + 1| + C \qquad \text{[Integrating]}$$

This is normal procedure in these types of examples. Normally you have to rewrite the integrand so that you can apply formula 8.42 . Similarly we have

$$\int \frac{5}{3x + 2} \, dx = \frac{5}{3} \ln|3x + 2| + C$$

$$\int \frac{x + 1}{x^2 + 2x + 5} \, dx = \frac{1}{2} \ln|x^2 + 2x + 5| + C$$

The last result might be more difficult to follow. Differentiating $x^2 + 2x + 5$ with respect to x gives $2x + 2$ and so rewriting the integrand we have

$$\frac{x + 1}{x^2 + 2x + 5} = \frac{1}{2} \frac{2x + 2}{x^2 + 2x + 5} \, dx \qquad \left[\text{Because } \frac{1}{2}[2x + 2] = x + 1\right]$$

Thus

$$\int \frac{x + 1}{x^2 + 2x + 5} \, dx = \frac{1}{2} \int \frac{2x + 2}{x^2 + 2x + 5} \, dx$$

$$= \frac{1}{2} \ln|x^2 + 2x + 5| + C$$

Example 12

Determine $\int \dfrac{t^2}{t^3 - 1} \, dt$

Solution

Differentiating $t^3 - 1$ gives us $3t^2$, but we are missing the 3 in the numerator. It is close, so let's try using the $\int \dfrac{du}{u}$ formula. Let $u = t^3 - 1$ and differentiating:

$$\frac{du}{dt} = 3t^2 \text{ which gives } dt = \frac{du}{3t^2}$$

Example 12 *continued*

Substituting these gives

$$\int \frac{t^2}{t^3 - 1}\, dt = \int \left(\frac{t^2}{u}\right)\frac{du}{3t^2} \underset{\text{cancelling } t^2}{=} \frac{1}{3}\int \frac{du}{u} = \frac{1}{3}\ln|u| + C = \frac{1}{3}\underbrace{\ln|t^3 - 1|}_{\text{replacing } u = t^3 - 1} + C$$

If you know that the derivative of the denominator is three times the numerator, then take out 1/3 and integrate the remaining function of the form $\dfrac{f'(x)}{f(x)}$.

SUMMARY

An important integral well worth remembering is

8.42
$$\int \frac{f'(x)}{f(x)}\, dx = \ln|f(x)| + C$$

Exercise 8(c)

Solutions at end of book. Complete solutions available at www.palgrave.com/engineering/singh

1 Obtain **a** $\displaystyle\int \sin(7x + 1)dx$

   b $\displaystyle\int \cos(7x + 1)dx$

2 Find **a** $\displaystyle\int \cos(\omega t)dt$ **b** $\displaystyle\int \cos(\omega t + \theta)dt$

3 Obtain **a** $\displaystyle\int \sin(\omega t)dt$ **b** $\displaystyle\int \sin(\omega t)\, d(\omega t)$

4 Determine **a** $\displaystyle\int \frac{2x}{x^2 - 1}\, dx$

   b $\displaystyle\int \frac{3x^2 - 6x}{x^3 - 3x^2 + 1}\, dx$

5 Show that $\displaystyle\int \cot(x)dx = \ln|\sin(x)| + C$

6 Show that

   a $\displaystyle\int \tanh(x)\, dx = \ln|\cosh(x)| + C$

   b $\displaystyle\int \coth(x)\, dx = \ln|\sinh(x)| + C$

7 Find **a** $\displaystyle\int \frac{dt}{7t - 1}$ **b** $\displaystyle\int \frac{t^3}{t^4 - 1}\, dt$

   c $\displaystyle\int \frac{t^2}{5 - t^3}\, dt$

8 Determine **a** $\displaystyle\int e^{11x + 5}\, dx$ **b** $\displaystyle\int e^{-2x + 1000}\, dx$

🚗 Questions 9 to 12 belong to the field of [*mechanics*].

Exercise **8(c) continued**

9 The vertical velocity component, v, of a projectile is defined as

$$v = \int (-g)\mathrm{d}t$$

where g is the constant acceleration due to gravity. Given that at $t = 0$, $v = v_0$, show that

$$v = v_0 - gt$$

10 The velocity, v, of a projectile is given by

$$v = \int (-g)\mathrm{d}t$$

Given that when $t = 0$, $v = u$, find an expression for v in terms of t.

11 The position, s, of a particle moving along a straight line is given by

$$s = \int 10(30t + 1)^{-1/2}\mathrm{d}t$$

Given that when $t = 0$, $s = \dfrac{2}{3}$, show that

$$s = \frac{2}{3}(30t + 1)^{1/2}$$

12 The velocity, v, of a particle is given by

$$v = \int (-6t)\mathrm{d}t$$

For the initial condition that when $t = 0$, $v = 48$ m/s, find

i an expression for the velocity, v

ii the time taken, t, for the velocity to become $v = 0$

13 [*fluid mechanics*]   The pressure, P, and density, ρ, of an airstream in an isentropic flow are related by

$$\frac{P}{\rho^\gamma} = k$$

where k is a constant and γ is the specific heat constant. Find $\int \dfrac{\mathrm{d}P}{\rho}$.

SECTION D **Applications of integration**

By the end of this section you will be able to:
▶ evaluate integrals for engineering examples

In this section we examine some engineering examples using the definite integral.

D1 **Fundamental theorem of calculus**

The Fundamental Theorem of Calculus states that differentiation and integration are inverse processes.

We defined the area under the curve $y = f(x)$ between the limits $x = a$ and $x = b$ (Fig. 8) as

8.43
$$\int_a^b f(x)\mathrm{d}x \Bigg|_{x=a}^{x=b} = \int_a^b f(x)\mathrm{d}x$$

Remember 8.43 is called the **definite integral**.

If $f(x) = F(x)$ then

8.44
$$\int_a^b f(x)\mathrm{d}x = F(b) - F(a)$$

Stating this formula in words:

$$\int_a^b f(x)\mathrm{d}x = \text{[the value of the integral}$$

at $x = b$] − [the value of the integral at $x = a$]

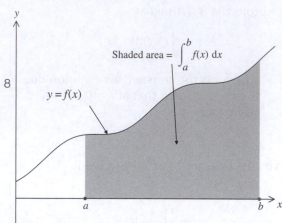

Fig. 8

Shaded area = $\displaystyle\int_a^b f(x)\,\mathrm{d}x$

$y = f(x)$

D2 **Engineering examples**

Let's examine some engineering examples.

Example 13 *mechanics*

The displacement, s, of an object is given by

$$s = \int_0^3 (t^4 + t)\mathrm{d}t$$

Evaluate s.

Solution

We integrate and then substitute $t = 3$ and $t = 0$:

$$s = \int_0^3 (t^4 + t)\mathrm{d}t$$

$$= \left[\frac{t^5}{5} + \frac{t^2}{2} \right]_0^3 \qquad [\text{Integrating by } 8.1]$$

$$= \underbrace{\left(\frac{3^5}{5} + \frac{3^2}{2} \right)}_{\text{substituting } t = 3} - \underbrace{\left(\frac{0^5}{5} + \frac{0^2}{2} \right)}_{\text{substituting } t = 0} = 53.1$$

$$s = 53.1 \text{ m}$$

8.1 $\displaystyle\int u^n \mathrm{d}u = \frac{u^{n+1}}{n+1}$

Example 14 *thermodynamics*

The change in specific enthalpy, Δh, in J/kg, is given by

$$\Delta h = \int_{150}^{300} (2.1 + (7 \times 10^{-3})T)\,\mathrm{d}T$$

Evaluate Δh.

Solution

Our limits of integration are $T = 300$ and $T = 150$:

$$\Delta h = \int_{150}^{300} (2.1 + (7 \times 10^{-3})T)\,\mathrm{d}T$$

$$= \left[2.1T + \frac{(7 \times 10^{-3})T^2}{2} \right]_{150}^{300} \quad [\text{Integrating by } \boxed{8.1}]$$

$$= \underbrace{\left((2.1 \times 300) + \frac{(7 \times 10^{-3})300^2}{2} \right)}_{\text{substituting } T = 300} - \underbrace{\left((2.1 \times 150) + \frac{(7 \times 10^{-3})150^2}{2} \right)}_{\text{substituting } T = 150}$$

$$= 945 - 393.75 = 551.25$$

$$\Delta h = 551.25 \text{ J/kg}$$

 Example 15 *mechanics*

Consider a deep well as shown in Fig. 9 on the next page. Let us assume we are trying to calculate how much rope we must attach to our bucket in order to reach the water.

A stone is dropped down the well and a splash is heard after **two** seconds. The acceleration a of the stone is $a = 9.8$ m/s². Determine

i A formula for the velocity v given that $a = \dfrac{\mathrm{d}v}{\mathrm{d}t}$.

ii A formula for the depth x given that $v = \dfrac{\mathrm{d}x}{\mathrm{d}t}$.

iii The depth of the well.

Assume at $t = 0$, the velocity $v = 0$ and depth $x = 0$.

$\boxed{8.1}$ $\displaystyle\int u^n\,\mathrm{d}u = u^{n+1}/n + 1$

Example 15 *continued*

Solution

i Substituting $a = 9.8$ into

$a = \dfrac{\mathrm{d}v}{\mathrm{d}t}$ gives $\dfrac{\mathrm{d}v}{\mathrm{d}t} = 9.8$.   Fig. 9

How can we find v from $\dfrac{\mathrm{d}v}{\mathrm{d}t} = 9.8$?

The inverse of differentiation is integration so we can integrate both sides to get

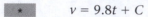 $v = 9.8t + C$

When $t = 0$ we have velocity $v = 0$. Substituting these into [*] yields

$$0 = (9.8 \times 0) + C \text{ so that } C = 0$$

Therefore the velocity is equal to $v = 9.8t$.

ii We are given that $v = \dfrac{\mathrm{d}x}{\mathrm{d}t}$. **How do we find the depth x?**

Integrate $v = \dfrac{\mathrm{d}x}{\mathrm{d}t} = 9.8t$

$$x = \frac{9.8}{2}t^2 + D, \quad x = 4.9t^2 + D$$

We call the new constant of integration D because we have already used C.

Substituting $t = 0$, $x = 0$ into $x = 4.9t^2 + D$ we have

$$0 = 4.9(0)^2 + D \text{ so that } D = 0$$

Substituting $D = 0$ into $x = 4.9t^2 + D$ gives

$$x = 4.9t^2$$

iii Substituting our given value of $t = 2$ into this formula gives the depth of the well

$$x = 4.9(2)^2 = 19.6$$

The depth of the well is therefore 19.6 m.

Sometimes in a definite integral we substitute a letter or a symbol rather than a number in the limits of integration. This results in an expression.

Example 16 *mechanics*

The moment of inertia, I, of a rod of length $2r$ is given by

$$I = \int_{-r}^{r} \frac{mx^2}{2r} \, \mathrm{d}x$$

where m is the mass of the rod and x is the distance from the axis. Show that $I = \dfrac{mr^2}{3}$.
(The mass, m, and length, r, are constant.)

Example 16 *continued*

Solution

Taking out the $m/2r$ gives

$$I = \frac{m}{2r}\int_{-r}^{r} x^2 dx = \frac{m}{2r}\left[\frac{x^3}{3}\right]_{x=-r}^{x=r} = \frac{m}{2r}\left[\frac{r^3-(-r)^3}{3}\right] = \frac{m}{2r}\left(\frac{2r^3}{3}\right) = \frac{mr^2}{3} \quad [\text{Cancelling}]$$

The next example is difficult because it involves trigonometric identities, integration, substitution and some unfamiliar symbols. It is more complicated than previous examples.

Example 17 *electrical principles*

The power, P, in a circuit is given by

$$P = \frac{\omega R}{2\pi}\int_0^{2\pi/\omega}(i^2)dt$$

where $i = I\sin(\omega t)$, R is resistance and ω is angular frequency. Show that

$$P = \frac{I^2 R}{2}$$

Solution

We have $i = I\sin(\omega t)$, therefore $i^2 = [I\sin(\omega t)]^2 = I^2\sin^2(\omega t)$. Substituting $i^2 = I^2\sin^2(\omega t)$

into $P = \frac{\omega R}{2\pi}\int_0^{2\pi/\omega}(i^2)dt$ gives

$$P = \frac{\omega R}{2\pi}\int_0^{2\pi/\omega}I^2\sin^2(\omega t)\,dt = \frac{\omega R I^2}{2\pi}\int_0^{2\pi/\omega}\sin^2(\omega t)dt \quad [\text{Taking out the constant } I^2]$$

The difficulty in this example is to find $\int_0^{2\pi/\omega}\sin^2(\omega t)dt$.

How do we integrate the $\sin^2$ term?

We need to avoid $\sin^2$ by finding another way of expressing it.

Example 17 *continued*

We use the trigonometric identity

4.68 $$\sin^2(x) = \frac{1}{2}\left[1 - \cos(2x)\right]$$

$$\sin^2(\omega t) = \frac{1}{2}\left[1 - \cos(2\omega t)\right]$$

Therefore we have

$$P = \frac{\omega R I^2}{2\pi} \int_0^{2\pi/\omega} \sin^2(\omega t)\, dt = \frac{\omega R I^2}{2\pi} \int_0^{2\pi/\omega} \frac{1}{2}\left[1 - \cos(2\omega t)\right] dt$$

$$= \frac{\omega R I^2}{4\pi} \int_0^{2\pi/\omega} \left[1 - \cos(2\omega t)\right] dt \quad \left[\text{Taking out } \frac{1}{2}\right]$$

$$= \frac{\omega R I^2}{4\pi} \left[t - \frac{\sin(2\omega t)}{2\omega}\right]_0^{2\pi/\omega} \quad \left[\text{Integrating using } \int \cos(kt)dt = \frac{\sin(kt)}{k}\right]$$

$$= \frac{\omega R I^2}{4\pi} \left\{\left(\underbrace{\frac{2\pi}{\omega} - \frac{\sin[2\omega\,(2\pi/\omega)]}{2\omega}}_{\text{substituting } t\, =\, 2\pi/\omega}\right) - \left[\underbrace{0 - \frac{\sin(0)}{2\omega}}_{\text{substituting } t\, =\, 0}\right]\right\}$$

$$= \frac{\omega R I^2}{4\pi} \left[\left(\frac{2\pi}{\omega} - \underbrace{\frac{\sin(4\pi)}{2\omega}}_{\substack{=\ 0 \text{ because} \\ \sin(4\pi)\, =\, 0}}\right) - 0\right]$$

$$= \frac{\omega R I^2}{4\pi} \left[\frac{2\pi}{\omega}\right] = \frac{I^2 R}{2} \quad\quad\quad \text{[Cancelling]}$$

$$P = \frac{I^2 R}{2}$$

Note that $\int \cos(kt)dt = \dfrac{\sin(kt)}{k}$ cropped up in the above example. Remember this is one of the three integrals that you should learn, as was suggested in the previous section.

Example 18 *reliability engineering*

The failure density function, $f(t)$, for a class of electronic components is given by

$$f(t) = 0.2e^{-0.2t}$$

The hazard function, $h(t)$, is defined as

$$h(t) = \frac{f(t)}{1 - \displaystyle\int_0^t f(x)dx}$$

Determine $h(t)$.

Solution

We are given $f(t)$, but what is $f(x)$ equal to?

Since $f(x)$ is a function of x, we simply replace the t with x, thus $f(x) = 0.2e^{-0.2x}$.

$$\int_0^t f(x)dx = 0.2\int_0^t e^{-0.2x}dx \quad \text{[Taking out 0.2]}$$

$$= 0.2\left[\frac{e^{-0.2x}}{-0.2}\right]_0^t \quad \left[\text{Integrating using } \int e^{kx}dx = \frac{e^{kx}}{k}\right]$$

$$= -\left[e^{-0.2x}\right]_0^t \quad \text{[Cancelling 0.2's and taking out the minus sign]}$$

$$= -\left(\underbrace{e^{-0.2t}}_{\text{substituting } x=t} - \underbrace{e^{-(0.2\times 0)}}_{\text{substituting } x=0}\right)$$

$$= -(e^{-0.2t} - 1) = -e^{-0.2t} + 1 \quad \text{[Remember } e^0 = 1\text{]}$$

$$\int_0^t f(x)dx = 1 - e^{-0.2t}$$

Simplifying the denominator in $h(t)$ gives

$$1 - \int_0^t f(x)dx = 1 - (1 - e^{-0.2t})$$

$$= 1 - 1 + e^{-0.2t}$$

$$= e^{-0.2t}$$

Replacing $1 - \displaystyle\int_0^t f(x)dx$ with $e^{-0.2t}$ in the denominator of the hazard function, $h(t)$,

gives

$$h(t) = \frac{0.2\, e^{-0.2t}}{e^{-0.2t}} = 0.2 \quad \text{[Cancelling } e^{-0.2t}\text{]}$$

Remember the variable x in $\displaystyle\int_a^b f(x)dx$ is called the **dummy variable** since x can be replaced by any letter because the value of the integral is the same. That is

$$\int_a^b f(x)dx = \int_a^b f(t)dt = \int_a^b f(\xi)d\xi = \cdots \text{ (A definite integral is a function of the limits **only**.)}$$

SUMMARY

We can use integration to solve engineering problems.

Exercise **8(d)**

Solutions at end of book. Complete solutions available at www.palgrave.com/engineering/singh

1 A building is designed to incorporate an archway as shown in Fig. 10. The opening needs to be glazed. Determine the area of glass needed.

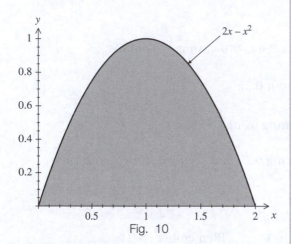

Fig. 10

🚗 Questions 2 to 6, inclusive, belong to the field of [*mechanics*].

2 A particle moves along a horizontal line with displacement, s, given by

$$s = \int_0^2 (t^2 - 2t)\,dt$$

Determine s.

3 The work done, W, to stretch a spring from its natural length to an extension of 0.5 m is given by

$$W = \int_0^{0.5} 100x\,dx$$

Evaluate W.

4 The force, F, required to compress a spring is given by

$$F = 1000x + 50x^3$$

where x is the displacement from its unstretched length. The work done, W, to compress a spring by 0.3 m is given by

$$W = \int_0^{0.3} F\,dx$$

Determine W.

5 The distance, s, travelled by a train on a straight track in the first two seconds is given by

$$s = \int_0^2 20(1 - e^{-t})\,dt$$

Find s.

6 The distance, s, covered by a vehicle between the 5th and 6th second is given by

$$s = \int_5^6 10e^{0.4t}\,dt$$

Evaluate s.

7 🌉 [*structures*]   A beam of length 6 m has a uniform distributed load, w, given by

$$w = 800 + \frac{1}{2}x^3$$

where x is the distance along the beam. The total load, P (N), and the moment about the origin, R (N m), are given by

$$P = \int_0^6 w\,dx \text{ and } R = \int_0^6 (wx)\,dx$$

Determine P and R.

Exercise **8(d) continued**

Solutions at end of book. Complete solutions available at www.palgrave.com/engineering/singh

8 [thermodynamics]   The specific heat, c, of steam is given by

$$c = \frac{3000 + T - 100}{1300}$$

The enthalpy change, Δh, is given by

$$\Delta h = \int_{150}^{500} 1.5c \, dT$$

Evaluate Δh.

9 [mechanics]   The time taken, t (hours), for a vehicle to reach a speed of 120 km/h with an initial speed of 80 km/h is given by

$$t = \int_{80}^{120} \frac{dv}{600 - 3v}$$

where v is velocity (km/h). Determine t.

10 [thermodynamics]   A gas in a cylinder obeys the law $PV^{1.25} = 1789$. The work done, W, is given by

$$W = \int_{0.01}^{0.1} P dV$$

Determine W.

Questions 11 to 15, inclusive, are in the field of [mechanics].

11 The moment of inertia, I, of a disc of radius r and mass m is given by

$$I = \int_0^r \left(\frac{2mx^3}{r^2} \right) dx$$

where x is the distance from an axis of rotation. Show that $I = \frac{mr^2}{2}$.

12 The moment of inertia, I, of an annulus (rubber ring) of inner radius a, outer radius b and mass m is given by

$$I = \int_a^b \frac{2mx^3}{b^2 - a^2} \, dx$$

where x is the distance from the axis of rotation. Show that $I = \frac{1}{2}m(a^2 + b^2)$.

13 The moment of inertia, I, of a rod of mass m and length $2r$ is given by

$$I = \int_0^{2r} \frac{mx^2}{2r} \, dx$$

where x is the distance from the axis of rotation. Show that $I = \frac{4mr^2}{3}$.

14 The moment of inertia, I, of a rod of length l and mass m is given by

$$I = \int_0^l \frac{mx^2 \sin^2(\theta)}{l} \, dx$$

where x is the distance along the rod and θ is the angle made between the rod and axis of rotation. Show that

$$I = \frac{ml^2 \sin^2(\theta)}{3}$$

15 The moment of inertia, J, of a disc of radius r and mass m is defined as

$$J = \int_0^r \frac{2m}{r^2} x^3 \, dx$$

Evaluate J.

*16 [fluid mechanics]   The momentum per unit time, M, for a fluid flow through a pipe of radius r is given by

$$M = k \int_0^r x^{2/7}(r - x) dx$$

where x is the distance from the pipe wall and k is a constant. Evaluate M.

17 [fluid mechanics]   The rate of flow, Q, of a fluid at a radius r in a pipe of radius R is given by

$$Q = \int_0^R (2\pi ur) dr$$

where u is the mean velocity, which is constant. Evaluate Q.

18 [signal processing]   The average voltage, v_{avg}, of a waveform with period T is given by

$$v_{avg} = \frac{1}{T} \int_0^T V \cos(\omega t + \theta) dt$$

Exercise **8(d) continued**

Solutions at end of book. Complete solutions available at www.palgrave.com/engineering/singh

where V is the peak voltage, ω is the angular frequency and θ is the phase. Evaluate v_{avg}.

19 [signal processing]   The mean, $\bar{x}$, of a signal over a period T is given by

$$\bar{x} = \frac{1}{T}\int_0^T a\cos(\omega t + \phi)\,dt$$

where a is the amplitude, ω is the angular frequency and ϕ is the phase. Determine $\bar{x}$.

Questions 20 to 26, inclusive, are in the field of [electrical principles].

20 The average value of current, I_{AV}, for a rectified alternating current is given by

$$I_{AV} = \frac{\omega}{\pi}\int_0^{\pi/\omega} i\,dt$$

where $i = I\sin(\omega t)$, ω is angular frequency, i is instantaneous current, I is peak current and t is time. Show that

$$I_{AV} = \frac{2}{\pi}I$$

21 The energy, W, stored in a capacitor of capacitance C charged with voltage V, is defined by

$$W = \int_0^V CV\,dV$$

Show that $W = \dfrac{1}{2}CV^2$.

22 A resistor shaped in the form of an annulus has an outer radius b and inner radius a and its resistance R is given by

$$R = \int_a^b \frac{\rho\,dr}{2\pi r}$$

where $a < r < b$ and ρ is the (constant) resistance per unit area. Evaluate R.

23 The voltage, v, across a capacitor is given by

$$v = \frac{1}{10 \times 10^{-6}}\int_0^t 100e^{-(5\times10^3)t}\,dt$$

Evaluate v.

24 The magnetising force, F, of a circular coil of radius r and at a distance d from the centre is given by

$$F = \int_0^{2\pi} \frac{I\,r\sin(\beta)}{4\pi(d^2 + r^2)}\,d\phi \quad \left[\frac{I\,r\sin(\beta)}{4\pi(d^2 + r^2)} \text{ is a constant}\right]$$

where I is the current carried by the coil, ϕ and β are angles. Evaluate F.

25 The magnetising force, F, of an infinite wire is given by

$$F = \int_0^{\pi} \frac{I\sin(\theta)}{4\pi r}\,d\theta$$

where r is perpendicular distance, θ is angle and I is current. Evaluate F.

26 The potential difference, V, between two concentric spheres of radii a and b is given by

$$V = \int_a^b \frac{Q}{4\pi\varepsilon_0 r^2}\,dr \text{ where } a < r < b \ (r \neq 0)$$

ε_0 is the permittivity constant and Q is charge. Show that

$$V = \frac{Q}{4\pi\varepsilon_0}\left(\frac{1}{a} - \frac{1}{b}\right) \text{ (where } a \neq 0, b \neq 0)$$

Questions 27 to 30, inclusive, are in the field of [reliability engineering].

27 The hazard function, $h(t)$, for a component is given by

$$h(t) = t^2 - 4t + 9 \quad (0 < t \leq 1)$$

Solutions at end of book. Complete solutions available at www.palgrave.com/engineering/singh

The failure density function, $f(t)$, is defined as

$$f(t) = h(t)\exp\left[-\int_0^t h(x)\,dx\right]$$

where exp is the exponential function. Determine $f(t)$.

28 The hazard function, $h(t)$, for a component is given by

$$h(t) = 3 - t \quad (0 < t \le 2)$$

Determine $f(t)$ where $f(t)$ is as defined in question **27**.

29 The failure density function, $f(t)$, for a set of components is given by

$$f(t) = 0.02(10 - t) \quad (0 < t \le 10)$$

The reliability function, $R(t)$, is defined as

$$R(t) = 1 - \int_0^t f(x)\,dx$$

and the hazard function, $h(t)$, is defined as

$$h(t) = \frac{f(t)}{R(t)}$$

Determine $R(t)$ and $h(t)$.

30 The hazard function, $h(t)$, is given by

$$h(t) = (2 \times 10^{-3})t^{-1/2}$$

The reliability function, $R(t)$, is defined as

$$R(t) = \exp\left[-\int_0^t h(x)\,dx\right]$$

Find $R(t)$.

⚙ **Questions 31 to 34, inclusive, are in the field of [thermodynamics].**

31 The work done, W, by a gas on a piston is given by

$$W = \int_{v_1}^{v_2} \frac{C}{V}\,dV$$

where C is a constant and V is volume. Show that $W = C\ln\left(\dfrac{v_2}{v_1}\right)$.

32 The work done, W, by a gas on a piston is given by

$$W = \int_{P_1}^{P_2} CP^{-1/n}\,dP$$

where P is pressure and C is constant. Evaluate W.

33 A gas expands according to the law

$$P_1 V_1^{k} = P_2 V_2^{k} = C$$

where P is pressure, V is volume, k and C are constants. The work done, W, by the gas is given by

$$W = \int_{V_1}^{V_2} CV^{-k}\,dV$$

Show that $W = \dfrac{P_1 V_1 - P_2 V_2}{k - 1}$.

34 A gas obeys the law

$$PV^{1.5} = C \quad \text{(constant)}$$

If the work done, $W = \displaystyle\int_{V_1}^{V_2} P\,dV$ then show that

$$W = 2C\left[\frac{1}{\sqrt{V_1}} - \frac{1}{\sqrt{V_2}}\right]$$

*35 🛩 [aerodynamics]   The drag coefficient, C_F, is defined as

$$C_F = k\int_0^1 \left(\frac{x}{L}\right)^{-\frac{1}{2}} d\left(\frac{x}{L}\right)$$

where k is a constant, L and x are lengths. Show that $C_F = 2k$.

SECTION E **Integration by parts**

By the end of this section you will be able to:
▶ understand the integration by parts formula
▶ apply the integration by parts formula
▶ apply the formula to engineering applications

This section might seem like a leap in sophistication from previous sections. You may initially find it difficult to understand the notation of the integration by parts formula.

E1 **Integration by parts formula**

The product rule for differentiation is given by

$$6.31 \qquad \frac{\mathrm{d}}{\mathrm{d}x}(uv) = u'v + uv'$$

where u and v are functions of x. The dash notation, ', represents derivatives. Integrating both sides of 6.31 :

$$uv = \int u'v\,\mathrm{d}x + \int uv'\mathrm{d}x$$

Rearranging yields

$$8.45 \qquad \int uv'\mathrm{d}x = uv - \int u'v\,\mathrm{d}x$$

8.45 is called the **integration by parts** formula. It is used to integrate **some** product of functions. If you have never seen this formula you might be confused with the symbols and notation. The v on the Right-Hand Side is the integral of v', that is

$$v = \int v'\mathrm{d}x$$

In general, the verbal form of the integration by parts formula is 'differentiate one (u) and integrate the other, (v')'.

Example 19

Find $\int xe^x\mathrm{d}x$.

Solution

Let $u = x$ and $v' = e^x$. Then to use the integration by parts formula, 8.45 , we need to differentiate u and integrate v':

$$u' = 1 \qquad v = \int e^x\mathrm{d}x = e^x$$

Example 19 *continued*

Don't worry about the constant (C) of integration, we add it on at the end. Substituting these into 8.45 :

$$\int xe^x dx = uv - \int vu' dx$$

$$= xe^x - \int (e^x)(1) dx$$

$$= xe^x - \int e^x dx$$

We have $\int xe^x dx = xe^x - \int e^x dx$.   Notice that we still need to integrate e^x, the last term on the Right-Hand Side. Hence

$$\int xe^x dx = xe^x - \underset{=\int e^x dx}{\underline{e^x}} + C$$

$$= e^x(x - 1) + C \qquad [\text{Taking out } e^x]$$

Observe that in the application of 8.45 , we still have an integral that we need to determine, $\int vu' dx$.

? **What is** $\dfrac{d}{dx}\left[e^x(x - 1) + C \right]$**?**

By inspecting **Example 19** we have

$$\frac{d}{dx}[e^x(x - 1) + C] = xe^x$$

because when we integrate xe^x we obtain $e^x(x - 1) + C$. Remember that differentiation is the reverse process of integration.

? **Why did we choose $u = x$ and $v' = e^x$ in Example 19?**

If we had selected u and v' the other way round, that is $u = e^x$ and $v' = x$ then the integration of xe^x becomes very complicated. Applying 8.45 gives

$$\int xe^x dx = e^x \frac{x^2}{2} - \int \frac{x^2}{2} e^x dx$$

The integral on the Right-Hand Side is more complicated than the integral of the given function, xe^x.

8.45 $\int uv' dx = uv - \int vu' dx$

? **In general, how do we know what to take *u* and *v′* to be?**

From previous experience we can establish some sort of ranking. Take *u* to be the function in the following order:

1 $\ln(x)$

2 x^n

3 e^{kx} where k is a constant

? **So if we want to find $\int x\ln(x)dx$, what is our *u* and *v′*?**

We take $u = \ln(x)$ and $v′ = x$ because $\ln(x)$ has priority over x. You can use this list to determine your *u* and *v′* of the given function. In general, you choose your *u* and *v′* such that the Right-Hand Side integral, $\int vu′dx$ of 8.45 , should be simpler than the original integral, $\int uv′dx$.

Example 20

Determine $\int x\ln(x)\,dx$

Solution

Let

$$u = \ln(x) \qquad\qquad v′ = x$$

$$u′ = \frac{1}{x} \quad \text{[Differentiating]} \quad v = \int x dx = \frac{x^2}{2} \qquad\qquad \text{[Integrating]}$$

Substituting these into the integration by parts formula 8.45 :

$$\int x\ln(x)dx = uv - \int u′v dx$$

$$= \ln(x)\left(\frac{x^2}{2}\right) - \int \left(\frac{1}{x}\right)\left(\frac{x^2}{2}\right)dx$$

$$= \left(\frac{x^2}{2}\right)\ln(x) - \frac{1}{2}\int x dx \quad \left[\text{Cancelling } x\text{'s and taking out } \frac{1}{2}\right]$$

$$= \left(\frac{x^2}{2}\right)\ln(x) - \frac{1}{2}\left(\frac{x^2}{2}\right) + C$$

$$= \frac{x^2}{2}\left(\ln(x) - \frac{1}{2}\right) + C \quad \left[\text{Taking out } \frac{x^2}{2}\right]$$

8.45 $\int uv′\,dx = uv - \int u′v\,dx$

E2 **Engineering applications of integration by parts**

Example 21 *electrical principles*

The energy, W, of an inductor is given by

$$* \qquad W = \int t\cos(t)\, dt$$

Determine W.

Solution

? **How do we integrate $t\cos(t)$?**

Since we have a product, $t \times \cos(t)$, we try using the integration by parts formula 8.45 .

? **What is u and v'?**

By using our list we see that t is on the list while $\cos(t)$ is **not**. Hence

$$u = t \quad \text{and} \quad v' = \cos(t)$$

Differentiating u and integrating v' gives

$$u' = 1 \qquad v = \int \cos(t)\, dt = \sin(t)$$

Putting these into 8.45 gives

$$\int t\cos(t)dt = uv - \int u'v\, dt$$

$$= t\sin(t) - \int (1)\sin(t)\, dt$$

$$= t\sin(t) - \int \sin(t)\, dt$$

$$= t\sin(t) - \underbrace{\left[-\cos(t) \right]}_{\text{by } 8.7} + C$$

$$W = \int t\cos(t)\, dt = t\sin(t) + \cos(t) + C$$

8.7 $\displaystyle\int \sin(t)dt = -\cos(t)$ 8.45 $\displaystyle\int uv'\, dx = uv - \int u'v\, dx$

Example 22 *mechanics*

The acceleration, $\ddot{x}$, of a particle is given by

$$\ddot{x} = te^{-t}$$

Find the velocity, $v = \dot{x}$, for the initial condition $t = 0$, $v = 0$.

Solution

$\dot{x}$ is obtained from $\ddot{x}$ by integrating $\ddot{x}$. Hence

$$v = \dot{x} = \int te^{-t}\mathrm{d}t$$

How do we integrate this function te^{-t}?

Use the integration by parts formula 8.45 :

$$u = t \qquad\qquad\qquad w' = e^{-t}$$

$$u' = 1 \quad \text{[Differentiating]} \qquad w = \int e^{-t}\mathrm{d}t = -e^{-t} \quad \text{[Integrating]}$$
$$\underset{\text{by } 8.41}{}$$

We use w because v represents velocity in this example.

Applying 8.45 :

$$v = \int te^{-t}\mathrm{d}t = uw - \int u'w\ \mathrm{d}t$$

$$= (t)(-e^{-t}) - \int (1)(-e^{-t})\mathrm{d}t$$

$$= -te^{-t} + \int e^{-t}\mathrm{d}t \qquad \text{[Simplifying]}$$

$$v = -te^{-t} - e^{-t} + C \qquad \text{[Integrating } e^{-t}\text{]}$$

Substituting the given conditions $t = 0$, $v = 0$ into $v = -te^{-t} - e^{-t} + C$ yields

$$0 = 0 - 1 + C \qquad \text{[Remember that } e^0 = 1\text{]}$$
$$C = 1$$

Hence putting $C = 1$ and taking out the common factor $-e^{-t}$ we have

$$v = -e^{-t}(t + 1) + 1 \text{ or } v = 1 - e^{-t}(t + 1)$$

For definite integrals the integration by parts formula is given by

8.46 $$\qquad \int_a^b uv'\mathrm{d}x = [uv]_a^b - \int_a^b u'v\mathrm{d}x$$

8.41 $$\int e^{kt+m}\ \mathrm{d}t = e^{kt+m}/k$$ 8.45 $$\int uv'\ \mathrm{d}x = uv - \int u'v\ \mathrm{d}x$$

Example 23 *electrical principles*

The energy stored, W, in an inductor is given by

$$W = \int_0^1 (10 \times 10^{-3})te^{-2t}dt$$

Find the exact value of W.

Solution

By taking out the constant, (10×10^{-3}), we have

$$W = (10 \times 10^{-3})\int_0^1 te^{-2t}dt$$

How do we find the definite integral $\int_0^1 te^{-2t}dt$?

Since the function is a product, we try the integration by parts formula for definite integrals, 8.46 . Differentiate u and integrate v':

$$u = t \qquad v' = e^{-2t}$$

$$u' = 1 \qquad v = \int e^{-2t}dt = -\frac{e^{-2t}}{2} \qquad \left[\text{By } 8.41 \right]$$

Thus, applying 8.46 with $a = 0$ and $b = 1$:

$$\int_0^1 te^{-2t}dt = [uv]_0^1 - \int_0^1 u'\,v\,dt$$

$$= \left[-\frac{te^{-2t}}{2}\right]_0^1 - \int_0^1\left(-\frac{e^{-2t}}{2}\right)dt$$

$$= -\frac{1}{2}\left(\underbrace{1e^{-2}}_{\substack{\text{substituting}\\ t=1}} - \underbrace{0}_{\substack{\text{substituting}\\ t=0}}\right) + \frac{1}{2}\int_0^1 e^{-2t}dt$$

We still need to integrate the last term on the Right-Hand Side of ★ :

$$\int_0^1 e^{-2t}dt \underset{\text{by } 8.41}{=} \left[\frac{e^{-2t}}{-2}\right]_0^1 = -\frac{1}{2}\left(e^{-2} - e^0\right) = -\frac{1}{2}\left(e^{-2} - 1\right)$$

8.41 $\displaystyle\int e^{(kt+m)}dt = e^{kt+m}/k$ 8.46 $\displaystyle\int_a^b uv'dt = [uv]_a^b - \int_a^b u'vdt$

> ### Example 23 *continued*
>
> Substituting this into ▪ * gives:
>
> $$\int_0^1 te^{-2t}dt = -\frac{1}{2}e^{-2} + \frac{1}{2}\left[-\frac{1}{2}\left(e^{-2} - 1\right)\right]$$
>
> $$= \frac{1}{2}\left[-e^{-2} - \frac{e^{-2}}{2} + \frac{1}{2}\right] \quad \left[\text{Taking out } \frac{1}{2} \text{ and opening the round brackets}\right]$$
>
> $$= \frac{1}{2}\left[-\frac{3e^{-2}}{2} + \frac{1}{2}\right] \quad \text{[Simplifying]}$$
>
> $$= \frac{1}{4}\left[1 - 3e^{-2}\right] \quad \left[\text{Taking out } \frac{1}{2}\right]$$
>
> The exact value of W is evaluated by putting this into ▪ † :
>
> $$W = \frac{10 \times 10^{-3}}{4}\left[1 - 3e^{-2}\right] = (2.5 \times 10^{-3})[\, 1 - 3e^{-2}] \text{ (joule)}$$

Sometimes we need to apply the integration by parts formula, 8.46 , twice as the following example shows.

> ### Example 24 *electrical principles*
>
> The energy stored, W, in an inductor is given by
>
> $$W = \int_0^2 (t^2 - 2t)e^{-2t}dt$$
>
> Evaluate W.
>
> Solution
>
> We need to use 8.46 . **What is u and v' in this formula?**
>
> By examining the priority list we let
>
> $$u = t^2 - 2t \qquad v' = e^{-2t}$$
>
> $$u' = 2t - 2 \qquad v = \int e^{-2t}dt = \frac{e^{-2t}}{-2} \qquad \left[\text{Using } \int e^{kt}\,dt = \frac{e^{kt}}{k}\right]$$
>
> Substituting these into 8.46 with $a = 0$ and $b = 2$ gives
>
> $$W = \int_0^2 (t^2 - 2t)e^{-2t}dt = [uv]_0^2 - \int_0^2 u'v\,dt$$
>
> $$= \left[(t^2 - 2t)\left(\frac{e^{-2t}}{-2}\right)\right]_0^2 - \int_0^2 (2t - 2)\left(\frac{e^{-2t}}{-2}\right)dt$$
>
> $$= \underbrace{\left((2^2 - (2 \times 2))\frac{e^{-4}}{-2} - 0\right)}_{= 0} + \frac{1}{2}\int_0^2 2(t - 1)e^{-2t}dt$$
>
> $$W = \int_0^2 (t - 1)e^{-2t}dt \qquad \text{[Cancelling 2's]}$$

8.46 $\displaystyle\int_a^b uv'\,dt = [uv]_a^b - \int_a^b u'v\,dt$

Example 24 *continued*

How can we find this integral?

Use 8.46 again since we have a product of $t - 1$ and e^{-2t}. Let

$$u = t - 1 \qquad v' = e^{-2t}$$

$$u' = 1 \qquad v = \int e^{-2t} dt = \frac{e^{-2t}}{-2}$$

Note that these u's and v's are different from those above. Putting these into the formula:

$$W = [uv]_0^2 - \int_0^2 u'\, v\, dt$$

$$= \left[(t - 1)\frac{e^{-2t}}{-2} \right]_0^2 - \int_0^2 (1)\frac{e^{-2t}}{-2}\, dt$$

$$= -\frac{1}{2}\left[(t - 1)e^{-2t} \right]_0^2 + \frac{1}{2}\int_0^2 e^{-2t} dt \qquad \left[\text{Taking out } -\frac{1}{2} \right]$$

$$= -\frac{1}{2}\left(\underbrace{(2 - 1)e^{-4}}_{\text{substituting } t = 2} - \underbrace{(0 - 1)e^0}_{\text{substituting } t = 0} \right) + \frac{1}{2}\int_0^2 e^{-2t} dt$$

$$= -\frac{1}{2}(e^{-4} + 1) + \frac{1}{2}\left[\frac{e^{-2t}}{-2} \right]_0^2$$

$$= -\frac{1}{2}(e^{-4} + 1) - \frac{1}{4}\left(\underbrace{e^{-4}}_{\text{substituting } t=2} - \underbrace{e^{0}}_{\text{substituting } t=0} \right)$$

$$= -\frac{1}{2}(e^{-4} + 1) - \frac{1}{4}(e^{-4} - 1) = -0.264 \qquad \text{[Evaluating]}$$

$$W = -0.264\,\text{J} \quad \text{(3 d.p.)}$$

The integration by parts formula **cannot** be applied to all products of functions. For example, to find $\int \cos(2x)\sin(3x)dx$ we would **not** use the integration by parts formula.

SUMMARY

Let u and v be functions of x, then

8.45 $\qquad \int uv'dx = uv - \int vu'dx$

8.45 is called the integration by parts formula and gives the integral of a product of functions.

Take u to be the function in the following order:
1 $\ln(x)$
2 x^n
3 e^{kx} where k is a constant

Exercise **8(e)**

Solutions at end of book. Complete solutions available at www.palgrave.com/engineering/singh

1 Determine the following integrals:

a $\displaystyle\int 2xe^x dx$ **b** $\displaystyle\int t\sin(t)\,dt$

2 Find **a** $\displaystyle\int q\cos(3q)\,dq$ **b** $\displaystyle\int s^2\ln(s)\,ds$

3 Obtain $\displaystyle\int te^{2t}\,dt$.

4 Show that

$$\int p\sqrt{1+p}\,dp = \frac{2}{15}(1+p)^{3/2}\,[3p-2] + C$$

5 By writing $\ln(x) = 1\times\ln(x)$, show that

$$\int \ln(x)\,dx = x[\ln(x)-1] + C$$

6 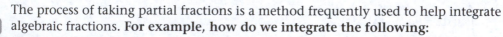 [*electrical principles*]   The current, i, through an inductor with inductance L is given by

$$i = \frac{1}{L}\int_0^t v\,dt$$

where v is the voltage across the inductor. For a circuit with $L = 10\times10^{-3}$ H and $v = 5te^{-t}$, find i.

7 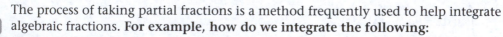 [*electrical principles*]   The energy, w, of an inductor with inductance L is given by

$$w = \frac{1}{2L}\left(\int_0^t v\,dt\right)^2$$

where v is the voltage across the inductor. For $L = 10$ mH and $v = t\cos(t)$, find w.

8 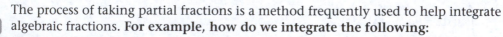 [*electrical principles*]   The current, i, through an inductor with inductance L is given by

$$i = \frac{1}{L}\int_0^1 v\,dt$$

For $L = (1\times10^{-3})$ H and $v = t^2e^{-t}$, find i.

SECTION F **Algebraic fractions**

By the end of this section you will be able to:
▶ understand what is meant by the term partial fraction
▶ find partial fractions of various expressions
▶ understand the procedure for finding partial fractions
▶ find partial fractions of improper fractions

The process of taking partial fractions is a method frequently used to help integrate algebraic fractions. **For example, how do we integrate the following:**

$$\int \frac{x}{(x+2)(x+1)}dx?$$

We can't use any of the techniques discussed earlier. We are compelled to express $\dfrac{x}{(x+2)(x+1)}$ as partial fractions and then integrate the result. This section could have been placed in the earlier **Chapters 1** and **2** on algebra, however partial fractions is a difficult topic and requires greater algebraic skills.

F1 **Partial fractions**

Consider the following example:

$$\frac{7}{12} = \frac{1}{4} + \frac{1}{3}$$

We say that $\frac{1}{4}$ and $\frac{1}{3}$ are partial fractions of $\frac{7}{12}$. The two or more fractions which create a single fraction are called **partial fractions**.

Consider the following algebraic expression:

$$\frac{x}{(x + 2)(x + 1)} = \frac{2}{x + 2} - \frac{1}{x + 1}$$

? **What are the partial fractions of $\frac{x}{(x + 2)(x + 1)}$?**

Clearly they are $\frac{2}{x + 2}$ and $\frac{-1}{x + 1}$.

In this section we study the processes of splitting a fraction such as $\frac{x}{(x + 2)(x + 1)}$ into its partial fractions.

? **How do we split $\frac{x}{(x + 2)(x + 1)}$ into its partial fractions?**

First we write this as

$$\dagger \qquad \frac{x}{(x + 2)(x + 1)} = \frac{A}{x + 2} + \frac{B}{x + 1}$$

where A and B are constants which we need to find.

? **How do we know that $\frac{x}{(x + 2)(x + 1)}$ breaks into $\frac{A}{x + 2} + \frac{B}{x + 1}$?**

Well, if we add the fractions

$$\frac{A}{x + 2} + \frac{B}{x + 1}$$

then we obtain a common denominator of $(x + 2)(x + 1)$. Thus

$$\frac{A}{x + 2} + \frac{B}{x + 1} = \frac{A(x + 1) + B(x + 2)}{(x + 2)(x + 1)}$$

So all we are left to achieve is to find the constants A and B so that the numerator gives an x.

$$\dagger\dagger \qquad \frac{x}{(x + 2)(x + 1)} = \frac{A(x + 1) + B(x + 2)}{(x + 2)\ (x + 1)}$$

Since on both sides of the equal sign we have a common denominator, the numerator must be the same, that is

$$* \qquad x = A(x + 1) + B(x + 2)$$

This can also be attained by multiplying both sides of ⚹⚹ by $(x + 2)(x + 1)$.

? **But how can we find A and B?**

We can choose values of x and substitute these values into ▮ . The evaluations of A and B are simpler if we select our x values such that some of the terms on the Right-Hand Side of ▮ vanish. For example, choosing $x = -1$ gives

$$-1 = A(-1 + 1) + B(-1 + 2)$$
$$-1 = 0 + B$$

Hence $B = -1$.

? **How can we find A?**

By choosing $x = -2$ and substituting into ▮ :

$$-2 = A(-2 + 1) + B(-2 + 2)$$
$$-2 = A(-1) + 0$$

Hence $A = 2$.

? **Why did we choose x = −2?**

Because this removes the B term of ▮ .

Putting $A = 2$ and $B = -1$ into † gives the identity

$$\frac{x}{(x + 2)(x + 1)} = \frac{2}{x + 2} - \frac{1}{x + 1}$$

We have broken the single fraction, $\dfrac{x}{(x + 2)(x + 1)}$, into the difference of two fractions, $\dfrac{2}{x + 2} - \dfrac{1}{x + 1}$. This is what we started with at the beginning of this discussion.

For the appropriate values of the constants, the partial fractions become an identity; that is, both sides are equal for all values of x.

Let's investigate another example.

Example 25

Resolve

$$\frac{2x}{x^2 - x - 2}$$

into partial fractions.

Solution

We first try to place the denominator in the form of two bracketed terms, that is to factorize the denominator $x^2 - x - 2$. This factorizes into

$$x^2 - x - 2 = (x + 1)(x - 2)$$

Therefore

$$\frac{2x}{x^2 - x - 2} = \frac{2x}{(x + 1)(x - 2)} = \frac{A}{x + 1} + \frac{B}{x - 2}$$

Multiply both sides by $(x + 1)(x - 2)$:

$$2x = \frac{A(x + 1)(x - 2)}{(x + 1)} + \frac{B(x + 1)(x - 2)}{(x - 2)}$$
$$= A(x - 2) + B(x + 1) \quad \text{[Cancelling]}$$

We have

▮ $\qquad 2x = A(x - 2) + B(x + 1)$

Example 25 *continued*

? **Which values of x should we select to find A and B?**

Putting $x = 2$ into ⬛ * removes the A term:

$$2 \times 2 = A(2 - 2) + B(2 + 1)$$
$$4 = 3B$$
$$B = \frac{4}{3}$$

To find A, put $x = -1$ into ⬛ * which removes the B term:

$$2 \times (-1) = A(-1 - 2) + B(0)$$
$$-2 = -3A$$
$$A = \frac{2}{3}$$

Putting $A = \frac{2}{3}$ and $B = \frac{4}{3}$ into

$$\frac{2x}{x^2 - x - 2} = \frac{A}{x + 1} + \frac{B}{x - 2}$$

gives

⬛ **
$$\frac{2x}{x^2 - x - 2} = \frac{\frac{2}{3}}{x + 1} + \frac{\frac{4}{3}}{x - 2}$$

Consider the first term on the Right-Hand Side of ⬛ ** :

$$\frac{\frac{2}{3}}{x + 1} = \frac{2}{3} \div (x + 1) = \frac{2}{3} \div \frac{(x + 1)}{1} = \frac{2}{3} \times \frac{1}{(x + 1)}$$

$$= \frac{2 \times 1}{3 \times (x + 1)} = \frac{2}{3(x + 1)}$$

Similarly the second term on the Right-Hand Side of ⬛ ** becomes

$$\frac{\frac{4}{3}}{x - 2} = \frac{4}{3(x - 2)}$$

Combining these two results we have the identity

$$\frac{2x}{x^2 - x - 2} = \frac{2}{3(x + 1)} + \frac{4}{3(x - 2)}$$

$$= \frac{2}{3}\left[\frac{1}{x + 1} + \frac{2}{x - 2}\right] \qquad \left[\text{Taking out } \frac{2}{3}\right]$$

Consider a single algebraic fraction

8.47
$$\frac{f(x)}{g(x)}$$

? where $g(x)$ factorizes and is of a higher degree polynomial than $f(x)$. **What do we mean by a higher degree polynomial?**

For example, if $g(x)$ is a cubic (containing x^3) then $f(x)$ is at most a quadratic (containing x^2). See Table 3.

TABLE 3	Polynomial	Degree	Name
	$x - 5$	1	linear
	$x^2 + x + 1$	2	quadratic
	$x^3 - x^2$	3	cubic
	$x^4 + x - 3$	4	quartic

In order to write 8.47 , $\dfrac{f(x)}{g(x)}$, as a sum of partial fractions, we first factorize the denominator, $g(x)$, and then use one of the following rules.

Each linear factor of the denominator has partial fractions of the form $\dfrac{A}{ax + b}$. Thus

8.48
$$\frac{f(x)}{(ax + b)(cx + d)} = \frac{A}{ax + b} + \frac{B}{cx + d}$$

Each repeated linear factor of the denominator has partial fractions of the form $\dfrac{A}{ax + b} + \dfrac{B}{(ax + b)^2}$. Thus

8.49
$$\frac{f(x)}{(ax + b)^2} = \frac{A}{ax + b} + \frac{B}{(ax + b)^2}$$

Each quadratic factor of the denominator has the partial fractions of the form $\dfrac{Ax + B}{ax^2 + bx + c}$.

Hence

8.50
$$\frac{f(x)}{ax^2 + bx + c} = \frac{Ax + B}{ax^2 + bx + c}$$

A combination of linear and quadratic factors gives

8.51
$$\frac{f(x)}{(ax^2 + bx + c)(dx + e)} = \frac{Ax + B}{ax^2 + bx + c} + \frac{C}{dx + e}$$

For the appropriate values of A, B and C, 8.48 to 8.51 are identities.

These identities look horrendous, but once we do a few examples you will become familiar with these identities. The procedure for finding the partial fractions of

$$8.47 \qquad \frac{f(x)}{g(x)}$$

is

1 Factorize the denominator, $g(x)$, as far as possible.
2 Write 8.47 as one of the general partial fractions by choosing the appropriate identity from 8.48 to 8.51 .
3 Find the values of the unknown constants A, B, C, etc.

Let's do an example.

Example 26

Express $\dfrac{3t^2 - t + 1}{(t^2 - t + 3)(t - 1)}$ as partial fractions.

Solution

The denominator does not factorize further into simpler factors. Each factor of the denominator, $t^2 - t + 3$ and $t - 1$, produces a partial fraction. **Which identity, 8.48 to 8.51 , is appropriate for**

$$\frac{3t^2 - t + 1}{(t^2 - t + 3)(t - 1)}?$$

Since we have quadratic and linear factors we use 8.51 :

$$† \qquad \frac{3t^2 - t + 1}{(t^2 - t + 3)(t - 1)} = \frac{At + B}{t^2 - t + 3} + \frac{C}{t - 1}$$

What do we need to find?

The constants A, B and C. **How can we find A, B and C?**

Multiplying both sides of † by $(t^2 - t + 3)(t - 1)$ we have

$$†† \qquad 3t^2 - t + 1 = (At + B)(t - 1) + C(t^2 - t + 3)$$

To remove the first term on the Right-Hand Side of †† we substitute $t = 1$ into †† :

$$3 - 1 + 1 = 0 + C(1^2 - 1 + 3)$$
$$3 = 3C \text{ gives } C = 1$$

Putting $C = 1$ into †† yields

$$* \qquad 3t^2 - t + 1 = (At + B)(t - 1) + (t^2 - t + 3)$$

How can we find A and B?

We can now substitute other values of t such as $t = 0$, but it is generally easier to **equate coefficients**. Conventionally we start to equate coefficients of the highest power first. So we first equate the number of t^2 on the left of * to the number of t^2 on the right of * .

Example 26 *continued*

We need to expand the Right-Hand Side of [*] because $At \times t$ gives At^2 and so the number of t^2 on the right is $A + 1$.

How many t^2 are on the Left-Hand Side of [*]?

3. Thus equating coefficients of t^2 we have

$$3 = A + 1 \text{ therefore } A = 2$$

Equating coefficients of t in [*] by expanding the Right-Hand Side of [*] gives

$$At \times (-1) = -At \text{ and } B \times t = Bt \text{ and } -t$$
$$-1 = -A + B - 1 \qquad\qquad [\text{Equating } t\text{'s of } [*]]$$

Substituting $A = 2$ yields

$$-1 = -2 + B - 1$$
$$-1 = -3 + B \text{ gives } B = 2$$

So we have $A = 2$, $B = 2$ and $C = 1$. Substituting these into [†] displays the partial fractions

$$\frac{3t^2 - t + 1}{(t^2 - t + 3)(t - 1)} = \frac{2t + 2}{t^2 - t + 3} + \frac{1}{t - 1}$$

F2 Improper fractions

If in the algebraic fraction

$$\frac{f(x)}{g(x)}$$

$f(x)$ is a polynomial of higher, or same, degree compared with $g(x)$ then $\frac{f(x)}{g(x)}$ is called an **improper fraction**. The degree of a polynomial, $f(x)$, is the highest power of x contained in the polynomial.

To find the partial fractions of an improper fraction, $\frac{f(x)}{g(x)}$, we first divide $f(x)$ by $g(x)$ by applying long division. If this division results in a polynomial $q(x)$ with remainder $R(x)$ then we can write

8.52 $$\frac{f(x)}{g(x)} = q(x) + \frac{R(x)}{g(x)}$$

where the remainder $R(x)$ is a polynomial of lower degree than $g(x)$. We can now treat $\frac{R(x)}{g(x)}$ in the usual way for partial fractions.

Let's first do an example on arithmetic long division. If you are not familiar with arithmetic long division, then you will need to revise this using the example at the top of the next page before continuing any further on this subsection.

Long division

Applying long division to $200 \div 15$

$\begin{array}{r} 0 \\ 15\overline{)200} \end{array}$	15 does **not** go into 2, therefore we place a 0 above the 2.
$\begin{array}{r} 01 \\ 15\overline{)200} \\ -15 \\ \hline 5 \end{array}$	However, 15 will go into the first two digits which are 20. Since $$15 \times 1 = 15$$ we place the 1 next to the 0 in the top row because 15 goes into 20 once. We subtract 15 from 20 to give us a remainder of 5.
$\begin{array}{r} 01 \\ 15\overline{)200} \\ -15\downarrow \\ \hline 50 \end{array}$	15 does **not** go into 5, so we bring down the 0 from the divisor to make 50.
$\begin{array}{r} 013 \\ 15\overline{)200} \\ -15\downarrow \\ \hline 50 \\ -45 \\ \hline 5 \end{array}$	Does 15 go into 50? Yes, $15 \times 3 = 45$. We place a 3 next to the 1 in the top row because three lots of 15 go into 50. Subtracting 45 from 50 leaves a remainder of 5. We know 15 will not go into 5 and there are no more numbers to bring down, hence 15 will go into 200 exactly 13 times with a remainder of 5.

We can write this result as:

$$\frac{200}{15} = 13 + \frac{5}{15}$$

Example 27

a Express $\dfrac{x^3}{x^2 - 4}$ in partial fractions.

b Divide $x^3 + 2x^2 - 3x + 1$ by $x - 1$.

Solution

a Since x^3 is of degree 3 and $x^2 - 4$ is of degree 2 we first apply long division.

Long division of algebraic expressions is a similar procedure to long division of numbers. We need to divide x^3 by $x^2 - 4$:

$$x^2 - 4\overline{)x^3}$$

What factor do we need to multiply $x^2 - 4$ by in order to get x^3?

x because $x(x^2 - 4) = x^3 - 4x$. We place the multiple x on the top line of the above as:

$$\begin{array}{r} x \\ x^2 - 4\,\overline{)x^3 + 0} \\ \underline{x^3 - 4x} \end{array} \quad \longleftarrow x(x^2 - 4) = x^3 - 4x \text{ goes here}$$

Now subtract $x^3 - 4x$ from the above algebraic expression:

Example 27 *continued*

$$x^2 - 4 \overline{\smash{)}\, x^3 + 0} \quad \frac{x}{}$$

$$\frac{-(x^3 - 4x)}{0 + 4x} \quad [\text{Subtracting } x^3 - 4x \text{ from } x^3]$$

Using **8.52** with $q(x) = x$ and $R(x) = 4x$ we have

†
$$\frac{x^3}{x^2 - 4} = x + \frac{4x}{x^2 - 4}$$

8.52 puts the expression $\dfrac{x^3}{x^2 - 4}$ into partial fractions, but we can cut it into even more

partial fractions because the denominator, $x^2 - 4$, factorizes. Thus

$$x^2 - 4 = x^2 - 2^2 = (x - 2)(x + 2) \quad [\text{Remember} \quad a^2 - b^2 = (a - b)(a+b)]$$

? Which identity, **8.48** to **8.51**, can we use to place $\dfrac{4x}{(x - 2)(x + 2)}$ into partial fractions?

Use

8.48
$$\frac{f(x)}{(ax + b)(cx + d)} = \frac{A}{ax + b} + \frac{B}{cx + d}$$

††
$$\frac{4x}{(x - 2)(x + 2)} = \frac{A}{x - 2} + \frac{B}{x + 2}$$

Multiplying both sides by $(x - 2)(x + 2)$ gives

*
$$4x = A(x + 2) + B(x - 2)$$

What are we trying to find?

The values of the constants A and B.

? **What should we substitute for x into** ***** **to obtain A and B?**

Putting $x = 2$ into ***** gives

$$(4 \times 2) = A(2 + 2) + 0$$
$$8 = 4A \text{ which gives } A = 2$$

Putting $x = -2$ into ***** gives

$$[4 \times (-2)] = 0 + B(-2 - 2)$$
$$-8 = -4B \text{ which gives } B = 2$$

Substituting $A = 2$, $B = 2$ into **††** yields

$$\frac{4x}{(x - 2)(x + 2)} = \frac{2}{x - 2} + \frac{2}{x + 2} \underset{\substack{\text{taking out the} \\ \text{common factor 2}}}{\equiv} 2\left[\frac{1}{x - 2} + \frac{1}{x + 2}\right]$$

Putting $\dfrac{4x}{x^2 - 4} = 2\left[\dfrac{1}{x - 2} + \dfrac{1}{x + 2}\right]$ into **†** gives the identity

$$\frac{x^3}{x^2 - 4} = x + 2\left[\frac{1}{x - 2} + \frac{1}{x + 2}\right]$$

Example 27 *continued*

b Applying our long division process we have

$$x - 1{\overline{\smash{\big)}\,x^3 + 2x^2 - 3x + 1}}$$

? **What do we need to multiply $x - 1$ by in order to get $x^3 + 2x^2 - 3x + 1$?**

x^2 because $x^2(x - 1) = x^3 - x^2$. We place the multiple x^2 on the top line:

$$
\begin{array}{r}
x^2 \\
x - 1{\overline{\smash{\big)}\,x^3 + 2x^2 - 3x + 1}} \\
\underline{-(x^3 - x^2)} \\
0 + 3x^2
\end{array}
$$
[Subtracting $x^3 - x^2$ from $x^3 + 2x^2 - 3x + 1$]

? **What is our next step?**

? Bring down the $- 3x$ so that we have $3x^2 - 3x$. **What multiple of $x - 1$ goes into $3x^2 - 3x$?**

$3x$ because $3x(x - 1) = 3x^2 - 3x$. We have

$$
\begin{array}{r}
x^2 + 3x \\
x - 1{\overline{\smash{\big)}\,x^3 + 2x^2 - 3x + 1}} \\
\underline{-(x^3 - x^2)} \quad\downarrow \\
3x^2 - 3x \\
\underline{-(3x^2 - 3x)} \\
0 - 0
\end{array}
$$
[Subtracting $3x^2 - 3x$ from $3x^2 - 3x$]

Finally $x - 1$ will **not** go into the 1 on the Right-Hand Side and we have **no** more expressions to bring down, therefore we have a remainder of 1. Hence we write

$$\frac{x^3 + 2x^2 - 3x + 1}{x - 1} = x^2 + 3x + \frac{1}{x - 1}$$

SUMMARY

The procedure for finding partial fractions of

8.47 $\dfrac{f(x)}{g(x)}$

where $g(x)$ factorizes and is of a higher degree than $f(x)$ is

1 factorize the denominator

2 write 8.47 as one of the general partial fractions determined by 8.48 to 8.51

3 find the values of the unknown constants A, B, C, etc.

If $f(x)$ is of the same or higher degree than $g(x)$ in 8.47, then this is called an improper fraction and we need to apply long division.

Solutions at end of book. Complete solutions available at www.palgrave.com/engineering/singh

Exercise **8(f)**

1 Express the following in partial fractions:

a $\dfrac{3x + 4}{(x + 1)(x + 2)}$ **b** $\dfrac{2t}{t^2 - 1}$

c $\dfrac{2s + 7}{s^2 + s - 2}$ **d** $\dfrac{-12u - 13}{(2u + 1)(u - 3)}$

2 Resolve the following into partial fractions:

a $\dfrac{x^2}{x + 1}$ **b** $\dfrac{x^5 - 2x^2}{x^2 - 1}$

3 Resolve the following into partial fractions:

a $\dfrac{4t^2 + t - 3}{(t^2 + t - 1)(t - 1)}$

b $\dfrac{z + 1}{(z - 1)^2}$

*c $\dfrac{2x^3 + 3x^2 + 5x + 2}{(x^2 + x + 1)^2}$

SECTION G **Integration of algebraic fractions**

By the end of this section you will be able to:

▶ integrate $\dfrac{f(x)}{g(x)}$ by using partial fractions

G1 **Integration by partial fractions**

In this section we integrate $\dfrac{f(x)}{g(x)}$ by expressing this in its partial fractions and then integrating. Generally you will discover that once we have found the partial fractions then to integrate we employ formula **8.42** :

8.42
$$\int \frac{f'(x)}{f(x)} \, dx = \ln|f(x)| + C$$

Example 28

Determine $\displaystyle\int \frac{s}{(s + 2)(s + 1)} \, ds$

Solution

First we express $\dfrac{s}{(s + 2)(s + 1)}$ in partial fractions. This was obtained in **Section F1** on page 442 with x in place of s:

$$\frac{s}{(s + 2)(s + 1)} = \frac{2}{s + 2} - \frac{1}{s + 1}$$

> **Example 28** *continued*
>
> **?** **How do we integrate** $\dfrac{s}{(s+2)(s+1)}$ **with respect to s?**
>
> $$\int \frac{s}{(s+2)(s+1)}\, ds = \int \left(\frac{2}{s+2} - \frac{1}{s+1} \right) ds$$
>
> $$= \int \frac{2ds}{s+2} - \int \frac{ds}{s+1} \qquad \text{[Splitting integrand]}$$
>
> $$= 2\int \frac{ds}{s+2} - \int \frac{ds}{s+1} \qquad \text{[Taking out 2]}$$
>
> $$= 2\ln|s+2| - \ln|s+1| + C \quad \text{[Integrating]}$$
> $$\text{by } 8.42$$
>
> $$= \ln(|s+2|^2) - \ln|s+1| + C$$
> $$\text{by } 5.14$$
>
> $$= \ln\left| \frac{(s+2)^2}{s+1} \right| + C \qquad \left[\text{By } 5.13 \right]$$

Note that to simplify the result of integration we apply the laws of logarithms. This is generally the case for integration by partial fractions and so make sure that the laws of logarithms, 5.12 to 5.14, are second nature to you.

> **Example 29**
>
> Find $\displaystyle\int \frac{2x}{x^2 - x - 2}\, dx$
>
> **Solution**
>
> We have already placed the integrand, $\dfrac{2x}{x^2-x-2}$, in partial fractions in **Example 25** on page 443, thus
>
> $$\frac{2x}{x^2-x-2} = \frac{2}{3}\left[\frac{1}{x+1} + \frac{2}{x-2} \right]$$
>
> Carrying out the integration by splitting the integrand and taking out the constant, 2/3, we have
>
> $$\int \frac{2x}{x^2-x-2}\, dx = \frac{2}{3}\int \left[\frac{1}{x+1} + \frac{2}{x-2} \right] dx$$
>
> $$= \frac{2}{3}\left[\int \frac{1}{x+1}\, dx + \int \frac{2}{x-2}\, dx \right]$$
>
> $$= \frac{2}{3}\left[\ln|x+1| + 2\ln|x-2| \right] + C \quad \text{[Integrating by using } 8.42 \text{]}$$
>
> $$= \frac{2}{3}\left[\ln|x+1| + \ln|x-2|^2 \right] + C$$
> $$\text{by } 5.14$$

5.13 $\ln(A) - \ln(B) = \ln(A/B)$ 5.14 $n\ln(A) = \ln(A^n)$

The next example is long but not difficult. First we need to find the partial fractions and then integrate.

🚗 | **Example 30** *mechanics*

The velocity, v, of an object falling in air at time t is given by

$$t = \int_0^v \frac{dv}{9 - 0.25v^2}$$

By using partial fractions, show that

$$t = \frac{1}{3}\ln\left|\frac{3 + 0.5v}{3 - 0.5v}\right|$$

Solution

First we need to factorize the denominator, $9 - 0.25v^2$:

$$9 - 0.25v^2 = 3^2 - (0.5v)^2$$
$$= (3 - 0.5v)(3 + 0.5v) \qquad [\text{Using } a^2 - b^2 = (a-b)(a+b)]$$

By placing into partial fractions we have

† $\qquad \dfrac{1}{9 - 0.25v^2} = \dfrac{1}{(3 - 0.5v)(3 + 0.5v)} \underset{\text{by } 8.48}{=} \dfrac{A}{3 - 0.5v} + \dfrac{B}{3 + 0.5v}$

Multiplying both sides by $(3 - 0.5v)(3 + 0.5v)$ gives

★ $\qquad 1 = A(3 + 0.5v) + B(3 - 0.5v)$

? | **How can we find A and B?**

Put $v = -6$ into **★** because this will remove the A term and therefore we can find B:

$$1 = A[3 + 0.5(-6)] + B[3 - 0.5(-6)]$$
$$1 = 0 + 6B$$

Thus $B = \dfrac{1}{6}$. Putting $v = 6$ into **★** gives

$$1 = A[3 + (0.5 \times 6)] + 0$$
$$1 = 6A$$

So $A = \dfrac{1}{6}$. Substituting $A = \dfrac{1}{6}$ and $B = \dfrac{1}{6}$ into **†** gives the partial fractions

$$\frac{1}{9 - 0.25v^2} = \frac{\frac{1}{6}}{3 - 0.5v} + \frac{\frac{1}{6}}{3 + 0.5v}$$

8.48 $\qquad \dfrac{f(v)}{(av + b)(cv + d)} = \dfrac{A}{av + b} + \dfrac{B}{cv + d}$

Example 30 *continued*

$$= \frac{1}{6}\left[\frac{1}{3-0.5v} + \frac{1}{3+0.5v}\right] \quad \left[\text{Taking out } \frac{1}{6}\right]$$

Considering the given integral we have

$$t = \int_0^v \frac{dv}{9-0.25v^2} = \frac{1}{6}\int_0^v\left(\frac{1}{3-0.5v} + \frac{1}{3+0.5v}\right)dv$$

†† $$t = \frac{1}{6}\left[\int_0^v \frac{dv}{3-0.5v} + \int_0^v \frac{dv}{3+0.5v}\right] \quad \text{[Splitting integrand]}$$

How do we find the first integral on the Right-Hand Side, $\int_0^v \dfrac{dv}{3-0.5v}$?

We can spot that differentiating the denominator gives -0.5. If we want to use

8.42 $$\int \frac{f'(v)}{f(v)}\,dv = \ln|f(v)| + C$$

then we need to have the derivative of the denominator on the numerator.
We can rewrite the integrand as follows:

$$\frac{1}{3-0.5v} = \frac{1}{-0.5}\left(\frac{-0.5}{3-0.5v}\right) = -2\left(\frac{-0.5}{3-0.5v}\right) \quad \left[\text{Because } \frac{1}{-0.5} = -2\right]$$

Evaluating the integral:

$$\int_0^v \frac{dv}{3-0.5v} = -2\int_0^v \frac{-0.5}{3-0.5v}\,dv$$

$$= -2\left[\underbrace{\ln|3-0.5v|}_{\text{by } 8.42}\right]_0^v$$

$$= -2\big[\ln|3-0.5v| - \ln(3)\big] \quad \text{[Substituting the limits]}$$

Similarly

$$\int_0^v \frac{dv}{3+0.5v} = \frac{1}{0.5}\int_0^v \frac{0.5\,dv}{3+0.5v}$$

$$= \frac{1}{0.5}\left[\ln|3+0.5v|\right]_0^v \quad \text{[Integrating]}$$

$$= 2\big[\ln|3+0.5v| - \ln(3)\big] \quad \text{[Substituting the limits]}$$

Substituting these results into †† produces

$$t = \frac{1}{6}\left[-2\big[\ln|3-0.5v| - \ln(3)\big] + 2\big[\ln|3+0.5v| - \ln(3)\big]\right]$$

$$= \frac{2}{6}\left[\ln(3) - \ln|3-0.5v| + \ln|3+0.5v| - \ln(3)\right] \quad \text{[Taking out 2]}$$

$$= \frac{1}{3}\left[\ln|3+0.5v| - \ln|3-0.5v|\right]$$

8.42 $$\int \frac{f'(v)}{f(v)}\,dv = \ln|f(v)|$$

Example 30 *continued*

$$= \frac{1}{3} \ln \underbrace{\left| \frac{3 + 0.5v}{3 - 0.5v} \right|}_{\text{by } 5.13} \quad \text{[Simplifying]}$$

There is an effortless way of obtaining the above. We can use

8.30
$$\int \frac{du}{a^2 - u^2} = \frac{1}{2a} \ln \left| \frac{a + u}{a - u} \right|$$

(this is given in Table 2 on page 410). Thus

$$\int_0^v \frac{dv}{9 - 0.25v^2} = \int_0^v \frac{dv}{3^2 - (0.5v)^2}$$

Let $u = 0.5v$, then

$$\frac{du}{dv} = 0.5 \text{ gives } dv = \frac{du}{0.5} = 2du$$

We can first evaluate the integral without the limits:

$$\int \frac{dv}{3^2 - (0.5v)^2} = \int \frac{2du}{3^2 - u^2} = 2\int \frac{du}{3^2 - u^2}$$

$$= 2\left[\frac{1}{2 \times 3} \ln \left| \frac{3 + u}{3 - u} \right| \right] \quad \left[\text{By } 8.30 \right]$$

$$= \frac{1}{3} \ln \left| \frac{3 + u}{3 - u} \right| \quad \text{[Cancelling 2s]}$$

$$= \frac{1}{3} \ln \left| \frac{3 + 0.5v}{3 - 0.5v} \right| \quad \text{[Remember that } u = 0.5v\text{]}$$

Using the limits of integration:

$$\int_0^v \frac{dv}{9 - 0.25v^2} = \frac{1}{3} \left[\ln \left| \frac{3 + 0.5v}{3 - 0.5v} \right| \right]_0^v$$

$$= \frac{1}{3} \left[\ln \left| \frac{3 + 0.5v}{3 - 0.5v} \right| - \ln \left| \frac{3}{3} \right| \right] \quad \text{[Substituting the limits]}$$

$$= \frac{1}{3} \ln \left| \frac{3 + 0.5v}{3 - 0.5v} \right| \quad \left[\text{Because } \ln \left| \frac{3}{3} \right| = \ln|1| = 0 \right]$$

SUMMARY

Integration by partial fractions involves first putting the function into partial fractions and then integrating. The most important integral formula for these fractions is 8.42 . Generally we need to use the laws of logarithms to simplify.

5.13 $\ln(A) - \ln(B) = \ln(A/B)$ 8.30 $\int \frac{du}{a^2 - u^2} = \frac{1}{2a} \ln \left| \frac{a + u}{a - u} \right|$ 8.42 $\int \frac{f'(v)}{f(v)} dv = \ln|f(v)|$

Exercise 8(g)

Solutions at end of book. Complete solutions available at www.palgrave.com/engineering/singh

1 Determine

a $\displaystyle\int \frac{3c + 4}{(c + 1)(c + 2)}\, dc$

b $\displaystyle\int \frac{2\lambda}{\lambda^2 - 1}\, d\lambda$

c $\displaystyle\int \frac{2a + 7}{a^2 + a - 2}\, da$

d $\displaystyle\int \frac{-12y - 13}{(2y + 1)(y - 3)}\, dy$

2 Determine

a $\displaystyle\int \frac{4p^2 + p - 3}{(p^2 + p - 1)(p - 1)}\, dp$

b $\displaystyle\int \frac{z + 1}{(z - 1)^2}\, dz$

3 Evaluate $\displaystyle\int_1^2 \frac{5z^2}{(z^2 + 1)(2z - 1)}\, dz$

4 🚗 [*mechanics*] The velocity, v, of an object moving in a medium at time t is given by

$$t = \int_{10}^{100} \frac{dv}{v(2v + 1)}$$

Evaluate t.

5 🚗 [*mechanics*] The velocity, v, of an object falling in air is given by

$$v = \int \frac{dv}{1 - (kv)^2}$$

where k is a non-zero constant. By using partial fractions, show that

$$v = \frac{1}{2k}\ln\left|\frac{1 + kv}{1 - kv}\right| + C$$

*6 Show that $\displaystyle\int_0^1 \frac{5 + 2x - x^2}{(x^2 + 1)(x + 1)}\, dx = \pi$.

7 i Express $\dfrac{x^3 + 1}{x^2 + 3x + 2}$ in partial fractions.

ii Determine $\displaystyle\int \frac{x^3 + 1}{x^2 + 3x + 2}\, dx$

SECTION H **Integration by substitution revisited**

By the end of this section you will be able to:
▶ evaluate integrals by substitution
▶ integrate $k\big[f(t)\big]^n f'(t)$

H1 **Integration by substitution**

In **Section C** of this chapter we found indefinite integrals (integrals without limits) by using a substitution. **Why did we use a substitution?**

The integral of interest was not among the list of standard integrals of Table 2. Additionally, the evaluation of the integral became elementary once we used the substitution.

Since then we have developed other methods of integration such as integration by parts and by partial fractions. In this section we examine integrals with limits which demand a substitution. Let's do an example.

Example 31

Evaluate $\displaystyle\int_1^4 \frac{t}{\sqrt{3t^2+1}}\, dt$.

Solution

Can you identify a substitution that we can use to evaluate this integral?

One choice might be to let $u = 3t^2 + 1$ because differentiating this results in

$$\frac{du}{dt} = 6t$$

which gives

$$\frac{du}{6t} = dt$$

Therefore we might be able to cancel the t's once substitution has taken place.

We also need to substitute new values for the limits $t = 1$ and $t = 4$. Why?

Because we are integrating with respect to a new variable, $u = 3t^2 + 1$.

When $t = 1$, $\quad u = (3 \times 1^2) + 1 = 4$

When $t = 4$, $\quad u = (3 \times 4^2) + 1 = 49$

Substituting these new limits and $u = 3t^2 + 1$, $\;dt = \dfrac{du}{6t}$, we have

$$\int_1^4 \left(\frac{t}{\sqrt{3t^2+1}}\right) dt = \int_{u=4}^{u=49} \left(\frac{t}{\sqrt{u}}\right)\frac{du}{6t}$$

$$= \frac{1}{6}\int_4^{49} \frac{du}{u^{1/2}} \qquad \left[\text{Cancelling } t\text{'s and taking out } \frac{1}{6}\right]$$

$$= \frac{1}{6}\int_4^{49} u^{-1/2}\, du$$

$$= \frac{1}{6}\left[\frac{u^{1/2}}{1/2}\right]_4^{49} \underset{\substack{\text{substituting}\\\text{the limits}}}{=} \frac{1}{3}\left[49^{1/2} - 4^{1/2}\right] = \frac{5}{3}$$

Notice that the t's in the above example cancel, and for this reason the integration becomes straightforward. In general, to integrate a function of the type

8.53 $\qquad k\,[f(t)]^n\, f'(t) \qquad$ (where $n \neq -1$ and k is a constant)

with respect to t, we use the substitution $u = f(t)$.

8.53 might look terrifying, but it is only asserting that if you have a function, $f(t)$, to the power n and its derivative multiplied by a constant in the form of 8.53 , then use the substitution $u = f(t)$. For **Example 31**:

$$f(t) = 3t^2 + 1, \;\; f'(t) = 6t, \;\; k = \frac{1}{6} \text{ and } n = -\frac{1}{2}$$

$$\frac{1}{6}(3t^2 + 1)^{-1/2}\, 6t = \frac{t}{\sqrt{3t^2 + 1}}$$

Let's explain this further by trying another example.

Example 32

Determine

$$\int \frac{t}{(t^2 + 1)^5} \, dt$$

Solution

If we let $u = t^2 + 1$, differentiating gives

$$\frac{du}{dt} = 2t$$

(In relation to $\boxed{8.53}$, we have $f(t) = t^2 + 1$ and $f'(t) = 2t$.)

Rearranging $\dfrac{du}{dt} = 2t$ gives

$$dt = \frac{du}{du/dt} = \frac{du}{2t}$$

Remember that we need to replace the dt in the original integral because we are

integrating with respect to u. Thus, substituting $u = t^2 + 1$ and $dt = \dfrac{du}{2t}$ gives

$$\int \left(\frac{t}{(t^2 + 1)^5} \right) dt = \int \left(\frac{t}{u^5} \right) \frac{du}{2t}$$

$$= \frac{1}{2} \int \frac{du}{u^5} \quad \left[\text{Cancelling } t\text{'s and taking out } \frac{1}{2} \right]$$

$$= \frac{1}{2} \int u^{-5} du \quad \left[\text{Writing } \frac{1}{u^5} = u^{-5} \right]$$

$$= \frac{1}{2} \left(\frac{u^{-4}}{-4} \right) + C = -\frac{u^{-4}}{8} + C = C - \frac{1}{8u^4}$$

We have

$$\int \frac{t}{(t^2 + 1)^5} \, dt = C - \frac{1}{8(t^2 + 1)^4} \quad \left[\text{Substituting } u = t^2 + 1 \text{ into } C - \frac{1}{8u^4} \right]$$

H2 Some trigonometric substitutions

The method of substitution is important when integrating trigonometric functions.

Example 33

Evaluate $\displaystyle\int_0^{\frac{\pi}{2}} \left(\cos(\alpha) \sqrt{\sin(\alpha)} \right) d\alpha$

Example 33 *continued*

Solution

When we differentiate $\sin(\alpha)$ we obtain $\cos(\alpha)$.

Thus the function $\cos(\alpha)\sqrt{\sin(\alpha)}$ seems to be of the form of 8.53 . So use the substitution $u = \sin(\alpha)$. Differentiating gives

$$\frac{du}{d\alpha} = \cos(\alpha)$$

$$d\alpha = \frac{du}{\cos(\alpha)}$$

We also need to change the limits of integration:

When $\alpha = 0$, $u = \sin(0) = 0$ [Because $u = \sin(\alpha)$]

When $\alpha = \dfrac{\pi}{2}$, $u = \sin\left(\dfrac{\pi}{2}\right) = 1$

Replacing the limits and $u = \sin(\alpha)$, $d\alpha = \dfrac{du}{\cos(\alpha)}$, into the original integral, we have

$$\int_{\alpha=0}^{\alpha=\frac{\pi}{2}}\left(\cos(\alpha)\sqrt{\sin(\alpha)}\right)d\alpha = \int_{u=0}^{u=1}\left(\cos(\alpha)\sqrt{u}\right)\frac{du}{\cos(\alpha)}$$

$$= \int_0^1 \sqrt{u}\ du \quad [\text{Cancelling } \cos(\alpha)]$$

$$= \int_0^1 u^{1/2}du \quad [\text{Rewriting } \sqrt{u} = u^{1/2}]$$

$$= \left[\frac{u^{3/2}}{3/2}\right]_0^1 = \frac{2}{3}\left[u^{3/2}\right]_0^1 = \frac{2}{3}\left(1^{3/2} - 0^{3/2}\right) = \frac{2}{3}$$

$$\int_0^{\frac{\pi}{2}}\left(\cos(\alpha)\sqrt{\sin(\alpha)}\right)d\alpha = \frac{2}{3}$$

SUMMARY

To integrate a function of the type

8.53 $k[f(t)]^n\ f'(t)$ (where $n \neq -1$ and k is a constant)

with respect to t, use the substitution $u = f(t)$.

For a definite integral we also need to replace the limits of integration:

$$\int_{x=a}^{x=b} f(x)dx = \int_{u=c}^{u=d} g(u)du$$

Solutions at end of book. Complete solutions available at www.palgrave.com/engineering/singh

Exercise **8(h)**

1 Determine the following integrals:

a $\displaystyle\int b(b^2 - 3)^7 \mathrm{d}b$

b $\displaystyle\int (5s - 1)^9 \mathrm{d}s$

c $\displaystyle\int (3a^2 - 4a)(a^3 - 2a^2 + 6)^4 \,\mathrm{d}a$

d $\displaystyle\int 21q^2\sqrt{7q^3 - 5} \,\mathrm{d}q$

e $\displaystyle\int \frac{p^2 - 1}{\sqrt{p^3 - 3p}} \,\mathrm{d}p$

f $\displaystyle\int \frac{\alpha - 1}{(\alpha^2 - 2\alpha + 10)^2} \,\mathrm{d}\alpha$

2 🔧 [*reliability engineering*]   The mean time to failure, MTTF, in years, for a set of components is given by

$$\text{MTTF} = \int_0^5 (1 - 0.2t)^{1.5}\mathrm{d}t$$

Evaluate MTTF.

3 Determine $\displaystyle\int_0^1 xe^{-x^2}\mathrm{d}x$. [Consider $u = x^2$]

4 Evaluate the following integrals:

a $\displaystyle\int_0^{\frac{\pi}{2}} \sin(\theta)\,\sqrt{\cos(\theta)}\,\mathrm{d}\theta$

b $\displaystyle\int_0^{\pi} \sin(\theta)\,\cos^5(\theta)\mathrm{d}\theta$

What do you notice about your result for **a**?

5 Determine the following indefinite integrals:

a $\displaystyle\int \sec^7(\beta)\tan(\beta)\mathrm{d}\beta$

b $\displaystyle\int \tan^5(A)\sec^2(A)\mathrm{d}A$

c $\displaystyle\int \cot^3(A)\operatorname{cosec}(A)\mathrm{d}A$

[Consider $u = \operatorname{cosec}(A)$]

SECTION I **Trigonometric techniques for integration**

By the end of this section you will be able to:
► use trigonometric substitutions to integrate trigonometric functions
► show some integral results of Table 2

This is a challenging section which requires knowledge of trigonometric identities, integration and substitution. Go through each example very carefully.

I1 **Trigonometric substitutions**

? **How do we integrate $\cos^2(t)$ with respect to t?**

? **Can we use integration by parts because $\cos^2(t) = \cos(t) \times \cos(t)$?**

No, because if we use integration by parts, then we will end up in a vicious circle.

We need to substitute something for $\cos^2(t)$. In **Chapter 4** we had

4.67
$$\cos^2(t) = \frac{1}{2}\left[1 + \cos(2t)\right]$$

Of course this identity might **not** have occurred to you since it is among many trigonometric identities in **Chapter 4**. However you can derive this identity from the fundamental identity:

$$\cos(2t) = 2\cos^2(t) - 1$$

Example 34

Determine $\int \cos^2(t)\, dt$.

Solution

Using 4.67 we have

$$\int \cos^2(t)\, dt = \int \frac{1}{2}\left[1 + \cos(2t)\right]dt$$

$$= \frac{1}{2}\int\left[1 + \cos(2t)\right]dt \qquad \left[\text{Taking out } \frac{1}{2}\right]$$

$$= \frac{1}{2}\left[t + \frac{\sin(2t)}{2}\right] + C \qquad \left[\text{By } \int \cos(kt)\, dt = \frac{\sin(kt)}{k}\right]$$

For these types of trigonometric functions, we need to use the appropriate identity from Chapter 4. Let's investigate an engineering example.

Example 35 *electrical principles*

The voltage, v, and current, i, across a pure capacitor is given by

$$v = V\sin(\omega t) \text{ and } i = I\cos(\omega t)$$

where V is the peak voltage, I is the peak current, ω is the angular frequency and t is time. The average power, p, is given by

$$p = \frac{\omega}{2\pi}\int_0^{2\pi/\omega}(vi)dt$$

Show that $p = 0$.

Solution

Substituting $v = V\sin(\omega t)$ and $i = I\cos(\omega t)$ into p produces:

$$p = \frac{\omega}{2\pi}\int_0^{2\pi/\omega}\left[V\sin(\omega t)\, I\cos(\omega t)\right]dt$$

$$= \frac{\omega VI}{2\pi}\int_0^{2\pi/\omega}\left[\sin(\omega t)\cos(\omega t)\right]dt$$

Example 35 *continued*

The problem is how do we integrate $\sin(\omega t)\cos(\omega t)$?

Using

4.53
$$2\sin(A)\cos(A) = \sin(2A)$$

$$2\sin(\omega t)\cos(\omega t) = \sin(2\omega t)$$

$$\sin(\omega t)\cos(\omega t) = \frac{1}{2}\sin(2\omega t) \quad \text{[Dividing by 2]}$$

Substituting this into ⭐ :

$$p = \frac{\omega VI}{2\pi}\int_0^{2\pi/\omega} \frac{1}{2}\sin(2\omega t)dt$$

$$= \frac{\omega VI}{4\pi}\int_0^{2\pi/\omega} \sin(2\omega t)dt \qquad \left[\text{Taking out } \frac{1}{2}\right]$$

$$= \frac{\omega VI}{4\pi}\left[-\frac{\cos(2\omega t)}{2\omega}\right]_0^{2\pi/\omega} \qquad \left[\text{Using }\int \sin(kt)dt = -\frac{\cos(kt)}{k}\right]$$

$$= -\frac{\omega VI}{4\pi 2\omega}\left(\underbrace{\cos\left[2\omega\left(\frac{2\pi}{\omega}\right)\right]}_{\substack{\text{substituting } t = 2\pi/\omega}} - \underbrace{\cos(0)}_{\substack{\text{substituting} \\ t = 0}}\right)$$

$$= -\frac{VI}{8\pi}\left(\underbrace{1}_{=\cos(4\pi)} - 1\right) = 0$$

$$p = 0$$

You could also have done **Example 35** by using the substitution $u = \sin(\omega t)$. Try it!

There are more problems on trigonometric substitutions which are given in **Exercise 8(i).** You need to search for the appropriate trigonometric identity.

12 **Further trigonometric substitutions**

Sometimes the substitution is not as clear-cut as in the previous examples. For some standard integrals there are suggested substitutions given in Table 4.

	Formulae number	The function that needs integrating contains	Try
TABLE 4	8.54	$\sqrt{a^2 - u^2}$	$u = a\sin(\theta)$ or $u = a\cos(\theta)$
	8.55	$\sqrt{u^2 - a^2}$	$u = a\sec(\theta)$
	8.56	$a^2 + u^2$	$u = a\tan(\theta)$

? **What is the connection between the integrand, $\sqrt{a^2 - u^2}$, and the trigonometric substitution $u = a\sin(\theta)$ or $u = a\cos(\theta)$?**

? Consider the case where $a = 1$ and $u = x$. **What does the integration $\int\limits_0^1 \sqrt{1 - x^2}\,dx$ represent?**

This is a quarter of a circle of radius 1 shown shaded in Fig. 11 below:

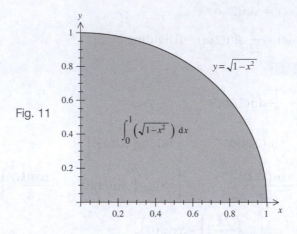

Fig. 11

We can write the integrand $\sqrt{1 - x^2}$ in terms of trigonometric functions because we can draw a right-angled triangle of radius 1 with the angle θ ranging between 0 and $\dfrac{\pi}{2}$ (or $90°$) as shown in Fig. 12 below:

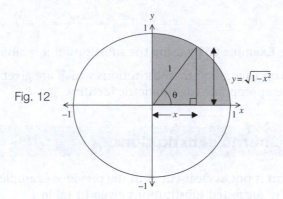

Fig. 12

By Pythagoras we have $y = \sqrt{1 - x^2}$ and from the definition of cosine we have

$$\cos(\theta) = \frac{adj}{hyp} = \frac{x}{1} = x \text{ or } x = \cos(\theta)$$

From the sine definition we have

$$\sin(\theta) = \frac{opp}{hyp} = \frac{\sqrt{1 - x^2}}{1} = \sqrt{1 - x^2} = y \text{ or } y = \sin(\theta)$$

$\sqrt{1 - x^2}$ simplifies to $\sin(\theta)$ which means we have got rid of the square root.

Let's do an example.

Example 36

Show that

$$\int \frac{du}{a^2 + u^2} = \frac{1}{a}\tan^{-1}\left(\frac{u}{a}\right) + C$$

This is result 8.26 of Table 2 on page 409.

Solution

Which substitution should we use?

By 8.56, let $u = a\tan(\theta)$.

We also need to replace the du. What is du?

Differentiating $u = a\tan(\theta)$ gives

$$\frac{du}{d\theta} \underset{\text{by } 6.8}{=} a\sec^2(\theta)$$

$$du = a\sec^2(\theta)d\theta$$

By substituting these into the given integral we have

$$\int \frac{du}{a^2 + u^2} = \int \frac{a\sec^2(\theta)d\theta}{a^2 + a^2\tan^2(\theta)}$$

$$= \int \frac{a\sec^2(\theta)d\theta}{a^2\underbrace{(1 + \tan^2(\theta))}_{=\,\sec^2(\theta)}}$$

$$= \frac{a}{a^2}\int \frac{\sec^2(\theta)}{\sec^2(\theta)}d\theta \quad \left[\text{Taking out } \frac{a}{a^2}\right]$$

$$= \frac{1}{a}\int d\theta \qquad \text{[Cancelling]}$$

Integrating $d\theta$ gives θ, hence

$$\boxed{*}\qquad \int \frac{du}{a^2 + u^2} = \frac{1}{a}\theta + C$$

What is θ?

From above we have

$$a\tan(\theta) = u$$

$$\tan(\theta) = \frac{u}{a}$$

6.8 $[\tan(\theta)]' = \sec^2(\theta)$

> ### Example 36 *continued*
>
> **[?]** **How do we find θ?**
>
> Take inverse tan, $\tan^{-1}$, of both sides:
>
> $$\theta = \tan^{-1}\left(\frac{u}{a}\right)$$
>
> Replacing $\theta = \tan^{-1}\left(\frac{u}{a}\right)$ in **[*]** displays the required result:
>
> $$\int \frac{du}{a^2 + u^2} = \frac{1}{a}\tan^{-1}\left(\frac{u}{a}\right) + C$$

Similarly we can show some of the other standard integrals of Table 2 by using the suggested substitution of Table 4.

In some cases you will be given the substitution to use, as the following example shows.

> ### Example 37
>
> By using the substitution $u = a\sinh(\theta)$, show that
>
> $$\int \frac{du}{\sqrt{a^2 + u^2}} = \sinh^{-1}\left(\frac{u}{a}\right) + C$$
>
> This is result **8.28** of Table 2.
>
> Solution
>
> Differentiating $u = a\sinh(\theta)$ gives
>
> $$\frac{du}{d\theta} = a\cosh(\theta) \quad \left[\text{By } \boxed{6.25}\right]$$
>
> $$du = a\cosh(\theta)d\theta$$
>
> Substituting $u = a\sinh(\theta)$ and $du = a\cosh(\theta)d\theta$ into the integral gives
>
> $$\int \frac{du}{\sqrt{a^2 + u^2}} = \int \frac{a\cosh(\theta)d\theta}{\sqrt{a^2 + (a\sinh(\theta))^2}}$$
>
> **[†]** $$= \int \frac{a\cosh(\theta)d\theta}{\sqrt{a^2 + a^2\sinh^2(\theta)}}$$
>
> The denominator $\sqrt{a^2 + a^2\sinh^2(\theta)}$ simplifies to
>
> $$\sqrt{a^2 + a^2\sinh^2(\theta)} = \sqrt{a^2\underbrace{[1 + \sinh^2(\theta)]}_{=\,\cosh^2(\theta)}} = \sqrt{a^2\cosh^2(\theta)} = a\cosh(\theta)$$

6.25 $[\sinh(\theta)]' = \cosh(\theta)$

Example 37 *continued*

Replacing this in gives

$$\int \frac{a\cosh(\theta)\mathrm{d}\theta}{a\cosh(\theta)} = \int \mathrm{d}\theta \quad [\text{Cancelling } a\cosh(\theta)]$$
$$= \theta + C \quad [\text{Integrating}]$$

Thus we have

$$* \qquad \int \frac{\mathrm{d}u}{\sqrt{a^2 + u^2}} = \theta + C$$

We need to replace θ. From the given substitution

$$a\sinh(\theta) = u$$

we have

$$\sinh(\theta) = \frac{u}{a}$$

Take inverse sinh, $\sinh^{-1}$, of both sides:

$$\theta = \sinh^{-1}\left(\frac{u}{a}\right)$$

We have our result by substituting this into $*$:

$$\int \frac{\mathrm{d}u}{\sqrt{a^2 + u^2}} = \sinh^{-1}\left(\frac{u}{a}\right) + C$$

SUMMARY

For integrating some trigonometric functions we select a relevant trigonometric identity and then integrate.
Some of the standard integrals of Table 2 can be established by using an appropriate substitution.

Exercise 8(i)

Solutions at end of book. Complete solutions available at www.palgrave.com/engineering/singh

Questions 1 to 5 are in the field of [*electrical principles*].

1 The average power, P, of an a.c. circuit is given by

$$P = \frac{\omega}{2\pi} \int_0^{\frac{2\pi}{\omega}} \left(i^2 R\right)\mathrm{d}t$$

where $i = I\sin(\omega t)$, ω is angular frequency, I is peak current, R is resistance and t is time. Show that

$$P = \frac{I^2 R}{2}$$

2 The voltage, v, produced by an electronic circuit is given by

$$v = 10\sin(\omega t)$$

The root mean square value, V_{RMS}, is defined as

$$V_{RMS} = \sqrt{\frac{\omega}{2\pi} \int_0^{\frac{2\pi}{\omega}} v^2 \,\mathrm{d}t}$$

Evaluate V_{RMS}.

Solutions at end of book. Complete solutions available at www.palgrave.com/engineering/singh

3 The average power, P, in an a.c. circuit is given by

$$P = \frac{\omega}{2\pi} \int_0^{2\pi/\omega} (vi)\mathrm{d}t$$

where $i = I\sin(\omega t)$ and $v = V\sin(\omega t)$.

(V and I are peak voltage and peak current respectively, ω is angular frequency and t is time.)

Show that $P = \dfrac{VI}{2}$.

4 The energy, W, of an inductance, L, is defined as

$$W = \frac{1}{2L}\left(\int_0^t V\mathrm{d}t + 1\right)^2$$

where V is the voltage across the inductor. For $L = 1 \times 10^{-3}$ henry and $V = \cos(t) - \sin(t)$, show that

$$W = 500[1 + \sin(2t)]$$

5 For $v = V\sin(\omega t)$ and $i = I\sin(\omega t + \phi)$, show that the power, P, in an a.c. circuit is given by

$$P = \frac{VI}{2}\cos(\phi)$$

where ϕ is the phase and P is as defined in question **3**.

*6 Show that

$$\int \frac{\mathrm{d}u}{u\sqrt{u^2 - a^2}} = \frac{1}{a}\sec^{-1}\left(\frac{u}{a}\right) + C.$$

7 Show that $\displaystyle\int \frac{\mathrm{d}u}{\sqrt{a^2 - u^2}} = \sin^{-1}\left(\frac{u}{a}\right) + C.$

Solutions at end of book. Complete solutions available at www.palgrave.com/engineering/singh

1 Integrate the following:

a $\displaystyle\int (3x - \sin(x))\mathrm{d}x$

b $\displaystyle\int_0^1 e^{3x}\,\mathrm{d}x$

2 Evaluate the following integrals:

a $\displaystyle\int 5x\cos(2x)\mathrm{d}x$, by parts

b $\displaystyle\int_1^2 x(x^2 + 5)^3\mathrm{d}x$, by substitution

Questions 1 and 2 are from University of Portsmouth, UK, 2007

3 Find the following indefinite integral:

$$\int\left(\frac{5\sin(2x)}{\cos(x)}\right)\mathrm{d}x$$

Arizona State University, USA (Review)

4 Evaluate the following integral:

$$\int \frac{[\ln(x)]^3}{x}\,\mathrm{d}x$$

North Carolina State University, USA, 2010

5 Find the integrals:

i $\displaystyle\int_{-1}^{2} (4x^3 - x + 1)\,\mathrm{d}x$

ii $\displaystyle\int_0^{\pi/4} \cos\left(2t - \frac{\pi}{4}\right)\mathrm{d}t$

iii $\displaystyle\int \frac{x}{x^2 + 1}\,\mathrm{d}x$

iv $\displaystyle\int \theta\sin(\theta)\,\mathrm{d}\theta$

v $\displaystyle\int \frac{3}{(x + 1)(x - 2)}\,\mathrm{d}x$

vi $\displaystyle\int_0^{1/2} \frac{1}{\sqrt{1 - x^2}}\,\mathrm{d}x$

6 Evaluate $I = \int_3^4 \dfrac{9x^2}{(x + 1)^2(2x - 1)}\, dx$.

Questions 5 and 6 are from University of Manchester, UK, 2009

7 Evaluate each of the following integrals:

a $\int \dfrac{2x^2 - 10x + 17}{x - 3}\, dx$

b $\int \dfrac{\cos^2(x)}{\sin(x)}\, dx$

c $\int x(3x - 7)^9 dx$

d $\int x \sec(x) \tan(x) dx$

8 Evaluate each of the following definite integrals:

a $\int_0^{\frac{\pi}{3}} 8 \cos^3(x) \sin(x) dx$

b $\int_1^4 \dfrac{\ln(x)}{\sqrt{x}}\, dx$

Questions 7 and 8 are from Memorial University of Newfoundland, Canada, 2010

The following questions from King's College London have been slightly edited.

9 Determine the integral $\int_0^{\ln(\pi)} e^x \sin(e^x) dx$.

10 Evaluate the integral $\int_1^e \ln(x) dx$.

Questions 9 and 10 are from King's College London, UK, 2008

11 Evaluate the definite integral:

$$\int_3^5 \dfrac{t + 5}{(t - 1)(t + 2)}\, dt$$

University of Aberdeen, UK, 2004

12 Find the value of the definite integrals:

i $\int_1^2 e^{x-1}\, dx$

ii $\int_{-1}^1 x^2 e^x\, dx$

Loughborough University, UK, 2008

*****13** Find $F'(-1)$ if $F(x) = \int_x^{x^2} \dfrac{1}{\sqrt{1 + t^2}}\, dt$.

University of Toronto, Canada, 2001

14 Consider the integral

$$I(x) = \int \dfrac{1}{x^2\sqrt{4 + x^2}}\, dx$$

Use the substitution $x = 2 \tan(\theta)$ to transform I to the following integral

$$I(\theta) = \dfrac{1}{4} \int \dfrac{\cos(\theta)}{\sin^2(\theta)}\, d\theta$$

Now use the substitution $u = \sin(\theta)$ to transform this integral to an integral $I(u)$ involving u.

Integrate $I(u)$ to show that

$$I(u) = -\dfrac{1}{4u} + c$$

Finally, express u in terms of x to obtain

$$I(x) = -\dfrac{\sqrt{x^2 + 4}}{4x} + c$$

University of Manchester, UK, 2007

15 Evaluate the following integral:

$$\int \cos \sqrt{x}\, dx$$

[*Hint:* You will need to use a substitition combined with another method of integration]

University of British Columbia, Canada, 2008

16 Evaluate the integral $\int e^{\sqrt{x}} \dfrac{dx}{\sqrt{x}}$.

17 Evaluate the integral $\int e^{\sqrt{x}}\, dx$.

Questions 16 and 17 are from University of California, Berkeley, USA, 2005

Miscellaneous exercise 8

Solutions at end of book. Complete solutions available at www.palgrave.com/engineering/singh

1 🚗 [mechanics]   The kinetic energy, KE, of a body is given by

$$\text{KE} = \int_0^v (mv)\,dv$$

where m is the mass of the body and v is the velocity. Evaluate KE.

2 🚗 [mechanics]   The work done, W, to stretch a spring by length l is given by

$$W = \int_0^l \frac{EAx}{L}\,dx \quad \left(\frac{EA}{L} \text{ is constant}\right)$$

where E is the modulus of elasticity, A is the cross-sectional area, L is the unstretched length and x is the displacement. Determine W.

3 🚗 [mechanics]   The work done, W, to stretch a spring from length x_1 to length x_2 is given by

$$W = \int_{x_1}^{x_2} (kx)\,dx$$

where k is the stiffness constant of the spring. Evaluate W.

4 🚗 [mechanics]   Figure 13 shows a belt wound partly around a pulley. The angle θ is given by

$$\theta = \int_{T_1}^{T_2} \frac{dT}{\mu T}$$

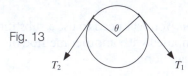

Fig. 13

where μ is the coefficient of friction between the pulley and the belt and T_1, T_2 are tensions as shown.
Show that

$$T_2 = T_1\, e^{\mu\theta}$$

5 🏗 [structures]   The work done, W (by external forces), on a cantilever beam of length L with uniformly distributed load q and a concentrated load p at the free end is given by

$$W = \int_0^L \frac{pq}{6EI}(2L^3\, x - 3L^2\, x^2 + x^4)\,dx$$

(EI is the flexural rigidity and x is the distance along the beam).
Show that

$$W = \frac{pqL^5}{30EI}$$

6 🪨 [materials]   The polar second moment of area, J, of a hollow circular shaft of outer diameter D_o and inner diameter D_i is given by

$$J = \int_{D_i/2}^{D_o/2} 2\pi r^3\,dr$$

Show that

$$J = \frac{\pi}{32}\left[D_o{}^4 - D_i{}^4\right]$$

7 Find $\displaystyle\int \frac{x^3}{x^2 - 4}\,dx$.

8 🌡 [thermodynamics]   The change in specific enthalpy, Δh, of a gas is given by

$$\Delta h = \int_{200}^{1000}\left(1.8 + (12 \times 10^{-3})T\right)dT$$

Evaluate Δh.

9 🌡 [thermodynamics]   The enthalpy change, Δh, is given by

$$\Delta h = \int_{T_1}^{T_2} C_p\,dT$$

where $C_p = R(a + bT + cT^2 + dT^3 + eT^4)$. ($R$, a, b, c, d and e are constants.) Find an expression for Δh.

10 Find $\displaystyle\int \frac{\cos(2x)}{\cos^2(x)\sin^2(x)}\,dx$.

11 📺 [*electronics*]   The potential, V, at the surface of a conductor at distance r from zero potential at distance z, is given by

$$V = -\int_z^r \frac{q}{2\pi\varepsilon_0 r}\, dr \quad (r \neq 0)$$

where q is the charge and ε_0 is the permittivity constant. Show that

$$V = \frac{q}{2\pi\varepsilon_0}\ln\left(\frac{z}{r}\right)$$

12 📟 [*thermodynamics*]   The work done, W, by a gas is given by

$$W = \int_{V_1}^{V_2} P\, dV$$

where P is the pressure, V is the volume, V_1 and V_2 are the initial and final volumes of the gas respectively. If the gas obeys the law $PV = C$, where C is constant, show that $W = PV\ln\left(\dfrac{V_2}{V_1}\right)$.

13 📟 [*thermodynamics*]   The work done, W, on a piston by a gas in a cylinder is given by $W = \int_{V_1}^{V_2} P\, dV$ where P is pressure and V is volume. If $PV^{1.32} = C$ (constant), find an expression for W.

*14 In Fig. 14:

i Evaluate the total area of the 100 rectangles under the graph $y = x^3$.

ii Determine $\displaystyle\int_0^1 x^3\, dx$.

iii Find the difference between the area of part **i** and $\displaystyle\int_0^1 x^3\, dx$. How can this difference be made smaller?

[*Hint*:

$$1^3 + 2^3 + 3^3 + \ldots + n^3 = \frac{n^2}{4}(n+1)^2$$

where n is a positive whole number.]

15 〜 [*fluid mechanics*]   The pressure, P, at a distance x of a fluid flowing around a sphere of radius r is given by

$$P = \int_{-\infty}^x k\left(\frac{1 + r^3/x^3}{x^4}\right) dx$$

where k is a constant. Evaluate P.

16 🌉 [*structures*]   A beam of length L has a uniform load w given by

$$w = w_0\sin\left(\frac{\pi x}{L}\right)$$

where w_0 is a constant and x is the distance along the beam. The total load P and reaction R are given by

$$P = \int_0^L w\, dx \quad \text{and} \quad R = \int_0^L (wx)\, dx$$

Show that **i** $P = \dfrac{2w_0 L}{\pi}$ and

ii $R = \dfrac{w_0 L^2}{\pi}$.

17 📺 [*electronics*]   In an RC (a resistor capacitor) network the charging voltage w and the instantaneous voltage v are related by $\dfrac{t}{RC} = \displaystyle\int_{-w}^v \frac{dv}{w - v}$.

Show that $v = w(1 - 2e^{-t/RC})$.

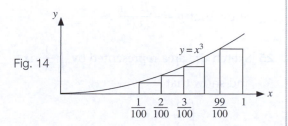

Fig. 14

$y = x^3$

18 [electrical principles]   The current, i, through an inductor of inductance $L = 10$ mH is given by

$$i = \frac{1}{L}\int_0^t v\,dx$$

where v represents voltage. The voltage v across the inductor is given by

$$v = -6e^{-(2\times10^3)x} + 10e^{-(8\times10^3)x}$$

Determine the current i.

19 [electrical principles]   The current, i, through an inductor of inductance L is given by

$$i = \frac{1}{L}\int_0^t v\,dt$$

where v is the voltage across the inductor. The energy, w, of an inductance L, is defined to be

$$w = \frac{1}{2}Li^2$$

For $L = 5$ mH and $v = 10\sin(100\pi t)$, find an expression for

i i **ii** w

20 [electromagnetism]   The flux, Φ, for a toroid (doughnut shaped object) of height h, inner radius a, and outer radius b, is given by

$$\Phi = \int_a^b \frac{\mu i N h}{2\pi r}\,dr \quad (r \neq 0)$$

where $a < r < b$, N is the number of turns, μ is the permeability constant and i is the current carried by the toroid coil.

If $L = N\dfrac{d\Phi}{di}$,

show that $L = \dfrac{\mu h N^2}{2\pi}\ln\left(\dfrac{b}{a}\right)$.

21 Evaluate $\displaystyle\int_0^1 \frac{4}{1+x^2}\,dx$.

*22 [mechanics]   The moment of inertia, I, of a triangular lamina (plane surface) of height h, base b and mass m is defined as

$$I = \int_0^h \frac{2m(h-y)y^2}{h^2}\,dy$$

where y is the distance from the axis of rotation. Show that $I = \dfrac{mh^2}{6}$.

*23 [structures]   A cantilever beam of length L, fixed at one end and deflected by a distance D at the free end has strain energy V given by

$$V = \frac{EI}{2}\int_0^L \left(\frac{d^2y}{dx^2}\right)^2 dx$$

where EI is the flexural rigidity. The deflection y at a distance x from the fixed end is given by

$$y = D\left[1 - \cos\left(\frac{\pi x}{2L}\right)\right]$$

Find V.

*24 [communication systems]   The total power, p, of an antenna of length L, carrying a peak current I at a distance r, is given by

$$p = \int_0^\pi \frac{\eta I^2 L^2 \pi \sin^3(\theta)d\theta}{4\lambda^2}$$

(λ = wavelength, η = space impedance and θ = angle).

Show that $p = \dfrac{\eta I^2 L^2 \pi}{3\lambda^2}$.

25 Sketch the area represented by $\displaystyle\int_1^e \frac{1}{x}\,dx$ and show that

$$\int_1^e \frac{1}{x}\,dx = 1$$

Miscellaneous exercise **8 continued**

Solutions at end of book. Complete solutions available at www.palgrave.com/engineering/singh

26 i By using a computer algebra system, or a graphical calculator, plot $y = (1 - x^2)^{1/2}$.

ii Evaluate $\int_{-1}^{1} (1 - x^2)^{1/2} dx$.

27 Show that

$$\int_{0}^{2} (4 - x^2)^{1/2} dx = \pi$$

28 By using the substitution $u = a \cosh(\theta)$, show that

$$\int \frac{du}{\sqrt{u^2 - a^2}} = \cosh^{-1}\left(\frac{u}{a}\right) + C$$

29 By substituting $u = a \tanh(\theta)$, show that

$$\int \frac{du}{a^2 - u^2} = \frac{1}{a}\tanh^{-1}\left(\frac{u}{a}\right) + C$$

*30 The Fourier coefficient, a_n, for a triangular waveform is given by

$$a_n = \frac{1}{\pi^2 \, n^2} \int_{0}^{2\pi} (\omega t) \cos(n\omega t) d(\omega t)$$

where n is a positive whole number. Show that $a_n = 0$.

For questions 31 and 32 use a computer algebra system or a graphical calculator.

31 ☒ [*thermodynamics*]   The rate of heat flow, q, is given by

$$q = 300 \int_{306}^{853} [3.7 - (1.9 \times 10^{-3})T$$
$$+ (4.7 \times 10^{-6})T^2 - (3.45 \times 10^{-11})T^3$$
$$+ (8.5 \times 10^{-13})T^4] dT$$

Evaluate q.

*32 ☒ [*thermodynamics*]   Determine the specific enthalpy change, Δh, for the following:

a $\Delta h = \int_{400}^{1000} \left[56 - \left(\frac{22 \times 10^3}{T^{0.75}} \right) \right.$

$$\left. + \left(\frac{116 \times 10^3}{T} \right) - \left(\frac{561 \times 10^3}{T^{1.5}} \right) \right] dT$$

b $\Delta h = \int_{400}^{1000} \left[81 - 18.66T^{0.25} \right.$

$$\left. + 0.54T^{0.75} - 0.04T \right] dT$$

c $\Delta h = \int_{400}^{1000} \left[143 - 58T^{0.25} \right.$

$$\left. + 8.3T^{0.5} - (37 \times 10^{-3})T \right] dT$$

d $\Delta h = \int_{400}^{1000} \left[(2.42 \times 10^{-6})T^2 \right.$

$$\left. - (41 \times 10^{-3})T + 3.05T^{0.5} - 3.7 \right] dT$$

(In each of these examples, T represents temperature and Δh is the specific enthalpy change of the gas being heated from 400 K to 1000 K. Also the units of Δh are kJ/kmol.)

CHAPTER 9
Engineering Applications of Integration

SECTION A **Trapezium rule**

By the end of this section you will be able to:
▶ understand the need for numerical methods
▶ apply the trapezium rule
▶ determine the discrepancy between the exact and approximate solutions

A1 **Introduction to numerical integration**

The advancement of technology has had a profound effect in engineering mathematics. The emphasis of many university engineering mathematics syllabi is now moving towards using numerical methods to solve problems rather than employing analytical techniques. The reason for this is that most realistic engineering problems require a numerical method because the problem cannot be solved analytically. **What do we mean by analytical techniques?** Analytical techniques are based on algebraic methods. All the integration examples we have done so far have involved analytical techniques such as integration by parts, substitution, partial fractions, etc. With these techniques you obtain an **exact** answer.

What does a numerical method involve?

Numerical method is based on arithmetic operations. This normally involves impractically lengthy calculations, so the use of technology is essential. Generally with numerical methods your final result is an **approximation**.

Why do we bother with an approximation when you can achieve an exact answer?

Most practical engineering problems cannot be solved by analytical methods. For example, the analysis of experimental data normally requires a numerical approach because a formula for the relationship between input and results in experiments simply isn't known – we don't have an $f(x)$ rule. Additionally, in many engineering problems an approximate solution is adequate. With the advancement of technology these numerical methods are becoming more important than the algebraic techniques.

Can you think of any examples of numerical methods?

The Newton–Raphson method for solving equations is an example of a numerical method. In **Sections A** and **B** we evaluate integrals with limits by using numerical methods such as the trapezium rule and Simpson's rule.

A2 **Trapezium rule**

What is the area represented by $\int_a^b y\,\mathrm{d}x$, **where** $y = f(x)$?

It is the area under the curve $y = f(x)$ and the x axis between a and b.

We can approximate this area, $\int_a^b y\,dx$, by fitting it with blocks of trapezia. Consider the shaded block of Fig. 1.

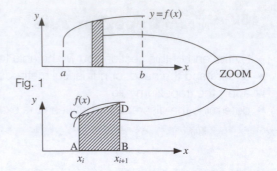

Fig. 1

The area under the curve $y = f(x)$ between x_i and x_{i+1} can be approximated by the area of the trapezium ABCD. **What is the area of the trapezium?**
It is equal to

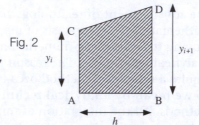

$$\frac{1}{2}\text{AB} \times (\text{AC} + \text{BD})$$

Let's rewrite $\frac{1}{2}\text{AB} \times (\text{AC} + \text{BD})$. If we use h as the distance AB and y_i as the value of $f(x_i)$, that is height AC, and y_{i+1} as the value of $f(x_{i+1})$, that is height BD (Fig. 2), then **what is the formula for the area of the trapezium ABCD?**

Fig. 2

$$\text{Area ABCD} = \frac{1}{2}h(y_i + y_{i+1})$$

Let's consider the whole area under $y = f(x)$ between a and b. Cut this area into n blocks of equal width h and height y_i at the point x_i (Fig. 3).

Fig. 3

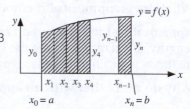

What is the area of the first block, between x_0 and x_1?

The heights are y_0 and y_1, therefore

$$\text{area of first block} = \frac{1}{2}h(y_0 + y_1)$$

What is the area of the second block between x_1 and x_2?

$$\frac{1}{2}h(y_1 + y_2)$$

The area of the nth block, between x_{n-1} and x_n is

$$\frac{1}{2}h(y_{n-1} + y_n)$$

How do we find the total area of these trapezia under the curve $y = f(x)$?

Add all these blocks together.

$$\text{Total area} = \underbrace{\frac{1}{2}h(y_0 + y_1)}_{\substack{\text{Area of first}\\\text{block}}} + \underbrace{\frac{1}{2}h(y_1 + y_2)}_{\substack{\text{Area of second}\\\text{block}}} + \frac{1}{2}h(y_2 + y_3) + \cdots + \underbrace{\frac{1}{2}h(y_{n-1} + y_n)}_{\text{Area of } n\text{th block}}$$

$$= \frac{1}{2}h\left[\underbrace{(y_0 + y_1)}_{} + \underbrace{(y_1 + y_2)}_{= 2y_1} + \underbrace{(y_2 + y_3)}_{= 2y_2} + \cdots + (y_{n-1} + y_n)\right]$$

taking out the
common factor of
$\frac{1}{2}h$

$$= \frac{1}{2}h\left[y_0 + 2y_1 + 2y_2 + \cdots + 2y_{n-1} + y_n\right]$$

Remember the area under the curve is represented by $\int_a^b y \, dx$. So by taking out the common factor of 2 from the middle terms we have:

9.1
$$\int_a^b y \, dx \approx \frac{h}{2}\left[y_0 + 2(y_1 + y_2 + \cdots + y_{n-1}) + y_n\right]$$

? **Why is there an approximation sign, $\approx$, in formula 9.1 ?**

Because near the edge of the curve we have the situation shown in Fig. 4.

Fig. 4 $f(x)$ ⟶ ◿ ⟵ Trapezium

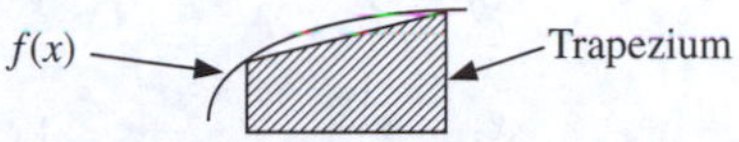

Generally the area under the curve, $\int_a^b y \, dx$,

cannot be evaluated exactly by the trapezia. Remember that y_i represents the height of the given function at the point x_i. This y_i is called the **ordinate**. Formula 9.1 can also be written as:

9.2
$$\text{Area} \approx \frac{\text{width of block}}{2}\left[(\text{first ordinate}) + (\text{last ordinate}) + 2(\text{remaining ordinates})\right]$$

These formulae, 9.1 and 9.2 , are known as the trapezium rule.

In **Examples 1** and **2** below, the analytical integration is a lot easier than it would be by applying the trapezium rule. However in these examples we can compare exact and approximate evaluations of the integral.

Example 1

i Evaluate $\int_0^1 x^2 \, dx$ by using the trapezium rule with four intervals.

ii Find the exact value of $\int_0^1 x^2 \, dx$.

iii Determine the discrepancy between the exact value and the value obtained by the trapezium rule, **i**.

Solution

i We need to apply

9.1
$$\int_b^a y \, dx \approx \frac{h}{2}\left[y_0 + 2(y_1 + y_2 + \cdots + y_{n-1}) + y_n\right]$$

Example 1 *continued*

What are the values of *a*, *b* and *h* for this example?

The lower limit $a = 0$ and therefore $b = 1$.
Remember that *h* is the uniform width
between 0 and 1. Since this is split into four
equal intervals, we have the situation shown
in Fig. 5.

Fig. 5

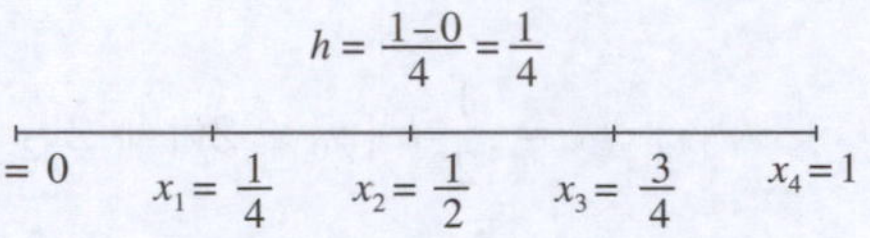

$$h = \frac{1-0}{4} = \frac{1}{4}$$

$x_0 = 0 \quad x_1 = \frac{1}{4} \quad x_2 = \frac{1}{2} \quad x_3 = \frac{3}{4} \quad x_4 = 1$

What is *y*?

$$y = x^2 \text{ (the function that needs to be integrated)}$$

What else do we require in order to use 9.1 ?

We need to obtain the values of $y_0, y_1, y_2, \ldots$ These are determined by *y* at $x_0, x_1, x_2, \ldots$
respectively. Thus substituting the *x* values into $y = x^2$ gives

$$x_0 = 0 \qquad\qquad y_0 = 0^2 = 0$$

$$x_1 = \frac{1}{4} \qquad\qquad y_1 = \left(\frac{1}{4}\right)^2 = \frac{1}{16}$$

$$x_2 = \frac{1}{2} \qquad\qquad y_2 = \left(\frac{1}{2}\right)^2 = \frac{1}{4}$$

$$x_3 = \frac{3}{4} \qquad\qquad y_3 = \left(\frac{3}{4}\right)^2 = \frac{9}{16}$$

$$x_4 = 1 \qquad\qquad y_4 = 1^2 = 1$$

Putting all these into the trapezium rule with $n = 4$ gives

$$\int_0^1 x^2 \, dx \approx \frac{h}{2} \left[y_0 + 2(y_1 + y_2 + y_3) + y_4 \right]$$

$$= \frac{1/4}{2} \left[0 + 2\left(\frac{1}{16} + \frac{1}{4} + \frac{9}{16} \right) + 1 \right] = \frac{11}{32}$$

With the trapezium rule, $\int_0^1 x^2 \, dx$ is $\frac{11}{32}$.

ii What is the exact value of $\int_0^1 x^2 \, dx$?

$$\int_0^1 x^2 \, dx \underset{\text{by } \boxed{8.1}}{=} \left[\frac{x^3}{3} \right]_0^1 = \left(\frac{1}{3} - 0 \right) = \frac{1}{3}$$

The exact value is 1/3.

iii The discrepancy between the exact and approximate (trapezium) value is

$$\frac{1}{3} - \frac{11}{32} = -\frac{1}{96}$$

8.1 $\int x^n \, dx = x^{n+1}/n + 1$

Figure 6 is the graphical representation of the trapezium rule with four intervals applied to $\int_0^1 x^2 \, dx$.

We can compare exact and approximate evaluations of the integral by determining the percentage error. The percentage error is given by

Fig. 6

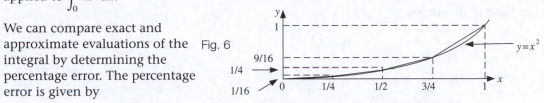

$y = x^2$

| 9.3 |

$$\% \text{ error} = \left(\frac{\text{exact} - \text{approximation}}{\text{exact}} \right) \times 100$$

$$\left(= \frac{\text{discrepancy}}{\text{exact}} \times 100 \right)$$

In **Example 1**:

$$\% \text{ error} = \frac{-1/96}{1/3} \times 100 = -3.125\%$$

The negative sign indicates that the exact value is smaller than the approximate value as seen in Fig. 6. In some engineering cases an error of more than 3% may be too large. **How can we achieve a better approximation with the trapezium rule?**

You could consider more intervals such as 8, 16, 32, etc.

Yes, but what penalty do we pay if we make an allowance for a large number of intervals?

More calculations are involved and of course in many cases it will not be feasible by hand. You need to use a graphical calculator or a computer algebra system such as MAPLE or numerical software such as MATLAB. Let's repeat **Example 1** using eight intervals, without a computer algebra system.

Example 2

Evaluate $\int_0^1 x^2 \, dx$ using the trapezium rule with eight intervals, and find the discrepancy from the exact value.

Solution

Since the interval 0 to 1 is sliced into eight equal blocks, the uniform width, h, is given by

$$h = \frac{1 - 0}{8} = \frac{1}{8}$$

In order to use the trapezium formula:

$$\int_a^b y \, dx \approx \frac{h}{2} \left[y_0 + 2(y_1 + y_2 + \cdots + y_{n-1}) + y_n \right]$$

Example 2 *continued*

we need to determine

$$x_0 = 0 \qquad\qquad y_0 = 0^2$$

$$x_1 = \frac{1}{8} \qquad\qquad y_1 = \left(\frac{1}{8}\right)^2$$

$$x_2 = \frac{2}{8} \qquad\qquad y_2 = \left(\frac{2}{8}\right)^2$$

$$\vdots \qquad\qquad\qquad \vdots$$

Thus we have

$$\int_0^1 x^2 \, dx \approx \frac{1/8}{2}\left\{0 + 2\left[\left(\frac{1}{8}\right)^2 + \left(\frac{2}{8}\right)^2 + \left(\frac{3}{8}\right)^2 + \left(\frac{4}{8}\right)^2 + \left(\frac{5}{8}\right)^2 + \left(\frac{6}{8}\right)^2 + \left(\frac{7}{8}\right)^2\right] + 1\right\}$$

$$= \frac{43}{128}$$

The exact value is $\frac{1}{3}$, therefore

$$\text{discrepancy} = \frac{1}{3} - \frac{43}{128} = -\frac{1}{384}$$

and

$$\% \text{ error} = -\frac{1/384}{1/3} \times 100 = -0.781\%$$

compared with a percentage error of more than -3% for four intervals.

Table 1 shows how the trapezium rule evaluation gets closer and closer to the exact value, $\frac{1}{3} = 0.33333\ldots$, as we consider more and more intervals.

TABLE 1	Number of intervals	4	8	16	32	64
	$\int_0^1 x^2 \, dx$ (trapezium rule (5 d.p.))	0.34375	0.33594	0.33398	0.33349	0.33337
	Percentage error	−3.125	−0.782	−0.194	−0.047	−0.011

The integral $\int_0^1 x^2 \, dx$ was comfortably evaluated by using an analytical method. However the trapezium rule (and other numerical methods) are effective when the solution cannot be evaluated, or is difficult to evaluate, analytically. Let's consider an example which involves experimental data.

A3 **Engineering applications**

 Example 3 *structures*

Fig. 7

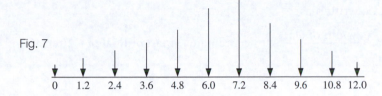

TABLE 2	x (m)	0	1.2	2.4	3.6	4.8	6.0	7.2	8.4	9.6	10.8	12.0
	w (kN/m)	2.6	3.8	5.6	7.1	9.3	13.7	15.0	10.7	8.1	6.2	3.5

Figure 7 and Table 2 show the loading on a beam of length 12 m and the corresponding figures for the loading in kN/m. The total force, F, exerted by the load is given by

$$F = \int_0^{12} w \, dx$$

Estimate F.

Solution

We cannot use analytical methods. Why not?

We are not given an algebraic function but a set of data values, so we are compelled to use a numerical method. For the trapezium rule:

9.1 $$\int_a^b y \, dx \approx \frac{h}{2}\left[y_0 + 2\left(y_1 + y_2 + \cdots + y_{n-1} \right) + y_n \right]$$

We have to find a, b, h, y_0, y_1, etc.

We know $a = 0$, $b = 12$ and $h = 1.2$, because h is the uniform width.

But what are the values of y_0, y_1, y_2, …?

These can be read off from the second row of Table 2:

$$y_0 = 2.6, \quad y_1 = 3.8, \quad y_2 = 5.6, \text{ etc.}$$

Thus we have

$$\int_0^{12} w \, dx \approx \frac{1.2}{2} \, [2.6 + 2\{3.8 + 5.6 + 7.1 + 9.3 + 13.7 + 15.0 + 10.7 + 8.1 + 6.2\} + 3.5]$$

$$= 99.06$$

The total force, $F \approx 99.06$ kN

SUMMARY

Generally numerical methods give an approximate solution. One numerical method to determine integrals with limits is the trapezium rule:

9.1
$$\int_a^b y \, dx \approx \frac{h}{2} \left[y_0 + 2(y_1 + y_2 + y_3 + \cdots) + y_n \right]$$

This can also be written as

9.2 $\text{Area} \approx \dfrac{\text{width of block}}{2} \left[(\text{first ordinate}) + (\text{last ordinate}) + 2(\text{remaining ordinates}) \right]$

Exercise 9(a)

Solutions at end of book. Complete solutions available at www.palgrave.com/engineering/singh

1 [mechanics]   A force, F, acting on a particle varies with time, t, according to the table below.

t (s)	0	0.5	1.0	1.5	2.0	2.5	3.0
F (N)	3.2	5.6	7.0	7.7	8.4	9.9	11.6

The impulse of this force is given by

$$\int_0^3 F \, dt$$

Find an approximate value for the impulse.

2 Use the trapezium rule with four equal intervals to find the approximate values of

a $\displaystyle\int_0^1 e^{-x^2} \, dx$ **b** $\displaystyle\int_0^{\pi/2} \sqrt{\cos(x)} \, dx$

3 i Apply the trapezium rule to find the approximate value of

$$\int_0^1 x^3 \, dx$$

with **a** four equal intervals
     b eight equal intervals.

ii Find the exact value of $\displaystyle\int_0^1 x^3 \, dx$.

iii Determine the percentage error between the exact value and the estimated values found in **i**.

iv Comment upon your results to **iii**.

4 [mechanics]   The velocity, v, of a model for a new ship is tested in an experiment for 5 seconds and has the values shown in the table below.

t (s)	0	1	2	3	4	5
v (m/s)	2.10	9.56	11.36	12.08	12.98	13.76

The distance travelled is given by

$$\int_0^5 v \, dt$$

Estimate the distance.

5 [fluid mechanics]   A river is 15 m wide. The depth of river is found in metres from one side of the embankment to the other with the results shown in the table below.

Dist. (m)	0	1.5	3.0	4.5	6.0	7.5
Depth (m)	0	1.04	1.65	3.10	4.66	4.12

Dist. (m)	9.0	10.5	12.0	13.5	15.0
Depth (m)	3.21	2.33	1.78	0.76	0

Given that the velocity of water is 2.05 m/s, find the approximate number of cubic metres of water flowing down the river per second.

[*Hint*: Volume of flow per second = cross-sectional area × velocity.]

Solutions at end of book. Complete solutions available at www.palgrave.com/engineering/singh

Exercise **9(a) continued**

6 [*electrical principles*]   A voltage, v, has the values at intervals of 0.1 s shown in the table below.

t (s)	0	0.1	0.2	0.3	0.4	0.5	0.6
v (volts)	4	3.92	3.86	3.77	3.61	3.52	3.41

The mean voltage, $\bar{v}$, is given by

$$\bar{v} = \frac{1}{0.6}\int_0^{0.6} v \, dt.$$ Find an approximate value for $\bar{v}$.

SECTION B **Further numerical integration**

By the end of this section you will be able to:
▶ apply Simpson's rule
▶ use a computer algebra system to determine definite integrals

B1 **Simpson's rule**

In this section we state the formula of another numerical method used to find the value of an integral with limits – Simpson's rule.

Simpson's rule is named after **Thomas Simpson** (1710–61) who was born in Leicestershire, England and was a private tutor in mathematics. He is primarily known for his work in numerical integration.

The formula is defined as

9.4 $$\int_a^b y \, dx \approx \frac{h}{3}\left[y_0 + 4(y_1 + y_3 + y_5 + \cdots) + 2(y_2 + y_4 + y_6 + \cdots) + y_n \right]$$

where h is the uniform width, $y_0, y_1, y_2, \cdots, y_n$ are the ordinates shown in Fig. 8 and n is the number of blocks, which must be **even**.

? **What do you notice about the ordinates** $y_0, y_1, \cdots, y_n$ **of formula** 9.4 **?**

Formula 9.4 can also be written as

Fig. 8

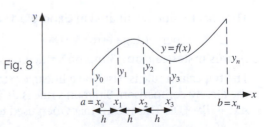

9.5 $$\text{Area} \approx \frac{h}{3}\left[\begin{array}{l} \text{(first ordinate)} + 4\text{(sum of odd ordinates)} \\ + \, 2\text{(sum of even ordinates)} + \text{(last ordinate)} \end{array} \right]$$

Note that to use formula 9.4 or 9.5 , the number of blocks, n, must be even.

Let's repeat **Example 1** with Simpson's rule.

Example 4

Evaluate $\int_0^1 x^2 \, dx$ using Simpson's rule with four intervals and find the percentage error from the exact value. Comment on your result for percentage error.

Solution

? **Can we use Simpson's rule?**

Yes, since we have an even number (4) of intervals. Putting $n = 4$ into Simpson's rule gives

> * $\int_0^1 x^2 \, dx \approx \dfrac{h}{3} \left[y_0 + 4(y_1 + y_3) + 2y_2 + y_4 \right]$

? **What is the value of the uniform width h?**

We require four equal intervals, therefore

$$h = \frac{1 - 0}{4} = \frac{1}{4}$$

? **What else do we need to find in order to use * ?**

The ordinates y_0, y_1, y_2, y_3 and y_4. With $y = x^2$ we have the same y values as those in **Example 1**:

$$x_0 = 0 \qquad\qquad y_0 = 0^2 = 0$$
$$x_1 = \frac{1}{4} \qquad\qquad y_1 = \left(\frac{1}{4}\right)^2 = \frac{1}{16}$$
$$x_2 = \frac{2}{4} = \frac{1}{2} \qquad\qquad y_2 = \left(\frac{1}{2}\right)^2 = \frac{1}{4}$$
$$x_3 = \frac{3}{4} \qquad\qquad y_3 = \left(\frac{3}{4}\right)^2 = \frac{9}{16}$$
$$x_4 = 1 \qquad\qquad y_4 = 1^2 = 1$$

Putting all these into * gives

$$\int_0^1 x^2 \, dx \approx \frac{h}{3} \left[y_0 + 4(y_1 + y_3) + 2y_2 + y_4 \right]$$
$$= \frac{1/4}{3} \left[0 + 4\left(\frac{1}{16} + \frac{9}{16}\right) + 2\left(\frac{1}{4}\right) + 1 \right] = \frac{1}{3}$$

The exact value (evaluated in **Example 1**) of $\int_0^1 x^2 \, dx = \dfrac{1}{3}$. So

$$\text{percentage error} = 0\%$$

? In this instance using Simpson's rule gives an exact value for the area. **How?**

The trapezium rule is based on a linear, or straight-line, equation, creating a series of steps under the arc. By comparison, Simpson's rule is derived in part from the quadratic equation for a curve $Ax^2 + Bx + C$. This means that when used to calculate the area under a quadratic function like $\int_0^1 x^2 \, dx$ it provides a perfect template. For equations of a higher order, its accuracy becomes increasingly compromised.

Example 5

Find the approximate value of

$$\int_0^{\pi/2} \sqrt{\cos(x)}\, dx$$

by using Simpson's rule with eight equal intervals and working to 4 d.p.

Solution

Applying Simpson with $n = 8$ we have

$$\int_0^{\pi/2} \sqrt{\cos(x)}\, dx \approx \frac{h}{3}\left[y_0 + 4\left(y_1 + y_3 + y_5 + y_7\right) + 2\left(y_2 + y_4 + y_6\right) + y_8\right] \quad *$$

What is the value of h?

$$h = \frac{\pi/2 - 0}{8} = \frac{\pi}{16}$$

We can find the y values, as shown in Table 3.

x	$\sqrt{\cos(x)} = y$
0	$\sqrt{\cos(0)} = 1 = y_0$
$\dfrac{\pi}{16}$	$\sqrt{\cos\left(\dfrac{\pi}{16}\right)} = 0.9903 = y_1$
$\dfrac{2\pi}{16} = \dfrac{\pi}{8}$	$\sqrt{\cos\left(\dfrac{\pi}{8}\right)} = 0.9612 = y_2$
$\dfrac{3\pi}{16}$	$\sqrt{\cos\left(\dfrac{3\pi}{16}\right)} = 0.9118 = y_3$
$\dfrac{4\pi}{16} = \dfrac{\pi}{4}$	$\sqrt{\cos\left(\dfrac{\pi}{4}\right)} = 0.8409 = y_4$
$\dfrac{5\pi}{16}$	$\sqrt{\cos\left(\dfrac{5\pi}{16}\right)} = 0.7454 = y_5$
$\dfrac{6\pi}{16} = \dfrac{3\pi}{8}$	$\sqrt{\cos\left(\dfrac{3\pi}{8}\right)} = 0.6186 = y_6$
$\dfrac{7\pi}{16}$	$\sqrt{\cos\left(\dfrac{7\pi}{16}\right)} = 0.4417 = y_7$
$\dfrac{8\pi}{16} = \dfrac{\pi}{2}$	$\sqrt{\cos\left(\dfrac{\pi}{2}\right)} = 0 = y_8$

TABLE 3

Substituting $h = \dfrac{\pi}{16}$ and the above y values into $*$ yields

Example 5 *continued*

$$\int_0^{\pi/2} \sqrt{\cos(x)}\,dx \approx \frac{h}{3}\left[y_0 + 4(y_1 + y_3 + y_5 + y_7) + 2(y_2 + y_4 + y_6) + y_8\right]$$

$$= \frac{\pi/16}{3}\left[1 + 4(0.9903 + 0.9118 + 0.7454 + 0.4417) + 2(0.9612 + 0.8409 + 0.6186) + 0\right]$$

$$= \frac{\pi}{48}\,[18.1982] = 1.1911$$

Example 6 *mechanics*

The height, s, of a rocket in the first 6 seconds is given by

$$s = (5 \times 10^3)\int_0^6 \ln\left(\frac{100}{100 - 4t}\right)dt - 180$$

i Evaluate $\int_0^6 \ln\left(\frac{100}{100 - 4t}\right)dt$ by using Simpson's rule with six equal intervals.

ii Determine s.

Solution

Let $y = \ln\left(\frac{100}{100 - 4t}\right)$

i Using

9.4 $$\int_a^b y\,dx \approx \frac{h}{3}\left[y_0 + 4(y_1 + y_3 + \cdots) + 2(y_2 + y_4 + \cdots) + y_n\right]$$

with six equal intervals, $n = 6$, and $dx = dt$ gives

* $$\int_0^6 y\,dt \approx \frac{h}{3}\left[y_0 + 4(y_1 + y_3 + y_5) + 2(y_2 + y_4) + y_6\right]$$

What is the value of the uniform width h?

$$h = \frac{6 - 0}{6} = 1$$

Thus $t_0 = 0$, $t_1 = 1$, $\cdots$, $t_6 = 6$.

What else do we require in order to use Simpson's rule?
We need to find $y_0, y_1, \cdots$ and y_6 by substituting the t values into $\ln\left[\frac{100}{100 - 4t}\right]$:

$$t_0 = 0 \qquad y_0 = \ln\left[\frac{100}{100 - (4 \times 0)}\right] = 0$$

$$t_1 = 1 \qquad y_1 = \ln\left[\frac{100}{100 - (4 \times 1)}\right] = 0.041$$

$$t_2 = 2 \qquad y_2 = \ln\left[\frac{100}{100 - (4 \times 2)}\right] = 0.083$$

🚗 Example 6 *continued*

$$t_3 = 3 \qquad\qquad y_3 = \ln\left[\frac{100}{100 - (4 \times 3)}\right] = 0.128$$

$$t_4 = 4 \qquad\qquad y_4 = \ln\left[\frac{100}{100 - (4 \times 4)}\right] = 0.174$$

$$t_5 = 5 \qquad\qquad y_5 = \ln\left[\frac{100}{100 - (4 \times 5)}\right] = 0.223$$

$$t_6 = 6 \qquad\qquad y_6 = \ln\left[\frac{100}{100 - (4 \times 6)}\right] = 0.274$$

Putting all this into   *   gives

$$\int_0^6 y\,dt \approx \frac{1}{3}\left[0 + 4(0.041 + 0.128 + 0.223) + 2(0.083 + 0.174) + 0.274\right] = 0.785$$

ii $s \approx [(5 \times 10^3) \times (0.785)] - 180 = 3745\ \text{m} = 3.75\ \text{km (3 s.f.)}$

Numerical methods are particularly useful if we have to integrate a function which is given by a set of experimental data.

Example 7

The shape of a piece of land is shown in Fig. 9. All dimensions are in metres.
By using Simpson's rule, estimate the area.

Solution

? **Can we use Simpson's rule? What do we need to check first?**

Fig. 9

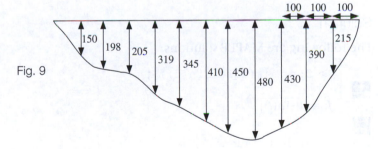

Check that there are an even number of blocks. In fact there are 12.

9.5 $\quad$ Area $\approx \dfrac{h}{3}\left[\begin{array}{l}\text{(first ordinate)} + 4\text{(sum of odd ordinates)} + \\ \qquad 2\text{(sum of even ordinates)} + \text{(last ordinate)}\end{array}\right]$

? **What is the value of the uniform width h?**

From Fig. 9, $h = 100$.

? **In order to use Simpson's rule, what else do we need to find?**

First ordinate = 0. Last ordinate = 0.

Even ordinates are 198, 319, 410, 480 and 390.

Example 7 *continued*

Odd ordinates are 150, 205, 345, 450, 430 and 215.

Substituting these into 9.5 gives

$$\text{Area} \approx \frac{100}{3}\left[\begin{aligned}&0 + 4(150 + 205 + 345 + 450 + 430 + 215)\\&\quad + 2(198 + 319 + 410 + 480 + 390) + 0\end{aligned}\right]$$

$$= 359\,133.33 \text{ m}^2 = 3.59 \times 10^5 \text{ m}^2 \text{ (3 s.f.)}$$

B2 Integration using a computer algebra system

We can use a computer algebra system to apply Simpson's rule or the trapezium rule with a large number of intervals. **Example 8** uses MAPLE to evaluate a definite integral with 100 intervals. Even if you don't have access to a computer algebra system, try following the example.

Example 8 *electrical principles*

The average value of an a.c. current waveform, i_{avg}, is given by

$$i_{avg} = \frac{1}{2\pi}\int_0^{2\pi} \sqrt{|\sin(t)|}\,dt$$

By using any appropriate mathematical software, determine an estimate for i_{avg} with 100 intervals using

a Simpson's rule   **b** the trapezium rule

Solution

The following are MAPLE solutions.

```
> f:=abs (sqrt(sin(t)));
```
$f: = \sqrt{|\sin(t)|}$

```
> with (student):
```

```
> i[avg]:=evalf((1/(2*Pi))*simpson(f,t=0 .. 2*Pi,100));
```
$i_{avg}: = 0.7619458331$

```
> i[avg]:=evalf((1/(2*Pi))*trapezoid(f,t=0 .. 2*Pi,100));
```
$i_{avg}: = 0.7606753615$

The next example shows how we can evaluate a difficult definite integral on a computer algebra system (MAPLE).

Example 9 *signal processing*

The probability, P, that a signal is greater than 0.1 volt is given by

$$P = \frac{5}{\sqrt{2\pi}} \int_{0.1}^{\infty} \exp\left(-\frac{25x^2}{2}\right) dx$$

Determine P.

Solution

Remember that exp is the exponential, e. Using MAPLE:

```
> g:=exp(-25*x^2/2);
g := e^(-25/2x²)
```

```
> f:=int(g, x=0.1 .. infinity);
f: = 0.1546777837
```

```
> evalf(5/sqrt(2*Pi)*f);
> 0.3085375386
```

We have $P = 0.3085$ (4 d.p.).

SUMMARY

Simpson's rule is

9.4 $$\int_a^b y\,dx \approx \frac{h}{3}\Big[y_0 + 4(y_1 + y_3 + y_5 + \cdots) + 2(y_2 + y_4 + \cdots) + y_n\Big]$$

9.5 $$\text{Area} \approx \frac{h}{3}\Big[(\text{first ordinate}) + 4(\text{sum of odd ordinates}) \\ + 2(\text{sum of even ordinates}) + (\text{last ordinate})\Big]$$

The above rules are particularly useful for functions which are given in experimental data form.

Normally it is more appropriate to use a computer algebra system to integrate difficult algebraic functions.

Exercise 9(b)

Solutions at end of book. Complete solutions available at www.palgrave.com/engineering/singh

1 Evaluate $\int_0^{\pi/2} \sin(t)\,dt$ by applying Simpson's rule with ten equal intervals. (Work to 3 d.p. but give the final answer to 4 d.p.) Determine the exact value of $\int_0^{\pi/2} \sin(t)\,dt$. Find the percentage error from the exact value for the above case.

Exercise **9(b) continued**

Solutions at end of book. Complete solutions available at www.palgrave.com/engineering/singh

2 Evaluate $\int_0^{\pi/3} \sqrt{\sin(x)}\,dx$ to 3 d.p., using six equal intervals.

Use a computer algebra system to find $\int_0^{\pi/3} \sqrt{\sin(x)}\,dx$ by applying

a the trapezium rule,   **b** Simpson's rule

with **i** 10 intervals,   **ii** 50 intervals,

  iii 100 intervals,   **iv** 1000 intervals.

3 🚗 [*mechanics*]   A force, F, acting on a particle varies with time according to the following experimental data:

t (s)	2.0	2.1	2.2	2.3
F (N)	230.0	202.6	188.3	178.7

t (s)	2.4	2.5	2.6	2.7
F (N)	172.4	168.0	165.6	163.4

t (s)	2.8	2.9	3.0	
F (N)	162.0	161.1	160.5	

The impulse of this force from 2 to 3 s is given by

$$\int_2^3 F\,dt$$

Estimate this impulse.

4 🏛 [*structures*]

x (m)	0	0.5	1.0	1.5	2.0
w (kN/m)	0	1.7	2.3	5.6	8.8

x (m)	2.5	3.0	3.5	4.0
w (kN/m)	7.3	4.9	3.2	2.5

The above table shows the loading, w, on a beam of length 4 m. The total force, F, exerted by the load is given by

$$F = \int_0^4 w\,dx$$

Estimate F.

5 📟 [*electrical principles*]   An a.c. current, i, varies with time, t, according to the following table:

t (s)	0	0.1	0.2	0.3	0.4	0.5	0.6
i (A)	0.0	0.25	0.31	0.43	0.37	0.20	0.0

The RMS current, i_{RMS}, is given by

$$i_{RMS} = \sqrt{\frac{1}{0.6}\int_0^{0.6} i^2\,dt}$$

Find an approximate value for i_{RMS}.

6 🌡 [*thermodynamics*]   The volume, V, and pressure, P, of a gas in a cylinder is measured in an experiment and the results are shown in the table below.

V (m³)	0.01	0.02	0.03
P (kPa)	560.73	237.86	140.33

V (m³)	0.04	0.05	0.06
P (kPa)	98.67	73.66	59.24

V (m³)	0.07	0.08	0.09
P (kPa)	48.73	42.31	35.20

The work done, W, by the gas on the face of the piston is given by

$$W = \int_{0.01}^{0.09} P\,dV$$

Find an approximate value for W.

Exercise **9(b) continued**

7 [fluid mechanics] A river is 25 m wide. The depth of river is measured at eleven equidistant points from one end, with the following results:

Dist. (m)	0	2.5	5.0	7.5
Depth (m)	0.3	1.78	2.05	2.96

Dist. (m)	10.0	12.5	15.0	17.5
Depth (m)	4.01	5.30	6.78	6.11

Dist. (m)	20.0	22.5	25.0
Depth (m)	4.35	1.76	0.4

Given that the velocity of water is 1.67 m/s, find the approximate number of cubic metres of water flowing down the river per second.

[Hint: Volume of flow per second = cross-sectional area × velocity.]

Use a computer algebra system for the remaining questions in this exercise.

8 [structures] The force, F, on a beam of length 5 m is given by

$$F = \int_0^5 \frac{e^x}{1 + \sqrt{x}}\,dx \quad \text{(where } x \text{ is distance)}$$

Evaluate F.

9 [signal processing] The probability, P, that a signal is greater than 3 V is given by

$$P = \sqrt{\frac{2}{\pi}} \int_3^\infty e^{-2x^2}\,dx$$

Evaluate P.

10 [signal processing] The Q function, $Q(x)$, in signal processing is defined as

$$Q(x) = \frac{1}{\sqrt{2\pi}} \int_x^\infty \exp\left(-\frac{t^2}{2}\right) dt$$

Evaluate $Q(1)$, $Q(2)$, $Q(3)$ and $Q(4)$.

11 [mechanics] The time taken, t, for an object falling in air to reach a velocity of 20 m/s is given by

$$t = \int_0^{20} \frac{dv}{9.8 - 0.1v^{1.23}}$$

Evaluate t with 100 intervals by using
a Simpson's rule **b** the trapezium rule

SECTION C **Engineering applications**

By the end of this section you will be able to:
▶ determine the area under a curve
▶ find the area between curves
▶ find the average value of a function
▶ find the root mean square (RMS) value of a function

C1 **Area under a curve**

? **How can we write the shaded area of Fig. 10 on the next page in terms of an integral?**

9.6 $$\text{Shaded area} = \int_a^b y\,dx$$

Let's examine this formula by doing an example.

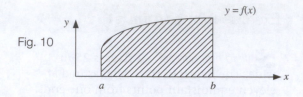

Fig. 10

Example 10

Find the area bounded by $\sin(x)$ and the x axis between $-\pi$ and π.

Solution

Using 9.6 we have the required area as:

$$\int_{-\pi}^{\pi} \sin(x)\, dx = \Big[-\cos(x)\Big]_{-\pi}^{\pi} \qquad \text{[By 8.7]}$$

$$= -[\cos(\pi) - \cos(-\pi)] \qquad \text{[Substituting]}$$

$$= -[-1 - (-1)] = 0$$

$$\int_{-\pi}^{\pi} \sin(x)\, dx = 0$$

This says that the area between $\sin(x)$ and the x axis (from $-\pi$ to π) is zero.

However Fig. 11 shows that there is an area $(A + B)$ between $\sin(x)$ and the x axis.

Fig. 11

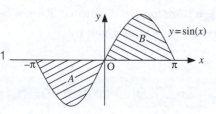

? **What's gone wrong with our calculation which evaluates the area to zero?**

Consider the following:

For area A we have

$$\int_{-\pi}^{0} \sin(x)\, dx = \Big[-\cos(x)\Big]_{-\pi}^{0}$$

$$= -[\cos(0) - \cos(-\pi)] = -[1 - (-1)] = -2$$

? **What does the negative sign in -2 for area A signify?**

The -2 for area A means that the area is two units below the x axis.
For area B we have

8.7 $\displaystyle\int \sin(x)\,dx = -\cos(x)$

Example 10 *continued*

$$\int_0^\pi \sin(x)dx = \left[-\cos(x)\right]_0^\pi = -\left[\cos(\pi) - \cos(0)\right] = -\left[-1 - 1\right] = 2$$

So the areas cancel each other out: $-2 + 2 = 0$

Thus the total area $= 2 + 2 = 4$.

The purpose of **Example 10** was to show the need to sketch a graph before finding the area. First we draw the graph and then we integrate each positive and negative area separately. If you get a negative answer then the area is below the x axis. Consider each area to be positive and then add to find the total area.

Example 11

Calculate the area between the curve $y = \dfrac{2}{x}$ and the line $y = 3 - x$.

Solution

The graph of $y = 3 - x$ is a straight line crossing the y axis at 3 and the x axis at 3.

The graph of $y = \dfrac{2}{x}$ is the normal reciprocal graph which is asymptotic to the x and y axes. (Refer back to **Chapter 2** if you need to remind yourself what these terms mean.)

? **Where do these graphs intersect?**

Intersection occurs where $y = \dfrac{2}{x}$ and $y = 3 - x$ are equal. Thus

$$3 - x = \frac{2}{x}$$

$$3x - x^2 = 2 \qquad \text{[Multiplying by } x]$$

Collecting on one side:

$$x^2 - 3x + 2 = 0 \qquad \text{[Quadratic equation]}$$

$$(x - 1)(x - 2) = 0 \qquad \text{[Factorizing]}$$

$$x = 1, \quad x = 2$$

? **What are the y values at $x = 1$ and $x = 2$?**

At $x = 1$, $y = \dfrac{2}{1} = 2$, the point (1, 2). At $x = 2$, $y = \dfrac{2}{2} = 1$, the point (2, 1).

Hence the points of intersection are (1, 2) and (2, 1). Thus we have the situation shown in Fig. 12.

Example 11 *continued*

?

How do we find the shaded area?

We can calculate the area below the
curve by integration and the area below
the line is a trapezium, ABCD. The
difference between the two is therefore
the shaded area.

Fig. 12

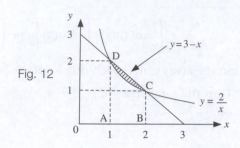

Area below $y = \dfrac{2}{x}$ and the x axis

between 1 and 2 is

$$\int_1^2 \frac{2}{x}\, dx = \left[\, 2\ln|x|\,\right]_1^2 \qquad \text{[By 8.2]}$$

$$= 2\left[\ln(2) - \ln(1)\right] = 2\ln(2)\ \text{(units)}^2$$

Area below $y = 3 - x$ and the x axis is the area of trapezium
ABCD (Fig. 13).

$$\text{Area ABCD} = \frac{1}{2}1(2 + 1) = \frac{3}{2}\ \text{(units)}^2$$

Fig. 13

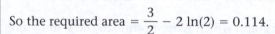

To find the required area we subtract the area below $y = \dfrac{2}{x}$
from the area of the trapezium.

So the required area $= \dfrac{3}{2} - 2\ln(2) = 0.114.$

The area between the line $y = 3 - x$ and $y = 2/x$ is 0.114 (units)^2.

C2 **Mean and root mean square (RMS) values**

?

What does the term *mean* represent?

Average. Average values of current, force and
other engineering quantities are important.
In this section we are interested in the
average value of a function. We can use what we
know about areas and curves to discover what the
mean value of a function is. Consider the function $y = f(x)$ between a and b (Fig. 14).

Fig. 14

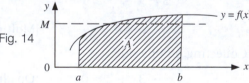

Let M be the mean value of the function, $y = f(x)$, between a and b. Then M is defined as the
y value such that

$$M(b - a) = \text{shaded area } A$$

$M(b - a)$ is equivalent in size to an area of a rectangle of width $b - a$ and height M (Fig. 15).

8.2 $\displaystyle\int dx/x = \ln|x|$

❓ How can we determine M?

We have

$$M(b - a) = \text{shaded area } A$$

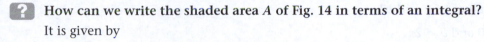

Fig. 15

Same as the shaded area of Fig. 14

Therefore transposing gives

† $$M = \frac{\text{shaded area } A}{b - a}$$

We can define the mean value, M, as

$$M = \frac{\text{area}}{\text{interval}}$$

❓ How can we write the shaded area A of Fig. 14 in terms of an integral?

It is given by

$$\text{shaded area } A = \int_a^b y \, dx$$

Putting this into † gives

$$M = \frac{\int_a^b y \, dx}{b - a} = \frac{1}{b - a} \int_a^b y \, dx \qquad \left[\text{Taking out } \frac{1}{b - a}\right]$$

So we have the mean value of the function $y = f(x)$ between $x = a$ and $x = b$ given by

9.7 $$\text{Mean value of } y = \frac{1}{b - a} \int_a^b y \, dx$$

Similarly we can define the root mean square, RMS, value of $y = f(x)$ between $x = a$ and $x = b$ as

$$\text{RMS} = \sqrt{\text{Mean value of } y^2}$$

$$(\text{RMS})^2 = \text{Mean value of } y^2 \quad [\text{Squaring both sides}]$$

❓ What is the mean value of y^2 between $x = a$ and $x = b$?

Using 9.7 :

$$\text{Mean value of } y^2 = \frac{1}{b - a} \int_a^b y^2 \, dx$$

Since $(\text{RMS})^2 = \text{Mean value of } y^2$:

9.8 $$(\text{RMS})^2 = \frac{1}{b - a} \int_a^b y^2 \, dx$$

and hence taking the square root gives

9.9 $$\text{RMS} = \sqrt{\frac{1}{b - a} \int_a^b y^2 \, dx}$$

Generally it is more appropriate to use 9.8 , evaluate $(\text{RMS})^2$ and then take the square root of the result to give the RMS value. A general error is that we forget to take the square root at the end.

The root mean square (RMS) value is important in electrical principles.

Example 12 *electrical principles*

Evaluate the RMS value of the current $i = 10\sin(t)$ over the interval 0 to 2π.

Solution

Which formula do we use?

9.8 $$(\text{RMS})^2 = \frac{1}{b - a} \int_a^b y^2 \, dx$$

and then take the square root of the result.

What are the values of a and b?

$$a = 0 \text{ and } b = 2\pi$$

What is y?

In this case $y = i = 10\sin(t)$ and we integrate with respect to t. Using 9.8 :

$$(\text{RMS})^2 = \frac{1}{2\pi - 0} \int_0^{2\pi} \left[10\sin(t) \right]^2 dt$$

$$= \frac{1}{2\pi} \int_0^{2\pi} 100\sin^2(t) \, dt \qquad \text{[Squaring the integrand]}$$

★ $$(\text{RMS})^2 \underset{\substack{\text{taking out the} \\ \text{constant } 100}}{=} \frac{100}{2\pi} \int_0^{2\pi} \sin^2(t) \, dt$$

How do we evaluate $\int_0^{2\pi} \sin^2(t) \, dt$?

We need to use the trigonometric identity

4.68 $$\sin^2(t) = \frac{1}{2} [1 - \cos(2t)]$$

Therefore

$$\int_0^{2\pi} \sin^2(t) \, dt = \int_0^{2\pi} \frac{1}{2} \left[1 - \cos(2t) \right] dt$$

$$= \frac{1}{2} \int_0^{2\pi} \left[1 - \cos(2t) \right] dt \quad \left[\text{Taking out } \frac{1}{2} \right]$$

$$= \frac{1}{2} \left[t - \frac{\sin(2t)}{2} \right]_0^{2\pi} \quad \left[\text{By} \int \cos(kt) \, dt = \frac{\sin(kt)}{k} \right]$$

$$= \frac{1}{2} \left\{ \left[2\pi - \frac{\sin[2(2\pi)]}{2} \right] - 0 \right\}$$

$$= \frac{1}{2} \left\{ [2\pi - 0] - 0 \right\} = \pi \quad \text{[Cancelling 2's]}$$

Example 12 *continued*

$$\int_0^{2\pi} \sin^2(t)\, dt = \pi$$

Substituting $\int_0^{2\pi} \sin^2(t)\, dt = \pi$ into ▮ * ▮ gives

$$(RMS)^2 = \frac{100}{2\pi}(\pi) = 50 \quad \text{[Cancelling } \pi\text{'s]}$$

? **Have we found the RMS value of $i = 10\sin(t)$?**

No, but we have established the $(RMS)^2$ value of $i = 10\sin(t)$.

? **How do we find the RMS value?**

Take the square root. Hence

$$RMS = \sqrt{50} = 7.07 \text{ A} \qquad \text{(2 d.p.)}$$

Example 13 *electrical principles*

Find the RMS value, i_{RMS}, of the current, i, shown in Fig. 16.

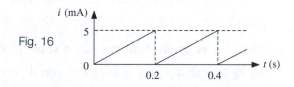

Fig. 16

Solution

To find the RMS value we first obtain $(i_{RMS})^2$ and then take the square root of our result.

Since the waveform repeats itself we only need to consider ONE cycle, say between $t = 0$ and $t = 0.2$.

? **What is the equation of the function, i, for t between 0 and 0.2?**

The waveform is a straight line going through 0 and having a gradient m given by

$$m = \frac{5}{0.2} = 25$$

Therefore using the straight-line equation, $i = mt + c$, we have

$$i = 25t \qquad \text{[Because } c = 0\text{]}$$

Substituting $a = 0$, $b = 0.2$, $y = i = 25t$ and $dx = dt$ into

9.8 $$(RMS)^2 = \frac{1}{b - a}\int_a^b y^2\, dx$$

gives

Example 13 *continued*

$$(i_{RMS})^2 = \frac{1}{0.2} \int_0^{0.2} (25t)^2 \, dt$$

$$= \frac{25^2}{0.2} \int_0^{0.2} t^2 \, dt \qquad \text{[Taking out } 25^2\text{]}$$

$$= 3125 \left[\frac{t^3}{3} \right]_0^{0.2} \qquad \text{[Integrating]}$$

$$= \frac{3125}{3} (0.2^3 - 0^3) = 8.33$$

Hence

$$i_{RMS} = \sqrt{8.33} = 2.89 \text{ mA} \qquad \text{(2 d.p.)}$$

Example 14 *mechanics*

The force, F, acting on a particle for the first 10 seconds is shown in Table 4.

TABLE 4	t (s)	0	1	2	3	4	5	6	7	8	9	10
	F (kN)	0	3.4	5.1	7.6	9.1	8.4	6.2	3.1	2.7	1.3	0.9

Estimate the mean value of the applied force, F.

Solution

What do we use to find the mean value of F?

Use **9.7** with $a = 0$, $b = 10$, $y = F$ and $dx = dt$:

9.7	Mean value of $y = \dfrac{1}{b-a} \displaystyle\int_a^b y \, dx$

$*$	Mean value of $F = \dfrac{1}{10} \displaystyle\int_0^{10} F \, dt$

How can we find $\displaystyle\int_0^{10} F \, dt$?

Since our F is in tabular form and goes up/down in steps we can only use a numerical method to evaluate $\displaystyle\int_0^{10} F \, dt$. Let's use Simpson's rule with an even number of entries $n = 10$, therefore

†	$\displaystyle\int_0^{10} F \, dt \approx \dfrac{h}{3} [F_0 + 4(F_1 + F_3 + F_5 + F_7 + F_9) + 2(F_2 + F_4 + F_6 + F_8) + F_{10}]$

What are the values of h, F_0, F_1, $\cdots$, F_{10}?

From Table 4 the uniform width $h = 1$. Using the second row of Table 4 we have $F_0 = 0$, $F_1 = 3.4$, $F_2 = 5.1$, $F_3 = 7.6$, $F_4 = 9.1$, $F_5 = 8.4$, $F_6 = 6.2$, $F_7 = 3.1$, $F_8 = 2.7$, $F_9 = 1.3$ and $F_{10} = 0.9$.

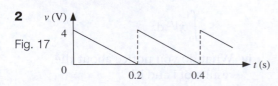

Example 14 *continued*

Substituting all this into † gives

$$\int_0^{10} F\, dt \approx \frac{1}{3}\, [0 + 4(3.4 + 7.6 + 8.4 + 3.1 + 1.3) + 2(5.1 + 9.1 + 6.2 + 2.7) + 0.9]$$

$$= 47.43$$

Is 47.43 the mean value of the force F?

No, we need to substitute this into * :

$$\text{Mean value of } F = \frac{1}{10}\,(47.43) = 4.74 \text{ kN} \quad (2 \text{ d.p.})$$

SUMMARY

The area between the graph and the x axis is given by

9.6 $$\text{area} = \int_a^b y\, dx$$

Remember, before finding areas under curves we must make a sketch of the graph(s) because areas under the x axis will otherwise be subtracted from areas above it and give the wrong answer.

Mean value of the function $y = f(x)$ between $x = a$ and $x = b$ is

9.7 $$\text{Mean value of } y = \frac{1}{b-a} \int_a^b y\, dx$$

Also root mean square, RMS, is

9.8 $$(\text{RMS})^2 = \frac{1}{b-a} \int_a^b y^2\, dx$$

9.9 $$\text{RMS} = \sqrt{\frac{1}{b-a} \int_a^b y^2 dx}$$

Exercise 9(c)

Solutions at end of book. Complete solutions available at www.palgrave.com/engineering/singh

1 Obtain the area bounded by $y = \sin(x)$ and the x axis between 0 and 2π.

Questions 2 to 10 belong to the field of [electrical principles].

2

v (V)

4

Fig. 17

0   0.2   0.4   ▶ t (s)

Determine the mean and RMS values of the voltage, v, shown in Fig. 17. (Consider ONE cycle.)

3 Show that the mean value of $i = I \sin(t)$ over the interval 0 to π is $2I/\pi$ where the constant I is the maximum current. Determine the mean value of $i = \sin(t)$ over this period.

Exercise **9(c) continued**

Solutions at end of book. Complete solutions available at www.palgrave.com/engineering/singh

4 Determine the mean value, over the interval 0 to 2π, of the following voltages:
i $v = 10\sin(t)$ **ii** $v = 10\cos(t)$

5 By using

| 9.7 | Mean value of $y = \dfrac{1}{b-a}\displaystyle\int_a^b y\,dx$

show that the mean value of the voltage $v = V\sin(\omega t)$ over the interval $\omega t = 0$ to $\omega t = 2\pi$ is 0 V.

6 Show that the RMS value, i_{RMS}, of the current $i = I\cos(t)$ over the interval 0 to 2π is given by

$$i_{RMS} = \frac{I}{\sqrt{2}}$$

where the constant I is the maximum current.

7 Evaluate the mean and RMS values of the voltage, v, over an interval 0 to 10 where

$$v = 10(1 - e^{-0.1t})$$

8

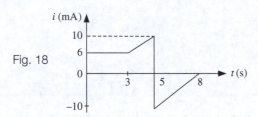

Fig. 18

Determine the mean value of the waveform in Fig. 18. Consider the waveform between 0 and 8 seconds.

*9 Determine the mean and RMS values of the voltage waveform, $v = \omega t \sin(\omega t)$ across the interval $\omega t = 0$ to $\omega t = \pi$. The form factor, f, is defined as

$$f = \frac{\text{RMS value}}{\text{Mean value}}$$

Determine f for the waveform v.

10 Find the mean value of the current, $i = \sqrt{\cos(\omega t)}$ for $\omega t = 0$ to $\omega t = \pi/2$. Note that we cannot integrate $\sqrt{\cos(\omega t)}$ analytically, so use Simpson's rule with four equal intervals.

11 🚗 [*mechanics*]   The force, F, acting on a particle for the first 10 seconds is given by:

t (s)	0	1	2	3	4	5
F (kN)	0	2.6	3.1	3.9	4.5	6.7

t (s)	6	7	8	9	10
F (kN)	7.1	4.5	3.1	2.0	1.5

Estimate the mean value of the applied force, F.

12

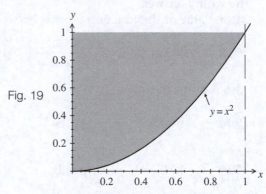

Fig. 19

i Evaluate the shaded area of Fig. 19 above.

ii Evaluate $\displaystyle\int_0^1 x^{1/2}\,dx$.

iii Draw the diagram of Fig. 19 and sketch the graph of $y = x^{1/2}$. Shade in the area represented by

$$\int_0^1 x^{1/2}\,dx$$

iv What do you notice about the results of **i** and **ii**?

SECTION D **Applications in mechanics**

By the end of this section you will be able to:
► use integration to find the displacement and velocity of a particle

D1 **Displacement, velocity and acceleration**

In **Chapter 7**, **Section D** (page 349) we stated that for a particle in straight-line motion the displacement, s, velocity, v, and acceleration, a, are related as follows:

$$v = \frac{ds}{dt}$$

$$a = \frac{dv}{dt} \left(= \frac{d^2s}{dt^2} \right)$$

In this section we carry out the inverse process, that is, we integrate both sides of these equations to give:

9.10 $$s = \int v \, dt$$

9.11 $$v = \int a \, dt$$

Remember that, when integrating, we obtain a constant of integration, C, and we can usually find the value of C if we are given some conditions associated with the problem, as the following example shows.

 Example 15 *mechanics*

The acceleration, a, of a particle is given by
$$a = \cos(3t)$$
At $t = 0$, $s = \frac{1}{3}$ and $\frac{ds}{dt} = 0$. Obtain an expression for the displacement, s.

Solution

We first obtain the velocity by using $v = \int a \, dt$ with $a = \cos(3t)$:

$$v = \int \cos(3t) \, dt$$

* $$= \frac{\sin(3t)}{3} + C \qquad \left[\text{Because} \int \cos kt \, dt = \frac{\sin kt}{k} \right]$$

Example 15 continued

To find C we use the initial condition, $t = 0$, $\dfrac{ds}{dt} = 0 = v$. Substituting these into ▮* :

$$0 = \frac{\sin(0)}{3} + C \text{ which gives } C = 0$$

Hence putting $C = 0$ into ▮* :

$$v = \frac{\sin(3t)}{3}$$

We integrate v to obtain the displacement, s:

$$s = \int \frac{\sin(3t)}{3} \, dt$$

▮**
$$= -\left(\frac{\cos(3t)}{9}\right) + D \qquad \left[\text{Because} \int \sin kt \, dt = -\frac{\cos kt}{k}\right]$$

The new constant of integration is called D because we have already used C for the first constant of integration in ▮* .

? **How do we find the value of D?**

Substituting the other condition $t = 0$, $s = 1/3$ into ▮** :

$$\frac{1}{3} = -\left(\frac{\cos(0)}{9}\right) + D$$

$$= -\frac{1}{9} + D \qquad\qquad [\text{Because } \cos(0) = 1]$$

$$\frac{1}{3} + \frac{1}{9} = D \text{ which gives } D = \frac{4}{9}$$

Substituting $D = 4/9$ into ▮** gives us the expression for s:

$$s = -\frac{\cos(3t)}{9} + \frac{4}{9} = \frac{1}{9}\left[4 - \cos(3t)\right] \qquad [\text{Factorizing}]$$

The given conditions in the above example are called the initial conditions. They give the values of displacement and velocity at $t = 0$.

Example 16 mechanics

A projectile is fired vertically upwards from the ground with an initial velocity of 50 m/s. Show that if the acceleration

$$a = -9.81$$

then the height, h, of the projectile is given by

$$h = 50t - 4.9t^2$$

Find the maximum height reached by the projectile and sketch the graph of h against t.

Example 16 *continued*

Solution

The acceleration

$$a = -9.81$$

(Negative sign because the projectile is fired against gravity.) Hence by $v = \int a\, dt$ where v is velocity we have

$$v = -\int 9.81\, dt = -9.81t + C \quad \text{[Integrating]}$$

When $t = 0$, $v = 50$ because the initial velocity is 50 m/s. Substituting these into $v = -9.81t + C$:

$$50 = 0 + C \text{ which gives } C = 50$$
$$v = -9.81t + 50$$

The height (or displacement), h, is obtained by using $h = \int v\, dt$:

$$h = \int(-9.81t + 50)dt$$
$$h = -4.9t^2 + 50t + D \qquad \text{[Integrating]}$$

When $t = 0$, $h = 0$ (the projectile is fired from the ground, $h = 0$) this gives $D = 0$:

    $$h = -4.9t^2 + 50t$$

How do we find the maximum height?

We use differentiation. The stationary point occurs at $\dfrac{dh}{dt} = 0$:

$$\frac{dh}{dt} = v = -9.81t + 50 = 0$$
$$9.81t = 50$$
$$t = \frac{50}{9.81} = 5.10$$

To check whether we have a maximum or a minimum at $t = 5.10$ we need to differentiate again:

$$\frac{d^2h}{dt^2} = a = -9.81 < 0 \quad \text{[Negative]}$$

At $t = 5.1$, h is maximum. The maximum height is obtained by substituting $t = 5.1$ into $*$:

$$h = -(4.9 \times 5.1^2) + (50 \times 5.1) = 127.55 \text{ m}$$

For a sketch we also need to find the values of t where $h = 0$:

$$50t - 4.9t^2 = 0$$
$$t(50 - 4.9t) = 0$$
$$t = 0, t = 10.2$$

Thus we have the situation shown in Fig. 20.

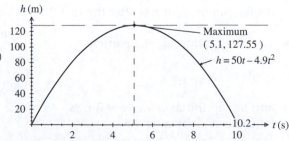

Fig. 20

SUMMARY

The displacement, s, velocity, v, and acceleration, a, are related by

9.10 $\qquad s = \int v \, dt$

9.11 $\qquad v = \int a \, dt$

Moreover if we are given initial conditions we can find the value of the integration constant C.

Exercise 9(d)

Solutions at end of book. Complete solutions available at www.palgrave.com/engineering/singh

 **All questions belong to the field of [*mechanics*].**

1 The acceleration, a, of a mass vibrating on a spring in simple harmonic motion is given by

$$a = -2.5 \cos(2\pi t)$$

Given that at $t = 0$, $v = 0$ and displacement $x = 0$, find an expression for displacement x.

2 The horizontal component of acceleration, a_x, of a particle moving with curvilinear motion is given by

$$a_x = t$$

Given that when $t = 0$, the velocity $v_x = 2$ and displacement $x = 3$, find an expression for displacement x.

3 The velocity, v, of an object is given by

$$v = 0.6t^2 + t$$

Determine the acceleration and displacement from rest after the first 2 s.

4 The acceleration, a, of a particle is given by

$$a = 4t^3$$

and has an initial velocity of 3 m/s. Find the velocity after 1.5 s.

5 A rocket is fired vertically upwards with an acceleration of 2.9 m/s². Find the height reached by the projectile after 2 minutes.

6 The table below shows the velocity, v, for the corresponding time, t, of a vehicle.

t (s)	0	25	50	75	100
v (m/s)	0	33	45	55	64
t (s)	125	150	175	200	
v (m/s)	72	77	80	82	

Find the approximate distance covered by the vehicle.

7 A particle has an acceleration, a, and at $t = 0$ the velocity $v = u$ and displacement $s = 0$. Show that

$$s = ut + \frac{1}{2}at^2$$

8 A stone is dropped vertically from a building of height 20 m with an acceleration, a, given by

$$a = 9.81$$

Calculate the time taken to reach the ground.

9 The acceleration, a, of a particle in simple harmonic motion is given by

$$a = -\cos(\omega t) + \sin(\omega t)$$

where ω is the angular frequency. Show that

$$\omega^2 x = \omega t + \cos(\omega t) - \sin(\omega t) - 1$$

given that x is the displacement and at $t = 0$ both $v = 0$ and $x = 0$.

SECTION E Miscellaneous applications of integration

By the end of this section you will be able to:
▶ use integration to find the work done in thermodynamics
▶ use integration to find the deflection of a beam
▶ use Maclaurin series to find certain integrals

This section is quite sophisticated compared to the previous ones. It applies integration to two fields of engineering – thermodynamics and structural mechanics. You should be able to follow the examples because they do **not** rely on previous knowledge of these two engineering disciplines. In the last subsection we use the Maclaurin series to find integrals of difficult integrands.

E1 Thermodynamics

Figure 21 shows a gas expanding in a cylinder according to the relation

$$PV^n = C \text{ (constant)}$$

The initial volume of gas is V_1 and the final volume is V_2. The work done, W, by the gas on the face of the piston is given by

Fig. 21

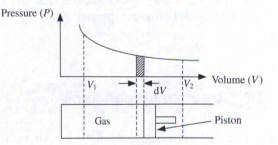

9.12
$$W = \int_{V_1}^{V_2} P dV$$

(This is the area under the curve shown in Fig. 21 between V_1 and V_2.)

 Example 17 *thermodynamics*

A gas expands in a cylinder according to the relation

$$PV^{1.3} = 1560$$

The initial volume of gas is 0.025 m³ and the final volume is 0.04 m³.

Evaluate the work done by the gas on the face of the piston.

Example 17 *continued*

Solution

Rearranging $PV^{1.3} = 1560$ to make P the subject gives

$$P = \frac{1560}{V^{1.3}} = 1560V^{-1.3} \qquad \left[\text{Remember } \frac{1}{a^n} = a^{-n}\right]$$

Using

9.12
$$W = \int_{V_1}^{V_2} P\,dV$$

with $V_1 = 0.025$, $V_2 = 0.04$, $P = 1560V^{-1.3}$ and W = work done gives

$$W = \int_{0.025}^{0.04} 1560V^{-1.3}\,dV$$

$$= 1560\int_{0.025}^{0.04} V^{-1.3}\,dV \qquad \text{[Taking out 1560]}$$

$$= 1560\left[\frac{V^{-0.3}}{-0.3}\right]_{0.025}^{0.04} \qquad \left[\text{By } \int V^n\,dV = \frac{V^{n+1}}{n+1}\right]$$

$$= -\frac{1560}{0.3}\left[(0.04)^{-0.3} - (0.025)^{-0.3}\right] \qquad \left[\text{Taking out } \frac{1}{-0.3}\right]$$

$$W = 2068.17\,J = 2.07\,kJ \ (J = \text{Joule})$$

If W is positive then this indicates that the gas is expanding, and a negative W signifies that the gas is being compressed.

E2 Structural mechanics

Figure 22 shows a beam of length L with a uniform distributed load of w per unit length.

In the general case, the deflection, y, of the beam at a distance x along the beam is related by

Fig. 22

9.13
$$\frac{d^2y}{dx^2} = \frac{M}{EI}$$

where M is the bending moment and EI is the flexural rigidity.

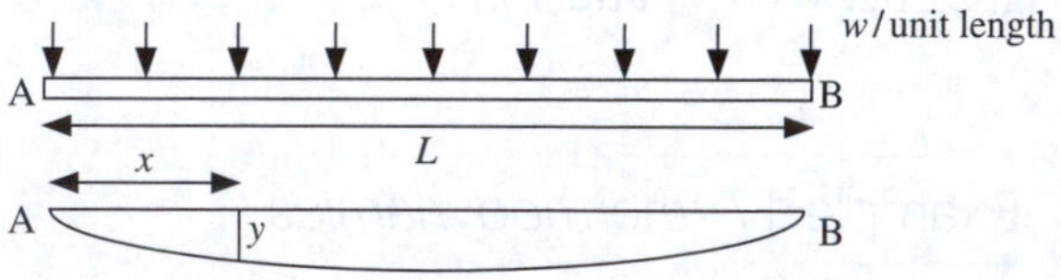

In this section we are interested in finding an expression for y, so we have to integrate both sides of 9.13 twice with respect to x. Generally the values of the (two) constants of integration are obtained by using the given conditions, as the following example shows.

Example 18 *structures*

The bending moment, M, of the beam shown in Fig. 22 is

$$M = \frac{w}{2}(x^2 - Lx)$$

Given the boundary conditions $x = 0$, $y = 0$ and $x = L$, $y = 0$ show that

$$y = \frac{w}{24EI}(x^4 - 2Lx^3 + L^3x)$$

Solution

Substituting $M = \frac{w}{2}(x^2 - Lx)$ into

9.13 $$\frac{d^2y}{dx^2} = \frac{M}{EI}$$

gives

$$\frac{d^2y}{dx^2} = \frac{w}{2EI}(x^2 - Lx)$$

Integrating this:

$$\frac{dy}{dx} = \frac{w}{2EI}\int (x^2 - Lx)\,dx \quad \left[\text{Taking out } \frac{w}{2EI}\right]$$

$$= \frac{w}{2EI}\left(\frac{x^3}{3} - \frac{Lx^2}{2}\right) + C \quad [\text{Integrating}]$$

To find y we integrate again:

$$y = \frac{w}{2EI}\left(\frac{x^4}{12} - \frac{Lx^3}{6}\right) + Cx + D$$

$$= \frac{w}{2EI}\left(\frac{x^4}{12} - \frac{2Lx^3}{12}\right) + Cx + D \quad \left[\text{Writing } \frac{1}{6} = \frac{2}{12}\right]$$

Taking out a common factor of $\frac{1}{12}$ from the brackets gives

* $$y = \frac{w}{24EI}(x^4 - 2Lx^3) + Cx + D$$

Substituting the condition $x = 0$, $y = 0$ into * :

$$0 = \frac{w}{24EI}[0] + C(0) + D$$

gives $D = 0$.

Substituting the other condition $x = L$, $y = 0$ into * :

$$0 = \frac{w}{24EI}[L^4 - 2LL^3] + CL + D$$

The square brackets become $-L^4$ because

$$L^4 - 2LL^3 = L^4 - 2L^4 = -L^4$$

Example 18 *continued*

Also we have already evaluated $D = 0$ therefore

$$0 = \frac{w}{24EI}\left[-L^4\right] + CL$$

Transposing for C:

$$C = \frac{w(L^4)}{24EI\,L} = \frac{wL^3}{24EI} \quad \text{[Cancelling L's]}$$

Putting $C = \dfrac{wL^3}{24EI}$ and $D = 0$ into ⬛ ★ gives the required result:

$$y = \frac{w}{24EI}\left(x^4 - 2Lx^3\right) + \frac{wL^3x}{24EI}$$

$$= \frac{w}{24EI}\left(x^4 - 2Lx^3 + L^3x\right) \quad \text{[Factorizing]}$$

There are many other applications of integration in engineering, such as finding moments of inertia, centroids and arc lengths for instance.

E3 Other applications of integration

Example 19

Find the first three non-zero terms of the Maclaurin series for $\cos(x^2)$, and use it to find an approximate value for

$$\int_0^1 \cos(x^2)\,dx$$

giving your answer to 4 d.p.

Solution

The Maclaurin series for $\cos(x)$ is

7.17 $\qquad \cos(x) = 1 - \dfrac{x^2}{2!} + \dfrac{x^4}{4!} - \ldots$

The Maclaurin series for $\cos(x^2)$ is found by replacing x with x^2 in **7.17**:

$$\cos(x^2) = 1 - \frac{(x^2)^2}{2!} + \frac{(x^2)^4}{4!} - \ldots$$

$$= 1 - \frac{x^4}{2!} + \frac{x^8}{4!} - \ldots \qquad \text{[Using the rules of indices } (a^m)^n = a^{m \times n}]$$

Example 19 *continued*

Integrating this function gives

$$\int_0^1 \cos(x^2)\mathrm{d}x = \int_0^1 \left(1 - \frac{x^4}{2!} + \frac{x^8}{4!} - \ldots\right)\mathrm{d}x$$

$$= \left[x - \frac{x^5}{5 \times 2!} + \frac{x^9}{9 \times 4!} - \ldots\right]_0^1 \qquad \text{[Integrating]}$$

$$= \left[1 - \frac{1^5}{10} + \frac{1^9}{216} - \ldots\right] - [0] = 0.9046$$

The answer to 4 d.p. is 0.9046.

SUMMARY

In this section we covered two areas of applications of integration, thermodynamics and structural mechanics. We also used the Maclaurin series to find integrals.
In thermodynamics the work done, W, by a gas on the face of a piston from an initial volume V_1 to a new volume V_2 is given by

9.12 $\qquad W = \int_{V_1}^{V_2} P \, \mathrm{d}V \qquad$ (P is pressure)

In structural mechanics the deflection, y, of a beam at a distance x is given by

9.13 $\qquad \dfrac{\mathrm{d}^2 y}{\mathrm{d}x^2} = \dfrac{M}{EI}$

where M = bending moment and EI = flexural rigidity.

Exercise 9(e)

Solutions at end of book. Complete solutions available at www.palgrave.com/engineering/singh

Questions 1 to 4 are in the field of [*thermodynamics*].

1 Find the work done on the face of a piston by a gas expanding from 0.01 m^3 to 0.1 m^3 according to $PV^{1.35} = 1500$.

2 The initial volume of a gas in a cylinder is 0.6 m^3. Find the work done by the gas if the final volume is 0.023 m^3 and the gas obeys the law

$$PV^{1.55} = 1789$$

3 A gas is compressed according to the relation

$$PV^{1.6} = 1675$$

If the volume changes from 0.15 m^3 to 0.037 m^3, find the work done by the gas.

Exercise **9(e) continued**

Solutions at end of book. Complete solutions available at www.palgrave.com/engineering/singh

4 A gas in a cylinder obeys the law

$$PV = C \quad \text{where } C \text{ is a constant}$$

Show that the work done, W, by the gas is given by

$$W = C \ln\left(\frac{V_2}{V_1}\right)$$

where V_1 and V_2 are the initial and final volumes of the gas respectively.

 Questions 5 to 8 are in the field of [structures].

5

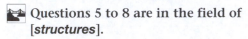

Fig. 23

Figure 23 shows a cantilever with a concentrated load P at a distance l from the fixed end. The bending moment, M, is given by

$$M = P(l - x)$$

At $x = 0$, $y = 0$ and $\frac{dy}{dx} = 0$. Determine y.

*6 Consider the cantilever beam of length L (Fig. 24).

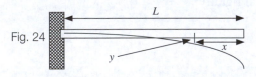

Fig. 24

Find the deflection y given $M = -\frac{wx^2}{2}$

(where w is the load per unit length)

and at $x = L$, $y = 0$ and $\frac{dy}{dx} = 0$.

*7 A cantilever beam of length L is subject to a uniform distributed load, w per unit

length, and the bending moment M is given by

$$M = -\frac{w}{2}(L - x)^2$$

Obtain an expression for y for the initial conditions at $x = 0$, $y = 0$ and $\frac{dy}{dx} = 0$.

*8

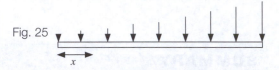

Fig. 25

Figure 25 shows a simply supported beam of length L carrying a load, wx, which increases uniformly. The bending moment, M, is given by

$$M = \frac{w}{6}(x^3 - L^2x)$$

For the conditions $x = 0$, $y = 0$ and $x = L$, $y = 0$ show that the deflection y is given by

$$y = \frac{w}{360EI}[3x^5 - 10L^2x^3 + 7L^4x]$$

9 Determine the first three non-zero terms of the series expansion of

$\sin(x)$ and determine $\displaystyle\int_0^1 \frac{\sin(x)}{x}\, dx$ to 3 d.p.

10 i Find the first four non-zero terms of the series expansion of e^{-x^2}.

ii Determine an approximation for

$$\int_0^1 e^{-x^2}\, dx \text{ correct to 3 d.p. using the}$$

series found in part **i**.

Solutions at end of book. Complete solutions available at www.palgrave.com/engineering/singh

Examination questions 9

1 Find the area of the region bounded by the curve $y = 3x - x^2$ and the x axis.

2 R is the region bounded by the curves $y = 3 - \sin(x)$, $y = \sin(x)$, $x = 0$ and $x = \pi$. Find the area of R.

Questions 1 and 2 are from University of Manchester, UK, 2009

3 Using Simpson's rule with 4 strips, find an approximate value of the integral

$$\int_0^1 e^{x^2}\, dx.$$

University of Manchester, UK, 2007

4 Write down the trapezoidal rule approximation T_3 for $\int_1^4 x \cos(\pi/x)dx$.

Leave your answer expressed as a sum involving cosines. [T_3 in this question means consider 3 subintervals].

University of British Columbia, Canada, 2008

5 Approximate $\int_0^2 \dfrac{x}{1 + x^2}\, dx$ using the trapezoidal rule with $n = 4$ subintervals.

North Carolina State University, USA, 2010

6 Find the area under the graph of $y = \ln x$ between $x = 1$ and $x = 3$.

University of Aberdeen, UK, 2004

7 Integrals like $\int \sin(x^2)\, dx$ arise in the diffraction of light but they cannot be evaluated exactly. Find the Maclaurin series for $\sin(x^2)$ up to and including the term in x^{10}, and use it to find an approximate value for

$$\int_0^1 \sin(x^2)\, dx$$

giving your answer to 4 d.p.

University of Surrey, UK, 2008

8 You are standing on the top of a building which is 256 ft. high. You drop a large orange pumpkin and watch as it falls helplessly to the ground.

 a Assume that the acceleration due to gravity is $s''(t) = -32$ ft./sec.^{2}.

   Derive the instantaneous velocity, $s'(t)$, and height (above ground), $s(t)$, formulas for this pumpkin.
 b In how many seconds will the pumpkin strike the ground?
 c What is the pumpkin's velocity as it strikes the ground?

University of California, Davis, USA

9 Consider the region bounded by the curves $y = -(x - 2)^2$ and $y = 4 - 2x$.

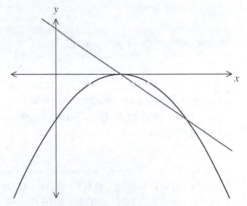

 a Express the area A of this region as an integral taken with respect to x.
 b Evaluate the integral to find A.

Memorial University of Newfoundland, Canada, 2010

10 Calculate an approximation to the integral

$$\int_1^3 \frac{x}{\sqrt{1 + x^2}}\, dx$$

using Simpson's rule by dividing the range into four equal intervals. You should work to an accuracy of 3 d.p. throughout. What is the approximate percentage error in the result (you may compute the integral directly or use the formula which approximates the error)?

University of Manchester, UK, 2008

11 Find the following:

a the average value of $f(x) = x^3 + x$ on the interval $1 \leq x \leq 3$

b the area of the region between the graphs of $f(x) = x^3$ and $g(x) = x$ for $0 \leq x \leq 2$.

University of Toronto, Canada, 2001

12 Find the first three non-zero terms in the power series representation in powers of x (i.e. the Maclaurin

series) for $\int_0^x t \cos(t^3) \, dt$.

13 Find the numbers b such that the average value of the function $f(x) = 3x^2 - 6x + 2$ on the interval $[0, b]$ is equal to 0.

University of British Columbia, Canada, 2007

14 Let $I = \int_0^1 \cos(x^2) \, dx$. Write down the first three non-zero terms obtained by using the Maclaurin series to estimate I, and explain why the error in using this estimate is less than 0.001.

University of British Columbia, Canada, 2008

Miscellaneous exercise 9

1 [thermodynamics]   The volume and pressure of a gas are measured in an experiment and give the results shown in the table below.

V (m³)	0.01	0.02	0.03	0.04	0.05
P (kPa)	946	358	203	136	99

V (m³)	0.06	0.07	0.08	0.09
P (kPa)	77	62	51	44

Find the approximate work done on the face of the piston by the gas to produce a volume change from 0.01 m³ to 0.09 m³.

2 [mechanics]   The following table shows the experimental results of a force, F, on a piston with displacement, x:

x (m)	0.025	0.05	0.075	0.1	0.125
F (kN)	10.06	4.23	2.55	1.78	1.34

x (m)	0.15	0.175	0.2	0.225
F (kN)	1.07	0.88	0.75	0.64

The work done, W, in moving the piston from 0.025 m to 0.225 m is given by

$$W = \int_{0.025}^{0.225} F \, dx$$

Find an approximate value for W.

3 [thermodynamics]   The specific heat, c, of steam at temperature T, is given in the following table:

T (K)	200	250	300
c (kJ/kg K)	3.57	3.63	3.70

T (K)	350	400	450
c (kJ/kg K)	3.75	3.80	3.86

Miscellaneous exercise **9 continued**   Solutions at end of book. Complete solutions available at www.palgrave.com/engineering/singh

The enthalpy change, Δh, is given by

$$\Delta h = \int_{200}^{450} c \, dT$$

Find an approximation for Δh.

4 [mechanics]   The height, h, of a projectile is given by

$$h = \int_{0}^{10}\left[5\ln\left|\frac{2}{2+3t}\right| + 9.81t\right]dt$$

By using Simpson's rule with four equal intervals, find an approximation for the height h.

5 [mechanics]   The time taken, t, for a particle to reach a velocity of 90 m/s with an initial velocity of 50 m/s is given by

$$t = \int_{50}^{90}\frac{dv}{5 - 2v^{0.11}}$$

where v is the velocity. Find an approximation for t.

6 [electrical principles]   An a.c. voltage waveform, v, is given by

$$v = \frac{\sin(t)}{t}$$

Find an approximation for the mean value of v over the interval 0.4 to 1. (You cannot find an analytical solution to the integral of $\sin(t)/t$, so use Simpson's rule with six equal intervals.)

7 [electrical principles]   An a.c. current, i, is given by

$$i = \sqrt{1 - 2\sin(t)}$$

Find an approximation for the mean value of i over the interval π to $3\pi/2$.

*8 [aerodynamics]   The area of an aerofoil is shown shaded in Fig. 26.

Fig. 26

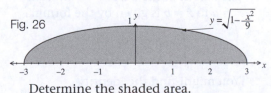
$$y = \sqrt{1 - \frac{x^2}{9}}$$

Determine the shaded area.

9 [electrical principles]   A voltage waveform, v, is given by

$$v = 5(1 - e^{-t/4}) \quad 0 < t < 20$$

Find the mean and RMS values of v.

10 [mechanics]   The acceleration, a, of a projectile fired vertically upwards is given by

$$a = \frac{10}{10 - t} + 9.8$$

where $0 \le t < 10$

a Show that the velocity, v, is given by

$$v = 10\ln\left(\frac{10}{10 - t}\right) + 9.8t$$

b Find an approximation for the height reached by the projectile after 6 seconds.

11 [thermodynamics]   Find the work done on the face of a piston by a gas expanding from 0.013 m³ to 0.09 m³ according to $PV^{1.41} = 1961$.

12 [structures]   The bending moment, M, of a beam is given by

$$M = 2.5(x - 5)^2 + 8(x - 3) - 5x$$

where x is the distance along the beam. Given that at $x = 0$, $y = 0$ and at $x = 5$, $y = 0$, find an expression for the deflection of the beam, y.

13 Use a computer algebra system or a graphical calculator to determine

$$\int_{0}^{1}\frac{x^4(1 - x)^4}{1 + x^2}\,dx$$

14 [mechanics]   Figure 27 shows the motion of a projectile.

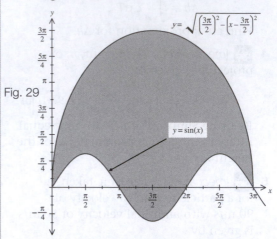

Fig. 27

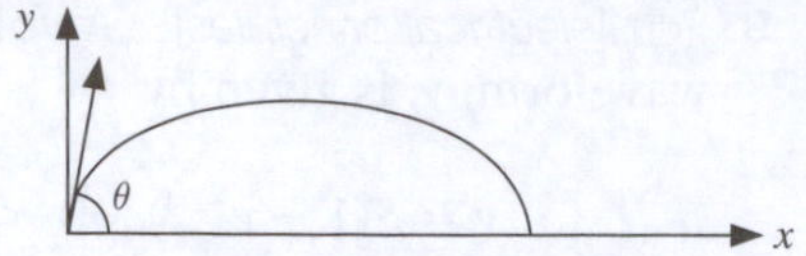

i The horizontal acceleration, $\dfrac{d^2x}{dt^2} = 0$.

For the initial conditions when

$t = 0$, $\dfrac{dx}{dt} = v\cos(\theta)$ and $x = 0$,

show that

$$x = [v\cos(\theta)]t$$

ii The vertical acceleration, $\dfrac{d^2y}{dt^2} = -g$.

For the initial conditions when

$t = 0$, both $\dfrac{dy}{dt} = v\sin(\theta)$ and $y = 0$,

show that

$$y = [v\sin(\theta)]t - \dfrac{gt^2}{2}$$

15 🖳 [*electrical principles*]   A voltage signal is given by

$$v = V_1\sin(\omega t) + V_3\sin(3\omega t)$$

where V_1, V_3 are the constant maximum voltages and ω is the angular frequency. Show that for the interval $t = 0$ to $t = \pi/\omega$:

$$(v_{\text{RMS}})^2 = \dfrac{V_1^2 + V_3^2}{2}$$

where v_{RMS} is the RMS value of v.

16 Show by applying integration that the area of the semicircle shown in Fig. 28 is $\dfrac{\pi}{2}$.

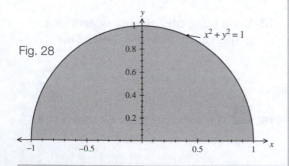

Fig. 28

17 Determine the shaded area shown in Fig. 29 below:

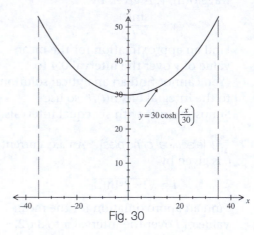

Fig. 29

18 Figure 30 shows a power cable hanging between the tops of two pylons of equal height, 70 m apart. The cable assumes the curve $y = 30\cosh\left(\dfrac{x}{30}\right)$ and is 30 m above ground level at its lowest point.

Fig. 30

Find the length of the cable, correct to 2s.f., between the pylons.
The length of the curve $y = f(x)$ from $x = a$ to $x = b$ is given by the formula

$$\int_a^b \sqrt{1 + \left(\dfrac{dy}{dx}\right)^2}\,dx$$

[You might find the identity $1 + \sinh^2(x) = \cosh^2(x)$ useful.]

CHAPTER 10

Complex Numbers

SECTION A **Arithmetic of complex numbers**

By the end of this section you will be able to:
▶ understand the term *complex number*
▶ add and subtract complex numbers
▶ evaluate and simplify powers of *j*
▶ multiply and divide two complex numbers
▶ use complex numbers to solve engineering problems

A1 **Introduction**

Complex numbers and complex functions are used in many engineering fields such as alternating current (a.c.) theory, vibration problems, fluid flow, elasticity, and many more. A particular application of complex numbers is in oscillations which include shaking of buildings, vibrations in cars and radio waves. The basic form of an oscillation is $A\cos(\omega t)$ where A is amplitude, ω is frequency and t is time. For simplifying calculations it is easier to replace $A\cos(\omega t)$ with a complex number form. The word 'complex' in this chapter does not mean complicated.

To understand complex numbers we use the following rules of indices:

10.1	$\sqrt{a \times b} = \sqrt{a} \times \sqrt{b} = a^{1/2} \times b^{1/2}$
10.2	$(a^m)^n = a^{m \times n}$
10.3	$a^m a^n = a^{m+n}$
10.4	$\dfrac{a^m}{a^n} = a^{m-n}$

Example 1

What is $\sqrt{-25}$ equal to?

Solution

If we try to evaluate this on a simple calculator we get an error message. **So if, in the course of trying to solve a mathematical problem, we needed to evaluate $\sqrt{-25}$, how would we go about it?**

There is **no** real number which when multiplied by itself gives -25:

$$5^2 = +25 \qquad\qquad (-5)^2 = +25$$

We can rewrite $-25 = 25 \times (-1)$ and then we have

$$\sqrt{-25} = \sqrt{25 \times (-1)} \underset{\text{by } 10.1}{=} \sqrt{25} \times \sqrt{(-1)}$$

$$= 5 \times \sqrt{(-1)}$$

$$= 5\sqrt{(-1)}$$

$$= 5\,j \text{ where } j = \sqrt{(-1)}$$

Example 1 *continued*

We introduce the number $j \left(=\sqrt{(-1)}\right)$ so that we can find the square root of a negative number.

Of course we also have $\sqrt{-25} = -5j$ because $\sqrt{25} = -5$ or 5.

? **How do we solve the equation $x^2 + 1 = 0$?**

The graph of $x^2 + 1$ is shown in Fig. 1.

Since $x^2 + 1$ does not cut the x axis there cannot be a real number which satisfies

Fig. 1

$$x^2 + 1 = 0$$

Let's rearrange this equation:

$$x^2 = -1$$

$$x = \pm \sqrt{(-1)} = \pm j \text{ (where } j = \sqrt{-1})$$

The numbers $+j$ and $-j$ are examples of complex numbers.

We let j (or i) $= \sqrt{-1}$. Engineers use j but mathematicians use i.

? **Why do engineers use j?**

Complex numbers are used in alternating current (a.c.) theory where i represents current.

We will use j to represent $\sqrt{-1}$.

Example 2

Solve
$$z^2 - 2z + 5 = 0$$

Solution

We need to use the quadratic formula

$$z = \frac{-b \pm \sqrt{b^2 - 4ac}}{2a}$$

? **What are the values of a, b and c in this example?**

$a = 1$, $b = -2$ and $c = 5$. Therefore

$$z = \frac{-(-2) \pm \sqrt{(-2)^2 - (4 \times 1 \times 5)}}{2 \times 1}$$

$$= \frac{2 \pm \sqrt{4 - 20}}{2}$$

$$z = \frac{2 \pm \sqrt{-16}}{2}$$

Example 2 *continued*

Previously we would have thought of this problem as unsolvable, because we end up needing to take the square root of a negative number (-16). This is where we can use complex numbers. Hence

$$\sqrt{-16} = \sqrt{16 \times (-1)} = \sqrt{16} \times \sqrt{-1} = 4\sqrt{-1} = 4j$$

Therefore

$$z = \frac{2 \pm \sqrt{-16}}{2}$$

$$= \frac{2 \pm j4}{2} = \frac{2}{2} \pm j\frac{4}{2} = 1 \pm j2$$

The roots are $z = 1 + j2$ and $z = 1 - j2$.

Imaginary numbers like j entered mathematics as a tool for solving quadratic and cubic equations. The German mathematician Gauss was the first to use complex numbers seriously.

Karl Friederick Gauss (1777–1855) is widely regarded as being one of the three greatest mathematicians of all time, the others being Archimedes and Newton. His talent as a mathematician was established as a young boy, when his teacher, as a means of occupying his pupils, instructed the class to find the sum of all the whole numbers from 1 to 100. Within minutes, Gauss handed his incredulous teacher the correct answer. He had realized that if he used 2 sets of whole numbers, and placed them end to end, the combination was simple.

	1	2	3	⋯	99	100
+	100	99	98	⋯	2	1
	101	101	101	⋯	101	101

Having established this, the solution was elementary:

$$\frac{100 \times 101}{2} = 5050$$

By the age of 11 Gauss could prove that $\sqrt{2}$ is irrational. At the age of 18 he constructed a regular 17-sided polygon with a compass and unmarked straight edge only. Gauss went to the world renowned centre for mathematics – Göttingen. Later in life Gauss took up a post at Göttingen and published papers in number theory, infinite series, algebra, astronomy and optics.

The unit of magnetic induction is named after Gauss.

A2 Complex numbers

A complex number is of the form $a + jb$ where a is called the **real part** and b is called the **imaginary part**.

For example, the complex number $3 + j9$ has real part = 3 and imaginary part = 9. In other books or in your notes $3 + j9$ might be written as $3 + 9j$. It doesn't make any difference!

Let z be a complex number then the real part of z is denoted by Re(z) and the imaginary part of z is denoted by Im(z). For example

$$\text{Re}(3 + j9) = 3 \qquad\qquad \text{Im}(3 + j9) = 9$$

A3 Addition and subtraction

For addition of complex numbers, add the real parts and add the imaginary parts:

10.5
$$(a + jb) + (c + jd) = \underbrace{a + c}_{\substack{\text{add the} \\ \text{real parts}}} + j\underbrace{(b + d)}_{\substack{\text{add the} \\ \text{imaginary parts}}}$$

Similarly for subtraction, subtract the real parts and subtract the imaginary parts:

10.6
$$(a + jb) - (c + jd) = \underbrace{a - c}_{\substack{\text{subtract the} \\ \text{real parts}}} + j\underbrace{(b - d)}_{\substack{\text{subtract the} \\ \text{imaginary parts}}}$$

Example 3

Evaluate

a $(2 + j7) + (3 + j12)$ **b** $(2 + j7) - (3 + j12)$ **c** $(3 + j4) - (3 - j4)$

Solution

a $(2 + j7) + (3 + j12) = (2 + 3) + j(7 + 12)$

$$= 5 + j19$$

b $(2 + j7) - (3 + j12) = (2 - 3) + j(7 - 12)$

$$= -1 + j(-5) = -1 - j5$$

c $(3 + j4) - (3 - j4) = (3 - 3) + j(4 - (-4))$

$$= 0 + j(4 + 4) = j8$$

A4 **Powers of j**

We can evaluate the first four powers of j by using the rules of indices:

$$j = \sqrt{-1}$$

$$j^2 = \left(\sqrt{-1}\right)^2 = (-1^{1/2})^2 = (-1)^1 = -1 \quad \text{[By 10.2]}$$

$$j^3 \underset{\text{by 10.3}}{=} j^2 \cdot j^1 = -1 \cdot j = -j$$

$$j^4 = (j^2)^2 = (-1)^2 = 1$$

These first four powers of j are important because we can write **all** the powers of j in terms of these first four powers:

10.7 $j = \sqrt{-1}$

10.8 $j^2 = -1$

10.9 $j^3 = -j$

10.10 $j^4 = 1$

Example 4

Simplify   **a** j^{12} **b** j^7 **c** j^{203}

Solution

We use the rules of indices and powers of j established above:

a $j^{12} \underset{\text{by 10.2}}{=} (j^4)^3 \underset{\text{by 10.10}}{=} (1)^3 = 1$

b $j^7 \underset{\text{by 10.3}}{=} j^4 \cdot j^3 = \underset{\text{by 10.10}}{1} \; j^3 \underset{\text{by 10.9}}{=} -j$

c $j^{203} \underset{\text{by 10.3}}{=} j^{200} \cdot j^3 = \underset{\text{by 10.2}}{(j^4)^{50}} \cdot j^3$

$$= (1)^{50}(-j) = -j$$

10.2 $(a^m)^n = a^{m \times n}$ 10.3 $a^m a^n = a^{m+n}$ 10.9 $j^3 = -j$ 10.10 $j^4 = 1$

? To simplify higher powers of j you only need to use the first four powers of j. **Which power of j is the most useful in simplifying powers of j?**

From **Example 4** you should notice that it is better to write most powers in terms of j^4 because $j^4 = 1$, and 1 to the power of any real number is also 1, that is $1^x = 1$. These, 10.7 to 10.10, four powers of j are very important and are well worth remembering, in particular $j^2 = -1$.

A5 Multiplication

The easiest way to multiply two complex numbers is to use First, Outside, Inside, Last (FOIL), as we used to expand brackets in **Chapter 1**:

10.11
$$(a + jb)(c + jd) = \underbrace{ac}_{F} + \underbrace{jad}_{O} + \underbrace{jbc}_{I} + \underbrace{j^2bd}_{L}$$

Collecting the j terms and using $j^2 = -1$ simplifies 10.11 to

$$(a + jb)(c + jd) = ac + j(ad + bc) - bd$$

$$= (ac - bd) + j(ad + bc) \quad \text{[Collecting real and imaginary parts]}$$

Example 5

Evaluate the following:

a $(2 + j3)\,(6 + j7)$ **b** $(2 + j3)\,(2 - j3)$

Solution

a $(2 + j3)(6 + j7) = \underbrace{(2 \times 6)}_{F} + \underbrace{j(2 \times 7)}_{O} + \underbrace{j(3 \times 6)}_{I} + \underbrace{j^2(3 \times 7)}_{L}$

$$= 12 + \underbrace{j14 + j18}_{= j32} + j^2 21$$

$$= 12 + j32 - 21 \quad \text{[Because } j^2 = -1\text{]}$$

$$= -9 + j32$$

b $(2 + j3)(2 - j3) = 4 - \underbrace{j6 + j6}_{= 0} - j^2 9$

$$= 4 - j^2 9$$

$$= 4 + 9 = 13 \quad \text{[Because } j^2 = -1\text{]}$$

A6 **Complex conjugate**

> ### Example 6
>
> Evaluate $(3 + j4)(3 - j4)$
>
> Solution
>
> $$(3 + j4)(3 - j4) = 9 \underbrace{- j12 + j12}_{= 0} - j^2 16 \qquad \text{[Applying FOIL]}$$
>
> $$= 9 - j^2 16$$
>
> $$= 9 + 16 \quad \text{[Remember that } j^2 = -1\text{]}$$
>
> $$= 25$$

Note that the answer to **Example 6** contains **no** j, that is multiplication of two complex numbers sometimes gives a real number. Here are some more examples of complex number multiplication in which the answer contains **no** j.

i $(5 - j8)(5 + j8) = 89$

ii $(1 - j)(1 + j) = 2$

iii $(7 + j5)(7 - j5) = 74$

? What do you notice about these complex numbers?

They are identical apart from the signs of the imaginary part.

A pair of complex numbers like these are called **conjugate complex numbers** and the product of these is always **real** (no j).

Notice: $(3 + j4)(3 - j4) = 3^2 + 4^2 = 25$

Similarly $(4 + j5)(4 - j5) = 4^2 + 5^2 = 41$

$$(2 - j6)(2 + j6) = 2^2 + 6^2 = 40$$

The complex conjugate of $a + jb$ is

10.12 $a - jb$

Also we have the fundamental identity $(a + jb)(a - jb) = a^2 + b^2$.

If z is a complex number, then we denote the **complex conjugate** of z by $\bar{z}$.

So if $z = 5 - j3$ then $\bar{z} = 5 + j3$.

A7 Division

Division of a complex number by a real number is straightforward. For example

$$\frac{3 - j4}{2} = \frac{3}{2} - j\frac{4}{2} = 1.5 - j2$$

? **But what is** $\dfrac{(5 - j3)}{3 + j4}$ **equal to?**

Example 7

Evaluate $z = \dfrac{(5 - j3)}{3 + j4}$.

Solution

We have just seen that dividing a complex number by a real number presents no problems, so if we could convert the denominator, $3 + j4$, into a real number then the problem can be solved.

? **How can we achieve this?**

Multiply by its complex conjugate $3 - j4$. We multiply numerator and denominator by $3 - j4$:

$$z = \frac{(5 - j3)(3 - j4)}{(3 + j4)(3 - j4)} = \frac{15 - j20 - j9 + j^2 12}{3^2 + 4^2} \quad \text{[Expanding]}$$

$$= \frac{15 - j29 + j^2 12}{25}$$

$$= \frac{15 - j29 - 12}{25} \quad \text{[Using } j^2 = -1\text{]}$$

$$= \frac{3 - j29}{25} = \frac{3}{25} - j\frac{29}{25} = 0.12 - j1.16$$

So to divide one complex number by another, we multiply numerator and denominator by the conjugate of the denominator. Hence

10.13
$$\frac{a + jb}{c + jd} = \frac{(a + jb)(c - jd)}{c^2 + d^2}$$

Remember that if we multiply a complex number by its conjugate, the result is a sum of squares, that is

$$(c + jd)(c - jd) = c^2 + d^2$$

That is why the denominator in 10.13 is $c^2 + d^2$.

A8 Applications of complex numbers

> ### Example 8 *electrical principles*

The admittance, Y, in a circuit is given by

$$Y = \frac{1}{Z}$$

where Z is the impedance of the circuit. If $Z = 100 - j25$, find Y.

Solution

This example demonstrates the process of dividing a real number by a complex number.

Substituting $Z = 100 - j25$ into $Y = \dfrac{1}{Z}$ gives

$$Y = \frac{1}{100 - j25}$$

To cut down the arithmetic we can take out a factor of 25 from the denominator because 25 is a common factor of 100 and -25:

$$Y = \frac{1}{25}\left[\frac{1}{4 - j}\right]$$

We need to use the rules of division.

What is the complex conjugate of $4 - j$?

$$4 + j$$

Multiplying numerator and denominator by the complex conjugate, $4 + j$, gives

$$\frac{1}{25}\left[\frac{1}{4 - j}\right] = \frac{1}{25}\left[\frac{4 + j}{4^2 + 1^2}\right]$$

$$= \frac{1}{25}\left[\frac{4 + j}{17}\right]$$

$$= \frac{4 + j}{25 \times 17}$$

$$= \frac{4 + j}{425} = \frac{4}{425} + j\frac{1}{425} = 0.0094 + j0.0024$$

Example 8 *continued*

Thus

$$Y = 0.0094 + j0.0024$$

Normally Z is in ohm and Y is in siemen, that is $Z = (100 - j25)$ ohm (Ω) and $Y = (0.0094 + j0.0024)$ siemen (S).

Example 9 *electrical principles*

The total impedance, Z, of a circuit containing an inductor with inductance L and a resistor with resistance R in series is given by

$$Z = R + j\,2\pi fL \quad \text{where } f = \text{frequency}$$

If $Z = (50 + j200)\ \Omega$ and $f = 50$ Hz, find R and L.

Solution

We have

$$R + j2\pi fL = 50 + j200$$

Equating real and imaginary parts gives

$$R = 50\ \Omega$$

$$2\pi fL = 200$$

$$L = \frac{200}{2\pi \times 50} = \frac{2}{\pi}\,\text{H} \qquad [\text{We are given } f = 50]$$

SUMMARY

A number $a + jb$ is called a complex number, where $j = \sqrt{(-1)}$.
To add and subtract complex numbers:

10.5 $\quad (a + jb) + (c + jd) = \underbrace{a + c}_{\substack{\text{add the}\\\text{real parts}}} + j\underbrace{(b + d)}_{\substack{\text{add the}\\\text{imaginary parts}}}$

10.6 $\quad (a + jb) - (c + jd) = \underbrace{a - c}_{\substack{\text{subtract the}\\\text{real parts}}} + j\underbrace{(b - d)}_{\substack{\text{subtract the}\\\text{imaginary parts}}}$

SUMMARY continued

Powers of j are given by

$$j^2 = -1, \quad j^3 = -j, \quad j^4 = 1$$

Multiplication of complex numbers is given by

10.11 $(a + jb)(c + jd) = \underset{F}{ac} + \underset{O}{jad} + \underset{I}{jbc} + \underset{L}{j^2bd}$

The complex conjugate of $a + jb$ is

10.12 $a - jb$

Division of complex numbers is given by

10.13 $\dfrac{a + jb}{c + jd} = \dfrac{(a + jb)(c - jd)}{c^2 + d^2}$

Exercise 10(a)

Solutions at end of book. Complete solutions available at www.palgrave.com/engineering/singh

1 Let $u = 2 + j3$ and $v = 5 + j8$. Express the following in the form $a + jb$ where a and b are real:

a $u + v$ **b** $u - v$
c uv **d** vu
e u/v **f** v/u
g $2u + v$ **h** u^2
i v^2 **j** $u^2 + v^2$

2 State the complex conjugate of the following:

$3 - j15, \quad 3 + j15, \quad j3, \quad j, \quad 0, \quad \pi$ and e.

3 If $z = 3 - j$ express $z^2 + 7z + 13$ in the form $a + jb$ where a and b are real.

4 Simplify j^{20}, j^9, j^{23}, j^{100} and j^{1007}.

5 **i** If $z = 2 + j$ find z^2, z^3, z^4 and $z^4 - 6z^2 + 25$.
ii Find one of the roots of $z^4 - 6z^2 + 25 = 0$.

6 [electrical principles]   Determine the admittance, $Y = \dfrac{1}{Z}$, in a circuit, given that $Z = (75 - j40) \, \Omega$.
(Ω = ohm is the SI unit of impedance.)

7 Assuming a and b are real, find them if

a $a + jb = 2 + j3$
b $a + jb = (2 + j3) - (2 - j3)$
c $a + jb = (1 + j)^2$
d $a + jb = \dfrac{1}{3 - j4}$
e $2a + jb = (1 + 2j)^2$
f $1.8a + j3.4b = \dfrac{7 + j5}{2 - j}$

8 Express the following in the form $a + jb$ where a and b are real:

a $-j5(3 + j4)$

b $\dfrac{3 + j4}{j5}$

c $(2 - j3)(1 + j)$

d $\left(\dfrac{3 + j4}{j5}\right) - \left(\dfrac{2 - j3}{1 - j}\right)$

9 [electrical principles]   If $Z_1 = (1 + j)$ ohm and $Z_2 = (3 + j4)$ ohm are impedances connected in parallel in a circuit, find the total impedance, Z_t, in the circuit given that

$$\frac{1}{Z_t} = \frac{1}{Z_1} + \frac{1}{Z_2}$$

10 [electrical principles]   A current, I, in a magnetically coupled circuit satisfies

$$20I + j100I = 200$$

Find the current I.

11 Find the exact values of z which satisfy the following equations:

a $z^2 + 2z + 26 = 0$

b $z^2 - 2z + 3 = 0$

c $3z^2 - 7z + 13 = 0$

By using a computer algebra system or a graphical calculator, plot on different axes the graphs $z^2 + 2z + 26$, $z^2 - 2z + 3$ and $3z^2 - 7z + 13$.

By looking at these graphs can you see why there are no real roots of these equations?

The remaining questions in this exercise are in the field of [electrical principles].

12 An a.c. voltage V across a circuit is given by $V = (3 + j6)$ volts. Given that

the impedance Z of the circuit is $(8 + j)\,\Omega$ find the current I where

$$I = \frac{V}{Z}$$

*13 Let I_p and I_s represent the primary and secondary currents through a mutually magnetically coupled circuit which satisfies the following equations:

$$10 = 4I_p + j6I_p - j4I_s$$
$$0 = 50I_s + j12I_s + 8I_s - j4I_p$$

Determine I_p and I_s. (If you have difficulty with this question try it after **Example 17**.)

*14 A resistance, R, and a capacitance, C, are connected in parallel. The impedance, Z, of the circuit is given by

$$\frac{1}{Z} = \frac{1}{R} + \frac{1}{X_c}$$

where $X_c = \dfrac{1}{j\omega C}$

(ω = angular frequency)

i Show that $Z = \dfrac{R}{1 + j\omega CR}$.

ii Find the real and imaginary parts of Z. ($\mathrm{Re}(Z)$ and $\mathrm{Im}(Z)$ respectively.)

SECTION B **Representation of complex numbers**

By the end of this section you will be able to:
► show how a complex number can be represented on an Argand diagram
► determine the polar form of a complex number
► convert from polar to rectangular form, and vice versa

B1 Complex plane

We know from our school days that a real number can be represented on a real line. For example, the number 5 can be represented as shown in Fig. 2a.

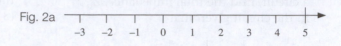

Fig. 2a

? **How can we represent a complex number such as 5 + j2?**

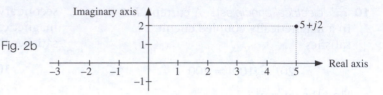

Fig. 2b

Consider the x–y plane, where the y axis represents the imaginary numbers and the x axis represents the real numbers. Then 5 + j2 is 5 across and 2 up (Fig. 2b).

Fig. 3

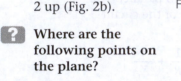

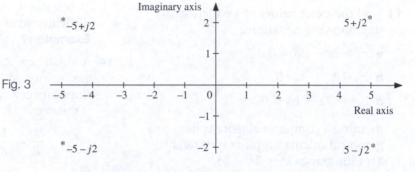

? **Where are the following points on the plane?**

a 5 − j2

b −5 + j2

c −5 − j2 (see Fig. 3)

This plane is called the **complex plane** or the **Argand diagram**.

A complex number x + jy can be represented by the point (x, y).

> The Argand diagram is named after the amateur French mathematician **Jean Robert Argand** (1768–1822). He also developed the concept of the modulus of a complex number.
>
> It was actually **Wessel** (1745–1818) who first developed the concept of a complex plane, but he published his work in Danish and as a result is **not** credited with this work.

Figure 4 shows the x–y plane where the x axis is the **real axis**, and the y axis is the **imaginary axis**.

Fig. 4

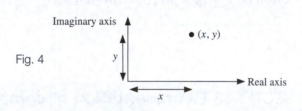

A complex number, written x + jy, is said to be in rectangular form because x and y represent the rectangular co-ordinates of the complex number x + jy (Fig. 5).

Fig. 5

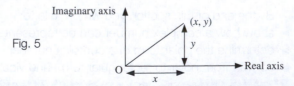

B2 Polar form of complex numbers

A complex number $z = x + jy$ can also be represented by its distance r from the origin and the angle θ it makes with the positive real axis. This is known as the **polar form** representation of z. You will see in Section C why this kind of representation is useful in that it allows us to multiply and

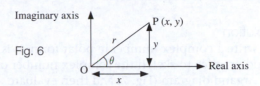

Fig. 6

divide complex numbers in a more straightforward manner than discussed earlier. First though, we need to familiarize ourselves with the method of conversion to polar form.

By examining Fig. 6 and using trigonometry we have

$$x = r\cos(\theta) \text{ and } y = r\sin(\theta)$$

$$z = x + jy = r\cos(\theta) + j\,r\sin(\theta)$$

$$= r[\cos(\theta) + j\sin(\theta)] \quad \text{[Taking out } r]$$

So the polar form of $z = x + jy$ can be written in terms of the angle θ, and length of the hypotenuse r.

10.14 $z = r[\cos(\theta) + j\sin(\theta)]$ or $r\angle\theta$

where r is the length OP and θ is the angle shown in Fig. 6. $r\angle\theta$ is a shorthand notation for $r[\cos(\theta) + j\sin(\theta)]$.

To calculate r we use Pythagoras:

10.15 $r = \sqrt{x^2 + y^2}$

and to calculate θ we use tangent definition $\tan(\theta) = \dfrac{y}{x}$:

10.16 $\theta = \tan^{-1}\left(\dfrac{y}{x}\right)$

We plot the complex number on an Argand diagram to check that the angle θ lies in the correct quadrant.

The length r is called the **modulus** of z, denoted by $|z|$ and

$$|z| = \sqrt{x^2 + y^2}$$

θ is called the **argument** of z, denoted by $\arg(z)$ and

$$\arg(z) = \tan^{-1}\left(\dfrac{y}{x}\right)$$

If θ satisfies $-180° < \theta \leq 180°$, then θ is called the **principal argument**. (This convention gives slightly easier arithmetic and ensures that we only have **one** value for θ.)

We need to be **careful** in determining the angle θ as you will see in **Example 11**.

Example 10

Express $2 + j3$ in polar form.

Solution
To write a complex number in polar form, it is good practice to sketch the complex number on an Argand diagram (Fig. 7) and then evaluate the modulus and argument.

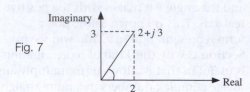

Fig. 7

Using $|x + jy| = \sqrt{x^2 + y^2}$ with $x = 2$ and $y = 3$, the modulus is

$$|2 + j3| = \sqrt{2^2 + 3^2} = \sqrt{13}$$

Using $\arg(x + jy) = \tan^{-1}\left(\dfrac{y}{x}\right)$ with $x = 2$ and $y = 3$ the argument is

$$\arg(2 + j3) = \tan^{-1}\left(\frac{3}{2}\right) = 56.31° \quad \text{[By calculator]}$$

In polar form, $2 + j3$ is (Fig. 8):

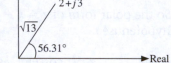

Fig. 8

$$\sqrt{13}\left[\cos(56.31°) + j\sin(56.31°)\right] \text{ or } \sqrt{13}\angle 56.31°$$

Example 11

Find the modulus and the principal argument of $-1 + j2$.

Solution
We first sketch $-1 + j2$ on an Argand diagram (Fig. 9).

Let r be the modulus and $\arg(-1 + j2)$ be the argument of $-1 + j2$. We have

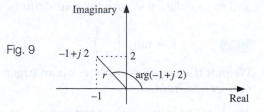

Fig. 9

$$r = |-1 + j2| \underset{\text{by } 10.15}{=} \sqrt{(-1)^2 + 2^2} = \sqrt{5}$$

$$\text{angle} \underset{\text{by } 10.16}{=} \tan^{-1}\left(\frac{2}{-1}\right) = -63.43° \quad \text{[By calculator]}$$

This angle, $-63.43°$, is illustrated in Fig. 10.

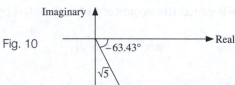

Fig. 10

But this does **not** correspond to the location of the complex number $-1 + j2$ in Fig. 9.

10.15 $|x + jy| = \sqrt{x^2 + y^2}$ **10.16** $\theta = \tan^{-1}(y/x)$

Example 11 *continued*

Why is this?

Your calculator does not recognize the significance of $\left(\dfrac{-2}{1}\right)$ as opposed to $\left(\dfrac{2}{-1}\right)$.

It deals with them as arithmetic fractions and reads both as $\left(-\dfrac{2}{1}\right)$. The same is true of

$\left(\dfrac{-2}{-1}\right)$, which it reads as $\left(\dfrac{2}{1}\right)$.

Fig. 11

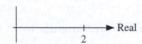

We can perform a simple check by plotting the given values on an Argand diagram (Fig. 11), and comparing the result to the angle suggested by the calculator. If it appears in the opposing quadrant, add 180°. Hence

$$\arg(-1 + j2) = (-63.43°) + 180° = 116.57°$$

$-1 + j2$ in polar form is

$$\sqrt{5}[\cos(116.57°) + j\sin(116.57°)] \text{ or } \sqrt{5}\angle116.57°$$

Note that the calculator did not in this example give the correct angle. It is important that you make a sketch of the complex number first.

How do we write 2 in polar form?

Plotting 2 on an Argand diagram:

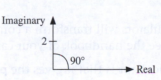

Thus the modulus of 2 is 2 and the argument is 0°, therefore $2 = 2\angle0°$.

How do we write $j2$ in polar form?

On an Argand diagram:

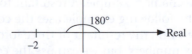

The modulus of $j2$ is 2 and the argument is 90°:

$$j2 = 2\angle90° \quad \text{[Polar form]}$$

How do we write −2 in polar form?

− 2 can be represented by

The modulus of −2 is 2 and the argument is 180°:

$$-2 = 2\angle180° \quad \text{[Polar form]}$$

? **How do we write −j2 in polar form?**

On an Argand diagram:

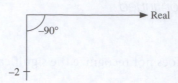

The modulus of − j2 is 2 and the argument is −90°:

$$- j2 = 2\angle(-90°) \qquad \text{[Polar form]}$$

Let's do an example from polar to rectangular. Converting back is very simple.

Example 12

Express $2\angle 53°$ in rectangular form, $x + jy$ (x and y are real).

Solution

Putting $r = 2$ and $\theta = 53°$ into $r[\cos(\theta) + j\sin(\theta)]$ gives

$$2[\cos(53°) + j\sin(53°)] = 2(0.6 + j\,0.8)$$
$$= (2 \times 0.6) + j(2 \times 0.8)$$
$$= 1.2 + j1.6$$

When transforming from polar to rectangular form you can avoid making a sketch because the angle is given.

Most basic scientific calculators will transform a complex number from rectangular to polar or polar to rectangular. See the handbook of your calculator for instructions.

One advantage of the calculator is that it gives the **principal argument**.

For example, transforming $3.1 - j\,6.5$ gives a modulus of 7.2 and argument of $-64.5°$. We will often use the shorthand notation $7.2\angle(-64.5°)$ rather than $7.2[\cos(-64.5°) + j\sin(-64.5°)]$.

Use your calculator to convert from polar to rectangular form, or vice versa.

Next we examine a mechanics example. A resultant force is the force which is equivalent to two or more forces. The following example uses the conversion from polar to rectangular form, and vice versa, to find the resultant of three forces. It is important that you realize forces are not complex numbers, but we can use the concept of conversion (polar ↔ rectangular) to find the resultant force. The example can also be tackled by using vectors (**Chapter 12**).

🚗 Example 13 *mechanics*

Figure 12 shows (coplanar) forces F_1, F_2 and F_3 acting at a common point O. Find the magnitude and direction of the resultant force, $F = F_1 + F_2 + F_3$.

Fig. 12

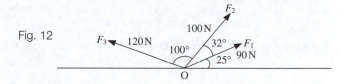

Solution

The forces F_1, F_2 and F_3 can be written in polar form as

$$F_1 = 90\underline{/25°}, F_2 = 100\underline{/(32° + 25°)} \text{ and } F_3 = 120\underline{/(100° + 32° + 25°)}$$

To add these forces we need to write them in rectangular form. Use your calculator to place each number into rectangular form:

$$F_1 = 90\underline{/25°} = 81.57 + j38.04$$

$$F_2 = 100\underline{/(32° + 25°)} = 100\underline{/57°} = 54.46 + j83.87$$

$$F_3 = 120\underline{/(100° + 57°)} = 120\underline{/157°} = -110.46 + j46.89$$

$$\text{Resultant force} = \underbrace{81.57 + j38.04}_{F_1} + \underbrace{54.46 + j83.87}_{F_2} + \underbrace{-110.46 + j46.89}_{F_3}$$

$$= (81.57 + 54.46 - 110.46) + j(38.04 + 83.87 + 46.89)$$

adding the real and imaginary parts

$$= 25.57 + j168.8$$

$$= 170.73\underline{/81.39°} \quad \text{[By calculator]}$$

The resultant force, F, has a modulus of 170.73 N at an angle of 81.39°, as shown in Fig. 13.

Fig. 13

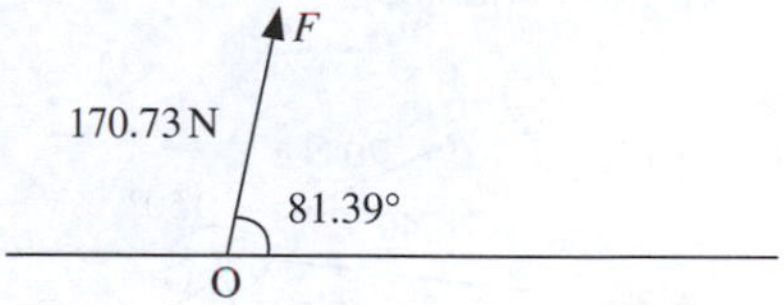

SUMMARY

If $z = x + jy$ then

10.15 $|z| = \sqrt{x^2 + y^2}$ [modulus]

10.16 $\arg(z) = \tan^{-1}\left(\dfrac{y}{x}\right)$ [argument]

SUMMARY continued

We can rewrite z as

$$r[\cos(\theta) + j\sin(\theta)] \text{ or } r\underline{/\theta}$$

where $r = |z|$ and $\theta = \arg(z)$.

Exercise 10(b)

Solutions at end of book. Complete solutions available at www.palgrave.com/engineering/singh

1 Label the following complex numbers on an Argand diagram:

a $2 + j3$ **b** $-1-j$ **c** $-1+j$
d $-j2$ **e** -3

2 Without using a calculator, express the following in polar form, giving the principal argument:

a $1+j$ **b** $-1+j$ **c** $\sqrt{3} - j$
d $-1 - j\sqrt{3}$ **e** j **f** jx
g -3 **h** 5

3 Express the following in rectangular form:

a $2\underline{/30°}$ **b** $5\underline{/23°}$ **c** $2\underline{/(-45°)}$
d $1.4\underline{/(-130°)}$

4 🚗 [*mechanics*]

Fig. 14

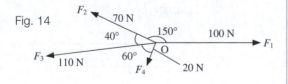

Figure 14 shows coplanar forces F_1, F_2, F_3 and F_4 acting at a common point O. Find the magnitude and direction of the resultant force F where

$$F = F_1 + F_2 + F_3 + F_4$$

5 🚗 [*mechanics*]

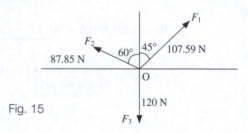

Fig. 15

Figure 15 shows three coplanar forces, F_1, F_2 and F_3, acting at a common point O. Find the resultant force $F = F_1 + F_2 + F_3$. From your solution, what can you conclude about F_1, F_2 and F_3?

SECTION C Multiplication and division in polar form

By the end of this section you will be able to:
► use the multiplication and division rules for complex numbers in polar form
► apply the polar form of complex numbers to engineering problems

C1 **Multiplication and division in polar form**

Example 14

Let $z = r[\cos(A) + j\sin(A)]$ and $w = q[\cos(B) + j\sin(B)]$ be any two complex numbers (Fig. 16). Show that

$$zw = rq[\cos(A + B) + j\sin(A + B)]$$

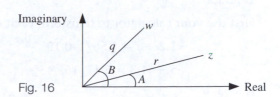

Fig. 16

Solution

Multiplying the two complex numbers gives

$$zw = rq\{[\cos(A) + j\sin(A)] \times [\cos(B) + j\sin(B)]\}$$

$$= rq\{\cos(A)\cos(B) + j\sin(B)\cos(A) + j\sin(A)\cos(B) + \underbrace{j^2}_{=-1}\sin(A)\sin(B)\}\ \left[\text{Using FOIL}\right]$$

$$= rq\{\underbrace{[\cos(A)\cos(B) - \sin(A)\sin(B)]}_{\text{real part}} + j\underbrace{[\cos(A)\sin(B) + \sin(A)\cos(B)]}_{\text{imaginary part}}\}$$

$$= rq\{\underbrace{\cos(A + B)}_{\text{by }4.39} + j\underbrace{\sin(A + B)}_{\text{by }4.37}\}$$

Hence if $z = r[\cos(A) + j\sin(A)]$ and $w = q[\cos(B) + j\sin(B)]$ then

Multiplication

10.17 $zw = rq[\cos(A + B) + j\sin(A + B)]$ or $r\angle A \times q\angle B = rq\angle(A + B)$

Division

10.18 $\dfrac{z}{w} = \dfrac{r}{q}[\cos(A - B) + j\sin(A - B)]$ or $\dfrac{r\angle A}{q\angle B} = \dfrac{r}{q}\angle(A - B)$

You are asked to show result 10.18 in **Exercise 10(c)**.

? **Can you see the advantages of putting a complex number in polar form?**

  i 10.17 shows that multiplication of complex numbers is just a matter of **multiplying** their **moduli** and **adding** their **arguments**.
  ii Similarly 10.18 shows that division of complex numbers is simply a matter of **dividing** their **moduli** and **subtracting** their **arguments**.

Generally it is better to use complex numbers in polar form if you want to multiply or to divide two or more complex numbers.

4.37 $\sin(A + B) = \sin(A)\cos(B) + \cos(A)\sin(B)$
4.39 $\cos(A + B) = \cos(A)\cos(B) - \sin(A)\sin(B)$

Example 15

Express $(-1.2 + j)(-5 + j1.3)$ in polar form.

Solution

First use your calculator to transform each number into polar form:

$$-1.2 + j = 1.56 \underline{/140.19°}$$

$$-5 + j1.3 = 5.17 \underline{/165.43°}$$

Hence

$$(-1.2 + j)(-5 + j1.3) = (1.56 \underline{/140.19°}) \times (5.17 \underline{/165.43°})$$

$$= 1.56 \times 5.17 \underline{/(140.19° + 165.43°)} \quad [\text{By } \boxed{10.17}]$$

$$= 8.07 \underline{/305.62°}$$

Since the question does not ask for the principal argument we can leave the argument as 305.62°.

C2 **Engineering applications**

Example 16 *electrical principles*

An a.c. voltage, $v = (7.6 - j6.8)$ volts, is applied across a circuit of impedance $z = (88 + j5)\,\Omega$. Find the current i through the circuit, given that $i = \dfrac{v}{z}$ (you may leave your answer in polar form).

Solution

Putting v and z into polar form via a calculator yields

$$7.6 - j6.8 = 10.20 \underline{/(-41.82°)}$$
$$88 + j5 = 88.14 \underline{/3.25°}$$

Hence

$$i = \frac{7.6 - j6.8}{88 + j5} = \frac{10.20 \underline{/(-41.82°)}}{88.14 \underline{/3.25°}}$$

$$\underset{\text{by } \boxed{10.18}}{=} \frac{10.20}{88.14} \underline{/(-41.82° - 3.25°)}$$

$$i = 0.12 \underline{/(-45.07°)} \text{ A} \quad (\text{A} = \text{amp})$$

$\boxed{10.17}$ $r \underline{/A} \times q \underline{/B} = rq \underline{/(A + B)}$ $\boxed{10.18}$ $\dfrac{r \underline{/A}}{r \underline{/B}} = \dfrac{r}{q} \underline{/(A - B)}$

The next example is particularly challenging. We need to solve simultaneous equations where the coefficients of the variables are complex numbers. Addition or subtraction of complex numbers is carried out in rectangular form, but multiplication and division are normally easier in polar form. Hence you will find that we interchange between polar ↔ rectangular as the arithmetic requires.

Example 17 *electrical principles*

Let I_1 and I_2 be the primary and secondary currents, respectively, of a magnetically coupled circuit. By Kirchhoff we obtain the following mesh equations:

$$20I_1 + j100I_1 - j20I_2 = 200$$

$$20I_2 + j100I_2 + 100I_2 - j20I_1 = 0$$

Determine, in polar form, the primary and secondary currents, I_1 and I_2 respectively, through the circuit.

Solution

Since 20 is a common factor of both equations we divide through by 20:

$$I_1 + j5I_1 - jI_2 = 10$$

$$I_2 + j5I_2 + 5I_2 - jI_1 = 0$$

Adding jI_1 to both sides of the second equation gives

$$I_2 + j5I_2 + 5I_2 = jI_1$$

$$(6 + j5)I_2 = jI_1 \quad \text{[Factorizing } I_2\text{]}$$

Dividing by $6 + j5$ gives

$$I_2 = \frac{jI_1}{6 + j5}$$

Substituting this for I_2 into the factorized first equation, $10 = (1 + j5)I_1 - jI_2$, gives

$$10 = (1 + j5)I_1 - j\frac{jI_1}{6 + j5}$$

$$= \left[1 + j5 - \frac{j^2}{6 + j5}\right]I_1 \quad \text{[Taking out } I_1\text{]}$$

Example 17 *continued*

$$= \left[1 + j5 + \frac{6 - j5}{6^2 + 5^2} \right] I_1 \quad \text{[Because } j^2 = -1 \text{ and the complex conjugate of } 6 + j5 \text{ is } 6 - j5\text{]}$$

$$= \left[1 + j5 + \frac{6 - j5}{61} \right] I_1$$

$$10 = \left[\frac{67 + j300}{61} \right] I_1 \quad \text{[Adding with common denominator]}$$

Transposing to make I_1 the subject results in

$$I_1 = \frac{61 \times 10}{[67 + j300]} = \frac{610 \angle 0°}{307.39 \angle 77.41°} \underset{\text{by } 10.18}{=} \frac{610}{307.39} \angle (0° - 77.41°)$$

$$I_1 = 1.98 \angle (-77.41°) \text{ A}$$

Substituting for I_1 into [*]:

$$I_2 = \frac{j}{6 + j5} 1.98 \angle (-77.41°)$$

$$= \frac{1 \angle 90° \times 1.98 \angle (-77.41°)}{7.81 \angle 39.81°} \quad \text{[Polar form]}$$

$$= \frac{1 \times 1.98}{7.81} \angle (90° + (-77.41°) - 39.81°) \quad \text{[By 10.17 and 10.18]}$$

$$I_2 = 0.25 \angle (-27.22°) \text{ A}$$

In a similar manner, try question **13** of **Exercise 10(a)**.

C3 Important angles

As you might expect, some angles crop up with more frequency than others. The sin, cos and tan of these angles, in degrees and radians, is given in Table 1 on the next page. Some of these results were shown in Chapter 4.

10.17 $r \angle A \times q \angle B = rq \angle (A + B)$ 10.18 $\dfrac{r \angle A}{r \angle B} = \dfrac{r}{q} \angle (A - B)$

TABLE 1		0	$\dfrac{\pi}{6}$	$\dfrac{\pi}{4}$	$\dfrac{\pi}{3}$	$\dfrac{\pi}{2}$	$\dfrac{2\pi}{3}$	$\dfrac{3\pi}{4}$	$\dfrac{5\pi}{6}$	π
	angle θ	0°	30°	45°	60°	90°	120°	135°	150°	180°
	$\sin(\theta)$	0	$\dfrac{1}{2}$	$\dfrac{1}{\sqrt{2}}$	$\dfrac{\sqrt{3}}{2}$	1	$\dfrac{\sqrt{3}}{2}$	$\dfrac{1}{\sqrt{2}}$	$\dfrac{1}{2}$	0
	$\cos(\theta)$	1	$\dfrac{\sqrt{3}}{2}$	$\dfrac{1}{\sqrt{2}}$	$\dfrac{1}{2}$	0	$-\dfrac{1}{2}$	$-\dfrac{1}{\sqrt{2}}$	$-\dfrac{\sqrt{3}}{2}$	-1
	$\tan(\theta)$	0	$\dfrac{1}{\sqrt{3}}$	1	$\sqrt{3}$	un-def	$-\sqrt{3}$	-1	$-\dfrac{1}{\sqrt{3}}$	0

(In Table 1 un-def means undefined.) Remember that the table shows the exact trigonometric values.

The argument of a complex number might be given in radians rather than degrees.

Example 18

Find the exact rectangular form of $\dfrac{1}{\sqrt{2}} \angle \left(\dfrac{\pi}{4}\right)$.

Solution

Using

$$r\angle\theta = r[\cos(\theta) + j\sin(\theta)]$$

with $r = \dfrac{1}{\sqrt{2}}$ and $\theta = \dfrac{\pi}{4}$ we have

$$\dfrac{1}{\sqrt{2}} \angle \left(\dfrac{\pi}{4}\right) = \dfrac{1}{\sqrt{2}}\left[\cos\left(\dfrac{\pi}{4}\right) + j\sin\left(\dfrac{\pi}{4}\right)\right] \underset{\text{by Table 1}}{=} \dfrac{1}{\sqrt{2}}\left[\dfrac{1}{\sqrt{2}} + j\left(\dfrac{1}{\sqrt{2}}\right)\right]$$

$$= \dfrac{1}{\sqrt{2}}\dfrac{1}{\sqrt{2}}\,[1 + j] \quad \left[\text{Taking out } \dfrac{1}{\sqrt{2}}\right]$$

$$= \dfrac{1}{2}[1 + j] \quad \left[\text{Because } \dfrac{1}{\sqrt{2}}\dfrac{1}{\sqrt{2}} = \dfrac{1}{2}\right]$$

SUMMARY

If $z = r\angle A$ and $w = q\angle B$ then

10.17 $zw = rq\angle(A + B)$ [Multiplication]

10.18 $\dfrac{z}{w} = \dfrac{r}{q}\angle(A - B)$ [Division]

Exercise 10(c)

Solutions at end of book. Complete solutions available at www.palgrave.com/engineering/singh

1 🔲 [*electrical principles*]   An a.c. voltage, $v = (2.9 + j6.2)$ volts, is applied across an impedance $Z = (8 + j1.9)$ ohms. Find the current, i, in polar form, through the circuit given that

$$i = \frac{v}{Z}$$

2 Find the modulus and argument of

$$z = \frac{2\angle 45° \times 2\angle(-60°)}{4\angle(-22°) \times 5\angle 33°}$$

3 Without using a calculator, express the following in rectangular form:

a $2\angle 30°$ **b** $2\angle\left(\dfrac{\pi}{6}\right)$ **c** $240\angle 0°$

*4 If $z = r\angle\theta$ show that $\bar{z} = r\angle(-\theta)$ ($\bar{z}$ is the complex conjugate of z).

5 Find $\bar{z}$ (complex conjugate of z) in polar form, if

a $z = 6\angle 45°$ **b** $z = 2\angle 131°$
c $z = 15\angle(-26°)$

*6 i Show that, if $z = a + jb$ such that $|z| = 1$ then

$$\frac{1}{z} = \bar{z}$$

ii If $z = \dfrac{1}{3}\left(\sqrt{8} + j\right)$ then determine

$|z|$ and $\dfrac{1}{z}$.

7 Let $z = r[\cos(\alpha) + j\sin(\alpha)]$ and $w = q[\cos(\beta) + j\sin(\beta)]$. Show that

$$\frac{z}{w} = \frac{r}{q}\left[\cos(\alpha - \beta) + j\sin(\alpha - \beta)\right].$$

8 🔲 [*electronics*]   An amplifier has a transfer function, T, given by

$$T = \frac{1000}{1 + j\omega(0.5 \times 10^{-3})}$$

where ω is the angular frequency. The gain and phase of the amplifier are given by the modulus and argument of T respectively. Find the gain and phase at an angular frequency of $\omega = 2 \times 10^3$ rad/s.

9 🔲 [*control engineering*]   A system has a transfer function, T, given by

$$T = \frac{50}{j\omega(1 + j0.01\omega)(1 + j0.2\omega)}$$

Evaluate the gain (modulus of T) and phase (argument of T) of the system for $\omega = 45$ rad/s.

10 🔲 [*electrical principles*]   By applying Kirchhoff's laws on a circuit we obtain the following mesh equations:

$$(6 + j2)I_1 + (2 + j2)I_2 = 10$$

$$(2 + j2)I_1 + 7I_2 = 10$$

Determine, in polar form, the currents I_1 and I_2.

SECTION D **Powers and roots of complex numbers**

By the end of this section you will be able to:
▶ apply de Moivre's theorem
▶ find roots of complex numbers
▶ plot these roots on an Argand diagram

D1 De Moivre's theorem

? Let $z = r\left[\cos(\theta) + j\sin(\theta)\right]$ be a complex number. **What are z^2 and z^3 equal to?**

$$z^2 = z \cdot z \underset{\text{by } \boxed{10.17}}{=} r \cdot r\left[\cos(\theta + \theta) + j\sin(\theta + \theta)\right]$$

$$z^2 = r^2\left[\cos(2\theta) + j\sin(2\theta)\right] \text{ or } r^2 \underline{/2\theta}$$

Similarly we find

$$z^3 = r^3\left[\cos(3\theta) + j\sin(3\theta)\right] \text{ or } r^3\underline{/3\theta}$$

? **What do you think z^n will be?**

$\boxed{10.19}$ $\qquad z^n = r^n\left[\cos(n\theta) + j\sin(n\theta)\right] \text{ or } (r\underline{/\theta})^n = r^n\underline{/n\theta}$

This is known as de Moivre's theorem. To use de Moivre's theorem we first place the complex number in polar form.

We also have

> **Abraham de Moivre** (1667–1754) was born in France but spent most of his life in England. This rule was known to de Moivre for the case when n is a whole number, but the result also holds for fractions.

$$\frac{1}{z^n} = z^{-n} = r^{-n}[\cos(-n\theta) + j\sin(-n\theta)]$$

$$= r^{-n}\left[\underset{\text{by } \boxed{4.51}}{\cos(n\theta)} + j(\underset{\text{by } \boxed{4.50}}{-\sin(n\theta))}\right] = r^{-n}\left[\cos(n\theta) - j\sin(n\theta)\right]$$

Thus we have

$\boxed{10.20}$ $\qquad z^{-n} = r^{-n}[\cos(n\theta) - j\sin(n\theta)]$

We will need to use the following definition of indices in this section:

$\boxed{10.21}$ $\qquad \sqrt[n]{z} = z^{1/n} \quad \left(\sqrt{z} = z^{1/2}\right)$

$\boxed{4.50}$ $\quad \sin(-x) = -\sin(x)$ $\qquad\qquad\qquad\qquad$ $\boxed{4.51}$ $\quad \cos(-x) = \cos(x)$

$\boxed{10.17}$ $\quad zw = rq[\cos(A + B) + j\sin(A + B)]$

Example 19

Evaluate $(1 - j)^5$ and plot the result on an Argand diagram.

Solution

First we place $(1 - j)$ in polar form:

$$1 - j = \sqrt{2}\angle(-45°)$$

$$= 2^{1/2}\angle(-45°) \qquad [\text{Remember } \sqrt{2} = 2^{1/2}]$$

Taking to the power of 5:

$$(1 - j)^5 = \left(2^{1/2}\angle(-45°)\right)^5$$

$$\underset{\text{by } 10.19}{=} (2^{1/2})^5 \ \angle\left(5\times(-45°)\right) \underset{\text{by } 10.2}{=} 2^{5/2} \ \angle(-225°)$$

Putting this, $2^{5/2}\angle(-225°)$, into rectangular form by using a calculator gives

$$(1 - j)^5 = 2^{5/2}\angle(-225°) = -4 + j4$$

The angle is measured in a clockwise direction, because we have a minus in front of 225, from the positive real axis (Fig. 17).

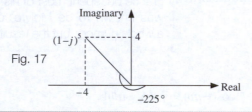

Fig. 17

The next example is a leap in sophistication compared to the above example. It uses algebraic substitution, rules of complex division and powers of complex numbers. Also we don't end up with simple whole numbers after substitution. The only other hiccup might be writing $j6.28 \times 10^{-4}$ in polar form. Remember that jy (y is real) on an Argand diagram (Fig. 18) is

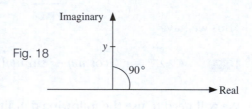

Fig. 18

$$jy = y\angle 90° \qquad [\text{Polar form}]$$

Similarly

$$j(6.28 \times 10^{-4}) = (6.28 \times 10^{-4})\angle 90° \quad [\text{Polar form}]$$

Remember that $j(6.28 \times 10^{-4}) = 0 + j(6.28 \times 10^{-4})$ which means there is **zero** real part to this complex number so it extends vertically up on the Argand diagram.

10.2 $(a^m)^n = a^{m \times n}$ 10.19 $(r\angle\theta)^n = r^n\angle n\theta$

Example 20 *electrical principles*

The characteristic impedance, Z_0, of a transmission line is given by

$$Z_0 = \left(\frac{R + j2\pi fL}{G + j2\pi fC} \right)^{1/2}$$

where $R = 5\ \Omega$, $L = 0.03 \times 10^{-3}$ H (henry), $C = 0.01 \times 10^{-6}$ F (farad), $G = 0$ siemen and $f = 10^4$ Hz. Determine Z_0 where $\mathrm{Re}(Z_0) \geq 0$.

Solution

In this example we use de Moivre's theorem with $n = 1/2$.

Substituting the given values into the numerator we have

$$R + j2\pi fL = 5 + j(2\pi \times 10^4 \times 0.03 \times 10^{-3})$$

$$= 5 + j1.88$$

$$= 5.34\underline{/20.61°} \qquad \text{[Converting to polar form]}$$

Substituting the given values into the denominator:

$$G + j2\pi fC = 0 + j(2\pi \times 10^4 \times 0.01 \times 10^{-6})$$

$$= j(6.28 \times 10^{-4})$$

$$= (6.28 \times 10^{-4})\underline{/90°} \qquad \text{[Converting to polar form]}$$

Substituting these $R + j2\pi fL = 5.34\underline{/20.61°}$ and $G + 2\pi fC = (6.28 \times 10^{-4})\underline{/90°}$ gives

$$Z_0 = \left(\frac{R + j2\pi fL}{G + j2\pi fC} \right)^{1/2}$$

$$= \left(\frac{5.34\underline{/20.61°}}{6.28 \times 10^{-4}\underline{/90°}} \right)^{1/2}$$

$$= \left(\frac{5.34}{6.28 \times 10^{-4}}\underline{/(20.61° - 90°)} \right)^{1/2} \qquad \text{[By 10.18]}$$

$$= \left(8503.18\underline{/(-69.39°)} \right)^{1/2}$$

$$= 8503.18^{1/2}\underline{/\left(\frac{1}{2} \times (-69.39°) \right)} \qquad \text{[By 10.19]}$$

$$= 92.21\underline{/(-34.70°)}$$

$$Z_0 = 92.21\underline{/(-34.70°)}\ \Omega = (75.81 - j52.49)\ \Omega \qquad \text{[Rectangular form]}$$

10.18 $\quad \dfrac{r\underline{/A}}{q\underline{/B}} = \dfrac{r}{q}\underline{/(A - B)}$ $\qquad$ **10.19** $\quad z^n = r^n\underline{/n\theta}$

D2 **Finding roots**

De Moivre's theorem can be used to find roots of complex numbers. The **Fundamental Theorem of Algebra** says that the complex number

$$z = (x + jy)^{1/n} \text{ has } \textbf{exactly } n \text{ roots } (x \text{ and } y \text{ are real numbers})$$

? **How many roots do $z = (1 + j)^{1/2}$, $z = (1 + j)^{1/5}$ and $z = (1 + j)^{1/12}$ have?**

2, 5 and 12 roots respectively.
The number $8^{1/3}$ has exactly three roots, namely 2, $2\angle120°$ and $2\angle240°$.

? **Why have we only one root for Example 20?**

Since in the question we have $\text{Re}(Z_0) \geq 0$ which means the real part of Z_0 is positive or zero, so we do not look for the other root. We will show later that the other root is $(-75.81 + j52.49)\,\Omega$.

We can represent any complex number in polar form in many different ways. For example, consider the real number 2 $= 2 + j0 = 2\angle0°$ (Fig. 19).

Fig. 19

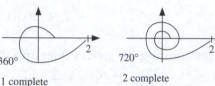

1 complete revolution

2 complete revolutions

$2 = 2\angle0°$ $2 = 2\angle360°$ $2 = 2\angle720°$

Hence we can make complete revolutions and still get back to the same point. **How else can we represent the complex number $4\angle30°$ in polar form?**

Fig. 20

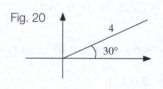

See Fig. 20. Thus $4\angle30° = 4\angle390°$.

We can use this concept to find **all** the roots of complex numbers, as the following example shows.

$4\angle30°$ $4\angle(360° + 30°) = 4\angle390°$

Example 21

Solve

$$z^2 - (2\sqrt{3} + j2) = 0$$

Solution

Adding $2\sqrt{3} + j2$ to both sides of the equation so that the complex number is on the Right-Hand Side:

$$z^2 = \left(2\sqrt{3} + j2\right) \text{ gives } z = \sqrt{2\sqrt{3} + j2}$$
$$\underset{\text{by } \boxed{10.21}}{=} \left(2\sqrt{3} + j2\right)^{1/2}$$

$\boxed{10.21}$ $\sqrt{z} = z^{1/2}$

Example 21 *continued*

How many roots does this equation have?

Two roots.

1st root. To use de Moivre's theorem we need to put the complex number, $2\sqrt{3} + j2$, into polar form (via a calculator or otherwise):

$$2\sqrt{3} + j2 = 4\angle 30° \quad \text{[Fig. 21]}$$

Applying de Moivre's theorem:

Fig. 21

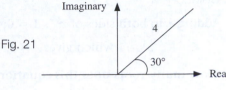

$$\left(2\sqrt{3} + j2\right)^{1/2} = (4\angle 30°)^{1/2}$$

$$= 4^{1/2}\angle\left(\frac{1}{2} \times 30°\right) \quad \text{[By 10.19]}$$

$$= 2\angle 15°$$

Hence one of the roots of $\left(2\sqrt{3} + j2\right)^{1/2}$ is $2\angle 15°$. **What about the other root?**

2nd root. If we make one complete revolution (360°) from $4\angle 30°$ we get back to the same point. Therefore

$$2\sqrt{3} + j2 \text{ is } 4\angle 30° \text{ or } 4\angle(30° + 360°) = 4\angle 390°$$

So for the second root we have

$$\left(2\sqrt{3} + j2\right)^{1/2} = \left(4\angle 390°\right)^{1/2}$$

$$\underset{\text{by } 10.19}{=} 4^{1/2}\angle\left(\frac{1}{2} \times 390°\right) = 2\angle 195°$$

So the **two** roots of $\left(2\sqrt{3} + j2\right)^{1/2}$ are $2\angle 15°$ and $2\angle 195°$ which can be represented on the Argand diagram as shown in Fig. 22.

In the above example, why not make 2, 3, 4, ... , etc. revolutions from $4\angle 30°$?

If we do make 2, 3, 4, ... , etc. revolutions we just reproduce one of the above two roots, $2\angle 15°$ or $2\angle 195°$.

Fig. 22

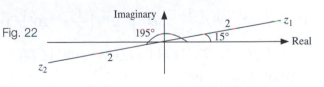

We can label the roots

$z_1 = 2\angle 15°$, $z_2 = 2\angle 195°$ and they satisfy the equation given in the question:

$$z^2 - \left(2\sqrt{3} + j2\right) = 0$$

10.19 $(r\angle\theta)^n = r^n\angle n\theta$

Example 22

Find all the roots of

$$z^5 - 1 = 0$$

and plot them on an Argand diagram.

Solution

Adding 1 to both sides of $z^5 - 1 = 0$:

$$z^5 = 1 \text{ which gives } z = 1^{1/5}$$

?
?   **How many roots does this equation have?**

Five roots. **Can you find any of the roots of this equation?**

1 is an obvious root ($1^5 - 1 = 0$)
which we need to place in polar
form in order to use de Moivre's
theorem. Since 1 is a real number
(no j), therefore:

Fig. 23

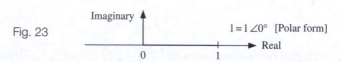

$1 = 1\angle 0°$ [Polar form]

$$1^{1/5} = (1\angle 0°)^{1/5} = 1\angle 0°   \text{[Fig. 23]}$$

?   So $1\angle 0°$ is a root of $z^5 = 1$. **Where are the other four roots?**

If we make one, two, three and four complete revolutions from $1\angle 0°$ we will end up at
the same point.

$1\angle 0°$, $1\angle 360°$ (one revolution), $1\angle(360° + 360°) = 1\angle 720°$ (two revolutions),
$1\angle(720° + 360°) = 1\angle 1080°$ (three revolutions) and $1\angle(1080° + 360°) = 1\angle 1440°$ (four
revolutions).

By applying de Moivre's theorem

10.19 $$(r\angle\theta)^n = r^n\angle n\theta$$

to each of these numbers with $n = \dfrac{1}{5}$ we obtain

$$1^{1/5} = 1\angle 0°, \ 1^{1/5}\angle\left(\frac{1}{5} \times 360°\right), \ 1^{1/5}\angle\left(\frac{1}{5} \times 720°\right), \ 1^{1/5}\angle\left(\frac{1}{5} \times 1080°\right) \text{ and } 1^{1/5}\angle\left(\frac{1}{5} \times 1440°\right)$$

$$= 1\angle 0°, \ 1\angle 72°, \ 1\angle 144°, \ 1\angle 216° \text{ and } 1\angle 288°$$

We label the roots
z_1, z_2, z_3, z_4 and z_5 and show
them on an Argand diagram
(Fig. 24).

Fig. 24

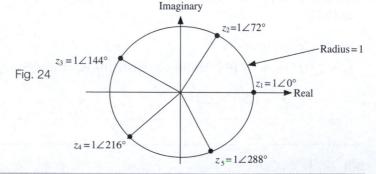

> Example 22 *continued*

Hence the five roots lie on a circle of radius 1 and are (Fig. 25)

$$z_1 = 1\underline{/0°}, \quad z_2 = 1\underline{/72°}, \quad z_3 = 1\underline{/144°}, \quad z_4 = 1\underline{/216°} \quad \text{and} \quad z_5 = 1\underline{/288°}$$

? **What do you notice about the size of each angle between the roots?**

They are all equal to 72°, that is

$$144° - 72° = 216° - 144° = \cdots = 72°$$

Hence a full revolution, 360°, is divided into five equal size angles of

$$\left(\frac{360}{5}\right)° = 72° \left(\text{or in radians } \frac{2\pi}{5}\right)$$

Fig. 25

These roots are called the five roots of unity.

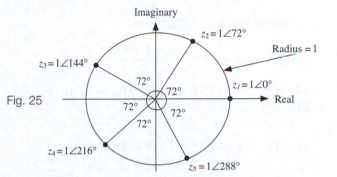

Also in **Example 21** we evaluated $\left(2\sqrt{3} + j2\right)^{1/2}$ and found that the **two** roots were equally spaced at angles of size

$$\left(\frac{360}{2}\right)° = 180° \left(\text{or } \frac{2\pi}{2} = \pi \text{ radians}\right)$$

For **Example 20** we found one of two roots

$$Z_0 = 92.21\underline{/(-34.70°)}$$

? **What is the value of the other root?**

Since Z_0 was the square root of a complex number we add $\left(\frac{360}{2}\right)° = 180°$:

$$92.21\underline{/(-34.70° + 180°)} = 92.21\underline{/145.3°}$$

In rectangular form this is

$$-75.81 + j52.49$$

In general, the n complex roots of $z = r\left[\cos(\theta) + j\sin(\theta)\right] = r\underline{/\theta}$ are equally spaced at angles of size $\left(\frac{360}{n}\right)° \left(\text{or } \frac{2\pi}{n} \text{ radians}\right)$. Hence the n roots of z are given by

10.22 $$z^{1/n} = r^{1/n}\underline{/\left(\frac{\theta + 360k°}{n}\right)} \quad \text{or} \quad z^{1/n} = r^{1/n}\underline{/\left(\frac{\theta + 2k\pi}{n}\right)}$$

where $k = 0, 1, 2, \ldots, n - 1$.

Example 23

Solve

$$z^6 - 1 = j\sqrt{3}$$

and represent the roots on an Argand diagram.

Solution

Adding 1 to both sides:

$$z^6 = 1 + j\sqrt{3} = 2\angle 60° \qquad \text{[Polar form]}$$

Putting $n = 6$ into 10.22 gives

⋆ $$z = 2^{1/6}\angle\left(\frac{60° + 360k°}{6}\right)$$

where $k = 0, 1, 2, 3, 4$ and 5. The roots are found by substituting these values of k into ⋆ :

$(k = 0)$ $z_1 = 2^{1/6}\angle\left(\dfrac{60°}{6}\right) = 2^{1/6}\angle 10°$

$(k = 1)$ $z_2 = 2^{1/6}\angle\left(\dfrac{60° + 360°}{6}\right) = 2^{1/6}\angle 70°$

$(k = 2)$ $z_3 = 2^{1/6}\angle\left(\dfrac{60° + (360 \times 2)°}{6}\right) = 2^{1/6}\angle 130°$

$(k = 3)$ $z_4 = 2^{1/6}\angle 190°$

$(k = 4)$ $z_5 = 2^{1/6}\angle 250°$

$(k = 5)$ $z_6 = 2^{1/6}\angle 310°$

These roots are equally spaced at angles of $\left(\dfrac{360}{6}\right)° = 60°$ on a circle of radius $2^{1/6}$ (Fig. 26).

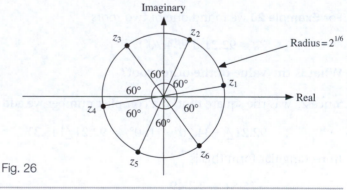

Fig. 26

Normally it is easier to find the first root, $r^{1/n}\angle\theta$, and then add $\left(\dfrac{360}{n}\right)°\left(\text{or } \dfrac{2\pi}{n} \text{ radians}\right)$ to the angle θ rather than using formula 10.22.

In **Example 23** we had $z_1 = 2^{1/16}\angle 10°$ and so for the other roots just add $60°$.

Hence

$$2^{1/6}\angle(10° + 60°), \; 2^{1/6}\angle(70° + 60°), \; 2^{1/6}\angle(130° + 60°), \ldots, 2^{1/6}\angle(250° + 60°)$$

? Also note that we do **not** need k values greater than 5. **Why?**

Each of the k values greater than 5 will reproduce one of the above six roots; $k = 6$ gives the same root as $k = 0$ and $k = 7$ gives the same root as $k = 1$, and so on.

SUMMARY

If $z = r\angle\theta$ then

10.19 $z^n = r^n\angle n\theta$

10.20 $z^{-n} = r^{-n}\left[\cos(n\theta) - j\sin(n\theta)\right]$

The n roots of z are given by

10.22 $z^{1/n} = r^{1/n}\angle\left(\dfrac{\theta + 360k°}{n}\right)$ or $z^{1/n} = r^{1/n}\angle\left(\dfrac{\theta + 2k\pi}{n}\right)$

where $k = 0, 1, 2, \ldots, n - 1$. Generally it is easier to find the first root, $r^{1/n}\angle\theta$, and then add $\left(\dfrac{360}{n}\right)°\left(\text{or } \dfrac{2\pi}{n} \text{ radians}\right)$ to θ.

Exercise **10(d)**

Solutions at end of book. Complete solutions available at www.palgrave.com/engineering/singh

1 By using your calculator to place the given number in polar form, or otherwise, evaluate the following, giving your answer in rectangular form:

a $(1 + j)^2$ **b** $(1 - j)^3$ **c** $(1 + j)^4$
d $(6 - j8)^7$ **e** $(-3 + j4)^7$ **f** $(-3 - j)^5$

2 Find all the roots of $z^4 - 1 = 0$.

3 Show that $(1 + j)^n = 2^{n/2}\angle\left(\dfrac{n\pi}{4}\right)$ and evaluate $(1 + j)^{12}$, giving your answer in rectangular form.

*4 [electrical principles]   A cable has the following constants: $R = 10\,\Omega$, $L = 0.1 \times 10^{-3}$ H, $G = 1 \times 10^{-6}$ siemen and $C = 1 \times 10^{-9}$ F. For $\omega = 10{,}000$ rad/s, determine:

i The characteristic impedance, Z_0, where

$$Z_0 = \sqrt{\dfrac{R + j\omega L}{G + j\omega C}}$$
$$(\text{Re}(Z_0) \geq 0)$$

ii The propagation coefficient, γ, given by

$$\gamma = \sqrt{(R + j\omega L)(G + j\omega C)}$$

$\text{Re}(\gamma) \geq 0$ and the units of γ are m^{-1}.

iii The wavelength, $\lambda = \dfrac{2\pi}{\beta}$, where

$\beta = \text{Im}(\gamma)$, (imaginary part of γ).

iv The velocity of propagation, $v = \dfrac{\omega}{\beta}$, where $\beta = \text{Im}(\gamma)$.

5 [control engineering]   A robot arm has a transfer function, G, given by

$$G = \dfrac{40}{(\omega + j\omega)^4}$$

where $\omega = $ angular frequency. Show that

$$G = -\dfrac{10}{\omega^4}$$

Exercise **10(d) continued**

Solutions at end of book. Complete solutions available at www.palgrave.com/engineering/singh

6 🔾 [*control engineering*]   A system has the following closed-loop transfer function:

$$G = \frac{K(z^4 + 16)}{z^3 + 64}$$

where K is a non-zero constant.
The **poles** of the system occur where the denominator of the transfer function is **zero**, that is $z^3 + 64 = 0$.
The **zeros** of the system occur where the numerator of the transfer function is **zero**, that is $z^4 + 16 = 0$.

i Determine the **poles** of the system and label (*) them on an Argand diagram.
ii Determine the **zeros** of the system and label (o) them on the same Argand diagram.

7 Let

$$z = \cos\left(\frac{\pi}{3}\right) + j\sin\left(\frac{\pi}{3}\right)$$

i Show that z is a root of $z^6 - 1 = 0$.
ii Determine $z^4 + z$.
iii Determine $z^6 + z^4 + z$.

8 Let a be a real number and satisfy

$$s^2 - (a + j\sqrt{3}a) = 0$$

Show that

$$s = \sqrt{\frac{a}{2}}\left(\sqrt{3} + j\right)$$

9 Simplify

$$\frac{(3 + j3)^3(1 - j)^4}{(1 + j\sqrt{3})^9}$$

*10 **i** Solve $t^2 - 2t + 4 = 0$.
ii Find all the roots of $z^6 - 2z^3 + 4 = 0$.

SECTION E **Exponential form of complex numbers**

By the end of this section you will be able to:
▶ convert a complex number from polar to exponential form
▶ convert from exponential to rectangular form
▶ apply the exponential form to engineering examples

E1 **Exponential form**

To write a complex number in exponential form, we need the argument of the complex number to be in **radians**. To convert from degrees → radians we use the following result:

10.23 $\theta° = \dfrac{\theta \times \pi}{180}$ radians

The exponential form of a complex number is

10.24 $r[\cos(\theta) + j\sin(\theta)] = re^{j\theta}$ or $r\angle\theta = re^{j\theta}$

where r = modulus and θ = argument in radians. You are asked to derive this result in **Exercise 10(e)**. You need to use the Maclaurin series expansion (**Chapter 7**) of $\sin(x)$, $\cos(x)$ and e^x to show 10.24. Remember that the Maclaurin series is only valid if x is in radians and that is why we need the argument of the complex number to be in radians.

Example 24

Write $2[\cos(60°) + j\sin(60°)]$ in exponential form.

Solution

What is wrong with $2[\cos(60°) + j\sin(60°)] = 2e^{j60°}$?

Remember that we need to put the argument 60° into radians. We have

$$\theta° = \frac{\theta \times \pi}{180}$$

$$60° = \frac{6\emptyset \times \pi}{18\emptyset} = \frac{\pi}{3}$$

Applying

10.24 $r[\cos(\theta) + j\sin(\theta)] = re^{j\theta}$

with $r = 2$ and $\theta = \pi/3$ gives

$$2[\cos(60°) + j\sin(60°)] = 2\left[\cos\left(\frac{\pi}{3}\right) + j\sin\left(\frac{\pi}{3}\right)\right] = 2e^{j\frac{\pi}{3}}$$

Properties of the complex exponential function are similar to those of the real exponential function. For example, if z_1 and z_2 are complex numbers then

$$e^{z_1 + z_2} = e^{z_1} e^{z_2}$$

and for a complex number z

$$(e^z)^n = e^{z \times n}$$

One advantage of using exponential form is that we do not need to remember special rules of the modulus and argument when multiplying, dividing and raising powers of complex numbers. We just use the laws of indices established earlier.

Example 25

Put $e^{2+j\frac{\pi}{4}}$ into rectangular form, $a + jb$.

Solution
Using the rules of indices we have

$$e^{2+j\frac{\pi}{4}} \underset{\text{by 10.3}}{=} e^2\, e^{j\frac{\pi}{4}}$$

$$\underset{\text{by 10.24}}{=} e^2\left[\cos\left(\frac{\pi}{4}\right) + j\sin\left(\frac{\pi}{4}\right)\right]$$

$$\underset{\text{by Table 1}}{=} e^2\left[\frac{1}{\sqrt{2}} + j\frac{1}{\sqrt{2}}\right] = \frac{e^2}{\sqrt{2}}\,[1+j] \quad \left[\text{Taking out } \frac{1}{\sqrt{2}}\right]$$

This format $\dfrac{e^2}{\sqrt{2}}[1+j]$ is neater than $\dfrac{e^2}{\sqrt{2}} + j\dfrac{e^2}{\sqrt{2}}$.

There are many additional identities associated with $re^{j\theta}$. For example, $re^{j\theta} = r[\cos(\theta) + j\sin(\theta)]$.
If $r = 1$ then

10.25 $e^{j\theta} = \cos(\theta) + j\sin(\theta)$

This is known as Euler's formula. This formula is one of the most famous results in mathematics. One of the most beautiful formula in mathematics is

$$e^{j\pi} + 1 = 0 \text{ or } e^{j\pi} = -1 \quad [\text{Means } 2.71828182846....^{\sqrt{-1}\,\times\,3.141592\,....} = -1]$$

because it shows the relationship between the numbers e, j, π, -1 and 0. Try proving the result by substituting $\theta = \pi$ into Euler's formula.

Example 26

Show that $e^{-j\theta} = \cos(\theta) - j\sin(\theta)$

Solution
Remember $e^{-j\theta} = e^{j(-\theta)}$ so we have

10.3 $a^{m+n} = a^m a^n$ 10.24 $re^{j\theta} = r[\cos(\theta) + j\sin(\theta)]$

Example 26 *continued*

$$e^{-j\theta} = \cos(-\theta) + j\sin(-\theta) \quad \text{[By Euler's formula]}$$

$$= \underbrace{\cos(\theta)}_{\text{by } 4.51} + j\,\underbrace{[-\sin(\theta)]}_{\text{by } 4.50}$$

$$= \cos(\theta) - j\sin(\theta)$$

This is another identity:

10.26 $e^{-j\theta} = \cos(\theta) - j\sin(\theta)$

Example 27 *electrical principles*

Express the following voltage

$$v = 20\sin(1000t - \phi) \text{ volts where } \phi = 30°$$

in complex exponential form.

Solution

For exponential form, we need to put 30° into radians:

$$30° \underset{\text{using } 10.23}{=} \frac{3\cancel{0} \times \pi}{18\cancel{0}} = \frac{\pi}{6}$$

Substituting $r = 20$ and $\theta = 1000t - \dfrac{\pi}{6}$ into $re^{j\theta} = r[\cos(\theta) + j\sin(\theta)]$ gives

$$20e^{j\left(1000t - \frac{\pi}{6}\right)} = 20\left[\cos\left(1000t - \frac{\pi}{6}\right) + j\sin\left(1000t - \frac{\pi}{6}\right)\right]$$

$$= 20\cos\left(1000t - \frac{\pi}{6}\right) + j\,\underbrace{20\sin\left(1000t - \frac{\pi}{6}\right)}_{=v}$$

We only want the j part because v is just the imaginary part of $20e^{j\left(1000t - \frac{\pi}{6}\right)}$ and so we can write this as

$$v = 20\sin\left(1000t - \frac{\pi}{6}\right) = \text{Imaginary part of } 20e^{j\left(1000t - \frac{\pi}{6}\right)}$$

Normally this is denoted as

$$v = \text{Im}\left[20e^{j\left(1000t - \frac{\pi}{6}\right)}\right] \text{V} \quad (\text{V} = \text{volt})$$

4.50 $\sin(-x) = -\sin(x)$ **4.51** $\cos(-x) = \cos(x)$ **10.23** $\theta° = (\theta \times \pi)/180 \text{ radians}$

Example 28 *control engineering*

A system transfer function, T, is given by

$$T = \frac{20 \times 6.32e^{j0.66}}{\left(3.87e^{j\frac{\pi}{2}}\right)(4.89e^{j0.91})(4e^{j1.32})}$$

Simplify T.

Solution

We can simplify T as an ordinary algebraic expression by using the rules of indices.
Writing

$$T = \frac{20 \times 6.32e^{j0.66}}{3.87 \times 4.89 \times 4e^{\underbrace{j\left(\frac{\pi}{2}+0.91+1.32\right)}_{\text{by } 10.3}}}$$

$$= \frac{20 \times 6.32}{3.87 \times 4.89 \times 4}\underbrace{e^{j\left[0.66-\left(\frac{\pi}{2}+0.91+1.32\right)\right]}}_{\text{by } 10.4}$$

$$= 1.67e^{-j3.14}$$

E2 De Moivre's theorem in exponential form

Let $z = r[\cos(\theta) + j\sin(\theta)] = re^{j\theta}$ then

$$z^n = (re^{j\theta})^n = r^n e^{jn\theta}$$

$$= r^n[\cos(n\theta) + j\sin(n\theta)]$$

which is **de Moivre's theorem**, 10.19.

Note: $z^{-n} = (re^{j\theta})^{-n} = r^{-n}[\cos(n\theta) - j\sin(n\theta)]$

E3 Oscillations

Consider a simple harmonic oscillator whose motion is illustrated in Fig. 27 on the next page. You can think of $P(t)$ as a complex number going around in a circle with angular velocity ω radians per second. We can use complex number notation to describe $P(t)$. **How?**

Well $P(t)$ is given by

$$P(t) = A\angle(\omega t) = A\big[\cos(\omega t) + j\sin(\omega t)\big] = Ae^{j\omega t}$$

10.3 $a^m a^n = a^{m+n}$ 10.4 $\dfrac{a^m}{a^n} = a^{m-n}$

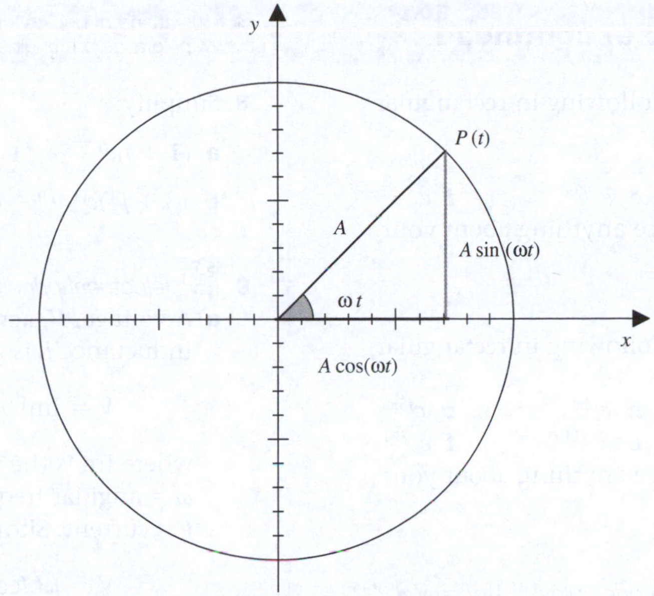

Fig. 27

The vertical and horizontal displacements of $P(t)$ are given by $A \sin(\omega t)$ and $A \cos(\omega t)$ respectively (see Fig. 27). Hence we can write these as

$$\text{Vertical displacement} = \text{Im}\left(Ae^{j\omega t}\right)$$

$$\text{Horizontal displacement} = \text{Re}\left(e^{j\omega t}\right)$$

In many engineering and physics problems the calculations are easier with the exponential form as stated above.

SUMMARY

The exponential form of a complex number is

10.24 $\qquad r[\cos(\theta) + j \sin(\theta)] = re^{j\theta}$

where r is the modulus and θ is the argument in radians.

Exercise 10(e)

Solutions at end of book. Complete solutions available at www.palgrave.com/engineering/singh

(Throughout this exercise x and y are real)

1 Express the following in complex exponential form:

  a $10\angle\pi$ **b** $10[\cos(180°) + j\sin(180°)]$

  c $j2$ **d** $1 + j$ **e** $\sqrt{3} - j$

2 Express the following in rectangular form, $x + jy$:

  a $e^{1+j\frac{\pi}{2}}$ **b** $e^{\pi(1+j)}$ **c** $2e^{2+j\frac{\pi}{6}}$

  d $e^{1000-j\pi/4}$

Exercise **10(e) continued**

Solutions at end of book. Complete solutions available at www.palgrave.com/engineering/singh

3 Express the following in rectangular form, $x + jy$:

   a $e^{j2\pi}$ **b** $e^{j4\pi}$ **c** $e^{j6\pi}$

   d $e^{-j2\pi}$ **e** $e^{-j8\pi}$ **f** $e^{-j20\pi}$

Do you notice anything about your results?

4 Express the following in rectangular form, $x + jy$:

   a $e^{j\pi}$ **b** $e^{j3\pi}$ **c** $e^{j5\pi}$

   d $e^{-j\pi}$ **e** $e^{-j11\pi}$ **f** $e^{-j21\pi}$

Do you notice anything about your results?

5 🔲 [*electrical principles*]   If $v = Ve^{j(\omega t + \alpha)}$ and $i = Ie^{j(\omega t + \beta)}$ then show that the

impedance $z\left(= \dfrac{v}{i}\right)$ is given by

$$z = \frac{V}{I}\left[\cos(\alpha - \beta) + j\sin(\alpha - \beta)\right]$$

6 🔲 [*control engineering*]   A system transfer function, G, is given by

$$G = \frac{10}{5\,e^{j\frac{\pi}{2}}\,e^{j\frac{\pi}{3}}}$$

Simplify G.

7 🔲 [*control engineering*]   A system has an open-loop transfer function

$$G = \frac{100}{10e^{j\frac{\pi}{2}}1.02e^{j0.2}1.8e^{j0.99}}$$

Simplify G.

8 Simplify:

   a $(1 + j)e^{j\frac{\pi}{4}} + (1 - j)e^{-j\frac{\pi}{4}}$

  *b $(3 + j4)e^{(1+j)t} + (3 - j4)e^{(1-j)t}$ (t is real)

9 🔲 [*electrical principles*]

   a A voltage, V, across an inductor with inductance L is given by

$$V = \text{Im}[\omega L\, Ie^{j(\omega t + \pi/2)}]$$

where Im is the imaginary part, ω = angular frequency, t = time and I = current. Show that

$$V = \omega LI\cos(\omega t)$$

   b A voltage, V, across a capacitor with capacitance C is given by

$$V = \text{Im}\left[\frac{I}{\omega C}e^{j(\omega t - \pi/2)}\right]$$

Show that

$$V = -\frac{I}{\omega C}\cos(\omega t)$$

(ω, I and t are as in part **a**).

10 Let $z = \ln(3) + j\pi/2$ and a, b be real such that

$$a + jb = e^{z}$$

Determine a and b.

11 Show the identity

$$re^{j\theta} = r[\cos(\theta) + j\sin(\theta)]$$

[*Hint*: Consider the Maclaurin series expansion of e^{x}, $\cos(x)$ and $\sin(x)$]

Examination questions 10   Solutions at end of book. Complete solutions available at www.palgrave.com/engineering/singh

The complex number j is denoted by i in some of the following questions.

1 Given that $z_1 = 4 + j2$, $z_2 = 3 - j$ find

   i $3z_1 + 2z_2$ **ii** $z_1 z_2$

University of Portsmouth, UK, 2007

2 Given $z_1 = 2 + j$, $z_2 = 3 - j4$, find

   a $z_1 z_2$ **b** Both values of $\sqrt{z_1}$

University of Portsmouth, UK, 2006

3 a Find the **four** solutions to the equation

$$z^4 = -2 - j2\sqrt{3}$$

   and display them on an Argand diagram.

  b Evaluate without using a table or a calculator $\dfrac{(\cos 53° + j \sin 53°)^3}{\cos 24° + j \sin 24°}$

Cork Institute of Technology, Ireland, 2007

4 a Solve the quadratic equation $x^2 - 6x + 13 = 0$, giving your answers in the form $x = a + bi$.

  b Write down $z = e^{i\theta}$ in terms of trigonometric functions. Deduce the value of $e^{i\pi/4}$.

  c Continuing with $z = e^{i\theta}$, state de Moivre's theorem giving z^n in terms of $\cos(n\theta)$ and $\sin(n\theta)$. Show that $z^n + \dfrac{1}{z^n} = a \cos(n\theta)$, find the value of a.

University of Manchester, UK, 2009

5 a If $a = 3 + i$, $b = 2 - 3i$, $c = -1 + 4i$ find (assume * means conjugate in this question):

   i $(a + b)c$ **ii** $(bc)^*$ **iii** b/a

   iv $|ab|$ **v** $\mathrm{Re}(ac)$

   vi Write a, b and c in polar form and hence determine the modulus and argument of $\left(\dfrac{ac}{b}\right)$.

  b Use de Moivre's theorem to find the 3 complex roots: $(3 - i)^{1/3}$.

Loughborough University, UK, 2007

6 Determine all complex 3rd roots of $8i$.

Colorado State University, USA, 2009

7 Determine all complex numbers z such that $z^5 = -32$.

Colorado State University, USA, 2008

*8 a Let $z = 3 + 5i$ and $w = 3e^{3\pi i/4}$. Find zw, $\dfrac{z}{w}$ and $z + w$ writing your answers in both Cartesian and polar form.

  b Solve the equation $z^2 + 2\bar{z} = -3$, where $\bar{z}$ denotes the complex conjugate of z.

  c Let $z = 1 + \sqrt{3}i$. Write z exactly in polar form. Find z^{10} and all values of $\sqrt[3]{z}$.

  d Find a formula for $\cos(3\theta)$ in terms of $\cos(\theta)$ and $\sin(\theta)$.

  [*Hint*: Use de Moivre's theorem to write $(\cos(\theta) + i \sin(\theta))^3$ in two different ways.]

University of Manchester, UK, 2008

The next question has been slightly edited.

9 Find the argument arg(z) of the complex number $z = -10e^{-i\pi/10}$.

King's College London, UK, 2008

10 a Let $z = 2 - j$ and $w = 3 + 2j$ be two complex numbers. Calculate and simplify each of the following:

$$\bar{z}, \quad z + w, \quad zw, \quad \frac{1}{w}, \quad |z|$$

b Express $z = -2 - 2\sqrt{3}j$ in modulus argument form. Hence find two complex numbers w such that $w^2 = -2 - 2\sqrt{3}j$. Give your answers exactly in modulus-argument form.

University of Aberdeen, UK, 2003

11 a Express the following complex numbers in the form $x + jy$ where x and y are real:

$$(2 + 3j)(1 - j), \quad \frac{2 + 3j}{1 - j}$$

b Let $z = -2 - 2\sqrt{3}j$. Give the modulus and exact (principal) argument of z.

c Express the complex number -25 in modulus-argument form. Hence find all solutions w to the equation

$$w^4 + 25 = 0$$

and mark them on an Argand diagram.

University of Aberdeen, UK, 2004

12 Answer the following questions showing your work.

a Write the complex number $1 - i$ in polar form (in the form $re^{i\theta}$, $r > 0$ and $-\pi \le \theta \le \pi$).

b Express $e^{i\pi}$ in the form $a + bi$, a and b are real.

c Simplify $\left(\frac{1}{\sqrt{2}} + i\frac{1}{\sqrt{2}} \right)^{60}$. (Write as $a + bi$, a and b are real.)

d Compute the cube roots of 8. You may leave your answer in polar form.

Colorado State University, USA, 2006

---

1 Mark on an Argand diagram the points z and $\bar{z}$ (complex conjugate of z) if

a $z = 2 + j3$ **b** $z = 1 - j$ **c** $z = j$

What do you notice?

2 Express the following in $x + jy$ form:

a $(1 + j)^{10}$ **b** $\left(-1 - j\sqrt{3} \right)^7$

c $\dfrac{\cos(5\theta) + j\sin(5\theta)}{\cos(\theta) + j\sin(\theta)}$

3 Find x and y if

a $x + jy = \dfrac{1}{e^{j\theta}}$ **b** $x + jy = \dfrac{1}{1 + e^{j\theta}}$

4 Evaluate the following:

a $\sqrt{2}e^{j\pi/4}$ **b** $-e^{-j\pi/6}$

*5 Let z be any complex number. Show that

$$z\bar{z} = |z|^2$$

6 [mechanics]

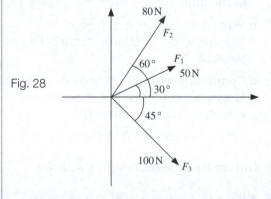

Fig. 28

Figure 28 shows three coplanar forces F_1, F_2 and F_3. Find the resultant magnitude and direction of F where

$$F = (F_1 + F_2 + F_3) - 155.8 \underline{/8.7°}$$

📶 **Questions 7 to 9 are in the field of [*electrical principles*].**

7 In a three-phase supply, $V_R = 240\ V$ and $z = (10 + j10)\ \Omega$. Find the current I_R through z $\left(I_R = \dfrac{V_R}{z}\right)$.

Find I_Y $\left(I_Y = \dfrac{V_Y}{z}\right)$ and I_B $\left(I_B = \dfrac{V_B}{z}\right)$

for the following data:

$V_Y = 240\underline{/120°}\ V$ with $z = (5 + j8)\ \Omega$
$V_B = 240\underline{/240°}\ V$ with $z = (2 + j9)\ \Omega$

The neutral line current, I_N, is given by

$$I_N = I_R + I_Y + I_B$$

Determine I_N.

8 For a series circuit with resistance R, inductance L and capacitance C, the total impedance z is given by

$$z = R + j\omega L + \frac{1}{j\omega C} \quad \text{where}$$

$$\omega = \text{angular frequency}$$

Resonance occurs when $\text{Im}(z) = 0$. Show that the resonant angular frequency, ω_r, is given by $\omega_r = \dfrac{1}{\sqrt{LC}}$.

*9 A cable of length L has the following impedances:

$$z_{oc} = 600e^{j\frac{\pi}{3}}$$
        (Open-circuit impedance)

$$z_{sc} = 400e^{-j\frac{\pi}{3}}$$
        (Short-circuit impedance)

Given that

$$\tanh(\gamma L) = \sqrt{\left(\frac{z_{sc}}{z_{oc}}\right)}$$

where γ is complex (propagation coefficient), show that

$$\tanh(\gamma L) = \frac{1}{\sqrt{6}}(1 - j\sqrt{3})$$

🌬 **Questions 10 to 14 belong to the field of [*control engineering*].**

10 Let $T(s) = \dfrac{K}{s^2 + 4s + K}$ (K is real) be the transfer function of a system. Let $\sigma \pm j\omega$ be the roots of $s^2 + 4s + K = 0$ (poles of the system). The natural angular frequency ω_n satisfies

$$\omega_n^2 = \sigma^2 + \omega^2$$

Find the value of K required for the natural angular frequency $\omega_n = 5$ rad/s.

11 A control system has a transfer function, $\dfrac{\theta_0}{\theta_1}$, given by

$$\frac{\theta_0}{\theta_1} = \frac{1000}{200 + 2j\omega + (j\omega)^2}$$

where ω is the angular velocity.

Determine the gain $\left(\left|\dfrac{\theta_0}{\theta_1}\right|\right)$ and phase $\left(\arg\left(\dfrac{\theta_0}{\theta_1}\right)\right)$ of the function for $\omega = 10$ rad/s.

12 A system is said to be **stable** if all the poles of the transfer function lie within the unit circle, $|z| < 1$. The system is said to be **critically stable** if it has a pole on the unit circle, $|z| = 1$ (Fig. 29).

Fig. 29

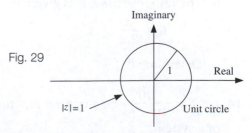

Determine whether the systems with the following transfer functions are stable or critically stable:

a $G(z) = \dfrac{10}{3z^2 + 2z + 1}$

b $G(z) = \dfrac{z - 1}{z^3 + 1}$

(Poles occur where the denominator of the transfer function is zero.)

13 A system with feedback has a transfer function, $G(s)$, given by

$$G(s) = \frac{10}{(s + 1 + j)(s + 1 - j) + 10}$$

(s is complex)

i Show that the roots of the characteristic equation

$(s + 1 + j)(s + 1 - j) + 10 = 0$

are $s = -1 \pm j\sqrt{11}$.

Let $\sigma = \text{Re}(s)$ and $\omega = \text{Im}(s)$. Find σ and ω. For the remaining question use these values of σ and ω.

ii The undamped natural frequency, ω_n, of the system is given by

$$\omega_n = \sqrt{\sigma^2 + \omega^2}$$

Determine ω_n.

iii The damping ratio, ζ, of the system is given by

$$\zeta = \frac{|\sigma|}{\omega_n}$$

Determine ζ.

iv The response time, t_r, is given by

$$t_r = \frac{\pi}{2\omega_n}$$

Determine t_r.

14 Let $T = \dfrac{10}{10 + j\omega}$ be a transfer function of a system.

i Evaluate the gain ($|T|$) and phase, $\arg(T)$, for $\omega = 1$, 10 and 100 rad/s.

ii Sketch, on different axes, the graphs of $|T|$ and $\arg(T)$ as a function of ω.

15 [electronics]   The gain of a phase shifter circuit consisting of an op-amp, with resistance R and capacitance C, is given by $|A|$ where

$$A = \frac{1 - R/X_c}{1 + R/X_c} \text{ where } X_c = \frac{1}{j\omega C}$$

Show that $|A| = 1$.

16 [electronics]   The transfer function, G, of a Butterworth low-pass filter is derived from the relationship

$$G = \frac{1}{1 - s^6}$$

where s is complex.
Find the poles of the filter, that is the values of s where $1 - s^6 = 0$, and label them on an Argand diagram.

17 [acoustics]   The specific acoustic impedance, z, is given by

$$z = \frac{jp\omega}{\left(\dfrac{1}{r} + jk\right)}$$

where r is the radial distance, p is the acoustic pressure and k is the wavenumber. Show that:
Specific acoustic resistance,

$$\text{Re}(z) = \frac{p\omega k r^2}{1 + k^2 r^2}$$

Specific acoustic reactance,

$$\text{Im}(z) = \frac{p\omega r}{1 + k^2 r^2}$$

18 [fluid mechanics]   The complex potential function $w = \phi + j\psi$ for a certain flow is given by

$$w = 3z e^{j\frac{\pi}{3}}$$

Miscellaneous exercise **10 continued**   Solutions at end of book. Complete solutions available at www.palgrave.com/engineering/singh

where $z = x + jy$. Determine the stream, ψ, and potential, ϕ, functions in terms of x and y.

19 We know from 10.19 that

$$\cos(3\theta) + j\sin(3\theta) = \left[\cos(\theta) + j\sin(\theta)\right]^3$$

By expanding the right-hand side, show that

$$\cos(3\theta) = 4\cos^3(\theta) - 3\cos(\theta)$$

and $\sin(3\theta) = 3\sin(\theta) - 4\sin^3(\theta)$

20 Let $z = \cos(\theta) + j\sin(\theta)$. Show that

i $z - \dfrac{1}{z} = 2j\sin(\theta)$

ii $z^3 - \dfrac{1}{z^3} = 2j\sin(3\theta)$

iii $z^5 - \dfrac{1}{z^5} = 2j\sin(5\theta)$

iv By using the binomial theorem, or otherwise, show that

$$\left(z - \frac{1}{z}\right)^5 = \left(z^5 - \frac{1}{z^5}\right) - 5\left(z^3 - \frac{1}{z^3}\right)$$
$$+ 10\left(z - \frac{1}{z}\right)$$

v Show that

$$\sin^5(\theta) = \frac{1}{16}\Big[\sin(5\theta) - 5\sin(3\theta)$$
$$+ 10\sin(\theta)\Big]$$

21 By 10.19 we have

$$\cos(4\theta) + j\sin(4\theta) = \left[\cos(\theta) + j\sin(\theta)\right]^4$$

By expanding the Right-Hand Side and by equating the real and imaginary parts, show that

$$\cos(4\theta) = \cos^4(\theta) - 6\cos^2(\theta)\sin^2(\theta)$$
$$+ \sin^4(\theta)$$

and

$$\sin(4\theta) = 4\cos^3(\theta)\sin(\theta) - 4\cos(\theta)\sin^3(\theta)$$

CHAPTER 11
Matrices

SECTION A **Manipulation of matrices**

By the end of this section you will be able to:
▶ understand the arithmetic of matrices
▶ perform transformations using matrices

The next two chapters (Matrices and Vectors) belong to a field of mathematics called linear algebra.

Linear algebra is a fundamental area of mathematics and is arguably one of the most powerful mathematical tools ever developed. It is a core topic of study within fields as diverse as: business, economics, engineering, physics, computer science, ecology, sociology, demography and genetics.

For an example of linear algebra at work, one needs to look no further than Google's search engine, which relies upon linear algebra to rank the results of a search with respect to relevance.

A1 **Introduction**

Matrices are used in control theory, electrical principles, vibrations, structural analysis, as well as many other fields. Also many engineering problems can be written in terms of simultaneous equations. If there are a large number of equations then in most cases it is easier to use matrices rather than the process of elimination discussed in **Chapter 2.**

Arthur Cayley (1821–95), an English lawyer who became a mathematician, was the first person to develop matrices as we know them today. He graduated in 1842 from Trinity College, Cambridge. In 1858 he published *Memoir on the Theory of Matrices* which contained the definition of a matrix, matrix addition, subtraction, etc. Cayley thought matrices were of no practical use, just a convenient notation. He could not have been more wrong. Matrices are used today in areas including economics, science, engineering, medicine and statistics.

Another application of matrices is in image processing. Many displays form images by lighting up tiny dots, called pixels, on their screens. Figure 1 (on the next page) shows a symbol represented by lighting up the appropriate pixels.

If we let 1 = 'pixel on' and 0 = 'pixel off' then we can represent this in matrix form as

$$\begin{pmatrix} 1 & 1 & 1 & 1 & 1 \\ 0 & 0 & 1 & 0 & 0 \\ 0 & 0 & 1 & 0 & 0 \\ 1 & 1 & 1 & 1 & 1 \end{pmatrix}$$

This is an example of a 4 × 5 (verbally stated as '4 by 5') matrix.

A matrix is an array of numbers enclosed in a bracket. It is used to store information. For example

$\begin{pmatrix} 12 & 6 \\ 3 & 7 \end{pmatrix}$ is a 2 × 2 matrix (2 rows and

2 columns) and is called a square matrix.

Fig. 1

$\begin{pmatrix} 3 & 2 \\ 7 & 6 \\ 5 & 9 \\ 2 & 1 \end{pmatrix}$ is a 4 × 2 matrix (4 rows and 2 columns) and is

not a square matrix.

An example of a 2 × 4 matrix is

$$\begin{pmatrix} 3 & 7 & 5 & 2 \\ 2 & 6 & 9 & 1 \end{pmatrix}$$

Note that when we say 2 × 4 matrix we state the number of **rows first** and then the number of columns. 2 × 4 ('2 by 4') refers to the **size** of the matrix.

Matrices is the plural of matrix. A common notation for matrices is:

$$\begin{array}{cc} & \text{Column 1} \quad \text{Column 2} \\ \text{Row 1} & \begin{pmatrix} a_{11} & a_{12} \\ a_{21} & a_{22} \end{pmatrix} = \mathbf{A} \\ \text{Row 2} & \end{array}$$

where a_{12} is the element in the first row and second column. Generally in this chapter, bold capital letters will denote matrices, as seen above.

A2 Arithmetic of matrices

To add matrices, we add the corresponding locations in each matrix. For example:

$$\begin{pmatrix} 2 & 5 \\ 4 & 1 \end{pmatrix} + \begin{pmatrix} 3 & 9 \\ 5 & 7 \end{pmatrix} = \begin{pmatrix} 2 + 3 & 5 + 9 \\ 4 + 5 & 1 + 7 \end{pmatrix} = \begin{pmatrix} 5 & 14 \\ 9 & 8 \end{pmatrix}$$

Similarly for subtraction, we subtract the corresponding locations in each matrix:

$$\begin{pmatrix} 2 & 5 \\ 4 & 1 \end{pmatrix} - \begin{pmatrix} 3 & 9 \\ 5 & 7 \end{pmatrix} = \begin{pmatrix} 2 - 3 & 5 - 9 \\ 4 - 5 & 1 - 7 \end{pmatrix} = \begin{pmatrix} -1 & -4 \\ -1 & -6 \end{pmatrix}$$

Matrices must be of the **same** size in order to add or subtract.

A scalar is a number which increases or decreases each entry of a matrix. The following are examples of scalar multiplication:

$$10 \begin{pmatrix} 3 & 1 \\ 7 & -9 \end{pmatrix} = \begin{pmatrix} 3 \times 10 & 1 \times 10 \\ 7 \times 10 & -9 \times 10 \end{pmatrix} = \begin{pmatrix} 30 & 10 \\ 70 & -90 \end{pmatrix}$$

and

$$\frac{1}{5} \begin{pmatrix} 20 & -10 \\ 5 & -35 \end{pmatrix} = \begin{pmatrix} 20/5 & -10/5 \\ 5/5 & -35/5 \end{pmatrix} = \begin{pmatrix} 4 & -2 \\ 1 & -7 \end{pmatrix}$$

Multiply each number in the matrix by the scalar. Also

$$7\begin{pmatrix} -1 \\ 5 \\ 3 \\ 9 \end{pmatrix} = \begin{pmatrix} -7 \\ 35 \\ 21 \\ 63 \end{pmatrix}$$

Next we examine the multiplication of matrices.

Row times column

Matrix multiplication is carried out by row times column. For example:

$$(1 \quad 2)\begin{pmatrix} 3 \\ 4 \end{pmatrix} = (1 \times 3) + (2 \times 4) = 11$$

In general we have

$$(a \quad b)\begin{pmatrix} c \\ d \end{pmatrix} = (a \times c) + (b \times d) = ac + bd$$

For 2×2 matrices:

$$\begin{pmatrix} a & b \\ c & d \end{pmatrix}\begin{pmatrix} e & f \\ g & h \end{pmatrix} = \begin{pmatrix} ae + bg & af + bh \\ ce + dg & cf + dh \end{pmatrix}$$

Applying this to the above matrices gives

$$\begin{pmatrix} 2 & 1 \\ 3 & 7 \end{pmatrix} \times \begin{pmatrix} 4 & 9 \\ 5 & 8 \end{pmatrix} = \begin{pmatrix} (2 \times 4) + (1 \times 5) & (2 \times 9) + (1 \times 8) \\ (3 \times 4) + (7 \times 5) & (3 \times 9) + (7 \times 8) \end{pmatrix}$$

$$= \begin{pmatrix} 13 & 26 \\ 47 & 83 \end{pmatrix} \quad ✓$$

The following is **not** how you multiply matrices:

$$\begin{pmatrix} 2 & 1 \\ 3 & 7 \end{pmatrix} \times \begin{pmatrix} 4 & 9 \\ 5 & 8 \end{pmatrix} \neq \begin{pmatrix} 2 \times 4 & 1 \times 9 \\ 3 \times 5 & 7 \times 8 \end{pmatrix} \qquad \text{[Not equal]} \quad ✗$$

We can **only** multiply matrices if the number of **columns** of the **first** matrix **equals** the number of **rows** of the **second** matrix.

$$\begin{array}{cc} & \text{Col. 1} \quad \text{Col. 2} \end{array}$$

$$\begin{array}{c} \text{Row 1} \\ \text{Row 2} \end{array}\begin{pmatrix} 2 & 1 \\ 3 & 7 \end{pmatrix} \times \begin{pmatrix} 4 & 9 \\ 5 & 8 \end{pmatrix} = \begin{pmatrix} \text{Row 1} \times \text{Col. 1} & \text{Row 1} \times \text{Col. 2} \\ \text{Row 2} \times \text{Col. 1} & \text{Row 2} \times \text{Col. 2} \end{pmatrix}$$

Example 1

Evaluate the following:

a $\begin{pmatrix} 2 & 3 \\ -1 & 5 \end{pmatrix}\begin{pmatrix} 2 \\ 1 \end{pmatrix}$
b $(3 \quad 4)\begin{pmatrix} 2 & 3 \\ -1 & 5 \end{pmatrix}$
c $(1 \quad 0)\begin{pmatrix} 3 \\ 4 \end{pmatrix}$

d $\begin{pmatrix} 2 & 3 & 6 \\ 1 & 5 & 7 \end{pmatrix}\begin{pmatrix} 3 & 7 \\ 4 & 2 \\ 1 & 3 \end{pmatrix}$
e $(1 \quad 2 \quad 3)\begin{pmatrix} 4 \\ 5 \\ 6 \end{pmatrix}$
f $\begin{pmatrix} 2 & 3 & 6 \\ 1 & 5 & 7 \end{pmatrix}\begin{pmatrix} 3 & 7 \\ 4 & 2 \end{pmatrix}$

> ## Example 1 continued
>
> ### Solution
>
> Matrix multiplication is row × column.
>
> **a** $\begin{pmatrix} 2 & 3 \\ -1 & 5 \end{pmatrix}\begin{pmatrix} 2 \\ 1 \end{pmatrix} = \begin{pmatrix} (2 \times 2) + (3 \times 1) \\ (-1 \times 2) + (5 \times 1) \end{pmatrix} = \begin{pmatrix} 7 \\ 3 \end{pmatrix}$
>
> **b** $(3 \quad 4)\begin{pmatrix} 2 & 3 \\ -1 & 5 \end{pmatrix} = \left((3 \times 2) + (4 \times (-1)) \quad (3 \times 3) + (4 \times 5)\right)$
>
> $\qquad\qquad = (2 \quad 29)$
>
> **c** $(1 \quad 0)\begin{pmatrix} 3 \\ 4 \end{pmatrix} = \left((1 \times 3) + (0 \times 4)\right) = (3) = 3$
>
> **d** $\begin{pmatrix} 2 & 3 & 6 \\ 1 & 5 & 7 \end{pmatrix}\begin{pmatrix} 3 & 7 \\ 4 & 2 \\ 1 & 3 \end{pmatrix} = \begin{pmatrix} (2 \times 3) + (3 \times 4) + (6 \times 1) & (2 \times 7) + (3 \times 2) + (6 \times 3) \\ (1 \times 3) + (5 \times 4) + (7 \times 1) & (1 \times 7) + (5 \times 2) + (7 \times 3) \end{pmatrix}$
>
> $\qquad\qquad\qquad = \begin{pmatrix} 24 & 38 \\ 30 & 38 \end{pmatrix}$
>
> **e** $(1 \quad 2 \quad 3)\begin{pmatrix} 4 \\ 5 \\ 6 \end{pmatrix} = \left((1 \times 4) + (2 \times 5) + (3 \times 6)\right) = (32) = 32$
>
> **f** $\begin{pmatrix} 2 & 3 & 6 \\ 1 & 5 & 7 \end{pmatrix}\begin{pmatrix} 3 & 7 \\ 4 & 2 \end{pmatrix}$. Since the number of columns of the first matrix is 3 and number
> of rows of the second matrix is 2, we **cannot** multiply these matrices.

A3 Applications to transformations

Computer graphics and games are based on transformations. For example, a computer animated sequence is based on modelling surfaces of connecting triangles. The computer stores the vertices of the triangle in its memory and then certain operations such as rotations, translations, reflections and enlargements are carried out. These operations are stored as matrices.

A transformation is a function like rotation, reflection and translation as shown in Fig. 2:

Fig. 2

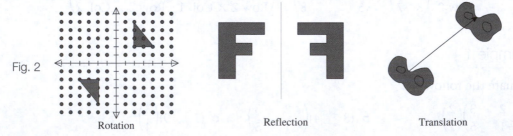

Rotation   Reflection   Translation

The figure shows various two-dimensional transformations. These transformations can be represented by a matrix.

For example, consider the vertices of a triangle $P(2, \ 0)$, $Q(2, \ 3)$ and $R(0, \ 0)$ shown in Fig. 3 on the next page.

This can be represented in matrix form with the co-ordinates of the point P as the entries in the first column, the co-ordinates of the point Q as entries in the second column and the co-ordinates of the point R as entries in the last column. Let **A** denote this matrix then

$$\begin{array}{ccc} P & Q & R \end{array}$$
$$\mathbf{A} = \begin{pmatrix} 2 & 2 & 0 \\ 0 & 3 & 0 \end{pmatrix}$$

Fig. 3

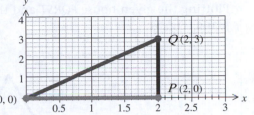

The five corners of a house shown in Fig. 4 can be represented by a matrix **H** given by:

$$\begin{array}{ccccc} P & Q & R & S & T \end{array}$$
$$\mathbf{H} = \begin{pmatrix} 1 & 1 & 1.5 & 2 & 2 \\ 2 & 4 & 5 & 4 & 2 \end{pmatrix}$$

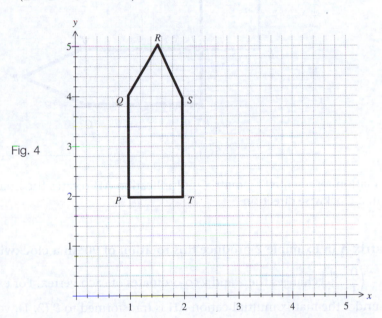

Fig. 4

Example 2

Let $\mathbf{A} = \begin{pmatrix} 0 & 1 \\ 1 & 0 \end{pmatrix}$ and $\mathbf{H} = \begin{pmatrix} 1 & 1 & 1.5 & 2 & 2 \\ 2 & 4 & 5 & 4 & 2 \end{pmatrix}$ be the matrix representing $PQRST$ shown in Fig. 4 above. Determine the image of the house under the transformation represented by the matrix multiplication **AH**. Illustrate and describe what effect this transformation has.

Solution

Let the image of the corners P, Q, R, S and T be denoted by P', Q', R', S' and T' respectively. Then the matrix multiplication **AH** yields:

$$\begin{array}{ccccc} P & Q & R & S & T \end{array} \qquad \begin{array}{ccccc} P' & Q' & R' & S' & T' \end{array}$$
$$\mathbf{AH} = \begin{pmatrix} 0 & 1 \\ 1 & 0 \end{pmatrix}\begin{pmatrix} 1 & 1 & 1.5 & 2 & 2 \\ 2 & 4 & 5 & 4 & 2 \end{pmatrix} = \begin{pmatrix} 2 & 4 & 5 & 4 & 2 \\ 1 & 1 & 1.5 & 2 & 2 \end{pmatrix}$$

Example 2 *continued*

Plotting the given house *PQRST* and the transformed house *P'Q'R'S'T'* on the same plane (Fig. 5):

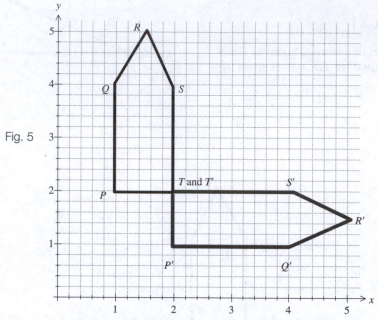

Fig. 5

The transformation represented by matrix multiplication **AH** rotates the house 90° about the point *T* in a clockwise direction.

Why does matrix A in Example 2 produce this rotation of 90° in a clockwise direction?

The matrix $\mathbf{A} = \begin{pmatrix} 0 & 1 \\ 1 & 0 \end{pmatrix}$ changes the *x* and *y* co-ordinates of each vertex. For example, the vertex *P*(1, 2) under the matrix multiplication **AH** is transformed to *P'*(2, 1), vertex *Q*(1, 4) is transformed to *Q'*(4, 1), vertex *R*(1.5, 5) is transformed to *R'*(5, 1.5), etc.

Example 3

Let $\mathbf{A} = \begin{pmatrix} 2 & 2 & 0 \\ 0 & 3 & 0 \end{pmatrix}$ be the matrix representing the triangle *PQR* shown in Fig. 3 on page 365. Determine the image of the triangle under the transformation represented by 2A. Illustrate and describe what effect this transformation has.

Solution

Carrying out the scalar multiplication we have:

$$\begin{array}{ccc} P & Q & R \end{array} \qquad \begin{array}{ccc} P' & Q' & R' \end{array}$$
$$2\mathbf{A} = 2\begin{pmatrix} 2 & 2 & 0 \\ 0 & 3 & 0 \end{pmatrix} = \begin{pmatrix} 4 & 4 & 0 \\ 0 & 6 & 0 \end{pmatrix}$$

Example 3 *continued*

Illustrating the given triangle *PQR* and the transformed triangle *P'Q'R'* (Fig. 6):

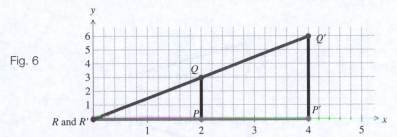

Fig. 6

What effect does scalar multiplication 2A have on the initial triangle *PQR*?

This scalar multiplication increases the length of sides of the triangle by a factor of 2. This means that 2**A** doubles the length of each side of the initial triangle.

? **What would be the image of the initial triangle *PQR* under the scalar multiplication of $\frac{1}{2}$ A?**

It scales each side of the triangle to half its initial size.

Example 4

Let **A** be the matrix representing the triangle with the vertices $P(2, 2)$, $Q(2, 5)$ and $R(4, 2)$.

Write the matrix **A** and determine the image of the triangle under the

transformation represented by matrix multiplication **BA** where $\mathbf{B} = \begin{pmatrix} 1 & 0 \\ 0 & -1 \end{pmatrix}$.

Illustrate and describe what effect this transformation has.

Solution

We are given the vertices of the triangle $P(2, 2)$, $Q(2, 5)$ and $R(4, 2)$, therefore

$$\begin{array}{ccc} P & Q & R \end{array}$$
$$\mathbf{A} = \begin{pmatrix} 2 & 2 & 4 \\ 2 & 5 & 2 \end{pmatrix}$$

Let the images of points of the given triangle *P*, *Q* and *R* under the transformation **BA** be given by *P'*, *Q'* and *R'*. We have

$$\begin{array}{cccccc} & P & Q & R & & P' & Q' & R' \end{array}$$
$$\mathbf{BA} = \begin{pmatrix} 1 & 0 \\ 0 & -1 \end{pmatrix} \begin{pmatrix} 2 & 2 & 4 \\ 2 & 5 & 2 \end{pmatrix} = \begin{pmatrix} 2 & 2 & 4 \\ -2 & -5 & -2 \end{pmatrix}$$

Example 4 *continued*

Plotting the given triangle $P(2, 2)$, $Q(2, 5)$ and $R(4, 2)$ and the transformed triangle $P'(2, -2)$, $Q'(2, -5)$ and $R'(4, -2)$ results in Fig. 7.

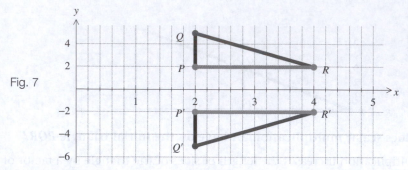

Fig. 7

The effect of transformation **B** is to reflect the triangle PQR in the x (horizontal) axis.

The matrix **B** changes the $+y$ co-ordinate of each vertex to $-y$.

Example 5

Determine the image of the triangle PQR given in **Example 4** under the transformation represented by **CA** where $\mathbf{C} = \begin{pmatrix} 0 & -1 \\ 1 & 0 \end{pmatrix}$.

Illustrate and describe what effect this transformation has.

Solution

The transformation representing the matrix multiplication **CA** is given by

$$\mathbf{CA} = \begin{pmatrix} 0 & -1 \\ 1 & 0 \end{pmatrix} \overset{\begin{matrix} P & Q & R \end{matrix}}{\begin{pmatrix} 2 & 2 & 4 \\ 2 & 5 & 2 \end{pmatrix}} = \overset{\begin{matrix} P' & Q' & R' \end{matrix}}{\begin{pmatrix} -2 & -5 & -2 \\ 2 & 2 & 4 \end{pmatrix}}$$

We can illustrate this transformation as shown in Fig. 8 below:

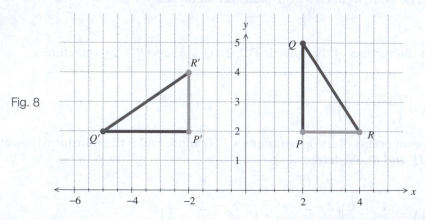

Fig. 8

The matrix multiplication **CA** produces a rotation of 90° anticlockwise about the origin.

We can also combine transformations which are normally called **composite** transformations. Composite transformations such as **CB(A)** means that we **first** carry out the transformation B on A and **then** perform transformation C on the result of **BA**.

Example 6

Determine the image of the triangle *PQR* (represented by matrix A) given in **Example 4** above under the composite transformations represented by

i BC(A)   ii CB(A)

where the matrices **B** and **C** are given in **Examples 4** and **5** above, respectively.

Illustrate and describe what effect each composite transformation has. **What do you notice about your results?**

Solution

i The transformation representing the matrix multiplication **BC(A)** is given by

$$\mathbf{BC(A)} = \begin{pmatrix} 1 & 0 \\ 0 & -1 \end{pmatrix}\begin{pmatrix} 0 & -1 \\ 1 & 0 \end{pmatrix}\begin{pmatrix} 2 & 2 & 4 \\ 2 & 5 & 2 \end{pmatrix}$$

$$= \begin{pmatrix} 1 & 0 \\ 0 & -1 \end{pmatrix}\overbrace{\begin{pmatrix} \overset{P'}{-2} & \overset{Q'}{-5} & \overset{R'}{-2} \\ 2 & 2 & 4 \end{pmatrix}}^{} = \overbrace{\begin{pmatrix} \overset{P''}{-2} & \overset{Q''}{-5} & \overset{R''}{-2} \\ -2 & -2 & -4 \end{pmatrix}}^{}$$

Triangle rotated by 90° anticlockwise as seen in **Example 5**   Reflection in the *x* axis

We can illustrate this composite transformation as shown in Fig. 9a below:

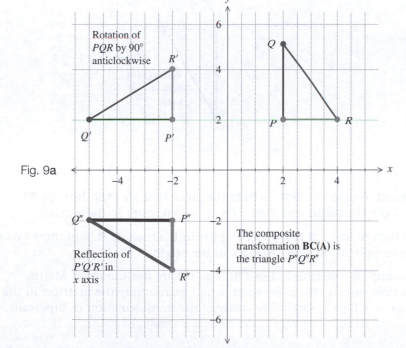

Rotation of *PQR* by 90° anticlockwise

Reflection of *P'Q'R'* in *x* axis

The composite transformation **BC(A)** is the triangle *P"Q"R"*

Fig. 9a

Example 6 *continued*

ii Similarly we have

$$\mathbf{CB(A)} = \begin{pmatrix} 0 & -1 \\ 1 & 0 \end{pmatrix}\begin{pmatrix} 1 & 0 \\ 0 & -1 \end{pmatrix}\begin{pmatrix} 2 & 2 & 4 \\ 2 & 5 & 2 \end{pmatrix}$$

$$= \begin{pmatrix} 0 & -1 \\ 1 & 0 \end{pmatrix} \underbrace{\begin{pmatrix} \overset{P'}{2} & \overset{Q'}{2} & \overset{R'}{4} \\ -2 & -5 & -2 \end{pmatrix}}_{\substack{\text{Reflection of triangle in } x \text{ axis} \\ \text{as seen in } \textbf{Example 4}}} = \underbrace{\begin{pmatrix} \overset{P''}{2} & \overset{Q''}{5} & \overset{R''}{2} \\ 2 & 2 & 4 \end{pmatrix}}_{\text{Rotation of } 90° \text{ anticlockwise}}$$

Figure 9b below illustrates the composite transformations:

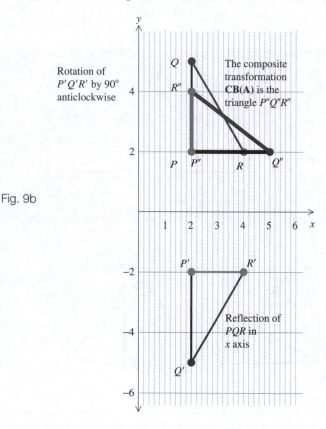

Fig. 9b

BC(A) is the transformation of **first** rotating the given triangle *PQR* through 90° anticlockwise about the origin and then reflecting your result in the *x* axis.

CB(A) is the transformation of **first** reflecting the initial triangle *PQR* in the *x* axis and then rotating through 90° anticlockwise about the origin.

Note that the composite transformations **BC** and **CB** are **not** the same. Matrix multiplication does **not** give the same result when we multiply the matrices in the other order, that is, **BC ≠ CB** [not equal]. The **order** of the transformation is important.

$\mathbf{BC} \neq \mathbf{CB}$:

$$\mathbf{BC} = \begin{pmatrix} 1 & 0 \\ 0 & -1 \end{pmatrix}\begin{pmatrix} 0 & -1 \\ 1 & 0 \end{pmatrix} = \begin{pmatrix} 0 & -1 \\ -1 & 0 \end{pmatrix}$$

$$\mathbf{CB} = \begin{pmatrix} 0 & -1 \\ 1 & 0 \end{pmatrix}\begin{pmatrix} 1 & 0 \\ 0 & -1 \end{pmatrix} = \begin{pmatrix} 0 & 1 \\ 1 & 0 \end{pmatrix}$$

Note also that the matrix representing the transformation that we apply first is the second matrix in the product and vice versa. This is because we apply the transformation to an object whose vertices are the columns of a matrix **A**. So assume **B** is the matrix representing the first transformation and **C** is the matrix representing the second matrix. After applying the first transformation to **A** we have **BA** and then we apply the second transformation to this to get **C(BA)**. Because matrix multiplication is associative, this is the same as **(CB)A**. Consequently the composite transformation is represented by **CB**.

? **Can we apply transformations to three dimensions?** Yes. It is difficult to visualize this in a book but we move on to this in **Section C**.

A4 Computation of matrices

? **Can we write the computation of matrices in compact form?** Yes, for example, let **A** and **B** be matrices, then **A** + **B** means add the matrices **A** and **B**. The computation 2**A** means multiply every entry of the matrix **A** by 2.

? **How can the tedious nature of some of the calculations be avoided?** A software package such as MATLAB can be used for computing matrices. In fact, MATLAB is short for 'Matrix Laboratory'. It is a very useful tool and can be used to eliminate the tedium of lengthy calculations. Many mathematical software packages exist, such as MAPLE and MATHEMATICA, but MATLAB is particularly useful for matrices. In the present release of MATLAB the matrices are entered with square brackets rather than round ones.

For example, in MATLAB to write the matrix $\mathbf{A} = \begin{pmatrix} 1 & 2 \\ 3 & 4 \end{pmatrix}$ we enter **A** = [1 2; 3 4] after the command prompt >>. The semicolon indicates the end of the row. The output shows:

A =

   1 2
   3 4

Example 7

Let

$$\mathbf{A} = \begin{pmatrix} 1 & 2 & 5 \\ 4 & 6 & 9 \end{pmatrix}, \mathbf{B} = \begin{pmatrix} 2 & 5 & 6 \\ 1 & 7 & 2 \\ 9 & 6 & 1 \end{pmatrix} \text{ and } \mathbf{C} = \begin{pmatrix} 5 & 4 & 9 \\ 7 & 4 & 0 \\ 6 & 9 & 8 \end{pmatrix}$$

Compute the following:

a **B** + **C** **b** 3**B** + 2**C** **c** **A** + **B** **d** **AB** **e** **BA**

Example 7 *continued*

?

What do you notice about your results to parts d and e?

a Adding the corresponding entries gives

$$\mathbf{B} + \mathbf{C} = \begin{pmatrix} 2 & 5 & 6 \\ 1 & 7 & 2 \\ 9 & 6 & 1 \end{pmatrix} + \begin{pmatrix} 5 & 4 & 9 \\ 7 & 4 & 0 \\ 6 & 9 & 8 \end{pmatrix} = \begin{pmatrix} 2+5 & 5+4 & 6+9 \\ 1+7 & 7+4 & 2+0 \\ 9+6 & 6+9 & 1+8 \end{pmatrix} = \begin{pmatrix} 7 & 9 & 15 \\ 8 & 11 & 2 \\ 15 & 15 & 9 \end{pmatrix}$$

To use MATLAB enter the matrices **B** and **C** after the command prompt >>. Separate each matrix by a comma. Then enter **B** + **C** and the output should give the same result as above.

b We have

$$3\mathbf{B} + 2\mathbf{C} = 3\begin{pmatrix} 2 & 5 & 6 \\ 1 & 7 & 2 \\ 9 & 6 & 1 \end{pmatrix} + 2\begin{pmatrix} 5 & 4 & 9 \\ 7 & 4 & 0 \\ 6 & 9 & 8 \end{pmatrix}$$

$$= \begin{pmatrix} 3 \times 2 & 3 \times 5 & 3 \times 6 \\ 3 \times 1 & 3 \times 7 & 3 \times 2 \\ 3 \times 9 & 3 \times 6 & 3 \times 1 \end{pmatrix} + \begin{pmatrix} 2 \times 5 & 2 \times 4 & 2 \times 9 \\ 2 \times 7 & 2 \times 4 & 2 \times 0 \\ 2 \times 6 & 2 \times 9 & 2 \times 8 \end{pmatrix}$$ $\begin{bmatrix} \text{Multiplying each} \\ \text{entry of matrix B by} \\ \text{3 and of matrix C by 2} \end{bmatrix}$

$$= \begin{pmatrix} 6 & 15 & 18 \\ 3 & 21 & 6 \\ 27 & 18 & 3 \end{pmatrix} + \begin{pmatrix} 10 & 8 & 18 \\ 14 & 8 & 0 \\ 12 & 18 & 16 \end{pmatrix}$$ $\begin{bmatrix} \text{Simplifying each} \\ \text{entry} \end{bmatrix}$

$$= \begin{pmatrix} 6+10 & 15+8 & 18+18 \\ 3+14 & 21+8 & 6+0 \\ 27+12 & 18+18 & 3+16 \end{pmatrix} = \begin{pmatrix} 16 & 23 & 36 \\ 17 & 29 & 6 \\ 39 & 36 & 19 \end{pmatrix}$$ $\begin{bmatrix} \text{Adding the} \\ \text{corresponding} \\ \text{entries} \end{bmatrix}$

In MATLAB, multiplication is carried out by using the command *. Once the matrices **B** and **C** have been entered you do not need to enter them again. Enter 3 * **B** + 2 * **C** to check the above evaluation.

?

c **A** + **B** is impossible because the matrices are of different sizes. **What sizes are matrix A and matrix B?**

$\mathbf{A} = \begin{pmatrix} 1 & 2 & 5 \\ 4 & 6 & 9 \end{pmatrix}$ is a 2×3 matrix and $\mathbf{B} = \begin{pmatrix} 2 & 5 & 6 \\ 1 & 7 & 2 \\ 9 & 6 & 1 \end{pmatrix}$ is a 3×3 matrix, therefore we

cannot add (or subtract) these matrices.

If we try this in MATLAB we receive a message saying 'Matrix dimensions must agree'.

d We have

$$\mathbf{AB} = \begin{pmatrix} 1 & 2 & 5 \\ 4 & 6 & 9 \end{pmatrix} \times \begin{pmatrix} 2 & 5 & 6 \\ 1 & 7 & 2 \\ 9 & 6 & 1 \end{pmatrix}$$

$$= \begin{pmatrix} (1 \times 2) + (2 \times 1) + (5 \times 9) & (1 \times 5) + (2 \times 7) + (5 \times 6) & (1 \times 6) + (2 \times 2) + (5 \times 1) \\ (4 \times 2) + (6 \times 1) + (9 \times 9) & (4 \times 5) + (6 \times 7) + (9 \times 6) & (4 \times 6) + (6 \times 2) + (9 \times 1) \end{pmatrix}$$

Example 7 *continued*

$$= \begin{pmatrix} 2+2+45 & 5+14+30 & 6+4+5 \\ 8+6+81 & 20+42+54 & 24+12+9 \end{pmatrix} \text{ [Simplifying]}$$

$$= \begin{pmatrix} 49 & 49 & 15 \\ 95 & 116 & 45 \end{pmatrix}$$

e **BA** is impossible because the number of columns in the left-hand matrix, **B**, is **not** equal to the number of rows in the right-hand matrix, **A**. **How many columns does matrix B have?**

$$B = \begin{pmatrix} 2 & 5 & 6 \\ 1 & 7 & 2 \\ 9 & 6 & 1 \end{pmatrix} \text{ has 3 columns. How many rows does matrix A have?}$$

Matrix $A = \begin{pmatrix} 1 & 2 & 5 \\ 4 & 6 & 9 \end{pmatrix}$ has 2 rows. Since the number of columns of matrix **B**, 3, does **not** match the number of rows of matrix **A**, 2, we **cannot** evaluate **BA**. As obvious as this may seem, many students get this wrong. Watch out for this!

If we were to try this in MATLAB we would receive an error message saying 'Inner matrix dimensions must agree'.

Notice again that the matrix multiplication **AB** does **not** equal the matrix multiplication **BA**. In fact, as we have just established, in this case **BA** cannot even be evaluated, while **AB** is computed in part **d** above. So matrix multiplication is **not** the same as multiplying two real numbers. In matrix multiplication the **order** of the multiplication **does** matter.

SUMMARY

When multiplying two matrices it is row × column. If **A** and **B** represent matrices then in general

$$AB \neq BA$$

Exercise **11(a)**

Solutions at end of book. Complete solutions available at www.palgrave.com/engineering/singh

You may like to check your numerical solutions by using MATLAB or any other appropriate software. Some calculators can also perform matrix calculations.

1 For $A = \begin{pmatrix} 1 & 2 \\ 3 & -1 \end{pmatrix}$, $B = \begin{pmatrix} 6 & -1 \\ 5 & 3 \end{pmatrix}$ and

$C = \begin{pmatrix} -1 \\ 1 \end{pmatrix}$, evaluate the following:

a A + B   **b** B + A   **c** A + A + A

d 3A + 2B   **e** A + C   **f** B + C

g AC   **h** BC   **i** 5A − 7BC

j 3AC − 2BC

2 Write down the matrix representing the ABCDEFGH in Fig. 10 below:

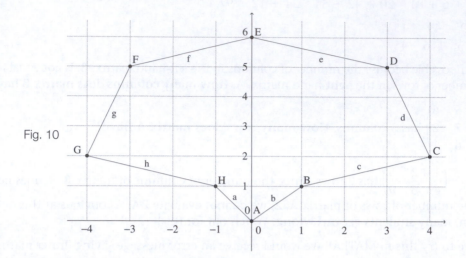

Fig. 10

3 Let the top of a table be given by the co-ordinates $P(1, 2)$, $Q(2, 2)$, $R(2, 4)$ and $S(1, 4)$. Write down the matrix **A** which represents this table top *PQRS*.

Let

$$\mathbf{B} = \begin{pmatrix} 4 & 4 & 4 & 4 \\ 0 & 0 & 0 & 0 \end{pmatrix}, \mathbf{C} = \begin{pmatrix} 0 & 1 \\ -1 & 0 \end{pmatrix} \text{ and}$$

$$\mathbf{D} = \begin{pmatrix} -1 & 0 \\ 0 & 1 \end{pmatrix}$$

Determine the image of the table top under the following transformation and illustrate the effect that each transformation has:

a A − B   **b** 3A   **c** CA   **d** DA
e CD(A)   **f** DC(A)

4 Let $\mathbf{A} = \begin{pmatrix} 1 & -1 \\ 1 & -1 \end{pmatrix}$ and find $\mathbf{A}^2 = \mathbf{A} \times \mathbf{A}$.

[A^2 in MATLAB is evaluated by the command A^2.] Comment upon your result. Can you get the same result when multiplying two non-zero real numbers?

5 Evaluate $\begin{pmatrix} 5 & -1 & -2 \\ 10 & -2 & -4 \\ 15 & -3 & -6 \end{pmatrix}\begin{pmatrix} 1 & 1 & 3 \\ 1 & -1 & -1 \\ 2 & 3 & 8 \end{pmatrix}$.

Comment upon your result. Can you get the same result when multiplying two non-zero real numbers?

SECTION B **Applications**

By the end of this section you will be able to:
► apply matrices to solve equations
► evaluate the determinant of a 2 × 2 matrix
► obtain the inverse matrix

B1 Determinant of a matrix

Example 8

Consider a triangle given by the co-ordinates $P(0, 0)$, $Q(2, 0)$ and $R(0, 3)$ (the shaded triangle shown in Fig. 11 below). Let the matrix **A** represent this triangle PQR and determine the image of this triangle under the transformation given by **BA** where

$$\mathbf{B} = \begin{pmatrix} 2 & 0 \\ 3 & 4 \end{pmatrix}.$$

By illustrating this transformation, determine the areas of the triangle PQR and the transformed triangle $P'Q'R'$. **How does this transformation B change the size of the area?**

Solution

We are given co-ordinates $P(0, 0)$, $Q(2, 0)$ and $R(0, 3)$, therefore $\mathbf{A} = \begin{pmatrix} 0 & 2 & 0 \\ 0 & 0 & 3 \end{pmatrix}$.

Evaluating the matrix multiplication **BA**:

$$\begin{matrix} & P & Q & R & & P' & Q' & R' \\ \mathbf{BA} = & \begin{pmatrix} 2 & 0 \\ 3 & 4 \end{pmatrix} & \begin{pmatrix} 0 & 2 & 0 \\ 0 & 0 & 3 \end{pmatrix} & = & \begin{pmatrix} 0 & 4 & 0 \\ 0 & 6 & 12 \end{pmatrix} \end{matrix}$$

These triangles are plotted in Fig. 11 below:

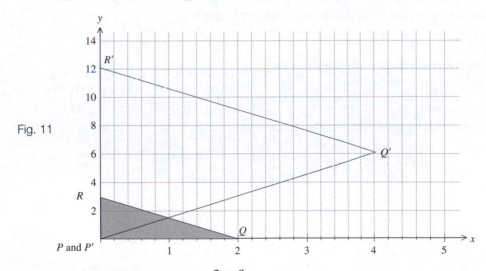

Fig. 11

The area of shaded triangle $PQR = \dfrac{2 \times 3}{2} = 3$ and the area of the large triangle

$P'Q'R' = \dfrac{12 \times 4}{2} = 24$.

The transformation **B** increases the area by a factor of 8 (because, of course, 24/3 = 8).

The factor 8 is called the **determinant** of the matrix **B** and describes the increase in size that **B** imposes on **A**. The determinant of the matrix **B** is calculated by:

$$\begin{pmatrix} 2 & 0 \\ 3 & 4 \end{pmatrix} = (2 \times 4) - (3 \times 0) = 8$$

In general, the determinant of a 2 × 2 matrix, $\begin{pmatrix} a & b \\ c & d \end{pmatrix}$, is given by

11.1 $\qquad \det\begin{pmatrix} a & b \\ c & d \end{pmatrix} = ad - cb$

$\det\begin{pmatrix} a & b \\ c & d \end{pmatrix}$ is used to represent the determinant. Another notation is $\begin{vmatrix} a & b \\ c & d \end{vmatrix}$. Note that the

determinant is a number, **not** a matrix. **Why?** Because the transformation $\begin{pmatrix} a & b \\ c & d \end{pmatrix}$ changes

area by a **factor** $(ad - bc)$. For example

$$\det\begin{pmatrix} 5 & 4 \\ -2 & 1 \end{pmatrix} = (5 \times 1) - (-2 \times 4) = 13$$

The determinant is also used to find the inverse of a matrix which will be defined in Section **B3**.

B2 Identity matrix

Example 9

Determine the image of the triangle PQR (represented by **A**) in **Example 8** on the previous page

under the transformations represented by the matrix multiplication **IA** where $\mathbf{I} = \begin{pmatrix} 1 & 0 \\ 0 & 1 \end{pmatrix}$.
Illustrate and describe what effect this transformation has.

Solution

Carrying out the matrix multiplication **IA** gives

$$\mathbf{IA} = \begin{pmatrix} 1 & 0 \\ 0 & 1 \end{pmatrix}\begin{pmatrix} 0 & 2 & 0 \\ 0 & 0 & 3 \end{pmatrix} = \begin{pmatrix} 0 & 2 & 0 \\ 0 & 0 & 3 \end{pmatrix}$$

We can plot this as shown in the Fig. 12 below:

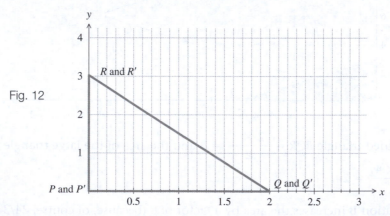

Fig. 12

The transformation given by the matrix multiplication **IA** does **not** change the given triangle PQR. This means that the triangle PQR remains fixed under this transformation.

Now evaluate

$$\begin{pmatrix} 5 & 3 \\ 1 & 2 \end{pmatrix}\begin{pmatrix} 1 & 0 \\ 0 & 1 \end{pmatrix} \text{ and } \begin{pmatrix} 6 & 7 \\ 2 & 3 \end{pmatrix}\begin{pmatrix} 1 & 0 \\ 0 & 1 \end{pmatrix}$$

? **What do you notice about your results?**

$$\begin{pmatrix} 5 & 3 \\ 1 & 2 \end{pmatrix}\begin{pmatrix} 1 & 0 \\ 0 & 1 \end{pmatrix} = \begin{pmatrix} 5 & 3 \\ 1 & 2 \end{pmatrix} \text{ and } \begin{pmatrix} 6 & 7 \\ 2 & 3 \end{pmatrix}\begin{pmatrix} 1 & 0 \\ 0 & 1 \end{pmatrix} = \begin{pmatrix} 6 & 7 \\ 2 & 3 \end{pmatrix}$$

The resultant matrix in each case is unchanged by this operation. In general, we have:

$$\begin{pmatrix} a & b \\ c & d \end{pmatrix}\begin{pmatrix} 1 & 0 \\ 0 & 1 \end{pmatrix} = \begin{pmatrix} a & b \\ c & d \end{pmatrix}$$

where $\begin{pmatrix} 1 & 0 \\ 0 & 1 \end{pmatrix}$ is called the identity matrix and is denoted by **I**, $\mathbf{I} = \begin{pmatrix} 1 & 0 \\ 0 & 1 \end{pmatrix}$.

The identity matrix, **I**, for matrices is analogous to the number 1 for numbers.

B3 Inverse of a matrix

Let **A** be the general 2 × 2 (square) matrix:

$$\mathbf{A} = \begin{pmatrix} a & b \\ c & d \end{pmatrix}$$

The inverse matrix of **A** is denoted by $\mathbf{A}^{-1}$ and has the property

11.2 $\mathbf{A}^{-1} \times \mathbf{A} = \mathbf{I}$

The notation $\mathbf{A}^{-1}$ is the inverse matrix and not A to the index -1. $\left[\mathbf{A}^{-1} \neq \dfrac{1}{\mathbf{A}}, \text{ not equal}\right]$

In transformations, this means that if the transformation **A** is applied to a particular figure then the transformation $\mathbf{A}^{-1}$ **undoes** A so the net result of the combined transformation $\mathbf{A}^{-1}\mathbf{A}$ is to leave the figure untouched, and this is why $\mathbf{A}^{-1}\mathbf{A} = \mathbf{I}$.

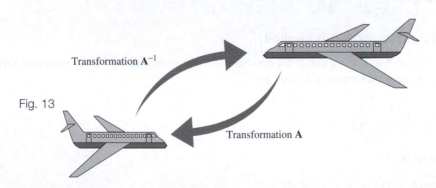

Transformation $\mathbf{A}^{-1}$

Fig. 13

Transformation **A**

In a computer game if a transformation **A** is applied to an object as shown in Fig. 13, then the transformation $\mathbf{A}^{-1}$ undoes A and so the net result of $\mathbf{A}^{-1}\mathbf{A} = \mathbf{I}$ is to leave the object unchanged.

If the determinant of A is equal to 1 then A^{-1} is given by

11.3 $A^{-1} = \begin{pmatrix} d & -b \\ -c & a \end{pmatrix}$ [provided that det A = 1]

Remember this only works if det A = 1.

Example 10

For $A = \begin{pmatrix} 3 & 7 \\ 2 & 5 \end{pmatrix}$, find A^{-1}.

Solution

First we obtain the determinant by using

11.1 $\det\begin{pmatrix} a & b \\ c & d \end{pmatrix} = ad - cb$

$\det\begin{pmatrix} 3 & 7 \\ 2 & 5 \end{pmatrix} = (3 \times 5) - (2 \times 7) = 1$

Since the determinant of **A** is 1 we can find A^{-1} by using   11.3 , that is, interchanging numbers 3 and 5 and putting a negative sign in front of the other numbers:

$A^{-1} = \begin{pmatrix} 5 & -7 \\ -2 & 3 \end{pmatrix}$

We can confirm the result by checking the matrix multiplication $A^{-1}A = I$:

$$A^{-1}A = \begin{pmatrix} 5 & -7 \\ -2 & 3 \end{pmatrix}\begin{pmatrix} 3 & 7 \\ 2 & 5 \end{pmatrix} = \begin{pmatrix} 1 & 0 \\ 0 & 1 \end{pmatrix} = I$$

? **What are the limitations of this method?**

The determinant of **A** has to equal 1. Of course there are going to be many matrices where the determinant does **not** equal 1. Consider

$A = \begin{pmatrix} 11 & 4 \\ 7 & 3 \end{pmatrix}$

The determinant is

$\det A = (11 \times 3) - (7 \times 4) = 5$

Therefore we **cannot** use $A^{-1} = \begin{pmatrix} d & -b \\ -c & a \end{pmatrix}$ to find A^{-1}.

? **However if we try using this, what matrix do we obtain?**

Call this matrix **B**. Interchanging the numbers 3 and 11 and placing a negative sign in front of the other numbers:

$$\mathbf{B} = \begin{pmatrix} 3 & -4 \\ -7 & 11 \end{pmatrix}$$

Multiplying the two matrices gives

$$\mathbf{BA} = \begin{pmatrix} 3 & -4 \\ -7 & 11 \end{pmatrix}\begin{pmatrix} 11 & 4 \\ 7 & 3 \end{pmatrix} = \begin{pmatrix} 5 & 0 \\ 0 & 5 \end{pmatrix}$$

Of course $\begin{pmatrix} 5 & 0 \\ 0 & 5 \end{pmatrix} \neq \begin{pmatrix} 1 & 0 \\ 0 & 1 \end{pmatrix} = \mathbf{I}$ but it can be made into the identity matrix.

? **How?**

By multiplying with a scalar of 1/5:

$$\frac{1}{5}\begin{pmatrix} 5 & 0 \\ 0 & 5 \end{pmatrix} = \begin{pmatrix} 1 & 0 \\ 0 & 1 \end{pmatrix} = \mathbf{I}$$

? **Where did the number 5 appear previously in this discussion?**

It is the determinant of the matrix **A**. Since $\frac{1}{5}$ **B** multiplied by **A** gives the identity matrix, so $\frac{1}{5}$ **B** must be the inverse of **A**. Thus

$$\mathbf{A}^{-1} = \frac{1}{5}\mathbf{B} = \frac{1}{5}\begin{pmatrix} 3 & -4 \\ -7 & 11 \end{pmatrix}$$

This means that when we apply the transformation **B** on **A** the area is increased by a factor of 5 (because det**B** = 5) and so to find the inverse we need to divide by 5.

In general, if $\mathbf{A} = \begin{pmatrix} a & b \\ c & d \end{pmatrix}$ then the inverse matrix, $\mathbf{A}^{-1}$, is defined as

11.4 $\mathbf{A}^{-1} = \dfrac{1}{\det\mathbf{A}}\begin{pmatrix} d & -b \\ -c & a \end{pmatrix} = \dfrac{1}{ad - cb}\begin{pmatrix} d & -b \\ -c & a \end{pmatrix}$ provided that det**A** ≠ 0

? If the determinant of **A** is zero then matrix **A** does not have an inverse. **Why not?**
Because the det of inverse matrix $\mathbf{A}^{-1}$ has 1/det**A** and if det**A** = 0 then we cannot have $\mathbf{A}^{-1}$. We can demonstrate this by using transformations.

Consider the image of the triangle *PQR* (represented by matrix A) given in **Example 8** on page 575, under the transformation **BA** where $\mathbf{B} = \begin{pmatrix} 3 & 2 \\ 6 & 4 \end{pmatrix}$. Multiplying the matrices we have:

$$\begin{array}{ccccccc} & P & Q & R & P' & Q' & R' \end{array}$$
$$\mathbf{BA} = \begin{pmatrix} 3 & 2 \\ 6 & 4 \end{pmatrix}\begin{pmatrix} 0 & 2 & 0 \\ 0 & 0 & 3 \end{pmatrix} = \begin{pmatrix} 0 & 6 & 6 \\ 0 & 12 & 12 \end{pmatrix}$$

The transformation **B** on **A** is illustrated in Fig. 14:

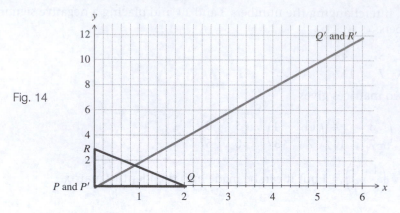

Fig. 14

Under the transformation **B** the triangle represented by matrix **A** becomes a line $P'Q'R'$ which means it has **no** area (Fig. 14). The transformation **B** collapses the area of the triangle to zero so that matrix **B** **cannot** have an inverse because there **no** way back to the original object. The area is increased by a factor of **zero**, therefore the determinant of **B** is zero $(\det(\mathbf{B}) = 0)$ and we can confirm this by evaluating using the formula:

$$\det\begin{pmatrix} 3 & 2 \\ 6 & 4 \end{pmatrix} = (3 \times 4) - (6 \times 2) = 0$$

Example 11

Let

$$\mathbf{A} = \begin{pmatrix} 7 & 9 \\ 5 & 7 \end{pmatrix}$$

Find $\mathbf{A}^{-1}$.

Solution

First we have to find $\det\begin{pmatrix} 7 & 9 \\ 5 & 7 \end{pmatrix}$. By

11.1 $$\det\begin{pmatrix} a & b \\ c & d \end{pmatrix} = ad - cb$$

$$\det\begin{pmatrix} 7 & 9 \\ 5 & 7 \end{pmatrix} = (7 \times 7) - (5 \times 9) = 4$$

With $\det\mathbf{A} = 4$:

$$\mathbf{A}^{-1} = \frac{1}{\det\mathbf{A}}\begin{pmatrix} d & -b \\ -c & a \end{pmatrix}$$

$$= \frac{1}{4}\begin{pmatrix} 7 & -9 \\ -5 & 7 \end{pmatrix}$$

$$= \begin{pmatrix} 7/4 & -9/4 \\ -5/4 & 7/4 \end{pmatrix}$$

Example 11 *continued*

To check that $\mathbf{A}^{-1}$ is indeed the inverse matrix of $\mathbf{A}$, we carry out the following matrix multiplication:

$$\mathbf{A}^{-1} \times \mathbf{A} = \frac{1}{4}\begin{pmatrix} 7 & -9 \\ -5 & 7 \end{pmatrix}\begin{pmatrix} 7 & 9 \\ 5 & 7 \end{pmatrix}$$

$$= \frac{1}{4}\begin{pmatrix} 4 & 0 \\ 0 & 4 \end{pmatrix}$$

$$= \begin{pmatrix} 1 & 0 \\ 0 & 1 \end{pmatrix} = \mathbf{I}$$

Thus $\mathbf{A}^{-1}$ is the inverse matrix of $\mathbf{A}$.

It's good practice to check your inverse matrix.

B4 Solving linear equations

Many problems in engineering and science are solved by using systems of simultaneous linear equations. Linear equations are equations in which the variables have an index of 1 or 0. For example, if you apply Hooke's law to a spring then this leads to a system of linear equations. If you apply Kirchhoff's law to a circuit which only contains resistors then the result is a set of linear equations. Linear equations arise in many different fields such as engineering, physics, computer science, technology, business and economics.

Generally a complex scientific or engineering problem can be reduced to a linear system of simultaneous equations.

We can use matrices to solve simultaneous equations, as the following example shows.

Example 12 *electrical principles*

By applying Kirchhoff's law to a circuit we obtain the following equations:

$$7i_1 + 9i_2 = 3$$
$$5i_1 + 7i_2 = 1$$

where i_1 and i_2 represent currents. Find the values of i_1 and i_2.

Solution

The product of matrices is row × column so the above equations can be written in matrix form as

$$\begin{pmatrix} 7 & 9 \\ 5 & 7 \end{pmatrix}\begin{pmatrix} i_1 \\ i_2 \end{pmatrix} = \begin{pmatrix} 3 \\ 1 \end{pmatrix}$$

Note that if we multiply the contents of the first two sets of brackets, using the row by column method of matrix multiplication, we get back

Example 12 *continued*

$$7i_1 + 9i_2 = 3$$
$$5i_1 + 7i_2 = 1$$

Writing equations in this format is why the multiplication rule of matrices is useful.

Let $\mathbf{A} = \begin{pmatrix} 7 & 9 \\ 5 & 7 \end{pmatrix}$. Therefore we have

$$\boxed{\quad * \quad} \qquad \mathbf{A}\begin{pmatrix} i_1 \\ i_2 \end{pmatrix} = \begin{pmatrix} 3 \\ 1 \end{pmatrix}$$

Since we want to find i_1 and i_2 we need to obtain

$$\begin{pmatrix} i_1 \\ i_2 \end{pmatrix} = \begin{pmatrix} ? \\ ? \end{pmatrix}$$

We need to remove the A from the Left-Hand Side. How?

Multiply both sides of $\boxed{*}$ by the inverse matrix, $\mathbf{A}^{-1}$, which gives

$$\underbrace{\mathbf{A}^{-1}\mathbf{A}}_{=\mathbf{I}}\begin{pmatrix} i_1 \\ i_2 \end{pmatrix} = \mathbf{A}^{-1}\begin{pmatrix} 3 \\ 1 \end{pmatrix}$$
$$\mathbf{I}\begin{pmatrix} i_1 \\ i_2 \end{pmatrix} = \mathbf{A}^{-1}\begin{pmatrix} 3 \\ 1 \end{pmatrix}$$

Remember that **I** is the identity matrix, so on the Left-Hand Side we only have

$\begin{pmatrix} i_1 \\ i_2 \end{pmatrix}$. Thus $\begin{pmatrix} i_1 \\ i_2 \end{pmatrix} = \mathbf{A}^{-1}\begin{pmatrix} 3 \\ 1 \end{pmatrix}$.

We need to find the inverse matrix, $\mathbf{A}^{-1}$, using

11.4
$$\begin{pmatrix} a & b \\ c & d \end{pmatrix}^{-1} = \frac{1}{ad - cb}\begin{pmatrix} d & -b \\ -c & a \end{pmatrix}$$

We have

$$\mathbf{A}^{-1} = \begin{pmatrix} 7 & 9 \\ 5 & 7 \end{pmatrix}^{-1} = \frac{1}{(7 \times 7) - (5 \times 9)}\begin{pmatrix} 7 & -9 \\ -5 & 7 \end{pmatrix} = \frac{1}{4}\begin{pmatrix} 7 & -9 \\ -5 & 7 \end{pmatrix}$$

Substituting this value for $\mathbf{A}^{-1}$ into the above $\begin{pmatrix} i_1 \\ i_2 \end{pmatrix} = \mathbf{A}^{-1}\begin{pmatrix} 3 \\ 1 \end{pmatrix}$ gives

$$\begin{pmatrix} i_1 \\ i_2 \end{pmatrix} = \frac{1}{4}\begin{pmatrix} 7 & -9 \\ -5 & 7 \end{pmatrix}\begin{pmatrix} 3 \\ 1 \end{pmatrix} = \frac{1}{4}\begin{pmatrix} (7 \times 3) + (-9 \times 1) \\ (-5 \times 3) + (7 \times 1) \end{pmatrix} = \frac{1}{4}\begin{pmatrix} 12 \\ -8 \end{pmatrix}$$

Multiplying each element of the last matrix by 1/4:

$$\begin{pmatrix} i_1 \\ i_2 \end{pmatrix} = \begin{pmatrix} 12/4 \\ -8/4 \end{pmatrix} = \begin{pmatrix} 3 \\ -2 \end{pmatrix}$$

Hence $i_1 = 3$ A (amp) and $i_2 = -2$ A. To check this solution is correct we substitute these into our given equations:

$$7i_1 + 9i_2 = 3 \qquad\qquad 7(3) + 9(-2) = 3$$
$$5i_1 + 7i_2 = 1 \qquad\qquad 5(3) + 7(-2) = 1$$

Normally an $n \times 1$ matrix is called a **column vector** and is denoted by a bold small case letter (n is any positive whole number).

The above equations are examples of linear equations

$$7i_1 + 9i_2 = 3$$
$$5i_1 + 7i_2 = 1$$

and can be written in matrix form as

$$\mathbf{Au} = \mathbf{b}$$

where $\mathbf{A} = \begin{pmatrix} 7 & 9 \\ 5 & 7 \end{pmatrix}$, $\mathbf{u} = \begin{pmatrix} i_1 \\ i_2 \end{pmatrix}$ and $\mathbf{b} = \begin{pmatrix} 3 \\ 1 \end{pmatrix}$. (In this case $\mathbf{u}$ and $\mathbf{b}$ are column vectors.)

For the general case

$$\mathbf{Au} = \mathbf{b}$$

where $\mathbf{A}$ is a 2×2 matrix with inverse $\mathbf{A}^{-1}$, $\mathbf{u} = \begin{pmatrix} x \\ y \end{pmatrix}$ (x and y are the variables whose values we are trying to find) and $\mathbf{b}$ is the value of the equations. To solve these equations we need

to find the values of x and y which satisfy these equations, hence we need to find $\mathbf{u} = \begin{pmatrix} x \\ y \end{pmatrix}$.
This can be determined by

11.5	$\mathbf{u} = \mathbf{A}^{-1}\mathbf{b}$

To solve simultaneous equations we can also use the following algorithm:

1 Write the given equations in matrix form: $\mathbf{Au} = \mathbf{b}$.
2 Find $\mathbf{A}^{-1}$.
3 The solution $\mathbf{u}$ is given by $\mathbf{u} = \mathbf{A}^{-1}\mathbf{b}$.

Example 13

Solve
$$2x + 3y = 4$$
$$10x + 4y = 9$$

Solution

Following the above algorithm.

1 We write the given simultaneous equations in matrix form:

$$\begin{pmatrix} 2 & 3 \\ 10 & 4 \end{pmatrix}\begin{pmatrix} x \\ y \end{pmatrix} = \begin{pmatrix} 4 \\ 9 \end{pmatrix}$$

2 Let $\mathbf{A} = \begin{pmatrix} 2 & 3 \\ 10 & 4 \end{pmatrix}$ so we need to find $\mathbf{A}^{-1}$ by using

11.4	$\begin{pmatrix} a & b \\ c & d \end{pmatrix}^{-1} = \dfrac{1}{ad - cb}\begin{pmatrix} d & -b \\ -c & a \end{pmatrix}$

$$\begin{pmatrix} 2 & 3 \\ 10 & 4 \end{pmatrix}^{-1} = \frac{1}{8 - 30}\begin{pmatrix} 4 & -3 \\ -10 & 2 \end{pmatrix} = -\frac{1}{22}\begin{pmatrix} 4 & -3 \\ -10 & 2 \end{pmatrix}$$

Example 13 *continued*

3 Hence

$$\begin{pmatrix} x \\ y \end{pmatrix} = A^{-1} \begin{pmatrix} 4 \\ 9 \end{pmatrix}$$

$$= -\frac{1}{22} \begin{pmatrix} 4 & -3 \\ -10 & 2 \end{pmatrix} \begin{pmatrix} 4 \\ 9 \end{pmatrix} = -\frac{1}{22} \begin{pmatrix} -11 \\ -22 \end{pmatrix} = \begin{pmatrix} 1/2 \\ 1 \end{pmatrix}$$

Therefore $x = 1/2$, $y = 1$. Check for yourself that $x = 1/2$, $y = 1$ do indeed satisfy the given simultaneous equations.

SUMMARY

The determinant of a matrix $\begin{pmatrix} a & b \\ c & d \end{pmatrix}$ is defined as

11.1 $\det \begin{pmatrix} a & b \\ c & d \end{pmatrix} = ad - cb$

The inverse matrix, A^{-1}, has the property

11.2 $A^{-1}A = I$

where $I = \begin{pmatrix} 1 & 0 \\ 0 & 1 \end{pmatrix}$ is the identity matrix.

The inverse matrix, A^{-1}, of $A = \begin{pmatrix} a & b \\ c & d \end{pmatrix}$ is defined as

11.4 $A^{-1} = \dfrac{1}{ad - cb} \begin{pmatrix} d & -b \\ -c & a \end{pmatrix}$ $(\det A \neq 0)$

Solving simultaneous equations of the form

$$Au = b$$

is given by

11.5 $u = A^{-1}b$

Exercise **11(b)**

Solutions at end of book. Complete solutions available at www.palgrave.com/engineering/singh

1 Compute detA and detB for the following:

a $A = \begin{pmatrix} 1 & 3 \\ 5 & 7 \end{pmatrix}$, $B = \begin{pmatrix} 1 & 5 \\ 3 & 7 \end{pmatrix}$

b $A = \begin{pmatrix} -1 & 2 \\ 5 & 3 \end{pmatrix}$, $B = \begin{pmatrix} -1 & 5 \\ 2 & 3 \end{pmatrix}$

c $A = \begin{pmatrix} \cos(-\pi) & \cos(\pi) \\ \sin(-\pi) & \sin(\pi) \end{pmatrix}$,

   $B = \begin{pmatrix} \cos(-\pi) & \sin(-\pi) \\ \cos(\pi) & \sin(\pi) \end{pmatrix}$

What do you notice about matrices A and B and your results for detA and detB?

Exercise **11(b) continued**

Solutions at end of book. Complete solutions available at www.palgrave.com/engineering/singh

2 For $A = \begin{pmatrix} a & b \\ c & d \end{pmatrix}$ and $B = \begin{pmatrix} a & c \\ b & d \end{pmatrix}$, show that $\det A = \det B$.

3 Evaluate **i** $\det A$ **ii** $\det B$ **iii** $\det A \times \det B$ **iv** $\det(AB)$ for the following:

a $A = \begin{pmatrix} 1 & 3 \\ 5 & 6 \end{pmatrix}$, $B = \begin{pmatrix} 3 & 7 \\ -1 & 5 \end{pmatrix}$

b $A = \begin{pmatrix} -1 & 10 \\ 170 & 1.5 \end{pmatrix}$, $B = \begin{pmatrix} -30 & -9.3 \\ 61 & -1.9 \end{pmatrix}$

c $A = \begin{pmatrix} 5 & -3 \\ 2.2 & -5.6 \end{pmatrix}$, $B = \begin{pmatrix} -7.1 & -2.1 \\ -3.5 & -12.2 \end{pmatrix}$

What do you notice about your results to parts **iii** and **iv**.

4 Find the inverse of the following matrices:

a $A = \begin{pmatrix} 9 & 2 \\ 13 & 3 \end{pmatrix}$ **b** $A = \begin{pmatrix} 17 & 7 \\ 12 & 5 \end{pmatrix}$

c $A = \begin{pmatrix} 1 & 0 \\ 0 & 1 \end{pmatrix}$

5 Find the inverse matrix of the following:

a $A = \begin{pmatrix} 5 & 4 \\ 3 & 1 \end{pmatrix}$ **b** $A = \begin{pmatrix} 3 & 6 \\ 7 & 8 \end{pmatrix}$

c $A = \begin{pmatrix} 7 & 1 \\ 14 & 2 \end{pmatrix}$

The next two questions are in the field of [*electrical principles*].

6 We obtain the following equations from an electrical circuit:

$$2i_1 + 3i_2 = 5$$
$$6i_1 + 7i_2 = 10$$

where i_1 and i_2 represent currents. Write these equations in matrix form and solve for i_1 and i_2.

*7 By applying Kirchhoff's law to a circuit we obtain the following equations:

$$30i_1 - 10i_2 = 12$$
$$-10i_1 + 35i_2 = 5$$

where i_1 and i_2 represent currents. Find the exact values of i_1 and i_2.

8 Solve the following equations:

a $\quad x + 3y = 14$
$$2x - 3y = -8$$
b $-2x - 5y = -3$
$$-5x - 2y = 3$$
c $\quad 4x + 6y = -1$
$$7x - \frac{3}{4}y = 2$$

SECTION C **3 × 3 matrices**

By the end of this section you will be able to:
▶ evaluate the determinant of a 3 × 3 matrix
▶ find the inverse of a 3 × 3 matrix
▶ solve equations using the inverse matrix method

This section may seem rather abstract compared to the previous two because it is much harder to visualize three-dimensional transformations.

C1 Determinants

If

$$A = \begin{pmatrix} a & b & c \\ d & e & f \\ g & h & i \end{pmatrix}$$

is a general 3×3 matrix then

11.6 $\det A = a\left[\det\begin{pmatrix} e & f \\ h & i \end{pmatrix}\right] - b\left[\det\begin{pmatrix} d & f \\ g & i \end{pmatrix}\right] + c\left[\det\begin{pmatrix} d & e \\ g & h \end{pmatrix}\right]$

The determinant of a 3×3 matrix is the increase in **volume** associated with three-dimensional transformation.

11.6 gives the determinant of a 3×3 matrix. The simplest way of evaluating the above determinant is **not** to remember all the details of 11.6 but to use the following procedure:

1 Delete the row and column containing the entry a:

$$\begin{pmatrix} (a) & & \\ & e & f \\ & h & i \end{pmatrix}$$

then evaluate the determinant of the remaining matrix $\begin{pmatrix} e & f \\ h & i \end{pmatrix}$.

2 Delete the row and column containing the entry b:

$$\begin{pmatrix} & (b) & \\ d & & f \\ g & & i \end{pmatrix}$$

then evaluate the determinant of the remaining matrix $\begin{pmatrix} d & f \\ g & i \end{pmatrix}$.

3 Delete the row and column containing the entry c:

$$\begin{pmatrix} & & (c) \\ d & e & \\ g & h & \end{pmatrix}$$

then evaluate the determinant of the remaining matrix $\begin{pmatrix} d & e \\ g & h \end{pmatrix}$.

4 Combine these steps as in 11.6 to obtain the determinant of the above matrix.

A general aid in finding the determinant is to place your thumb over the number and ignore the horizontal row and vertical column containing that number.

Example 14

Evaluate the determinant of

$$A = \begin{pmatrix} 1 & 7 & 5 \\ 3 & 2 & -7 \\ 6 & -8 & 9 \end{pmatrix}$$

Solution

By using steps 1–4 and 11.6 we have

$$\det A = 1\det\begin{pmatrix} 2 & -7 \\ -8 & 9 \end{pmatrix} - 7\det\begin{pmatrix} 3 & -7 \\ 6 & 9 \end{pmatrix} + 5\det\begin{pmatrix} 3 & 2 \\ 6 & -8 \end{pmatrix}$$

$$\underset{\text{by } 11.1}{=} \left[(2 \times 9) - (-8 \times (-7))\right] - 7\left[(3 \times 9) - (6 \times (-7))\right] + 5\left[(3 \times (-8)) - (6 \times 2)\right]$$

$$= -701$$

C2 Cofactors

A general 3×3 matrix is $\begin{pmatrix} a & b & c \\ d & e & f \\ g & h & i \end{pmatrix}$. The determinant of the remaining matrix after

deleting the rows and columns of an entry is called the **minor** of that entry. For example, in the above case

$$\det\begin{pmatrix} e & f \\ h & i \end{pmatrix} \text{ is the minor of } a$$

$$\det\begin{pmatrix} d & f \\ g & i \end{pmatrix} \text{ is the minor of } b$$

? **What is the minor of c?**

$$\det\begin{pmatrix} d & e \\ g & h \end{pmatrix}$$

? **What is the minor of e?**

Deleting the rows and columns containing the entry e we have

$$\begin{pmatrix} a & & c \\ & (e) & \\ g & & i \end{pmatrix}$$

Hence $\det\begin{pmatrix} a & c \\ g & i \end{pmatrix}$ is the minor of e.

11.1 $\det\begin{pmatrix} a & b \\ c & d \end{pmatrix} = ad - cb$

Example 15

Evaluate the minor of −1 in

$$\begin{pmatrix} 3 & 5 & 7 \\ -1 & 2 & 3 \\ -4 & 4 & -9 \end{pmatrix}$$

Solution

After deleting the row and column containing −1

$$\begin{pmatrix} & 5 & 7 \\ (-1) & & \\ & 4 & -9 \end{pmatrix}$$

we obtain the matrix $\begin{pmatrix} 5 & 7 \\ 4 & -9 \end{pmatrix}$. The minor of −1 is

$$\det\begin{pmatrix} 5 & 7 \\ 4 & -9 \end{pmatrix} \underset{\text{by } 11.1}{=} [5 \times (-9)] - [4 \times 7] = -73$$

The **cofactor** of −1 in **Example 15** is −(−73) = 73. The cofactor of an entry is the minor with a **place sign**. The place sign is found from the rule:

$$\begin{matrix} + & - & + \\ - & + & - \\ + & - & + \end{matrix}$$

The first entry of a matrix has a positive place sign and then the place signs alternate.

Example 16

Find the cofactor of 5 of the matrix in **Example 15**.

Solution

The minor of 5 is

$$\det\begin{pmatrix} -1 & 3 \\ -4 & -9 \end{pmatrix} \underset{\text{by } 11.1}{=} [-1 \times (-9)] - [(-4) \times 3] = 21$$

According to the rule, the place sign is negative, so the cofactor of 5 is −21.

Note that the **minor** of 5 is 21 but the **cofactor** is −21 because the position of 5 in the matrix gives it a negative place sign.

Hence the cofactor is a minor with a place sign.

In **Section C1** we calculated the determinant of a 3 × 3 matrix using formula   11.6 .

11.1 $\det\begin{pmatrix} a & b \\ c & d \end{pmatrix} = ad - cb$

We can rewrite a 3 × 3 matrix in terms of its cofactors, that is, if **A** is a general 3 × 3 matrix:

$$\mathbf{A} = \begin{pmatrix} a & b & c \\ d & e & f \\ g & h & i \end{pmatrix}$$

We can also use a 3 × 3 cofactor matrix to calculate its determinant using:

11.7 $\det\mathbf{A} = a(\text{cofactor of } a) + b(\text{cofactor of } b) + c(\text{cofactor of } c)$

We can find the determinant of a matrix by expanding along **any** of the rows or **any** of the columns. For example, the formula for expanding along the second row is

$$\det\mathbf{A} = d(\text{cofactor of } d) + e(\text{cofactor of } e) + f(\text{cofactor of } f)$$

? **What is the formula for expanding along the third row?**

$$\det\mathbf{A} = g(\text{cofactor of } g) + h(\text{cofactor of } h) + i(\text{cofactor of } i)$$

We can also expand along any column. Expansion along the first column is

$$\det\mathbf{A} = a(\text{cofactor of } a) + d(\text{cofactor of } d) + g(\text{cofactor of } g)$$

and so on.

Example 17

Find the determinant of

$$\mathbf{A} = \begin{pmatrix} -1 & -6 & 3 \\ 5 & 6 & -7 \\ -2 & 0 & 1 \end{pmatrix}$$

Solution

Since there is a zero in the last row it is easier to expand along this row:

$$\det\begin{pmatrix} -1 & -6 & 3 \\ 5 & 6 & -7 \\ -2 & 0 & 1 \end{pmatrix} = -2\det\begin{pmatrix} -6 & 3 \\ 6 & -7 \end{pmatrix} - 0\det\begin{pmatrix} -1 & 3 \\ 5 & -7 \end{pmatrix} + 1\det\begin{pmatrix} -1 & -6 \\ 5 & 6 \end{pmatrix}$$

$$\underset{\text{by } \boxed{11.1}}{=} -2(42 - 18) - 0 + (-6 + 30)$$

$$= -24$$

Note that the middle term on the Right-Hand Side is zero and so we just need to evaluate the other two terms. We could also have found the determinant of **A** by expanding along the second column because this also contains (the same) zero.

11.1 $\det\begin{pmatrix} a & b \\ c & d \end{pmatrix} = ad - cb$

Example 17 *continued*

If a row or column contains zero(s) then it is easier to expand along that row or column.

We can also obtain the determinant of a 4 × 4, 5 × 5, 6 × 6, etc. matrix but it becomes very laborious to do this by just using pen and paper. In these cases it is more convenient to use a graphical calculator or a computer algebra system.

C3 Cofactor matrix

Let **C** be the new matrix consisting of the cofactors of the general matrix **A**. If

$$\mathbf{A} = \begin{pmatrix} a & b & c \\ d & e & f \\ g & h & i \end{pmatrix} \quad \text{then} \quad \mathbf{C} = \begin{pmatrix} A & B & C \\ D & E & F \\ G & H & I \end{pmatrix}$$

where A is the cofactor of a, B is the cofactor of b, etc. **C** is called the cofactor matrix and it is used in finding the inverse matrix.

Example 18

Find the cofactor matrix **C** of

$$\mathbf{A} = \begin{pmatrix} 1 & -1 & 5 \\ 3 & 9 & 7 \\ -2 & 1 & 0 \end{pmatrix}$$

Solution

Cofactor of the first entry, 1, is

$$\det\begin{pmatrix} 9 & 7 \\ 1 & 0 \end{pmatrix} \underset{\text{by } \mathbf{11.1}}{=} [(9 \times 0) - (1 \times 7)] = -7$$

Cofactor of -1 is

$$\underset{\substack{\text{minus}\\\text{place sign}}}{=}\det\begin{pmatrix} 3 & 7 \\ -2 & 0 \end{pmatrix} \underset{\text{by } \mathbf{11.1}}{=} -[(3 \times 0) - (-2 \times 7)] = -14$$

Cofactor of 5 is

$$\det\begin{pmatrix} 3 & 9 \\ -2 & 1 \end{pmatrix} \underset{\text{by } \mathbf{11.1}}{=} [(3 \times 1) - (-2 \times 9)] = 21$$

Cofactor of 3 is

$$\underset{\substack{\text{minus}\\\text{place sign}}}{=}\det\begin{pmatrix} -1 & 5 \\ 1 & 0 \end{pmatrix} \underset{\text{by } \mathbf{11.1}}{=} -[(-1 \times 0) - (1 \times 5)] = 5$$

11.1 $\det\begin{pmatrix} a & b \\ c & d \end{pmatrix} = ad - cb$

Example 18 *continued*

Cofactor of 9 is

$$\det\begin{pmatrix} 1 & 5 \\ -2 & 0 \end{pmatrix} \underset{\text{by }11.1}{\equiv} [(1 \times 0) - (-2 \times 5)] = 10$$

Cofactor of 7 is

$$\underset{\substack{\text{minus} \\ \text{place sign}}}{-}\det\begin{pmatrix} 1 & -1 \\ -2 & 1 \end{pmatrix} \underset{\text{by }11.1}{\equiv} -[(1 \times 1)-(-2 \times (-1))] = 1$$

Cofactor of −2 is

$$\det\begin{pmatrix} -1 & 5 \\ 9 & 7 \end{pmatrix} \underset{\text{by }11.1}{\equiv} [(-1 \times 7) - (9 \times 5)] = -52$$

Cofactor of 1 (the 1 on the bottom row of the given matrix) is

$$\underset{\substack{\text{minus} \\ \text{place sign}}}{-}\det\begin{pmatrix} 1 & 5 \\ 3 & 7 \end{pmatrix} \underset{\text{by }11.1}{\equiv} -[(1 \times 7)-(3 \times 5)] = 8$$

Cofactor of 0 is

$$\det\begin{pmatrix} 1 & -1 \\ 3 & 9 \end{pmatrix} \underset{\text{by }11.1}{\equiv} [(1 \times 9) - (3 \times (-1))] = 12$$

Hence by collecting these together and placing them in the corresponding position gives the cofactor matrix:

$$\mathbf{C} = \begin{pmatrix} -7 & -14 & 21 \\ 5 & 10 & 1 \\ -52 & 8 & 12 \end{pmatrix}$$

C4 Inverse of a matrix

In order to find the inverse of a 3 × 3 matrix, we must define some terms. The **transpose** of a matrix is evaluated by interchanging the rows and columns. Let $\mathbf{A}$ be a matrix then the transpose is denoted by $\mathbf{A}^T$. The first row of $\mathbf{A}$ becomes the first column of $\mathbf{A}^T$, the second row of $\mathbf{A}$ becomes the second column of $\mathbf{A}^T$ and so on. If

$$\mathbf{A} = \begin{pmatrix} a & b & c \\ d & e & f \\ g & h & i \end{pmatrix} \text{ then } \mathbf{A}^T = \begin{pmatrix} a & d & g \\ b & e & h \\ c & f & i \end{pmatrix}$$

11.1 $\det\begin{pmatrix} a & b \\ c & d \end{pmatrix} = ad - cb$

Example 19

Find A^T and C^T of **Example 18**.

Solution

We have

$$A = \begin{pmatrix} 1 & -1 & 5 \\ 3 & 9 & 7 \\ -2 & 1 & 0 \end{pmatrix} \quad C = \begin{pmatrix} -7 & -14 & 21 \\ 5 & 10 & 1 \\ -52 & 8 & 12 \end{pmatrix}$$

Interchanging rows and columns gives

$$A^T = \begin{pmatrix} 1 & 3 & -2 \\ -1 & 9 & 1 \\ 5 & 7 & 0 \end{pmatrix} \quad C^T = \begin{pmatrix} -7 & 5 & -52 \\ -14 & 10 & 8 \\ 21 & 1 & 12 \end{pmatrix}$$

We can find the transpose of **any** matrix, it doesn't need to be a square matrix. For example, if

$$A = \begin{pmatrix} 1 & 2 & 3 \\ 4 & 5 & 6 \end{pmatrix} \text{ then } A^T = \begin{pmatrix} 1 & 4 \\ 2 & 5 \\ 3 & 6 \end{pmatrix}$$

The transpose of a matrix is an important concept in matrix theory.

The **adjoint** of a square matrix A, denoted adjA, is defined as

11.8　　$\text{adj}A = C^T$

where C is the matrix of cofactors of A. For **Example 19**:

$$\text{adj}A = C^T = \begin{pmatrix} -7 & 5 & -52 \\ -14 & 10 & 8 \\ 21 & 1 & 12 \end{pmatrix}$$

The **inverse** of any square matrix A, denoted by A^{-1}, is given by

11.9　　$A^{-1} = \dfrac{1}{\det A}(\text{adj}A)$　　[Provided that $\det A \neq 0$]

and it is defined as the matrix which satisfies

$$A^{-1} \times A = I$$

where I is the identity matrix.

Example 20

Determine A^{-1} of

$$A = \begin{pmatrix} 1 & -1 & 5 \\ 3 & 9 & 7 \\ -2 & 1 & 0 \end{pmatrix}$$

Solution

Since $A^{-1} = \dfrac{1}{\det A}(\text{adj}A)$ we need to find adjA and detA.

What is adjA equal to?

Because the matrix A is the same matrix A as in **Examples 18** and **19**, we have already evaluated adjA ($= \mathbf{C}^T$, that is, the cofactor matrix transposed). Hence from above we have

$$\text{adj}A = \begin{pmatrix} -7 & 5 & -52 \\ -14 & 10 & 8 \\ 21 & 1 & 12 \end{pmatrix} \quad [= \mathbf{C}^T]$$

We only need to find detA. Expanding along the **bottom row** of the given A because it contains a 0:

$$\det A = -2\det\begin{pmatrix} -1 & 5 \\ 9 & 7 \end{pmatrix} - 1\det\begin{pmatrix} 1 & 5 \\ 3 & 7 \end{pmatrix} + 0\det\begin{pmatrix} 1 & -1 \\ 3 & 9 \end{pmatrix}$$

$$\underset{\text{by } \boxed{11.1}}{=} -2[-7 - 45] - 1[7 - 15] + 0$$

$$= 112$$

Substituting these into $\boxed{11.9}$ gives

$$A^{-1} = \frac{1}{\det A}(\text{adj}A) = \frac{1}{112}\begin{pmatrix} -7 & 5 & -52 \\ -14 & 10 & 8 \\ 21 & 1 & 12 \end{pmatrix}$$

Example 21

Check that A^{-1} is indeed the inverse matrix of A in **Example 20**.

Solution

We need to check that

$$A^{-1} \times A = I = \begin{pmatrix} 1 & 0 & 0 \\ 0 & 1 & 0 \\ 0 & 0 & 1 \end{pmatrix}$$

$\boxed{11.1}$ $\det\begin{pmatrix} a & b \\ c & d \end{pmatrix} = ad - cb$

Example 21 *continued*

I is the identity matrix of a 3 × 3 matrix, with 1's along the main diagonal and zeros elsewhere. Also remember that multiplication of matrices is row × column.

$$\mathbf{A}^{-1} \times \mathbf{A} = \frac{1}{112}\begin{pmatrix} -7 & 5 & -52 \\ -14 & 10 & 8 \\ 21 & 1 & 12 \end{pmatrix}\begin{pmatrix} 1 & -1 & 5 \\ 3 & 9 & 7 \\ -2 & 1 & 0 \end{pmatrix}$$

$$= \frac{1}{112}\begin{pmatrix} 112 & 0 & 0 \\ 0 & 112 & 0 \\ 0 & 0 & 112 \end{pmatrix} = \begin{pmatrix} 1 & 0 & 0 \\ 0 & 1 & 0 \\ 0 & 0 & 1 \end{pmatrix} = \mathbf{I}$$

The identity matrix is a square matrix with 1's along the diagonal and zeros elsewhere.

Example 22 *electrical principles*

By applying Kirchhoff's law to a circuit we obtain the following equations:

$$i_1 + 2i_2 + i_3 = 4$$
$$3i_1 - 4i_2 - 2i_3 = 2$$
$$5i_1 + 3i_2 + 5i_3 = -1$$

where i_1, i_2 and i_3 represent currents. By using the inverse matrix method, find the values of i_1, i_2 and i_3.

Solution

Writing the above equations in matrix form we obtain

$$\begin{pmatrix} 1 & 2 & 1 \\ 3 & -4 & -2 \\ 5 & 3 & 5 \end{pmatrix}\begin{pmatrix} i_1 \\ i_2 \\ i_3 \end{pmatrix} = \begin{pmatrix} 4 \\ 2 \\ -1 \end{pmatrix}$$

Let $\mathbf{A} = \begin{pmatrix} 1 & 2 & 1 \\ 3 & -4 & -2 \\ 5 & 3 & 5 \end{pmatrix}$, then we have

$$\mathbf{A}\begin{pmatrix} i_1 \\ i_2 \\ i_3 \end{pmatrix} = \begin{pmatrix} 4 \\ 2 \\ -1 \end{pmatrix}$$

Multiplying both sides by the inverse matrix, $\mathbf{A}^{-1}$, gives

$$\begin{pmatrix} i_1 \\ i_2 \\ i_3 \end{pmatrix} = \mathbf{A}^{-1}\begin{pmatrix} 4 \\ 2 \\ -1 \end{pmatrix}$$

🧩 Example 22 *continued*

❓ We need to find A^{-1}. **How do we obtain A^{-1}?**

We need $\det A$ because $A^{-1} = \dfrac{1}{\det A}(\text{adj}A)$. Thus

$$\det\begin{pmatrix} 1 & 2 & 1 \\ 3 & -4 & -2 \\ 5 & 3 & 5 \end{pmatrix} = \det\begin{pmatrix} -4 & -2 \\ 3 & 5 \end{pmatrix} - 2\det\begin{pmatrix} 3 & -2 \\ 5 & 5 \end{pmatrix} + \det\begin{pmatrix} 3 & -4 \\ 5 & 3 \end{pmatrix}$$

$$\underset{\text{by }\boxed{11.1}}{=} -14 - 50 + 29$$

$$\det A = -35$$

❓ **What else do we need to find?**

adjA

adjA is the cofactor matrix **C** transposed. Evaluating the cofactors of each entry horizontally we obtain the following:

Cofactor of 1 (first entry) is

$$\det\begin{pmatrix} -4 & -2 \\ 3 & 5 \end{pmatrix} \underset{\text{by }\boxed{11.1}}{=} (-4 \times 5)-(3 \times (-2)) = -14$$

Cofactor of 2 is

$$-\det\begin{pmatrix} 3 & -2 \\ 5 & 5 \end{pmatrix} \underset{\text{by }\boxed{11.1}}{=} -[(3 \times 5) - (5 \times (-2))] = -25$$

Cofactor of 1 is

$$\det\begin{pmatrix} 3 & -4 \\ 5 & 3 \end{pmatrix} = 29$$

Cofactor of 3 (in the second row) is

$$-\det\begin{pmatrix} 2 & 1 \\ 3 & 5 \end{pmatrix} = -7$$

Cofactor of -4 is

$$\det\begin{pmatrix} 1 & 1 \\ 5 & 5 \end{pmatrix} = 0$$

Cofactor of -2 is

$$-\det\begin{pmatrix} 1 & 2 \\ 5 & 3 \end{pmatrix} = 7$$

Cofactor of 5 (the left 5 in the bottom row) is

$$\det\begin{pmatrix} 2 & 1 \\ -4 & -2 \end{pmatrix} = 0$$

$\boxed{11.1}$ $\det\begin{pmatrix} a & b \\ c & d \end{pmatrix} = ad - cb$

Example 22 *continued*

Cofactor of 3 (in the last row) is

$$-\det\begin{pmatrix} 1 & 1 \\ 3 & -2 \end{pmatrix} = 5$$

Cofactor of 5 (the right 5 in the bottom row) is

$$\det\begin{pmatrix} 1 & 2 \\ 3 & -4 \end{pmatrix} = -10$$

Thus the cofactor matrix $\mathbf{C} = \begin{pmatrix} -14 & -25 & 29 \\ -7 & 0 & 7 \\ 0 & 5 & -10 \end{pmatrix}$

Transposing (rows ↔ columns) this matrix gives the adjoint:

$$\text{adj}\mathbf{A} = \mathbf{C}^{\mathrm{T}} = \begin{pmatrix} -14 & -7 & 0 \\ -25 & 0 & 5 \\ 29 & 7 & -10 \end{pmatrix}$$

Substituting this and $\det\mathbf{A} = -35$ into 11.9

$$\mathbf{A}^{-1} = \frac{1}{\det\mathbf{A}}(\text{adj}\mathbf{A}) = \frac{1}{-35}\begin{pmatrix} -14 & -7 & 0 \\ -25 & 0 & 5 \\ 29 & 7 & -10 \end{pmatrix}$$

Substituting this into ✱ yields

$$\begin{pmatrix} i_1 \\ i_2 \\ i_3 \end{pmatrix} = \mathbf{A}^{-1}\begin{pmatrix} 4 \\ 2 \\ -1 \end{pmatrix}$$

$$= \frac{1}{-35}\begin{pmatrix} -14 & -7 & 0 \\ -25 & 0 & 5 \\ 29 & 7 & -10 \end{pmatrix}\begin{pmatrix} 4 \\ 2 \\ -1 \end{pmatrix}$$

$$= -\frac{1}{35}\begin{pmatrix} [-14 \times 4] + [-7 \times 2] + [0 \times (-1)] \\ [-25 \times 4] + [0 \times 2] + [5 \times (-1)] \\ [29 \times 4] + [7 \times 2] + [-10 \times (-1)] \end{pmatrix} = -\frac{1}{35}\begin{pmatrix} -70 \\ -105 \\ 140 \end{pmatrix}$$

$$\begin{pmatrix} i_1 \\ i_2 \\ i_3 \end{pmatrix} = -\frac{1}{35}\begin{pmatrix} -70 \\ -105 \\ 140 \end{pmatrix}$$

$$= \begin{pmatrix} -70/(-35) \\ -105/(-35) \\ 140/(-35) \end{pmatrix} = \begin{pmatrix} 2 \\ 3 \\ -4 \end{pmatrix}$$

Hence $i_1 = 2$ A, $i_2 = 3$ A and $i_3 = -4$ A.

You can apply the inverse matrix method as described in **Example 22** to solve 3, 4, 5, . . . simultaneous equations. It might seem long and tedious but generally solving three or more equations is more laborious unless you use a graphical calculator or a computer algebra system.

SUMMARY

The determinant of a 3×3 matrix is

$$\boxed{11.6} \quad \det\begin{pmatrix} a & b & c \\ d & e & f \\ g & h & i \end{pmatrix} = a\left[\det\begin{pmatrix} e & f \\ h & i \end{pmatrix}\right] - b\left[\det\begin{pmatrix} d & f \\ g & i \end{pmatrix}\right] + c\left[\det\begin{pmatrix} d & e \\ g & h \end{pmatrix}\right]$$

The cofactor of a in $\boxed{11.6}$ is the determinant of the matrix after deleting the row and column containing a with a place sign.
The cofactor matrix is the matrix of cofactors and the adjoint is defined as the cofactor matrix transposed.

The inverse of a square matrix A is

$$\boxed{11.9} \quad \mathbf{A}^{-1} = \frac{1}{\det\mathbf{A}}(\text{adj}\mathbf{A}) \quad [\text{Provided that } \det\mathbf{A} \neq 0]$$

Exercise 11(c)

Solutions at end of book. Complete solutions available at www.palgrave.com/engineering/singh

1 Calculate the determinants of the following matrices:

a $\mathbf{A} = \begin{pmatrix} 1 & 3 & -1 \\ 2 & 0 & 5 \\ -6 & 3 & 1 \end{pmatrix}$

b $\mathbf{B} = \begin{pmatrix} 2 & -10 & 11 \\ 5 & 3 & -4 \\ 7 & 9 & 12 \end{pmatrix}$

c $\mathbf{C} = \begin{pmatrix} -12 & 9 & -5 \\ 3 & 9 & 1 \\ -7 & 2 & -2 \end{pmatrix}$

2 Find $\mathbf{A}^T$ (A transposed) for the following:

a $\mathbf{A} = \begin{pmatrix} -2 & 3 \\ 1 & 5 \end{pmatrix}$ **b** $\mathbf{A} = \begin{pmatrix} 1/3 & -2/5 \\ -3/7 & \pi \end{pmatrix}$

c $\mathbf{A} = \begin{pmatrix} 7 & 3 & 4 \\ 2 & 6 & 1 \\ -3 & -3 & 1 \end{pmatrix}$

d $\mathbf{A} = \begin{pmatrix} 1.17 & 1.36 \\ 9.39 & -1.45 \\ 2.11 & 5.20 \end{pmatrix}$

e $\mathbf{A} = \begin{pmatrix} -7 & 2 & 1 & 5 \\ 3 & 6 & -4 & 7 \\ 8 & 3 & -3 & 5 \\ -4 & 6 & 7 & 0 \end{pmatrix}$

f $\mathbf{A} = \begin{pmatrix} 1 & -3 & 7 \\ -9 & 4 & 6 \\ -4 & 2 & 8 \\ 9 & 19 & 11 \end{pmatrix}$

3 Show that

$$\det\begin{pmatrix} \mathbf{i} & \mathbf{j} & \mathbf{k} \\ 7 & 3 & -2 \\ 4 & 2 & 7 \end{pmatrix} = 25\mathbf{i} - 57\mathbf{j} + 2\mathbf{k}$$

4 Find the values of t so that

$$\det\begin{pmatrix} 1 & 0 & 3 \\ 5 & t & -7 \\ 3 & 9 & t-1 \end{pmatrix} = 0$$

5 Find the cofactor matrix C and $\mathbf{C}^T$ of

$$\mathbf{A} = \begin{pmatrix} 1 & 0 & 5 \\ -2 & 3 & 7 \\ 6 & -1 & 0 \end{pmatrix}$$

6 [electrical principles]   By applying Kirchhoff's law and Ohm's law to a circuit, we obtain the following equations:

$$3i_1 - 5i_2 + 3i_3 = 7.5$$
$$2i_1 + i_2 - 7i_3 = -17.5$$
$$-10i_1 + 4i_2 + 5i_3 = 16$$

where i_1, i_2 and i_3 represent currents. By using the inverse matrix method, find the values of i_1, i_2 and i_3.

Solutions at end of book. Complete solutions available at www.palgrave.com/engineering/singh

Exercise **11(c) continued**

7 Find the determinant of the following:

a $A = \begin{pmatrix} 2 & 3 & 5 \\ 0 & 0 & 6 \\ 1 & 5 & 3 \end{pmatrix}$

b $B = \begin{pmatrix} 6 & 7 & 1 \\ 1 & 3 & 2 \\ 0 & 1 & 5 \end{pmatrix}$

c $C = \begin{pmatrix} 1 & 5 & 1 \\ 0 & 3 & 7 \\ 0 & 2 & 9 \end{pmatrix}$

d $D = \begin{pmatrix} 9 & 5 & 1 \\ 13 & 0 & 2 \\ 11 & 0 & 3 \end{pmatrix}$

For the following use a computer algebra system or a graphical calculator.

8 Evaluate the determinant of

a $A = \begin{pmatrix} 1 & 2 & 3 & 4 \\ 5 & 6 & 7 & 8 \\ 9 & 10 & 11 & 12 \\ 13 & 14 & 15 & 16 \end{pmatrix}$

b $B = \begin{pmatrix} -1.10 & 4.23 & 2.67 & 7.45 & 9.62 \\ 19.61 & 6.40 & 3.12 & 11.89 & 2.36 \\ -17.5 & -9.73 & 5.23 & 8.54 & 2.51 \\ 6.19 & 2.91 & 17.64 & 8.93 & 8.98 \\ 3.98 & 11.84 & 4.78 & 9.85 & 3.22 \end{pmatrix}$

SECTION D **Gaussian elimination**

By the end of this section you will be able to:
► understand the process of Gaussian elimination
► apply the Gaussian elimination procedure to a system

Solving three (or more) simultaneous equations by the inverse matrix method can be a lengthy task, as we observed in the previous section. The Gaussian elimination procedure is often a more straightforward method for solving simultaneous linear equations.

D1 **Gaussian elimination**

The Gaussian elimination procedure (named after the great German mathematician **K.F. Gauss**) is often used in computer software to solve systems of linear equations.

For square matrices, MATLAB uses Gaussian elimination to reduce a linear system to an appropriate format.

Example 23

Solve the following linear equations using the Gaussian elimination procedure:

$$\begin{aligned} x - 3y + 5z &= -9 \\ 2x - y - 3z &= 19 \\ 3x + y + 4z &= -13 \end{aligned}$$

Example 23 *continued*

Solution

This can be written in matrix form as

$$\begin{pmatrix} 1 & -3 & 5 \\ 2 & -1 & -3 \\ 3 & 1 & 4 \end{pmatrix} \begin{pmatrix} x \\ y \\ z \end{pmatrix} = \begin{pmatrix} -9 \\ 19 \\ -13 \end{pmatrix}$$

What are we trying to find?

The values of x, y and z which satisfy the given equations.

If we can somehow transform the above matrix to

11.10
$$\begin{matrix} \text{Row 1} \\ \text{Row 2} \\ \text{Row 3} \end{matrix} \begin{pmatrix} * & * & * \\ 0 & * & * \\ 0 & 0 & A \end{pmatrix} \begin{pmatrix} x \\ y \\ z \end{pmatrix} = \begin{pmatrix} * \\ * \\ B \end{pmatrix}$$

then we can find z. **How?**

Look at the last row of the 3×3 matrix in 11.10 and multiply it out with the adjacent

column vector, $\begin{pmatrix} x \\ y \\ z \end{pmatrix}$. (An $n \times 1$ matrix is generally called a column vector.) We obtain

$$(0 \times x) + (0 \times y) + (A \times z) = B$$
$$Az = B \text{ gives } z = \frac{B}{A}$$

Therefore from the last row we can find z.

Look at row 2. On the Left-Hand Side we get an equation in y and z. From above we know the value of $z = B/A$, so we can substitute $z = B/A$ and also obtain y. Similarly from the first row we can find x by substituting the values of y and z.

The matrix in 11.10 shows that we need to achieve 0's in the position shown in order to create equations with fewer unknowns so that by substitution we can find the answers.

Let's try this method for the matrix in question:

$$\begin{pmatrix} 1 & -3 & 5 \\ 2 & -1 & -3 \\ 3 & 1 & 4 \end{pmatrix} \begin{pmatrix} x \\ y \\ z \end{pmatrix} = \begin{pmatrix} -9 \\ 19 \\ -13 \end{pmatrix}$$

We need to enter the Right-Hand Side within the matrix:

$$\begin{matrix} \text{Row 1} \\ \text{Row 2} \\ \text{Row 3} \end{matrix} \quad \begin{matrix} R_1 \\ R_2 \\ R_3 \end{matrix} \begin{pmatrix} 1 & -3 & 5 & -9 \\ 2 & -1 & -3 & 19 \\ 3 & 1 & 4 & -13 \end{pmatrix}$$

This is called an **augmented** matrix. R_1, R_2 and R_3 just refer to row 1, row 2 and row 3 respectively. We need to obtain the matrix in 11.10 from the above matrix, therefore we

---

Example 23 *continued*

require zeros in the bottom left-hand corner of this matrix; that is, the 2 from the second row must become zero and the 3 and 1 from the bottom row must become zero. Hence $2 \to 0$, $3 \to 0$ and $1 \to 0$.

How is this achieved?

Recall from our work with simultaneous equations in **Chapter 1**, that we can multiply an entire equation by a non-zero constant, and take one equation away from another. In terms of matrices this means that we can multiply a whole row by a non-zero constant and take one row away from another.

How do we get 0 in place of 2?

To get 0 in place of 2, we multiple row 1, R_1, by 2 and subtract from row 2, R_2; that is, we carry out the row operation $R_2 - 2R_1$:

$$\begin{array}{c} R_1 \\ R_2{}^* = R_2 - 2R_1 \\ R_3 \end{array} \left(\begin{array}{ccc|c} 1 & -3 & 5 & -9 \\ 2 - 2(1) & -1 - [2 \times (-3)] & -3 - (2 \times 5) & 19 - (2 \times (-9)) \\ 3 & 1 & 4 & -13 \end{array} \right)$$

We call the new middle row $R_2{}^*$ say. Simplifying the middle row gives

$$\begin{array}{c} R_1 \\ R_2{}^* \\ R_3 \end{array} \left(\begin{array}{ccc|c} 1 & -3 & 5 & -9 \\ 0 & 5 & -13 & 37 \\ 3 & 1 & 4 & -13 \end{array} \right)$$

Where else do we need a 0?

We need to get a 0 in place of 3 in the bottom row. **How?**

We multiply the top row R_1 by 3 and subtract from the bottom row R_3; that is, we carry out the row operation, $R_3 - 3R_1$:

$$\begin{array}{c} R_1 \\ R_2{}^* \\ R_3{}^* = R_3 - 3R_1 \end{array} \left(\begin{array}{ccc|c} 1 & -3 & 5 & -9 \\ 0 & 5 & -13 & 37 \\ 3 - 3(1) & 1 - (3 \times (-3)) & 4 - (3 \times 5) & -13 - (3 \times (-9)) \end{array} \right)$$

We can call $R_3{}^*$ the new bottom row of this matrix. (Of course you can use other symbols such as small r for the new rows, that is, r_2 and r_3.)

Simplifying the arithmetic in the entries gives:

$$\begin{array}{c} R_1 \\ R_2{}^* \\ R_3{}^* \end{array} \left(\begin{array}{ccc|c} 1 & -3 & 5 & -9 \\ 0 & 5 & -13 & 37 \\ 0 & 10 & -11 & 14 \end{array} \right)$$

Note that we only need to convert the 10 into 0 in the bottom row. **How do we get a 0 in place of 10?**

Example 23 *continued*

We can **only** use the bottom two rows, R_2^* and R_3^*, otherwise we get a non-zero number in place of zeros already established. We execute $R_3^* - 2R_2^*$ because

$$10 - (2 \times 5) = 0 \text{ (gives a 0 in place of 10)}$$

Therefore we have:

$$
\begin{array}{c}
R_1 \\
R_2^* \\
R_3^{**} = R_3^* - 2R_2^*
\end{array}
\left(
\begin{array}{ccc|c}
1 & -3 & 5 & -9 \\
0 & 5 & -13 & 37 \\
0 - (2 \times 0) & 10 - (2 \times 5) & -11 - [2 \times (-13)] & 14 - (2 \times 37)
\end{array}
\right)
$$

which simplifies to

$$
\begin{array}{c}
 \\
(\dagger) \\
 \\
\end{array}
\begin{array}{c}
R_1 \\
R_2^* \\
R_3^{**}
\end{array}
\begin{array}{ccc}
x & y & z \\
\end{array}
\left(
\begin{array}{ccc|c}
1 & -3 & 5 & -9 \\
0 & 5 & -13 & 37 \\
0 & 0 & 15 & -60
\end{array}
\right)
$$

From the bottom row R_3^{**} we have:

$$15z = -60 \text{ which gives } z = -\frac{60}{15} = -4$$

How do we find the other two unknowns x and y?

By expanding the middle row R_2^* of $(\dagger)$ we have:

$$5y - 13z = 37$$

We can find y by substituting $z = -4$ into this:

$$5y - 13(-4) = 37 \quad [\text{Substituting } z = -4]$$

$$5y + 52 = 37 \text{ implies that } 5y = -15, \text{ therefore } y = -\frac{15}{5} = -3$$

How can we find the last unknown x?

By expanding the first row R_1 of $(\dagger)$ we have:

$$x - 3y + 5z = -9$$

Substituting $y = -3$ and $z = -4$ into this:

$$x - [3 \times (-3)] + [5 \times (-4)] = -9$$

$$x + 9 - 20 = -9 \text{ which gives } x = 2$$

Hence our solution to the linear system is $x = 2$, $y = -3$ and $z = -4$. Check that this solution is correct by substituting these values into the given equations:

$$
\begin{array}{ll}
x - 3y + 5z = -9 & 2 - 3(-3) + 5(-4) = -9 \\
2x - y - 3z = 19 & 2(2) - (-3) - 3(-4) = 19 \\
3x + y + 4z = -13 & 3(2) + (-3) + 4(-4) = -13
\end{array}
$$

The above process is called **Gaussian elimination**. We can:

1 interchange any two rows or columns (apart from the augmented column)
2 multiply any row by a non-zero scalar
3 add or subtract rows from each other.

The aim of the Gaussian elimination process is to produce the 'triangular' matrix shown in 11.10 , that is, zeros in the lower triangle.

When interchanging two columns we need to be careful. Consider the matrix used in **Example 23**:

$$\begin{matrix} x & y & z \\ \begin{pmatrix} 1 & -3 & 5 \\ 2 & -1 & -3 \\ 3 & 1 & 4 \end{pmatrix} \end{matrix}$$

Remember that column 1 represents the coefficients of x, column 2 represents the coefficients of y and column 3 represents the coefficients of z. So when interchanging any of these columns we need to remember the variable associated with that column. Interchanging column 1 and column 3 in the above gives

$$\begin{matrix} z & y & x \\ \begin{pmatrix} 5 & -3 & 1 \\ -3 & -1 & 2 \\ 4 & 1 & 3 \end{pmatrix} \end{matrix}$$

 Example 24 *mechanics*

A pulley system gives the following equations

$$\ddot{x}_1 + \ddot{x}_2 = 0$$
$$2\ddot{x}_1 = 20 - T$$
$$5\ddot{x}_2 = 50 - T$$

where $\ddot{x}_1, \ddot{x}_2$ represent acceleration and T represents tension in the rope. Determine $\ddot{x}_1, \ddot{x}_2$ and T by using Gaussian elimination.

Solution

Rearranging the above equations gives

$$\ddot{x}_1 + \ddot{x}_2 \qquad = 0$$
$$2\ddot{x}_1 \qquad + T = 20$$
$$5\ddot{x}_2 + T = 50$$

Putting this in matrix form:

$$\begin{pmatrix} 1 & 1 & 0 \\ 2 & 0 & 1 \\ 0 & 5 & 1 \end{pmatrix} \begin{pmatrix} \ddot{x}_1 \\ \ddot{x}_2 \\ T \end{pmatrix} = \begin{pmatrix} 0 \\ 20 \\ 50 \end{pmatrix}$$

Example 24 *continued*

In augmented matrix form we have

$$\begin{array}{c} R_1 \\ R_2 \\ R_3 \end{array} \left(\begin{array}{ccc|c} 1 & 1 & 0 & 0 \\ 2 & 0 & 1 & 20 \\ 0 & 5 & 1 & 50 \end{array} \right)$$

To get a 0 in place of 2 we execute $R_2 - 2R_1$:

$$\begin{array}{c} R_1 \\ R_2^* = R_2 - 2R_1 \\ R_3 \end{array} \left(\begin{array}{ccc|c} 1 & 1 & 0 & 0 \\ 0 & -2 & 1 & 20 \\ 0 & 5 & 1 & 50 \end{array} \right)$$

We need to get 0 in place of 5 in the bottom row, R_3. How?

Carrying out the row operation $R_3 + \dfrac{5}{2}R_2^*$:

$$\begin{array}{c} R_1 \\ R_2^* \\ R_3^* = R_3 + \dfrac{5}{2}R_2^* \end{array} \left(\begin{array}{ccc|c} 1 & 1 & 0 & 0 \\ 0 & -2 & 1 & 20 \\ 0 & 0 & 7/2 & 100 \end{array} \right)$$

From the bottom row, R_3^*, we can now conclude

$$\frac{7}{2}T = 100 \text{ which gives } T = \frac{200}{7}\text{ N}$$

Substituting $T = \dfrac{200}{7}$ into the middle row, R_2^*, we have

$$-2\ddot{x}_2 + \frac{200}{7} = 20$$

Solving

$$\ddot{x}_2 = \frac{30}{7}\text{ m/s}^2$$

Similarly from the first row, R_1, we obtain $\ddot{x}_1 = -\dfrac{30}{7}\text{ m/s}^2$.

Check for yourself that these values do indeed satisfy the original equations.

Example 25 *electrical principles*

The mesh equations of a circuit are given by:

$$\begin{aligned} i_1 + 3i_2 + 2i_3 &= 13 \\ 4i_1 + 4i_2 - 3i_3 &= 3 \\ 5i_1 + i_2 + 2i_3 &= 13 \end{aligned}$$

Determine the values of i_1, i_2 and i_3.

Example 25 *continued*

Solution

How can we find the unknowns i_1, i_2 and i_3?

We use **Gaussian elimination with back substitution**.

Let's try this method for the given equations. The augmented matrix is:

$$
\begin{array}{cc}
\text{Row 1} & R_1 \\
\text{Row 2} & R_2 \\
\text{Row 3} & R_3
\end{array}
\left(\begin{array}{ccc|c}
1 & 3 & 2 & 13 \\
4 & 4 & -3 & 3 \\
5 & 1 & 2 & 13
\end{array}\right)
$$

We need to convert this matrix so that there are zeros in the bottom left-hand corner triangle of this matrix, that is, first 4 in the second row needs to become 0, and 5 and 1 from the bottom row also need to become 0. Hence $4 \to 0$, $5 \to 0$ and $1 \to 0$.

To get 0 in place of the first 4 in the middle row we multiply row 1, R_1, by 4 and take it away from row 2, R_2, that is, $R_2 - 4R_1$. To get 0 in place of 5 in the bottom row we multiply row 1, R_1, by 5 and take away from row 3, R_3, that is, $R_3 - 5R_1$. Combining the two row operations, $R_2 - 4R_1$ and $R_3 - 5R_1$, we have

$$
\begin{array}{c}
R_1 \\
R_2{}^\dagger = R_2 - 4R_1 \\
R_3{}^\dagger = R_3 - 5R_1
\end{array}
\left(\begin{array}{ccc|c}
1 & 3 & 2 & 13 \\
4-(4\times1) & 4-(4\times3) & -3-(4\times2) & 3-(4\times13) \\
5-(5\times1) & 1-(5\times3) & 2-(5\times2) & 13-(5\times13)
\end{array}\right)
$$

We call the new rows 2 and 3 by $R_2{}^\dagger$ and $R_3{}^\dagger$ respectively. This simplifies to

$$
\begin{array}{c}
R_1 \\
R_2{}^\dagger \\
R_3{}^\dagger
\end{array}
\left(\begin{array}{ccc|c}
1 & 3 & 2 & 13 \\
0 & -8 & -11 & -49 \\
0 & -14 & -8 & -52
\end{array}\right)
$$

The new row 3 is called $R_3{}^\dagger$ say.

We have nearly obtained the required matrix with zeros in the bottom left-hand corner. We need a 0 in place of -14 in the new bottom row, $R_3{}^\dagger$. We can **only** use the second and third rows, $R_2{}^\dagger$ and $R_3{}^\dagger$. **Why?**

Because if we use the first and second rows, R_1 and $R_2{}^\dagger$, we will get a non-zero number in place of the zero already established in $R_2{}^\dagger$. Similarly this holds for R_1 and $R_3{}^\dagger$.

How can we obtain 0 in place of -14?

$$
R_3{}^\dagger - \frac{14}{8}R_2{}^\dagger \text{ because } -14 - \left[\frac{14}{8}\times(-8)\right] = 0 \qquad \text{[gives 0 in place of } -14]
$$

Therefore

$$
\begin{array}{c}
R_1 \\
R_2{}^\dagger \\
R_3{}^{\dagger\dagger} = R_3{}^\dagger - \frac{14}{8}R_2{}^\dagger
\end{array}
\left(\begin{array}{ccc|c}
1 & 3 & 2 & 13 \\
0 & -8 & -11 & -49 \\
0 & -14-\left[\frac{14}{8}\times(-8)\right] & -8-\left[\frac{14}{8}\times(-11)\right] & -52-\left[\frac{14}{8}\times(-49)\right]
\end{array}\right)
$$

Example 25 *continued*

which simplifies to

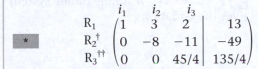

$$
\begin{array}{c}
 & i_1 & i_2 & i_3 & \\
R_1 \\
R_2^{\dagger} \\
R_3^{\dagger\dagger}
\end{array}
\left(
\begin{array}{ccc|c}
1 & 3 & 2 & 13 \\
0 & -8 & -11 & -49 \\
0 & 0 & 45/4 & 135/4
\end{array}
\right)
$$

This time we have called the bottom row $R_3^{\dagger\dagger}$. From this row, $R_3^{\dagger\dagger}$, we have

$$\frac{45}{4} i_3 = \frac{135}{4} \text{ which gives } i_3 = \frac{135}{45} = 3$$

How can we find the unknown i_2?

Examining the second row $R_2^{\dagger}$ in * we have

$$-8i_2 - 11i_3 = -49$$

We have established that $i_3 = 3$, therefore substituting $i_3 = 3$ gives

$$
\begin{aligned}
-8i_2 - 11(3) &= -49 \\
-8i_2 - 33 &= -49 \\
-8i_2 &= -49 + 33 = -16 \text{ which yields } i_2 = (-16)/(-8) = 2
\end{aligned}
$$

So far we have $i_2 = 2$ and $i_3 = 3$. **How can we find the last unknown i_1?**

By expanding the first row, R_1, of * we have

$$i_1 + 3i_2 + 2i_3 = 13$$

Substituting our values already found, $i_2 = 2$ and $i_3 = 3$, we have

$$
\begin{aligned}
i_1 + (3 \times 2) + (2 \times 3) &= 13 \\
i_1 + 6 + 6 &= 13 \text{ which gives } i_1 = 1
\end{aligned}
$$

Hence $i_1 = 1$, $i_2 = 2$ and $i_3 = 3$ is our solution. You can check that this solution is correct by substituting these values, $i_1 = 1$, $i_2 = 2$ and $i_3 = 3$, into the given equations.

SUMMARY

The Gaussian elimination procedure obtains the following triangular matrix

11.10
$$
\begin{pmatrix}
* & * & * \\
0 & * & * \\
0 & 0 & *
\end{pmatrix}
$$

and then uses back substitution to solve the given linear equations.

Exercise **11(d)**

Solutions at end of book. Complete solutions available
at www.palgrave.com/engineering/singh

1 By using Gaussian elimination, solve

   a $x + 2y - 3z = 3$
       $2x - y - z = 11$
       $3x + 2y + z = -5$

   b $2x + y + 2z = 10$
       $x + 4y - 3z = 0$
       $3x + 6y - z = 12$

   c $x + y + 2z = 1$
       $2x + 3y + 2z = 2$
       $5x + 2y + 8z = 4$

2 [electrical principles]   Mesh equations
of a circuit are given by

$$(9 \times 10^3)i_1 - (3 \times 10^3)i_2 = 10$$
$$-(3 \times 10^3)i_1 + (13 \times 10^3)i_2 = 0$$

By using Gaussian elimination,
solve these equations for the currents i_1
and i_2.

[*Hint*: Work in kilo (lose the 10^3), then
the answer comes out automatically in
mA (10^{-3}) for i_1 and i_2.]

3 By using Gaussian elimination solve

$$10x + y - 5z = 18$$
$$-20x + 3y + 20z = 14$$
$$5x + 3y + 5z = 9$$

4 [*mechanics*]   A simple pulley system
gives the equations

$$\ddot{x}_1 = T - g$$
$$2\ddot{x}_2 = T - 2g$$
$$\ddot{x}_1 + \ddot{x}_2 = 0$$

where $\ddot{x}_1$, $\ddot{x}_2$ represent acceleration, T is the
tension in the rope and g is the accelera-
tion due to gravity. Determine $\ddot{x}_1$, $\ddot{x}_2$ and T.

5 [*mechanics*]   A simple pulley system
gives the equations

$$\ddot{x}_1 + \ddot{x}_2 = 0$$
$$-3\ddot{x}_1 + T = 3g$$
$$-2\ddot{x}_2 + T = 2g$$

where $\ddot{x}_1$, $\ddot{x}_2$ represent acceleration, T is the
tension in the rope and g is the accelera-
tion due to gravity. Determine $\ddot{x}_1$, $\ddot{x}_2$ and T.

6 [electrical principles]   The mesh
equations of a circuit are given by

$$(4 \times 10^3)I_1 - (3 \times 10^3)I_2 = 10$$
$$-(3 \times 10^3)I_1 + (18 \times 10^3)I_2$$
$$-(10 \times 10^3)I_3 = 0$$
$$-(10 \times 10^3)I_2 + (23 \times 10^3)I_3 = -15$$

By using Gaussian elimination, solve
these equations for I_1, I_2 and I_3.

[See *Hint* for question **2**.]

SECTION E **Linear equations**

By the end of this section you will be able to:
▶ investigate equations which can be written in the form **Au** = **b**
▶ understand that a solution to **Au** = **b** might still exist even if **A**$^{-1}$
   does not exist
▶ understand the concept of an infinite number of solutions

E1 Properties of linear equations

Linear equations are equations whose variables have an index of 1 or 0. So far we have looked at linear equations of the form

$$2x + 3y + 5z = 9$$
$$x + 2y + 7z = 13$$
$$4x + 11y + 3z = -1$$

These are called a system of linear equations.

? **How can we write these in matrix form?**

$$\begin{pmatrix} 2 & 3 & 5 \\ 1 & 2 & 7 \\ 4 & 11 & 3 \end{pmatrix} \begin{pmatrix} x \\ y \\ z \end{pmatrix} = \begin{pmatrix} 9 \\ 13 \\ -1 \end{pmatrix}$$

Also this can be written as

$$\mathbf{Au} = \mathbf{b} \quad \text{where } \mathbf{A} = \begin{pmatrix} 2 & 3 & 5 \\ 1 & 2 & 7 \\ 4 & 11 & 3 \end{pmatrix}, \mathbf{u} = \begin{pmatrix} x \\ y \\ z \end{pmatrix} \text{ and } \mathbf{b} = \begin{pmatrix} 9 \\ 13 \\ -1 \end{pmatrix}.$$

? **How do we find the values of x, y and z?**

We find the inverse matrix $\mathbf{A}^{-1}$ and multiply both sides by $\mathbf{A}^{-1}$ to obtain

$$\underbrace{\mathbf{A}^{-1} \mathbf{A}}_{=\mathbf{I}} \mathbf{u} = \mathbf{A}^{-1}\mathbf{b}$$
$$\mathbf{Iu} = \mathbf{A}^{-1}\mathbf{b} \qquad [\text{I is the identity matrix}]$$
$$\mathbf{u} = \mathbf{A}^{-1}\mathbf{b}$$

This solution, $\mathbf{u}$, is unique. In earlier sections we used inverse matrix, $\mathbf{A}^{-1}$, to solve linear equations. The solutions of simultaneous equations depend on the determinant of matrix A which we examine next.

We need to find $\mathbf{A}^{-1}$ for this method to work. We can only find $\mathbf{A}^{-1}$ if $\det \mathbf{A} \neq 0$ because $\mathbf{A}^{-1} = \dfrac{1}{\det \mathbf{A}}(\text{adj}\mathbf{A})$ and of course we **cannot** divide by 0.

In general if **A** is an $n \times n$ matrix, **u** and **b** are $n \times 1$ matrices ($n \times 1$ matrices are called vectors) such that

11.11	$\mathbf{Au} = \mathbf{b}$ with $\det \mathbf{A} \neq 0$

then we have a unique solution, that is, we have a unique **u** which satisfies $\mathbf{Au} = \mathbf{b}$.

? **However if $\det \mathbf{A} = 0$ then what is the inverse matrix, $\mathbf{A}^{-1}$, of A?**

We **cannot** find $\mathbf{A}^{-1}$ because $\mathbf{A}^{-1}$ is defined with $\det \mathbf{A}$ in the denominator.

Hence the inverse matrix, $\mathbf{A}^{-1}$, does **not** exist.

? **How can we solve linear equations when $\mathbf{A}^{-1}$ does not exist? Is it possible?**

In the next example we examine these questions.

Example 26

Solve for x and y:

a $2x + 3y = 7$
  $2x + 3y = 12$

b $2x + 10y = 10$
  $6x + 30y = 30$

Solution

a Writing equations **a** in matrix form:

$$\begin{pmatrix} 2 & 3 \\ 2 & 3 \end{pmatrix}\begin{pmatrix} x \\ y \end{pmatrix} = \begin{pmatrix} 7 \\ 12 \end{pmatrix}$$

Let $A = \begin{pmatrix} 2 & 3 \\ 2 & 3 \end{pmatrix}$, then

$$\det A = (2 \times 3) - (2 \times 3) = 0 \qquad \text{[By 11.1]}$$

? **What is A^{-1}?**

Hence we do not have an inverse matrix, A^{-1}.

? **So what are the values of x and y?**

Look at these equations:

* $\quad\quad 2x + 3y = 7$
** $\quad\quad 2x + 3y = 12$

? **What do you notice?**

The coefficients of x and y are the same but the Right-Hand Side values are different, 7 and 12. Clearly they are **inconsistent**. Moreover if we subtract ** from * then we obtain $0 = -5$ which is clearly incorrect. Thus there is **no** solution to these simultaneous equations. The graphs of these equations are shown in Fig. 15.

Fig. 15

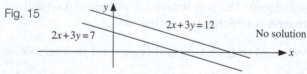

$2x+3y=7$ $2x+3y=12$ No solution

We have parallel lines therefore the lines don't cross, which explains why there is no solution to these simultaneous equations.

b Writing the equations **b** in matrix form:

$$\begin{pmatrix} 2 & 10 \\ 6 & 30 \end{pmatrix}\begin{pmatrix} x \\ y \end{pmatrix} = \begin{pmatrix} 10 \\ 30 \end{pmatrix}$$

Let $A = \begin{pmatrix} 2 & 10 \\ 6 & 30 \end{pmatrix}$ then

$$\det A = (2 \times 30) - (6 \times 10) = 0 \qquad \text{[By 11.1]}$$

11.1 $\det\begin{pmatrix} a & b \\ c & d \end{pmatrix} = ad - cb$

Example 26 *continued*

? What is A⁻¹?

? Again A⁻¹ does not exist. **Are there any solutions to these equations?**

? Yes, try $x = 0$, $y = 1$. **Do these equations have any other solutions?**

Many:

$$x = 2, \quad y = 3/5$$
$$x = 3, \quad y = 2/5$$
$$x = -1, y = 6/5$$
$$\vdots \qquad \vdots$$

These equations have an infinite number of solutions. They can be found as follows:

The equations are

† $2x + 10y = 10$

†† $6x + 30y = 30$

From **†** we have

$$10y = 10 - 2x$$
$$y = \frac{10 - 2x}{10} = \frac{10}{10} - \frac{2}{10}x = 1 - \frac{x}{5} \qquad \text{[Cancelling]}$$

Thus we have $y = 1 - \dfrac{x}{5}$.

Let $x = a$, where a is any real number, then

$$y = 1 - \frac{a}{5}$$

Putting $a = 1$ we obtain $x = 1$, $y = 4/5$.

Putting $a = 2$ we obtain $x = 2$, $y = 3/5$ and so on.

? **What do you notice about equations b?**

$$2x + 10y = 10$$
$$6x + 30y = 30$$

The second equation is three times the first equation. The graphs of these are shown in Fig. 16.

Since the lines lie on top of each other, we have an infinite number of solutions.

Fig. 16

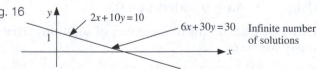

Infinite number of solutions

If we consider the general n simultaneous equations written in matrix form as

11.12 $\mathbf{Au} = \mathbf{b}$ [A is a square matrix, $\mathbf{b} \neq \mathbf{0}$]

with detA = 0, then the inverse matrix, A^{-1}, does not exist. Furthermore 11.12 satisfies one of the following:

i there is no solution (there is no **u** satisfying **Au** = **b**), or

ii there are an infinite number of solutions (there are an infinite number of **u**'s which satisfy **Au** = **b**).

In **Example 26a**: $A = \begin{pmatrix} 2 & 3 \\ 2 & 3 \end{pmatrix}$, $\mathbf{u} = \begin{pmatrix} x \\ y \end{pmatrix}$ and $\mathbf{b} = \begin{pmatrix} 7 \\ 12 \end{pmatrix}$ and there were **no** solutions.

In **Example 26b**: $A = \begin{pmatrix} 2 & 10 \\ 6 & 30 \end{pmatrix}$, $\mathbf{u} = \begin{pmatrix} x \\ y \end{pmatrix}$ and $\mathbf{b} = \begin{pmatrix} 10 \\ 30 \end{pmatrix}$ and we had an **infinite** number

of solutions of the form: $x = a$, $y = 1 - \dfrac{a}{5}$ where a is any real number.

In both cases we had detA = 0.

Now we look at equations of the form

$$\mathbf{Au} = \mathbf{0} \text{ with detA} = 0$$

Note that the Right-Hand Side is zero. Consider the equations

| * | $2x + 10y = 0$ |
| ** | $6x + 30y = 0$ |

From * we have $x = -5y$, that is, x is -5 times the y value. Hence $y = 1$, $x = -5$ is a solution to both equations * and ** . The following are also solutions of the simultaneous equations:

$$y = 2, \ x = -10$$
$$y = 3, \ x = -15$$
$$y = 4, \ x = -20$$
$$\vdots \quad \vdots$$

There are an infinite number of solutions. The equations can be written in matrix form as

$$\begin{pmatrix} 2 & 10 \\ 6 & 30 \end{pmatrix}\begin{pmatrix} x \\ y \end{pmatrix} = \begin{pmatrix} 0 \\ 0 \end{pmatrix}$$

or   **Au** = **0**

where $A = \begin{pmatrix} 2 & 10 \\ 6 & 30 \end{pmatrix}$, $\mathbf{u} = \begin{pmatrix} x \\ y \end{pmatrix}$, $\mathbf{0} = \begin{pmatrix} 0 \\ 0 \end{pmatrix}$ and detA = 0.

In general if

11.13   **Au** = **0** with detA = 0

then 11.13 has an infinite number of solutions. That is an infinite number of **u**'s that satisfy **Au** = **0**.

Now consider equations of the form

$$\mathbf{Au} = \mathbf{0} \text{ with detA} \ne 0$$

? **Does A^{-1} exist?**

Yes. Multiply both sides by A^{-1}

$$A^{-1}Au = A^{-1}0 = 0$$

$$\underset{\substack{\text{because} \\ A^{-1}A = 1}}{I}\ u = 0$$

$$u = 0 \qquad \text{[Because } Iu = u]$$

The solution $u = 0$ means $x = 0$, $y = 0$, $z = 0$, . . . This is known as the **trivial** solution.

In general if

11.14	$Au = 0$ with $\det A \neq 0$

then there is only the trivial solution, $x = 0$, $y = 0$, $z = 0$, ...

Summarizing the above, we have four cases of

∗	$Au = b$

11.11	If $\det A \neq 0$ then ∗ has a unique solution

11.12	If $b \neq 0$, $\det A = 0$ then either ∗ has no solution or there are an infinite number of solutions

11.13	If $b = 0$, $\det A = 0$ then ∗ has an infinite number of solutions

11.14	If $b = 0$, $\det A \neq 0$ then ∗ only has the trivial solution $u = 0$; $x = 0$, $y = 0$, $z = 0$, . . .

Note that for the case with $b = 0$ we have non-trivial solutions only if $\det A = 0$.

The next three examples look into the various types of solution outlined in 11.11 to 11.14 .

Example 27

Find all the solutions of

$$x + 2y + z = 4$$
$$3x - 4y - 2z = 2$$
$$5x + 3y + 5z = -1$$

Solution

These equations can be written in matrix form as

$$Au = b$$

where $A = \begin{pmatrix} 1 & 2 & 1 \\ 3 & -4 & -2 \\ 5 & 3 & 5 \end{pmatrix}$, $u = \begin{pmatrix} x \\ y \\ z \end{pmatrix}$ and $b = \begin{pmatrix} 4 \\ 2 \\ -1 \end{pmatrix}$. The equations were solved in

Example 22 with the variables i_1, i_2 and i_3 instead of x, y and z, hence

∗	$x = 2$, $y = 3$ and $z = -4$

Since $\det A = -35 \neq 0$ (evaluated in **Example 22**), we can conclude from 11.11 that these solutions are unique, that is $x = 2$, $y = 3$ and $z = -4$ are the only values which satisfy the given equations.

Example 28

Find all the values of x, y and z that satisfy

$$2x + 3y + 4z = 0$$
$$5x + 7y + 3z = 0$$
$$7x + 3y + 11z = 0$$

Solution

Putting these in matrix form gives

$$\begin{pmatrix} 2 & 3 & 4 \\ 5 & 7 & 3 \\ 7 & 3 & 11 \end{pmatrix} \begin{pmatrix} x \\ y \\ z \end{pmatrix} = \begin{pmatrix} 0 \\ 0 \\ 0 \end{pmatrix}$$

This is of the form $\mathbf{Au} = \mathbf{b}$ where $\mathbf{A} = \begin{pmatrix} 2 & 3 & 4 \\ 5 & 7 & 3 \\ 7 & 3 & 11 \end{pmatrix}$, $\mathbf{u} = \begin{pmatrix} x \\ y \\ z \end{pmatrix}$ and $\mathbf{b} = \begin{pmatrix} 0 \\ 0 \\ 0 \end{pmatrix}$.

We first obtain det$\mathbf{A}$:

$$\det \begin{pmatrix} 2 & 3 & 4 \\ 5 & 7 & 3 \\ 7 & 3 & 11 \end{pmatrix} = 2\det \begin{pmatrix} 7 & 3 \\ 3 & 11 \end{pmatrix} - 3\det \begin{pmatrix} 5 & 3 \\ 7 & 11 \end{pmatrix} + 4\det \begin{pmatrix} 5 & 7 \\ 7 & 3 \end{pmatrix}$$

$$= 2[(7 \times 11) - (3 \times 3)] - 3[(5 \times 11) - (7 \times 3)] + 4[(5 \times 3) - (7 \times 7)]$$

$$= -102$$

Since det$\mathbf{A} = -102 \neq 0$ we can conclude by 11.14 that we only have the trivial solution:

$$x = 0, y = 0 \text{ and } z = 0$$

Example 29

Find the values of λ that give a non-trivial solution of

$$(\mathbf{A} - \lambda\mathbf{I})\mathbf{u} = \mathbf{0}$$

where $\mathbf{A} = \begin{pmatrix} 2 & 3 \\ 7 & 1 \end{pmatrix}$, $\mathbf{u} = \begin{pmatrix} x \\ y \end{pmatrix}$ and $\mathbf{I} = \begin{pmatrix} 1 & 0 \\ 0 & 1 \end{pmatrix}$ is the identity matrix.

Solution

This is quite a difficult example to follow. It requires use of some algebra as well as matrix manipulation.

By 11.13 we have a non-trivial solution if

⋆ $$\det(\mathbf{A} - \lambda\mathbf{I}) = 0$$

Example 29 *continued*

Substituting the given information:

$$\det(\mathbf{A} - \lambda\mathbf{I}) = \det\left[\begin{pmatrix} 2 & 3 \\ 7 & 1 \end{pmatrix} - \lambda\begin{pmatrix} 1 & 0 \\ 0 & 1 \end{pmatrix}\right]$$

$$= \det\left[\begin{pmatrix} 2 & 3 \\ 7 & 1 \end{pmatrix} - \begin{pmatrix} \lambda & 0 \\ 0 & \lambda \end{pmatrix}\right]$$

$$= \det\begin{pmatrix} 2-\lambda & 3 \\ 7 & 1-\lambda \end{pmatrix}$$

$$\underset{\text{by } 11.1}{=} (2 - \lambda)(1 - \lambda) - (7 \times 3)$$

$$= 2 - 3\lambda + \lambda^2 - 21 = \lambda^2 - 3\lambda - 19$$

By ***** we equate this to zero:

$$\lambda^2 - 3\lambda - 19 = 0 \qquad \text{[Quadratic equation]}$$

? **How do we find λ?**

By using the quadratic formula

1.16 $$\lambda = \frac{-b \pm \sqrt{b^2 - 4ac}}{2a}$$

with $a = 1$, $b = -3$ and $c = -19$:

$$\lambda = \frac{3 \pm \sqrt{9 - (4 \times 1 \times (-19))}}{2}$$

$$= \frac{3 \pm \sqrt{85}}{2}$$

$$\lambda_1 = \frac{3 + \sqrt{85}}{2}, \lambda_2 = \frac{3 - \sqrt{85}}{2}$$

Different λ's are normally denoted by λ_1 and λ_2.

These λ are called the **eigenvalues** of the matrix A. Eigenvalues are the subject of the next section.

These terms *eigenvalue* and *eigenvector* are derived from the German word *Eigenwerte* which means 'proper value'. The word *eigen* is pronounced 'i-gun' (the eigenvalues are usually represented by the Greek letter lambda λ).

SUMMARY

A system of linear equations can be written as

* $\mathbf{Au} = \mathbf{b}$

It has the following properties:

11.11   If detA ≠ 0 then ***** has a unique solution

11.1 $\det\begin{pmatrix} a & b \\ c & d \end{pmatrix} = ad - cb$

11.12	If $\mathbf{b} \neq 0$, det$A = 0$ then either $\boxed{*}$ has no solution or there are an infinite number of solutions
11.13	If $\mathbf{b} = 0$, det$A = 0$ then $\boxed{*}$ has an infinite number of solutions
11.14	If $\mathbf{b} = 0$, det$A \neq 0$ then $\boxed{*}$ only has the trivial solution $x = 0, y = 0, z = 0, \ldots$

Exercise 11(e)

Solutions at end of book. Complete solutions available at www.palgrave.com/engineering/singh

1 Solve:
$$x + 3y + 5z = 0$$
$$7x - y + z = 0$$
$$2x + 3y - 8z = 0$$

2 Solve and find whether the solution is unique:
$$2x + 3y = 5$$
$$4x + 5y = 9$$

3 Find x and y which satisfy:
$$2x + y = 0$$
$$4x + 2y = 0$$
Is the solution unique?

4 Test whether the solutions of questions **1, 4** and **5** of **Exercise 11(d)** are unique.

5 Find the values of λ that give a non-trivial solution of

$$(\mathbf{A} - \lambda\mathbf{I})\mathbf{u} = \mathbf{0}$$

for

a $\mathbf{A} = \begin{pmatrix} -1 & 1 \\ 9 & -1 \end{pmatrix}$ **b** $\mathbf{A} = \begin{pmatrix} 1 & 3 \\ -6 & 5 \end{pmatrix}$

$\left[\mathbf{u} = \begin{pmatrix} x \\ y \end{pmatrix} \text{ and } \mathbf{I} \text{ is the identity matrix} \right]$

SECTION F **Eigenvalues and eigenvectors**

By the end of this section you will be able to:
▶ understand the terms eigenvalues and eigenvectors
▶ find the eigenvalues and eigenvectors of 2×2 matrices
▶ find the eigenvalues and eigenvectors of 3×3 matrices
▶ examine vibrations of buildings

This is a challenging section because it assumes a thorough understanding of matrices and it also requires a proficiency in algebra on topics such as expansion of brackets, solving equations and substitution.

F1 **Introduction**

Example 30

If $\mathbf{A} = \begin{pmatrix} 4 & -2 \\ 1 & 1 \end{pmatrix}$ and $\mathbf{u} = \begin{pmatrix} 2 \\ 1 \end{pmatrix}$ then evaluate $\mathbf{Au}$.

Example 30 *continued*

Solution

Multiplying the matrix and column vector:

$$\mathbf{Au} = \begin{pmatrix} 4 & -2 \\ 1 & 1 \end{pmatrix}\begin{pmatrix} 2 \\ 1 \end{pmatrix} = \begin{pmatrix} 6 \\ 3 \end{pmatrix} = 3\begin{pmatrix} 2 \\ 1 \end{pmatrix} = 3\mathbf{u}$$

In geometric terms, this means that the matrix **A** scalar multiples the vector **u** by 3 (Fig. 17):

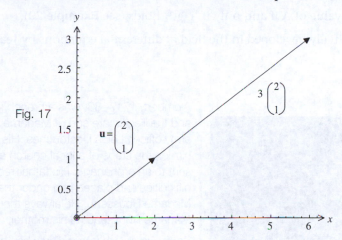

Fig. 17

Note that we have **Au** = 3**u**.

In general terms, this can be stated as

⬛*   **Au** = λ**u**

where **A** is a square matrix, **u** is a column vector and λ is a scalar (a number). **Can you think of a vector, u, which satisfies** ⬛* **?**

        u = 0

This is called the trivial solution: $x = 0$, $y = 0$, $z = 0$ . . .

Au = λ**u** is only interesting if **u** ≠ **0**, otherwise we always have the trivial solution. For **u** ≠ **0**, λ is called an **eigenvalue** of the matrix **A** and **u** is called an **eigenvector** belonging to λ.

In this section the problem is that we need to find the eigenvalue, λ, and the corresponding eigenvector, **u**, of a square matrix **A**. **How can we find λ?**

We have

        Au = λ**u**

†   **Au** − λ**u** = 0

Remember the properties of the identity matrix, **I**:

        u = **Iu** and so λ**u** = λ**Iu**

† therefore becomes

        Au − λ**Iu** = 0

Factorizing we obtain

11.15 $\qquad (A - \lambda I)\mathbf{u} = \mathbf{0}$

? **Are there any solutions of 11.15?**

Yes, the trivial solution $\mathbf{u} = \mathbf{0}$. However for eigenvalues and eigenvectors, $\mathbf{u} \neq \mathbf{0}$.

? **What does matrix A need to satisfy in order to have a non-trivial solution to $(A - \lambda I)\mathbf{u} = \mathbf{0}$?**

As discussed in **Example 29** of the previous section, for non-trivial solution

11.16 $\qquad \det(A - \lambda I) = 0$

Hence λ is an eigenvalue of A if and only if 11.16 holds (see **Example 29**).

Eigenvalues were initially developed in the field of differential equations by **Jean D'Alembert**.

Alembert, 1717–83, was a French mathematician and the illegitimate son of Madame Tencin and an army officer, Louis Destouches. His mother left him on the steps of a local church and he was sent to an orphanage. His father recognized his difficulties and placed him under the care of Madame Rousseau. He always thought of Madame Rousseau as his mother.

Sadly, Alembert's father died when he was only 9 years old and his father's family had to look after his financial situation so that he could be educated.

In 1735 Alembert graduated and thought a career in law would suit him, but his real thirst and enthusiasm were for mathematics which he studied in his spare time. Three years later he did qualify as an advocate but did not pursue a career in this field, choosing mathematics instead.

For most of his life he worked for the Paris Academy of Science and the French Academy.

He was well known to have a short fuse and often argued with his contemporaries.

F2 Finding eigenvalues and eigenvectors

Example 31

Find the eigenvalues of $A = \begin{pmatrix} 2 & 0 \\ 1 & 3 \end{pmatrix}$.

Example 31 *continued*

Solution

We need to find λ in $\det(A - \lambda I) = 0$. First we obtain $A - \lambda I$:

$$A - \lambda I = \begin{pmatrix} 2 & 0 \\ 1 & 3 \end{pmatrix} - \lambda \begin{pmatrix} 1 & 0 \\ 0 & 1 \end{pmatrix} \qquad \left[I = \text{identity matrix} = \begin{pmatrix} 1 & 0 \\ 0 & 1 \end{pmatrix} \right]$$

$$= \begin{pmatrix} 2 & 0 \\ 1 & 3 \end{pmatrix} - \begin{pmatrix} \lambda & 0 \\ 0 & \lambda \end{pmatrix}$$

$$= \begin{pmatrix} 2 - \lambda & 0 \\ 1 & 3 - \lambda \end{pmatrix}$$

Substituting this into $\det(A - \lambda I)$ gives

$$\det(A - \lambda I) = \det \begin{pmatrix} 2 - \lambda & 0 \\ 1 & 3 - \lambda \end{pmatrix} \underset{\text{by } \boxed{11.1}}{=} (2 - \lambda)(3 - \lambda) - 0$$

For eigenvalues we equate the determinant to zero:

$$(2 - \lambda)(3 - \lambda) = 0$$

$$\lambda = 2, \lambda = 3 \qquad\qquad \text{[Solving]}$$

Example 32

Find the eigenvector corresponding to $\lambda = 2$ and $\lambda = 3$ for A in **Example 31**.

Solution

Let **u** be the eigenvector corresponding to $\lambda = 2$. Substituting $A = \begin{pmatrix} 2 & 0 \\ 1 & 3 \end{pmatrix}$ and $\lambda = 2$ into

$\boxed{11.15}$ $[A - \lambda I]\mathbf{u} = 0$

gives

$$\left[\begin{pmatrix} 2 & 0 \\ 1 & 3 \end{pmatrix} - \begin{pmatrix} 2 & 0 \\ 0 & 2 \end{pmatrix} \right] \mathbf{u} = 0$$

$$\begin{pmatrix} 0 & 0 \\ 1 & 1 \end{pmatrix} \mathbf{u} = 0 \qquad\qquad \text{[Simplifying]}$$

Remember the zero column vector, $0 = \begin{pmatrix} 0 \\ 0 \end{pmatrix}$. Let $\mathbf{u} = \begin{pmatrix} x \\ y \end{pmatrix}$ and so we have

$$\begin{pmatrix} 0 & 0 \\ 1 & 1 \end{pmatrix} \begin{pmatrix} x \\ y \end{pmatrix} = \begin{pmatrix} 0 \\ 0 \end{pmatrix}$$

Multiplying out gives

$$0 + 0 = 0$$
$$x + y = 0$$

$\boxed{11.1}$ $\det \begin{pmatrix} a & b \\ c & d \end{pmatrix} = ad - cb$

Example 32 *continued*

Remember that the eigenvector cannot be zero; $\mathbf{u} \neq \mathbf{0}$ therefore at least one of the values, x or y, must be non-zero. From the second equation we have

$$x + y = 0$$
$$x = -y$$

Solutions are

$$x = 1, \ y = -1$$
$$x = 2, \ y = -2$$
$$x = \pi, \ y = -\pi$$

We have an infinite number of solutions, so for the time being we need to write the eigenvector $\mathbf{u}$ in general terms. **How?**

Let $x = a$ then $y = -a$ where $a \neq 0$ and is any real number.

The eigenvectors corresponding to $\lambda = 2$ are

$$\mathbf{u} = \begin{pmatrix} a \\ -a \end{pmatrix} = a \begin{pmatrix} 1 \\ -1 \end{pmatrix} \text{ where } a \neq 0$$

Similarly we find the general eigenvector $\mathbf{v}$ corresponding to $\lambda = 3$. Putting $\lambda = 3$ into $[\mathbf{A} - \lambda \mathbf{I}]\mathbf{v} = \mathbf{0}$ gives

$$\left[\begin{pmatrix} 2 & 0 \\ 1 & 3 \end{pmatrix} - \begin{pmatrix} 3 & 0 \\ 0 & 3 \end{pmatrix} \right] \mathbf{v} = \mathbf{0}$$

$$\begin{pmatrix} -1 & 0 \\ 1 & 0 \end{pmatrix} \mathbf{v} = \mathbf{0}$$

By writing $\mathbf{v} = \begin{pmatrix} x \\ y \end{pmatrix}$ [different x and y from those above] and $\mathbf{0} = \begin{pmatrix} 0 \\ 0 \end{pmatrix}$ we obtain

$$\begin{pmatrix} -1 & 0 \\ 1 & 0 \end{pmatrix}\begin{pmatrix} x \\ y \end{pmatrix} = \begin{pmatrix} 0 \\ 0 \end{pmatrix}$$

Multiplying out:

$$-x + 0 = 0$$
$$x + 0 = 0$$

From these equations we must have $x = 0$. **What is y equal to?**

We can choose y to be any real number apart from zero. Thus

$$y = a \text{ where } a \neq 0$$

The general eigenvector of $\lambda = 3$ is found by substituting $x = 0$ and $y = a$:

$$\mathbf{v} = \begin{pmatrix} 0 \\ a \end{pmatrix} = a \begin{pmatrix} 0 \\ 1 \end{pmatrix}$$

Note that for one eigenvalue there are infinitely many eigenvectors. We can choose any one of these eigenvectors for the corresponding eigenvalue. By putting $a = 1$, an eigenvector can be selected. In **Example 32**:

$$\mathbf{u} = \begin{pmatrix} 1 \\ -1 \end{pmatrix} \text{ is an eigenvector corresponding to } \lambda = 2$$

$$\mathbf{v} = \begin{pmatrix} 0 \\ 1 \end{pmatrix} \text{ is an eigenvector corresponding to } \lambda = 3$$

? **What does all this mean?**

It means that we must have $\mathbf{Au} = \lambda_1\mathbf{u}$ and $\mathbf{Av} = \lambda_2\mathbf{v}$ which in this case is

$$\mathbf{Au} = 2\mathbf{u} \text{ and } \mathbf{Av} = 3\mathbf{v}$$

Plotting these eigenvectors and the effect of multiplying by the matrix A gives:

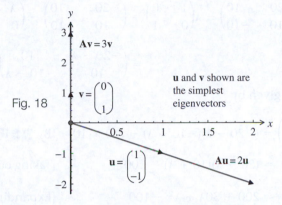

Fig. 18

The effect on the eigenvectors of multiplying by the matrix A is to produce scalar multiples of itself, as you can see in Fig. 18.

Eigenvectors are non-zero vectors which are transformed to scalar multiples of themselves by a square matrix A.

Generally the values of the elements of the eigenvectors are scaled to the smallest possible size of whole numbers.

The resulting equation from $\det(\mathbf{A} - \lambda\mathbf{I}) = 0$ is called the characteristic equation.

You need to be very careful with the algebra when expanding brackets. For example, expanding brackets such as $(-1 - \lambda)(-2 - \lambda)$ can lead to confusion with the minus signs.

However if you factor out the two minus signs then it should be easier to expand this:

$$(-1 - \lambda)(-2 - \lambda) = {\underset{= \,+}{=}}(1 + \lambda)(2 + \lambda) = (1 + \lambda)(2 + \lambda)$$

F3 **Applications to vibrations**

Example 33

Consider a two-storey building subject to earthquake oscillations, as shown in Fig. 19.

Fig. 19

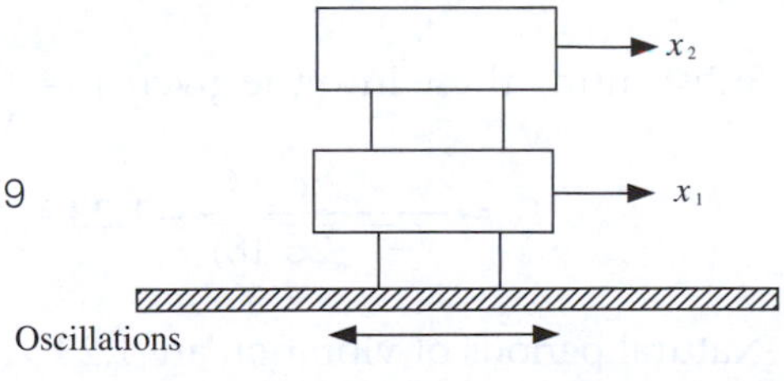

Oscillations

The equations of motion are expressed as

$\ddot{\mathbf{x}} = \mathbf{Ax}$ where $\mathbf{x} = \begin{pmatrix} x_1 \\ x_2 \end{pmatrix}$.

The period, T, of natural vibrations is given by

$T = \dfrac{2\pi}{\sqrt{-\lambda}}$ where λ is the eigenvalue of the matrix A. Find the period(s) if $\mathbf{A} = \begin{pmatrix} -20 & 10 \\ 10 & -10 \end{pmatrix}$.

Example 33 *continued*

Solution

Substituting the given matrix **A** into **A** − λ**I** yields

$$\mathbf{A} - \lambda\mathbf{I} = \begin{pmatrix} -20 & 10 \\ 10 & -10 \end{pmatrix} - \lambda\begin{pmatrix} 1 & 0 \\ 0 & 1 \end{pmatrix} = \begin{pmatrix} -20 & 10 \\ 10 & -10 \end{pmatrix} - \begin{pmatrix} \lambda & 0 \\ 0 & \lambda \end{pmatrix}$$

$$= \begin{pmatrix} -20 - \lambda & 10 \\ 10 & -10 - \lambda \end{pmatrix}$$

Therefore det(**A** − λ**I**) is given by

$$\det\begin{pmatrix} -20 - \lambda & 10 \\ 10 & -10 - \lambda \end{pmatrix} = (-20 - \lambda)(-10 - \lambda) - (10 \times 10) \quad \text{[By } \boxed{11.1}\text{]}$$

$$= (20 + \lambda)(10 + \lambda) - 100 \qquad \text{[Taking out two minus signs]}$$

$$= 200 + 30\lambda + \lambda^2 - 100 \qquad \text{[Expanding]}$$

$$= \lambda^2 + 30\lambda + 100$$

Equating this to zero gives the quadratic equation

$$\lambda^2 + 30\lambda + 100 = 0$$

? **How do we solve this quadratic equation?**

Apply the formula

$\boxed{1.16}$ $\qquad \lambda = \dfrac{-b \pm \sqrt{b^2 - 4ac}}{2a}$

with $a = 1$, $b = 30$ and $c = 100$:

$$\lambda = \frac{-30 \pm \sqrt{(-30)^2 - (4 \times 1 \times 100)}}{2} = \frac{-30 \pm \sqrt{500}}{2}$$

$$= -26.18, \, -3.82$$

The two eigenvalues are $\lambda_1 = -26.18$ and $\lambda_2 = -3.82$.

Substituting these into the given $T = \dfrac{2\pi}{\sqrt{-\lambda}}$:

$$T_1 = \frac{2\pi}{\sqrt{-(-26.18)}} = 1.23 \text{ s}, \quad T_2 = \frac{2\pi}{\sqrt{-(-3.82)}} = 3.21 \text{ s}$$

Natural periods of vibration are 1.23 s (2 d.p.) and 3.21 s (2 d.p.).

$\boxed{11.1}$ $\det\begin{pmatrix} a & b \\ c & d \end{pmatrix} = ad - cb$

F4 **3 × 3 Matrices**

Example 34

Determine the eigenvalues of

$$A = \begin{pmatrix} 1 & 0 & 4 \\ 0 & 4 & 0 \\ 3 & 5 & -3 \end{pmatrix}$$

Solution

We have

$$A - \lambda I = \begin{pmatrix} 1 & 0 & 4 \\ 0 & 4 & 0 \\ 3 & 5 & -3 \end{pmatrix} - \begin{pmatrix} \lambda & 0 & 0 \\ 0 & \lambda & 0 \\ 0 & 0 & \lambda \end{pmatrix} = \begin{pmatrix} 1-\lambda & 0 & 4 \\ 0 & 4-\lambda & 0 \\ 3 & 5 & -3-\lambda \end{pmatrix}$$

It's easier to remember that $A - \lambda I$ is actually A with $-\lambda$ associated with each entry along the leading diagonal (from top left to bottom right). We need to evaluate $\det(A - \lambda I)$. **What is the easiest way to evaluate this?**

From the properties of determinants we know that it will be easier to evaluate the determinant along the second row, containing the elements 0, $4 - \lambda$ and 0. **Why?**

Because it has two zeros, so we do not have to evaluate the 2 × 2 determinants associated with these zeros.

$$\det(A - \lambda I) = \det\begin{pmatrix} 1-\lambda & 0 & 4 \\ 0 & 4-\lambda & 0 \\ 3 & 5 & -3-\lambda \end{pmatrix}$$

$$= (4-\lambda)\left[\det\begin{pmatrix} 1-\lambda & 4 \\ 3 & -3-\lambda \end{pmatrix}\right] \quad \text{[Expanding second row]}$$

$$= (4-\lambda)\left[(1-\lambda)(-3-\lambda) - (3 \times 4)\right] \quad \text{[By } \boxed{11.1} \text{]}$$

$$= (4-\lambda)\left[(\lambda-1)(3+\lambda) - 12\right] \quad \text{[Taking out minus signs]}$$

$$= (4-\lambda)[3\lambda + \lambda^2 - 3 - \lambda - 12]$$

$$= (4-\lambda)[\lambda^2 + 2\lambda - 15] \quad \text{[Simplifying]}$$

$$= (4-\lambda)(\lambda+5)(\lambda-3) \quad \text{[Factorizing]}$$

This is equal to zero because for eigenvalues $\det(A - \lambda I) = 0$:

$$(4-\lambda)(\lambda+5)(\lambda-3) = 0$$

$$\lambda=4, \quad \lambda=-5, \quad \lambda=3$$

$\boxed{11.1}$ $\det\begin{pmatrix} a & b \\ c & d \end{pmatrix} = ad - cb$

Example 35

Find the eigenvectors associated with $\lambda = 3$ for the matrix **A** in **Example 34**.

Solution

Substituting $\lambda = 3$ and $\mathbf{A} = \begin{pmatrix} 1 & 0 & 4 \\ 0 & 4 & 0 \\ 3 & 5 & -3 \end{pmatrix}$ into $(\mathbf{A} - \lambda \mathbf{I})\mathbf{u} = \mathbf{0}$ gives (subtract 3 from the leading diagonal):

$$(\mathbf{A} - \lambda \mathbf{I})\mathbf{u} = \begin{pmatrix} 1-3 & 0 & 4 \\ 0 & 4-3 & 0 \\ 3 & 5 & -3-3 \end{pmatrix}\mathbf{u} = \mathbf{0}$$

where **u** is the eigenvector corresponding to $\lambda = 3$. Remember that $\mathbf{0} = \begin{pmatrix} 0 \\ 0 \\ 0 \end{pmatrix}$ and let $\mathbf{u} = \begin{pmatrix} x \\ y \\ z \end{pmatrix}$, so we have

$$\begin{pmatrix} -2 & 0 & 4 \\ 0 & 1 & 0 \\ 3 & 5 & -6 \end{pmatrix}\begin{pmatrix} x \\ y \\ z \end{pmatrix} = \begin{pmatrix} 0 \\ 0 \\ 0 \end{pmatrix}$$

Expanding this gives

†	$-2x + 0 + 4z = 0$
††	$0 + y + 0 = 0$
†††	$3x + 5y - 6z = 0$

From middle equation †† we know that $y = 0$. From top equation † we have

$$2x = 4z \text{ which gives } x = 2z$$

If $z = 1$ then $x = 2$; or more generally if $z = a$ then $x = 2a$ where $a \neq 0$.

The general eigenvector $\mathbf{u} = \begin{pmatrix} 2a \\ 0 \\ a \end{pmatrix} = a\begin{pmatrix} 2 \\ 0 \\ 1 \end{pmatrix}$ where $a \neq 0$ corresponds to $\lambda = 3$.

The eigenvectors corresponding to $\lambda = 4$, $\lambda = -5$ are part of **Exercise 11(f)**, question **4**.

F5 Eigenspace

Note that for the $\lambda = 3$ in the above **Example 35** we have an infinite number of eigenvectors by substituting various non-zero values of a:

$$\mathbf{u} = \begin{pmatrix} 2 \\ 0 \\ 1 \end{pmatrix} \text{ or } \mathbf{u} = \begin{pmatrix} 4 \\ 0 \\ 2 \end{pmatrix} \text{ or } \mathbf{u} = \begin{pmatrix} 1 \\ 0 \\ 1/2 \end{pmatrix} \text{ or } \mathbf{u} = \begin{pmatrix} -4 \\ 0 \\ -2 \end{pmatrix} \text{ etc.}$$

These solutions are given by **all** the points (apart from $x = y = z = 0$) on the line shown below in the Fig. 20:

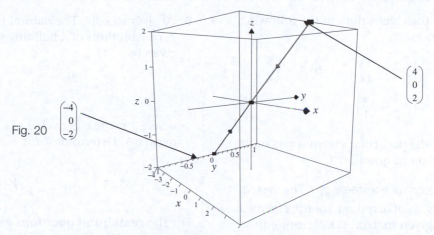

Fig. 20

In general, if **A** is a square matrix and λ is an eigenvalue of **A** with an eigenvector **u** then every scalar multiple (apart from 0) of the vector **u** is also an eigenvector belonging to the eigenvalue λ.

This space is called an **eigenspace** of λ and is denoted by E_λ.

For example, the eigenspace relating to **Example 32** for the eigenvalue $\lambda = 2$ is the eigenvector $\mathbf{u} = a\begin{bmatrix} 1 \\ -1 \end{bmatrix}$ and for $\lambda = 3$ the eigenvector $\mathbf{v} = a\begin{bmatrix} 0 \\ 1 \end{bmatrix}$, as shown below in Fig. 21:

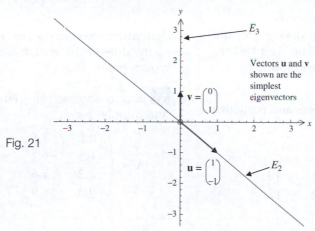

Fig. 21

SUMMARY

A scalar λ is called an eigenvalue of a square matrix **A** if it satisfies

11.15 $\qquad (\mathbf{A} - \lambda \mathbf{I})\mathbf{u} = 0$

and the corresponding **u** ($\neq \mathbf{0}$) is called an eigenvector.

We find λ by solving the resulting equation from

11.16 $\qquad \det(\mathbf{A} - \lambda \mathbf{I}) = 0$

Exercise **11(f)**

1 Find the eigenvalues of the following matrices:

a $A = \begin{pmatrix} 7 & 3 \\ 0 & -4 \end{pmatrix}$ **b** $A = \begin{pmatrix} 5 & -2 \\ 4 & -1 \end{pmatrix}$

c $A = \begin{pmatrix} -1 & 4 \\ 2 & 1 \end{pmatrix}$

2 Find the particular eigenvectors of the matrices of question **1**.

3 [control engineering] The system poles, λ, of a system are the eigenvalues of a given matrix **A**. Determine the system poles for

a $A = \begin{pmatrix} -1 & 1 \\ -2 & 1 \end{pmatrix}$

b $A = \begin{pmatrix} 1 & 2 \\ -4 & -3 \end{pmatrix}$

c $A = \begin{pmatrix} 5 & -5 \\ 5 & -2 \end{pmatrix}$

4 Obtain the general eigenvectors corresponding to $\lambda = 4$ and $\lambda = -5$ for the matrix in **Example 34**. (The eigenvector corresponding to $\lambda = 3$ was found in **Example 35**.)

5 Determine the eigenvalues and particular eigenvectors of

$$A = \begin{pmatrix} 5 & 0 & 0 \\ -9 & 4 & -1 \\ -6 & 2 & 1 \end{pmatrix}$$

6 [vibrations] The natural period, T, of vibrations of a building is given by

$$T = \frac{2\pi}{\sqrt{-\lambda}}$$

where λ is the eigenvalue of a given matrix **A**. Determine T for

$$A = \begin{pmatrix} -15 & 5 \\ 10 & -10 \end{pmatrix}$$

For the remaining questions use a computer algebra system or a graphical calculator.

7 [control engineering] The state matrix, **A**, of a system is given by

$$A = \begin{pmatrix} 3.16 & 2.87 & 1.61 \\ 4.93 & 12.03 & 2.64 \\ 5.12 & 3.84 & 4.20 \end{pmatrix}$$

Determine the eigenvalues. The eigenvalues of the system are called system poles.

8 [control engineering] Find the system poles for

$$A = \begin{pmatrix} -0.2 & 5.6 & 7.1 & 3.3 \\ 1.5 & -6.1 & -6.3 & -7.0 \\ -2.2 & 3.6 & 4.8 & 8.3 \\ -9.3 & 8.9 & 6.4 & -6.1 \end{pmatrix}$$

SECTION G **Diagonalization**

By the end of this section you will be able to:
▶ diagonalize a matrix
▶ find powers of matrices

G1 Diagonal matrices

? **What is meant by a diagonal matrix?**

A diagonal matrix is an n by n matrix where **all** entries to both sides of the leading diagonal are zero. (The leading diagonal is the diagonal from top left to bottom right of the matrix.)

? **Can you think of an example of a diagonal matrix?**

The identity matrix $\mathbf{I} = \begin{pmatrix} 1 & 0 & 0 \\ 0 & 1 & 0 \\ 0 & 0 & 1 \end{pmatrix}$. Another example is $\begin{pmatrix} 1 & 0 & 0 \\ 0 & 2 & 0 \\ 0 & 0 & 3 \end{pmatrix}$.

In this section we look at the diagonalization problem for matrices which is the following:

? **For a square matrix A, is there a matrix P such that $\mathbf{P}^{-1}\mathbf{AP}$ is a diagonal matrix?**

This process of converting any n by n matrix into a diagonal matrix is called **diagonalization**.

In this section we will show that diagonalizing a matrix is equivalent to finding the eigenvalues and corresponding eigenvectors of the matrix. Actually we will demonstrate that the eigenvalues are the leading diagonal entries in the diagonal matrix.

In the next subsection we state a procedure for diagonalization.

G2 Algorithm for diagonalization

Example 36

Let $\mathbf{A} = \begin{pmatrix} 1 & 0 \\ 1 & 2 \end{pmatrix}$ and $\mathbf{P} = \begin{pmatrix} 0 & -1 \\ 1 & 1 \end{pmatrix}$. Determine $\mathbf{P}^{-1}\mathbf{AP}$.

Solution

? **What do we need to find first?**

The inverse of the matrix $\mathbf{P}$, denoted $\mathbf{P}^{-1}$ is given by:

$$\mathbf{P}^{-1} = \begin{pmatrix} 0 & -1 \\ 1 & 1 \end{pmatrix}^{-1} = \frac{1}{0-(-1)}\begin{pmatrix} 1 & 1 \\ -1 & 0 \end{pmatrix} \quad \left[\begin{pmatrix} a & b \\ c & d \end{pmatrix}^{-1} = \frac{1}{ad-bc}\begin{pmatrix} d & -b \\ -c & a \end{pmatrix} \right]$$

$$= \begin{pmatrix} 1 & 1 \\ -1 & 0 \end{pmatrix}$$

Carrying out the matrix multiplication

$$\mathbf{P}^{-1}\mathbf{AP} = \begin{pmatrix} 1 & 1 \\ -1 & 0 \end{pmatrix}\begin{pmatrix} 1 & 0 \\ 1 & 2 \end{pmatrix}\begin{pmatrix} 0 & -1 \\ 1 & 1 \end{pmatrix}$$

$$= \begin{pmatrix} 2 & 2 \\ -1 & 0 \end{pmatrix}\begin{pmatrix} 0 & -1 \\ 1 & 1 \end{pmatrix} = \begin{pmatrix} 2 & 0 \\ 0 & 1 \end{pmatrix}$$

Note that $\mathbf{P}^{-1}\mathbf{AP} = \begin{pmatrix} 2 & 0 \\ 0 & 1 \end{pmatrix}$ is a diagonal matrix. (The entries of the leading diagonal, 2 and 1, are the eigenvalues of the matrix **A**.)

11.17 The procedure for diagonalizing an n by n matrix **A** with eigenvalues $\lambda_1, \lambda_2, \lambda_3, \cdots$ and λ_n is:

1 Determine the eigenvectors of the given matrix **A**. Call these $\mathbf{p}_1, \mathbf{p}_2, \mathbf{p}_3, \ldots$ and $\mathbf{p}_n$.
2 Form the matrix **P** by having the eigenvectors of step **1**, $\mathbf{p}_1, \mathbf{p}_2, \mathbf{p}_3, \ldots$ and $\mathbf{p}_n$, as its columns. That is, let

$$\mathbf{P} = (\mathbf{p}_1 \vdots \mathbf{p}_2 \vdots \mathbf{p}_3 \vdots \cdots \vdots \mathbf{p}_n) \text{ provided } \det(\mathbf{P}) \neq 0$$

If $\det(\mathbf{P}) = 0$ then we **cannot** diagonalize matrix **A**.

3 The diagonal matrix $\mathbf{D} = \mathbf{P}^{-1}\mathbf{AP}$ will have the eigenvalues $\lambda_1, \lambda_2, \lambda_3, \cdots$ and λ_n along its leading diagonal, that is

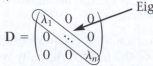

Eigenvalues of matrix **A**

$$\mathbf{D} = \begin{pmatrix} \lambda_1 & 0 & 0 \\ 0 & \ddots & 0 \\ 0 & 0 & \lambda_n \end{pmatrix}$$

4 It is good practice to check that matrices **P** and **D** actually work. For matrices of size **greater than** 2 by 2 the evaluation of the inverse matrix $\mathbf{P}^{-1}$ can be a tedious task. To avoid this evaluation, check that $\mathbf{PD} = \mathbf{AP}$. **Why?**

Because left-multiplying the above $\mathbf{D} = \mathbf{P}^{-1}\mathbf{AP}$ by **P** gives

$$\mathbf{PD} = \mathbf{P}(\mathbf{P}^{-1}\mathbf{AP})$$
$$= (\mathbf{PP}^{-1})\mathbf{AP} = \mathbf{IAP} = \mathbf{AP} \quad [\text{Remember } \mathbf{PP}^{-1} = \mathbf{I}]$$

Hence it is enough to check that $\mathbf{PD} = \mathbf{AP}$.

Next we illustrate this diagonalization process. We will **not** find the eigenvalues and eigenvectors in the examples below because you should be familiar with determining these by now. However you will need to find these for the questions in **Exercise 11(g)**.

Example 37

Determine the matrix **P** which diagonalizes the matrix $\mathbf{A} = \begin{pmatrix} 1 & 4 \\ 2 & 3 \end{pmatrix}$ given that the eigenvalues of this matrix are $\lambda_1 = -1$ and $\lambda_2 = 5$ with the corresponding eigenvectors $\mathbf{u} = \begin{pmatrix} -2 \\ 1 \end{pmatrix}$ and $\mathbf{v} = \begin{pmatrix} 1 \\ 1 \end{pmatrix}$ respectively.

Solution

Step 1:

We have been given the eigenvectors in the question.

Step 2:

The matrix $\mathbf{P} = (\mathbf{u} \vdots \mathbf{v}) = \begin{pmatrix} -2 & 1 \\ 1 & 1 \end{pmatrix}$ where **u** and **v** are the given eigenvectors.

> **Example 37** *continued*
>
> *Step 3:*
>
> Before we write down the diagonal matrix we need to make sure that the order of eigenvalues corresponds to the order of the eigenvectors given in matrix **P**. We have **P** = (**u** ⋮ **v**) therefore our diagonal matrix **D** is given by
>
> $$\mathbf{D} = \begin{pmatrix} -1 & 0 \\ 0 & 5 \end{pmatrix}$$
>
> *Step 4:*
>
> We need to confirm that this matrix **P** does indeed diagonalize the given matrix **A**. **How?**
>
> By checking that **PD = AP**:
>
> $$\mathbf{PD} = \begin{pmatrix} -2 & 1 \\ 1 & 1 \end{pmatrix}\begin{pmatrix} -1 & 0 \\ 0 & 5 \end{pmatrix} = \begin{pmatrix} 2 & 5 \\ -1 & 5 \end{pmatrix}$$
>
> $$\mathbf{AP} = \begin{pmatrix} 1 & 4 \\ 2 & 3 \end{pmatrix}\begin{pmatrix} -2 & 1 \\ 1 & 1 \end{pmatrix} = \begin{pmatrix} 2 & 5 \\ -1 & 5 \end{pmatrix}$$
>
> This confirms that the matrix **P** does diagonalize the given matrix **A**.

Notice that the eigenvalues, $\lambda_1 = -1$ and $\lambda_2 = 5$, of the given matrix **A** are entries along the leading diagonal in the matrix **D**. They occur in the order λ_1 and then λ_2 because the matrix **P** is created by **P** = (**u** ⋮ **v**) where **u** is the eigenvector belonging to λ_1 and **v** is the eigenvector belonging to the other eigenvalue λ_2. **What would be the diagonal matrix if the matrix P was created by interchanging u and v, that is, P = (v ⋮ u)?**

Our diagonal matrix would be **D** = $\begin{pmatrix} 5 & 0 \\ 0 & -1 \end{pmatrix}$.   [See **Exercise 11(g)**]

The process of finding the inverse matrix $\mathbf{P}^{-1}$ of a 3 by 3 matrix can be a tedious task as you will have noticed in **Section 11C**. For the next example, check your result by showing that **PD = AP**.

> **Example 38**
>
> For the matrix A = $\begin{pmatrix} 1 & -2 & 3 \\ 0 & 2 & 5 \\ 0 & 0 & 3 \end{pmatrix}$, determine the matrix **P** which diagonalizes matrix **A** given that the eigenvalues of **A** are $\lambda_1 = 1$, $\lambda_2 = 2$ and $\lambda_3 = 3$ with corresponding eigenvectors **u** = $\begin{pmatrix} 1 \\ 0 \\ 0 \end{pmatrix}$, **v** = $\begin{pmatrix} -2 \\ 1 \\ 0 \end{pmatrix}$ and **w** = $\begin{pmatrix} -7 \\ 10 \\ 2 \end{pmatrix}$ respectively.
>
> Check that **P** does indeed diagonalize the given matrix **A**.

Example 38 *continued*

Solution

Step 1 and Step 2:

We have been given the eigenvalues and corresponding eigenvectors of the matrix A.
What is our matrix P equal to?

We have

$$P = (u \vdots v \vdots w) = \begin{pmatrix} 1 & -2 & -7 \\ 0 & 1 & 10 \\ 0 & 0 & 2 \end{pmatrix} \quad \left[\text{Given } u = \begin{pmatrix} 1 \\ 0 \\ 0 \end{pmatrix}, v = \begin{pmatrix} -2 \\ 1 \\ 0 \end{pmatrix} \text{ and } w = \begin{pmatrix} -7 \\ 10 \\ 2 \end{pmatrix} \right]$$

Step 3:

The diagonal matrix has the eigenvalues $\lambda_1 = 1$, $\lambda_2 = 2$ and $\lambda_3 = 3$ along the leading
diagonal, that is

$$P^{-1}AP = D = \begin{pmatrix} 1 & 0 & 0 \\ 0 & 2 & 0 \\ 0 & 0 & 3 \end{pmatrix}$$

Step 4:

Checking that we have the correct **P** and **D** matrices by showing PD = AP:

$$PD = \begin{pmatrix} 1 & -2 & -7 \\ 0 & 1 & 10 \\ 0 & 0 & 2 \end{pmatrix} \begin{pmatrix} 1 & 0 & 0 \\ 0 & 2 & 0 \\ 0 & 0 & 3 \end{pmatrix} = \begin{pmatrix} 1 & -4 & -21 \\ 0 & 2 & 30 \\ 0 & 0 & 6 \end{pmatrix}$$

$$AP = \begin{pmatrix} 1 & -2 & 3 \\ 0 & 2 & 5 \\ 0 & 0 & 3 \end{pmatrix} \begin{pmatrix} 1 & -2 & -7 \\ 0 & 1 & 10 \\ 0 & 0 & 2 \end{pmatrix} = \begin{pmatrix} 1 & -4 & -21 \\ 0 & 2 & 30 \\ 0 & 0 & 6 \end{pmatrix}$$

Hence this confirms that the matrix **P** does indeed diagonalize the given matrix **A**.

In general a diagonal matrix is easier to work with because if you are multiplying, solving a
system of equations or finding eigenvalues, it is always preferable to have a diagonal matrix.

However **not** all matrices can be diagonalized. For example, $A = \begin{pmatrix} 2 & 2 \\ 0 & 2 \end{pmatrix}$ has eigenvalues

$\lambda_1 = 2$ and $\lambda_2 = 2$ with corresponding eigenvectors $u = \begin{pmatrix} 1 \\ 0 \end{pmatrix}$ and $v = \begin{pmatrix} 1 \\ 0 \end{pmatrix}$. **What is matrix P**
equal to?

$$P = (u \vdots v) = \begin{pmatrix} 1 & 1 \\ 0 & 0 \end{pmatrix}$$

What is the determinant of matrix P?

$$\det(P) = \det\begin{pmatrix} 1 & 1 \\ 0 & 0 \end{pmatrix} = (1 \times 0) - (0 \times 1) = 0$$

We have $\det(P) = 0$ therefore we **cannot** diagonalize matrix A.
Next we look at the applications of diagonalization.

G3 **Powers of matrices**

? We can use the above diagonalization process to find powers of matrices. **What does diagonalization have to do with powers of matrices?**

If **A** is a square matrix which is diagonalizable then there is a matrix **P** such that $\mathbf{P^{-1}AP = D}$ where **D** is a diagonal matrix. It can be shown that

$$\mathbf{A}^m = \mathbf{PD}^m\mathbf{P}^{-1} \text{ where } m \text{ is a positive whole number}$$

? **We can use this formula to find $\mathbf{A}^m$ but how do we determine $\mathbf{D}^m$?**

The matrix $\mathbf{D}^m$ is a diagonal matrix with the leading diagonal entries raised to the power m, that is

$$\text{If } \mathbf{D} = \begin{pmatrix} a_1 & 0 & 0 \\ 0 & \ddots & 0 \\ 0 & 0 & a_n \end{pmatrix} \text{ then } \mathbf{D}^m = \begin{pmatrix} a_1^{\,m} & 0 & 0 \\ 0 & \ddots & 0 \\ 0 & 0 & a_n^{\,m} \end{pmatrix}$$

We can apply this to evaluate powers of the square matrix **A** because multiplying diagonal matrices or finding powers of diagonal matrices is much easier. (We only need to find the powers of the entries on the **leading diagonal**.)

In general to find $\mathbf{A}^m$ we have to multiply m copies of the matrix **A** and this is a tedious and laborious task. It is less demanding if we use the above, that is $\mathbf{A}^m = \mathbf{PD}^m\mathbf{P}^{-1}$ even though we need to find **P**, $\mathbf{P}^{-1}$ and **D** which is no easy task in itself. This formula means that if you want to work out $\mathbf{A}^m$ without working out lower powers, then using the diagonal matrix is efficient.

? Note that in the above subsection when we diagonalized a matrix we could avoid the calculation of $\mathbf{P}^{-1}\,\mathbf{AP}$. **Why?**

Because we check that $\mathbf{PD = AP}$ which means that $\mathbf{P^{-1}AP}$ is a diagonal matrix with eigenvalues along the leading diagonal. However in evaluating $\mathbf{A}^m$ we need to find $\mathbf{P}^{-1}$ because

$$\mathbf{A}^m = \mathbf{PD}^m\mathbf{P}^{-1}$$

Example 39

Let $\mathbf{A} = \begin{pmatrix} 1 & -2 & 3 \\ 0 & 2 & 5 \\ 0 & 0 & 3 \end{pmatrix}$. Determine $\mathbf{A}^5$ given that

$$\mathbf{P} = \begin{pmatrix} 1 & -2 & -7 \\ 0 & 1 & 10 \\ 0 & 0 & 2 \end{pmatrix} \text{ and } \mathbf{P}^{-1} = \frac{1}{2}\begin{pmatrix} 2 & 4 & -13 \\ 0 & 2 & -10 \\ 0 & 0 & 1 \end{pmatrix}$$

Example 39 *continued*

Solution

The given matrix **A** is the same as the matrix **A** given in previous **Example 38**. The

diagonal matrix is $\mathbf{D} = \mathbf{P}^{-1}\mathbf{AP} = \begin{pmatrix} 1 & 0 & 0 \\ 0 & 2 & 0 \\ 0 & 0 & 3 \end{pmatrix}$. How do we find $\mathbf{A}^5$?

By applying the above proposition $\mathbf{A}^m = \mathbf{P}\mathbf{D}^m\mathbf{P}^{-1}$ with $m = 5$:

$$\mathbf{A}^5 = \mathbf{P}\mathbf{D}^5\mathbf{P}^{-1}$$

Substituting the matrices $\mathbf{P} = \begin{pmatrix} 1 & -2 & -7 \\ 0 & 1 & 10 \\ 0 & 0 & 2 \end{pmatrix}$, $\mathbf{D} = \begin{pmatrix} 1 & 0 & 0 \\ 0 & 2 & 0 \\ 0 & 0 & 3 \end{pmatrix}$ and

$\mathbf{P}^{-1} = \dfrac{1}{2}\begin{pmatrix} 2 & 4 & -13 \\ 0 & 2 & -10 \\ 0 & 0 & 1 \end{pmatrix}$ into $\mathbf{A}^5 = \mathbf{P}\mathbf{D}^5\mathbf{P}^{-1}$ gives

$\mathbf{A}^5 = \mathbf{P}\mathbf{D}^5\mathbf{P}^{-1}$

$= \begin{pmatrix} 1 & -2 & -7 \\ 0 & 1 & 10 \\ 0 & 0 & 2 \end{pmatrix}\begin{pmatrix} 1 & 0 & 0 \\ 0 & 2 & 0 \\ 0 & 0 & 3 \end{pmatrix}^5 \dfrac{1}{2}\begin{pmatrix} 2 & 4 & -13 \\ 0 & 2 & -10 \\ 0 & 0 & 1 \end{pmatrix}$

$= \dfrac{1}{2}\begin{pmatrix} 1 & -2 & -7 \\ 0 & 1 & 10 \\ 0 & 0 & 2 \end{pmatrix}\begin{pmatrix} 1^5 & 0 & 0 \\ 0 & 2^5 & 0 \\ 0 & 0 & 3^5 \end{pmatrix}\begin{pmatrix} 2 & 4 & -13 \\ 0 & 2 & -10 \\ 0 & 0 & 1 \end{pmatrix}$ $\left[\text{Taking } \dfrac{1}{2} \text{ to the front}\right]$

$= \dfrac{1}{2}\begin{pmatrix} 1 & -2 & -7 \\ 0 & 1 & 10 \\ 0 & 0 & 2 \end{pmatrix}\begin{pmatrix} 1 & 0 & 0 \\ 0 & 32 & 0 \\ 0 & 0 & 243 \end{pmatrix}\begin{pmatrix} 2 & 4 & -13 \\ 0 & 2 & -10 \\ 0 & 0 & 1 \end{pmatrix}$ $\begin{bmatrix}\text{Replacing} \\ 1^5 = 1, 2^5 = 32 \text{ and } 3^5 = 243\end{bmatrix}$

$= \dfrac{1}{2}\begin{pmatrix} 1 & -64 & -1701 \\ 0 & 32 & 2430 \\ 0 & 0 & 486 \end{pmatrix}\begin{pmatrix} 2 & 4 & -13 \\ 0 & 2 & -10 \\ 0 & 0 & 1 \end{pmatrix}$ $\begin{bmatrix}\text{Multiplying the first} \\ \text{two matrices on the left}\end{bmatrix}$

$= \dfrac{1}{2}\begin{pmatrix} 2 & -124 & -1074 \\ 0 & 64 & 2110 \\ 0 & 0 & 486 \end{pmatrix} = \begin{pmatrix} 1 & -62 & -537 \\ 0 & 32 & 1055 \\ 0 & 0 & 243 \end{pmatrix}$ $\begin{bmatrix}\text{Multiplying by} \\ \text{scalar } 1/2\end{bmatrix}$

You may check this final result by using any appropriate software.

By examining the diagonal matrix **D** you can find the determinant of the initial matrix **A** by multiplying the entries on the leading diagonal, because the determinant of **A** is the product of the eigenvalues which are the entries on the leading diagonal, that is

11.18 $\det(\mathbf{A}) = \lambda_1 \lambda_2 \cdots \lambda_n$

Clearly you don't need to diagonalize, you just need to find the eigenvalues.

SUMMARY

If an n by n matrix $\mathbf{A}$ is diagonalizable with $\mathbf{P}^{-1}\mathbf{A}\mathbf{P} = \mathbf{D}$ where $\mathbf{D}$ is a diagonal matrix then $\mathbf{A}^m = \mathbf{P}\mathbf{D}^m\mathbf{P}^{-1}$.

Exercise 11(g)

Solutions at end of book. Complete solutions available at www.palgrave.com/engineering/singh

In this exercise you may check your numerical answers using mathematical software.

1 For the following matrices find:

 i The eigenvalues and corresponding eigenvectors.
 ii Matrices $\mathbf{P}$ and $\mathbf{D}$ such that $\mathbf{D} = \mathbf{P}^{-1}\mathbf{A}\mathbf{P}$ where $\mathbf{D}$ is a diagonal matrix.

a $\mathbf{A} = \begin{bmatrix} 1 & 0 \\ 0 & 2 \end{bmatrix}$ **b** $\mathbf{A} = \begin{bmatrix} 1 & 1 \\ 1 & 1 \end{bmatrix}$

c $\mathbf{A} = \begin{bmatrix} 3 & 0 \\ 4 & 4 \end{bmatrix}$ **d** $\mathbf{A} = \begin{bmatrix} 2 & 2 \\ 1 & 3 \end{bmatrix}$

2 For the matrices in question 1 find $\mathbf{A}^5$.

3 For the following matrices determine:

 i The eigenvalues and corresponding eigenvectors.
 ii Matrices $\mathbf{P}$ and $\mathbf{D}$ where $\mathbf{P}$ is an invertible (has an inverse) matrix and $\mathbf{D} = \mathbf{P}^{-1}\mathbf{A}\mathbf{P}$ is a diagonal matrix.
 iii Determine $\mathbf{A}^4$ in each case by using the results of parts **i** and **ii**.
 To find $\mathbf{P}^{-1}$ you may use MATLAB.

a $\mathbf{A} = \begin{bmatrix} 1 & 0 & 0 \\ 0 & 2 & 0 \\ 0 & 0 & 3 \end{bmatrix}$

b $\mathbf{A} = \begin{bmatrix} -1 & 4 & 0 \\ 0 & 4 & 3 \\ 0 & 0 & 5 \end{bmatrix}$

Examination questions 11

Solutions at end of book. Complete solutions available at www.palgrave.com/engineering/singh

In this exercise you may check your numerical answers using mathematical software.

1 Let $\mathbf{A} = \begin{pmatrix} 1 & 2 \\ 3 & 4 \end{pmatrix}$ and $\mathbf{B} = \begin{pmatrix} 5 & 6 \\ 7 & 8 \end{pmatrix}$.
Determine

 i $(\mathbf{A} - \mathbf{B})(\mathbf{A} + \mathbf{B})$

 ii $\mathbf{A}^2 - \mathbf{B}^2$

University of Hertfordshire, UK, 2008

2 Given the matrix

$$\mathbf{A} = \frac{1}{7}\begin{pmatrix} 3 & -2 & -6 \\ -2 & 6 & -3 \\ -6 & -3 & -2 \end{pmatrix}$$

a Compute $\mathbf{A}^2$ and $\mathbf{A}^3$.
b Based on these results, determine the matrices $\mathbf{A}^{-1}$ and $\mathbf{A}^{2004}$.

University of Wisconsin, USA, 2004

3 Let $\mathbf{A}$, $\mathbf{B}$ and $\mathbf{C}$ be matrices defined by

$$\mathbf{A} = \begin{pmatrix} 1 & 0 & -1 \\ 1 & 1 & 1 \\ 1 & 2 & 3 \end{pmatrix}, \quad \mathbf{B} = \begin{pmatrix} 1 & 1 \\ 1 & 2 \\ 0 & -1 \\ 0 & 1 \end{pmatrix},$$

$$\mathbf{C} = \begin{pmatrix} 1 & -1 & 1 & 0 \\ -2 & 1 & 2 & 3 \end{pmatrix}$$

Which of the following are defined?

$\mathbf{A}^T$, $\mathbf{A}\mathbf{B}$, $\mathbf{B} + \mathbf{C}$, $\mathbf{A} - \mathbf{B}$, $\mathbf{C}\mathbf{B}$, $\mathbf{B}\mathbf{C}^T$, $\mathbf{A}^2$

Compute those matrices which are defined.

Jacobs University, Germany, 2002

4 a If $\mathbf{A} = \begin{pmatrix} 1 & 2 \\ 3 & 4 \end{pmatrix}$ and $\mathbf{B} = \begin{pmatrix} 0 & 1 \\ -1 & 0 \end{pmatrix}$,

 compute $\mathbf{A}^2$, $\mathbf{B}^2$, $\mathbf{A}\mathbf{B}$ and $\mathbf{B}\mathbf{A}$.

Examination questions **11 continued**   Solutions at end of book. Complete solutions available at www.palgrave.com/engineering/singh

b If $A = \begin{pmatrix} a & b \\ c & d \end{pmatrix}$ and $B = \begin{pmatrix} e & f \\ g & h \end{pmatrix}$, compute $AB - BA$.

Queen Mary, University of London, UK, 2006

5 Let $A = \begin{pmatrix} \dfrac{1}{3} & \dfrac{1}{3} \\ \dfrac{1}{3} & \dfrac{1}{3} \end{pmatrix}$. Determine **i** A^2 **ii** A^3

University of Hertfordshire, UK, 2007

6 Let A be the 3×3 matrix determined by

$$A\begin{pmatrix} 0 \\ 1 \\ 1 \end{pmatrix} = \begin{pmatrix} -1 \\ 0 \\ 2 \end{pmatrix}, \quad A\begin{pmatrix} 1 \\ 0 \\ 1 \end{pmatrix} = \begin{pmatrix} 0 \\ -1 \\ 2 \end{pmatrix},$$

$$A\begin{pmatrix} 1 \\ 1 \\ 0 \end{pmatrix} = \begin{pmatrix} 1 \\ 1 \\ 2 \end{pmatrix}$$

Find A.

Columbia University, New York, USA, 2006

The next question has been slightly edited.

7 Find the determinant of the matrix

$$A = \begin{pmatrix} 0 & 1 & 5 \\ 3 & -6 & 9 \\ 2 & 6 & 1 \end{pmatrix}$$

University of Ottawa, Canada, 2004

8 Compute the inverse of the matrix

$$A = \begin{pmatrix} 2 & -1 & -4 \\ -1 & 1 & 2 \\ -1 & 1 & 3 \end{pmatrix}$$

Is A diagonalizable? Give reasons for your answer.

University of New Brunswick, Canada, 2000

9 Find adjA, det A and hence A^{-1} when

$$A = \begin{pmatrix} 1 & -1 \\ -1 & 1 \end{pmatrix}$$

and when

$$A = \begin{pmatrix} 3 & -4 & 1 \\ 1 & -1 & 3 \\ 2 & -2 & 5 \end{pmatrix}$$

University of Sussex, UK, 2008

10 If $A = \begin{pmatrix} 1 & 1 \\ 0 & 3 \end{pmatrix}$

a Find all the eigenvalues of A.
b Find a non-singular matrix Q and a diagonal matrix D such that $Q^{-1}AQ = D$ (that is, $A = QDQ^{-1}$).
c For the matrix A find A^5. (Non-singular means the matrix Q has an inverse.)

Purdue University, USA, 2006

11 Show that the eigenvalues and corresponding eigenvectors of

$$A = \begin{pmatrix} 1 & 1 & 1 \\ 1 & 1 & 1 \\ 1 & 1 & 1 \end{pmatrix} \text{ are given by}$$

$\lambda_1 = 0, \mathbf{u} = \begin{pmatrix} 1 \\ 1 \\ -2 \end{pmatrix}, \lambda_2 = 0, \mathbf{v} = \begin{pmatrix} 1 \\ -1 \\ 0 \end{pmatrix},$

and $\lambda_3 = 3, \mathbf{w} = \begin{pmatrix} 1 \\ 1 \\ 1 \end{pmatrix}$

University of Hertfordshire, UK, 2009

12 a Find the eigenvalues and corresponding eigenvectors of $A = \begin{pmatrix} 6 & 2 \\ 2 & 3 \end{pmatrix}$.

b Determine which of the following vectors

$$\vec{v} = \begin{pmatrix} 1 \\ 1 \\ 1 \\ -1 \end{pmatrix}, \quad \vec{w} = \begin{pmatrix} 1 \\ -1 \\ -1 \\ 2 \end{pmatrix}$$

is an eigenvector of the matrix

$$\begin{pmatrix} 1 & 1 & 1 & 1 \\ 1 & -1 & 1 & -1 \\ 1 & 1 & -1 & -1 \\ 1 & -1 & -1 & 1 \end{pmatrix} \text{ and find the}$$

corresponding eigenvalue.

University of New Brunswick, Canada, 2004

13 a Find the eigenvalues and corresponding eigenvectors for

$$A = \begin{pmatrix} 1 & 1 & -1 \\ 1 & 1 & 1 \\ 1 & 1 & 1 \end{pmatrix}$$

b Is A diagonalizable? Give reasons for your answer.

University of New Brunswick, Canada, 2000

1 Evaluate the following:

i $\begin{pmatrix} 5 & 0 \\ 0 & 5 \end{pmatrix}\begin{pmatrix} 3 & 0 \\ 0 & 3 \end{pmatrix}$

ii $\begin{pmatrix} 6 & 0 \\ 0 & 6 \end{pmatrix}\begin{pmatrix} -7 & 0 \\ 0 & -7 \end{pmatrix}$

iii $\begin{pmatrix} a & 0 \\ 0 & b \end{pmatrix}\begin{pmatrix} c & 0 \\ 0 & d \end{pmatrix}$

2 Find a, b, c and d which satisfy

$$\begin{pmatrix} 1 & 2 \\ 3 & 4 \end{pmatrix}\begin{pmatrix} a & b \\ c & d \end{pmatrix} = \begin{pmatrix} 1 & 0 \\ 0 & 1 \end{pmatrix}$$

3 If $A = \begin{pmatrix} 2 & 0 \\ 0 & 2 \end{pmatrix}$, find A^2, A^3 and A^{10}.

4 i Find x and y which satisfy

$$\begin{pmatrix} 5 & 6 \\ 4 & 5 \end{pmatrix}\begin{pmatrix} x \\ y \end{pmatrix} = \begin{pmatrix} 1 \\ 0 \end{pmatrix}$$

ii Find a and b which satisfy

$$\begin{pmatrix} 5 & 6 \\ 4 & 5 \end{pmatrix}\begin{pmatrix} a \\ b \end{pmatrix} = \begin{pmatrix} 0 \\ 1 \end{pmatrix}$$

iii Obtain the inverse of $\begin{pmatrix} 5 & 6 \\ 4 & 5 \end{pmatrix}$.

5 a For $A = \begin{pmatrix} 2 & 1 \\ 0 & 3 \end{pmatrix}$ show that

$A^2 - 5A + 6I = 0$.

b For $A = \begin{pmatrix} 2 & 3 \\ 4 & 1 \end{pmatrix}$ show that

$A^2 - 3A - 10I = 0$.

6 Evaluate

$$\begin{pmatrix} 5 & -1 & -2 \\ 10 & -2 & -4 \\ 15 & -3 & -6 \end{pmatrix}\begin{pmatrix} 1 & 1 & 3 \\ 1 & -1 & -1 \\ 2 & 3 & 8 \end{pmatrix}$$

What do you notice?
Could the same thing happen when multiplying two non-zero numbers?

7 Evaluate the following:

$$\det\begin{bmatrix} 1 & 15 & -6 & 7 & 77 \\ 0 & 2 & -973 & 533 & 207 \\ 0 & 0 & 3 & -66 & 1057 \\ 0 & 0 & 0 & 4 & 855 \\ 0 & 0 & 0 & 0 & 0 \end{bmatrix}$$

8 [electrical principles]   By using Kirchhoff's laws on a circuit we obtain the equations

$$100I_1 - 30I_2 - 40I_3 = -8$$
$$-30I_1 + 140I_2 - 60I_3 = -2$$
$$-40I_1 - 60I_2 + 170I_3 = 10$$

Write these equations in matrix form and solve for I_1, I_2 and I_3 by using Gaussian elimination.

9 [electrical principles]   The equations of a circuit are given by

$$9i_1 \quad -3i_2 \quad -5i_3 \quad = \quad 12 \times 10^{-3}$$
$$-3i_1 \quad +16.5i_2 \quad -0.5i_3 \quad = \quad 0$$
$$-5i_1 \quad -0.5i_2 \quad +20.5i_3 \quad = \quad 15 \times 10^{-3}$$

Solve for i_1, i_2 and i_3 by using matrices. Is the solution unique?

*10 [vibrations]   Consider the two-storey building shown in Fig. 22.

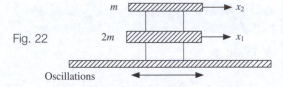

Fig. 22

Oscillations

The equations of motion are expressed as

$$\begin{pmatrix} 3k/2m & k/2m \\ k/m & k/m \end{pmatrix} \begin{pmatrix} x_1 \\ x_2 \end{pmatrix} = \begin{pmatrix} 0 \\ 0 \end{pmatrix}$$

where k is a negative constant and m is mass. Find the period, T, given by

$$T = \frac{2\pi}{\sqrt{-\lambda}} \quad \text{where } \lambda \text{ is an eigenvalue.}$$

*11 [control engineering]   A system is described by the equations

$$\dot{x}_1 = 2x_1 + 2x_2 + u_1$$
$$\dot{x}_2 = -x_1 + 5x_2 + u_2$$
$$y = x_1 + 4x_2$$

Write these equations in the form

$$\dot{x} = Ax + Iu$$
$$y = Cx$$

where $x = \begin{pmatrix} x_1 \\ x_2 \end{pmatrix}$, $\dot{x} = \begin{pmatrix} \dot{x}_1 \\ \dot{x}_2 \end{pmatrix}$ is the time derivative, $u = \begin{pmatrix} u_1 \\ u_2 \end{pmatrix}$ and I is the identity matrix.

i Determine the eigenvalues (system poles) of A.

ii Determine the corresponding general eigenvectors.

12 [control engineering]   The transfer function, TF, of a system is given by

$$TF = C(sI - A)^{-1}B$$

where A is the state matrix, B is the input matrix, C is the output matrix and I is the identity matrix.

For

$$A = \begin{pmatrix} 0 & -1 \\ 3 & -2 \end{pmatrix}, B = \begin{pmatrix} 1 \\ 0.6 \end{pmatrix} \text{ and }$$
$$C = (1 \;\; 0)$$

obtain TF.

For the remaining questions use a computer algebra system or a graphical calculator.

*13 For a system with

$$A = \begin{pmatrix} -5 & 2 & 1 \\ -6 & 5 & 9 \\ 7 & -1 & -4 \end{pmatrix}, B = \begin{pmatrix} -1 \\ 3 \\ 8 \end{pmatrix},$$
$$k_0 = 5, k_1 = -3 \text{ and } k_2 = 9$$

find the feedback matrix, F, by using Ackermann's formula:

$$F = (0 \;\; 0 \;\; 1)C_T^{-1} p(A)$$

where $C_T = (B \quad AB \quad A^2B)$ and $p(A) = A^3 + k_2A^2 + k_1A + k_0I$
(I is the 3 × 3 identity matrix).

14 By using $p(A) = A^3 + k_2A^2 + k_1A + k_0I$, find $p(A)$ for

a $A = \begin{pmatrix} 6 & 3 & -11 \\ -2 & 15 & 3 \\ 7 & 9 & 12 \end{pmatrix}$,

$$k_2 = 5, k_1 = -7 \text{ and } k_0 = 6$$

b $A = \begin{pmatrix} 9.31 & -6.11 & 2.30 \\ -2.75 & 3.93 & -6.51 \\ -8.99 & 12.23 & -1.75 \end{pmatrix}$,

$k_2 = -6.73$, $k_1 = 5.29$ and $k_0 = 6$

15 The transfer function, TF, of a system is given by

$$TF = C(sI - A)^{-1}B$$

where A is the state matrix, B is the input matrix and C is the output matrix.

For

$$A = \begin{pmatrix} 2.2 & 3.1 & 6.7 & -2.7 \\ 1.1 & -2.8 & 5.6 & -5.5 \\ 3.4 & 2.7 & -1.9 & -4.6 \\ -7.6 & 8.2 & 1.5 & 4.2 \end{pmatrix},$$

$$B = \begin{pmatrix} -1.1 \\ 2.3 \\ 1.7 \\ -4.1 \end{pmatrix} \quad \text{and}$$

$$C = (9.1 \quad 6.2 \quad 4.7 \quad 5.5), \text{ find TF.}$$

Vectors

SECTION A **Vector representation**

By the end of this section you will be able to:
▶ understand the terms vector and scalar
▶ apply vector addition and subtraction
▶ apply vectors to problems in mechanics

A1 **Vectors and scalars**

? A vector is a quantity that has **size** (magnitude) and **direction**. Examples of vectors are velocity, acceleration, force and displacement. A force of 10 N (newtons) upwards is a vector. **So what are scalars?**

A scalar is a quantity that has size but no direction. Examples of scalars are mass, length, volume, speed and temperature.

? **How do we write down vectors and scalars and how can we distinguish between them?**

A vector **from** O **to** A is denoted by $\overrightarrow{OA}$ or written in bold typeface **a** (Fig. 1). A scalar is denoted by a, **not** in bold, so that we can distinguish between vectors and scalars.

Fig. 1

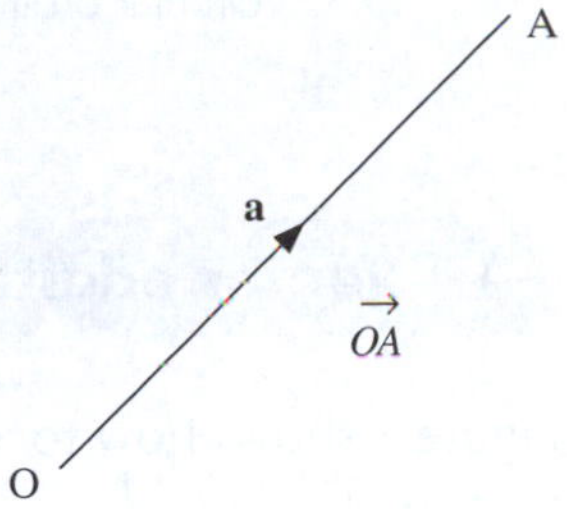

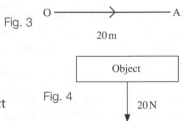

The vectors shown in Fig. 2 are given below in the notation outlined:

a is $\overrightarrow{CD}$ **b** is $\overrightarrow{DC}$ **c** is $\overrightarrow{EF}$

d is $\overrightarrow{HG}$ **e** is $\overrightarrow{IJ}$

Two vectors are **equivalent** if they have the same direction and magnitude. In Fig. 2, $\overrightarrow{HG} = \overrightarrow{IJ}$ because $\overrightarrow{HG}$ and $\overrightarrow{IJ}$ have the same direction and magnitude (length) but only differ in position.

Fig. 2

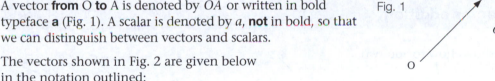

Also note that the direction of the arrow gives the direction of the vector, that is $\overrightarrow{CD}$ is **different** from $\overrightarrow{DC}$. The magnitude (or length) of the vector $\overrightarrow{AB}$ is denoted by $|\overrightarrow{AB}|$.

Fig. 3

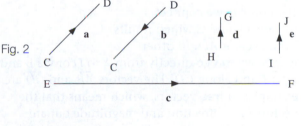

There are many examples of vectors in engineering, particularly in dynamics:

a A displacement of 20 m to the horizontal right of an object from O to A (Fig. 3).

Fig. 4

b A force on an object acting vertically downwards (Fig. 4).

c The velocity and acceleration of an object thrown vertically upwards, as illustrated in Fig. 5. The acceleration is the acceleration due to gravity, 9.8 m/s².

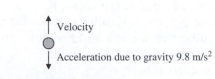

Fig. 5 Ball in upwards motion

d Figure 6 shows the various directions of the velocity and acceleration of a particle rotating in a circle. In each case the velocity and acceleration have magnitude and direction so they can be represented by vectors. The acceleration **a** illustrated in Fig. 6 is acting towards the centre of the circle.

The velocity of P is in the direction of the tangent at P.

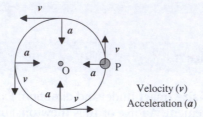

Velocity (*v*)
Acceleration (*a*)

Fig. 6 A ball on a piece of string being swung round the point O. (The expected velocity and direction of the ball in the absence of the string is indicated by the vectors **v**, but at every point P, the string forces the ball into a circular orbit, denoted by the vectors **a**.)

A2 Vector addition

Figure 7 shows how to add two vectors $\overrightarrow{AB}$ and $\overrightarrow{CD}$.

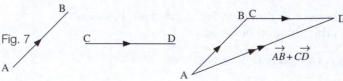

To add vectors $\overrightarrow{AB}$ and $\overrightarrow{CD}$ we connect the ends B and C which results in the vector $\overrightarrow{AD}$. In other words, we can go directly from A to D or via B and C, $\overrightarrow{AB}$ and along $\overrightarrow{CD}$. The vectors $\overrightarrow{AB}$ and $\overrightarrow{CD}$ are examples of **free** vectors, which means that the vectors have direction and magnitude but are not at a fixed position.

Consider a robot which moves an object from A to B and then moves it to C. Thus the object can be moved from A to C directly or via B as shown in Fig. 8.

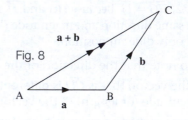

Consider the addition of vectors **a** and **b** in Fig. 9.

We shift vector **b** to the end of vector **a** and the result is going from O to the end of vector **b**. Generally **a** + **b** is called the resultant vector of **a** and **b** and is represented by the line with two arrows.
Let's do an example.

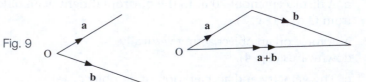

Example 1 *mechanics*

Two forces $\mathbf{F_1}$ and $\mathbf{F_2}$ are applied to an object O. Determine the magnitude and direction of the resultant force, $\mathbf{R} = \mathbf{F_1} + \mathbf{F_2}$.

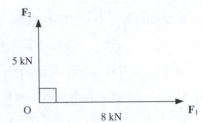

Fig. 10

Solution

Adding vectors $\mathbf{F_1}$ and $\mathbf{F_2}$ in Fig. 10 gives $\mathbf{R}$ (Fig. 11).

How can we obtain the magnitude (i.e. the length) of the vector R?

We can use Pythagoras:

$$|\mathbf{R}| = \sqrt{5^2 + 8^2} = 9.43 \text{ kN}$$

Fig. 11

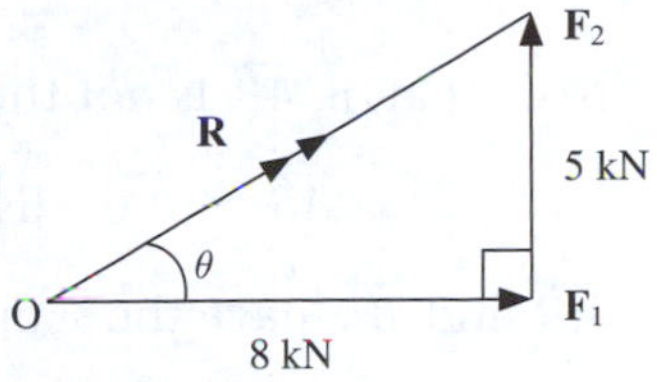

We also need to find the direction of $\mathbf{R}$ which is given by the angle θ in Fig. 11. By using trigonometry we have

$$\tan(\theta) = \frac{\text{opposite}}{\text{adjacent}} = \frac{5}{8}$$

Therefore taking inverse tan gives

$$\theta = \tan^{-1}\left(\frac{5}{8}\right) = 32.01° \quad \text{[By calculator]}$$

The resultant force, $\mathbf{R}$, has a magnitude of 9.43 kN and an angle of 32.01° as shown.

A3 **Subtraction of vectors**

We define $\overrightarrow{OA} - \overrightarrow{OB}$ as $\overrightarrow{OA} + \left(-\overrightarrow{OB}\right)$.

Example 2

By considering vectors $\overrightarrow{OA}$ and $\overrightarrow{OB}$ as in Fig. 12, show on a different diagram the vectors

i $\overrightarrow{OA} - \overrightarrow{OB}$ and ii $\overrightarrow{OB} - \overrightarrow{OA}$ iii Find $\overrightarrow{OA} - \overrightarrow{OA}$.

Fig. 12

Solution

i $-\overrightarrow{OB}$ is the vector with the same length as $\overrightarrow{OB}$ but is in the opposite direction. In fact $-\overrightarrow{OB}$ is the vector $\overrightarrow{BO}$ (Fig. 13).

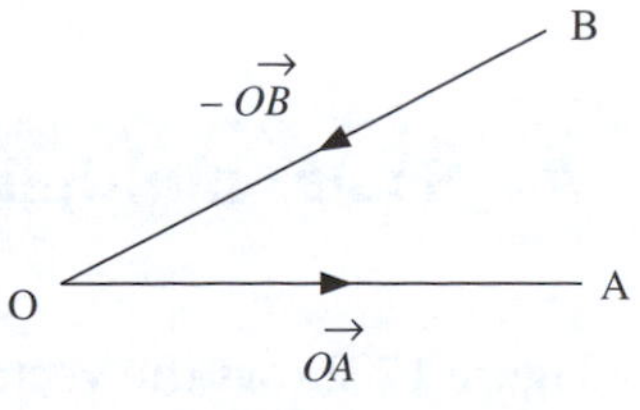

Therefore $\overrightarrow{OA} - \overrightarrow{OB}$ is given by

$$\overrightarrow{OA} - \overrightarrow{OB} = \overrightarrow{BA} \quad \text{(Fig. 14)}$$

Fig. 13

We can justify this by using algebra:

$$\overrightarrow{OA} - \overrightarrow{OB} = \overrightarrow{OA} + (-\overrightarrow{OB})$$
$$= -\overrightarrow{OB} + \overrightarrow{OA}$$
$$= \overrightarrow{BO} + \overrightarrow{OA} = \overrightarrow{BA}$$

Fig. 14

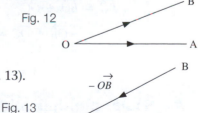

Example 2 *continued*

ii Similarly $-\overrightarrow{OA}$ is the vector $\overrightarrow{AO}$. Hence

$$\overrightarrow{OB} - \overrightarrow{OA} = \overrightarrow{AB} \quad \text{(Fig. 15)}$$

iii Clearly $\overrightarrow{OA} - \overrightarrow{OA} = 0$ [Zero vector]

The zero vector is denoted by **0**, and has zero magnitude and direction.

Fig. 15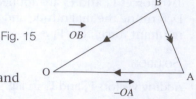

Note that, if $\overrightarrow{AB}$ is not the zero vector then

$$\overrightarrow{AB} \neq \overrightarrow{BA} \quad \text{[Not equal vectors]}$$

$\overrightarrow{AB}$ and $\overrightarrow{BA}$ have the same magnitude but opposite directions, in fact

$$\overrightarrow{AB} = -\overrightarrow{BA}$$

Example 3

Write **i** $\overrightarrow{AB}$ **ii** $\overrightarrow{BA}$ and **iii** $\overrightarrow{BC}$ in terms of **a**, **b** and **c** (see Fig. 16).

Solution

i $\overrightarrow{AB} = \overrightarrow{AO} + \overrightarrow{OB}$

$$= -\overrightarrow{OA} + \overrightarrow{OB} = -\mathbf{a} + \mathbf{b} = \mathbf{b} - \mathbf{a}$$

Fig. 16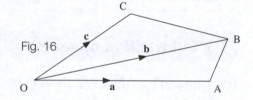

ii The vector $\overrightarrow{BA}$ is given by

$$\overrightarrow{BA} = \overrightarrow{BO} + \overrightarrow{OA}$$

$$= -\overrightarrow{OB} + \overrightarrow{OA} = -\mathbf{b} + \mathbf{a} = \mathbf{a} - \mathbf{b}$$

iii The vector $\overrightarrow{BC}$ is

$$\overrightarrow{BC} = \overrightarrow{BO} + \overrightarrow{OC}$$

$$= -\overrightarrow{OB} + \overrightarrow{OC} = -\mathbf{b} + \mathbf{c} = \mathbf{c} - \mathbf{b}$$

A4 **Scalar multiplication**

Figure 17 shows the vector **a** and its scalar multiples $2\mathbf{a}$, $-2\mathbf{a}$ and $\frac{1}{2}\mathbf{a}$. Notice that $2\mathbf{a}$ has the same direction as vector **a** but is twice as long,

Fig. 17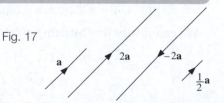

and $\frac{1}{2}$**a** is half the length of **a**. Also -2**a** has the same magnitude as 2**a** but is in the opposite direction.

Example 4

Figure 18 shows a triangle OAB with M as the midpoint of the line AB. Determine $\overrightarrow{OM}$ in terms of **a** and **b**.

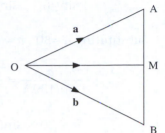

Fig. 18

Solution

Since M is the midpoint of BA we have

> † $\qquad \overrightarrow{OM} = \overrightarrow{OB} + \frac{1}{2}\overrightarrow{BA}$

We have $\overrightarrow{OB} = $ **b** but we need to express $\overrightarrow{BA}$ in terms of **a** and **b**.

$$\overrightarrow{BA} = \overrightarrow{BO} + \overrightarrow{OA}$$

$$= -\overrightarrow{OB} + \overrightarrow{OA}$$

$$= \overrightarrow{OA} - \overrightarrow{OB} = \mathbf{a} - \mathbf{b}$$

Substituting $\overrightarrow{OB} = $ **b** and $\overrightarrow{BA} = \mathbf{a} - \mathbf{b}$ into † gives

$$\overrightarrow{OM} = \mathbf{b} + \frac{1}{2}(\mathbf{a} - \mathbf{b})$$

$$= \mathbf{b} + \frac{1}{2}\mathbf{a} - \frac{1}{2}\mathbf{b}$$

$$= \frac{1}{2}\mathbf{b} + \frac{1}{2}\mathbf{a} = \frac{1}{2}(\mathbf{a} + \mathbf{b}) \quad \left[\text{Taking out } \frac{1}{2}\right]$$

A5 Applications of vectors in mechanics

Example 5 *mechanics*

Find the magnitude, $|\mathbf{R}|$, of the resultant force **R** shown in Fig. 19.

Solution

What is angle *B* equal to?

$\qquad$ angle $B = 180° - (20° + 30°) = 130°$

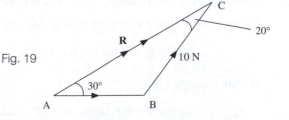

Fig. 19

Example 5 *continued*

To find $|\mathbf{R}|$ we use the sine rule. Let a and b represent the sides opposite angles A and B respectively. Thus $a = 10$ and $b = |\mathbf{R}|$. We need to find the length b by using

$$\frac{b}{\sin(B)} = \frac{a}{\sin(A)}$$

Substituting $A = 30°$, $a = 10$ and $B = 130°$:

$$\frac{b}{\sin(130°)} = \frac{10}{\sin(30°)} = 20$$

$$b = 20 \times \sin(130°) = 15.32$$

Thus $|\mathbf{R}| = 15.32$ N.

Example 6 *mechanics*

The point C of Fig. 20 shows a particle rotating in a circle with the spin velocity represented by $\overrightarrow{CA}$ and the forward velocity represented by $\overrightarrow{CB}$. The resultant velocity, **v**, is vertically downwards. Find **v** if $|\overrightarrow{CB}| = 5$ m/s.

Fig. 20

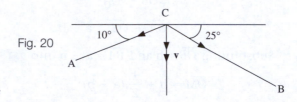

Solution

We shift $\overrightarrow{CA}$ to the end of $\overrightarrow{CB}$ to make the triangle shown in Fig. 21.

To obtain $\overrightarrow{CX}$ of Fig. 21 we use the sine rule:

$$\frac{x}{\sin(X)} = \frac{b}{\sin(B)}$$

Fig. 21

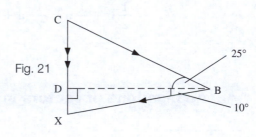

where x and b are the sides opposite angles X and B respectively.

Considering the triangle BDX, what is angle X equal to?

Recall that the sum of the angles inside a triangle add up to 180°, Thus

$$\text{angle } X = 180° - (90° + 10°) = 80°$$

In our triangle CBX:

$$\text{angle } X = 80°, \text{ angle } B = 10° + 25° = 35° \text{ and } x = 5 \text{ because } |\overrightarrow{CB}| = 5$$

Example 6 *continued*

By substituting these values into $\dfrac{x}{\sin(X)} = \dfrac{b}{\sin(B)}$ we have

$$\frac{5}{\sin(80°)} = \frac{b}{\sin(35°)}$$

$$b = \frac{5}{\sin(80°)} \times \sin(35°) = 2.912$$

Therefore the velocity **v** = 2.91 m/s vertically downwards.

SUMMARY

A vector has direction and magnitude, and a scalar has magnitude but no direction. Vectors are basic tools used in engineering.

Exercise 12(a)

Solutions at end of book. Complete solutions available at www.palgrave.com/engineering/singh

1 [mechanics]

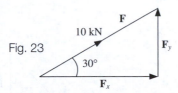

Fig. 22

Figure 22 shows two forces $\mathbf{F}_1$ and $\mathbf{F}_2$ acting on an object O. Determine the resultant force, $\mathbf{R} = \mathbf{F}_1 + \mathbf{F}_2$.

2 [mechanics]

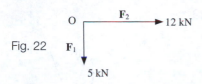

Fig. 23

Determine the horizontal, $\mathbf{F}_x$, and the vertical, $\mathbf{F}_y$, components of the force **F** shown in Fig. 23.

3 [mechanics]   A force of 12 000 N is applied to a body at an angle of 45° to the horizontal. Determine the vertical and horizontal components of the force.

4 [mechanics]

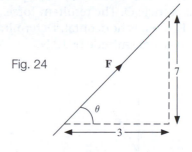

Fig. 24

Figure 24 shows a force, **F**, of 10 kN applied at an angle θ. Obtain the horizontal and vertical components of the force.

5

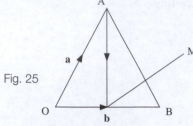

Fig. 25

In the triangle OAB shown in Fig. 25, the point M is the midpoint of OB. Determine AM in terms of **a** and **b**.

6

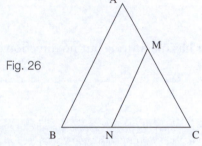

Fig. 26

Figure 26 shows a triangle ABC where M is the midpoint of AC and N is the midpoint of BC. Show that

$$\overrightarrow{MN} = \frac{1}{2}\overrightarrow{AB}.$$

🚗 **The remaining questions are in the field of [*mechanics*].**

7 Figure 27 shows forces $\mathbf{F}_1$ and $\mathbf{F}_2$ exerted on a particle O. The resultant force, $\mathbf{F} = \mathbf{F}_1 + \mathbf{F}_2$, is horizontal. Determine **F** if $\mathbf{F}_2$ has a magnitude of 10 N.

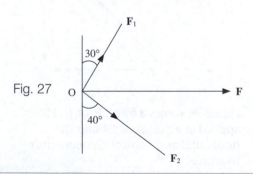

Fig. 27

8 Figure 28 shows forces $\mathbf{F}_1$ and $\mathbf{F}_2$ applied to an object O. The resultant force **R** is vertical.

Fig. 28

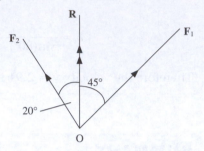

Determine the magnitude of **R** if $\mathbf{F}_1$ has a magnitude of 12 kN.

9 Figure 29 shows forces $\mathbf{F}_1$ and $\mathbf{F}_2$ of magnitude 7 kN and 20 kN respectively. Determine the resultant force.

Fig. 29

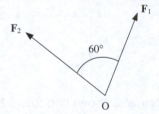

10 Determine the resultant force, $\mathbf{F}_1 + \mathbf{F}_2$, of the forces $\mathbf{F}_1$ and $\mathbf{F}_2$ shown in Fig. 30.

Fig. 30

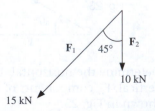

SECTION B **Vectors in Cartesian co-ordinates**

By the end of this section you will be able to:
▶ write vectors of two dimensions in **i** and **j** notation
▶ use a calculator to find the resultant vector
▶ examine engineering applications

B1 **Vectors in two dimensions**

We define vectors **i** and **j** in the *x–y* plane as the unit vectors in the *x* and *y* directions respectively (Fig. 31). The term 'unit vector' means that the magnitude (size) of the vector is equal to 1. So **i** is one unit in the direction of the *x* axis and **j** is one unit in the direction of the *y* axis. The vectors **i** and **j** are shown in Fig. 31.

Fig. 31

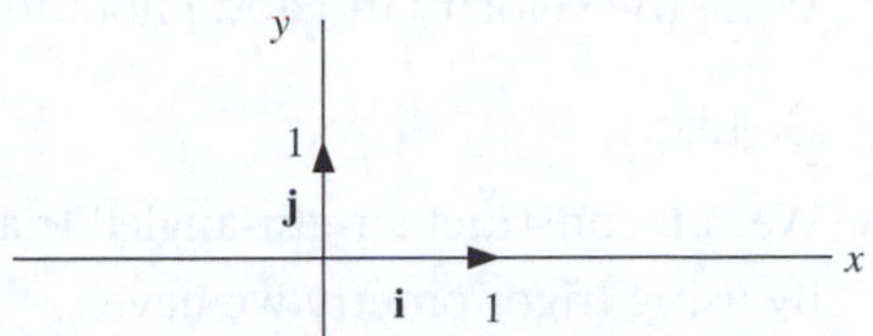

Do not confuse the unit vector **j** with the complex number $\sqrt{-1}$ of Chapter 10. The unit vector **j** is denoted by bold typeface.

Consider the vector $\overrightarrow{OA}$ shown in Fig. 32.

To get from O to A we can go 3 units along the *x* axis and 5 units up the *y* axis. We can write this as

$$3\mathbf{i} + 5\mathbf{j} \text{ or } \overrightarrow{OA} = 3\mathbf{i} + 5\mathbf{j}$$

Fig. 32

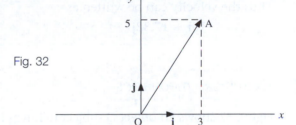

Write the vectors $\overrightarrow{OA}, \overrightarrow{OB}$ and $\overrightarrow{OC}$, shown in Fig. 33, in terms of the unit vectors **i** and **j**.

The vector $\overrightarrow{OA}$ is 7 units across and 10 units up, so

$$\overrightarrow{OA} = 7\mathbf{i} + 10\mathbf{j}$$

Similarly

Fig. 33

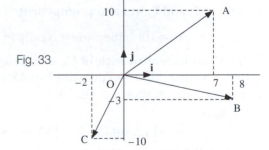

$$\overrightarrow{OB} = 8\mathbf{i} - 3\mathbf{j}$$

Also the vector $\overrightarrow{OC}$ is 2 units to the left and 10 units down, therefore

$$\overrightarrow{OC} = -2\mathbf{i} - 10\mathbf{j}$$

Generally a vector from the origin, O, to a point, P, is called the position vector of P and is denoted by $\overrightarrow{OP}$. Moreover the vector $\overrightarrow{OP}$ from the origin to the point P = (*x*, *y*) can be written as

12.1 $$\overrightarrow{OP} = x\mathbf{i} + y\mathbf{j}$$

We can solve problems in mechanics by using this new notation. The advantage of this approach is that we can find the resultant vector of two or more vectors without using the sine and cosine rules. Consider the following examples.

The displacement, **s**, of an object by 5 m is shown in Fig. 34.

This displacement **s** can be represented in **i** and **j** notation as

$$s = 5i + 0j = 5i$$

Fig. 34

Example 7 *mechanics*

A projectile is fired with an initial velocity of 20 m/s and makes an angle of 60° with the horizontal (Fig. 35).

Write the velocity in **i** and **j** notation.

Fig. 35

Solution

We can construct a right-angled triangle (Fig. 36).

By using trigonometry we have

$$adj = hyp \times \cos(\theta) = 20\cos(60°) = 10$$

$$opp = hyp \times \sin(\theta) = 20\sin(60°) = 17.32$$

Fig. 36

Thus the velocity can be written as

$$10i + 17.32j$$

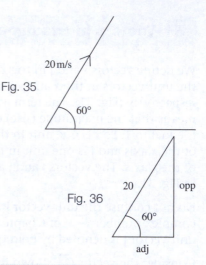

Example 8 *mechanics*

Figure 37 shows an object O subject to forces $\mathbf{F}_1$ and $\mathbf{F}_2$. Find the resultant force **R**.

Solution

Fig. 37

Writing $\mathbf{F}_1$ and $\mathbf{F}_2$ in **i** and **j** components:

$$\mathbf{F}_1 = 10i \quad \text{[Because it is only in the i direction]}$$

The 12 kN force at an angle of 45° is made up of $12\cos(45°)$ in the **i** direction (the adjacent) and $12\sin(45°)$ in the **j** direction (the opposite), as shown in Fig. 38.

We have

$$\mathbf{F}_2 = \left[12\cos(45°)\right]i + \left[12\sin(45°)\right]j$$

Fig. 38

Adding the two vectors gives

$$\mathbf{R} = \mathbf{F}_1 + \mathbf{F}_2$$

$$= 10i + 12\cos(45°)i + 12\sin(45°)j$$

$$= \left[10 + 12\cos(45°)\right]i + 12\sin(45°)j \quad \text{[Collecting like terms]}$$

$$= 18.485i + 8.485j \quad \text{[Evaluating]}$$

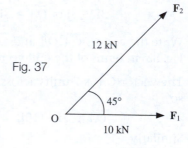

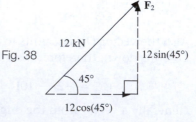

Example 8 *continued*

Hence the vector **R** is 18.485 units in the horizontal direction and 8.485 units up (Fig. 39).

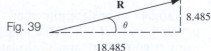

Fig. 39

We can find the magnitude (length) by using Pythagoras:

$$|\mathbf{R}| = \sqrt{18.485^2 + 8.485^2} = 20.34 \text{ kN} \qquad [\text{Scalar}]$$

Now we need to find the direction of **R**. **What is the value of angle θ?**

By trigonometry

$$\tan(\theta) = \frac{\text{opposite}}{\text{adjacent}} = \frac{8.485}{18.485}$$

Taking inverse tan gives

$$\theta = \tan^{-1}\left(\frac{8.485}{18.485}\right) = 24.66°$$

The resultant force **R** has a magnitude of 20.34 kN at an angle of 24.66° inclined to the horizontal. Remember **R** is a vector but $|\mathbf{R}|$ is only concerned with its magnitude, **not** its direction, so is a scalar.

Example 9 *mechanics*

Figure 40 shows three forces $\mathbf{F}_1$, $\mathbf{F}_2$ and $\mathbf{F}_3$ applied to an object O. Determine the resultant force **R**.

Solution

Fig. 40

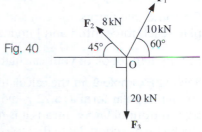

We can measure each angle from the positive horizontal axis. Therefore the 8 kN force is at an angle of 135° (=180° − 45°). Putting each force into its **i** and **j** components by using trigonometry gives

$$\mathbf{F}_1 = [10\cos(60°)]\mathbf{i} + [10\sin(60°)]\mathbf{j}$$

$$\mathbf{F}_2 = [8\cos(135°)]\mathbf{i} + [8\sin(135°)]\mathbf{j}$$

$$\mathbf{F}_3 = -20\mathbf{j}$$

Adding the forces gives the resultant force, **R**:

$$\mathbf{R} = \mathbf{F}_1 + \mathbf{F}_2 + \mathbf{F}_3$$

$$= [10\cos(60°)]\mathbf{i} + [10\sin(60°)]\mathbf{j} + [8\cos(135°)]\mathbf{i} + [8\sin(135°)]\mathbf{j} - 20\mathbf{j}$$

$$= [10\cos(60°) + 8\cos(135°)]\mathbf{i} + [10\sin(60°) + 8\sin(135°) - 20]\mathbf{j}$$

$$= -0.657\mathbf{i} - 5.683\mathbf{j}$$

The magnitude of the resultant force, $|\mathbf{R}|$, is found by using Pythagoras:

$$|\mathbf{R}| = \sqrt{(-0.657)^2 + (-5.683)^2} = 5.72 \text{ kN}$$

Example 9 *continued*

The angle can be determined by plotting $-0.657\mathbf{i} - 5.683\mathbf{j}$ and then using trigonometry and a calculator.

So **R** has magnitude 5.72 kN and is at an angle of $\theta = 83.41°$, as shown in Fig. 41.

Fig. 41

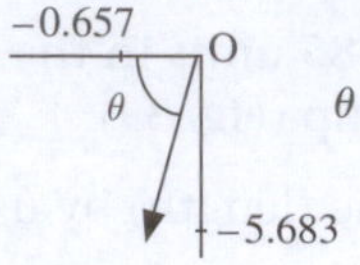

$$\theta = \tan^{-1}\left(\frac{5.683}{0.657}\right) = 83.41°$$

B2 Use of calculator

The calculations in examples can be simplified by using a calculator.

Consider **Example 9.** We can also write forces $\mathbf{F}_1$, $\mathbf{F}_2$ and $\mathbf{F}_3$ in polar form, that is

$$\mathbf{F}_1 = 10\angle 60°$$

Measuring the angle anticlockwise from the positive horizontal axis:

$$\mathbf{F}_2 = 8\angle 135°$$

Measuring the angle clockwise from the positive horizontal axis:

$$\mathbf{F}_3 = 20\angle(-90°)$$

$10\angle 60°$ means the vector has a magnitude of 10 units and is at an angle of 60°. We used this polar form notation for complex numbers in Chapter 10.

On some calculators you can add these forces directly in polar form, in others you need to place each force into **i** and **j** notation and then carry out the addition.

See the handbook of your calculator.

Solving **Example 9** on the calculator gives the answer in polar form as $5.72 \angle -96.59°$. The minus sign in front of 96.59° means it is measured in a clockwise direction. Note that the calculator gives the angle measured from the positive horizontal axis.

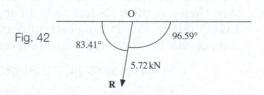

Fig. 42

The angle 96.59° in Fig. 42 is the same as the angle 83.41° in Fig. 41.

The resultant vector, **R**, has a magnitude of 5.72 kN and an angle of 96.59° as shown in Fig. 42.

It is important to note that this use of the calculator only works for vectors in two dimensions, and we cannot use this technique for three-dimensional vectors, which are discussed in the next section.

SUMMARY

Unit vectors in the direction of the *x* and *y* axes are denoted by **i** and **j** respectively.
We apply trigonometry to find the resultant vector.
We can also use a calculator to find the resultant vector.

Exercise **12(b)**   Solutions at end of book. Complete solutions available
at www.palgrave.com/engineering/singh

🚗 **All questions are in the field of [*mechanics*].**

1

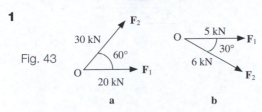

Fig. 43

a   **b**

Figure 43, **a** and **b**, shows forces
F_1 and F_2 applied to an object O.

Determine the resultant force,
$R = F_1 + F_2$, in each case.

2 The following forces have units
of kN:

$$F_1 = 4i - 6j$$
$$F_2 = 7i - 12j$$

Determine the magnitude of the force
$F = 31F_1 - 3F_2$.

3

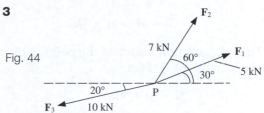

Fig. 44

Figure 44 shows forces F_1, F_2 and F_3
acting on a particle P. Determine
the resultant force **R**.

4 Determine the resultant force of the
forces shown in Fig. 45.

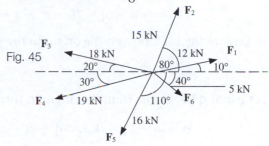

Fig. 45

SECTION C **Three-dimensional vectors**

By the end of this section you will be able to:
► write vectors of three dimensions in **i**, **j** and **k** notation
► find the magnitude of a vector in three dimensions
► evaluate the unit vector

C1 **Vectors in three dimensions**

The x–y plane can be extended to cover three dimensions by
including a third axis called the z axis. This axis is at right
angles to the other two, x and y, axes. The position of a point
in three dimensions is given by three co-ordinates (x, y, z). Real
life is three dimensional, hence many engineering problems
require the application of three dimensions.

Let's examine vectors in three dimensions. We define **i**, **j** and **k** as
the unit vectors in the x, y and z directions respectively (Fig. 46).

Fig. 46

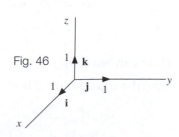

Consider the vector $\vec{OA}$ in Fig. 47.

To get from O to A we can go 3 units along the x axis, 5 units along the y axis and then 7 units up the z axis. Therefore

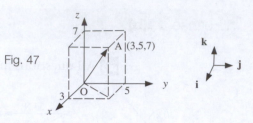

Fig. 47

$$\vec{OA} = 3\mathbf{i} + 5\mathbf{j} + 7\mathbf{k}$$

Thus the position vector of the point with co-ordinates (3, 5, 7) is

$$3\mathbf{i} + 5\mathbf{j} + 7\mathbf{k}$$

? **How would you write the vectors from the origin to the points:**

$$\mathbf{B} = (-3, 5, 7), \ \ \mathbf{C} = (1, 1, -1) \text{ and } \mathbf{D} = (-5, -11, 23)?$$

We have

$$\vec{OB} = -3\mathbf{i} + 5\mathbf{j} + 7\mathbf{k}, \ \ \vec{OC} = \mathbf{i} + \mathbf{j} - \mathbf{k} \text{ and } \vec{OD} = -5\mathbf{i} - 11\mathbf{j} + 23\mathbf{k}$$

In general the vectors from the origin to the point A = (x, y, z) are given by

$$\vec{OA} = x\mathbf{i} + y\mathbf{j} + z\mathbf{k}$$

Let **p** and **q** be vectors from the origin in three-dimensional space such that

$$\mathbf{p} = a\mathbf{i} + b\mathbf{j} + c\mathbf{k} \text{ and } \mathbf{q} = d\mathbf{i} + e\mathbf{j} + f\mathbf{k}$$

Then the addition of **p** and **q** is defined by

12.2	$\mathbf{p} + \mathbf{q} = (a + d)\mathbf{i} + (b + e)\mathbf{j} + (c + f)\mathbf{k}$

All you need to do is add the **i**, **j** and **k** components separately.

Consider the vector $\mathbf{r} = a\mathbf{i} + b\mathbf{j} + c\mathbf{k}$.

? **What is the magnitude (length) of the vector r?**

Figure 48 shows the vector **r**. By Pythagoras, we can find the length OA^2:

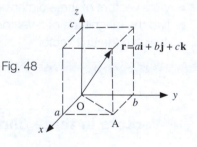

Fig. 48

$$OA^2 = a^2 + b^2$$

The length of the vector $|\mathbf{r}|$ is found by applying Pythagoras again:

$$|\mathbf{r}|^2 = OA^2 + c^2$$
$$= a^2 + b^2 + c^2$$

? **How can we find $|\mathbf{r}|$?**

Take the square root of both sides:

$$|\mathbf{r}| = \sqrt{a^2 + b^2 + c^2}$$

Hence the magnitude, $|\mathbf{r}|$, of a vector

$$\mathbf{r} = a\mathbf{i} + b\mathbf{j} + c\mathbf{k}$$

is given by

12.3 $$|\mathbf{r}| = \sqrt{a^2+b^2+c^2}$$

In two-dimensional vectors we always found the direction of the vector by evaluating the angle that the vector made with the horizontal axis. In three-dimensional vectors there is **no one** angle between a three-dimensional vector and the horizontal axis or any other axes which specifies the direction.

 Example 10 *mechanics*

Three forces $\mathbf{F}_1$, $\mathbf{F}_2$ and $\mathbf{F}_3$ (measured in newton) act on a particle. Obtain the magnitude of the resultant force, $\mathbf{R} = \mathbf{F}_1 + \mathbf{F}_2 + \mathbf{F}_3$, in each of the following:

a $\mathbf{F}_1 = 5\mathbf{i} - \mathbf{j} - \mathbf{k}$, $\mathbf{F}_2 = 3\mathbf{i} + \mathbf{j}$, $\mathbf{F}_3 = 2\mathbf{i} - 7\mathbf{j} + 3\mathbf{k}$
b $\mathbf{F}_1 = 2\mathbf{i} - 3\mathbf{j} + 2\mathbf{k}$, $\mathbf{F}_2 = \mathbf{i}$, $\mathbf{F}_3 = \mathbf{i} + 3\mathbf{j} - \mathbf{k}$
c $\mathbf{F}_1 = 7\mathbf{i} - \mathbf{j}$, $\mathbf{F}_2 = 2\mathbf{i} + \mathbf{j} - \mathbf{k}$, $\mathbf{F}_3 = 7\mathbf{i} - \mathbf{j} - 21\mathbf{k}$

Solution

a Adding the forces, $\mathbf{F}_1 + \mathbf{F}_2 + \mathbf{F}_3$, gives

$$\mathbf{R} = (5\mathbf{i} - \mathbf{j} - \mathbf{k}) + (3\mathbf{i} + \mathbf{j}) + (2\mathbf{i} - 7\mathbf{j} + 3\mathbf{k})$$
$$= (5 + 3 + 2)\mathbf{i} + (-1 + 1 - 7)\mathbf{j} + (-1 + 0 + 3)\mathbf{k} \quad \text{[Collecting like terms]}$$
$$= 10\mathbf{i} - 7\mathbf{j} + 2\mathbf{k}$$

To find the magnitude we use $|\mathbf{r}| = \sqrt{a^2+b^2+c^2}$ with $a = 10$, $b = -7$ and $c = 2$:

$$|\mathbf{R}| = \sqrt{10^2+(-7)^2 + 2^2} = 12.37 \text{ N} \ (2 \text{ d.p.})$$

b We have

$$\mathbf{R} = (2\mathbf{i} - 3\mathbf{j} + 2\mathbf{k}) + \mathbf{i} + (\mathbf{i} + 3\mathbf{j} - \mathbf{k})$$
$$= (2 + 1 + 1)\mathbf{i} + (-3 + 0 + 3)\mathbf{j} + (2 + 0 - 1)\mathbf{k}$$
$$= 4\mathbf{i} + \mathbf{k} \qquad \text{[No } \mathbf{j} \text{ term]}$$

The magnitude is

$$|\mathbf{R}| = \sqrt{4^2+0^2+1^2} = 4.12 \text{ N} \ (2 \text{ d.p.})$$

c Similarly we have

$$\mathbf{R} = (7\mathbf{i} - \mathbf{j}) + (2\mathbf{i} + \mathbf{j} - \mathbf{k}) + (7\mathbf{i} - \mathbf{j} - 21\mathbf{k})$$
$$= (7 + 2 + 7)\mathbf{i} + (-1 + 1 - 1)\mathbf{j} + (0 - 1 - 21)\mathbf{k}$$
$$= 16\mathbf{i} - \mathbf{j} - 22\mathbf{k}$$

and the magnitude is

$$|\mathbf{R}| = \sqrt{16^2+(-1)^2+(-22)^2} = 27.22 \text{ N} \quad (2 \text{ d.p.})$$

C2 **Unit vectors**

Unit vectors are vectors which have a magnitude of 1. If a vector, **r**, has a magnitude of 5 then the unit vector, denoted by **u**, in the direction of **r** is given by

$$\mathbf{u} = \frac{1}{5}\mathbf{r} \qquad \left[\text{Sometimes this is denoted by } \hat{\mathbf{r}} = \frac{1}{5}\mathbf{r} \right]$$

For a general vector, **r**, we have the situation shown in Fig. 49.

The unit vector, **u**, along the direction of a vector **r** is given by

12.4 $\qquad \mathbf{u} = \dfrac{\mathbf{r}}{|\mathbf{r}|} \quad \text{or} \quad \hat{\mathbf{r}} = \dfrac{\mathbf{r}}{|\mathbf{r}|}$

provided that **r** is not the zero vector.

A unit vector is a vector which has a magnitude of 1.

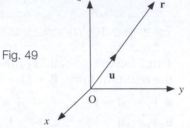

Fig. 49

? **What is the unit vector, u, along the direction of the vector r = $a\mathbf{i} + b\mathbf{j} + c\mathbf{k}$?**

The magnitude, $|\mathbf{r}|$, of the vector $\mathbf{r} = a\mathbf{i} + b\mathbf{j} + c\mathbf{k}$ is given by

$$|\mathbf{r}| = \sqrt{a^2 + b^2 + c^2}$$

Thus

12.5 $\qquad \mathbf{u} = \dfrac{1}{\sqrt{a^2 + b^2 + c^2}}(a\mathbf{i} + b\mathbf{j} + c\mathbf{k})$

Let's find the unit vector, **u** (or $\hat{\mathbf{r}}$), in the direction of the vector $\mathbf{r} = 2\mathbf{i} - 2\mathbf{j} + \mathbf{k}$.

The magnitude is $|\mathbf{r}| = \sqrt{2^2 + (-2)^2 + 1^2} = 3$. Therefore the unit vector is given by

$$\mathbf{u} = \frac{1}{3}(2\mathbf{i} - 2\mathbf{j} + \mathbf{k})$$

? **What is the magnitude (size) of the vector u = $\frac{1}{3}(2\mathbf{i} - 2\mathbf{j} + \mathbf{k})$?**

It's 1 because it is a unit vector as $|\mathbf{u}| = \dfrac{1}{3}|\mathbf{r}| = \dfrac{1}{3}(3) = 1$.

Example 11

Determine the unit vector in the direction of $\overrightarrow{AB}$ where A = (1, 3, 5) and B = (7, 1, 2).

Solution

In Fig. 50:

$$\overrightarrow{AB} = \overrightarrow{AO} + \overrightarrow{OB}$$

Fig. 50

$$\overrightarrow{AB} = -\overrightarrow{OA} + \overrightarrow{OB} = \overrightarrow{OB} - \overrightarrow{OA}$$

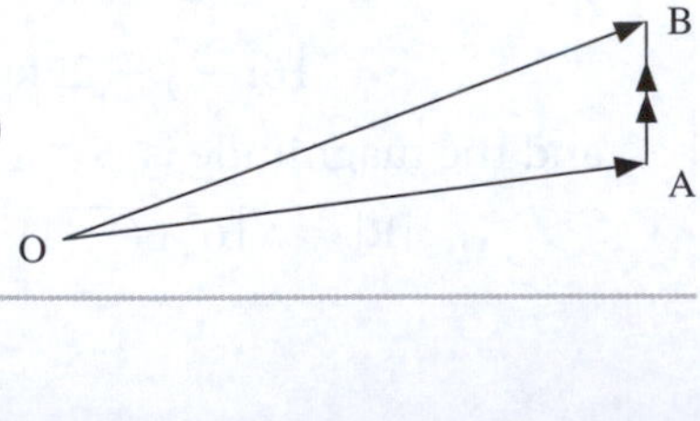

Example 11 *continued*

The vector, $\overrightarrow{OA}$, from the origin to A = (1, 3, 5) is written as

$$\overrightarrow{OA} = \mathbf{i} + 3\mathbf{j} + 5\mathbf{k}$$

Similarly for B = (7, 1, 2) we have

$$\overrightarrow{OB} = 7\mathbf{i} + \mathbf{j} + 2\mathbf{k}$$

Substituting these into [⋆] gives

$$\overrightarrow{AB} = \overrightarrow{OB} - \overrightarrow{OA}$$

$$= (7\mathbf{i} + \mathbf{j} + 2\mathbf{k}) - (\mathbf{i} + 3\mathbf{j} + 5\mathbf{k})$$

$$= (7 - 1)\mathbf{i} + (1 - 3)\mathbf{j} + (2 - 5)\mathbf{k}$$

$$= 6\mathbf{i} - 2\mathbf{j} - 3\mathbf{k}$$

The unit vector in the direction of $\overrightarrow{AB} = 6\mathbf{i} - 2\mathbf{j} - 3\mathbf{k}$ is determined by

12.5

$$\mathbf{u} = \frac{1}{\sqrt{a^2 + b^2 + c^2}} (a\mathbf{i} + b\mathbf{j} + c\mathbf{k})$$

$$\text{unit vector} = \frac{1}{\sqrt{6^2 + (-2)^2 + (-3)^2}} (6\mathbf{i} - 2\mathbf{j} - 3\mathbf{k})$$

$$= \frac{1}{7} (6\mathbf{i} - 2\mathbf{j} - 3\mathbf{k})$$

Example 12 *mechanics*

Determine the components of the force, **F**, which have a combined magnitude of 28 kN in the direction of the vector $\overrightarrow{AB}$ where

$$A = (1, 3, 5) \text{ and } B = (7, 1, 2)$$

Solution

We first obtain the unit vector in the direction of $\overrightarrow{AB}$ and then multiply the result by 28 kN because the unit vector gives the direction of $\overrightarrow{AB}$ and 28 kN gives the magnitude. **How do we find the unit vector?**

A = (1, 3, 5) and B = (7, 1, 2) are the same co-ordinates as those of **Example 11**, so the unit vector = $\frac{1}{7}$ (6**i** − 2**j** − 3**k**). Thus

$$\mathbf{F} = 28 \times \frac{1}{7} (6\mathbf{i} - 2\mathbf{j} - 3\mathbf{k})$$

$$= 4 (6\mathbf{i} - 2\mathbf{j} - 3\mathbf{k})$$

$$= (24\mathbf{i} - 8\mathbf{j} - 12\mathbf{k}) \text{ kN}$$

SUMMARY

Unit vectors in the direction of the x, y and z axes are denoted by **i**, **j** and **k** respectively. The magnitude, $|\mathbf{r}|$, of a vector $\mathbf{r} = a\mathbf{i} + b\mathbf{j} + c\mathbf{k}$ is

| 12.3 | $$|\mathbf{r}| = \sqrt{a^2 + b^2 + c^2}$$ |

The unit vector, **u** (or $\hat{\mathbf{r}}$), in the direction of the non-zero vector $\mathbf{r} = a\mathbf{i} + b\mathbf{j} + c\mathbf{k}$ is given by

| 12.5 | $$\mathbf{u} = \frac{1}{\sqrt{a^2 + b^2 + c^2}} (a\mathbf{i} + b\mathbf{j} + c\mathbf{k})$$ |

Exercise 12(c)

Solutions at end of book. Complete solutions available at www.palgrave.com/engineering/singh

🚗 **All questions belong to the field of [*mechanics*].**

1 Forces F_1, F_2, F_3 and F_4 act on a particle so that it is in equilibrium, that is $F_1 + F_2 + F_3 + F_4 = 0$. Determine F_4 for each of the following:

   a $F_1 = 2\mathbf{i} - \mathbf{j} - 3\mathbf{k}$, $F_2 = 7\mathbf{i} - 7\mathbf{j} - 7\mathbf{k}$, $F_3 = 9\mathbf{i} - 6\mathbf{j} - \mathbf{k}$

   b $F_1 = 61\mathbf{i} - 64\mathbf{j} - 89\mathbf{k}$, $F_2 = 93\mathbf{i} - 98\mathbf{j}$, $F_3 = 22\mathbf{i} - 41\mathbf{j} + 43\mathbf{k}$

2 Forces F_1, F_2, F_3 and F_4 all act through the same point in a body such that the body is in equilibrium. Obtain the force F_1 given that $F_2 = 3\mathbf{i} + \mathbf{j} - \mathbf{k}$, $F_3 = -2\mathbf{i} - 3\mathbf{j} - 10\mathbf{k}$ and $F_4 = -\mathbf{i} - 5\mathbf{j} + 3\mathbf{k}$.

3 Forces
$F_1 = 2\mathbf{i} - \mathbf{j} + 10\mathbf{k}$, $\quad F_2 = 12\mathbf{i} - 3\mathbf{j} + 22\mathbf{k}$,
$F_3 = 7\mathbf{i} + \mathbf{j} - 7\mathbf{k}$, $\quad F_4 = -9\mathbf{i} - 4\mathbf{j} + 5\mathbf{k}$

   are applied on a particle. Determine $F = F_1 + F_2 + F_3 + F_4$ in terms of **i**, **j** and **k**.

4 Determine the resultant, $F_1 + F_2 + F_3 + F_4$, of the following forces:
$F_1 = \mathbf{i} - \mathbf{j} + 3\mathbf{k}$
F_2 has a magnitude of 26 N in the direction of $12\mathbf{i} + 5\mathbf{k}$.

F_3 has a magnitude of 18 N in the direction of $2\mathbf{i} - 2\mathbf{j} + \mathbf{k}$.
F_4 has a magnitude of 12 N in the direction of $4\mathbf{i} - 4\mathbf{j} - 2\mathbf{k}$.

5 A force F of magnitude 36 kN is applied in the direction of the vector $\overrightarrow{AB}$ where $A = (7, 5, 14)$ and $B = (-1, 1, 15)$. Find the force F.

6 Find the force F which has a magnitude of 11.3 kN in the direction of the vector $\overrightarrow{AB}$ where $A = (1.1, -2.6, -4.3)$ and $B = (2.2, 5.6, 7.1)$.

7 Let **A**, **B** and **C** have co-ordinates $(-16, 30, 19)$, $(-12, -17, -23)$ and $(-1, 27, -31)$ respectively. Let F_{AB} be the force applied in the direction of $\overrightarrow{AB}$ and F_{AC} be the force applied in the direction of $\overrightarrow{AC}$. Given that the magnitude of F_{AB} is 850 kN and the magnitude of F_{AC} is 700 kN, determine

$$F = F_{AB} + F_{AC}$$

Also find the magnitude of **F**.

SECTION D **Scalar products**

By the end of this section you will be able to:
▶ find scalar products
▶ examine engineering applications of a scalar product

D1 **Scalar product**

Figure 51 shows a constant force **F** applied to an object and as a result the object moves from A to B. The work done by the force is the product of the magnitude of the force in the line of action, $|\mathbf{F}|\cos(\theta)$, and the distance the object has moved,

$|\overrightarrow{AB}|$. Hence

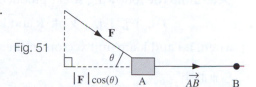

Fig. 51

$$\text{Work done} = |\mathbf{F}|\cos(\theta)|\overrightarrow{AB}|$$

This $|\mathbf{F}|\cos(\theta)|\overrightarrow{AB}|$ gives the scalar product of the force vector **F** and displacement vector $\overrightarrow{AB}$. Thus the work done is a scalar quantity.

When the angle θ between the force **F** and distance $\overrightarrow{AB}$ is 0 then

$$\text{Work done} = |\mathbf{F}|\cos(\theta)|\overrightarrow{AB}| = |\mathbf{F}||\overrightarrow{AB}| \qquad [\text{Because } \cos(0) = 1]$$

This is the least possible force used to push the object because we are pushing in the same direction as we would like the object to move.

In general if a constant force, **F**, moves a particle from A to B through a displacement vector **r**, then the work done by the force **F** is given by

$$\text{Work done} = \mathbf{F} \cdot \mathbf{r} = |\mathbf{F}||\mathbf{r}|\cos(\theta)$$

$\mathbf{F} \cdot \mathbf{r}$ is called the scalar product or the dot product of **F** and **r**. We will return to this application of scalar product later in this section.

First let's consider the more general case of a scalar product of any two vectors, **p** and **q**.

Figure 52 shows two vectors **p** and **q** making an angle θ. The scalar product of **p** and **q**, denoted by $\mathbf{p} \cdot \mathbf{q}$, is defined by

12.6 $$\mathbf{p} \cdot \mathbf{q} = |\mathbf{p}||\mathbf{q}|\cos(\theta)$$

where $|\mathbf{p}|$ is the magnitude of **p** and $|\mathbf{q}|$ is the magnitude of **q**.

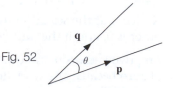

Fig. 52

The dot in $\mathbf{p} \cdot \mathbf{q}$ represents the scalar product (or dot product) of **p** and **q**. Note that $\mathbf{p} \cdot \mathbf{q}$ is a scalar quantity, that is a vector combined with another vector in this way gives a scalar result.

The scalar product has the following properties:

| 12.7 | $\mathbf{p} \cdot \mathbf{q} = \mathbf{q} \cdot \mathbf{p}$ |

| 12.8 | $(\mathbf{p} + \mathbf{q}) \cdot \mathbf{r} = \mathbf{p} \cdot \mathbf{r} + \mathbf{q} \cdot \mathbf{r}$ |

| 12.9 | $\mathbf{p} \cdot (\mathbf{q} + \mathbf{r}) = \mathbf{p} \cdot \mathbf{q} + \mathbf{p} \cdot \mathbf{r}$ |

| 12.10 | $k(\mathbf{p} \cdot \mathbf{q}) = k\mathbf{p} \cdot \mathbf{q} = \mathbf{p} \cdot k\mathbf{q}$ [k is a constant] |

We will apply these properties in evaluating a scalar product of two vectors in component form.

Example 13

Determine the following scalar products:

$$\mathbf{i} \cdot \mathbf{i}, \ \mathbf{i} \cdot \mathbf{j}, \ \mathbf{j} \cdot \mathbf{j}, \ \mathbf{j} \cdot \mathbf{k}, \ \mathbf{k} \cdot \mathbf{k} \text{ and } \mathbf{k} \cdot \mathbf{i}$$

where $\mathbf{i}$, $\mathbf{j}$ and $\mathbf{k}$ are unit vectors in the x, y and z directions respectively.

Solution

Since $\mathbf{i}$, $\mathbf{j}$ and $\mathbf{k}$ are unit vectors, they all have a magnitude of 1. Furthermore $\mathbf{i}$, $\mathbf{j}$ and $\mathbf{k}$ are at an angle of 90° to each other (Fig. 53).

We use the formula

| 12.6 | $\mathbf{p} \cdot \mathbf{q} = |\mathbf{p}||\mathbf{q}| \cos(\theta)$ |

The angle between $\mathbf{i}$ and $\mathbf{i}$ is 0° therefore

$$\mathbf{i} \cdot \mathbf{i} = |\mathbf{i}||\mathbf{i}| \underbrace{\cos(0°)}_{=1} = 1 \times 1 \times 1 = 1$$

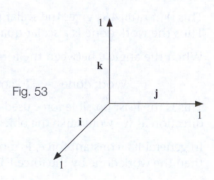

Fig. 53

Similarly $\mathbf{j} \cdot \mathbf{j} = |\mathbf{j}||\mathbf{j}| \cos(0°) = 1$ and
$\mathbf{k} \cdot \mathbf{k} = |\mathbf{k}||\mathbf{k}| \cos(0°) = 1$.

Also

$$\mathbf{i} \cdot \mathbf{j} = |\mathbf{i}||\mathbf{j}| \underbrace{\cos(90°)}_{=0} = 1 \times 1 \times 0 = 0$$

$$\mathbf{j} \cdot \mathbf{k} = |\mathbf{j}||\mathbf{k}| \cos(90°) = 0$$

$$\mathbf{k} \cdot \mathbf{i} = |\mathbf{k}||\mathbf{i}| \cos(90°) = 0$$

We have

| 12.11 | $\mathbf{i} \cdot \mathbf{i} = 1, \mathbf{j} \cdot \mathbf{j} = 1$ and $\mathbf{k} \cdot \mathbf{k} = 1$ |
| 12.12 | $\mathbf{i} \cdot \mathbf{j} = 0, \mathbf{j} \cdot \mathbf{k} = 0$ and $\mathbf{k} \cdot \mathbf{i} = 0$ |

Notice that the scalar product of two different unit vectors which are at right angles to each other is 0, and if the unit vectors are the same then the scalar product is 1.

Any two vectors $\mathbf{p}$ and $\mathbf{q}$ which are perpendicular to each other will have a scalar product of **zero** because the work done by a force perpendicular to displacement is 0 (see Fig. 54a on the following page).

This is why $\mathbf{i} \cdot \mathbf{j} = 0$, $\mathbf{j} \cdot \mathbf{k} = 0$ and $\mathbf{k} \cdot \mathbf{i} = 0$.

Any two vectors **p** and **q** which are in parallel will have a scalar product $|\mathbf{p}||\mathbf{q}|$ (see Fig. 54b).

This is why $\mathbf{i}\cdot\mathbf{i} = |\mathbf{i}||\mathbf{i}| = 1 \times 1 = 1$,
$\mathbf{j}\cdot\mathbf{j} = |\mathbf{j}||\mathbf{j}| = 1 \times 1 = 1$ and $\mathbf{k}\cdot\mathbf{k} = |\mathbf{k}||\mathbf{k}| = 1 \times 1 = 1$.   Fig. 54

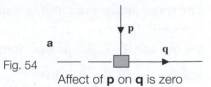

Affect of **p** on **q** is zero

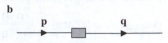

Example 14

Let **p** and **q** be vectors such that

$$\mathbf{p} = a\mathbf{i} + b\mathbf{j} + c\mathbf{k}$$
$$\mathbf{q} = d\mathbf{i} + e\mathbf{j} + f\mathbf{k}$$

Show that the scalar product of **p** and **q** is given by

$$\mathbf{p}\cdot\mathbf{q} = ad + be + cf$$

Solution

Expanding as algebraic brackets gives

$$\mathbf{p}\cdot\mathbf{q} = (a\mathbf{i} + b\mathbf{j} + c\mathbf{k})\cdot(d\mathbf{i} + e\mathbf{j} + f\mathbf{k})$$
$$= (ad)\mathbf{i}\cdot\mathbf{i} + (ae)\mathbf{i}\cdot\mathbf{j} + (af)\mathbf{i}\cdot\mathbf{k} + (bd)\mathbf{j}\cdot\mathbf{i} + (be)\mathbf{j}\cdot\mathbf{j} + (bf)\mathbf{j}\cdot\mathbf{k} + (cd)\mathbf{k}\cdot\mathbf{i}$$
$$+ (ce)\mathbf{k}\cdot\mathbf{j} + (cf)\mathbf{k}\cdot\mathbf{k}$$

Since the scalar product of two different unit vectors which are at right angles to each other is 0, the only terms that do not vanish in the above are the ones containing $\mathbf{i}\cdot\mathbf{i}$, $\mathbf{j}\cdot\mathbf{j}$ and $\mathbf{k}\cdot\mathbf{k}$, and all these are equal to 1. Thus

$$\mathbf{p}\cdot\mathbf{q} = ad + be + cf$$

Example 14 shows that the scalar product of two vectors is defined as

12.13 $(a\mathbf{i} + b\mathbf{j} + c\mathbf{k})\cdot(d\mathbf{i} + e\mathbf{j} + f\mathbf{k}) = ad + be + cf$

To find the scalar product of two vectors given in component form you multiply the corresponding coefficients in the **i**, **j** and **k** directions and then add them together.

Example 15

Determine $(2\mathbf{i} - 3\mathbf{j} + \mathbf{k})\cdot(-5\mathbf{i} + \mathbf{j} - 4\mathbf{k})$

Solution

Applying

$$(a\mathbf{i} + b\mathbf{j} + c\mathbf{k})\cdot(d\mathbf{i} + e\mathbf{j} + f\mathbf{k}) = ad + be + cf$$

gives

$$(2\mathbf{i} - 3\mathbf{j} + \mathbf{k})\cdot(-5\mathbf{i} + \mathbf{j} - 4\mathbf{k}) = [2 \times (-5)] + [(-3) \times 1] + [1 \times (-4)] = -17$$

Notice that the scalar product of two vectors gives a scalar, -17.

The scalar product can also be implemented on a calculator. (See the handbook of your calculator.)

We can use the scalar product formula

12.6 $\mathbf{p} \cdot \mathbf{q} = |\mathbf{p}||\mathbf{q}|\cos(\theta)$

to find the angle θ between vectors $\mathbf{p}$ and $\mathbf{q}$. Rearranging 12.6 gives

12.14 $\cos(\theta) = \dfrac{\mathbf{p} \cdot \mathbf{q}}{|\mathbf{p}||\mathbf{q}|}$

where $\mathbf{p}$ and $\mathbf{q}$ are not the zero vectors.

Remember $\mathbf{p} \cdot \mathbf{q}$ is the work done by the force $\mathbf{p}$ in moving an object through a displacement $\mathbf{q}$ as illustrated in part **a** of Fig. 55.

If $\mathbf{p}$ and $\mathbf{q}$ are parallel, then the angle between them is 0°. If $\mathbf{p}$ and $\mathbf{q}$ are in opposite directions, then the angle between them is 180°. For example, when a ball is thrown vertically upwards then the angle between the velocity and acceleration is 180° as the ball is moving up, while gravity is attempting to accelerate it vertically downwards as shown in part **b** of Fig. 55.

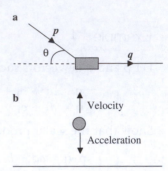

Fig. 55 Ball in upwards motion

D2 Applications of scalar product

Example 16 *mechanics*

The velocity, $\mathbf{v}$, and acceleration, $\mathbf{a}$, of a particle are given by
$$\mathbf{v} = 18\mathbf{i} - 6\mathbf{j} + \mathbf{k}$$
$$\mathbf{a} = 8\mathbf{i} + 4\mathbf{j} + \mathbf{k}$$
Determine the angle between the velocity, $\mathbf{v}$, and acceleration, $\mathbf{a}$.

Solution

We have

† $\cos(\theta) = \dfrac{\mathbf{v} \cdot \mathbf{a}}{|\mathbf{v}||\mathbf{a}|}$

We use

12.3 $|a\mathbf{i} + b\mathbf{j} + c\mathbf{k}| = \sqrt{a^2 + b^2 + c^2}$

to find the magnitude of the velocity and acceleration:

$$|\mathbf{v}| = |18\mathbf{i} - 6\mathbf{j} + \mathbf{k}| = \sqrt{18^2 + (-6)^2 + 1^2} = 19$$
$$|\mathbf{a}| = |8\mathbf{i} + 4\mathbf{j} + \mathbf{k}| = \sqrt{8^2 + 4^2 + 1^2} = 9$$
$$\mathbf{v} \cdot \mathbf{a} = (18\mathbf{i} - 6\mathbf{j} + \mathbf{k}) \cdot (8\mathbf{i} + 4\mathbf{j} + \mathbf{k}) \quad \text{[Scalar product]}$$
$$= (18 \times 8) + [(-6) \times 4] + [1 \times 1] = 121$$

by 12.13

12.13 $(a\mathbf{i} + b\mathbf{j} + c\mathbf{k}) \cdot (d\mathbf{i} + e\mathbf{j} + f\mathbf{k}) = ad + be + cf$

Example 16 *continued*

Substituting $\mathbf{v} \cdot \mathbf{a} = 121$, $|\mathbf{v}| = 19$ and $|\mathbf{a}| = 9$ into ▢† gives

$$\cos(\theta) = \frac{121}{(19 \times 9)}$$

Taking inverse cos gives the angle

$$\theta = \cos^{-1}\left(\frac{121}{(19 \times 9)}\right) = 44.96°$$

The angle between velocity and acceleration is 44.96°.

As mentioned earlier in this section, the scalar product gives the work done by a force $\mathbf{F}$. If a constant force $\mathbf{F}$ moves a particle through a displacement vector $\mathbf{r}$, then the work done by the force $\mathbf{F}$ is given by

▢12.15   Work done $= \mathbf{F} \cdot \mathbf{r}$

where $\mathbf{F} \cdot \mathbf{r}$ is the scalar product of $\mathbf{F}$ and $\mathbf{r}$.

 Example 17 *mechanics*

Determine the work done by a force, $\mathbf{F}$, in moving an object through a displacement $\mathbf{r}$ in each of the following:

a $\mathbf{F} = 3\mathbf{i} - \mathbf{j} - \mathbf{k}$, $\mathbf{r} = \mathbf{i} - \mathbf{j} - \mathbf{k}$
b $\mathbf{F} = 6\mathbf{i} + 2\mathbf{j} - 6\mathbf{k}$, $\mathbf{r} = 9\mathbf{i} + 6\mathbf{j} + \mathbf{k}$
c $\mathbf{F} = 37\mathbf{i} + 35\mathbf{j} - \mathbf{k}$, $\mathbf{r} = 61\mathbf{i} + 64\mathbf{j} + 74\mathbf{k}$

Solution

We need to use

▢12.15   Work done $= \mathbf{F} \cdot \mathbf{r}$

to find the work done for each of **a**, **b** and **c**.

a $(3\mathbf{i} - \mathbf{j} - \mathbf{k}) \cdot (\mathbf{i} - \mathbf{j} - \mathbf{k}) = (3 \times 1) + (-1 \times (-1)) + (-1 \times (-1))$
$= 5 \text{ J}$ (J = joule is the SI unit of work)

b $(6\mathbf{i} + 2\mathbf{j} - 6\mathbf{k}) \cdot (9\mathbf{i} + 6\mathbf{j} + \mathbf{k}) = (6 \times 9) + (2 \times 6) + (-6 \times 1)$
$= 60 \text{ J}$

c $(37\mathbf{i} + 35\mathbf{j} - \mathbf{k}) \cdot (61\mathbf{i} + 64\mathbf{j} + 74\mathbf{k}) = (37 \times 61) + (35 \times 64) + (-1 \times 74)$
$= 4423 \text{ J}$

D3 **Differentiation of vectors**

Remember $\dot{x}$ means the time derivative, $\dfrac{dx}{dt}$, that is differentiate x with respect to t.

Suppose a particle moves in two dimensions, then its position can be defined by the x, y co-ordinates which in turn are functions of time, t. The position vector, $\mathbf{r}$, of a particle can be specified by

$$\mathbf{r} = x\mathbf{i} + y\mathbf{j}$$

where x and y are functions of time, t. The velocity, $\mathbf{v}$, and acceleration, $\mathbf{a}$, of a particle can be found by differentiating $\mathbf{r}$ with respect to t, that is

$$\mathbf{v} = \dot{\mathbf{r}} = \dot{x}\mathbf{i} + \dot{y}\mathbf{j} \quad \text{[First derivative]}$$

$$\mathbf{a} = \ddot{\mathbf{r}} = \ddot{x}\mathbf{i} + \ddot{y}\mathbf{j} \quad \text{[Second derivative]}$$

Remember $\dot{\mathbf{r}} = \dfrac{d\mathbf{r}}{dt}$, $\dot{x} = \dfrac{dx}{dt}$, $\dot{y} = \dfrac{dy}{dt}$, $\ddot{x} = \dfrac{d^2x}{dt^2}$ and $\ddot{y} = \dfrac{d^2y}{dt^2}$.

Example 18 *mechanics*

An object is dropped and its position vector, $\mathbf{r}$, is given by

$$\mathbf{r} = (t^4 - 3t^3)\mathbf{i} + (6t^2 - 2t)\mathbf{j}$$

Find the velocity, $\mathbf{v}$, and acceleration, $\mathbf{a}$, and the angle between $\mathbf{v}$ and $\mathbf{a}$ at $t = 1$.

Solution

Differentiating $\mathbf{r}$ with respect to t:

$$\mathbf{r} = (t^4 - 3t^3)\mathbf{i} + (6t^2 - 2t)\mathbf{j}$$

$$\mathbf{v} = (4t^3 - 9t^2)\mathbf{i} + (12t - 2)\mathbf{j} \quad \text{[Differentiating]}$$

Differentiating again:

$$\mathbf{a} = (12t^2 - 18t)\mathbf{i} + 12\mathbf{j}$$

Substituting $t = 1$ into $\mathbf{v}$ and $\mathbf{a}$ yields

$$\mathbf{v} = (4 - 9)\mathbf{i} + (12 - 2)\mathbf{j}$$

$$= -5\mathbf{i} + 10\mathbf{j}$$

$$\mathbf{a} = (12 - 18)\mathbf{i} + 12\mathbf{j}$$

$$= -6\mathbf{i} + 12\mathbf{j}$$

To find the angle between the velocity and acceleration we need to use

$$\cos(\theta) = \dfrac{\mathbf{a} \cdot \mathbf{v}}{|\mathbf{a}||\mathbf{v}|}$$

★

Example 18 *continued*

We have

$$|\mathbf{v}| = |-5\mathbf{i} + 10\mathbf{j}| = \sqrt{(-5)^2 + 10^2} = \sqrt{125}$$

$$|\mathbf{a}| = |-6\mathbf{i} + 12\mathbf{j}| = \sqrt{(-6)^2 + 12^2} = \sqrt{180}$$

$$\mathbf{a} \cdot \mathbf{v} = (-6\mathbf{i} + 12\mathbf{j}) \cdot (-5\mathbf{i} + 10\mathbf{j})$$
$$= [-6 \times (-5)] + (12 \times 10)$$
$$= 150$$

Substituting these into ⬛ * ⬛ gives

$$\cos(\theta) = \frac{150}{\sqrt{180} \sqrt{125}} = 1$$

Taking inverse cos yields

$$\theta = \cos^{-1}(1) = 0°$$

The angle between velocity and acceleration is 0°. **What does this 0° signify?**

Since we have an angle of 0° between velocity and acceleration then these vectors are parallel (or the same vector). We have

$$\mathbf{v} = -5\mathbf{i} + 10\mathbf{j} \quad \text{and} \quad \mathbf{a} = -6\mathbf{i} + 12\mathbf{j}$$

Observe that $\mathbf{a} = 1.2\mathbf{v}$ which means $\mathbf{a}$ has the same direction as $\mathbf{v}$ but the magnitude is 1.2 times larger.

SUMMARY

Let $\mathbf{p}$ and $\mathbf{q}$ be vectors with angle θ between them. Then the scalar product $\mathbf{p} \cdot \mathbf{q}$ is given by

12.6 $$\mathbf{p} \cdot \mathbf{q} = |\mathbf{p}||\mathbf{q}| \cos(\theta)$$

Also

12.13 $$(a\mathbf{i} + b\mathbf{j} + c\mathbf{k}) \cdot (d\mathbf{i} + e\mathbf{j} + f\mathbf{k}) = ad + be + cf$$

The scalar product is applied in mechanics to find the work done:

12.15   Work done $= \mathbf{F} \cdot \mathbf{r}$

Solutions at end of book. Complete solutions available at www.palgrave.com/engineering/singh

1 Determine the scalar products of the following:

a $(3\mathbf{i} + \mathbf{j} + \mathbf{k}) \cdot (5\mathbf{i} + \mathbf{j} + 7\mathbf{k})$

b $(-\mathbf{i} + \mathbf{j} - \mathbf{k}) \cdot (3\mathbf{i} + 2\mathbf{j} - \mathbf{k})$

c $(10\mathbf{i} - 15\mathbf{j} + \mathbf{k}) \cdot (-21\mathbf{i} - \mathbf{k})$

2 🚗 [*mechanics*] A force $\mathbf{F} = 100\mathbf{i} + 220\mathbf{j} - 250\mathbf{k}$ moves an object through a displacement $\mathbf{s} = 150\mathbf{i} + 200\mathbf{j}$. Find the work done by the force $\mathbf{F}$.

3 🚗 [*mechanics*] A force $\mathbf{F} = 5\mathbf{i} + 3\mathbf{j} - 2\mathbf{k}$ is applied to an object which moves from A = (1, 1, 1) to B = (5, −1, 2). Determine the work done by the force $\mathbf{F}$.

4 🚗 [*mechanics*] A force of magnitude 20 N is applied to an object in the direction $3\mathbf{i} + \mathbf{j} + \mathbf{k}$. Determine the work done by the force in moving the object from A = (3, −1, 5) to B = (−1, 7, 10).

5 Let $\mathbf{r} = x\mathbf{i} + y\mathbf{j} + z\mathbf{k}$ be a vector. Determine

   i $\mathbf{r} \cdot \mathbf{r}$ **ii** $|\mathbf{r}|^2$

where $|\mathbf{r}|$ is the magnitude of $\mathbf{r}$.

What do you notice about your results?

6 🚗 [*mechanics*] The velocity, $\mathbf{v}$, and acceleration, $\mathbf{a}$, of a particle are given by

$$\mathbf{v} = 2\mathbf{i} + 3\mathbf{j}, \ \mathbf{a} = 2\mathbf{i} + 2\mathbf{j}$$

Find the angle between these vectors.

7 🚗 [*mechanics*] The position vector, $\mathbf{r}$, of a particle moving in curvilinear motion is given by

$$\mathbf{r} = (t^3 - t^2)\mathbf{i} + t^4\mathbf{j}$$

  i Find expressions for $\mathbf{v} = \dot{\mathbf{r}}$ and $\mathbf{a} = \ddot{\mathbf{r}}$.
  ii Determine the angle between $\mathbf{v}$ and $\mathbf{a}$ for $t = 1$.
     ($\mathbf{v}$ is velocity and $\mathbf{a}$ is acceleration.)

SECTION E **Vector products**

By the end of this section you will be able to:
▶ find vector products
▶ examine engineering applications of vector products

E1 **Vector product**

In the previous **Section D** we defined the scalar product of two vectors denoted by $\mathbf{p} \cdot \mathbf{q}$. Here we examine the vector or cross product of the vectors, denoted by $\mathbf{p} \times \mathbf{q}$.

The vector product of vectors $\mathbf{p}$ and $\mathbf{q}$, denoted by $\mathbf{p} \times \mathbf{q}$, is defined as the vector of size

$$|\mathbf{p}||\mathbf{q}|\sin(\theta)$$

in the direction 90° to the plane containing the vectors **p** and **q**. We have

12.16 $\qquad \mathbf{p} \times \mathbf{q} = |\mathbf{p}||\mathbf{q}|\sin(\theta)\mathbf{u}$

where θ is the angle between the
vectors **p** and **q**, and **u** is the unit
vector in the direction of $\mathbf{p} \times \mathbf{q}$
shown in part **a** of Fig. 56. The
vector product $\mathbf{p} \times \mathbf{q}$ is a **vector
quantity**, that is the cross product
of a vector combined with a
vector gives a vector. The direction
of the vector $\mathbf{p} \times \mathbf{q}$ is determined by the direction of a
right-handed screw turned from **p** to **q** as shown
in part **b** of Fig. 56.

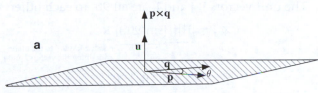

Note that the vector $\mathbf{q} \times \mathbf{p}$ is in the direction of a
right-handed screw turned from **q** to **p** (part **c** of Fig. 56).

? **What do you notice about $\mathbf{p} \times \mathbf{q}$ and $\mathbf{q} \times \mathbf{p}$?**

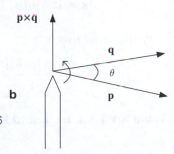
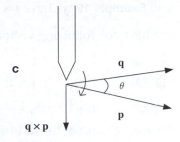

Fig. 56

$$\mathbf{p} \times \mathbf{q} \neq \mathbf{q} \times \mathbf{p} \quad \text{[Not equal]}$$

In fact

12.17 $\qquad \mathbf{p} \times \mathbf{q} = -(\mathbf{q} \times \mathbf{p})$

The vector product also has the following properties:

12.18 $\qquad \mathbf{p} \times (\mathbf{q} + \mathbf{r}) = (\mathbf{p} \times \mathbf{q}) + (\mathbf{p} \times \mathbf{r})$

12.19 $\qquad k(\mathbf{p} \times \mathbf{q}) = (k\mathbf{p}) \times \mathbf{q} = \mathbf{p} \times (k\mathbf{q})$
$\qquad$ [k is a constant]

12.20 $\qquad$ If $\mathbf{p} = 0$ then $\mathbf{p} \times \mathbf{q} = 0$

Example 19

Determine the following vector products:

$$\mathbf{i} \times \mathbf{i}, \ \mathbf{i} \times \mathbf{j}, \ \mathbf{k} \times \mathbf{i}, \ \mathbf{j} \times \mathbf{j}, \ \mathbf{j} \times \mathbf{k} \text{ and } \mathbf{k} \times \mathbf{k}$$

where **i**, **j** and **k** are unit vectors in the directions of
the x, y and z axes respectively.

Solution

Plot the unit vectors, as shown in Fig. 57.

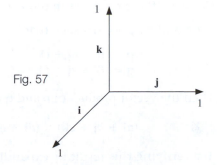

Fig. 57

? **Clearly $\mathbf{i} \times \mathbf{i} = 0$. Why?**

Because the angle between **i** and **i** is 0° and
$\sin(0°) = 0$. Thus

$$\mathbf{i} \times \mathbf{i} = |\mathbf{i}||\mathbf{i}|\sin(0°)\mathbf{j} = 0$$

Example 19 *continued*

Similarly $j \times j = 0$ and $k \times k = 0$.

The unit vectors i, j and k are all 90° to each other. We have

$$i \times j = \underbrace{|i|}_{=1}\underbrace{|j|}_{=1} \underbrace{[\sin(90°)]}_{=1} k$$

because k is 90° to the plane containing the vectors i and j

$$= (1)(1)(1)k = k$$

Likewise

$$k \times i = \underbrace{|k|}_{=1}\underbrace{|i|}_{=1} \underbrace{[\sin(90°)]}_{=1}j = j$$

By the same token:

$$j \times k = \underbrace{|j|}_{=1}\underbrace{|k|}_{=1} \underbrace{[\sin(90°)]}_{=1}i = i$$

? What are $j \times i$, $i \times k$ and $k \times j$ equal to?

By

12.17 $p \times q = -(q \times p)$

and **Example 19** we have $j \times i = -k$, $i \times k = -j$ and $k \times j = -i$.

We have the following vector products of the unit vectors:

12.21 $i \times i = 0$, $j \times j = 0$ and $k \times k = 0$

12.22 $i \times j = k$, $j \times k = i$ and $k \times i = j$

12.23 $j \times i = -k$, $k \times j = -i$ and $i \times k = -j$

We can remember the last two formulae by using:

Moving clockwise you get $i \times j = k$, $j \times k = i$ and $k \times i = j$. Going anticlockwise you get $j \times i = -k$, $i \times k = -j$ and $k \times j = -i$.

Compare these results with those given for scalar products in 12.11 and 12.12 .

Let p and q be vectors such that

$$p = ai + bj + ck$$
$$q = di + ej + fk$$

Then the vector product of p and q can be shown to be

  * $(ai + bj + ck) \times (di + ej + fk) = (bf - ce)i + (cd - af)j + (ae - bd)k$

Try verifying this result by expanding brackets and using the preceding results for i, j and k.

Rather than applying this formula to find the vector product we can use a determinant as a device for evaluating the vector product:

12.24 $$\mathbf{p} \times \mathbf{q} = (a\mathbf{i} + b\mathbf{j} + c\mathbf{k}) \times (d\mathbf{i} + e\mathbf{j} + f\mathbf{k}) = \det\begin{pmatrix} \mathbf{i} & \mathbf{j} & \mathbf{k} \\ a & b & c \\ d & e & f \end{pmatrix}$$

This can be shown by expanding the determinant:

$$\det\begin{pmatrix} \mathbf{i} & \mathbf{j} & \mathbf{k} \\ a & b & c \\ d & e & f \end{pmatrix} = \mathbf{i}\left[\det\begin{pmatrix} b & c \\ e & f \end{pmatrix}\right] - \mathbf{j}\left[\det\begin{pmatrix} a & c \\ d & f \end{pmatrix}\right] + \mathbf{k}\left[\det\begin{pmatrix} a & b \\ d & e \end{pmatrix}\right]$$

$$= \mathbf{i}(bf - ec) - \mathbf{j}(af - dc) + \mathbf{k}(ae - db) \quad [\text{By } 11.1]$$

$$= (bf - ec)\mathbf{i} + (dc - af)\mathbf{j} + (ae - db)\mathbf{k}$$

$$= (a\mathbf{i} + b\mathbf{j} + c\mathbf{k}) \times (d\mathbf{i} + e\mathbf{j} + f\mathbf{k}) \quad [\text{By } ★]$$

Hence we evaluate this determinant to find the vector product.

Example 20

Obtain the vector product of

$$\mathbf{r} = 5\mathbf{i} + 7\mathbf{j} + 6\mathbf{k}$$

$$\mathbf{s} = 3\mathbf{i} - \mathbf{j} + \mathbf{k}$$

Solution

Using the determinant in 12.24 to find $\mathbf{r} \times \mathbf{s}$ gives

$$(5\mathbf{i} + 7\mathbf{j} + 6\mathbf{k}) \times (3\mathbf{i} - \mathbf{j} + \mathbf{k}) = \det\begin{pmatrix} \mathbf{i} & \mathbf{j} & \mathbf{k} \\ 5 & 7 & 6 \\ 3 & -1 & 1 \end{pmatrix}$$

$$= \mathbf{i}\left[\det\begin{pmatrix} 7 & 6 \\ -1 & 1 \end{pmatrix}\right] - \mathbf{j}\left[\det\begin{pmatrix} 5 & 6 \\ 3 & 1 \end{pmatrix}\right] + \mathbf{k}\left[\det\begin{pmatrix} 5 & 7 \\ 3 & -1 \end{pmatrix}\right]$$

$$= \mathbf{i}(7 + 6) - \mathbf{j}[(5 \times 1) - (3 \times 6)] + \mathbf{k}[(5 \times (-1)) - (3 \times 7)]$$
by 11.1
$$= 13\mathbf{i} + 13\mathbf{j} - 26\mathbf{k}$$

We can also calculate the vector product of $\mathbf{r}$ and $\mathbf{s}$ of **Example 20** on a calculator.
Of course $\mathbf{s} \times \mathbf{r} = -13\mathbf{i} - 13\mathbf{j} + 26\mathbf{k}$ (that is $-(\mathbf{r} \times \mathbf{s})$).

11.1 $\det\begin{pmatrix} a & b \\ c & d \end{pmatrix} = ad - cb$

E2 **Applications of vector product**

Next we look at an application of vector products in mechanics.

A force, **F**, is applied to an object, making it rotate around a fixed axis as shown in Fig. 58 (for example, a spinning top). A moment of force is the tendency of a force to turn or rotate an object. The moment of a vector is a generalization of the moment of force. The moment vector, **M**, of a force, **F**, about the origin O is defined as

Fig. 58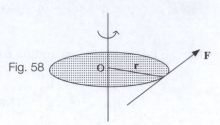

12.25 $\mathbf{M} = \mathbf{r} \times \mathbf{F}$

where **r** is the position vector from O of any point on the line of action of **F**.

Note that the moment, **M**, is a vector. (The moment vector is called the torque in many books on mechanics.)

? **What is the direction of the vector M?**

Fig. 59

It is 90° to the plane containing the vectors **r** and **F** (Fig. 59).

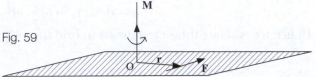

Example 21 *mechanics*

Determine the moment vector, **M**, about the origin, O, of a force $\mathbf{F} = 10\mathbf{i} - \mathbf{j} + 2\mathbf{k}$, passing through the point with position vector $\mathbf{r} = 2\mathbf{i} - 3\mathbf{j} + 5\mathbf{k}$.

Solution

By 12.25 we have

$$\mathbf{M} = \mathbf{r} \times \mathbf{F}$$
$$= (2\mathbf{i} - 3\mathbf{j} + 5\mathbf{k}) \times (10\mathbf{i} - \mathbf{j} + 2\mathbf{k})$$

$$\underset{\text{by } 12.24}{=} \det \begin{pmatrix} \mathbf{i} & \mathbf{j} & \mathbf{k} \\ 2 & -3 & 5 \\ 10 & -1 & 2 \end{pmatrix}$$

$$= \mathbf{i}\left[\det\begin{pmatrix} -3 & 5 \\ -1 & 2 \end{pmatrix}\right] - \mathbf{j}\left[\det\begin{pmatrix} 2 & 5 \\ 10 & 2 \end{pmatrix}\right] + \mathbf{k}\left[\det\begin{pmatrix} 2 & -3 \\ 10 & -1 \end{pmatrix}\right]$$

$$= -\mathbf{i} + 46\mathbf{j} + 28\mathbf{k} \quad \text{[Evaluating determinant]}$$

12.24   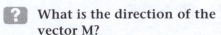 $(a\mathbf{i} + b\mathbf{j} + c\mathbf{k}) \times (d\mathbf{i} + e\mathbf{j} + f\mathbf{k}) = \det \begin{pmatrix} \mathbf{i} & \mathbf{j} & \mathbf{k} \\ a & b & c \\ d & e & f \end{pmatrix}$

The next example is more complicated than the previous examples. The complication is to determine vector moments about other points rather than just the origin.

Example 22 *mechanics*

Let A and B have position vectors
$$\mathbf{a} = 2\mathbf{i} - 3\mathbf{j} + \mathbf{k}$$
$$\mathbf{b} = 7\mathbf{i} - 4\mathbf{j} - \mathbf{k}$$
respectively. The force $\mathbf{F} = 3\mathbf{i} - \mathbf{j} + 5\mathbf{k}$ passes through the point A. Determine the vector moment of $\mathbf{F}$ about the point B.

Solution

Draw the vectors as in Fig. 60.
Find the resultant vector first, $\overrightarrow{BA}$:

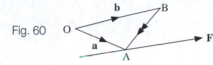

Fig. 60

$$\overrightarrow{BA} = \overrightarrow{OA} - \overrightarrow{OB}$$
$$= (2\mathbf{i} - 3\mathbf{j} + \mathbf{k}) - (7\mathbf{i} - 4\mathbf{j} - \mathbf{k}) = -5\mathbf{i} + \mathbf{j} + 2\mathbf{k}$$

Let $\mathbf{M}$ be the moment of $\mathbf{F}$, then by using $\boxed{12.25}$ and substituting $\overrightarrow{BA} = -5\mathbf{i} + \mathbf{j} + 2\mathbf{k}$, $\mathbf{F} = 3\mathbf{i} - \mathbf{j} + 5\mathbf{k}$ we have

$$\mathbf{M} = \overrightarrow{BA} \times \mathbf{F}$$

$$= (-5\mathbf{i} + \mathbf{j} + 2\mathbf{k}) \times (3\mathbf{i} - \mathbf{j} + 5\mathbf{k})$$

$$\underset{\text{by } \boxed{12.24}}{=} \det\begin{pmatrix} \mathbf{i} & \mathbf{j} & \mathbf{k} \\ -5 & 1 & 2 \\ 3 & -1 & 5 \end{pmatrix}$$

$$= \mathbf{i}\left[\det\begin{pmatrix} 1 & 2 \\ -1 & 5 \end{pmatrix}\right] - \mathbf{j}\left[\det\begin{pmatrix} -5 & 2 \\ 3 & 5 \end{pmatrix}\right] + \mathbf{k}\left[\det\begin{pmatrix} -5 & 1 \\ 3 & -1 \end{pmatrix}\right]$$

$$= \mathbf{i}(5 + 2) - \mathbf{j}(-25 - 6) + \mathbf{k}(5 - 3)$$

$$= 7\mathbf{i} + 31\mathbf{j} + 2\mathbf{k}$$

SUMMARY

The vector product of $\mathbf{p}$ and $\mathbf{q}$ is defined as

$\boxed{12.16}$ $$\mathbf{p} \times \mathbf{q} = |\mathbf{p}||\mathbf{q}|\sin(\theta)\,\mathbf{u}$$

$\boxed{12.24}$ $$(a\mathbf{i} + b\mathbf{j} + c\mathbf{k}) \times (d\mathbf{i} + e\mathbf{j} + f\mathbf{k}) = \det\begin{pmatrix} \mathbf{i} & \mathbf{j} & \mathbf{k} \\ a & b & c \\ d & e & f \end{pmatrix}$$

SUMMARY continued

where **u** is the unit vector in the direction of $\mathbf{p} \times \mathbf{q}$. Also

12.24 $(a\mathbf{i} + b\mathbf{j} + c\mathbf{k}) \times (d\mathbf{i} + e\mathbf{j} + f\mathbf{k}) = \det\begin{pmatrix} \mathbf{i} & \mathbf{j} & \mathbf{k} \\ a & b & c \\ d & e & f \end{pmatrix}$

The moment vector, **M**, about the origin is defined as

12.25 $\mathbf{M} = \mathbf{r} \times \mathbf{F}$

where **r** is the position vector from the origin of any point on the line of action of **F**.

Exercise 12(e)

Solutions at end of book. Complete solutions available at www.palgrave.com/engineering/singh

1 If $\mathbf{r} = 5\mathbf{i} + \mathbf{j} + 3\mathbf{k}$ and $\mathbf{s} = 2\mathbf{i} - 3\mathbf{j} - 4\mathbf{k}$, determine the vector products

  i $\mathbf{r} \times \mathbf{s}$ **ii** $\mathbf{s} \times \mathbf{r}$

2 🚗 [*mechanics*] Forces $\mathbf{F}_1$, $\mathbf{F}_2$ and $\mathbf{F}_3$ act at a point on a body with the corresponding position vectors:

$$\mathbf{F}_1 = 2\mathbf{i} + \mathbf{j} - 3\mathbf{k}, \; \mathbf{r}_1 = \mathbf{i} + 3\mathbf{j} + \mathbf{k}$$
$$\mathbf{F}_2 = 3\mathbf{i} - 4\mathbf{j} + 6\mathbf{k}, \; \mathbf{r}_2 = \mathbf{i} - \mathbf{j} + 2\mathbf{k}$$
$$\mathbf{F}_3 = \mathbf{i} + 10\mathbf{j} + 7\mathbf{k}, \; \mathbf{r}_3 = \mathbf{i} + 4\mathbf{j} + 2\mathbf{k}$$

Show that the body is in equilibrium, that is

$$(\mathbf{r}_1 \times \mathbf{F}_1) + (\mathbf{r}_2 \times \mathbf{F}_2) + (\mathbf{r}_3 \times \mathbf{F}_3) = 0$$

3 🚗 [*mechanics*] Forces $\mathbf{F}_1$, $\mathbf{F}_2$, $\mathbf{F}_3$ and $\mathbf{F}_4$ act at a point on a body with the corresponding position vectors:

$$\mathbf{F}_1 = 2\mathbf{i} - \mathbf{j} + \mathbf{k}, \qquad \mathbf{r}_1 = 3\mathbf{i} + \mathbf{j} + 5\mathbf{k}$$
$$\mathbf{F}_2 = 8\mathbf{i} - 6\mathbf{j} + 6\mathbf{k}, \quad \mathbf{r}_2 = 5\mathbf{i} - 2\mathbf{j} - \mathbf{k}$$
$$\mathbf{F}_3 = 4\mathbf{i} + 3\mathbf{k}, \qquad\quad \mathbf{r}_3 = \mathbf{i} - 6\mathbf{j} + 7\mathbf{k}$$
$$\mathbf{F}_4 = \mathbf{i} - 5\mathbf{j}, \qquad\quad\;\; \mathbf{r}_4 = \mathbf{i} + 6\mathbf{k}$$

Show that the body is in equilibrium, that is

$$(\mathbf{r}_1 \times \mathbf{F}_1) + (\mathbf{r}_2 \times \mathbf{F}_2)$$
$$+ (\mathbf{r}_3 \times \mathbf{F}_3) + (\mathbf{r}_4 \times \mathbf{F}_4) = 0$$

4 🚗 [*mechanics*] Obtain the moment, M, about the point B of a force, F, passing through the point A for the following:

  a $\mathbf{F} = 2\mathbf{i} - \mathbf{j} + \mathbf{k}$, A is (1, 2, 3), B is (5, 7, 9)

  b $\mathbf{F} = 4\mathbf{i} - 3\mathbf{j} - 6\mathbf{k}$, A is (1, −1, 0), B is (3, −1, −1)

5 🚗 [*mechanics*] The points R and S have position vectors

$$\mathbf{r} = \mathbf{i} - \mathbf{j} - \mathbf{k}, \; \mathbf{s} = 7\mathbf{i} + \mathbf{j} - 9\mathbf{k}$$

respectively. A force $\mathbf{F} = 14\mathbf{i} + 2\mathbf{j} - 18\mathbf{k}$ passes through R. Calculate the vector moment of F about the point S.

6 Let $\mathbf{r} = a\mathbf{i} + b\mathbf{j}$ and $\mathbf{s} = c\mathbf{i} + d\mathbf{j}$ be any vectors. Show that

$$\mathbf{r} \times \mathbf{s} = (ad - cb)\mathbf{k}$$

7 Let $\mathbf{r} = a\mathbf{i} + b\mathbf{j} + c\mathbf{k}$ and $\mathbf{s} = d\mathbf{i} + e\mathbf{j} + f\mathbf{k}$ be any vectors. Show that

$$\mathbf{r} \times \mathbf{s} = -(\mathbf{s} \times \mathbf{r})$$

Solutions at end of book. Complete solutions available at www.palgrave.com/engineering/singh

Examination questions 12

In this exercise you may check your numerical answers using mathematical software.

The first two questions have been slightly edited.

1 Given $r = 2i + 3j - 4k$ and $s = 4i - 5j - 2k$, find the vector cross product $r \times s$.

2 a Given force $F = 6i + 2j - 5k$, and displacement vector $r = 9i + 6j + k$, find

   i The work done by F on r
   ii The angle between F and r.

  b Given the position vector of an object in terms of time is

$$r = (t^3 - 3t^2)i + (6t^2 - 2t)j$$

   i Find the velocity and acceleration in terms of time
   ii Find the initial velocity and acceleration
   iii Find the time when the horizontal velocity = 0.

Questions 1 and 2 are from University of Portsmouth, UK, 2007

3 If $a = (3, 5, -2)$ and $b = (2, 4, 7)$, find $a \times b$.

University of Sussex, UK, 2007

4 An electric charge q with velocity v produces a magnetic field B given by:

$$B = \frac{\mu q}{4\pi} \cdot \frac{v \times r}{|r|^2}$$

Find B, in terms of μq, given $r = 3i + j - 2k$ and $v = i - 2j + 3k$.

University of Portsmouth, UK, 2006

5 Given the three vectors $a = 2i - j + k$, $b = 2i + 2j - k$ and $c = -i + 2j - k$ find

  i $a \times c$
  ii $a \cdot (c \times a)$
  iii $(a \times c) \times b$
  iv the value of x which makes the vector $(x, 1, 0)$ to be at $60°$ to b.

University of Manchester, UK, 2007

6 If $a = i + 2j - 4k$, $b = 2i + k$, $c = 3i - j - 5k$ find

  i $a - 2b + c$
  ii $b \cdot c$
  iii $b \times c$
  iv $b \cdot (a \times c) - a \cdot (c \times b)$
  v $|c| - |a|$
  vi a unit vector in the direction of b
  vii a unit vector perpendicular to a and b.

Loughborough University, UK, 2007

7 Find the cosine of the angle between the vectors $u \times v$ and w and between $u \times w$ and v where $u = (1, -1, 1)$, $v = (0, 1, 2)$, $w = (1, 0, 5)$.

Jacobs University, Germany, 2002

Miscellaneous exercise 12

Solutions at end of book. Complete solutions available at www.palgrave.com/engineering/singh

1 Find the exact unit vector in the direction of the vector $F = 5i + 7j + 12k$.

2 Let $a = 2i + 3j - k$, $b = 7i + 12k$, $c = -i - j - 6k$ and $d = 12i - j - 11k$. Determine:

  i $a + c$ **ii** $a + b + c + d$ **iii** $a \cdot b$
  iv $b \cdot a$ **v** $a \cdot (b + c)$
  vi $(c \cdot d)\hat{a}$ where $\hat{a}$ is the unit vector in the direction of a
  vii the magnitude of $a + b + c + d$.

🚗 **Questions 3 to 14 are in the field of [mechanics].**

3 Let the velocity, v, and acceleration, a, be given by

$$v = -4i - 8j + 4k \text{ and}$$
$$a = 2i + j + 4k$$

Find the angle between the velocity and acceleration.

4 The velocity, **v**, and acceleration, **a**, of a particle are given by

$$v = 6i + 2j, \quad a = 3i + 7j$$

Find the angle between **v** and **a**.

5 Determine the resultant force, **R**, of the forces shown in Fig. 61.

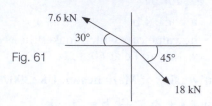

Fig. 61

6 Determine the resultant force, **R**, of the forces shown in Fig. 62.

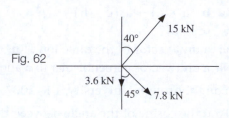

Fig. 62

7 Find the horizontal, F_x, and vertical, F_y, components of the force F shown in Fig. 63.

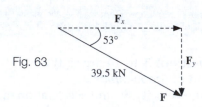

Fig. 63

8 Find the resultant force, $R = F_1 + F_2 + F_3 + F_4$, in each of the following cases:

a $F_1 = 2i - j + 7k$,
   $F_2 = -3i - 2j + 11k$,

   $F_3 = 61i + 64j + 19k$ and
   $F_4 = 3i + 8j + 11k$

b $F_1 = i + j + k$,
   $F_2 = 16i + 99j + 19k$,
   $F_3 = 47i + 84j + 88k$ and
   $F_4 = 22i + 41j + 3k$

9 A force $F = i + j + 3k$ moves an object through a displacement $s = 15i + 20j$. Find the work done by the force **F**.

10 A force $F = 2i - 7j + 12k$ acts on a particle P. Calculate the work done by F as the particle P moves from the point $A = (2, 3, 7)$ to the point $B = (7, -11, 37)$.

11 Determine the work done by a force F on a particle P moving from point A to point B where

a $F = 3i - j - 2k$,
   $A = (1, 7, 1)$, $B = (3, 4, -1)$

b $F = 2i + j - 7k$,
   $A = (3, 4, 3)$, $B = (6, 7, -17)$

12 A force $F = 5i - j - k$ passes through the point A whose position vector is $a = 3i + j + 8k$. Determine the vector moment, **M**, of F about the point B with position vector $b = 2i - 3j + 7k$. Also find the magnitude of **M**.

13 Determine the force F which has a magnitude of 39 kN in the direction of the vector $\overrightarrow{AB}$ where $A = (-1, 2, 3)$ and $B = (5, 1, -2)$.

14 Forces F_1, F_2 and F_3 act at a point in a body with the corresponding position vectors:

$F_1 = i + 3j - 6k$, $r_1 = 2i + 2j + 2k$
$F_2 = -7i + 2j + 5k$, $r_2 = 3i - 2j - k$
$F_3 = 2j + 3k$, $r_3 = 2i + 2j - 10k$

Show that the body is in equilibrium, that is

$$(\mathbf{r}_1 \times \mathbf{F}_1) + (\mathbf{r}_2 \times \mathbf{F}_2) + (\mathbf{r}_3 \times \mathbf{F}_3) = 0$$

15 Let $\mathbf{r} = a\mathbf{i} + b\mathbf{k}$ and $\mathbf{s} = c\mathbf{i} + d\mathbf{k}$ be any vectors. Determine the vector product $\mathbf{r} \times \mathbf{s}$.

16 Let $\mathbf{a} = 2\mathbf{i} - 3\mathbf{j} + x\mathbf{k}$ and $\mathbf{b} = -3\mathbf{i} + \mathbf{j} + x\mathbf{k}$ be vectors. Find the value(s) of x for which the vectors $\mathbf{a}$ and $\mathbf{b}$ are at right angles.

17 Consider the vectors

$$\mathbf{a} = 2\mathbf{i} + 2\mathbf{j} + \mathbf{k} \text{ and } \mathbf{b} = x\mathbf{i} + \mathbf{j} - \mathbf{k}$$

Find the value of x for which the angle between $\mathbf{a}$ and $\mathbf{b}$ is 45°.

CHAPTER 13
First Order Differential Equations

SECTION A **Solving differential equations**

By the end of this section you will be able to:
▶ find the order of a differential equation
▶ solve first order differential equations by using direct integration
▶ solve first order differential equations by separating variables
▶ solve engineering applications with initial conditions

In the next two chapters we discuss differential equations. The entire field of engineering and science – heat, light, sound, gravitation, magnetism, fluid flow, population dynamics and mechanics – can be described by differential equations. Other modern technologies such as radio, television, cars and aircraft all depend on the mathematics of differential equations.

A1 **Introduction**

What is a differential equation?

An equation which contains derivatives.

For example

$$3\frac{dy}{dx} + x = \cos(y) \text{ is a first order differential equation}$$

$$3\frac{d^2y}{dx^2} + \frac{dy}{dx} + x = \cos(y) \text{ is a second order differential equation}$$

$$3\frac{d^3y}{dx^3} + \frac{dy}{dx} + x = \cos(y) \text{ is a third order differential equation}$$

A differential equation can be classified according to its order. The **order** is the order of the highest derivative appearing in the differential equation.

Example 1

If

$$y = e^{5x}$$

Show that

$$\frac{dy}{dx} - 5y = 0$$

Solution

By using

| 6.11 | $\dfrac{d}{dx}(e^{kx}) = ke^{kx}$ |

Example 1 *continued*

on $y = e^{5x}$ we have

$$\frac{dy}{dx} = \frac{d}{dx}(e^{5x}) = 5e^{5x}$$

Substituting $y = e^{5x}$, so $5y = 5e^{5x}$ and we have

$$5e^{5x} - 5y = 5e^{5x} - 5e^{5x} = 0$$

Hence $\frac{dy}{dx} - 5y = 0$ holds.

Up until now we have dealt with the situation of **Example 1**, where we are given a solution, $y = e^{5x}$, and asked to verify that it satisfies the differential equation, $\frac{dy}{dx} - 5y = 0$.

In this chapter we are given the differential equation, $\frac{dy}{dx} - 5y = 0$, and we are asked to find a solution, $y = e^{5x}$. **If we are asked to solve $\frac{dy}{dx} - 5y = 0$ then what are we trying to find?**

A function of x, $y = y(x)$, which satisfies $\frac{dy}{dx} - 5y = 0$.

Note that x is the independent variable and y is the dependent variable.

Consider an investment P which earns an interest rate of r. These two quantities are related by the differential equation.

$$\frac{dP}{dt} = rP$$

This differential equation also describes an exponential growth or decaying system, such as radioactive decay and the current through a series RL circuit. This type of equation is an important first order differential equation which crops up in finance, engineering and science.

A2 **Direct integration**

If we are given the general first order differential equation

13.1 $$\frac{dy}{dx} = f(x) \qquad \text{[Function of } x\text{]}$$

then we can use direct integration to find y.

Example 2

Solve

$$\frac{dy}{dx} = 4x$$

Solution

We have

$$dy = 4x\,dx$$

because $dy = \dfrac{dy}{dx}\,dx$. Integrating gives

$$\int dy = \int 4x\,dx$$

$$y = \frac{4x^2}{2} + C$$

$$y = 2x^2 + C$$

$y = 2x^2 + C$ is called the general solution of $\dfrac{dy}{dx} = 4x$.

? Why is it called the general solution?

Because the solution has a constant, C, which does not have a particular value. The solution, $y = 2x^2 + C$, satisfies the differential equation, $\dfrac{dy}{dx} = 4x$, for any value of C.

We can sketch these solutions, $y = 2x^2 + C$, for different values of C.

Figure 1 shows the graphs of $y = 2x^2 + C$.

The differential equation, $\dfrac{dy}{dx} = 4x$, has an infinite number

of solutions which are called the **family of solutions.**

Fig. 1

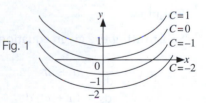

Example 3 *materials*

The bending moment, M, of a beam of length l is given by

$$\frac{dM}{dx} = w(x + l)$$

where w is the constant load. Find an expression for M in terms of x.

Example 3 *continued*

Solution

For the given differential equation, M is the dependent variable and x is the independent variable. We have

$$dM = w(x + l)\, dx$$
$$= (wx + wl)\, dx \qquad \text{[Expanding brackets]}$$

Integrating both sides:

$$\int dM = \int (wx + wl)\, dx$$

$$M = \frac{wx^2}{2} + wlx + C$$

This $M = \dfrac{wx^2}{2} + wlx + C$ is the general solution of the differential equation $\dfrac{dM}{dx} = w(x + l)$.

Example 4 *mechanics*

The velocity, $v = u + at$, of an object is defined as

$$\boxed{\;*\;} \qquad v = \frac{ds}{dt}$$

where s is the displacement, u is the initial velocity and a is the acceleration. Given that when $t = 0$, $s = 0$ show that

$$s = ut + \frac{1}{2}at^2$$

Solution

Substituting the given equation $v = u + at$ into $\boxed{\;*\;}$ yields

$$\frac{ds}{dt} = u + at$$
$$ds = (u + at)\, dt$$

Integrating:

$$\int ds = \int (u + at)\, dt$$

$$s = ut + \frac{at^2}{2} + C$$

Substituting the given conditions $t = 0$, $s = 0$:

$$0 = 0 + 0 + C \text{ gives } C = 0$$

Hence we have

$$s = ut + \frac{1}{2}at^2$$

Notice that $s = ut + \dfrac{1}{2}at^2$ is a **particular solution** of $\dfrac{ds}{dt} = u + at$ because we have a particular value for C (= 0). A particular solution is **one** member of the family of general solutions. Generally if we can find a value for C by using some given conditions then the solution is called a particular solution and the conditions are called the **initial conditions** or **boundary conditions** of the differential equation.

A3 Separating variables

A general first order differential equation of the form

13.2 $\dfrac{dy}{dx} = f(x)g(y)$

is called a separable equation because we can separate the variables by dividing through by $g(y)$. In 13.2 , $f(x)$ is a function of x only and $g(y)$ is a function of y only.

Example 5

Solve

$$\frac{dy}{dx} = (1 + x)(1 + y) \quad (y > -1)$$

Solution

Since

$$dy = \frac{dy}{dx}\,dx$$

substituting $\dfrac{dy}{dx} = (1 + x)(1 + y)$ gives

$$dy = (1 + x)(1 + y)\,dx$$

Dividing through by $1 + y$ so that the y's and dy's are on one side (separating variables) gives

$$\frac{dy}{1 + y} = (1 + x)\,dx$$

? **What are we trying to find?**

We need to find y as a function of x which satisfies the differential equation.

? Therefore we eliminate dy and dx. **How?**

Example 5 *continued*

We integrate both sides:

$$\int \frac{dy}{1+y} = \int (1+x)\,dx$$

$$\underbrace{\ln(1+y)}_{\text{by } 8.42} = x + \frac{x^2}{2} + C$$

? We need to extract y from the Left-Hand Side. **How?**

Take exponentials of both sides:

$$e^{\ln(1+y)} = e^{x+(x^2/2)+C}$$

By using the rules of indices and exponentials we have

$$\underbrace{(1+y)}_{\text{by } 5.17} = e^x e^{x^2/2} e^C$$

$$= Ae^x\, e^{x^2/2} \text{ where } A = e^C \text{ is a constant}$$

Subtracting 1 from both sides gives

$$y = Ae^x\, e^{x^2/2} - 1$$

Again this $y = Ae^x e^{x^2/2} - 1$ is the general solution of $\dfrac{dy}{dx} = (1+x)(1+y)$ because we do not have a particular value for A.

Sometimes it can be difficult to extract y from the resulting integration.

In the above example we have used exponential and logarithmic properties from **Chapter 5**. It is critical that you are comfortable in using the following properties of logs and exponentials because they are used throughout the first four sections of this chapter:

5.12 $\ln(A) + \ln(B) = \ln(AB)$

5.13 $\ln(A) - \ln(B) = \ln\left(\dfrac{A}{B}\right)$

5.14 $\ln(A^n) = n\ln(A)$

5.16 $\ln(e^x) = x$

5.17 $e^{\ln x} = x$

5.17 $e^{\ln(x)} = x$ **8.42** $\displaystyle\int \frac{f'(x)}{f(x)}\,dx = \ln|f(x)|$

> Example 6 *fluid mechanics*

The streamlines of a fluid flow are given by the first order differential equation

$$\frac{dy}{dx} = -\frac{x}{y}$$

Solve the differential equation.

Solution

Separating the variables of the given differential equation:

$$y\,dy = -x\,dx$$

$$\int y\,dy = \int -x\,dx$$

$$\frac{y^2}{2} = -\frac{x^2}{2} + C \qquad \text{[Integrating]}$$

Multiplying both sides by 2:

$$y^2 = -x^2 + 2C$$
$$y^2 + x^2 = r^2 \text{ where } r^2 = 2C$$

The equation

$$x^2 + y^2 = r^2$$

is an equation of a circle with centre origin and radius r. [In the above we chose $r^2 = 2C$ because r = radius.] By using this equation we can sketch the streamlines of the above differential equation.

[A streamline is a curve in the x–y plane given by $y = f(x)$.]

Here the streamlines are circles with centre at the origin.

Fig. 2

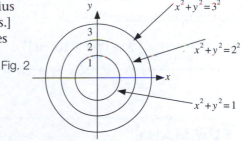

Figure 2 shows the streamlines for $2C = 1$, $2C = 4$ and $2C = 9$ where $2C = r^2$. There are an infinite number of these streamlines.

> Example 7 *fluid mechanics*

Consider a tank full of water which is being drained out through an outlet. The height H (m) of water in the tank at time t (s) is given by

$$\frac{dH}{dt} = -(2.8 \times 10^{-3})\sqrt{H}$$

Given that when $t = 0$, $H = 4$, find an expression for H in terms of t.

Example 7 *continued*

Solution

Separating variables:

$$\frac{dH}{\sqrt{H}} = -(2.8 \times 10^{-3})\, dt$$

$$H^{-1/2}\, dH = -(2.8 \times 10^{-3})\, dt \qquad [\text{Because } 1/\sqrt{H} = H^{-1/2}\,]$$

Integrating:

$$\int H^{-1/2}\, dH = -\int (2.8 \times 10^{-3})\, dt$$

$$\underbrace{\frac{H^{1/2}}{1/2}}_{\text{by } 8.1} = -(2.8 \times 10^{-3})t + C$$

$$2H^{1/2} = -(2.8 \times 10^{-3})t + C$$

Substituting the given initial condition $t = 0$, $H = 4$ yields

$$2 \times 4^{1/2} = 0 + C \text{ which gives } C = 4$$

Substituting $C = 4$ and dividing by 2 gives

$$H^{1/2} = \frac{4 - (2.8 \times 10^{-3})t}{2}$$

$$= 2 - (1.4 \times 10^{-3})t$$

Squaring both sides:

$$H = \left[2 - (1.4 \times 10^{-3})t\right]^2$$

SUMMARY

Differential equations are equations which contain a derivative. Solving a differential equation containing $\frac{dy}{dx}$ involves finding an expression for the dependent variable, y, in terms of the independent variable, x. Two methods were discussed in this section – direct integration and separation of variables.

8.1 $\int x^n\, dx = \dfrac{x^{n+1}}{n+1}$

Exercise **13(a)**

Solutions at end of book. Complete solutions available at www.palgrave.com/engineering/singh

≈ **Questions 1 to 4 are in the field of [*fluid mechanics*].**

1 The pressure, p, on an object under a fluid of density ρ (Greek letter rho) is given by

$$\frac{dp}{dz} = -\rho g$$

where z represents depth and g is the acceleration due to gravity. Find an expression for p.

2 Consider a fluid rotating about a vertical axis. The pressure p at radius r from the axis is given by

$$\frac{dp}{dr} = -\rho \omega^2 r$$

where ρ is the density and ω is the angular velocity of the fluid. Find an expression for p.

3 The streamlines of a fluid flow are given by

$$\frac{dy}{dx} = C$$

where C is a constant. Plot the streamline that passes through the origin and $C = 5$.

4 The streamlines of a fluid flow are given by

$$\frac{dy}{dx} = e^x$$

Solve the differential equation and sketch the streamlines.

🚗 **Questions 5 to 7 are in the field of [*mechanics*].**

5 The acceleration, a, of an object is defined as

$$a = \frac{dv}{dt}$$

where v is the velocity of the object at time t. Given that when $t = 0$, $v = u$, and the acceleration, a, is constant, show that $v = u + at$.

6 The acceleration, a, of a particle is given by

$$a = 5 - 3t$$

Given that the initial displacement at $t = 0$ is -2.1 m and the initial velocity is 8 m/s, find expressions for the velocity and displacement.

[*Hint*: We have, $a = \dfrac{dv}{dt}$, $v = \dfrac{ds}{dt}$ where s is the displacement]

7 An object rotates with angular acceleration, α, defined as

$$\alpha = \frac{d\omega}{dt}$$

where ω is the angular velocity. Given that when $t = 0$, $\omega = \omega_0$, show that

$$\omega = \omega_0 + \alpha t$$

Also the angular velocity ω is defined as

$$\omega = \frac{d\theta}{dt}$$

where θ is the angular displacement. Given that when $t = 0$, $\theta = 0$ show that

$$\theta = \omega_0 t + \frac{1}{2}\alpha t^2$$

≈ **Questions 8 to 10 belong to [*fluid mechanics*].**

8 The streamlines of a fluid flow are given by

$$\frac{dy}{dx} = -\frac{x}{y}$$

Show that $x^2 + y^2 = A$ where A is a constant. Sketch the streamlines for $A = 1$, 25 and 100.

9 The streamlines of a fluid flow are given by

$$\frac{dy}{dx} = -\frac{y}{x} \quad (y > 0 \text{ and } x > 0)$$

Show that $y = \dfrac{A}{x}$ and sketch on the same axes the streamlines for $A = 1$, 5 and 8 (A is a constant).

10 The streamlines of a fluid flow are given by the first order differential equation

$$\frac{dy}{dx} = \frac{y + 1}{x + 1} \quad (y > -1 \text{ and } x > -1)$$

Show that $y = Ax + A - 1$ where A is a constant.

11 🌡 [*heat transfer*]   The temperature gradient, $\dfrac{d\theta}{dx}$, of a slab of thickness t and with thermal conductivity k is given by

$$\frac{d\theta}{dx} = C$$

where C is a constant and θ is a function of x.

By using the conditions $\theta(0) = \theta_1$, $\theta(t) = \theta_2$ and Fourier's law

$$Q = -kA\frac{d\theta}{dx}$$

show that

$$Q = -kA\left(\frac{\theta_2 - \theta_1}{t}\right)$$

[$\theta(0) = \theta_1$ means that when x is 0 then $\theta = \theta_1$, and $\theta(t) = \theta_2$ means that when x is t then $\theta = \theta_2$.]

12 ✈ [*aerodynamics*]   The pressure, $p > 0$, of a gas in isothermal condition is given by

$$\frac{dp}{dz} = -\frac{pmg}{RT}$$

where z is the altitude, T (constant) is the temperature, m is the molar mass and R is a gas constant. Show that

$$p = Ae^{-\frac{mg}{RT}z}$$

(where A is a constant).

*13 ✈ [*aerodynamics*]   The pressure, p, of the atmosphere at an altitude z is given by

$$\frac{dp}{dz} = -kp^{\frac{1}{\gamma}} \quad (k \neq 0)$$

where γ is the specific heat constant ($\gamma > 1$) and k is a constant. Show that

$$z = \frac{\gamma p^{\frac{\gamma-1}{\gamma}}}{k(1 - \gamma)} + A$$

(where A is a constant).

SECTION B **Using the integrating factor**

By the end of this section you will be able to:
▶ understand the integrating factor method for solving first order differential equations
▶ apply the integrating factor method to examples

B1 **Integrating factor**

Example 8 is on differentiation and should ideally be placed in **Chapter 6** but we establish a result in this example which is used to explain the integrating factor.

Example 8

Find $\dfrac{d}{dx}(ye^{Px})$ where y is a function of x and P is a constant.

Solution

We can differentiate ye^{Px} by using the product rule:

6.31 $\dfrac{d}{dx}(uv) = u'v + uv'$

with

$$u = y \qquad\qquad v = e^{Px}$$

$$u' = \dfrac{dy}{dx} \qquad\qquad v' = Pe^{Px} \qquad\qquad \text{[Differentiating]}$$

Substituting these into 6.31 gives

$$\dfrac{d}{dx}(ye^{Px}) = \dfrac{dy}{dx}e^{Px} + Pye^{Px}$$

Suppose we have a general first order differential equation of the form

13.3 $\dfrac{dy}{dx} + Py = Q(x)$

which we need to solve, that is find y which satisfies the differential equation. Multiply both sides of 13.3 by e^{Px}:

$$\dfrac{dy}{dx}e^{Px} + Pye^{Px} = e^{Px}\,Q(x)$$

? **What do you notice about the Left-Hand Side?**

By the result of **Example 8,** the Left-Hand Side $\dfrac{dy}{dx}e^{Px} + Pye^{Px} = \dfrac{d}{dx}(ye^{Px})$. Hence we have

$$\frac{d}{dx}(ye^{Px}) = e^{Px}Q(x)$$

? **How can we obtain y from this?**

Integrate both sides with respect to x.

Integration removes the derivative, $\dfrac{d}{dx}$, on the Left-Hand Side:

$$ye^{Px} = \int e^{Px}Q(x)\,dx$$

Since P is constant, $Px = \int P\,dx$. Thus replacing Px with $\int P\,dx$ we have

$$ye^{\int P dx} = \int e^{\int P dx}Q(x)\,dx$$

$e^{\int Pdx}$ is called the integrating factor (IF). Summarizing the above:
The solution of the differential equation

13.3 $\dfrac{dy}{dx} + Py = Q(x)$

is given by

13.4 $y(\text{IF}) = \int\big[(\text{IF})Q(x)\big]\,dx$ where IF $= e^{\int Pdx}$

In the above we illustrated the integrating factor method for a constant, P, but it may be generalized to cover P as a function of x, that is, $P = P(x)$. We still obtain formula 13.4 . We will not go over the same treatment for a non-constant P but just apply 13.4 in the appropriate case.

Example 9

Solve

$$\frac{dy}{dx} + 3y = e^x$$

with the initial condition that when $x = 0$, $y = 0$.

Solution

The given equation is of the same format as 13.3 , so we can use 13.4 with $P = 3$:

$$\text{IF} = e^{\int Pdx} = e^{\int 3dx} = e^{3x}$$

Applying $y(\text{IF}) = \int\big[(\text{IF})Q(x)\big]\,dx$ with IF $= e^{3x}$ and $Q(x) = e^x$ we have

Example 9 *continued*

$$ye^{3x} = \int [e^{3x}\, e^x]\, dx$$

$$= \int [e^{3x+x}]\, dx$$

$$= \int e^{4x}\, dx$$

$$= \frac{e^{4x}}{4} + C \qquad \left[\text{Integrating by } \int e^{kx}\, dx = \frac{e^{kx}}{k} \right]$$

Substituting the initial condition $x = 0$, $y = 0$ into $ye^{3x} = \dfrac{e^{4x}}{4} + C$ gives

$$0 = \frac{1}{4} + C \text{ and so } C = -\frac{1}{4} \qquad [\text{Remember that } e^0 = 1]$$

Putting $C = -1/4$:

$$ye^{3x} = \frac{e^{4x}}{4} - \frac{1}{4}$$

Remember we need to find y. **How?**

Divide through by e^{3x} and then use the rules of indices to simplify:

$$y = \frac{e^{4x}}{4e^{3x}} - \frac{1}{4e^{3x}}$$

$$= \frac{e^{4x-3x}}{4} - \frac{e^{-3x}}{4} \qquad \left[\text{Using } \frac{a^m}{a^n} = a^{m-n} \text{ and } \frac{1}{a^n} = a^{-n} \right]$$

$$= \frac{e^x}{4} - \frac{e^{-3x}}{4}$$

$$= \frac{1}{4}(e^x - e^{-3x}) \qquad \left[\text{Taking out } \frac{1}{4} \right]$$

B2 Engineering applications

In this section we examine electrical engineering applications. In electrical principles R, L and C are constants representing resistance, inductance and capacitance respectively. E represents the e.m.f. and $v = v(t)$, $i = i(t)$ represent voltage and current respectively at time t.

Example 10 *electrical principles*

The first order differential equation

$$RC\frac{dv}{dt} + v = Ee^{-\frac{t}{RC}}$$

describes a series RC circuit. By using the initial condition $v(0) = 1$, show that

$$v = e^{-\frac{t}{RC}}\left(\frac{Et}{RC} + 1\right)$$

Solution

Divide the given differential equation by RC:

$$\frac{dv}{dt} + \frac{v}{RC} = \frac{E}{RC}e^{-\frac{t}{RC}}$$

For this differential equation, can we use the integrating factor method?

Yes, because we can write the differential equation as

$$\frac{dv}{dt} + Pv = Q(t)$$

where $P = \dfrac{1}{RC}$ because $\dfrac{v}{RC} = \dfrac{1}{RC}v$ and $Q(t) = \dfrac{E}{RC}e^{-\frac{t}{RC}}$

With $P = \dfrac{1}{RC}$ and integrating with respect to t:

$$\text{IF} = e^{\int P dt} = e^{\int \frac{1}{RC} dt} = e^{\frac{t}{RC}}$$

Applying

13.4
$$v(\text{IF}) = \int \left[(\text{IF})Q(t) \right] dt$$

with $\text{IF} = e^{t/RC}$ and $Q(t) = \dfrac{E}{RC}e^{-t/RC}$ we have

$$ve^{\frac{t}{RC}} = \int \left(\frac{E}{RC}e^{\frac{t}{RC}}e^{-\frac{t}{RC}} \right) dt$$

$$= \frac{E}{RC} \int \left(e^{\frac{t}{RC} - \frac{t}{RC}} \right) dt \qquad \left[\text{Taking out } \frac{E}{RC} \text{ and using rules of indices} \right]$$

$$= \frac{E}{RC} \int \underset{=1}{(e^{0})}\, dt$$

$$= \frac{E}{RC} \int dt$$

$$ve^{\frac{t}{RC}} = \frac{Et}{RC} + D \qquad\qquad \text{[Integrating]}$$

> ### Example 10 *continued*

where D is the constant of integration.

The initial condition $v(0) = 1$ means that when $t = 0$, $v = 1$. Substituting this condition, $t = 0$, $v = 1$, gives

$$1 = 0 + D \qquad \text{[Remember that } e^0 = 1\text{]}$$

Hence $D = 1$ and therefore we have

$$ve^{\frac{t}{RC}} = \frac{Et}{RC} + 1$$

Dividing through by $e^{t/RC}$ to find v on its own gives

$$v = \frac{1}{e^{\frac{t}{RC}}}\left(\frac{Et}{RC} + 1\right)$$

$$= e^{-\frac{t}{RC}}\left(\frac{Et}{RC} + 1\right)$$

So we have the required result.

In the above example we used the initial condition $v(0) = 1$ which means when $t = 0$, $v = 1$. In general the initial condition $y(x_0) = y_0$ means when $x = x_0$ then $y = y_0$. It is important that you understand this notation because it is used extensively throughout this chapter.

SUMMARY

The solution of the general first order differential equation of the form

13.3
$$\frac{dy}{dx} + Py = Q(x)$$

where $P = P(x)$ is given by

13.4
$$y(\text{IF}) = \int \big[(\text{IF})Q(x)\big]\,dx \quad \text{where IF} = e^{\int P dx}$$

Exercise 13(b)

Solutions at end of book. Complete solutions available at www.palgrave.com/engineering/singh

1 Solve the differential equation

$$\frac{dy}{dx} + 2y = e^{-x}$$

with the initial condition $x = 0$, $y = 0$.

2 Solve the differential equation

$$\frac{dy}{dx} - y = e^{2x}$$

3 Solve the differential equation

$$(1 - x^2)\frac{dy}{dx} - 2xy = 1 \quad (x < 1)$$

with the initial condition $x = 0$, $y = 0$.

Exercise **13(b) continued**

Solutions at end of book. Complete solutions available at www.palgrave.com/engineering/singh

🚗 Questions 4 and 5 belong to the field of [*mechanics*].

4 The velocity, v, of an object of mass m in free fall can be described by the first order differential equation

$$m\frac{dv}{dt} = mg - kv$$

where k is a positive constant and g is the acceleration due to gravity. For the initial condition, when $t = 0$, $v = 0$, show by using the integrating factor method that

$$v = \frac{mg}{k}\left(1 - e^{-\frac{k}{m}t}\right)$$

5 The velocity, v, of an object is given by the differential equation

$$\frac{dv}{dt} = \frac{1}{t}(1 - v) \qquad (t > 0)$$

By using the integrating factor method, determine an expression for v with the initial condition that when $t = 1$, $v = 0$.

What is the terminal velocity (the velocity as $t \to \infty$)?

Sketch the graph of v versus t for $t > 0$.

▦ **The remaining questions are in the field of [*electrical principles*].**

6 By applying Kirchhoff's voltage law to a series *RL* circuit we obtain the first order differential equation

$$L\frac{di}{dt} + Ri = E(t)$$

where $E(t)$ is the applied voltage and $i = i(t)$ is the current through the circuit.

i Show that $i = \dfrac{e^{-\frac{R}{L}t}}{L}\int\left(E(t)e^{\frac{R}{L}t}\right)dt$.

ii If $E(t)$ is constant, $E(t) = E$, and $i(0) = i_0$ then show that

$$i = \frac{E}{R} + \left(i_0 - \frac{E}{R}\right)e^{-\frac{R}{L}t}$$

iii Sketch the graph of i versus t (assuming $i_0 < E/R$). What value does i tend to as t becomes large?

7 A circuit consists of a resistor of resistance R, and a capacitor of capacitance C, connected in series, and is described by the first order differential equation

$$RC\frac{dv}{dt} + v = E$$

where E is the constant e.m.f. and v is the voltage across the capacitor. Given that $v(0) = 0$, show by using the integrating factor method that

$$v = E(1 - e^{-t/(RC)})$$

8 By applying Kirchhoff's voltage law to a series *RL* circuit we obtain the differential equation

$$(1 \times 10^{-3})\frac{di}{dt} + (3 \times 10^3)i = 10e^t$$

where $i = i(t)$ is the current through the circuit. Given the initial condition $i(0) = 0$, determine an expression for the current i.

9 The charge, q, on a capacitor of capacitance C, of a series *RC* circuit is given by

$$R\frac{dq}{dt} + \frac{q}{C} = Ve^{-t}$$

Show that when $t = 0$, $q = 0$ gives

$$q = \frac{VC}{1 - RC}\left(e^{-t} - e^{-\frac{t}{RC}}\right)$$

SECTION C **Applications to electrical principles**

By the end of this section you will be able to:
► model electrical circuits which give rise to a first order differential equation
► solve the resulting first order differential equation
► use a computer algebra system to solve a first order differential equation

C1 **Electrical laws**

Below are some of the laws regarding basic electrical components such as a resistor, a capacitor and an inductor. Kirchhoff's voltage law says:

In a closed path

13.5   sum of the voltage rises = sum of the voltage drops

With reference to Fig. 3, Kirchhoff's law says

$$v = v_1 + v_2 + v_3$$ Fig. 3

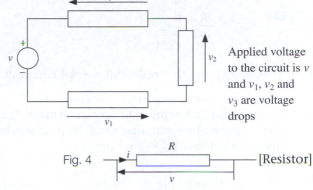

Applied voltage to the circuit is v and v_1, v_2 and v_3 are voltage drops

The voltage, v, across a resistor of resistance R (Fig. 4), is given by

13.6 $v = iR$ (Ohm's law)

The voltage, v, across an inductor of inductance L (Fig. 5), is given by

Fig. 4 ── [Resistor]

13.7 $$v = L\frac{di}{dt}$$ Fig. 5 ── [Inductor]

The voltage, v, across a capacitor of capacitance C (Fig. 6), is related to the current by

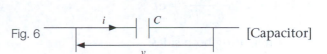

Fig. 6 ── [Capacitor]

13.8 $$i = C\frac{dv}{dt}$$

Remember that R, L and C are positive constants.

Next we use these rules to form differential equations of electrical circuits.

The following example is complicated because it involves a number of concepts such as forming and solving a differential equation, using algebraic techniques and graphing functions.

C2 **Electrical circuits**

Example 11 *electrical principles*

Figure 7 shows a resistor of resistance R, connected in series with an inductor of inductance L, and an applied constant voltage E.

i Obtain a first order differential equation for the current i at time t.

ii Solve this differential equation for the initial condition, when $t = 0$, $i = 0$.

iii What is the value of i as $t \to \infty$?

iv Sketch the graph of i versus t for $t \geq 0$.

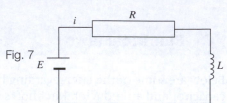

Fig. 7

Solution

i By applying Kirchhoff's law to the circuit of Fig. 7 we have:

$$E = (\text{voltage drop across resistance, } R) + (\text{voltage drop across inductance, } L)$$

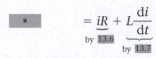

$$\underset{\text{by } \underline{13.6}}{= \underbrace{iR}} + \underset{\text{by } \underline{13.7}}{\underbrace{L\dfrac{di}{dt}}}$$

This, ⬛ * , is a first order differential equation.

ii What does 'solve' mean in this context?

It means find an expression for i in terms of E, R, L and t. This differential equation can be solved by separating variables as discussed in **Section 13A**. By subtracting iR from both sides of ⬛ * we have

$$L\frac{di}{dt} = E - iR$$

Separating variables:

$$L\frac{di}{E - iR} = dt$$

Integrating both sides:

$$L\int \frac{di}{E - iR} = \int dt$$

†
$$\underset{\text{by } \underline{8.42}}{\underbrace{-\frac{L}{R}\ln(E - iR)}} = t + C \qquad [\text{Because the derivative of } E - iR \text{ is } -R]$$

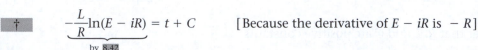

| $\underline{13.6}$ | $v = iR$ | $\underline{13.7}$ | $v = L di/dt$ | $\underline{8.42}$ | $\displaystyle\int \frac{f'(x)}{f(x)}\,dx = \ln|f(x)|$ |

Example 11 *continued*

Substituting the given initial condition $t = 0$, $i = 0$ into † :

$$-\frac{L}{R}\ln(E) = C$$

Placing $C = -\dfrac{L}{R}\ln(E)$ into † gives

$$-\frac{L}{R}\ln(E - iR) = t - \frac{L}{R}\ln(E)$$

$$-t = \frac{L}{R}\left[\ln(E - iR) - \ln(E)\right]$$

Using the properties of logs inside the square brackets and multiplying both sides by $\dfrac{R}{L}$:

$$-\frac{Rt}{L} = \underbrace{\ln\left(\frac{E - iR}{E}\right)}_{\text{by } 5.13}$$

Taking exponentials of both sides:

$$e^{-\frac{Rt}{L}} = e^{\ln\left(\frac{E-iR}{E}\right)} \underset{\text{by } 5.17}{=} \frac{E - iR}{E}$$

$$Ee^{-\frac{Rt}{L}} = E - iR \quad \text{[Multiplying by } E\text{]}$$

Rearranging gives

$$iR = E - Ee^{-\frac{Rt}{L}}$$

$$= E\left(1 - e^{-\frac{Rt}{L}}\right) \quad \text{[Taking out } E\text{]}$$

Dividing through by R yields i:

$$i = \frac{E}{R}\left(1 - e^{-\frac{Rt}{L}}\right)$$

iii Since R and L are positive, we have $\dfrac{R}{L} > 0$, $e^{-\frac{Rt}{L}} \to 0$ as $t \to \infty$ and so $i \to \dfrac{E}{R}$.

iv At $t = 0$, $i = 0$ therefore the graph goes through the origin and is asymptotic to the line $i = \dfrac{E}{R}$ (Fig. 8).

Fig. 8

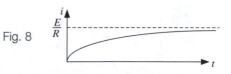

The next example uses the application of a computer algebra system such as MAPLE.

5.13 $\ln(A) - \ln(B) = \ln\left(\dfrac{A}{B}\right)$ 5.17 $e^{\ln(x)} = x$

Example 12 *electrical principles*

Figure 9 shows a series *RL* circuit with an
applied voltage, $E = 240(1 - e^{-t/3})$.

Fig. 9

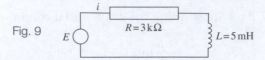

i Obtain a differential equation giving the
current *i* at time *t*.

ii Solve the differential equation for the initial condition, when $t = 0$, $i = 0$.

iii Plot the graph of *i* versus *t*. [Use MAPLE for parts **ii** and **iii**.]

Solution

i By applying Kirchhoff's law we have

E = (voltage drop across resistor) + (voltage drop across inductor)

$$= \underbrace{iR}_{\text{by }\boxed{13.6}} + \underbrace{L\frac{di}{dt}}_{\text{by }\boxed{13.7}}$$

Substituting the given values [remember that k = kilo = 10^3 and
m = milli = 10^{-3}], $R = 3 \times 10^3$, $L = 5 \times 10^{-3}$, $E = 240(1 - e^{-t/3})$ into
the differential equation:

$$240(1 - e^{-t/3}) = (3 \times 10^3)i + (5 \times 10^{-3})\frac{di}{dt}$$

ii We solve this first order differential equation, by using MAPLE:

```
> de_1:=(3*10^3)*i(t)+(5*10^(-3))*diff(i(t),t)=240
  *(1-exp(-t/3));
```

$$de\_1: = 3000\, i(t) + \frac{1}{200}\left(\frac{\partial}{\partial t}i(t)\right) = 240 - 240\, e^{(-1/3t)}$$

```
> soln:=dsolve({de_1,i(0)=0},i(t));
```

$$soln: = i(t) = \frac{2}{25} - \frac{144000}{1799999}\, e^{(-1/3t)} + \frac{2}{44999975}\, e^{(-600000t)}$$

$\boxed{13.6}$ $v = IR$ $\boxed{13.7}$ $v = L\, di/dt$

Example 12 *continued*

iii

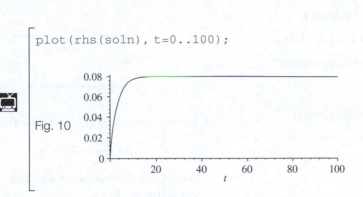

```
plot(rhs(soln), t=0..100);
```

Fig. 10

In [*] of **Example 12**, the extreme right term is called the transient term. Transient is a brief disturbance in a circuit. Because it happens to have a large negative exponent, it dies away very rapidly, and cannot be shown on the scale of Fig. 10. The first two terms show the time variation of the applied voltage. The current is eventually E/R, or $2/25 = 0.08$ in this case, which is shown in Fig. 10.

SUMMARY

13.6 The voltage, v, across a resistor of resistance R, is given by $v = iR$ (Ohm's law)

13.7 The voltage, v, across an inductor of inductance L, is given by $v = L\dfrac{di}{dt}$

13.8 The voltage, v, across a capacitor of capacitance C, is related to the current by $i = C\dfrac{dv}{dt}$

Exercise **13(c)**

Solutions at end of book. Complete solutions available at www.palgrave.com/engineering/singh

All questions belong to the field of [*electrical principles*].

1 Figure 11 shows a series RL circuit.

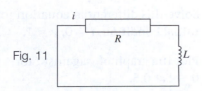

Fig. 11

i Show that
$$\frac{di}{dt} = -\frac{iR}{L}$$
where i represents the current through the circuit.

ii Show that for the initial condition, $t = 0$ s, $i = 1$ A, then
$$i = e^{-\frac{R}{L}t}$$

Solutions at end of book. Complete solutions available at www.palgrave.com/engineering/singh

iii Sketch the graph of i against t.

iv What is the value of i at $t = \tau$ where $\tau = \dfrac{L}{R}$ (τ is called the time constant for this circuit)?

2 Figure 12 shows a series RC circuit.

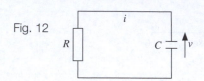

Fig. 12

i Show that the differential equation relating the voltage, v, across the capacitor of capacitance C at time t is given by

$$\frac{dv}{dt} = -\frac{v}{RC}$$

ii Show that for the initial condition when $t = 0$, $v = E\ (>0)$ that

$$v = Ee^{-t/RC}$$

iii Sketch the graph of v against t for $t \geq 0$.

iv What is the value of v at $t = \tau$ where $\tau = RC$ (τ is called the time constant)?

3 Figure 13 shows a series RC circuit with a constant applied e.m.f. $= E$.

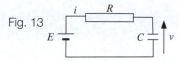

Fig. 13

Show that the differential equation of this circuit is given by

$$\frac{dv}{dt} = \frac{1}{RC}(E - v)$$

where v is the voltage across the capacitor. Solve this differential equation for the initial condition, $t = 0$, $v = 0$.

4

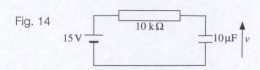

Fig. 14

i For the circuit of Fig. 14 show that

$$0.1\frac{dv}{dt} = 15 - v$$

where v is the voltage across the capacitor at time t.

ii Solve this differential equation for the initial condition, when $t = 0$, $v = 0$.

iii Sketch the graph of v against t for $t \geq 0$.

iv What percentage of the final voltage is gained at each of the following stages: $t = 0.1, 0.2$ and 0.3? (0.1 is the time constant for this circuit and 0.2, 0.3 are 2 and 3 time constants respectively.)

Use a computer algebra system for the remaining questions.

5

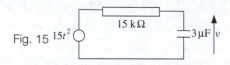

Fig. 15

i Show that the differential equation of the circuit in Fig. 15 is given by

$$0.045\frac{dv}{dt} = 15t^2 - v$$

ii Solve this differential equation for the initial condition, $t = 0$, $v = 1$.

iii Plot the graph of v against t for $0 \leq t \leq 0.5$.

Exercise **13(c) continued**

Solutions at end of book. Complete solutions available at www.palgrave.com/engineering/singh

6 Figure 16 shows a series resistor–inductor circuit.

Fig. 16

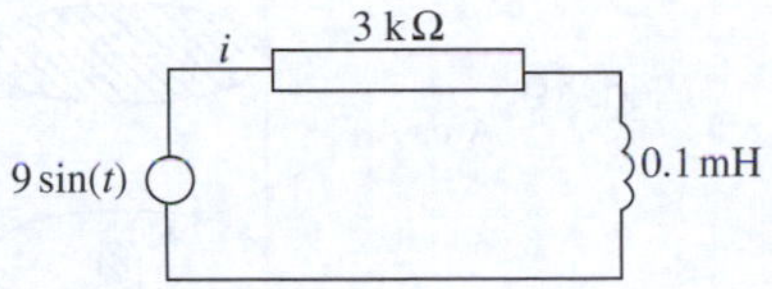

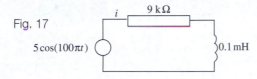

$9\sin(t)$ $3\,k\Omega$ i $0.1\,mH$

  i Show that
  $$\frac{di}{dt} = (90 \times 10^3)\sin(t) - (30 \times 10^6)i$$

  where i is the current through the circuit.

  ii Solve for the initial condition,
  $t = 0$, $i = 1 \times 10^{-3}$.

  iii Plot, on different axes, the graphs of i against t for

   a $0 \le t \le 20$

   b $0 \le t \le 2 \times 10^{-7}$.

   Comment upon your results.

7

Fig. 17

$5\cos(100\pi t)$ $9\,k\Omega$ i $0.1\,mH$

  i For the circuit of Fig. 17, show that
  $$\frac{di}{dt} = (50 \times 10^3)\cos(100\pi t) - (90 \times 10^6)i$$

  ii Solve the differential equation for the initial condition, $t = 0$, $i = 0$.

  iii Plot the graph of i against t. Choose a range that includes 5 complete cycles. Also plot the range $0 \le t \le 10^{-7}$.

  Comment upon your results.

8

Fig. 18

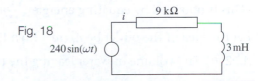

$240\sin(\omega t)$ $9\,k\Omega$ i $3\,mH$

  i Show that the differential equation of Fig. 18 is given by
  $$\frac{di}{dt} = (80 \times 10^3)\sin(\omega t) - (3 \times 10^6)i$$

  ii Solve this differential equation for the initial condition, $t = 0$, $i = 0$.

  iii Plot the graph of i against ωt for $\omega t \ge 0$. Initially choose $\omega = 10\,000$ and $\omega = 10^6$. Plot about 5 to 10 complete cycles. From the results you get, choose other values of ω which give interesting results.

SECTION D **Further engineering applications**

By the end of this section you will be able to:
▶ apply first order differential equations to other fields of engineering such as fluid mechanics and heat transfer
▶ understand Torricelli's law and derive and solve the relevant first order differential equation
▶ apply Newton's law of cooling

D1 Torricelli's law (fluid mechanics)

Figure 19 shows a water tank with an outlet of
cross-sectional area A_o (m²) and water in the tank
filled to a height of h (m). Torricelli's law states
that the velocity, v (m/s), of water exiting
through the outlet is given by

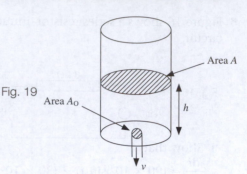

Fig. 19

$$v = \sqrt{2gh}$$

where g is the acceleration due to
gravity

(This is obtained by equating energies, $\frac{1}{2}mv^2 = mgh$ and solving for v.)

Let the area of the outlet be A_o (m²) then the volume of water leaving the tank per second is
$A_o\sqrt{2gh}$. The volume of water leaving the tank in time Δt is

$$\left(A_o\sqrt{2gh}\right)\Delta t$$

Hence the change of volume, ΔV, of water in the tank in time Δt is given by

$$\Delta V = -\left(A_o\sqrt{2gh}\right)\Delta t$$

where the negative sign indicates that the volume V is decreasing. We have

$$\frac{\Delta V}{\Delta t} = -\left(A_o\sqrt{2gh}\right)$$

Taking limits gives

13.9 $$\frac{dV}{dt} = -A_o\sqrt{2gh}$$

If the volume of water in the tank can be written as Ah where A is the constant
cross-sectional area of the tank then

$$\frac{dV}{dt} = \frac{d}{dt}(Ah) = A\frac{dh}{dt} \qquad [A \text{ is constant}]$$

Substituting $\dfrac{dV}{dt} = A\dfrac{dh}{dt}$ into 13.9 gives

$$A\frac{dh}{dt} = -A_o\sqrt{2gh}$$

Dividing both sides by A:

13.10 $$\frac{dh}{dt} = -\frac{A_o}{A}\sqrt{2gh}$$

We take g to be 9.81 m/s².

Example 13 *fluid mechanics*

A cubical tank of side 3 m is filled with water to a height of h (m). The water is drained through a circular hole of diameter 0.1 m.

  i Obtain the differential equation relating the height h of water at time t.
  ii Solve this differential equation for the initial condition, $t = 0$, $h = 2$.
  iii How long (in minutes) does it take to empty the tank from 2 m full?
  iv By using a computer algebra system or a graphical calculator, plot the graph of h against t for $t \geq 0$. Comment upon your graph.

Solution

  i The outlet area, A_o, is a circle of diameter 0.1 m and so the radius $r = 0.1/2$:

$$A_o = \pi \times r^2$$

$$= \pi \times \left(\frac{0.1}{2}\right)^2$$

$$= \pi \times 0.05^2 = (2.5 \times 10^{-3})\pi$$

Since we have a cubical tank of side 3 m, therefore

$$\text{cross-sectional area of tank } A = 3^2 = 9$$

Substituting $g = 9.81$, $A_o = (2.5 \times 10^{-3})\pi$ and $A = 9$ into

13.10
$$\frac{dh}{dt} = -\frac{A_o}{A}\sqrt{2gh}$$

$$\frac{dh}{dt} = -\frac{(2.5 \times 10^{-3})\pi}{9}\sqrt{2 \times 9.81 \times h}$$

$$= -\frac{(2.5 \times 10^{-3})\pi\sqrt{2 \times 9.81}}{9}\sqrt{h}$$

$$= -(3.87 \times 10^{-3})h^{1/2} \qquad \text{[Evaluating and writing } \sqrt{h} = h^{1/2}\text{]}$$

  ii The remaining evaluation is very similar to **Example 7**. Separating variables of

$$\frac{dh}{dt} = -(3.87 \times 10^{-3})h^{1/2}$$

gives

$$\frac{dh}{h^{1/2}} = -(3.87 \times 10^{-3})\ dt$$

Integrating:

$$\int h^{-1/2}dh = \int -(3.87 \times 10^{-3})\ dt \qquad \text{[Because } 1/h^{1/2} = h^{-1/2}\text{]}$$

$$\frac{h^{1/2}}{1/2} = -(3.87 \times 10^{-3})t + C \qquad \text{[Integrating]}$$

$$2h^{1/2} = -(3.87 \times 10^{-3})t + C$$

Substituting the given initial condition, $t = 0$, $h = 2$:

$$2(2)^{1/2} = 0 + C$$

Example 13 *continued*

Putting $C = 2(2)^{1/2}$:

$$2h^{1/2} = -(3.87 \times 10^{-3})t + 2(2)^{1/2}$$

$$h^{1/2} = -(1.935 \times 10^{-3})t + 2^{1/2} \qquad \text{[Dividing by 2]}$$

How can we find h?

Square both sides:

$$\dagger \qquad h = \left[-(1.935 \times 10^{-3})t + 2^{1/2}\right]^2$$

iii What are we trying to find?

The value of t when $h = 0$ because the tank is empty at $h = 0$.

Substituting $h = 0$ and then taking the square root of $\boxed{\dagger}$ gives

$$0 = -(1.935 \times 10^{-3})t + 2^{1/2}$$

$$t = \frac{2^{1/2}}{(1.935 \times 10^{-3})} = 731\,\text{s} \quad (\text{s} = \text{seconds})$$

What is t in minutes (mins)?

$$t = \frac{731}{60} = 12.18 \text{ mins}$$

It takes just over 12 mins to drain the water in the tank.

iv We plot $h = \left[-(1.935 \times 10^{-3})t + 2^{1/2}\right]^2$ against t using MAPLE:

```
> h:=t->(((-1.935*10^(-3))*t+sqrt(2))^2);
  h: = t → (-.001935000000 t + √2)²
```

```
> plot(h(t), t=0..731);
```

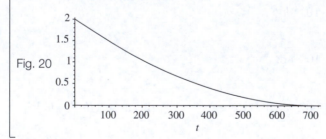

Fig. 20

The graph of Fig. 20 is only valid for $0 \le t \le 731$. When $t > 731$ the tank is empty and hence the height of the water in the tank is **zero**. Moreover the expression plotted is only valid for the tank which is 2 m full.

From the graph it is clear that the height of the water is decreasing with time t.

We can use MAPLE to solve the differential equation in **Example 13** and find how long it takes to empty the tank. It is in forming the differential equation [part **i**] where we cannot use MAPLE.

D2 **Newton's law of cooling (heat transfer)**

Newton's law of cooling states that the rate at which a body cools is proportional to the difference in temperature between the body and its surrounding environment. If $\theta = \theta(t)$ is the temperature of the body at time t and T is the constant surrounding temperature then

$$\frac{d\theta}{dt} \propto \theta - T \quad (\propto \text{ means directly proportional to})$$

Note that θ is a function of time t. We have

13.11 $$\frac{d\theta}{dt} = k(\theta - T)$$

where k is called the constant of proportionality.

Example 14 *heat transfer*

A body is exposed to a constant temperature of 280 K. After 1 minute the temperature of the body is 350 K and after 5 minutes it is 310 K. Find an expression for the temperature θ at time t. Sketch the graph of θ against t for $t \geq 0$. Comment upon your graph.

Solution
Using

13.11 $$\frac{d\theta}{dt} = k(\theta - T)$$

with $T = 280$ gives

$$\frac{d\theta}{dt} = k(\theta - 280)$$

Separating variables:

$$\frac{d\theta}{\theta - 280} = k\,dt$$

Integrating both sides:

$$\int \frac{d\theta}{\theta - 280} = \int k\,dt$$

$$\underbrace{\ln(\theta - 280)}_{\text{by } 8.42} = kt + C$$

Since t is measured in seconds we have $t = 1 \times 60 = 60$ and $\theta = 350$.

Substituting these values into $\ln(\theta - 280) = kt + C$ gives

---

8.42 $$\int \frac{f'(\theta)}{f(\theta)}\,d\theta = \ln|f(\theta)|$$

Example 14 *continued*

† $\ln(350 - 280) = 60k + C$

We are also given $t = 5 \times 60 = 300$ and $\theta = 310$. Similarly we have

†† $\ln(310 - 280) = 300k + C$

Solving the simultaneous equations † and †† gives $k = -3.53 \times 10^{-3}$ and $C = 4.46$.

Hence substituting these values into $\ln(\theta - 280) = kt + C$ yields

$$\ln(\theta - 280) = -(3.53 \times 10^{-3})t + 4.46$$

Taking exponentials of both sides to find θ we have

$$\theta - 280 = e^{-(3.53 \times 10^{-3})t + 4.46} \qquad [\text{Because } e^{\ln(\theta - 280)} = \theta - 280]$$

$$\theta = 280 + e^{4.46}e^{-(3.53 \times 10^{-3})t}$$

$$\theta = 280 + \underbrace{86.49e}_{= e^{4.46}}{}^{-(3.53 \times 10^{-3})t}$$

For sketching the graph we need to find the value of θ at $t = 0$:

$$\theta = 280 + 86.49e^{0}$$

$$= 280 + 86.49 = 366.49 \qquad [\text{Because } e^{0} = 1]$$

Also as $t \to \infty$, $e^{-(3.53 \times 10^{-3})t} \to 0$, thus $\theta \to 280$.

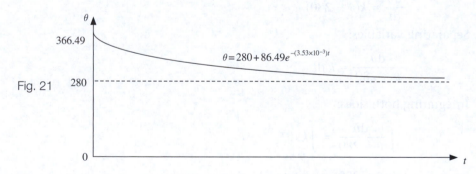

Fig. 21

The graph in Fig. 21 shows that the temperature of the body will eventually reach very close to the temperature of the surroundings at 280 K. This is because $86.49e^{-(3.53 \times 10^{-3})t}$ is the **transient term** and decays to zero as t gets large.

SUMMARY

The differential equation describing the tank filled to a height h (m) at time t is given by

Fig. 22

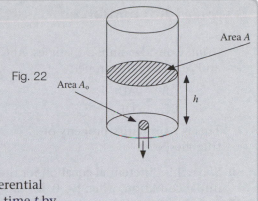

13.10 $\dfrac{\mathrm{d}h}{\mathrm{d}t} = -\dfrac{A_{\mathrm{o}}}{A} \sqrt{2gh}$

where A_{o} and A are the cross-sectional areas of the outlet and the tank respectively. Newton's law of cooling gives a first order differential equation which describes the temperature θ at time t by

13.11 $\dfrac{\mathrm{d}\theta}{\mathrm{d}t} = k(\theta - T)$

where k is a constant and T is the constant surrounding temperature.

Exercise 13(d)

Solutions at end of book. Complete solutions available at www.palgrave.com/engineering/singh

Plot the graphs in this exercise by using a computer algebra system or a graphical calculator.

 Questions 1 to 4 are in the field of [*fluid mechanics*].

1 A cubical tank of side 2 m is filled with water to a height of h (m). The water is drained through a circular hole of diameter 0.2 m.
   i Obtain a differential equation giving the height h at any time t.
   ii Given that when $t = 0$, $h = 1.3$, solve this differential equation to give an expression for h in terms of t.
   iii How long (in min) does it take to empty the tank if it is 1.3 m full?
   iv Plot the graph of h against t for $t \geq 0$. Comment upon your graph.

2 A cylindrical tank of diameter 1.5 m contains water of height h. The water is drained through a square hole of length 0.05 m.

   i Obtain a differential equation representing the height h at any time t.
   ii Given that when $t = 0$, $h = 1.6$, solve this differential equation to give an expression for h in terms of t.
   iii How long (in min) does it take to empty the tank if it is 1.6 m full?
   iv Plot the graph of h against t for $t \geq 0$.

3

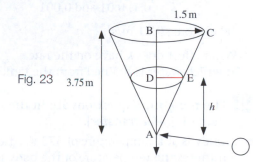

Fig. 23

Figure 23 shows a conical tank containing water of height h. The outlet is a circular hole of diameter 0.05 m.
   i Obtain the differential equation relating the height h of water at time t.

Exercise **13(d) continued**

Solutions at end of book. Complete solutions available at www.palgrave.com/science/engineering/singh

[*Hint*: Consider similar triangles, ABC and ADE. Use the property

$$\frac{AB}{BC} = \frac{AD}{DE}$$

This is a similarity property of triangles]

ii Solve this differential equation for the initial condition, when $t = 0$, $h = 3$.

iii How long (in min) does it take for the tank to empty if it is initially 3 m full?

iv Plot the graph of h against t for $t \geq 0$. Comment upon your graph.

v How long does it take to empty the tank if it is 3.8 m full?

4 The general differential equation giving the height, h, of water in a tank at time t is given by

$$\frac{dh}{dt} = -k\sqrt{2gh}$$

where k is a constant. Solve this differential equation for the initial condition, $h(0) = 1$.

Plot the graphs of the solutions of height, h, against time, t, for

$$k = 0.1, 0.01 \text{ and } 0.001$$

Take g to be 9.81 m/s².

What effect does k have on the rate at which water is drained from the tank?

🌡 **The remaining questions are in the field of [*heat transfer*].**

5 A body is at a temperature of 373 K. After 5 minutes the temperature of the body is 330 K. Find an expression for $\theta = \theta(t)$ given that the constant surrounding temperature is 300 K. Sketch the graph of θ against t for $t \geq 0$.

What does θ tend to as $t \to \infty$?

6 An object is initially at 400 K, and the constant surrounding temperature is 300 K. Determine an expression that gives the temperature $\theta = \theta(t)$ at time t.

7 Newton's law of cooling gives

$$\frac{d\theta}{dt} = k(\theta - T)$$

where θ is the temperature at time t, T is the constant surrounding temperature and k is a constant. Given that $\theta(0) = T_0$, show that

$$\theta = (T_0 - T)e^{kt} + T$$

8 By applying Newton's law of cooling to an object we obtain

$$\frac{d\theta}{dt} = k(\theta - 320)$$

where θ is the temperature at time t and k is a constant. Given that when $t = 0$, $\theta = 348$ K, find an expression for θ. Plot, on the same axes, the graphs of θ against t for $t \geq 0$ and $k = -0.001$, -0.01 and -0.1.

How does the temperature, θ, change with k?

*9 Stefan's law of radiation gives the rate of change of temperature, $\dfrac{d\theta}{dt}$, of a body by

$$\frac{d\theta}{dt} = k(\theta^4 - T^4)$$

where k is a constant, θ $(= \theta(t))$ is the temperature of the body at time t and T is the constant ambient temperature. Show, by solving the differential equation, that

$$\frac{1}{4T^3}\left[\ln\left(\frac{\theta - T}{\theta + T}\right) - 2\tan^{-1}\left(\frac{\theta}{T}\right)\right] = kt + C$$

where C is the constant of integration.

SECTION E **Euler's numerical method**

By the end of this section you will be able to:
► apply Euler's numerical formula
► use a computer algebra system to find the analytical and numerical solutions of differential equations
► use a spreadsheet to find numerical solutions

Up until now we have solved differential equations by employing analytical methods such as separating variables and the integrating factor. However many differential equations **cannot** be solved by using these analytical techniques. Sometimes it is difficult, or impossible, to find a solution to a given differential equation using these methods. With the advancement in technology it is becoming easier to use numerical methods. These numerical methods involve using a repetitive process, hence the use of computers is beneficial in eliminating the repetitive calculations. Each repetitive mathematical process is called an **iteration**.

You can use a spreadsheet, such as EXCEL, or mathematical software, such as MAPLE, MATLAB or MATHEMATICA, to perform these processes.

In this chapter we cover three main numerical methods: Euler's, improved Euler's and 4th order Runge–Kutta. There are others but we will confine ourselves to these three. In this section we examine Euler's method.

E1 **Euler's method**

Euler's method approximates the solution of the differential equation by using tangent lines. Remember that the solution of a differential equation is a function of x which we can plot on a graph.

Fig. 24

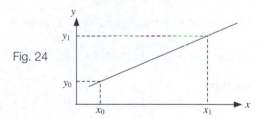

The gradient, m, of the line in the graph of Fig. 24 is given by

$$m = \frac{y_1 - y_0}{x_1 - x_0}$$

$$y_1 - y_0 = m(x_1 - x_0)$$

Adding y_0 to both sides makes y_1 the subject:

† $$y_1 = y_0 + m(x_1 - x_0)$$

We use this result,   †   , in deriving Euler's numerical formula.

Suppose we want to solve the first order differential equation

$$\frac{\mathrm{d}y}{\mathrm{d}x} = f(x, y)$$

which satisfies the initial condition, at $x = x_0$ we have $y = y_0$. The notation $f(x, y)$ means a function of x and y. **What do we mean by 'solve'?**

Ideally we need to find a solution of the form, y = function of x.

Suppose we know the solution, $y = g(x)$, and it has the graph shown in Fig. 25.

We know that $g(x_0) = y_0$
because this is the given
initial condition.

Fig. 25   Approximating
by a tangent

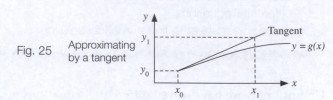

**Can we find the equation
of the tangent line at
(x_0, y_0) shown in Fig. 25?**

It is a straight line so we can use the above $y_1 = y_0 + m(x_1 - x_2)$ † . **How do we find the gradient, m?**

Remember the derivative gives the gradient. Hence the gradient at $x = x_0$ is $\dfrac{dy}{dx}$. **What is $\dfrac{dy}{dx}$ equal to at $x = x_0$?**

From the given differential equation we have

$$\frac{dy}{dx} = f(x, y)$$

and at $x = x_0$

$$\frac{dy}{dx} = f(x_0, y_0) = m \qquad \text{[Gradient]}$$

That means by the equation of a tangent

† $$y_1 = y_0 + m(x_1 - x_0)$$

we have

$$y_1 = y_0 + f(x_0, y_0)[x_1 - x_0] \qquad \text{[Substituting for } m]$$

Thus y_1 is an approximate value of the actual solution, $g(x_1)$.

We can repeat this procedure to find y_2 and y_3, the y values at x_2 and x_3 respectively. Moreover we can similarly show that

$$y_2 = y_1 + f(x_1, y_1)[x_2 - x_1]$$

and

$$y_3 = y_2 + f(x_2, y_2)[x_3 - x_2]$$

What would be the formula for y_{n+1}?

* $$y_{n+1} = y_n + f(x_n, y_n)[x_{n+1} - x_n]$$

If we assume a uniform step size of h (Fig. 26)

then $x_{n+1} - x_n = h$ and replacing this in formula * we have

13.12 $$y_{n+1} = y_n + h[f(x_n, y_n)]$$

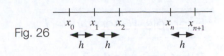

Fig. 26

or

$$y_{n+1} = y_n + hK_1 \qquad \text{where } K_1 = f(x_n, y_n)$$

13.12 is known as Euler's numerical formula. This formula approximates the actual solution by tangent lines. By using this formula we can find an approximate solution to a differential equation. Note that

$$x_{n+1} = x_n + h \qquad\qquad y_{n+1} = y_n + h[f(x_n, y_n)]$$

which means

$$(\text{new } x) = (\text{old } x) + h \qquad (\text{new } y) = (\text{old } y) + h[f(\text{old } x, \text{ old } y)]$$

We use the old x and y values to find the new x and y values. By repeating this procedure we obtain the collection of lines shown in Fig. 27.

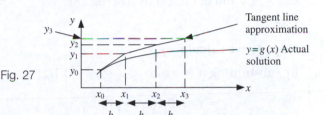

Fig. 27

? **How can we obtain a more accurate solution?**

Well, the smaller the step size, h, the better the approximation to the actual solution. Let's do an example.

Example 15

Consider the differential equation

$$\frac{dy}{dx} = x^2 + y^2 \text{ with } y(0) = 1 \quad [\text{which means that when } x = 0 \text{ then } y = 1]$$

Use Euler's method to find the approximate values of y at $x = 0.1, 0.2$ and 0.3 with step size $h = 0.1$.

Solution

Putting $h = 0.1$ into formula **13.12** gives

$$y_{n+1} = y_n + 0.1[f(x_n, y_n)]$$

? **What is $f(x_n, y_n)$?**

Remember that $f(x_n, y_n)$ is the given function of x and y:

$$f(x_n, y_n) = x_n^2 + y_n^2$$

Substituting this into y_{n+1}:

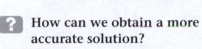 $$y_{n+1} = y_n + 0.1(x_n^2 + y_n^2)$$

Step 1:

Starting with $n = 0$ gives

$$y_1 = y_0 + 0.1(x_0^2 + y_0^2)$$

Example 15 *continued*

?

What are x_0 and y_0?

These values are given by the initial condition, $y(0) = 1$, which means that $x_0 = 0$ and $y_0 = 1$. Substituting these into y_1 gives

$$y_1 = y_0 + 0.1(x_0^2 + y_0^2)$$
$$= 1 + 0.1(0^2 + 1^2) = 1.1$$

The y_1 value at $x = 0.1$ is 1.1.

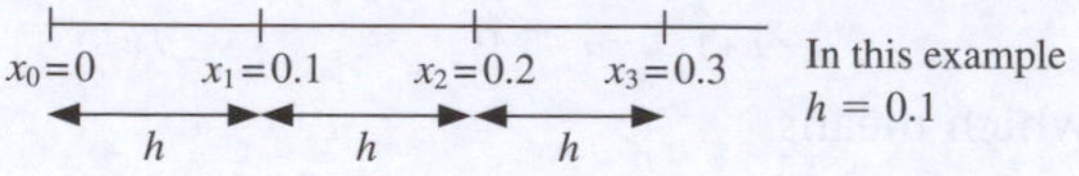

Fig. 28

$$\begin{array}{cccc} x_0=0 & x_1=0.1 & x_2=0.2 & x_3=0.3 \end{array}$$

In this example $h = 0.1$

We need to evaluate y_2 and y_3, the y values at $x = 0.2$ and 0.3 respectively (Fig. 28).

Step 2:

?

How do we calculate y_2?

By substituting $n = 1$ into $y_{n+1} = y_n + 0.1(x_n^2 + y_n^2)$, which gives

$$y_2 = y_1 + 0.1(x_1^2 + y_1^2)$$

?

What are x_1 and y_1 equal to?

$$x_1 = 0.1 \text{ and } y_1 = 1.1 \qquad \text{[Evaluated above]}$$

Substituting these into y_2 gives

$$y_2 = y_1 + 0.1(x_1^2 + y_1^2)$$
$$= 1.1 + 0.1(0.1^2 + 1.1^2) = 1.222$$

Step 3:

Similarly, substituting $n = 2$ into $\boxed{\;*\;}$ gives y_3:

$$y_3 = y_2 + 0.1(x_2^2 + y_2^2)$$

?

What are x_2 and y_2 equal to?

$$x_2 = 0.2 \text{ and } y_2 = 1.222$$

Putting these into y_3 gives

$$y_3 = y_2 + 0.1(x_2^2 + y_2^2)$$
$$= 1.222 + 0.1(0.2^2 + 1.222^2) = 1.375\ 328\ 4$$

We have $y_1 = 1.1$, $y_2 = 1.222$ and $y_3 = 1.375\ 328\ 4$, and these are the approximate values of y at $x = 0.1$, 0.2 and 0.3 respectively.

Note that the **old** values of x and y are used to find the **new** values. For example, to find y_3 we need to use x_2 and y_2.

We can apply MAPLE or any other computer algebra system to find the exact solution of **Example 15** and determine the error in using Euler's method at y_1, y_2 and y_3. The commands for using MAPLE are given in the next example. If you find difficulty in following these commands, have a look at the help menu. You can copy and paste examples from the help menu and make the appropriate changes for your example.

Example 16

By using a computer algebra system, solve the differential equation

$$\frac{dy}{dx} = x^2 + y^2 \quad \text{with } y(0) = 1$$

Determine y at $x = 0.1$, $x = 0.2$ and $x = 0.3$, and plot your solution. Compare your answers with y_1, y_2 and y_3 respectively, obtained in **Example 15**.

Solution

By using MAPLE we have:

```
> de_1:=diff(y(x),x)=(x^2)+(y(x)^2);
```

$$de\_1 := \frac{\partial}{\partial x}y(x) = x^2 + y(x)^2$$

```
> soln:=dsolve({de_1,y(0)=1},y(x)):
> evalf(subs(x=0.1,soln));
```

y(.1) = 1.111463377

```
> evalf(subs(x=0.2,soln));
```

y(.2) = 1.253016934

```
> evalf(subs(x=0.3,soln));
```

y(.3) = 1.439670778

```
> e_soln:=dsolve({de_1,y(0)=1},y(x),type=numeric,
method=classical,output=array([0.1,0.2,0.3]),stepsize=0.1);
```

$$e\_soln := \begin{bmatrix} [x,y(x)] \\ \begin{bmatrix} .1 & 1.10000000000000008 \\ .2 & 1.22200000000000020 \\ .3 & 1.37532840000000034 \end{bmatrix} \end{bmatrix}$$

```
> with(plots):
> p_1:=plots[odeplot](e_soln,[x,y(x)],linestyle=6):
> p_2:=plot(rhs(soln),x=0.1..0.3):
> display({p_1,p_2});
```

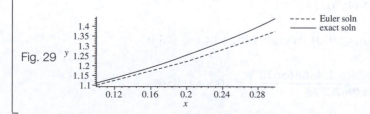

Fig. 29

Figure 29 shows the Euler solution and the actual solution; notice the discrepancy. We explore this discrepancy by evaluating the percentage error, % error, which is defined as

13.13 $\% \text{ error} = \dfrac{(\text{actual value}) - (\text{approximate value})}{\text{actual value}} \times 100$

The percentage error for y_3 for **Example 16** is

$$\left(\frac{1.439670778 - 1.3753284}{1.439670778} \right) \times 100 = 4.47\%$$

Note that the actual value in this example is only correct to 9 decimal places.

So Euler's method gives an error of more than 4%.

? **How can we reduce this error?**

By considering a smaller value of h.

? **Are there any penalties if we consider a smaller value of h?**

We will need more iterations to evaluate y_1, y_2 and y_3. We would need 30 iterations to evaluate y at $x = 0.3$ if $h = 0.01$ and $x_0 = 0$.

? **How many iterations are required for y at $x = 0.3$ if $h = 0.001$ and $x_0 = 0$?**

300. Thus to do justice to Euler's method, or any other numerical method for differential equations, we need to use mathematical software or write a procedure in a programmable calculator. Nobody would want to perform 300 iterations by hand.

By using MAPLE with different values of h we obtain Table 1 for **Example 16**.

x	$h = 0.1$	$h = 0.01$	$h = 0.001$	Solution (correct to 9 d.p.)
0.1	1.1	1.110129764	1.111327762	1.111463377
0.2	1.222	1.249332251	1.252641232	1.253016934
0.3	1.3753284	1.431804205	1.438865618	1.439670778

TABLE 1

The percentage error for $h = 0.01$ in evaluating y at $x = 0.3$ is

$$\left(\frac{1.439670778 - 1.431804205}{1.439670778} \right) \times 100 = 0.55\%$$

The percentage error for $h = 0.001$ in evaluating y at $x = 0.3$ is

$$\left(\frac{1.439670778 - 1.438865618}{1.439670778} \right) \times 100 = 0.056\%$$

Note how the error decreases as h gets smaller.

Most students are more familiar with spreadsheets than with a computer algebra system. Finding numerical solutions of differential equations is a straightforward task for spreadsheets such as EXCEL. These numerical solutions are evaluated in a worksheet document. A worksheet is an array of cells labelled A1, A2, ..., B666, etc.

We will not give detailed instructions on how to use a spreadsheet but just assume that the reader is familiar with the process.

Table 2 shows an EXCEL spreadsheet for **Example 16**:

$$\frac{dy}{dx} = x^2 + y^2 \quad \text{with } y(0) = 1$$

and $h = 0.1$ for $x = 0.1$ to $x = 1$.

x_n	y_n	$K_1 (=x_n^2 + y_n^2)$	y_{n+1}
0	1	1	1.1
0.1	1.1	1.22	1.222
0.2	1.222	1.533284	1.3753284
0.3	1.3753284	1.98152821	1.57348122
0.4	1.57348122	2.63584315	1.83706554
0.5	1.83706554	3.62480978	2.19954651
0.6	2.19954651	5.19800487	2.719347
0.7	2.719347	7.88484811	3.50783181
0.8	3.50783181	12.944884	4.80232022
0.9	4.80232022	23.8722794	7.18954816
1	7.18954816	52.6896027	12.4585084

TABLE 2

The y_{n+1} column shows the y values by using Euler's method with the step size of 0.1.

SUMMARY

Euler's numerical method gives an approximate solution to the first order differential equation

$$\frac{dy}{dx} = f(x, y) \quad \text{with } y(x_0) = y_0$$

This is obtained by using the formula

13.12 $y_{n+1} = y_n + h\left[f(x_n, y_n)\right]$

The smaller the value of h the better the accuracy, but then we need more iterations.

1 Consider the differential equation

$$\frac{dy}{dx} = x + y^2 \quad \text{with } y(0) = 1$$

Use Euler's numerical formula with $h = 0.1$ to determine approximate solutions of y at $x = 0.1, 0.2$ and 0.3.

2 Consider the initial value problem

$$\frac{dy}{dx} = x^2 + y^2 \quad \text{with } y(0) = 1$$

Use Euler's numerical formula with $h = 0.1$ to determine the approximate solution of y at $x = 0.5$ (work to 7 d.p.).

3 For the differential equation

$$\frac{dy}{dt} = t \quad \text{with } y(0) = 0$$

complete Table 3.

TABLE 3	t	Euler y with $h = 1$	Exact $y = 0.5t^2$
	0	0	0
	1		
	2		
	3		
	4		
	5		

4 [mechanics]   The velocity, v, of an object falling under gravity is given by

$$\frac{dv}{dt} = 10 - 0.1v^2 \quad \text{with } v(0) = 0$$

Determine $v(0.3)$ by using Euler's numerical formula with $h = 0.05$ (work to 5 d.p.).

Use a computer algebra system or a spreadsheet for the remaining questions.

5 [mechanics]   The velocity, v, of a sphere falling under gravity is given by

$$\frac{dv}{dt} = 10 - v^2 \quad \text{with } v(0) = 0$$

i Use Euler's numerical formula with $h = 0.1$ and $h = 0.01$ to obtain approximate values of v at $t = 0.5, 1, 1.5$ and 2.
ii Determine the solution of the given differential equation.
iii Plot, on the same axes, the graphs for $h = 0.1$, $h = 0.01$ and the solution of part **ii**.

6 Consider the initial value problem

$$\frac{dy}{dx} = \sqrt{x + y} \quad \text{with } y(0) = 0$$

Complete Table 4.

TABLE 4	x	0	0.1	0.2	0.3	0.4	0.5
	Euler y with $h = 0.1$						
	Euler y with $h = 0.01$						
	Euler y with $h = 0.005$						

7 Consider the differential equation

$$\frac{dx}{dt} = t^2 + xe^{-t} \text{ with } x(0.1) = -1$$

Solutions at end of book. Complete solutions available at www.palgrave.com/engineering/singh

Exercise **13(e) continued**

i Determine the exact solution of the given differential equation.
ii Use Euler's formula with $h = 0.25$ and $h = 0.005$ to obtain approximate values of x at $t = 0.5, 1, 1.5, 2$ and 2.5.

iii Plot, on the same axes, the graphs for $h = 0.25$, $h = 0.005$ and the solution in part **i**.
iv Determine the percentage error in x for $h = 0.25$ and $h = 0.005$ at $t = 2$.

SECTION F **Improved Euler's method**

By the end of this section you will be able to:
► use the improved Euler's formula
► compare errors between the improved Euler's and Euler's methods
► use a computer algebra system and a spreadsheet to evaluate and compare solutions

F1 **Improved Euler's method (Heun's method)**

Say we want to solve the general first order differential equation

$$\frac{dy}{dx} = f(x, y) \text{ with } y(x_n) = y_n \qquad \left[\text{when } x = x_n \text{ then } y = y_n\right]$$

using a numerical method. With Euler's method we have the situation shown in Fig. 30.

Euler's numerical formula from **Section E** is

Fig. 30

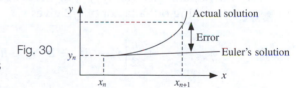

13.12 $$y_{n+1} = y_n + h\left[f(x_n, y_n)\right]$$

and at $x = x_n$ we have $f(x_n, y_n) = \dfrac{dy}{dx} = y'_n$. This is the slope at $x = x_n$. Thus substituting $f(x_n, y_n) = y'_n$ into 13.12 yields

* $$y_{n+1} = y_n + hy'_n$$

Euler's method involves finding the slope, y'_n, at the beginning of the interval, $[x_n, x_{n+1}]$. Generally this leads to a large error unless h is very small.

? **Where can we consider the slope in order to reduce this error?**

We consider two slopes, one at the beginning and the other at the end of the interval. Then taking the average of these two slopes gives an improved estimate of the slope for that interval.

Let y'_n be the slope at x_n, and y'_{n+1} be the slope at x_{n+1}. We can replace the slope, y'_n, by the average slope

$$\frac{y'_n + y'_{n+1}}{2}$$

We thus have the situation shown in Fig. 31.

Fig. 31

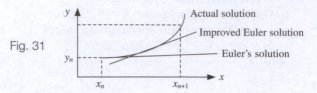

Replacing y'_n with $\dfrac{y'_n + y'_{n+1}}{2}$ in

Euler's numerical formula, ⬛ * , gives

†

$$y_{n+1} = y_n + h\underbrace{\left(\frac{y'_n + y'_{n+1}}{2}\right)}_{\text{average slope}}$$

We need to replace the y'_{n+1} term on the Right-Hand Side because we can only find y'_{n+1} by using the old x and y values (x_n and y_n). We have

$$y'_{n+1} = f(x_{n+1}, y_{n+1}) = f\left(x_{n+1}, \underbrace{y_n + hy'_n}_{\text{by } *}\right)$$

Remember that $x_{n+1} = x_n + h$. Substituting this, $x_{n+1} = x_n + h$, into y'_{n+1}:

$$y'_{n+1} = f(x_n + h, \ y_n + hy'_n)$$

We have now established (new) y'_{n+1} in terms of the nth (old) iteration. Replacing this term in † and taking out a factor of $1/2$ gives

13.14

$$y_{n+1} = y_n + \frac{h}{2}\left[y'_n + f(x_n + h, \ y_n + hy'_n)\right]$$

13.14 is called the improved Euler's formula or Heun's formula.

This formula is complicated compared to Euler's formula of **Section E**.

Example 17

For the differential equation

$$\frac{dy}{dx} = x^2 + y^2 \quad \text{with } y(0) = 1$$

find the approximate values of y at $x = 0.1$ and $x = 0.2$ by using the improved Euler's method with $h = 0.1$ (work to 6 d.p.).

Example 17 *continued*

Solution

Applying formula 13.14 with $h = 0.1$ gives

$$y_{n+1} = y_n + \frac{0.1}{2}\left[y_n' + f(x_n + 0.1, \ y_n + 0.1y_n')\right]$$

† $\quad y_{n+1} = y_n + 0.05\left[y_n' + f(x_n + 0.1, \ y_n + 0.1y_n')\right]$

Step 1:

What are we trying to find?

First we need to obtain y_1, the y value at $x = 0.1$ (Fig. 32).

Fig. 32

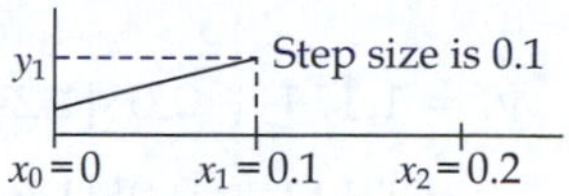

Step size is 0.1

y_1

$x_0 = 0 \quad x_1 = 0.1 \quad x_2 = 0.2$

Substituting $n = 0$ into † gives

$$y_1 = y_0 + 0.05[y_0' + f(x_0 + 0.1, \ y_0 + 0.1y_0')]$$

What are the values of x_0 and y_0?

These are the given initial condition, $y(0) = 1$, which means $x_0 = 0$ and $y_0 = 1$.

Substituting these into y_1:

* $\quad y_1 = 1 + 0.05[y_0' + f(0.1, \ 1 + 0.1y_0')]$

What is y_0' equal to?

From the question we have

$$y' = \frac{dy}{dx} = x^2 + y^2 = f(x, y)$$
$$y_0' = x_0^2 + y_0^2$$
$$= 0^2 + 1^2 = 1$$

Substituting $y_0' = 1$ into * gives

$$y_1 = 1 + 0.05\left[y_0' + f(0.1, 1 + 0.1y_0')\right]$$
$$= 1 + 0.05[1 + f(0.1, 1 + (0.1 \times 1))]$$
$$= 1 + 0.05[1 + f(0.1, 1.1)]$$
$$= 1 + 0.05\left[1 + \underbrace{(0.1^2 + 1.1^2)}_{\text{by using } f(x,y) = x^2 + y^2}\right] = 1.111$$

Step 2:

Next we determine y_2, the y value at $x = 0.2$. Substituting $n = 1$ into † gives

$$y_2 = y_1 + 0.05[y_1' + f(x_1 + 0.1, \ y_1 + 0.1y_1')]$$

What are the values of x_1 and y_1?

$$x_1 = x_0 + 0.1 = 0 + 0.1 = 0.1$$
$$y_1 = 1.111 \qquad \text{[Evaluated above]}$$

Example 17 *continued*

Putting these into $y_2 = y_1 + 0.05[y_1' + f(x_1 + 0.1, y_1 + 0.1y_1')]$ gives

✱✱ $y_2 = 1.111 + 0.05[y_1' + f(0.2, 1.111 + 0.1y_1')]$

What is the value of y_1'?

$$y_1' = x_1^2 + y_1^2 \quad \text{[Given function of } x, y]$$

$$= 0.1^2 + 1.111^2 = 1.244321$$

Substituting $y_1' = 1.244321$ into **✱✱** gives

$$y_2 = 1.111 + 0.05[1.244321 + f(0.2, \ 1.111 + (0.1 \times 1.244321))]$$

$$= 1.111 + 0.05[1.244321 + f(0.2, \ 1.235432)]$$

$$= 1.111 + 0.05\left[1.244321 + (0.2^2 + 1.235432^2)\right] \quad \text{[Using } f(x, y) = x^2 + y^2]$$

$$y_2 = 1.251531$$

We have calculated the following values: at $x = 0.1$, $y = 1.111$ and at $x = 0.2$, $y = 1.251531$.

Normally we would not work to so many significant figures but because of the effect of cumulative rounding errors, we must keep many more significant figures as guard digits.

The solutions to the differential equation of **Example 17** at $x = 0.1$ and $x = 0.2$ are

$$y(0.1) = 1.111463 \text{ and } y(0.2) = 1.253017$$

These were found by using MAPLE as shown in **Example 16**. The solutions are given correct to 6 d.p. Hence the percentage error at $x = 0.2$ is

$$\left(\frac{\text{actual value} - \text{approximate value}}{\text{actual value}}\right) \times 100 = \left(\frac{1.253017 - 1.251531}{1.253017}\right) \times 100 = 0.12\%$$

The error produced by the improved method is approximately 0.12%.

What is the size of the error developed by Euler's technique?

Euler's technique gave $y_2 = 1.222$ as evaluated in **Example 15**. The percentage error with this process at $x = 0.2$ is

$$\left(\frac{1.253017 - 1.222}{1.253017}\right) \times 100 = 2.48\%$$

Hence the improved method gives an error of 0.12% while Euler's numerical formula produces an error of 2.48%. For this example, Euler's formula gives an error of more than 20 times the error produced by the improved Euler's method. Generally there is a substantial reduction in error by using the improved approach.

Next we use MAPLE to compare exact, Euler's and improved Euler's solutions. You could use any other mathematical software such as MATHEMATICA, MATLAB, etc. Moreover you could employ a spreadsheet such as EXCEL to find the numerical solution.

Example 18

Consider the initial value problem

$$\frac{dy}{dx} = x^2 + y^2 \quad \text{with } y(0) = 1$$

i Solve this differential equation.

ii By employing Euler's numerical formula with $h = 0.1$, determine y at $x = 0.1$, 0.2, 0.3, 0.4 and 0.5.

iii By using the improved Euler's formula with $h = 0.1$, determine y at $x = 0.1, 0.2, 0.3, 0.4$ and 0.5.

iv Plot, on the same axes, the graphs for Euler's, the improved Euler's and the solution in part **i**.

Some commands might be difficult to follow, have a look at the help menu in MAPLE.

Solution

Using MAPLE we have

i

```
> de_18:=diff(y(x),x)=(y(x))^2+x^(2);
```

$$de\_18: = \frac{\partial}{\partial x}\, y(x) = y(x)^2 + x^2$$

```
> soln:= dsolve({de_18,y(0)=1},y(x)):
```

ii

```
> soln_Euler:=evalf(dsolve({de_18,y(0)=1},y(x),type=numeric,
  method=classical,output=array([0.1,0.2,0.3,0.4,0.5]),
  stepsize=0.1));
```

$$soln\_Euler : = \begin{bmatrix} [x, \ y(x)] \\ \begin{bmatrix} .1 & 1.100000000 \\ .2 & 1.222000000 \\ .3 & 1.375328400 \\ .4 & 1.573481221 \\ .5 & 1.837065536 \end{bmatrix} \end{bmatrix}$$

iii

```
> soln_IMPROVEul:=evalf(dsolve({de_18, y(0)=1}, y(x),
  type=numeric,method=classical[heunform],output=array([0.1,0.2,
  0.3,0.4,0.5]),stepsize=0.1));
```

$$Soln\_IMPROVEul : = \begin{bmatrix} [x, \ y(x)] \\ \begin{bmatrix} .1 & 1.111000000 \\ .2 & 1.251530674 \\ .3 & 1.436057423 \\ .4 & 1.688007332 \\ .5 & 2.048770723 \end{bmatrix} \end{bmatrix}$$

Example 18 *continued*

iv

```
> with(plots):
> p_1:=plots[odeplot](soln_Euler,[x,y(x)],linestyle=6,
  color=blue):
> p_2:=plots[odeplot](soln_IMPROVEul,[x,y(x)],linestyle=4,
  color=black):
> p_3:=plot(rhs(soln),x=0..0.5):
> display({p_1,p_2,p_3});
```

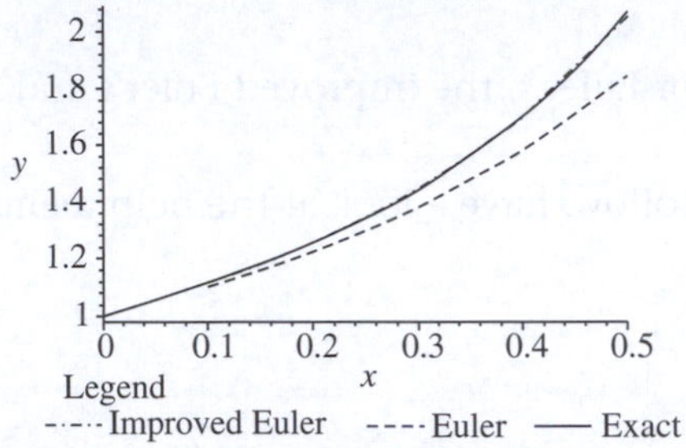

Fig. 33

Legend
----- Improved Euler   ---- Euler   —— Exact

The numerical results in the graph of Fig. 33 are plotted as points which are joined by straight lines.

The improved Euler's formula is sometimes written as

$$y_{n+1} = y_n + \frac{h}{2}\left[K_1 + K_2\right]$$

where $K_1 = f(x_n, y_n)$ and $K_2 = f(x_n + h, y_n + hK_1)$.

We can use a spreadsheet such as EXCEL to obtain the numerical solution of the above example: $\dfrac{dy}{dx} = x^2 + y^2$, $y(0) = 1$.

TABLE 5

x_n	y_n	$x_n + h$	K_1	$y_n + h*K_1$	K_2	y_{n+1}
0	1	0.1	1	1.1	1.22	1.111
0.1	1.111	0.2	1.244321	1.2354321	1.56629247	1.25153067
0.2	1.25153067	0.3	1.60632903	1.41216358	2.08420597	1.43605742
0.3	1.43605742	0.4	2.15226092	1.65128352	2.88673725	1.68800733
0.4	1.68800733	0.5	3.00936875	1.98894421	4.20589906	2.04877072
0.5	2.04877072	0.6	4.44746147	2.49351687	6.57762638	2.60002512
0.6	2.60002512	0.7	7.1201306	3.31203818	11.4595969	3.52901149
0.7	3.52901149	0.8	12.9439221	4.8234037	23.9052232	5.37146876
0.8	5.37146876	0.9	29.4926766	8.32073641	70.0446545	10.3483353
0.9	10.3483353	1	107.898044	21.1381397	447.820949	38.1342849
1	38.1342849	1.1	1455.22369	183.656654	33730.9765	1797.44429

The y_{n+1} column in Table 5 shows the approximate y values for $x = 0.1, 0.2, \ldots, 1.1$ using the improved Euler's method.

Look at the last figure in the y_{n+1} column, 1797.44429. This figure is too large. The reason is that there is a zero in the denominator of the analytical solution at about $x = 0.96$. If we plot the graph (in MAPLE) of the analytical solution between 0.9 and 1 we can see that there is a pole at $x = 0.96$.

Euler and other numerical methods step over the pole, so its effect could be missed. The moral is, don't believe everything your computer says. Euler's methods are intrinsically unstable. If you go far enough, the error in the Euler methods will always diverge. Runge–Kutta methods, discussed in the next section, are stable; errors are finite.

SUMMARY

The Improved Euler's method gives an approximate solution to the first order differential equation

$$\frac{dy}{dx} = f(x, y) \text{ with } y(x_n) = y_n$$

This is attained by using the formula

13.14 $\quad y_{n+1} = y_n + \dfrac{h}{2}[y'_n + f(x_n + h, \ y_n + hy'_n)]$

This formula produces a substantial reduction in error compared with Euler's numerical formula, 13.12 .

Exercise 13(f)

Solutions at end of book. Complete solutions available at www.palgrave.com/engineering/singh

1 Consider the differential equation

$$\frac{dy}{dx} = x + y \text{ with } y(0) = 2$$

Use the improved Euler's formula with $h = 0.1$ to obtain approximate solutions of y at $x = 0.1$ and $x = 0.2$.

2 For the initial value problem

$$\frac{dy}{dx} = x^2 + y^2, \quad y(0) = 1$$

approximate $y(0.3)$ by using the improved Euler's method with $h = 0.1$ (work to 6 d.p.).

3 For the differential equation

$$\frac{dy}{dx} = \ln|x + y|, \quad y(1.2) = 1$$

approximate $y(1.6)$ by employing the improved Euler's formula with $h = 0.2$ (work to 4 d.p.).

Use a computer algebra system or a spreadsheet for the remaining questions.

4 Consider the initial value problem

$$\frac{dx}{dt} = t^2 + xe^{-t} \text{ with } x(0.1) = -1$$

   i Solve this differential equation.
   ii Determine the approximate values of x at $t = 0.5, 1, 1.5, 2$ and 2.5 by using the Euler's numerical formula with $h = 0.25$.
   iii Determine the approximate values of the x at $t = 0.5, 1, 1.5, 2$ and 2.5 by using the improved Euler's formula with $h = 0.25$.
   iv Plot, on the same axes, the graphs of Euler's, the improved Euler's and the exact solutions.

Exercise **13(f) continued**

Solutions at end of book. Complete solutions available at www.palgrave.com/engineering/singh

5 Consider the initial value problem

$$\frac{dy}{dx} = x^3 + y^2, \ y(0) = -1$$

  i Solve this differential equation.

  ii Determine the approximate values of y at $x = 0.5, 1, 1.5$ and 2 by using Euler's numerical formula with $h = 0.01$.

  iii Determine the approximate values of y at $x = 0.5, 1, 1.5$ and 2 by using the improved Euler's formula with $h = 0.01$.

  iv Plot, on the same axes, the graphs of Euler's, the improved Euler's and the solution of part **i**.

6 Consider the initial value problem

$$\frac{dx}{dt} = 10 - x^2 \text{ with } x(0) = 0$$

  i Solve this differential equation.

  ii Determine the approximate values of x at $t = 0.1, 0.2, 0.3, 0.4$ and 0.5 by using Euler's numerical formula with $h = 0.09$.

  iii Determine the approximate values of x at $t = 0.1, 0.2, 0.3, 0.4$ and 0.5 by using the improved Euler's formula with $h = 0.09$.

  iv Plot, on the same axes, the graphs of Euler's, the improved Euler's and the exact solutions.

  v Determine the percentage error at $t = 0.5$ for Euler's and the improved Euler's methods. What do you notice?

SECTION G **Fourth order Runge–Kutta**

By the end of this section you will be able to:
▶ apply the 4th order Runge–Kutta method
▶ use a computer algebra system to find the Runge–Kutta solutions
▶ compare errors between Euler's, the improved Euler's and the Runge–Kutta methods

We discussed Euler's and the improved Euler's method in previous sections. In this part we examine another numerical method called 4th order Runge–Kutta. 1st, 2nd and 3rd order Runge–Kutta methods also exist but the 4th order is most commonly used and is one of the most accurate numerical methods.

Carle Runge was born in Germany in 1856. He studied mathematics at the University of Munich. He obtained a chair at Hannover and 18 years later he was offered a chair of Applied Mathematics at the University of Göttingen.

Martin Kutta was born in Poland in 1867 and also studied at the University of Munich. He became professor at Stuttgart in 1911.

The Runge–Kutta method was developed in 1901.

G1 **Fourth order Runge–Kutta**

The 4th order Runge–Kutta method is used to determine approximate solutions to a first order differential equation. Consider the **initial value problem**:

$$\frac{dy}{dx} = f(x, y) \text{ with } y(x_n) = y_n$$

Sometimes a differential equation with initial conditions is called an initial value problem.

The 4th order Runge–Kutta method is defined as

13.15 $$y_{n+1} = y_n + \frac{h}{6}[K_1 + 2(K_2 + K_3) + K_4]$$

where

$$K_1 = f(x_n, y_n)$$

$$K_2 = f\left(x_n + \frac{1}{2}h, \ y_n + \frac{1}{2}hK_1\right)$$

$$K_3 = f\left(x_n + \frac{1}{2}h, \ y_n + \frac{1}{2}hK_2\right)$$

$$K_4 = f(x_n + h, \ y_n + hK_3)$$

Example 19

Consider the differential equation

$$\frac{dy}{dx} = x^2 + y^2 \text{ with } y(0) = 1$$

Use the 4th order Runge–Kutta method to determine the approximate values of y at $x = 0.1$ and $x = 0.2$ with $h = 0.1$ (work to 6 d.p.).

Solution

We need to ascertain y_1 and y_2, the y values at $x = 0.1$ and $x = 0.2$ respectively. Substituting $n = 0, \ h = 0.1$ into formula **13.15** produces

* $$y_1 = y_0 + \frac{0.1}{6}\left[K_1 + 2(K_2 + K_3) + K_4\right]$$

For $n = 0$ we have $K_1 = f(x_0, y_0)$.

Step 1:

? **What are the values of x_0 and y_0?**

These are given by the initial condition, $y(0) = 1$, which means $x_0 = 0$ and $y_0 = 1$. Putting these into $K_1 = f(x_0, y_0)$ gives

† $$K_1 = f(0, 1)$$

? **What does f represent?**

This is the function of x and y on the Right-Hand Side of the given differential equation:

$$f(x,y) = x^2 + y^2$$

$$f(0,1) = 0^2 + 1^2 = 1$$

Example 19 *continued*

It follows from ⊞ † that $K_1 = 1$. We need to find K_2. Putting $h = 0.1$, $n = 0$ and $K_1 = 1$ into

$K_2 = f\left(x_n + \dfrac{1}{2}h,\ y_n + \dfrac{1}{2}hK_1\right)$ gives

$$K_2 = f\left(x_0 + \frac{1}{2}(0.1),\ y_0 + \frac{1}{2}[\,0.1 \times 1]\right)$$
$$= f(0 + 0.05,\ 1 + 0.05) \qquad [\text{Substituting } x_0 = 0 \text{ and } y_0 = 1]$$
$$= f(0.05,\ 1.05)$$
$$K_2 = 0.05^2 + 1.05^2 = 1.105$$

What else do we need to find to use $*$?

K_3 and K_4. Evaluating K_3 is similar to finding K_2. Substituting $h = 0.1$, $n = 0$ and $K_2 = 1.105$ into $K_3 = f\left(x_n + \dfrac{1}{2}h,\ y_n + \dfrac{1}{2}hK_2\right)$ gives

$$K_3 = f\left[x_0 + \frac{1}{2}(0.1),\ y_0 + \frac{1}{2}(0.1 \times 1.105)\right]$$
$$= f(0 + 0.05,\ 1 + 0.05525) \qquad [\text{Substituting } x_0 = 0 \text{ and } y_0 = 1]$$
$$= f(0.05,\ 1.05525)$$
$$K_3 = 0.05^2 + 1.05525^2 = 1.116053$$

How do we calculate K_4?

By placing $n = 0$, $h = 0.1$ and $K_3 = 1.116053$ into $K_4 = f(x_n + h,\ y_n + hK_3)$:

$$K_4 = f(x_0 + 0.1,\ y_0 + [\,0.1 \times 1.116053])$$
$$= f(0 + 0.1,\ 1 + 0.111605)$$
$$= f(0.1,\ 1.111605)$$
$$K_4 = 0.1^2 + 1.111605^2 = 1.245666$$

Substituting the calculated values, $y_0 = 1$, $K_1 = 1$, $K_2 = 1.105$, $K_3 = 1.116053$ and $K_4 = 1.245666$, into $*$ gives

$$y_1 = y_0 + \frac{0.1}{6}[K_1 + 2(K_2 + K_3) + K_4]$$
$$= 1 + \frac{0.1}{6}[1 + 2(1.105 + 1.116053) + 1.245666]$$
$$= 1.111463$$

Step 2:

Similarly we evaluate y_2. Putting $n = 1$ and $h = 0.1$ into 13.15 results in

∗∗ $\quad y_2 = y_1 + \dfrac{0.1}{6}[K_1 + 2(K_2 + K_3) + K_4]$

What is K_1 equal to in this case?

$$K_1 = f(x_1,\ y_1)$$

Example 19 *continued*

What are the values of x_1 and y_1?

$$x_1 = 0.1 \text{ and } y_1 = 1.111463 \qquad \text{[Evaluated above]}$$

Substituting these into $K_1 = f(x_1, y_1)$ yields

$$K_1 = f(0.1, 1.111463)$$

$$= 0.1^2 + 1.111463^2 = 1.245350$$

We need to determine K_2 next. Putting $n = 1$, $h = 0.1$ and $K_1 = 1.245350$ into

$$K_2 = f\left(x_n + \frac{1}{2}h, \; y_n + \frac{1}{2}hK_1\right) \text{ gives}$$

$$K_2 = f\left[x_1 + \frac{1}{2}(0.1), \; y_1 + \frac{1}{2}(0.1 \times 1.245350)\right]$$

$$= f(0.1 + 0.05, 1.111463 + 0.062268) \quad \text{[Substituting } x_1 = 0.1, y_1 = 1.111463]$$

$$= f(0.15, 1.173731)$$

$$K_2 = 0.15^2 + 1.173731^2 = 1.400144$$

Similarly by the definition of K_3 we have

$$K_3 = f\left[x_1 + \frac{1}{2}(0.1), \; y_1 + \frac{1}{2}(0.1 \times 1.400144)\right]$$

$$= f(0.1 + 0.05, 1.111463 + 0.070007)$$

$$= f(0.15, 1.181470)$$

$$K_3 = 0.15^2 + 1.181470^2 = 1.418371$$

Next we determine K_4:

$$K_4 = f[x_1 + 0.1, y_1 + (0.1 \times 1.418371)]$$

$$= f(0.1 + 0.1, 1.111463 + 0.141837)$$

$$= f(0.2, 1.2533)$$

$$K_4 = 0.2^2 + 1.2533^2 = 1.610761$$

Substituting $y_1 = 1.111463$, $K_1 = 1.245350$, $K_2 = 1.400144$, $K_3 = 1.418371$ and $K_4 = 1.610761$ into $\boxed{**}$ gives

$$y_2 = 1.111463 + \frac{0.1}{6}[1.245350 + 2(1.400144 + 1.418371) + 1.610761]$$

$$= 1.253015$$

The calculated values are $x = 0.1$, $y = 1.111463$ and at $x = 0.2$, $y = 1.253015$.

Next we compare the errors for the differential equation of **Example 19** between Euler's, the improved Euler's and the 4th order Runge–Kutta solutions.

The y values at $x = 0.2$ for Euler's, the improved Euler's and the 4th order Runge–Kutta solutions are 1.222, 1.251531 and 1.253015 respectively. The actual value (correct to 6 d.p.)

is 1.253017 as evaluated in **Example 16**. The Runge–Kutta value might well be the same as, or closer to, the actual value if we had performed the calculations to more decimal places. However, notice how close the Runge–Kutta approximation is to the actual value compared to Euler's and the improved Euler's.

We can use MAPLE to plot solutions of Euler's, the improved Euler's and the 4th order Runge–Kutta methods, as the next example shows.

Example 20

Consider the differential equation

$$\frac{dy}{dx} = x^2 + y^2 \text{ with } y(0) = 1$$

i Solve this differential equation.
ii Find y at $x = 0.2, 0.4, 0.6$ and 0.8 by using Euler's method with $h = 0.1$.
iii Repeat **ii** for the improved Euler's method with $h = 0.1$.
iv Repeat **ii** for the 4th order Runge–Kutta method with $h = 0.1$.
v Plot, on the same axes, the graphs of Euler's, the improved Euler's and the 4th order Runge–Kutta solutions, and the solution of part **i**.
What do you notice about your graphs?

Solution

Using Maple we have

i
```
> de_20:=diff(y(x),x)=(x^2)+(y(x)^2);
```
$$de\_20: = \frac{\partial}{\partial x} y(x) = x^2 + y(x)^2$$
```
> soln:=dsolve({de_20,y(0)=1},y(x)):
```

ii
```
> e_soln:=dsolve({de_20,y(0)=1}, y(x), type=numeric, method=
  classical, output=array ([0.2, 0.4, 0.6, 0.8]), stepsize=0.1);
```

$$e\_soln: = \begin{bmatrix} [x, y(x)] \\ \begin{bmatrix} .2 & 1.22200000000000020 \\ .4 & 1.57348122078465646 \\ .6 & 2.19954651435706426 \\ .8 & 3.50783181255390319 \end{bmatrix} \end{bmatrix}$$

iii
```
> improve_soln:=dsolve({de_20, y(0)=1},y(x),type=numeric,
  method=classical [heunform], output=array([.2,.4,.6,.8]),
  stepsize=0.1);
```

$$improve\_soln := \begin{bmatrix} [x, y(x)] \\ \begin{bmatrix} .2 & 1.25153067368552052 \\ .4 & 1.68800733199093366 \\ .6 & 2.60002511535832781 \\ .8 & 5.37146875547270320 \end{bmatrix} \end{bmatrix}$$

Example 20 *continued*

iv

```
> rk_soln:=dsolve({de_20,y(0)=1},y(x),type=numeric,
  method=classical[rk4], output=array([.2,.4,.6,.8]),
  stepsize=0.1);
```

$$rk\_soln := \begin{bmatrix} & [x, y(x)] \\ .2 & 1.25301517460353452 \\ .4 & 1.69609790372281698 \\ .6 & 2.64386019707933162 \\ .8 & 5.84201333527994748 \end{bmatrix}$$

v

```
> with(plots):p_1:=plots[odeplot](e_soln,[x,y(x)],
  linestyle=4,color=black):
```

```
> p_2:=plots[odeplot](improve_soln,[x,y(x)],linestyle=2,
  color=blue):
```

```
> p_3:=plots[odeplot](rk_soln,[x,y(x)],linestyle=6, color=coral):
```

```
> p_4:=plot(rhs(soln),x=0..0.8):
```

```
> display({p_1,p_2,p_3,p_4});
```

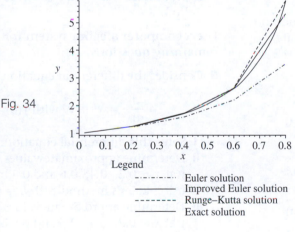

Fig. 34

Legend
- – · – · – · – · Euler solution
- ——————— Improved Euler solution
- – – – – – – – Runge–Kutta solution
- ——————— Exact solution

The Runge–Kutta solution seems to be the closest to the actual solution. The improved Euler's is a close second and Euler's solution increases in error as you move away from 0. Remember that the numerical results are plotted as points. So when comparing exact and numerical solutions, we should ignore the portions between the kinks in the lines.

We can also find the Runge–Kutta solution of differential equations by using a spreadsheet. Try the above example on the type of spreadsheet available to you.

SUMMARY

The fourth order Runge–Kutta solution of the differential equation

$$\frac{dy}{dx} = f(x, y) \text{ with } y(x_n) = y_n$$

is given by

13.5 $$y_{n+1} = y_n + \frac{h}{6}[K_1 + 2(K_2 + K_3) + K_4]$$

where

$$K_1 = f(x_n, y_n)$$

$$K_2 = f\left(x_n + \frac{1}{2}h,\ y_n + \frac{1}{2}hK_1\right)$$

$$K_3 = f\left(x_n + \frac{1}{2}h,\ y_n + \frac{1}{2}hK_2\right)$$

$$K_4 = f(x_n + h,\ y_n + hK_3)$$

Exercise **13(g)**

Solutions at end of book. Complete solutions available at www.palgrave.com/engineering/singh

1 Consider the differential equation

$$\frac{dy}{dx} = x + y \text{ with } y(0) = 2$$

Use the 4th order Runge–Kutta method with $h = 0.1$ to obtain approximate solutions of y at $x = 0.1$ and $x = 0.2$ (work to 6 d.p.).

2 For the initial value problem

$$\frac{dy}{dx} = \sqrt{x + y},\quad y(0) = 0$$

find $y(0.4)$ using the 4th order Runge–Kutta formula with $h = 0.2$ (work to 4 d.p.).

3 Consider the differential equation

$$\frac{dy}{dx} = \ln(x + y) \text{ with } y(1.2) = 1$$

Approximate $y(1.6)$ by employing the 4th order Runge–Kutta formula with $h = 0.2$ (work to 4 d.p.).

4 🚗 [*mechanics*]   The velocity, v, of an object dropped in air is given by

$$\frac{dv}{dt} = 9.8 - \frac{v^2}{10} \text{ with } v(0) = 0$$

Apply the 4th order Runge–Kutta method with $h = 1$ to find an approximate v at $t = 2$ s (work to 4 d.p.).

Use a computer alegbra system for the remaining questions.

5 Consider the differential equation

$$\frac{dy}{dx} = x^4 + y^2 \text{ with } y(0) = 1$$

  i Solve this differential equation.
  ii Determine approximate values of y at $x = 0.2,\ 0.4,\ 0.6$ and 0.8 by using Euler's method with $h = 0.05$.
  iii Determine approximate values of y at $x = 0.2,\ 0.4,\ 0.6$ and 0.8 by using the 4th order Runge–Kutta method with $h = 0.05$.
  iv Plot, on the same axes, the graphs of Euler's, Runge–Kutta and the exact solutions.

6 Consider the differential equation

$$\frac{dy}{dx} = e^y \sin(x),\ y(0) = -1$$

Solutions at end of book. Complete solutions available at www.palgrave.com/engineering/singh

Exercise **13(g) continued**

i Solve this differential equation.

ii Determine approximate values of y at $x = 1, 2, 3$ and 4 by using the improved Euler's method with $h = 0.2$.

iii Determine the approximate values of y at $x = 1, 2, 3$ and 4 by using the

4th order Runge–Kutta method with $h = 0.2$.

iv Plot, on the same axes, the graphs of the improved Euler's, the Runge–Kutta and the exact solutions.

Examination questions 13

1 Determine the solution of the differential equations

i $y \dfrac{dy}{dx} = 3x^2$ $\qquad y(0) = 1$

ii $\dfrac{dy}{dx} - y = e^{2x}$ $\qquad y(0) = 0$

Loughborough University, UK, 2009

2 The velocity v (m/s) of an object initially at rest (that is, $v = 0$ at $t = 0$) is given by

$$150 \frac{dv}{dt} = 350 - 15v$$

a Find an expression for v in terms of time t.

b Find the value of t when $v = 2$ m/s.

University of Portsmouth, UK, 2009

3 Dr Gourley likes to drink a cup of coffee before his 11 a.m. lecture to the engineering students. The coffee is at 200°F when poured at 10.15 a.m., and 15 minutes later it cools to 120°F in a room in which the temperature is 70°F. Dr Gourley will not drink the coffee until it has cooled to 90°F. At what time (to the nearest minute) will he start drinking it?

[Assume that the temperature T of the coffee satisfies $dT/dt = -k(T - 70)$ for some number k.]

University of Surrey, UK, 2002

4 i Use the integrating factor method to find the solution of the linear differential equation $\dfrac{dy}{dx} + \dfrac{y}{x} = x$ for which $y = 3$ at $x = 1$.

*ii Use the substitution $y = vx$ to find the general solution of the differential equation

$$\frac{dy}{dx} = \frac{y}{x - y}$$

University of Manchester, UK, 2008

5 Solve the problem $y' = e^{x + 2y}$.

University of California, Berkeley, USA 2005

6 Find $f(x)$ if $f'(x) = \dfrac{2x + \sqrt{x}}{x}$, $f(1) = 3$.

Arizona State University, USA (Review)

7 Use Euler's method with step size 0.5 to estimate $y(3)$, where $y(x)$ is the solution of the initial value problem: $y' = y - 2x$ $y(1) = 0$.

North Carolina State University, USA, 2010

8 Show that if $y = \ln(\cos x)$, then $\dfrac{dy}{dx} = -\tan(x)$. Use this result to find an integrating factor of the differential equation

$$\frac{dy}{dx} + 2y \tan x = \sin x$$

Hence solve the differential equation, given that $y(\pi/3) = 0$.

University of Sussex, UK, 2007

9 Solve the initial value problem

$$\frac{dy}{dt} - 2\tan(t)y = \tan(t),\ y\left(\frac{\pi}{4}\right) = 5.$$

Memorial University of Newfoundland, Canada, 2010

10 Find all real solutions to the differential equation $x\dfrac{dy}{dx} + 2y = \sin x$.

Princeton University, USA, 2006

11 The population of fish in a lake is m million, where $m = m(t)$ varies with time t (in years). The number of fish is currently 2 million.

  a Suppose m satisfies the logistic-growth equation

$$\frac{dm}{dt} = 16m\left(1 - \frac{m}{4}\right)$$

  When will the number of fish equal 3 million? You may use the fact that the general solution to the logistic-growth differential equation $y' = ky[1-(y/K)]$ is, $y = K/(1+Ae^{-kt})$ where A is a constant.

  b Suppose instead that (because of fishing by humans) m satisfies

$$\frac{dm}{dt} = 16m\left(1 - \frac{m}{4}\right) - 12$$

Will the population ever equal 3 million? You must give justification for your answer.

University of British Columbia, Canada, 2007

12 Torricelli's law states that

$$A(y)\frac{dy}{dt} = -a\sqrt{2gy}$$

where y is the depth of a fluid in a tank at time t, $A(y)$ is the cross-sectional area of the tank at height y above the exit hole, a is the cross-sectional area of the exit hole, and $g = 9.8$ m/s^2 is the acceleration due to gravity.

A container in the form of an inverted right circular cone of radius 1 m and height 3 m is full of water. A circular plug of radius 1 cm is pulled open at the bottom of the container. How long will it take for the container to become completely empty of all water?

University of Toronto, Canada, 2001

Miscellaneous exercise **13**

1 [*thermodynamics*]   In thermodynamics the relationship between mass, m, volume, V, and pressure, P, is given by

$$mCd(pV) = pVdm$$

where C is a constant. Show that

$$m = (pV)^C$$

(Assume the constant of integration to be zero.)

2 [*mechanics*]   The acceleration, a, of a vehicle is given by

$$a = v^2$$

where v is the velocity. Find an expression for v given that $a = \dfrac{dv}{dt}$ and when $t = 0$, $v = 10$.

3 [*mechanics*]   The acceleration, $a = \dfrac{dv}{dt}$, of a vehicle is given by

$$a = -kv^2$$

where k is a positive constant and v is the velocity. Given that when $t = 0$, $v = 0.1$, determine v as a function of time t.

By using a graphical calculator or a computer algebra system, plot graphs of v against t for $k = 0.1$, 1 and 10 ($0 \leq t \leq 10$). What effect does k have?

4 [car] [*mechanics*]   The velocity, v, of a vehicle during the application of brakes is given by

$$\frac{dv}{dt} = -kv$$

where t is time and k is a constant. If the velocity at time $t = 0$ is V_0, show that $v = V_0\, e^{-kt}$.

5 [thermo] [*thermodynamics*]   The PV relationship of an ideal gas is given by

$$\frac{dP}{dV} = -\frac{kP}{V}$$

where P is pressure, V is volume and k is a constant. Show that

$$PV^k = \text{constant}$$

6 [thermo] [*thermodynamics*]   The PV relationship of an ideal gas is given by

$$\frac{dP}{P} + k\frac{dV}{V} = 0$$

where P is pressure, V is volume and k is a constant. Show that

$$PV^k = \text{constant}$$

7 [electrical] [*electrical principles*]   The voltage, v, across an inductor of inductance L, is given by

$$v = L\frac{di}{dt}$$

where i is the current through the inductor. For $v = 240\sin(100t)$, $L = 1 \times 10^{-3}$ H and the initial condition $t = 0$, $i = 0$, show that

$$i = 2400\big[1 - \cos(100t)\big]$$

8 [electrical] [*electrical principles*]   A circuit consists of a resistor of resistance R, and a capacitor of capacitance C, in series, and is described by the differential equation

$$RC\frac{dV}{dt} + V = E$$

where E is the constant applied e.m.f. and V is the voltage across the capacitor. Show that, if $V(0) = 0$, then

$$V = E(1 - e^{-t/RC})$$

[car] **Questions 9 to 11 are in the field of [*mechanics*].**

9 The velocity, v, of an object initially at rest ($t = 0$, $v = 0$) is given by

$$100\,\frac{dv}{dt} = 200 - 10v$$

i Find an expression for v in terms of t.
ii Find v for large t ($t \to \infty$). (This v is called the terminal velocity.)
iii Sketch the graph of v against t for $t \geq 0$.

*10 An object of mass, m, falls from rest ($t = 0$, $v = 0$) and its velocity v is given by

$$m\,\frac{dv}{dt} = mg - kv \qquad \left(g > \frac{k}{m}\,v\right)$$

where k is a positive constant. Show that

$$v = \frac{mg}{k}\,(1 - e^{-kt/m})$$

Determine the (terminal) velocity as $t \to \infty$.

*11 By applying Newton's law to a parachutist falling from rest, we obtain the first order differential equation

$$m\,\frac{dv}{dt} = mg - kv^2 \qquad \left(v > \sqrt{\frac{mg}{k}}\,\right)$$

where m is mass, v is velocity, t is time, g is acceleration due to gravity and k is a constant. Show that

$$v = \sqrt{\frac{mg}{k}}\left(\frac{1 + Ae^{-2\beta t}}{1 - Ae^{-2\beta t}}\right)$$

where $\beta = \sqrt{\dfrac{gk}{m}}$ and A is a constant.

What is the value of the terminal velocity?

$$\left[\text{Hint: Use} \int \frac{du}{u^2 - a^2} = \frac{1}{2a}\ln\left|\frac{u-a}{u+a}\right|\right]$$

📠 Questions 12 to 14 are in the field of [**electrical principles**].

12 The current, i, through a series RL circuit is given by

$$L\frac{di}{dt} + Ri = t$$

Show that, if $t = 0$, $i = 0$, then

$$i = \frac{t}{R} + \frac{L}{R^2}\left(e^{-\frac{R}{L}t} - 1\right)$$

13

Fig. 35

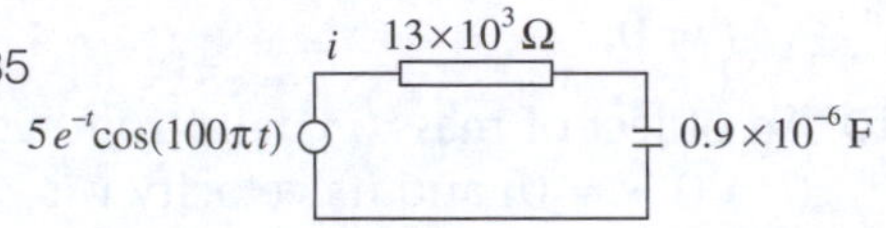

Figure 35 shows a voltage, $5e^{-t}\cos(100\pi t)$, applied to a circuit with a capacitance, (0.9×10^{-6}) F, in series with a resistance, (13×10^3) Ω. Show that

$$\frac{dv}{dt} = 427.35e^{-t}\cos(100\pi t) - 85.47v$$

where v is the voltage across the capacitor.

*14 By applying Kirchhoff's law to a circuit we obtain the differential equation

$$L\frac{di}{dt} + Ri = E\cos(\omega t)$$

where L is inductance, R is resistance, E is e.m.f., i is current, t is time and ω is angular frequency.

For the initial condition, when $t = 0$, $i = 0$, show that

$$i = \frac{E}{R^2 + \omega^2 L^2}\Big[R\cos(\omega t)$$
$$+ \omega L\sin(\omega t) - Re^{-Rt/L}\Big]$$

Use a computer algebra system for the remaining questions.

15 Consider the differential equation

$$\frac{dy}{dx} = x^2 + y^2 \text{ with } y(0.5) = 0$$

  i Solve this differential equation.
 ii Determine approximate values of y at $x = 1$, 1.2, 1.4, 1.6 and 1.8 by using Euler's method with $h = 0.1$.
iii Determine approximate values of y at $x = 1$, 1.2, 1.4, 1.6 and 1.8 by using the 4th order Runge–Kutta method with $h = 0.1$.
 iv Plot, on the same axes, the graphs of Euler's, Runge–Kutta and the exact solutions.
What do you notice about your results?

16 Consider the differential equation

$$\frac{dy}{dx} = x^2 + ye^x, \quad y(0) = -1$$

  i Solve this differential equation.
 ii Determine approximate values of y at $x = 0.2$, 0.4, 0.6 and 1.2 by using the improved Euler's method with $h = 0.2$.
iii Determine approximate values of y at $x = 0.2$, 0.4, 0.6 and 1.2 by using the 4th order Runge–Kutta method with $h = 0.2$.
 iv Plot, on the same axes, the graphs of the improved Euler's, the Runge–Kutta and the exact solutions.

17 📠 [*electrical principles*] Applying Kirchhoff's law to a circuit we obtain the differential equation

Miscellaneous exercise **13 continued**   Solutions at end of book. Complete solutions available at www.palgrave.com/engineering/singh

$$(2 \times 10^{-3})\frac{di}{dt} + (1 \times 10^{3})i = 10$$

where i is the current through the circuit.
Given the initial condition $i(0) = 0$:
i determine the current i
ii plot the graph of i against t between $t = 0$ and $t = 10$ μs ($\mu = 10^{-6}$).

18  [electrical principles]

Fig. 36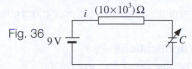

Figure 36 shows a variable capacitor of capacitance C, connected in series with a resistor of resistance 10 kΩ and a 9 V supply. The differential equation for this circuit is

$$(10 \times 10^{3})C\frac{dv}{dt} + v = 9$$

where v is the voltage across the capacitance, C.
i Solve this differential equation for the initial condition, $t = 0$, $v = 0$.
ii Substitute the following values of C into your solution:

$$C = 10 \times 10^{-6}, 20 \times 10^{-6}, 30 \times 10^{-6},$$
$$40 \times 10^{-6} \text{ and } 50 \times 10^{-6}$$

iii Plot, on the same axes, v against t for the above values of C and $0 \le t \le 1.5$.
iv What value does the voltage, v, tend to as $t \to \infty$? (This is called the 'final value'.)

v What effect does the change in capacitance, C, have on the voltage, v, across it?

19 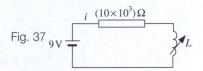 [electrical principles]

Fig. 37

Figure 37 shows a variable inductor of inductance L, connected in series with a resistor of value 10 kΩ (10×10^{3}) and an applied voltage of 9 V. The differential equation for this circuit is

$$(10 \times 10^{3})i + L\frac{di}{dt} = 9$$

i Solve this differential equation for the initial condition, $t = 0$, $i = 0$.
ii Substitute the following values of L into your solution:

$$L = \frac{1}{2}, \frac{1}{4}, \frac{1}{8}, \frac{1}{16} \text{ and } \frac{1}{32}$$

iii Plot, on the same axes, i against t for the values of L in part ii and $0 \le t \le 0.0001$.
iv What value does the current, i, tend to as $t \to \infty$? (This is called the 'final value'.)
v What effect does increasing the inductance, L, have on the current, i?

Second Order Linear Differential Equations

SECTION A **Homogeneous differential equations**

By the end of this section you will be able to:
► understand the term *homogeneous second order differential equation*
► state the relevant solution to second order differential equations by examining solutions of the corresponding characteristic equation

A1 **Introduction**

Second order differential equations are frequently found in many areas of engineering and science, such as vibrations, deflection of beams and the electrical quantities within *RLC* circuits.

A commonplace application is that of calculating an object's acceleration a in terms of its displacement x:

$$a = \frac{d^2x}{dt^2} \text{ where } t \text{ is time}$$

A demonstration of this is given in the following example.

Example 1 *mechanics*

Consider a deep well as shown in Fig. 1. A stone is dropped down the well and a splash is heard after **three** seconds. The acceleration $a = \dfrac{d^2x}{dt^2}$ of the stone is 9.8 m/s^2.

Determine the depth of the well. Assume at $t = 0$, the velocity $\dfrac{dx}{dt} = 0$ and depth $x = 0$.

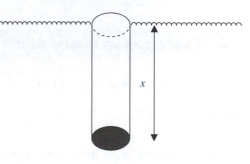

Fig. 1

Solution

We are given that $\dfrac{d^2x}{dt^2} = 9.8$ and we need to find a formula for depth x. **How?**

Integrate $\dfrac{d^2x}{dt^2} = 9.8$ twice. Integrating this once gives

$$† \qquad \frac{dx}{dt} = \int 9.8 \, dt = 9.8t + C$$

Example 1 *continued*

We are given that at $t=0$ the velocity $\dfrac{dx}{dt}=0$ therefore we can find the constant C. **How?**

By substituting this condition $\dfrac{dx}{dt}=0$ at $t=0$ into [†] :

$$(9.8 \times 0) + C = 0 \text{ implies that } C = 0$$

We have $\dfrac{dx}{dt} = 9.8t$. Integrating this again gives

$$x = \int 9.8t \, dt = \frac{9.8t^2}{2} + D$$

$$= 4.9t^2 + D$$

D is the constant of integration. **How can we find D?**

By substituting the other given condition, that at $t = 0$, $x = 0$:

$$(4.9 \times 0^2) + D = 0 \text{ gives } D = 0$$

Hence $x = 4.9t^2$. The splash is heard after 3 seconds therefore we substitute $t = 3$ into $x = 4.9t^2$ to find the depth of the well:

$$x = 4.9t^2 = 4.9 \times 3^2 = 44.1$$

Thus the depth of the well is 44.1 m.

$\dfrac{d^2x}{dt^2} = 9.8$ is an example of a **second order** differential equation. Second order differential equations may also contain the variables $\dfrac{dx}{dt}$ and x, as we discuss in the next subsection.

In solving a second order differential equation we obtain two arbitrary constants such as C and D as seen in **Example 1** above. We can calculate values for these constants if we have initial conditions.

A2 Homogeneous equations

Generally an equation of the form:

14.1 $$a\frac{d^2y}{dx^2} + b\frac{dy}{dx} + cy = f(x) \quad (a \neq 0)$$

is called a second order differential equation (2nd order DE), where a, b and c are constants and $f(x)$ is a function of x. Note the presence of $\dfrac{dy}{dx}$ and y in the general case. Also y is a function of the independent variable, x. **What does the term 'second order' mean?**

The highest order of the derivative in the differential equation is 2. If $f(x) = 0$ then **14.1** becomes

14.2 $$a\frac{d^2y}{dx^2} + b\frac{dy}{dx} + cy = 0$$

This, 14.2 , is called a **homogeneous** second order differential equation because the Right-Hand Side $f(x)$ is zero.

? **If we are asked to solve such an equation then what are we trying to find?**

We need to find y, a **function** of x, which satisfies 14.2 . Note that the solution is a function of x, that is $y = y(x)$.

? **We could begin by substituting the function $y = e^{mx}$ where m is unknown. Why?**

Because if you differentiate e^{mx} you end up with another e^{mx} term. Thus

$$\text{if } y = e^{mx} \text{ then } \frac{dy}{dx} = me^{mx}$$

Differentiating again: $\dfrac{d^2y}{dx^2} = m^2 e^{mx}$

Substituting these into

14.2 $a\dfrac{d^2y}{dx^2} + b\dfrac{dy}{dx} + cy = 0$

gives

$$am^2 e^{mx} + bme^{mx} + ce^{mx} = 0$$

$$e^{mx}(am^2 + bm + c) = 0 \quad \text{[Factorizing out } e^{mx}\text{]}$$

? **So we have $e^{mx} = 0$ or $am^2 + bm + c = 0$. What about $e^{mx} = 0$?**

For any real values of mx, $e^{mx} \neq 0$. (One of the properties of the exponential function is that it cannot equal zero.) Therefore we must have

$$am^2 + bm + c = 0$$

which is a quadratic equation. So if we solve this quadratic equation for m, we have a value for the 'm' in $y = e^{mx}$. All we need is a value for 'm', which we can find by solving $am^2 + bm + c = 0$. Since every quadratic equation is solvable, $y = e^{mx}$ is a solution of 14.2 provided that m satisfies

14.3 $am^2 + bm + c = 0$

This quadratic equation is called the **characteristic** or **auxiliary equation** of the differential equation

$$a\dfrac{d^2y}{dx^2} + b\dfrac{dy}{dx} + cy = 0$$

? **What is the characteristic equation of $2\dfrac{d^2y}{dx^2} + 3\dfrac{dy}{dx} + 7y = 0$?**

$$2m^2 + 3m + 7 = 0$$

The solution of 14.3 is given by the quadratic equation formula:

$$m = \frac{-b \pm \sqrt{b^2 - 4ac}}{2a}$$

or by factorizing the quadratic where appropriate.

? **How many different types of roots can result from a quadratic equation?**

Three:

    i   Real and different, m_1 and m_2
    ii Real and equal, $m_1 = m_2 = m$
    iii Complex, $m = \alpha \pm j\beta$

The theory of differential equations says that, if $y_1(x)$ and $y_2(x)$ are two independent solutions of the differential equation

$$a\frac{d^2y}{dx^2} + b\frac{dy}{dx} + cy = 0$$

then $Ay_1(x) + By_2(x)$ is also a solution (A and B are any real constants). As demonstrated in **Example 1** above there are two arbitrary constants in the solution of a second order differential equation. This is only true if the differential equation is linear, that is y has index 1 in the differential equation.

For the case of **i** real and different roots, m_1 and m_2, we have $y_1(x) = e^{m_1 x}$ and $y_2(x) = e^{m_2 x}$ as solutions of the differential equation. Hence

$$y = Ae^{m_1 x} + Be^{m_2 x}$$

is also a solution. This is called the **general solution** because we do not have specific values for the constants A and B.

Hence the solution, y, of the differential equation

$$a\frac{d^2y}{dx^2} + b\frac{dy}{dx} + cy = 0$$

depends on the type of root of the characteristic equation

$$am^2 + bm + c = 0$$

If the roots are real and different, m_1 and m_2, then

 14.4 $y = Ae^{m_1 x} + Be^{m_2 x}$

If the roots are real and equal, m_1 (or m_2), then

 14.5 $y = (A + Bx)e^{m_1 x}$

If the roots are complex, $m = \alpha \pm j\beta$, then

 14.6 $y = e^{\alpha x}\left[A\cos(\beta x) + B\sin(\beta x)\right]$

(A and B are constants in 14.4 to 14.6 .)

We will not show results 14.5 and 14.6 but just use them whenever appropriate.

Every homogeneous differential equation has a **trivial solution**, $y = 0$, because differentiating $y = 0$ once and twice still gives 0, therefore it satisfies the homogeneous differential equation. So if the question says solve a differential equation it means find the **non-trivial** solution.

Example 2

Solve

$$\frac{d^2y}{dx^2} + 2\frac{dy}{dx} - 3y = 0$$

Solution

We need to find y which satisfies the above equation. **What is the characteristic equation in this case?**

$$m^2 + 2m - 3 = 0$$
$$(m + 3)(m - 1) = 0 \quad \text{[Factorizing]}$$
$$m_1 = -3, \ m_2 = 1 \quad \text{[Solving]}$$

Hence the solution to the characteristic equation is two real and different roots, $m_1 = -3$ and $m_2 = 1$. Substituting these into

14.4 $y = Ae^{m_1 x} + Be^{m_2 x}$

gives

$$y = Ae^{-3x} + Be^x$$

This is the general solution of the given differential equation.

Example 3

Solve

$$\frac{d^2y}{dx^2} + 2\frac{dy}{dx} + y = 0$$

Solution

The characteristic equation is

$$m^2 + 2m + 1 = 0$$
$$(m + 1)^2 = 0 \quad \text{[Factorizing]}$$
$$m_1 = m_2 = -1 \quad \text{[Solving]}$$

The solution to the characteristic equation is two equal roots, $m_1 = m_2 = -1$. Substituting $m_1 = -1$ into

14.5 $y = (A + Bx)e^{m_1 x}$

gives the general solution

$$y = (A + Bx)e^{-x}$$

For **Example 4** we need to use the following result of complex numbers:

14.7 $\sqrt{-k^2} = \pm jk$

where k is any real number. For example, if $k = 2$ we have $\sqrt{-2^2} = \sqrt{-4} = \pm j\,2$

Example 4

Find the general solution of

$$\frac{d^2y}{dx^2} + \frac{dy}{dx} + y = 0$$

Solution

The characteristic equation is

$$m^2 + m + 1 = 0$$

To solve this quadratic equation we use the quadratic formula:

1.16 $m = \dfrac{-b \pm \sqrt{b^2 - 4ac}}{2a}$

Putting $a = 1$, $b = 1$ and $c = 1$ into 1.16 gives

$$m = \frac{-1 \pm \sqrt{1^2 - 4}}{2}$$

$$= \frac{-1 \pm \sqrt{-3}}{2} = -\frac{1}{2} \pm \frac{\sqrt{-3}}{2}$$

$$m = -\frac{1}{2} \pm j\,\frac{\sqrt{3}}{2} \quad \text{[Complex roots]}$$
$$\text{by } 14.7$$

Equating this with $m = \alpha \pm j\beta$ gives $\alpha = -\dfrac{1}{2}$ and $\beta = \dfrac{\sqrt{3}}{2}$.

Substituting this, $\alpha = -\dfrac{1}{2}$ and $\beta = \dfrac{\sqrt{3}}{2}$, into

14.6 $y = e^{\alpha x}\big[A\cos(\beta x) + B\sin(\beta x)\big]$

gives the general solution

$$y = e^{-x/2}\left[A\cos\left(\frac{\sqrt{3}}{2}x\right) + B\sin\left(\frac{\sqrt{3}}{2}x\right)\right]$$

We started this section with the differential equation

$$a\frac{d^2y}{dx^2} + b\frac{dy}{dx} + cy = f(x) \quad (a \neq 0)$$

A differential equation written in this format is said to be in **standard form**. Sometimes we have to rearrange a given differential equation to place it into standard form, for example

$$\frac{d^2y}{dx^2} + 3\frac{dy}{dx} + \cos(x) = 0$$

is not in standard form. **How do we write this in standard form?**

Subtract $\cos(x)$ from both sides:

$$\frac{d^2y}{dx^2} + 3\frac{dy}{dx} = -\cos(x)$$

The function of x, $f(x)$, is on the Right-Hand side of the equation.

SUMMARY

Solution of the linear homogeneous differential equation $a\dfrac{d^2y}{dx^2} + b\dfrac{dy}{dx} + cy = 0$ can be found using:

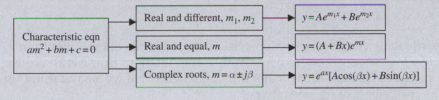

Characteristic eqn $am^2 + bm + c = 0$

Real and different, m_1, m_2 ⟶ $y = Ae^{m_1 x} + Be^{m_2 x}$

Real and equal, m ⟶ $y = (A + Bx)e^{mx}$

Complex roots, $m = \alpha \pm j\beta$ ⟶ $y = e^{\alpha x}[A\cos(\beta x) + B\sin(\beta x)]$

Exercise 14(a)

Solutions at end of book. Complete solutions available at www.palgrave.com/engineering/singh

1 Determine the displacement s in each of the following cases:

a $\dfrac{d^2s}{dt^2} = -9.8$ given that at $t = 0$, both

$\dfrac{ds}{dt} = 0$ and $s = 0$.

b $\dfrac{d^2s}{dt^2} = 9.8t$ given that at $t = 0$, both

$\dfrac{ds}{dt} = 0$ and $s = 0$.

c $\dfrac{d^2s}{dt^2} = a$ (a is acceleration) given that

at $t = 0$, both $\dfrac{ds}{dt} = u$ and $s = 0$

2 Find the general solutions of the following equations:

a $\dfrac{d^2y}{dx^2} + 5\dfrac{dy}{dx} + 6y = 0$

b $\dfrac{d^2y}{dx^2} + 4\dfrac{dy}{dx} + 4y = 0$

c $\dfrac{d^2y}{dx^2} - 2\dfrac{dy}{dx} + 4y = 0$

3 Solve:

a $\dfrac{d^2y}{dx^2} + 3\dfrac{dy}{dx} + 2y = 0$

Exercise **14(a) continued**

Solutions at end of book. Complete solutions available at www.palgrave.com/engineering/singh

b $\dfrac{d^2y}{dx^2} - 7\dfrac{dy}{dx} + 6y = 0$

c $\dfrac{d^2y}{dx^2} - 6\dfrac{dy}{dx} + 9y = 0$

d $\dfrac{d^2y}{dx^2} - 4\dfrac{dy}{dx} + 5y = 0$

e $\dfrac{d^2y}{dx^2} - 8\dfrac{dy}{dx} + 16y = 0$

4 Solve:

a $\dfrac{d^2y}{dx^2} + 6\dfrac{dy}{dx} + 9y = 0$

b $10\dfrac{d^2y}{dx^2} + 50\dfrac{dy}{dx} + 250y = 0$

c $-\dfrac{d^2y}{dx^2} - 3\dfrac{dy}{dx} + 8y = 0$

SECTION B **Engineering applications**

By the end of this section you will be able to:
▶ solve engineering applications of second order differential equations with initial conditions
▶ solve boundary value problems

This is a particularly challenging section because there are lots of examples from engineering which involve complicated symbols and notation. However the concepts are similar in nature to **Section A**, the major difference being that we are not using simple numbers as coefficients of the derivatives.

B1 **Particular homogeneous solutions**

In engineering applications we are normally given some initial conditions of the problem. For example, in finding the depth of a well in **Example 1** the velocity and depth were zero at time $t = 0$. These values are the initial conditions. The differential equation with initial conditions is called an **initial value problem**.

Given initial conditions we can evaluate unknowns A and B in the general solution of the equation, and this is then known as the **particular solution**.

Example 5 *electrical principles*

By applying Kirchhoff's voltage law to a circuit we obtain the second order differential equation

$$\dfrac{d^2i}{dt^2} + 7\dfrac{di}{dt} + 10i = 0$$

Example 5 *continued*

where i is the current and t is the time. Find the current i in terms of t for the initial conditions, when $t = 0$, both $i = 0$ and $\dfrac{di}{dt} = 3$.

Solution

The characteristic equation is

$$m^2 + 7m + 10 = 0$$
$$(m + 5)(m + 2) = 0 \quad \text{[Factorizing]}$$
$$m_1 = -5, \ m_2 = -2 \quad \text{[Solving]}$$

We have real and different roots, so putting these into

14.4 $\qquad i = Ae^{m_1 t} + Be^{m_2 t}$

gives the general solution

† $\qquad i = Ae^{-5t} + Be^{-2t}$

Now we use the given initial conditions to find values for A and B in † .

Substituting the given initial condition $t = 0$, $i = 0$ into †

$$Ae^0 + Be^0 = 0$$
$$A + B = 0 \quad \text{[Because } e^0 = 1\text{]}$$

Using the other given initial condition $t = 0$, $\dfrac{di}{dt} = 3$: to find $\dfrac{di}{dt}$ we need to differentiate $i = Ae^{-5t} + Be^{-2t}$:

$$\frac{di}{dt} = -5Ae^{-5t} - 2Be^{-2t} \quad \left[\text{Because } \frac{d}{dt}(e^{kt}) = ke^{kt} \right]$$
$$3 = -5Ae^0 - 2Be^0 = -5A - 2B$$

Thus we have two simultaneous equations:

$$A + B = 0$$
$$-5A - 2B = 3$$

From the first equation we have

$$A = -B$$

Substituting this into the second equation $-5A - 2B = 3$ gives

$$-5(-B) - 2B = 5B - 2B = 3B = 3$$

Therefore $B = 1$. Since $A = -B$ so $A = -1$.

Substituting $A = -1$ and $B = 1$ into $i = Ae^{-5t} + Be^{-2t}$ gives

$$i = -e^{-5t} + e^{-2t} = e^{-2t} - e^{-5t}$$

This is a particular solution since we have found particular values of A and B.

B2 Differential equation of the form $\dfrac{d^2y}{dx^2} + k^2y = 0$

Consider $\dfrac{d^2y}{dx^2} + k^2y = 0$ where k is a constant. Differential equations of this type arise in vibrational problems. One of the simplest cases is known as simple harmonic motion (SHM). Consider a particle, P, moving to and fro about its equilibrium position O (Fig. 2).

Fig. 2

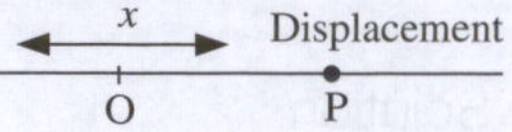

x Displacement

O   P

The equation of motion is given by

$$\frac{d^2x}{dt^2} + k^2x = 0$$

where x is displacement, k is a constant and t is time.

In mechanics the differential equation might be written as

$$\ddot{x} + k^2x = 0$$

where $\ddot{x}$ represents the second derivative, $\dfrac{d^2x}{dt^2}$.

The characteristic equation of the general second order differential equation $\dfrac{d^2y}{dx^2} + k^2y = 0$ is

$$m^2 + k^2 = 0$$
$$m^2 = -k^2$$
$$m = \sqrt{-k^2} \underset{\text{by } \boxed{14.7}}{\equiv} \pm jk \text{ hence } m = 0 \pm jk$$

Equating $m = 0 \pm jk$ with $m = \alpha \pm j\beta$ gives $\alpha = 0$ and $\beta = k$.

Putting $\alpha = 0$ and $\beta = k$ into

$$\boxed{14.6} \qquad y = e^{\alpha x}\big[A\cos(\beta x) + B\sin(\beta x)\big]$$

gives

$$y = \underset{=1}{\underline{e^0}}\big[A\cos(kx) + B\sin(kx)\big]$$

$$= A\cos(kx) + B\sin(kx)$$

Hence the general solution of the differential equation $\dfrac{d^2y}{dx^2} + k^2y = 0$ is

$$\boxed{14.8} \qquad y = A\cos(kx) + B\sin(kx)$$

$\boxed{14.7}$ $\sqrt{-k^2} = \pm jk$

(Ψ) Example 6 *vibrations*

Fig. 3

Mass

The motion of a spring–mass system (Fig. 3) is described by

$$\frac{d^2x}{dt^2} + 25x = 0$$

where x is displacement and t is time. Determine the particular solution for this differential equation with initial conditions, when $t = 0$, both $x = 1$ and $\dfrac{dx}{dt} = 10$.

Solution

Since $25 = 5^2$, the given differential equation can be written as

$$\frac{d^2x}{dt^2} + 5^2x = 0$$

Putting $k = 5$ into

14.8 $x = A\cos(kt) + B\sin(kt)$

gives the general solution

$$x = A\cos(5t) + B\sin(5t)$$

Substituting the initial condition, $t = 0$, $x = 1$, yields

$$1 = A\underbrace{\cos(0)}_{=1} + B\underbrace{\sin(0)}_{=0}$$

$$1 = A$$

To use the other initial condition we need to differentiate first:

$$x = A\cos(5t) + B\sin(5t)$$

$$\frac{dx}{dt} = \underbrace{-5A\sin(5t)}_{\text{by } 6.13} + \underbrace{5B\cos(5t)}_{\text{by } 6.12}$$

Substituting $t = 0$, $\dfrac{dx}{dt} = 10$ gives

$$10 = -5A\underbrace{\sin(0)}_{=0} + 5B\underbrace{\cos(0)}_{=1}$$

$$10 = 0 + 5B \text{ so it follows that } B = 2$$

Putting $A = 1$ and $B = 2$ into $x = A\cos(5t) + B\sin(5t)$ yields the particular solution

$$x = \cos(5t) + 2\sin(5t)$$

6.12 $[\sin(kt)]' = k\cos(kt)$ **6.13** $[\cos(kt)]' = -k\sin(kt)$

B3 **Boundary value problems**

If the given conditions are stated at different values of the (independent) variable, x, then we have **boundary conditions**. In **Example 7** which follows, the y values are given for two different values of x, $x = 0$ and $x = L$, so the given conditions are boundary conditions. Compare **Example 5** where we were given the following conditions, $i = 0$ and $\dfrac{di}{dt} = 3$, at the **same** value of the (independent) variable $t = 0$. Hence these are called initial conditions.

A differential equation associated with boundary conditions is called a **boundary value problem**.

The following example is more sophisticated than previous examples because it involves engineering symbols and properties of trigonometric functions. However the concepts are the same as above.

The next example is on structures and, as for all problems on structures, EI represents flexural rigidity (an object's tendency to bend) and $EI \neq 0$.

Example 7 *structures*

A column of length L is subject to a compressive load P (Fig. 4).

The displacement, y, at a distance x is given

by the solution of

$$EI\,\frac{d^2y}{dx^2} + Py = 0$$

Fig. 4 $P\rightarrow$ $\leftarrow P$ — Column (x, y)

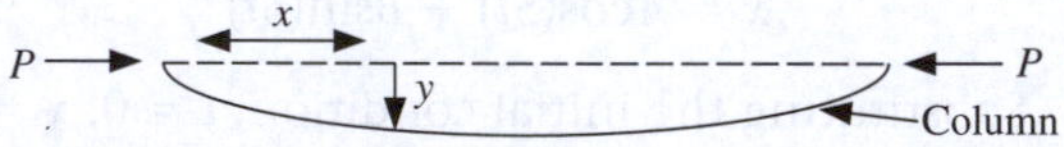

i Show that $y = A\cos(kx) + B\sin(kx)$ where $k = \sqrt{\dfrac{P}{EI}}$.

ii The critical load is the value of the load, P, that will cause buckling.
The critical load, P_{cr}, occurs for the boundary conditions $x = 0$, $y = 0$ and $x = L$, $y = 0$.
Show that the critical load is given by

$$P_{cr} = \frac{n^2\pi^2\,EI}{L^2}$$

where n is any whole number.

[*Hint*: Use the result $\sin(n\pi) = 0$ where n is any whole number.]

Solution

i Dividing the given differential equation by EI:

$$\frac{d^2y}{dx^2} + \frac{P}{EI}y = 0$$

Compare this with

$$\frac{d^2y}{dx^2} + k^2y = 0 \text{ where } k^2 = \frac{P}{EI}$$

Example 7 *continued*

Taking square roots gives $k = \sqrt{\dfrac{P}{EI}}$. Hence by 14.8

★ $y = A\cos(kx) + B\sin(kx)$

ii Substituting the given boundary condition, $x = 0, y = 0$, into ★ :

$$0 = A\underbrace{\cos(0)}_{=1} + B\underbrace{\sin(0)}_{=0} = A$$

We have $A = 0$. Using the other boundary condition, $x = L$, $y = 0$, gives

$$0 = 0 + B\sin(kL) \quad \text{so } B\sin(kL) = 0$$

$B \neq 0$, **why?**

If $B = 0$ then $y = 0$ because putting $A = 0$ and $B = 0$ into ★ gives the trivial solution, $y = 0$. Therefore we must have

$$\sin(kL) = 0$$

$$\underset{\text{by Hint}}{kL = n\pi} \text{ which gives } k = \frac{n\pi}{L} \quad \text{[Dividing by } L\text{]}$$

Squaring both sides:

$$k^2 = \left(\frac{n\pi}{L}\right)^2 = \frac{n^2\pi^2}{L^2}$$

Remember that $k^2 = \dfrac{P}{EI}$ and for the given boundary conditions, $P = P_{cr}$:

$$k^2 = \frac{P_{cr}}{EI} = \frac{n^2\pi^2}{L^2}$$

$$P_{cr} = \frac{n^2\pi^2\, EI}{L^2} \quad \text{[Multiplying by } EI\text{]}$$

This non-trivial solution is **only** valid at the critical loads. For loads below P_{cr} the only solution which satisfies the differential equation with the given boundary conditions is the trivial solution $y = 0$. That is, for loads less than $\pi^2 EI/L^2$ the beam does not buckle, it is only compressed. Buckling will not occur until the load equals or exceeds P_{cr}.

B4 Differential equation of the form $\dfrac{d^2y}{dx^2} - k^2y = 0$

Now consider the differential equation of the form

$$\frac{d^2y}{dx^2} - k^2y = 0$$

The characteristic equation is

$$m^2 - k^2 = 0$$
$$m^2 = k^2$$
$$m = \pm k \qquad \text{[Taking square root]}$$
$$m_1 = k, \ m_2 = -k \qquad \text{[Real and different]}$$

14.8 $x = A\cos(kt) + B\sin(kt)$

Putting these into

14.4 $y = Ae^{m_1 x} + Be^{m_2 x}$

gives

$y = Ae^{kx} + Be^{-kx}$

The general solution of the differential equation $\dfrac{d^2 y}{dx^2} - k^2 y = 0$ is

14.9 $y = Ae^{kx} + Be^{-kx}$

SUMMARY

If the characteristic equation is $m^2 + k^2 = 0$ then

14.8 $y = A\cos(kx) + B\sin(kx)$

If the characteristic equation is $m^2 - k^2 = 0$ then

14.9 $y = Ae^{kx} + Be^{-kx}$

If the initial and boundary conditions are given then A and B can be evaluated and we obtain a particular solution.

Exercise 14(b)

Solutions at end of book. Complete solutions available at www.palgrave.com/engineering/singh

1 [electrical principles]   Consider the second order differential equation

$$\frac{d^2 i}{dt^2} + 7\frac{di}{dt} + 10i = 0$$

where i represents current. Determine the current, i, in terms of t for the initial conditions, when $t = 0$, both

$$i = 0 \text{ and } \frac{di}{dt} = 6.$$

2 [vibrations]   For the following simple harmonic motions, state the general solution:

a $\dfrac{d^2 x}{dt^2} + 9x = 0$ **b** $\dfrac{d^2 x}{dt^2} + 16x = 0$

c $\dfrac{d^2 x}{dt^2} + 2x = 0$ **d** $\dfrac{d^2 x}{dt^2} + 5x = 0$

3 [vibrations]

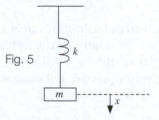

Fig. 5

Figure 5 shows a body of mass m attached to a spring of stiffness k. The displacement, x, is related by

$$\frac{d^2 x}{dt^2} + \frac{k}{m}x = 0$$

Show that $x = A\cos(\omega t) + B\sin(\omega t)$

where A, B are constants and $\omega = \sqrt{\dfrac{k}{m}}$.

Exercise **14(b) continued**

4 [electrical principles]   The voltage, v, of an *RLC* circuit is given by the second order differential equation

$$\frac{d^2v}{dt^2} + (4 \times 10^{-6})v = 0$$

with the initial conditions that when $t = 0$, both $v = 1$ and $\dfrac{dv}{dt} = 2 \times 10^{-3}$.

Show that

$$v = \sqrt{2}\cos\left[(2 \times 10^{-3})t - \frac{\pi}{4}\right]$$

[*Hint*: Use 4.75 to convert into a cosine term only.]

*5 [structures]   A load, P, is applied at both ends of a column of length L. The second order differential equation

$$EI\frac{d^2y}{dx^2} + Py = 0$$

gives the relationship between the deflection y and the distance x. Show that if $y = e$ (eccentricity) at both $x = L/2$ and $x = -L/2$ then

$$y = e\sec\left(\frac{kL}{2}\right)\cos(kx) \text{ where } k = \sqrt{P/EI}$$

$$\left[\text{Assume that } \sin\left(\frac{kL}{2}\right) \neq 0\right]$$

[*Hint*: Use the solution * to **Example 7**.]

Questions 6 to 8 are in the field of [*vibrations*].

6

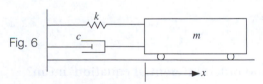

Fig. 6

Figure 6 shows a body of mass m attached to a spring of stiffness k and a viscous damper with a damping coefficient c. The displacement, x, is given by

$$\frac{d^2x}{dt^2} + 2\zeta\omega\frac{dx}{dt} + \omega^2x = 0$$

where $\omega = \sqrt{\dfrac{k}{m}}$ and the viscous damping factor $\zeta = \dfrac{c}{2m\omega}$.

Show that the solution of this differential equation for $\zeta > 1$ is given by

$$x = Ae^{r_1 t} + Be^{r_2 t}$$
$$(A \text{ and } B \text{ are constants})$$

where $r_1 = \omega\left(-\zeta + \sqrt{\zeta^2 - 1}\right)$ and

$$r_2 = -\omega\left(\zeta + \sqrt{\zeta^2 - 1}\right).$$

*7 The equation of motion for a vertical spring–mass system is given by

$$\ddot{x} + 2\zeta\omega\dot{x} + \omega^2x = 0$$

where ζ is the damping factor, ω is the angular frequency and x is the displacement of mass from its equilibrium position. When $t = 0$, both $x = 5$ and $\dot{x} = 0$. Find an expression for x in terms of t with $\zeta = 1$.

[*Hint*: Note that we have the same characteristic equation as in question **6**.]

Plot the graph of x versus ωt for $\omega t > 0$. (Use a graphical calculator or a computer algebra system for this part.)

*8 The equation of motion for a spring–mass system is given by

$$\ddot{x} + 2\zeta\omega\dot{x} + \omega^2 x = 0$$

where x is the displacement of mass from its equilibrium level, ζ is the damping factor and ω is the angular frequency. For $\zeta < 1$ and the initial conditions, when $t = 0$, both $x = 0$ and $\dot{x} = \beta\omega$ where $\beta = \sqrt{1 - \zeta^2}$, show that

$$x = e^{-\zeta\omega t}\sin(\beta\omega t)$$

Questions 9 to 11 are in the field of [*electrical principles*].
(R is resistance, L is inductance and C is capacitance)

*9 The voltage, v, in a parallel RLC circuit satisfies

$$C\frac{d^2v}{dt^2} + \frac{1}{R}\frac{dv}{dt} + \frac{v}{L} = 0$$

Find an expression for the voltage, v, in terms of t, R, L and C, for the following cases:

a $L = 4CR^2$ (Critically damped)
b $L > 4CR^2$ (Over damped)
c $L < 4CR^2$ (Under damped)

10 A capacitance $C = (1 \times 10^{-9})$ F, a resistance $R = 10 \times 10^3$ Ω and a variable inductance L are connected in parallel. The voltage, v, is governed

by the second order differential equation

$$C\frac{d^2v}{dt^2} + \frac{1}{R}\frac{dv}{dt} + \frac{v}{L} = 0$$

The inductance, L, is varied so that the roots of the characteristic equation become $-(50 \times 10^3) \pm j(30 \times 10^3)$. Determine this value of L given that $L < 0.4$ H.

*11 A circuit consisting of a resistor of resistance R, inductor of inductance L and capacitor of capacitance C has the following second order differential equation:

$$C\frac{d^2v}{dt^2} + \frac{1}{R}\frac{dv}{dt} + \frac{v}{L} = 0$$

By considering the characteristic equation $m^2 + 2\zeta\omega m + \omega^2 = 0$, find expressions for the natural frequency, ω, and damping ratio, ζ.

SECTION C **Non-homogeneous (inhomogeneous) differential equations**

By the end of this section you will be able to:
▶ understand the term *non-homogeneous*
▶ understand why we need a particular integral $Y(x)$
▶ use a trial function to find a particular integral
▶ evaluate general solutions of non-homogeneous second order differential equations

C1 **Non-homogeneous equations**

What does the term 'non-homogeneous second order differential equation' mean?

Differential equations of the type

$$a\frac{d^2y}{dx^2} + b\frac{dy}{dx} + cy = f(x) \text{ where } f(x) \neq 0$$

Remember that homogeneous is where $f(x) = 0$, therefore non-homogeneous is where $f(x) \neq 0$, does **not** equal zero. Another name for these equations is **inhomogeneous**.

Example 8

Solve

$$\boxed{*} \qquad \frac{d^2y}{dx^2} - \frac{dy}{dx} - 2y = 5e^{4x}$$

Solution

What are we trying to find?

We need to find y (a function of x) which satisfies $\boxed{*}$.

Let us assume for a moment that the Right-Hand Side is 0 then we have

$$\frac{d^2y}{dx^2} - \frac{dy}{dx} - 2y = 0$$

and the characteristic equation is

$$m^2 - m - 2 = 0$$
$$(m - 2)(m + 1) = 0 \qquad \text{[Factorizing]}$$
$$m_1 = 2, \; m_2 = -1 \qquad \text{[Solving]}$$

Since we have two real and different roots we substitute these, $m_1 = 2$ and $m_2 = -1$, into

$\boxed{14.4}$ $\qquad y = Ae^{m_1x} + Be^{m_2x}$

which gives

$$y_c = Ae^{2x} + Be^{-x}$$

(y_c denotes complementary function and will be explained later.)

If we substitute y_c into the Left-Hand Side of $\boxed{*}$ and then simplify, what will be the final answer?

ZERO

Why do you think we obtain zero?

Because we have just solved a homogeneous differential equation

$$\frac{d^2y}{dx^2} - \frac{dy}{dx} - 2y = 0$$

and not $\dfrac{d^2y}{dx^2} - \dfrac{dy}{dx} - 2y = 5e^{4x}$ [the given equation]

So there must be an extra term, function $Y = Y(x)$ (a function of x), to the existing homogeneous solution, $y_c = Ae^{2x} + Be^{-x}$.

What is this function Y going to be?

Y must satisfy the given differential equation

$$\frac{d^2Y}{dx^2} - \frac{dY}{dx} - 2Y = 5e^{4x}$$

Try the function $Y = Ce^{4x}$, where C is a constant. Why?

Because when we differentiate Ce^{4x} we still have a term in e^{4x} and on the Right-Hand Side we have $5e^{4x}$.

Example 8 *continued*

?

We need to find a value for the constant C. How can we find C?

Since

$$Y = Ce^{4x}$$

satisfies the given equation, we differentiate this:

$$\frac{dY}{dx} \underset{\text{by } 6.11}{\equiv} 4Ce^{4x}$$

$$\frac{d^2Y}{dx^2} \underset{\text{by } 6.11}{\equiv} 16Ce^{4x}$$

and substitute into

$$\frac{d^2Y}{dx^2} - \frac{dY}{dx} - 2Y = 5e^{4x}$$

which gives

$$16Ce^{4x} - 4Ce^{4x} - 2Ce^{4x} = 5e^{4x}$$
$$(16 - 4 - 2)Ce^{4x} = 5e^{4x} \qquad [\text{Taking out } Ce^{4x}]$$
$$10Ce^{4x} = 5e^{4x} \qquad [\text{Simplifying}]$$
$$10C = 5$$
$$C = \frac{5}{10} = \frac{1}{2}$$

Hence putting $C = \dfrac{1}{2}$ into $Y = Ce^{4x}$ gives $Y = \dfrac{e^{4x}}{2}$.

The general solution of the given differential equation

$$\frac{d^2y}{dx^2} - \frac{dy}{dx} - 2y = 5e^{4x}$$

is found by adding together the homogeneous solution $y_c = Ae^{2x} + Be^{-x}$ and the

particular integral $Y = \dfrac{e^{4x}}{2}$.

Therefore the general solution is $y = Ae^{2x} + Be^{-x} + \dfrac{e^{4x}}{2}$.

The homogeneous solution, y_c, is called the **complementary function** and Y is called the **particular integral**. To find the particular integral we use a trial function which depends on $f(x)$.

Summarizing the above:

The non-homogeneous differential equation

$$a\frac{d^2y}{dx^2} + b\frac{dy}{dx} + cy = f(x) \quad \text{where } f(x) \neq 0$$

is solved by:

1 finding the complementary function y_c (homogeneous solution)

$$a\frac{d^2y_c}{dx^2} + b\frac{dy_c}{dx} + cy_c = 0$$

6.11 $(e^{kx})' = ke^{kx}$

2 determining the particular integral Y by using a trial function (depends on $f(x)$)
3 adding together y_c and Y. The solution is

14.10 $y = y_c + Y$

Property **3** only holds if the given differential equation is **linear**, that is the power of y is 1, we do **not** have $\sqrt{y}$, y^2, etc. in the differential equation. Thus the general solution of a linear non-homogeneous differential equation is

14.10a   (complementary function) + (particular integral)

C2 Particular integrals

Example 9

Determine the particular integral, Y, of

$$\frac{d^2y}{dx^2} - \frac{dy}{dx} - 2y = 2x$$

Solution

Since on the Right-Hand Side we have a **linear** function, $f(x) = 2x$, we try the general **linear** function:

$$Y = ax + b$$

Do not confuse the coefficients a and b in Y with the coefficients of the differential equation, they are not the same.

This Y satisfies the given differential equation

 * $$\frac{d^2Y}{dx^2} - \frac{dY}{dx} - 2Y = 2x$$

Differentiating Y:

$$Y = ax + b$$
$$\frac{dY}{dx} = a$$
$$\frac{d^2Y}{dx^2} = 0$$

Substituting these into * gives

$$0 - a - 2(ax + b) = 2x$$
 ** $$-a - 2ax - 2b = 2x$$

? **We need to find the values of a and b. How?**

? We could equate coefficients of ** , thus equating coefficients of x. **What does this mean?**

Example 9 *continued*

The number of x's on the Left-Hand Side is equal to the number of x's on the Right-Hand Side of $\boxed{**}$. **How many x's are on the left of the equal sign in $\boxed{**}$?**

$- 2a$. **How many x's are on the right?**

2. Thus we have

$$-2a = 2 \text{ which gives } a = -1$$

Equating constants (putting $x = 0$ in $\boxed{**}$): $-a - 2b = 0$

$$-(-1) - 2b = 0 \qquad \text{[Because } a = -1]$$

$$1 - 2b = 0 \text{ which gives } b = \frac{1}{2}$$

Substituting $a = -1$ and $b = \dfrac{1}{2}$ into the trial function, $Y = ax + b$, gives the particular integral:

$$Y = -x + \frac{1}{2} = \frac{1}{2} - x$$

Example 10

Determine the particular integral, Y, of

$$\frac{d^2y}{dx^2} - \frac{dy}{dx} - 2y = 5x^2$$

Solution

We know the Right-Hand Side is a quadratic, $f(x) = 5x^2$, so what is Y?

A trial function, Y, is the general quadratic, that is

$$Y = ax^2 + bx + c$$

Y satisfies

$\boxed{*}$ $\qquad \dfrac{d^2Y}{dx^2} - \dfrac{dY}{dx} - 2Y = 5x^2$

So we need to differentiate Y to find a, b and c:

$$Y = ax^2 + bx + c$$
$$\frac{dY}{dx} = 2ax + b$$
$$\frac{d^2Y}{dx^2} = 2a$$

Substituting these into $\boxed{*}$ gives

Example 10 *continued*

$$\frac{d^2Y}{dx^2} - \frac{dY}{dx} - 2Y = 2a - (2ax + b) - 2(ax^2 + bx + c)$$

$$= 2a - 2ax - b - 2ax^2 - 2bx - 2c$$

$$= -2ax^2 + (-2a - 2b)x + (2a - b - 2c) \quad \text{[Factorizing]}$$

How can we find *a*, *b* and *c*?

One way is by equating coefficients. The above is equal to $5x^2$, thus equating coefficients of the highest power of *x* first:

$$-2ax^2 + (-2a - 2b)x + (2a - b - 2c) = 5x^2$$

x^2: $\qquad -2a = 5$ gives $a = -\dfrac{5}{2}$

x: $\qquad -2a - 2b = 0$

$\qquad\qquad -2b = 2a$

$$b = -a = -\left(-\frac{5}{2}\right) = \frac{5}{2} \quad \left[\text{Substituting } a = -\frac{5}{2}\right]$$

constants: $\qquad 2a - b - 2c = 0$

$$2\left(-\frac{5}{2}\right) - \frac{5}{2} - 2c = 0, \text{ and solving gives } c = -\frac{15}{4}$$

Substituting $a = -\dfrac{5}{2}$, $b = \dfrac{5}{2}$ and $c = -\dfrac{15}{4}$ into $Y = ax^2 + bx + c$ gives

$$Y = -\frac{5}{2}x^2 + \frac{5}{2}x - \frac{15}{4} = -\frac{10}{4}x^2 + \frac{10}{4}x - \frac{15}{4} \quad \text{[Common denominator]}$$

Taking out the common factor $-1/4$ gives the particular integral:

$$Y = -\frac{1}{4}(10x^2 - 10x + 15)$$

The particular integral, *Y*, of a general second order differential equation

$$L\frac{d^2y}{dx^2} + M\frac{dy}{dx} + Ny = f(x)$$

depends on $f(x)$. Table 1 shows trial functions that can be used to find the particular integral *Y*. (We are using the letters *L*, *M* and *N* for the coefficients of the derivatives so that you do not get confused with the letters *a*, *b* and *c* used for the trial functions in Table 1.)

TABLE 1

Reference number	$f(x)$	Trial function
14.11	A (constant)	C (constant)
14.12	$Ax + B$	$ax + b$
14.13	$Ax^2 + Bx + C$	$ax^2 + bx + c$
14.14	$A_n x^n + A_{n-1} x^{n-1} + \cdots + A_0$	$a_n x^n + a_{n-1} x^{n-1} + \cdots + a_0$
14.15	$A\cos(kx)$	$a\cos(kx) + b\sin(kx)$
14.16	$A\sin(kx)$	$a\cos(kx) + b\sin(kx)$
14.17	Ae^{kx}	Ce^{kx}

Note the trial function for the trigonometric functions $\cos(kx)$ and $\sin(kx)$. If $f(x) = A\cos(kx)$, then we consider the trial function $Y = a\cos(kx) + b\sin(kx)$ and **not** $Y = a\cos(kx)$ because the presence of first and second derivatives in the equation will **produce** sine and cosine terms from either sine **or** cosine. Also if $f(x)$ is an nth degree polynomial, 14.14, then a trial function is the general nth degree polynomial. For example, if $f(x) = x^3 + 1$ then our trial function is

$$Y = ax^3 + bx^2 + cx + d$$

Example 11 *electrical principles*

By applying Kirchhoff's voltage law to a circuit we obtain the following second order differential equation:

$$\frac{d^2i}{dt^2} + (12 \times 10^6)\frac{di}{dt} + (36 \times 10^{12})\, i = 5 \times 10^9$$

where i is the current through the circuit. Determine the current, i, at any time t.

Solution

What do we evaluate first?

Complementary function, i_c (place the Right-Hand Side equal to 0):

$$\frac{d^2i}{dt^2} + (12 \times 10^6)\frac{di}{dt} + (36 \times 10^{12})i = 0$$

The characteristic equation is

$$m^2 + (12 \times 10^6)\, m + (36 \times 10^{12}) = 0$$

Example 11 *continued*

To solve this characteristic equation we use the quadratic equation formula:

1.16 $m = \dfrac{-b \pm \sqrt{b^2 - 4ac}}{2a}$

with $a = 1$, $b = 12 \times 10^6$ and $c = 36 \times 10^{12}$ which gives

$$m = \frac{-(12 \times 10^6) \pm \sqrt{(12 \times 10^6)^2 - (4 \times 1 \times 36 \times 10^{12})}}{2 \times 1}$$

$$= \frac{-(12 \times 10^6)}{2} \pm \frac{1}{2}\underbrace{\sqrt{(144 \times 10^{12}) - (144 \times 10^{12})}}_{=\,0}$$

$$m = -(6 \times 10^6) \quad\text{[Equal roots]}$$

Since we have equal roots, $m = -6 \times 10^6$, of the characteristic equation so we substitute this into

14.5 $i = (A + Bt)e^{mt}$

which gives

$$i_c = (A + Bt)e^{-(6 \times 10^6)t}$$

Is this the solution of the given differential equation

$$\frac{d^2i}{dt^2} + (12 \times 10^6)\frac{di}{dt} + (36 \times 10^{12})i = 5 \times 10^9?$$

No, because the Right-Hand Side is **not** zero (i_c is the homogeneous solution).

Hence we need to find the particular integral I. Since $f(t) = 5 \times 10^9$, a constant, so by
14.11 a trial function is

$$I = C \text{ where } C \text{ is a constant}$$

$$\frac{dI}{dt} = 0 \text{ and } \frac{d^2I}{dt^2} = 0   \quad\text{[Differentiating a constant]}$$

So the first and second derivatives are zero. Substituting $I = C$ into the given differential equation

$$\frac{d^2I}{dt^2} + (12 \times 10^6)\frac{dI}{dt} + (36 \times 10^{12})I = 5 \times 10^9$$

yields

$$0 + 0 + (36 \times 10^{12})C = 5 \times 10^9$$

$$C = \frac{5 \times 10^9}{36 \times 10^{12}} = 1.39 \times 10^{-4}$$

Hence $I = C = 1.39 \times 10^{-4}$. Since $i = i_c + I$ so we have the general solution

$$i = (A + Bt)e^{-(6 \times 10^6)t} + (1.39 \times 10^{-4})$$

The next example might seem like a colossal leap from previous examples. The difficulty is the complicated notation and engineering symbols used. Don't be put off because the concepts are the same as above. Remember that $\ddot{x}$ is a second order time derivative.

Example 12 *vibrations*

An undamped spring–mass system is subject to a force, $F\sin(2t)$. The equation of motion is

$$m\ddot{x} = -kx + F\sin(2t) \qquad (k - 4m \neq 0)$$

where m is the mass, k is the stiffness of the spring, x is the displacement and t is the time. Show that

$$x = A\cos(\omega t) + B\sin(\omega t) + \frac{F}{k - 4m}\sin(2t) \text{ where } \omega = \sqrt{\frac{k}{m}}$$

Solution

By adding kx to both sides, the differential equation becomes

$$m\ddot{x} + kx = F\sin(2t)$$

The differential equation is now in standard form. (Standard form is covered at the end of **Section A**.)

Hence the complementary function x_c (Right-Hand Side = 0) is given by

$$m\ddot{x}_c + kx_c = 0$$
$$\ddot{x}_c + \frac{k}{m}x_c = 0 \quad \text{[Dividing by } m\text{]}$$

Thus the characteristic equation is

$$r^2 + \left(\sqrt{\frac{k}{m}}\right)^2 = 0 \quad \left[\text{Remember that } \frac{k}{m} = \left(\sqrt{\frac{k}{m}}\right)^2\right]$$

By 14.8 the complementary function is

$$x_c = A\cos(\omega t) + B\sin(\omega t) \text{ where } \omega = \sqrt{\frac{k}{m}}$$

Particular integral X:

Since the Right-Hand Side is equal to $F\sin(2t)$, by 14.16 we consider the trial function:

$$X = a\cos(2t) + b\sin(2t)$$

Differentiating gives

$$\dot{X} = \underbrace{-2a\sin(2t)}_{\text{by } 6.13} + \underbrace{2b\cos(2t)}_{\text{by } 6.12}$$

$$\ddot{X} = -4a\cos(2t) - 4b\sin(2t) \quad \left[\text{Differentiating}\right]$$
$$= -4\left[a\cos(2t) + b\sin(2t)\right] \quad \left[\text{Taking out } -4\right]$$

6.12 $[\sin(kt)]' = k\cos(kt)$ 6.13 $[\cos(kt)]' = -k\sin(kt)$

Example 12 *continued*

Substituting into

$$m\ddot{X} + kX = F\sin(2t)$$

gives

$$-4m\big[a\cos(2t) + b\sin(2t)\big] + k\big[a\cos(2t) + b\sin(2t)\big] = F\sin(2t)$$
$$(k - 4m)a\cos(2t) + (k - 4m)b\sin(2t) = F\sin(2t) \quad \text{[Factorizing]}$$

Equating coefficients of

$\cos(2t)$: $(k - 4m)a = 0$, which gives $a = 0$ because $k - 4m \neq 0$
$\sin(2t)$: $(k - 4m)b = F$

$$b = \frac{F}{k - 4m} \quad \text{[Dividing by } k - 4m\text{]}$$

Substituting $a = 0$ and $b = \dfrac{F}{k - 4m}$ into our trial function $X = a\cos(2t) + b\sin(2t)$ gives the particular integral

$$X = \frac{F}{k - 4m}\sin(2t)$$

The general solution, $x = x_c + X$, is

$$x = A\cos(\omega t) + B\sin(\omega t) + \frac{F}{k - 4m}\sin(2t) \quad \text{where } \omega = \sqrt{\frac{k}{m}}$$

Table 1 is a good guide for finding the particular integral, Y, but sometimes we need to adjust the trial function to find the particular integral, as the next example shows.

Example 13

Solve

$$\frac{d^2y}{dx^2} - \frac{dy}{dx} - 2y = 3e^{2x}$$

Solution

The complementary function, y_c, is the same as **Example 8** because we have the **same** characteristic equation. Hence

$$y_c = Ae^{2x} + Be^{-x}$$

We need to find the particular integral Y. By 14.17 of Table 1 $Y = Ce^{2x}$, because $f(x) = 3e^{2x}$. Let's find C:

$$\frac{dY}{dx} = 2Ce^{2x} \quad \left[\text{Differentiating by } \frac{d}{dx}(e^{kx}) = ke^{kx}\right]$$

$$\frac{d^2Y}{dx^2} = 4Ce^{2x} \quad \left[\text{Differentiating by } \frac{d}{dx}(e^{kx}) = ke^{kx}\right]$$

Example 13 *continued*

Substituting these into $\dfrac{d^2Y}{dx^2} - \dfrac{dY}{dx} - 2Y = 3e^{2x}$ gives

$$4Ce^{2x} - 2Ce^{2x} - 2Ce^{2x} = 3e^{2x}$$

$$(4 - 2 - 2)Ce^{2x} = 3e^{2x} \quad \text{[Taking out } Ce^{2x}\text{]}$$

$$0 = 3e^{2x}$$

This is impossible because $3e^{2x}$ cannot be zero.

?   **Where have we gone wrong in the above derivation?**

?   **Are you surprised by the result?**

?   You shouldn't be surprised because by taking the trial function, Y, to be Ce^{2x} was clearly going to give zero. **Why?**

Because Ae^{2x} is part of the complementary function, y_c, so Ce^{2x} is also part of the complementary function, that is part of the solution with $f(x) = 0$.

?   In the case when the trial function in Table 1 is part of the complementary function then we need to change our trial function. We cannot make a dramatic change since we need an e^{2x} term because $f(x) = 3e^{2x}$. **What is the simplest function we can try with an e^{2x} term?**

$$xe^{2x}$$

We try the function

$$Y = Cxe^{2x}$$

(Call **this** function Y from now on, rather than $Y = Ce^{2x}$, as above.)

?   **How do we differentiate $Y = Cxe^{2x}$?**

Use the product rule with the constant C outside:

$$u = x \qquad v = e^{2x}$$

$$u' = 1 \qquad v' = 2e^{2x}$$

$$\frac{dY}{dx} = \frac{d}{dx}(Cxe^{2x}) = [u'v + uv']C \qquad \text{[Taking out constant } C\text{]}$$

$$= [(1)e^{2x} + x(2e^{2x})]C$$

$$= [e^{2x} + 2xe^{2x}]C$$

Using the product rule again:

$$\frac{d^2Y}{dx^2} = [2e^{2x} + 2(e^{2x} + 2xe^{2x})]C$$

$$= \left[\underbrace{2e^{2x} + 2e^{2x}}_{=4e^{2x}} + 4xe^{2x}\right]C$$

$$= [4e^{2x} + 4xe^{2x}]C$$

Example 13 *continued*

Substituting these, $\dfrac{dY}{dx} = [e^{2x} + 2xe^{2x}]C$ and $\dfrac{d^2Y}{dx^2} = [4e^{2x} + 4xe^{2x}]C$, into the differential equation

$$\frac{d^2Y}{dx^2} - \frac{dY}{dx} - 2Y = 3e^{2x}$$

gives

$$[4e^{2x} + 4xe^{2x} - e^{2x} - 2xe^{2x} - 2xe^{2x}]C = 3e^{2x}$$

$$[4e^{2x} + 4xe^{2x} - e^{2x} - 2xe^{2x} - 2xe^{2x}]C = 3e^{2x} \qquad \text{[Expanding]}$$

$$\underbrace{(4 + 4x - 1 - 2x - 2x)}_{=\,3}Ce^{2x} = 3e^{2x} \qquad \text{[Taking out } e^{2x}]$$

$$3Ce^{2x} = 3e^{2x} \qquad \text{[Simplifying]}$$

$$3C = 3 \quad \text{so that } C = 1$$

Putting $C = 1$ into our trial function $Y = Cxe^{2x}$ gives $Y = xe^{2x}$.

Adding together $Y = xe^{2x}$ and $y_c = Ae^{2x} + Be^{-x}$ gives the general solution:

$$y = Ae^{2x} + Be^{-x} + xe^{2x}$$

The above example shows that if $f(x)$ is part of the complementary function then the particular integral Y is found by selecting the appropriate trial function from Table 1 and multiplying it by x.

This snag rarely occurs, but when it does, we can still find the solution of the differential equation using the method above.

SUMMARY

The solution of the linear non-homogeneous differential equation $a\dfrac{d^2y}{dx^2} + b\dfrac{dy}{dx} + cy = f(x)$ where $f(x) \neq 0$ can be found using:

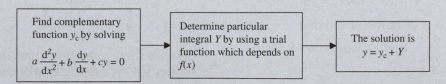

If $f(x)$ is already part of the complementary function, y_c, then choose the appropriate trial function from Table 1 and multiply by x.

Exercise **14(c)**

1 Determine the particular integrals of

$$\frac{d^2y}{dx^2} - \frac{dy}{dx} - 2y = f(x) \text{ with}$$

a $f(x) = 18$ **b** $f(x) = 2x + 3$
c $f(x) = 130\sin(3x)$ **d** $f(x) = \cos(x)$
e $f(x) = e^{3x}$

(Note that the Left-Hand Side of the differential equation is the same as **Example 8**.)

2 Find the general solutions of the following differential equations:

a $\dfrac{d^2y}{dx^2} + 3\dfrac{dy}{dx} + 2y = 6$

b $3\dfrac{d^2y}{dx^2} - 2\dfrac{dy}{dx} - y = 2x - 3$

c $\dfrac{d^2y}{dx^2} + 2\dfrac{dy}{dx} + y = 36e^{5x}$

d $\dfrac{d^2y}{dx^2} + 3\dfrac{dy}{dx} - 4y = -34\sin(x)$

⟨Ψ⟩ Questions 3 to 7 belong to the field of [vibrations].

3 The equation of motion of mass, m, on one end of a cantilever beam is given by

$$m\ddot{x} = mg - kx$$

where x is the displacement, k is a constant and g is acceleration due to gravity. Find an expression for x in terms of t. $\left(\text{Remember that } \ddot{x} = \dfrac{d^2x}{dt^2}.\right)$

4 A vehicle of mass $m = (1 \times 10^3)$ kg is brought to rest by a force F given by

$$F = (0.5 + 4x) \times 10^3$$

where x is distance. Given that $F = m\dfrac{d^2x}{dt^2}$, find an expression for x in terms of t.

*5

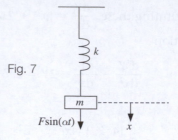

Fig. 7

Figure 7 shows an undamped spring–mass system subject to a force $F\sin(\alpha t)$. The equation of motion is

$$m\ddot{x} = -kx + F\sin(\alpha t) \quad \left(\alpha^2 \neq \frac{k}{m}\right)$$

where m is the mass, k is the stiffness of the spring, x is the displacement and t is the time. Show that

$$x = A\cos(\omega t) + B\sin(\omega t) + \frac{F}{k - m\alpha^2}\sin(\alpha t)$$

where $\omega = \sqrt{\dfrac{k}{m}}$

6 The spring–mass system of Fig. 7 is subject to a force $F\cos(\alpha t)$. Show that

$$x = A\cos(\omega t) + B\sin(\omega t) + \frac{F}{k - m\alpha^2}\cos(\alpha t)$$

where $\omega = \sqrt{\dfrac{k}{m}}$

*7

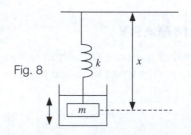

Fig. 8

Figure 8 shows a body of mass m suspended by a spring of stiffness k in

Exercise **14(c) continued**

Solutions at end of book. Complete solutions available at www.palgrave.com/engineering/singh

a fluid. The equation of motion is given by

$$m\ddot{x} = mg - k(x - L) - c\dot{x}$$

where x is the length shown, L is the unstretched length of the spring and c is a constant. Show that, if $c^2 < 4km$ then

$$x = e^{-\frac{c}{2m}t}\left[A\cos(\beta t) + B\sin(\beta t)\right] + \frac{mg}{k} + L$$

where $\beta = \frac{1}{2m}\sqrt{4km - c^2}$ and A, B are constants.

8 [electrical principles]   By applying Kirchhoff's law to a parallel *RLC* circuit (all three are in parallel with each other) we obtain the second order differential equation

$$CL\frac{d^2i}{dt^2} + \frac{L}{R}\frac{di}{dt} + i = I$$

where i is the current through the circuit and I is the d.c. current source applied to the circuit. The capacitance $C = 10 \times 10^{-9}$ F, inductance $L = 50 \times 10^{-3}$ H, resistance $R = 1 \times 10^{-3}\Omega$ and $I = 50 \times 10^{-3}$ A. Find an expression for i in terms of t.

9 Find the general solution of

$$\frac{d^2y}{dx^2} - \frac{dy}{dx} - 2y = 5e^{-x}$$

10 Determine the form of the trial function, Y, for the following:

a $\dfrac{d^2y}{dx^2} + 4\dfrac{dy}{dx} + 3y = 6e^{-3x}$

b $\dfrac{d^2y}{dx^2} + 9y = 6\cos(3x)$

*11 [structures]   A load, P, is applied to a column of length L. The differential equation relating the deflection y at a distance x along the column is given by

$$EI\frac{d^2y}{dx^2} + P\left[y + m\sin\left(\frac{\pi x}{L}\right)\right] = 0$$

$$\left(P \neq \frac{EI\pi^2}{L^2}\right)$$

where m is the initial maximum deflection. Show that

$$y = A\cos(kx) + B\sin(kx)$$

$$+ \frac{PL^2 m}{EI\pi^2 - PL^2}\sin\left(\frac{\pi x}{L}\right) \text{ where } k = \sqrt{\frac{P}{EI}}$$

SECTION D **Particular solutions**

By the end of this section you will be able to:
► find particular solutions of non-homogeneous equations

> ## D1 Particular solutions of non-homogeneous equations

In the previous section we only found **general solutions** of non-homogeneous differential equations. In **Example 8**

$$\frac{d^2y}{dx^2} - \frac{dy}{dx} - 2y = 5e^{4x}$$

the general solution was

$$y = Ae^{2x} + Be^{-x} + \frac{e^{4x}}{2}$$

If we are given initial (or boundary) conditions, then we can find values of the constants A and B. In this case y is called the **particular solution**. The names **particular solution** and **particular integral** can be a cause of confusion. These two terms are **not** the same and therefore **cannot** be interchanged. The particular integral is the term added to the complementary function to give a solution, while the particular solution is the solution of the differential equation which has particular values for the constants.

Example 14 *mechanics*

A vehicle is brought to rest by a buffer stop. By applying Newton's second law we obtain the second order differential equation

$$(2 \times 10^3)\ddot{x} + (18 \times 10^3)x + (5 \times 10^3) = 0$$

where x is the distance by which the buffer is compressed. Given the initial conditions, when $t = 0$, both $x = 0$ and $\dot{x} = 0$, find the expression for x in terms of t.

(Remember that $\dot{x}$ represents the first derivative with respect to time and $\ddot{x}$ represents the second derivative with respect to time.)

Solution

Dividing the given differential equation by 10^3 yields

$$2\ddot{x} + 18x + 5 = 0$$
$$2\ddot{x} + 18x = -5$$

*

The differential equation is now in standard form.

First we find the complementary function, x_c:

$$2\ddot{x}_c + 18x_c = 0 \qquad \text{[Right-Hand Side = 0]}$$

The characteristic equation is

$$2m^2 + 18 = 0$$

Dividing through by 2:

$$m^2 + 9 = 0$$
$$m^2 + 3^2 = 0$$

Example 14 *continued*

By 14.8 :

$$x_c = A\cos(3t) + B\sin(3t)$$

Now we need to find the particular integral, X:

Since $f(t) = -5$ is a constant, we use the trial function $X = C$ where C is a constant. Differentiating $X = C$ gives

$$\dot{X} = \ddot{X} = 0$$

Substituting into $2\ddot{X} + 18X = -5$ yields

$$18C = -5, \text{ giving } C = -\frac{5}{18}$$

which means

$$X = C = -\frac{5}{18}$$

The general solution is given by $x = x_c + X$, thus

** $$x = A\cos(3t) + B\sin(3t) - \frac{5}{18}$$

Substituting the initial condition, when $t = 0$, $x = 0$, into ****** gives

$$0 = A\underbrace{\cos(0)}_{=1} + B\underbrace{\sin(0)}_{=0} - \frac{5}{18}$$

$$0 = A - \frac{5}{18} \text{ giving } A = \frac{5}{18}$$

Using the other initial condition, when $t = 0$, $\dot{x} = 0$, means we need to differentiate ****** :

$$x = A\cos(3t) + B\sin(3t) - \frac{5}{18}$$

$$\dot{x} = \underbrace{-3A\sin(3t)}_{\text{by } 6.13} + \underbrace{3B\cos(3t)}_{\text{by } 6.12}$$

Putting $\dot{x} = 0$ and $t = 0$ gives

$$0 = -3A\underbrace{\sin(0)}_{=0} + 3B\underbrace{\cos(0)}_{=1}$$

$$0 = 0 + 3B \text{ which gives } B = 0$$

Substituting $A = \dfrac{5}{18}$ and $B = 0$ into ****** yields the particular solution:

$$x = \frac{5}{18}\cos(3t) - \frac{5}{18}$$

$$x = \frac{5}{18}(\cos(3t) - 1) \qquad \left[\text{Taking out } \frac{5}{18}\right]$$

6.12 $\dfrac{d}{dt}\big[\sin(kt)\big] = k\cos(kt)$ 6.13 $\dfrac{d}{dt}\big[\cos(kt)\big] = -k\sin(kt)$

14.8 If $m^2 + k^2 = 0$ then $x = A\cos(kt) + B\sin(kt)$

In the above example we have found the particular solution to a non-homogeneous differential equation.

? **What is the general solution of this differential equation?**

It is ****** , that is

**
$$x = A\cos(3t) + B\sin(3t) - \frac{5}{18}$$

Example 15 *electrical principles*

By applying Kirchhoff's voltage law to a series *RLC* circuit we obtain the differential equation

$$\frac{d^2i}{dt^2} + 3\frac{di}{dt} + 2i = 5e^{-3t}$$

Solve this differential equation for the initial conditions, when $t = 0$, both $i = 0$ and $\frac{di}{dt} = 0$.

Solution

For the complementary function we first consider the homogeneous equation by putting the Right-Hand Side equal to zero:

$$\frac{d^2i}{dt^2} + 3\frac{di}{dt} + 2i = 0$$

The characteristic equation is

$$m^2 + 3m + 2 = 0$$
$$(m + 2)(m + 1) = 0 \qquad \text{[Factorizing]}$$
$$m_1 = -2, \ m_2 = -1 \qquad \text{[Real and different]}$$

Substituting these into

14.4 $\quad i = Ae^{m_1t} + Be^{m_2t}$

gives the complementary function

$$i_c = Ae^{-2t} + Be^{-t}$$

For particular integral I: since $f(t) = 5e^{-3t}$ so our trial function is

$$I = Ce^{-3t}$$

Differentiating:

$$\frac{dI}{dt} = -3Ce^{-3t} \qquad \left[\text{By } \frac{d}{dt}(e^{kt}) = ke^{kt} \right]$$

$$\frac{d^2I}{dt^2} = 9Ce^{-3t}$$

Substituting into $\dfrac{d^2I}{dt^2} + 3\dfrac{dI}{dt} + 2I = 5e^{-3t}$ gives

$$9Ce^{-3t} + 3(-3Ce^{-3t}) + 2Ce^{-3t} = 5e^{-3t}$$

Example 15 *continued*

$$\underbrace{9Ce^{-3t} - 9Ce^{-3t}}_{=0} + 2Ce^{-3t} = 5e^{-3t}$$

$$2Ce^{-3t} = 5e^{-3t}$$

$$2C = 5 \text{ thus } C = 2.5$$

Substituting $C = 2.5$ into $I = Ce^{-3t}$ yields

$$I = 2.5e^{-3t}$$

The general solution, $i = i_c + I$, is

† $\quad i = \underbrace{Ae^{-2t} + Be^{-t}}_{= \, i_c} + \underbrace{2.5e^{-3t}}_{= I}$

To find A and B we use the given initial conditions:

Substituting $t = 0$, $i = 0$ into †

$$0 = Ae^0 + Be^0 + 2.5e^0$$

$$= A + B + 2.5 \qquad [\text{Because } e^0 = 1]$$

$$A + B = -2.5$$

Differentiating † by applying $(e^{kt})' = ke^{kt}$ yields

$$\frac{di}{dt} = -2Ae^{-2t} - Be^{-t} - 7.5e^{-3t}$$

Putting $t = 0$ and $\dfrac{di}{dt} = 0$ gives

$$0 = -2A - B - 7.5$$

$$-2A - B = 7.5$$

We have the simultaneous equations

$$A + B = -2.5$$

$$-2A - B = 7.5$$

Solving these gives

$$A = -5 \text{ and } B = 2.5$$

Substituting $A = -5$ and $B = 2.5$ into $i = Ae^{-2t} + Be^{-t} + 2.5e^{-3t}$ yields

$$i = -5e^{-2t} + 2.5e^{-t} + 2.5e^{-3t}$$

$$= 2.5(e^{-3t} + e^{-t}) - 5e^{-2t}$$

The next example is more sophisticated than the above examples. It uses unfamiliar symbols as coefficients rather than simple numbers. Again the technique of finding a particular solution is the same.

Example 16 *structures*

A load, P, is applied to each end of a column of length L and is also subject to a force, F, at $x = L/2$ (Fig. 9). The differential equation is given by

$$EI\frac{d^2y}{dx^2} + Py + \frac{F}{2}x = 0 \qquad 0 \le x \le \frac{L}{2}$$

where y is the deflection at a point x. If at $x = 0$, $y = 0$ and at $x = \frac{L}{2}$, $\frac{dy}{dx} = 0$ then show that

$$y = \frac{F}{2kP}\left[\sin(kx)\sec\left(\frac{kL}{2}\right) - kx\right] \quad \text{where } k = \sqrt{\frac{P}{EI}}$$

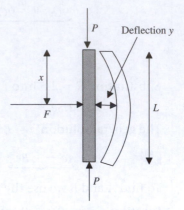

Fig. 9 Shows a column of length L with a load P applied at both ends and also subject to a force F

Solution

Rearranging the given differential equation to standard form:

$$EI\frac{d^2y}{dx^2} + Py = -\frac{Fx}{2}$$

Dividing through by EI:

$$\frac{d^2y}{dx^2} + \frac{P}{EI}y = -\frac{Fx}{2EI}$$

We found the complementary function y_c in **Example 7** on page 742:

$$y_c = A\cos(kx) + B\sin(kx) \quad \text{where } k = \sqrt{P/EI}$$

The particular integral Y: since the Right-Hand Side is a linear expression, our trial function is

$$Y = ax + b$$

$$\frac{dY}{dx} = a, \quad \frac{d^2Y}{dx^2} = 0 \qquad \text{[Differentiating]}$$

Substituting into $\frac{d^2Y}{dx^2} + \frac{P}{EI}Y = -\frac{F}{2EI}x$ gives

$$0 + \frac{P}{EI}(ax + b) = -\frac{F}{2EI}x$$

Equating coefficients of x: $\qquad \frac{P}{E\!I}a = -\frac{F}{2E\!I}$ which gives $a = -\frac{F}{2P}$ $\qquad$ [Cancelling EI]

Equating constants: $\qquad\qquad\quad \frac{Pb}{EI} = 0$, hence $b = 0$

Example 16 *continued*

Placing $a = -\dfrac{F}{2P}$ and $b = 0$ into $Y = ax + b$ yields $Y = -\dfrac{Fx}{2P}$.

Since $y = y_c + Y$, we have

$$y = \underbrace{A\cos(kx) + B\sin(kx)}_{= y_c} \underbrace{-\, \frac{Fx}{2P}}_{= Y}$$

We are given boundary conditions so we can find the constants A and B. Substituting the given boundary conditions, $x = 0$, $y = 0$:

$$0 = A + 0 - 0, \text{ hence } A = 0$$

Substituting $A = 0$ into $y = A\cos(kx) + B\sin(kx) - \dfrac{Fx}{2P}$ gives

$$y = B\sin(kx) - \frac{Fx}{2P}$$

Differentiating:

$$\frac{dy}{dx} = \underbrace{kB\cos(kx)}_{\text{by } 6.12} - \frac{F}{2P}$$

Substituting the other boundary condition, $x = \dfrac{L}{2}$, $\dfrac{dy}{dx} = 0$:

$$0 = kB\cos\left(\frac{kL}{2}\right) - \frac{F}{2P}$$

$$kB\cos\left(\frac{kL}{2}\right) = \frac{F}{2P}$$

$$B = \frac{F}{2kP\cos\left(\dfrac{kL}{2}\right)} \qquad \left[\text{Dividing by } k\cos\left(\frac{kL}{2}\right)\right]$$

$$= \frac{F}{2kP}\sec\left(\frac{kL}{2}\right) \qquad \left[\text{Remember that } \frac{1}{\cos(x)} = \sec(x)\right]$$

Substituting $B = \dfrac{F}{2kP}\sec\left(\dfrac{kL}{2}\right)$ into $y = B\sin(kx) - \dfrac{Fx}{2P}$ gives

$$y = \frac{F}{2kP}\sec\left(\frac{kL}{2}\right)\sin(kx) - \frac{Fx}{2P}$$

$$= \frac{F}{2kP}\sec\left(\frac{kL}{2}\right)\sin(kx) - \frac{Fkx}{2kP} \qquad \left[\text{Writing } \frac{Fx}{2P} = \frac{Fkx}{2kP}\right]$$

$$= \frac{F}{2kP}\left[\sec\left(\frac{kL}{2}\right)\sin(kx) - kx\right] \qquad \left[\text{Taking out } \frac{F}{2kP}\right]$$

6.12 $\dfrac{d}{dx}[\sin(kx)] = k\cos(kx)$

SUMMARY

Solution of the linear non-homogeneous differential equation $a\dfrac{d^2y}{dx^2} + b\dfrac{dy}{dx} + cy = f(x)$ with initial or boundary conditions can be found using:

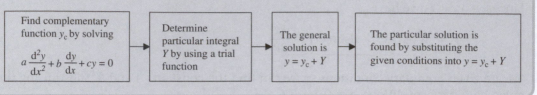

| Find complementary function y_c by solving $a\dfrac{d^2y}{dx^2} + b\dfrac{dy}{dx} + cy = 0$ | → | Determine particular integral Y by using a trial function | → | The general solution is $y = y_c + Y$ | → | The particular solution is found by substituting the given conditions into $y = y_c + Y$ |

Exercise **14(d)**   Solutions at end of book. Complete solutions available at www.palgrave.com/engineering/singh

1 [electrical principles]   The current, i, in a circuit is given by the second order differential equation

$$\frac{d^2i}{dt^2} + 8\frac{di}{dt} + 15i = 150$$

When $t = 0$, both $i = 0$ and $\dfrac{di}{dt} = 0$. Find the current, i, in terms of t.

2 [electrical principles]   By applying Kirchhoff's voltage law to a series RLC circuit we obtain the differential equation

$$\frac{d^2i}{dt^2} + 3\frac{di}{dt} + 2i = 5e^{-3t}$$

Solve this differential equation for the initial conditions, when $t = 0$, both $i = 0$ and $\dfrac{di}{dt} = 5$.

3 [mechanics]   By applying Newton's second law to a body of mass m at a distance x, we obtain

$$\ddot{x} + 9x = -18$$

Solve this differential equation for the initial conditions, when $t = 0$, both $x = 0$ and $\dot{x} = 3$.

4 [vibrations]

Fig. 10

The undamped mass–spring system in Fig. 10 is subject to an external force $F\sin(\alpha t)$. The equation of motion is

$$m\ddot{x} + kx = F\sin(\alpha t) \quad \left(\alpha \neq \sqrt{\frac{k}{m}}\right)$$

where x is the displacement of mass m from its equilibrium level and k is the spring stiffness. By assuming the trial function $X = C\sin(\alpha t)$ and the initial conditions, when $t = 0$, both $x = 0$ and $\dot{x} = 0$, show that

$$x = \frac{F}{\omega(k - m\alpha^2)}[\omega\sin(\alpha t) - \alpha\sin(\omega t)]$$

where $\omega = \sqrt{\dfrac{k}{m}}$

 The remaining questions are in the field of [structures].

14 ▶ Second Order Linear Differential Equations

Exercise 14(d) continued

Solutions at end of book. Complete solutions available at www.palgrave.com/engineering/singh

*5 A strut of length L is subject to a load P. The deflection y at a distance x is related by the second order differential equation

$$EI\frac{d^2y}{dx^2} + Py = Fx$$

where F is a force. By using the boundary conditions, at $x = 0$, $y = 0$ and at

$x = L$, $\dfrac{dy}{dx} = 0$, show that

$$y = \frac{F}{kP}\left[kx - \sec(kL)\sin(kx)\right]$$

$$\text{where } k = \sqrt{\frac{P}{EI}}$$

*6 A strut of length L is fixed at one end and is subject to a load P at the other end (free end). The differential equation relating deflection y at a distance x from the free end is given by

$$EI\frac{d^2y}{dx^2} + P(y - e - d) = 0$$

where e is the eccentricity and d is the deflection at the free end.

Using the initial conditions at $x = 0$, both $y = 0$ and $\dfrac{dy}{dx} = 0$, show that

$$y = (e + d)[1 - \cos(kx)] \text{ where } k = \sqrt{\frac{P}{EI}}$$

7 A load, P, is applied to each end of a strut of length L. The deflection y at a distance x is related by

$$EI\frac{d^2y}{dx^2} + Py = M$$

where M is the moment at each end of the strut.

By using the initial conditions, at $x = 0$, both $y = 0$ and $\dfrac{dy}{dx} = 0$, show that

$$y = \frac{M}{P}[1 - \cos(kx)] \text{ where } k = \sqrt{\frac{P}{EI}}$$

Examination questions 14

Solutions at end of book. Complete solutions available at www.palgrave.com/engineering/singh

1 Find the solution of the differential equation

$$\frac{d^2y}{dx^2} + 2\frac{dy}{dx} + 2y = 0$$

that satisfies the initial condition $y = 1$ and $\dfrac{dy}{dx} = 0$ at $x = 0$.

University of Manchester, UK, 2008

2 Solve the following differential equations for y:

i $\dfrac{d^2y}{dx^2} + 16y = 0$ subject to $y(0) = 3$ and $y'(0) = -2$

ii $\dfrac{d^2y}{dx^2} + 2\dfrac{dy}{dx} - 15y = 2e^{4x}$

University of Surrey, UK, 2009

3 A weight is attached to a spring which moves up and down, so that the equation of motion is given by:

$$\frac{d^2s}{dt^2} + 25s = 0, \text{ where } s \text{ (cm) is the}$$

extension of the spring at time t (seconds).

a If $s = 2$ and $\dfrac{ds}{dt} = 5$, when $t = 0$, find s in terms of t.

b Find s when $t = \pi/4$ seconds.

c Find the value of s, when the spring is initially at rest.

University of Portsmouth, UK, 2009

Examination questions **14 continued**

Solutions at end of book. Complete solutions available at www.palgrave.com/engineering/singh

4 Using the complementary function/ particular integral approach, or otherwise, determine the solution to:
$$\frac{d^2y}{dx^2} + y = 0.001x^2, \; y(0) = 0, \; y'(0) = 1.5$$
Loughborough University, UK, 2009

5 Find the general solution of the differential equation
$$\frac{d^2y}{dx^2} + \frac{dy}{dx} - 2y = -4x$$
Find the solution which satisfies $y(0) = 4, y'(0) = 5$.
University of Sussex, UK, 2007

6 a Solve the initial value problem $2y'' + 5y' + 3y = 0, \; y(0) = 3, \; y'(0) = -4$.
b Find the general solution of the differential equation $y'' - y' = \sin(2x)$.
University of British Columbia, Canada, 2008

7 Which of the following functions is the solution of the differential equation $y'' + y = \sin(x)$?
a $y = \dfrac{1}{2} x \sin(x)$
b $y = -\dfrac{1}{2} x \cos(x)$
North Carolina State University, USA, 2010

8 Find all the real solutions to the differential equation $\dfrac{d^2y}{dx^2} + \dfrac{dy}{dx} - 2y = e^{3x}$.
Princeton University, USA, 2006

9 Find the general solution to the differential equation
$$y'' + y = \cos(x), \; y(0) = 0, \; y'(0) = 5/2$$
University of California, Berkeley, USA, 2005

10 Use the method of undetermined coefficients to solve the equation
$$\frac{d^2y}{dt^2} + 9y = 9t^2 - 12\cos(3t)$$
Memorial University of Newfoundland, Canada, 2010

11 a Solve the initial value problem $\ddot{x} = -9x$ where $x = -1$ and $\dot{x} = 3$ when $t = 0$.
Express the solution in the form $x = R\sin(\omega t + \phi)$. State the amplitude and period of the oscillations and sketch the graph of x for $0 \le t \le 2\pi$.
b Find the general solution of the non-homogeneous ordinary differential equation:
$$\frac{d^2x}{dt^2} - 2\frac{dx}{dt} + 10x = 20t + 6$$
University of Aberdeen, UK, 2003

Miscellaneous exercise **14**

1 Find the general solutions of
a $\dfrac{d^2y}{dx^2} + 8\dfrac{dy}{dx} + 16y = 0$
b $\dfrac{d^2y}{dx^2} - 2\dfrac{dy}{dx} - 3y = 0$
c $\dfrac{d^2y}{dx^2} - 6\dfrac{dy}{dx} + 7y = 0$

2 Find the general solutions of the following:
a $\dfrac{d^2y}{dx^2} + 8\dfrac{dy}{dx} + 16y = 8$
b $\dfrac{d^2y}{dx^2} + 8\dfrac{dy}{dx} + 16y = 8x$
c $\dfrac{d^2y}{dx^2} + 8\dfrac{dy}{dx} + 16y = x^2 + x + 1$

Miscellaneous exercise **14 continued**　Solutions at end of book. Complete solutions available at www.palgrave.com/engineering/singh

3 [vibrations]　A shaft of length l with modulus of rigidity G has angular displacement θ, a function of time t, given by

$$\frac{d^2\theta}{dt^2} + \frac{GJ}{Il}\theta = 0$$

where I is moment of inertia and J is second moment of area. Show that

$$\theta = A\cos(\omega t) + B\sin(\omega t) \text{ where } \omega = \sqrt{\frac{GJ}{Il}}$$

4 [heat transfer]　The temperature, $T\,(=T(x))$, of a fin of perimeter P with cross-sectional area A is given by

$$\frac{d^2T}{dx^2} - \frac{hP}{kA}T = 0$$

where h is the convection heat transfer coefficient, k is the thermal conductivity and x is the distance along the fin. Find an expression for T in terms of x.

5 [vibrations]　An undamped spring–mass system is subject to a force $F\cos(\omega t)$ where $\omega = \sqrt{\dfrac{k}{m}}$. The equation of motion is given by

$$m\ddot{x} = -kx + F\cos(\omega t) \quad (\omega \neq 0)$$

where m is mass, k is stiffness of spring, x is displacement and t is time. Show that

$$x = A\cos(\omega t) + B\sin(\omega t) + \frac{Ft}{2m\omega}\sin(\omega t)$$

*6 [vibrations]

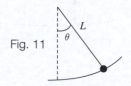

Fig. 11

The equation of motion of a simple pendulum of length L is given by

$$L\ddot{\theta} + g\theta = 0$$

where θ is the angle shown in Fig. 11, $\ddot{\theta} = \dfrac{d^2\theta}{dt^2}$, $\dot{\theta} = \dfrac{d\theta}{dt}$ and $\omega = \sqrt{\dfrac{g}{L}}$ (angular frequency). For the initial conditions $\theta = 1$ and $\dot{\theta} = \sqrt{3}\omega$ when $t = 0$, show that

$$\theta = 2\cos\left(\omega t - \frac{\pi}{3}\right)$$

and sketch the graph of θ versus ωt for $\omega t \geq 0$.

Questions 7 to 11 are in the field of [electrical principles].

7 A parallel LC circuit has the second order differential equation

$$C\frac{d^2v}{dt^2} + \frac{v}{L} = 0$$

If $\omega^2 = \dfrac{1}{LC}$, show that the voltage, v, can be written as

$$v = r\cos(\omega t - \alpha)$$

8 A second order differential equation for a parallel RLC circuit is given by

$$\boxed{*}\quad C\frac{d^2v}{dt^2} + \frac{1}{R}\frac{dv}{dt} + \frac{v}{L} = \frac{di}{dt}$$

i Determine the characteristic equation of $\boxed{*}$.

ii The characteristic equation can be written as

$$m^2 + 2\zeta\omega m + \omega^2 = 0$$

where ζ is the damping ratio and ω is the angular frequency. Critical damping occurs when $\zeta = 1$. We define the critical resistance, R_{cr}, as the resistance value which gives critical damping. Determine R_{cr} for the characteristic equation of part **i**.

Miscellaneous exercise **14 continued**   Solutions at end of book. Complete solutions available at www.palgrave.com/engineering/singh

9 A second order differential equation for an *RLC* series circuit is given by

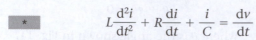

$$L\frac{d^2i}{dt^2} + R\frac{di}{dt} + \frac{i}{C} = \frac{dv}{dt}$$

i Determine the characteristic equation of ⁕ .
ii Determine the critical resistance, R_{cr}, where R_{cr} is defined as in question **8**.

10

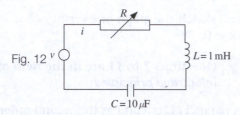

Fig. 12

Figure 12 shows a series *RLC* circuit. By applying Kirchhoff's voltage law, we obtain the second order differential equation

$$L\frac{d^2i}{dt^2} + R\frac{di}{dt} + \frac{i}{C} = \frac{dv}{dt}$$

i By using the solution to question **9**, determine the value of the critical resistance, R_{cr}.
ii For $R = R_{cr}$, obtain the homogeneous solution of the given second order differential equation.

11 The differential equation governing the voltage, *v*, of a parallel *RLC* circuit is given by

$$C\frac{d^2v}{dt^2} + \frac{1}{R}\frac{dv}{dt} + \frac{v}{L} = 0$$

The resistance $R = 500\ \Omega$, inductance $L = 100 \times 10^{-3}$ H and capacitance $C = 0.5 \times 10^{-6}$ F.

Determine an expression for the voltage, *v*, in terms of *t* with the initial conditions, when $t = 0$, both $v = 9$ and $\frac{dv}{dt} = 0$.

*12 🌡 [*heat transfer*]   Consider the differential equation of question **4**:

$$\frac{d^2T}{dx^2} - \frac{hP}{kA}T = 0 \qquad \left(\frac{hP}{kA} \neq 0\right)$$

Given the boundary conditions, at $x = 0$, $T = T_B$ (base temperature) and at $x = L$, $\frac{dT}{dx} = 0$, show that

$$T = T_B\frac{\cosh[m(L-x)]}{\cosh(mL)} \text{ where } m = \sqrt{\frac{hP}{kA}}$$

13 🌡 [*heat transfer*]   The second order differential equation of question **4**:

$$\frac{d^2T}{dx^2} - \frac{hP}{kA}T = 0 \qquad \left(\frac{hP}{kA} \neq 0\right)$$

has boundary conditions, at $x = 0$, $T = T_B$ and at $x = L$, $T = 0$. Show that

$$T = T_B\frac{\sinh[m(L-x)]}{\sinh(mL)} \text{ where } m^2 = \frac{hP}{kA}$$

Questions 14 to 17 belong to the field of [*structures*].

*14 Figure 13 shows an elastic strut with a load *P* applied at both ends with an eccentricity *e*. The relationship between the deflection *y* at a distance *x* is given by

$$EI\frac{d^2y}{dx^2} + Py = 0$$

Fig. 13

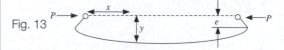

If $y = e$ at $x = 0$ and $x = L$, show that

$$y = e\left[\cos(kx) + \tan\left(\frac{kL}{2}\right)\sin(kx)\right]$$

where $k = \sqrt{\dfrac{P}{EI}}$

$$\left[Hint: \frac{1 - \cos(x)}{\sin(x)} = \tan\left(\frac{x}{2}\right)\right]$$

*15 A column of length L has a uniform load w and an external load P. The deflection, y, at a distance, x, is related by the second order differential equation

$$EI\frac{d^2y}{dx^2} + Py + \frac{w}{2}(Lx - x^2) = 0$$

i Show that

$$y = A\cos(kx) + B\sin(kx)$$
$$+ \frac{w}{2P}\left[x^2 - Lx - \frac{2}{k^2}\right]$$

where $k^2 = \dfrac{P}{EI}$

ii By using the boundary conditions, at $x = 0$, $y = 0$ and at $x = L$, $y = 0$, and the identity

$$\tan\left(\frac{x}{2}\right) = \frac{1 - \cos(x)}{\sin(x)}$$

show that

$$y = \frac{w}{Pk^2}\left[\tan\left(\frac{kL}{2}\right)\sin(kx) + \cos(kx)\right]$$
$$+ \frac{w}{2P}\left[x^2 - Lx - \frac{2}{k^2}\right]$$

*16 A load, P, is applied to a strut of length L which is fixed at one end and pinned at the other end. The deflection y is given by

$$EI\frac{d^2y}{dx^2} + Py + Fx + M = 0$$

where F is the applied force, M is the moment and x is distance. By

using the boundary conditions, at $x = 0$, $y = 0$ and at $x = L$, $y = 0$, show that

$$Py = M\cos(kx)$$
$$+ \left(FL\csc(kL) + M\tan\left(\frac{kL}{2}\right)\right)\sin(kx)$$
$$- Fx - M$$

where $k = \sqrt{\dfrac{P}{EI}}$. (You may use the Hint of question **14**.)

*17 A load, P, is applied to a column of length L fixed at both ends and with a uniform load q. The differential equation

$$EI\frac{d^2y}{dx^2} + Py = \frac{q}{2}(x^2 - xL) - M$$

gives the relationship between the deflection y and the distance x (M is the fixed end moment). At $x = 0$, both $y = 0$ and $\dfrac{dy}{dx} = 0$. Show that

$$y = \frac{1}{2P}\left[\frac{qL}{k}\sin(kx) + 2\left(\frac{q}{k^2} + M\right)\cos(kx)\right.$$
$$\left. + qx^2 - qLx - 2M - \frac{2q}{k^2}\right]$$

where $k^2 = \dfrac{P}{EI}$.

You may use a computer algebra system for question 18.

18 ⓨ [*vibrations*]   The equation of motion for a mass–spring system is given by

$$\ddot{x} + 6\dot{x} + 500x = 1000t + 400$$

Solve this differential equation for the initial conditions, when $t = 0$, both $x = 0$ and $\dfrac{dx}{dt} = 0$. Plot the graph of x versus t for $0 \le t \le 1$.

CHAPTER 15
Partial Differentiation

SECTION A **Partial derivatives**

By the end of this section you will be able to:
▶ distinguish between partial and ordinary differentiation
▶ understand the notation used for partial differentiation
▶ find the first and second partial derivatives

So far we have only differentiated functions of one variable $f(x)$, which means f is a function of x only. In this chapter we look at differentiating functions of two or more variables, $f(x, y)$, a function of x and y, or $f(x, y, z)$, a function of x, y and z. Most engineering problems involve two or more variables. For example:

a Thermodynamics

$$P = \frac{nRT}{V}$$

The pressure, P, of a gas in a cylinder will change with temperature, T, volume, V, and the number of moles of gas present, n (R is a constant).

b Electronics

$$\alpha = K\frac{BL}{\sqrt{V}}$$

The angular deflection, α, of a beam of electrons depends on magnetic field, B, length, L, and accelerating voltage, V (K is a constant).

c Materials

$$\sigma = K\left(\frac{r}{L}\right)^2$$

The axial stress, σ, varies with radius of gyration, r, and length, L (K is a constant).

A1 **Notation**

The volume, V, of a right circular cylinder is given by

$$V = \pi r^2 h$$

where r is the radius and h is the height.

If we change the radius from r to $r + \Delta r$ and keep the height h fixed, the rate of change in volume V is denoted by

$$\left[\frac{dV}{dr}\right]_{h \text{ constant}}$$

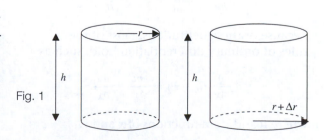
Fig. 1

This means the rate of change in volume V with respect to radius r, while keeping the height h constant. This is normally written as

$$\frac{\partial V}{\partial r}$$

Hence $\left[\dfrac{dV}{dr}\right]_{h\ \text{constant}} = \dfrac{\partial V}{\partial r}$. Also $\left[\dfrac{dV}{dh}\right]_{r\ \text{constant}} = \dfrac{\partial V}{\partial h}$

'Curly dees' are used to distinguish between partial and ordinary differentiation.

A function of two variables is written as $f(x, y)$, that is, f is a function of x and y. For example

$$f(x, y) = x^2 + e^{2y} + 2\cos(xy^2)$$

A function of two variables, $f(x, y)$, is graphed in a similar way to that of one variable, $f(x)$. However in this case we have a three-dimensional co-ordinate system where a point is identified by three co-ordinates (x, y, z).

The dependent variable, z, is plotted on the vertical axis and the two independent variables on the horizontal axes, as shown in Fig. 2. Notice that a function of two variables can be visualized as a surface, and that z is a function of x and y.

Fig. 2

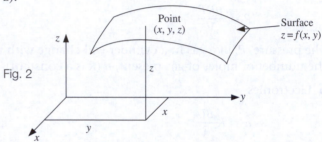

The partial derivative of $f(x, y)$ with respect to x is denoted by $\dfrac{\partial f}{\partial x}$, which means differentiate f with respect to x while keeping y fixed (or treating y as a constant) (Fig. 3, part **a**). Thus

$$\frac{\partial f}{\partial x} = \lim_{h \to 0} \frac{f(x + h, y) - f(x, y)}{h}$$ **a**

$\dfrac{\partial f}{\partial x}$ is stated verbally as 'partial dee f by dee x'.

Similarly $\dfrac{\partial f}{\partial y}$ is the partial derivative of $f(x, y)$ with respect to y, which means differentiate f with respect to y keeping x fixed (Fig. 3, part **b**). Hence

$$\frac{\partial f}{\partial y} = \lim_{h \to 0} \frac{f(x, y + h) - f(x, y)}{h}$$ **b**

$\dfrac{\partial f}{\partial x}$ = gradient of line in the direction of x, with y fixed (hence treat it as a constant in this case)

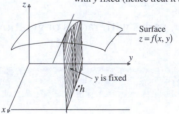

$\dfrac{\partial f}{\partial y}$ = gradient of line in the direction of y, with x fixed (hence treat it as a constant in this case)

Because of these definitions, most of the rules of ordinary differentiation hold, such as

$$\frac{\partial}{\partial x}(u + v) = \frac{\partial u}{\partial x} + \frac{\partial v}{\partial x}$$

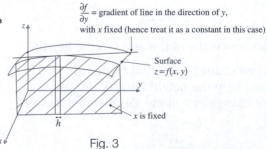

Fig. 3

where u and v are functions of x and y.

Example 1

Consider the function

$$f(x, y) = x^2 + 2xy^2 + y^2$$

Determine $\dfrac{\partial f}{\partial x}$ and $\dfrac{\partial f}{\partial y}$.

Solution

$\dfrac{\partial f}{\partial x}$ means differentiate f with respect to x and think of y as a constant. As with ordinary

differentiation you can partially differentiate each part and then add your results:

$$f(x, y) = x^2 + 2xy^2 + y^2$$

$$\frac{\partial f}{\partial x} = 2x + 2y^2 + \underset{\substack{\text{because } y \\ \text{is treated as} \\ \text{a constant}}}{0} = 2x + 2y^2$$

Similarly

$$\frac{\partial f}{\partial y} = \underset{\substack{\text{because } x \\ \text{is treated as} \\ \text{a constant}}}{0} + 4xy + 2y = 4xy + 2y$$

 ## Example 2 *materials*

The second moment of area, I, of a rectangle with dimensions b and d is given by

$$I = \frac{bd^3}{12}$$

Determine $\dfrac{\partial I}{\partial b}$ and $\dfrac{\partial I}{\partial d}$.

Solution

Just because the function is not in terms of x and y does not mean it is more difficult. We have

$$\frac{\partial I}{\partial b} = \frac{d^3}{12} \quad \text{and} \quad \frac{\partial I}{\partial d} = \frac{3bd^2}{12} = \frac{bd^2}{4}$$

 ## Example 3 *thermodynamics*

The pressure, P, of a gas is given by

$$P = \frac{KT}{V}$$

where K is a constant, T is temperature and V is volume. Find

$$\frac{\partial P}{\partial T} \quad \text{and} \quad \frac{\partial P}{\partial V}$$

Example 3 *continued*

Solution

We have

$$\frac{\partial P}{\partial T} = \frac{\partial}{\partial T}\left(\frac{KT}{V}\right) = \frac{K}{V}$$

We rewrite P as $P = \dfrac{KT}{V} = KTV^{-1}$. Differentiating partially with respect to V gives

$$\frac{\partial P}{\partial V} = -KTV^{-2}$$

$$= -\frac{KT}{V^2}$$

Example 4 *thermodynamics*

From the ideal gas equation, $PV = RT$, show that

$$\left(\frac{\partial P}{\partial T}\right)\left(\frac{\partial T}{\partial V}\right)\left(\frac{\partial V}{\partial P}\right) = -1$$

Solution

By rearranging the given equation, $PV = RT$, and partially differentiating we get

$$P = \frac{RT}{V} \text{ and } \frac{\partial P}{\partial T} = \frac{R}{V}$$

$$T = \frac{PV}{R} \text{ and } \frac{\partial T}{\partial V} = \frac{P}{R}$$

$$V = \frac{RT}{P} = RTP^{-1} \text{ and } \frac{\partial V}{\partial P} = -RTP^{-2} = -\frac{RT}{P^2}$$

Substituting these partial derivatives, $\dfrac{\partial P}{\partial T} = \dfrac{R}{V}$, $\dfrac{\partial T}{\partial V} = \dfrac{P}{R}$ and $\dfrac{\partial V}{\partial P} = -\dfrac{RT}{P^2}$, gives

$$\left(\frac{\partial P}{\partial T}\right)\left(\frac{\partial T}{\partial V}\right)\left(\frac{\partial V}{\partial P}\right) = \frac{R}{V}\frac{P}{R}\left(-\frac{RT}{P^2}\right)$$

$$= -\frac{RT}{VP} \qquad \text{[Cancelling]}$$

$$= -\frac{RT}{V\left(\dfrac{RT}{V}\right)} \qquad \left[\text{Because } P = \frac{RT}{V}\right]$$

$$= -1 \qquad \text{[Cancelling]}$$

A function of three variables is denoted by $f(x, y, z)$.

Example 5 *electronics*

The resistance, R, of a wire of length L and diameter D is given by

$$R = \frac{k\rho L}{D^2}$$

where k is constant and ρ is resistivity. Find $\dfrac{\partial R}{\partial L}$, $\dfrac{\partial R}{\partial \rho}$ and $\dfrac{\partial R}{\partial D}$.

Solution

Note that R is a function of ρ, L and D, that is, $R = f(\rho, L, D)$.

The first two, $\dfrac{\partial R}{\partial L}$ and $\dfrac{\partial R}{\partial \rho}$, are straightforward:

$$\frac{\partial R}{\partial L} = \frac{k\rho}{D^2} \quad \text{and} \quad \frac{\partial R}{\partial \rho} = \frac{kL}{D^2}$$

For $\dfrac{\partial R}{\partial D}$ we rewrite $R = \dfrac{k\rho L}{D^2} = k\rho L D^{-2}$, hence considering $k\rho L$ as a constant we differentiate

$$\frac{\partial R}{\partial D} = -2k\rho L D^{-3} = -\frac{2k\rho L}{D^3}$$

A2 **Higher derivatives**

Let f be a function of x and y then $\dfrac{\partial^2 f}{\partial x^2}$ means partially differentiate $\dfrac{\partial f}{\partial x}$ with respect to x.

Example 6

Let

$$f = f(x, y) = x^3 + y^2\cos(x)$$

Determine $\dfrac{\partial^2 f}{\partial x^2}$ and $\dfrac{\partial^2 f}{\partial y^2}$.

Solution

How can we find $\dfrac{\partial^2 f}{\partial x^2}$?

First find $\dfrac{\partial f}{\partial x}$ by differentiating f with respect to x and treating y as a constant:

$$\frac{\partial f}{\partial x} = 3x^2 - y^2\sin(x) \qquad \left[\text{Remember } [\cos(x)]' = -\sin(x)\right]$$

Example 6 *continued*

Note that $\dfrac{\partial f}{\partial x}$ is a function of x and y. We can partially differentiate this,

$\dfrac{\partial f}{\partial x} = 3x^2 - y^2 \sin(x)$, with respect to x to give $\dfrac{\partial^2 f}{\partial x^2}$:

$$\frac{\partial^2 f}{\partial x^2} = 6x - y^2 \cos(x)$$

To find $\dfrac{\partial^2 f}{\partial y^2}$ we need to first obtain $\dfrac{\partial f}{\partial y}$:

$$f = x^3 + y^2 \cos(x)$$
$$\frac{\partial f}{\partial y} = 0 + 2y\cos(x) = 2y\cos(x)$$

Partially differentiating this, $\dfrac{\partial f}{\partial y} = 2y\cos(x)$, with respect to y gives $\dfrac{\partial^2 f}{\partial y^2}$:

$$\frac{\partial^2 f}{\partial y^2} = 2\cos(x)$$

Hence we define $\dfrac{\partial^2 f}{\partial x^2}$ and $\dfrac{\partial^2 f}{\partial y^2}$ for $f = f(x, y)$, a general function of x and y, as

15.1 $$\frac{\partial^2 f}{\partial x^2} = \frac{\partial}{\partial x}\left(\frac{\partial f}{\partial x}\right)$$

15.2 $$\frac{\partial^2 f}{\partial y^2} = \frac{\partial}{\partial y}\left(\frac{\partial f}{\partial y}\right)$$

15.1 says we first find $\dfrac{\partial f}{\partial x}$ and then partially differentiate this again with respect to x to give

$\dfrac{\partial^2 f}{\partial x^2}$. In each case we treat y as a constant. **15.2** is similar with the variables interchanged.

In this chapter we will use the term *differentiate* to mean partially differentiate.

A3 Mixed partial derivatives

There are two other second derivatives, $\dfrac{\partial^2 f}{\partial x \partial y}$ and $\dfrac{\partial^2 f}{\partial y \partial x}$, which are defined as

15.3 $$\frac{\partial^2 f}{\partial x \partial y} = \frac{\partial}{\partial x}\left(\frac{\partial f}{\partial y}\right)$$

15.4 $$\frac{\partial^2 f}{\partial y \partial x} = \frac{\partial}{\partial y}\left(\frac{\partial f}{\partial x}\right)$$

15.3 says differentiate f with respect to y and then differentiate this result with respect to x. In each case we treat the other variables as constants. 15.4 is similar with variables interchanged.

Example 7

With $f = f(x, y)$ as in **Example 6**, find $\dfrac{\partial^2 f}{\partial x \, \partial y}$ and $\dfrac{\partial^2 f}{\partial y \, \partial x}$.

Solution

We established in **Example 6** that

$$\frac{\partial f}{\partial x} = 3x^2 - y^2 \sin(x)$$

Differentiating this with respect to y:

$$\frac{\partial^2 f}{\partial y \, \partial x} = \frac{\partial}{\partial y} \, [3x^2 - y^2 \sin(x)]$$
$$= 0 - 2y \sin(x)$$
$$= -2y \sin(x)$$

From **Example 6** we have $\dfrac{\partial f}{\partial y} = 2y \cos(x)$ and differentiate this with respect to x:

$$\frac{\partial^2 f}{\partial x \, \partial y} = \frac{\partial}{\partial x} \, [2y \cos(x)]$$
$$= -2y \sin(x)$$

? **What do you notice about your results to Example 7?**

15.5
$$\frac{\partial^2 f}{\partial x \, \partial y} = \frac{\partial^2 f}{\partial y \, \partial x}$$

The result, 15.5 , is usually but not always true. For all the functions we deal with we will assume it is true.

Many of the rules of ordinary differentiation can be used in a similar manner for partial differentiation. For example, the product and quotient rules are defined by

15.6
$$\frac{\partial}{\partial x} \, (uv) = v \, \frac{\partial u}{\partial x} + u \, \frac{\partial v}{\partial x} \qquad \text{[Product rule]}$$

15.7
$$\frac{\partial}{\partial x} \left(\frac{u}{v} \right) = \frac{v \, \dfrac{\partial u}{\partial x} - u \, \dfrac{\partial v}{\partial x}}{v^2} \qquad \text{[Quotient rule]}$$

where u and v are functions x, y, z, etc. Similarly we can find

$$\frac{\partial}{\partial y} \, (uv), \ \frac{\partial}{\partial y} \left(\frac{u}{v} \right), \ \frac{\partial}{\partial z} \, (uv), \ \text{etc.}$$

Moreover we can apply the differentiation table (Table 3 of **Chapter 6** on pages 291 and 292) by replacing the ordinary dee with a curly dee. For example

6.18
$$\frac{\mathrm{d}}{\mathrm{d}x} \, [\ln(u)] = \frac{1}{u} \, \frac{\mathrm{d}u}{\mathrm{d}x}$$

becomes $\dfrac{\partial}{\partial x} \, [\ln(u)] = \dfrac{1}{u} \, \dfrac{\partial u}{\partial x}$ where u is a function of x, y, z, etc.

≋ Example 8 *fluid mechanics*

For a fluid flow to be a potential flow it must satisfy Laplace's equation which is given by

$$\frac{\partial^2 \psi}{\partial x^2} + \frac{\partial^2 \psi}{\partial y^2} = 0$$

where $\psi = \psi(x, y)$ is the stream function. Show that if

$$\psi = \ln(x^2 + y^2)$$

then ψ is a potential flow.

Solution

Don't be put off by the Greek symbol, ψ (psi).

? **First we need to find $\dfrac{\partial \psi}{\partial x}$. How?**

Applying

$$\frac{\partial}{\partial x}\left[\ln(u)\right] = \frac{1}{u}\frac{\partial u}{\partial x}$$

with $u = x^2 + y^2$ we have

$$\frac{\partial}{\partial x}\left[\ln(x^2 + y^2)\right] = \frac{1}{x^2 + y^2}\frac{\partial}{\partial x}(x^2 + y^2)$$

$$= \frac{1}{x^2 + y^2}\,2x = \frac{2x}{x^2 + y^2}$$

Hence $\dfrac{\partial \psi}{\partial x} = \dfrac{2x}{x^2 + y^2}$. To find $\dfrac{\partial^2 \psi}{\partial x^2}$ we need to differentiate $\dfrac{\partial \psi}{\partial x}$. Applying the quotient rule, 15.7 , with

$$u = 2x \qquad v = x^2 + y^2$$
$$\frac{\partial u}{\partial x} = 2 \qquad \frac{\partial v}{\partial x} = 2x$$

we have

$$\frac{\partial^2 \psi}{\partial x^2} = \frac{\dfrac{\partial u}{\partial x}v - u\dfrac{\partial v}{\partial x}}{v^2}$$

$$= \frac{2(x^2 + y^2) - 2x(2x)}{(x^2 + y^2)^2} \qquad \text{[Substituting]}$$

$$= \frac{2x^2 + 2y^2 - 4x^2}{(x^2 + y^2)^2} \qquad \text{[Expanding numerator]}$$

$$= \frac{2y^2 - 2x^2}{(x^2 + y^2)^2} \qquad \text{[Simplifying numerator]}$$

Example 8 *continued*

We go through the same procedure for $\dfrac{\partial^2 \psi}{\partial y^2}$.

$$\psi = \ln(x^2 + y^2)$$

$$\frac{\partial \psi}{\partial y} = \frac{1}{x^2 + y^2} \frac{\partial}{\partial y}\left(x^2 + y^2\right) = \frac{1}{x^2 + y^2} 2y = \frac{2y}{x^2 + y^2}$$

Applying the quotient rule, 15.7 , with

$$u = 2y \qquad v = x^2 + y^2$$
$$\frac{\partial u}{\partial y} = 2 \qquad \frac{\partial v}{\partial y} = 2y$$

we have

$$\frac{\partial^2 \psi}{\partial y^2} = \frac{\dfrac{\partial u}{\partial y}\, v - u\, \dfrac{\partial v}{\partial y}}{v^2}$$

$$= \frac{2\,(x^2 + y^2) - 2y\,(2y)}{(x^2 + y^2)^2}$$

$$= \frac{2x^2 + 2y^2 - 4y^2}{(x^2 + y^2)^2} = \frac{2x^2 - 2y^2}{(x^2 + y^2)^2}$$

Adding both the second derivatives gives

$$\frac{\partial^2 \psi}{\partial x^2} + \frac{\partial^2 \psi}{\partial y^2} = \frac{2y^2 - 2x^2}{(x^2 + y^2)^2} + \frac{2x^2 - 2y^2}{(x^2 + y^2)^2}$$

$$= \frac{2y^2 - 2x^2 + 2x^2 - 2y^2}{(x^2 + y^2)^2}$$

$$= \frac{0}{(x^2 + y^2)^2} = 0$$

Hence $\psi = \ln(x^2 + y^2)$ is a potential flow.

$\Bigg($ Alternatively we can find $\dfrac{\partial^2 \psi}{\partial y^2}$ by noticing the symmetry of the expression $\psi = \ln(x^2 + y^2)$,

therefore $\dfrac{\partial^2 \psi}{\partial y^2} = \dfrac{2x^2 - 2y^2}{(x^2 + y^2)^2}$ because $\dfrac{\partial^2 \psi}{\partial x^2} = \dfrac{2y^2 - 2x^2}{(x^2 + y^2)^2}$. $\Bigg)$

Laplace's equation $\dfrac{\partial^2 \psi}{\partial x^2} + \dfrac{\partial^2 \psi}{\partial y^2} = 0$ is an example of a **partial differential equation**. Partial differential equations in engineering are used to describe the flow of water past submarines or the airflow past aircraft and cars.

Example 9

Let

$$f = f(x, y) = \cos(xy)$$

Determine $\dfrac{\partial^2 f}{\partial x^2}$, $\dfrac{\partial^2 f}{\partial y^2}$ and $\dfrac{\partial^2 f}{\partial x \partial y}$.

Solution

To find $\dfrac{\partial^2 f}{\partial x^2}$ we have first to obtain $\dfrac{\partial f}{\partial x}$:

$$f = \cos(xy)$$

$$\frac{\partial f}{\partial x} = -y\sin(xy) \quad \left[\text{Differentiating } \frac{\partial}{\partial x}\left[\cos(kx)\right] = -k\sin(kx)\right]$$

$$\frac{\partial^2 f}{\partial x^2} = -y^2\cos(xy) \quad \left[\text{Differentiating } \frac{\partial}{\partial x}\left[\sin(kx)\right] = k\cos(kx)\right]$$

Similarly we find $\dfrac{\partial^2 f}{\partial y^2}$:

$$f = \cos(xy)$$

$$\frac{\partial f}{\partial y} = -x\sin(xy) \quad \left[\text{Differentiating } \frac{\partial}{\partial y}\left[\cos(ky)\right] = -k\sin(ky)\right]$$

$$\frac{\partial^2 f}{\partial y^2} = -x^2\cos(xy) \quad \left[\text{Differentiating } \frac{\partial}{\partial y}\left[\sin(ky)\right] = k\cos(ky)\right]$$

To find the mixed derivative $\dfrac{\partial^2 f}{\partial x \partial y}$ means we need to differentiate $\dfrac{\partial f}{\partial y} = -x\sin(xy)$ with respect to x. Using the product rule we have

$$u = x \qquad v = \sin(xy)$$
$$u' = 1 \qquad v' = y\cos(xy)$$

$$\frac{\partial^2 f}{\partial x \partial y} = \frac{\partial}{\partial x}\left[-x\sin(xy)\right]$$

$$= -[u'v + uv'] \qquad\qquad [\text{Taking out minus sign}]$$
$$= -[(1)\sin(xy) + (x)y\cos(xy)] \qquad [\text{Product rule}]$$
$$= -[\sin(xy) + xy\cos(xy)]$$

SUMMARY

In this section we considered differentiating functions of two or more variables.

For example, let $f = f(x, y)$ then $\dfrac{\partial f}{\partial x}$ means differentiate f with respect to x while treating y as a constant. Higher derivatives are defined as

15.1 $$\frac{\partial^2 f}{\partial x^2} = \frac{\partial}{\partial x}\left(\frac{\partial f}{\partial x}\right)$$

SUMMARY continued

15.2
$$\frac{\partial^2 f}{\partial y^2} = \frac{\partial}{\partial y}\left(\frac{\partial f}{\partial y}\right)$$

15.3
$$\frac{\partial^2 f}{\partial x \partial y} = \frac{\partial}{\partial x}\left(\frac{\partial f}{\partial y}\right)$$

15.4
$$\frac{\partial^2 f}{\partial y \partial x} = \frac{\partial}{\partial y}\left(\frac{\partial f}{\partial x}\right)$$

We can modify the rules of ordinary differentiation by replacing 'dee' by a 'curly dee'.

Exercise 15(a)

Solutions at end of book. Complete solutions available at www.palgrave.com/engineering/singh

1 Find $\dfrac{\partial f}{\partial x}$ and $\dfrac{\partial f}{\partial y}$ for each of the following functions:

a $f(x, y) = x^2 + y^2$

b $f(x, y) = \sin(x) + \cos(y)$

c $f(x, y) = x^3 + y^3 + 3xy$

d $f(x, y) = 2y - \dfrac{1000}{x}$

2 [*materials*]   A beam is subject to a uniform load of w per unit length and a concentrated load P. The bending moment M at a distance x from one end is given by

$$M = Px + \frac{wx^3}{3}$$

Determine $\dfrac{\partial M}{\partial P}$ and $\dfrac{\partial M}{\partial x}$.

3 [*materials*]   Castigliano's theorem says the displacement, Δ, under a force P is given by

$$\Delta = \frac{\partial U}{\partial P}$$

where U is the internal strain energy of the body. For the following, find Δ:

a $U = \dfrac{P^2 L}{2AE}$ **b** $U = \dfrac{9P^2 L^3}{96EI}$

4 [*structures*]   Given the stress function, $\Omega(x, y)$, as (A and B are constants)

$$\Omega(x, y) = Ax^2 y^4 + Bx^4 y^2$$

find the stresses, σ_x and σ_y, where

$$\sigma_x = \frac{\partial^2 \Omega}{\partial y^2} \text{ and } \sigma_y = \frac{\partial^2 \Omega}{\partial x^2}$$

*5 [*fluid mechanics*]   The continuity equation (partial differential equation) is defined as

$$\frac{\partial u}{\partial x} + \frac{\partial v}{\partial y} = 0$$

where u, v are velocities of the fluid in the x, y directions respectively. Show that each of the following satisfy the continuity equation:

a $u = y^2 - x^2$, $v = 2xy$

b $u = \tan^{-1}\left(\dfrac{y}{x}\right)$, $v = \dfrac{1}{2}\ln(x^2 + y^2)$

c $u = \dfrac{-2y}{(1+x)^2 + y^2}$, $v = \dfrac{1 - x^2 - y^2}{(1+x)^2 + y^2}$

6 [*fluid mechanics*]   The stream function, $\psi(x, y)$, is related to the velocity components u and v of the fluid flow by

$$u = \frac{\partial \psi}{\partial y} \text{ and } v = -\frac{\partial \psi}{\partial x}$$

If $\psi = \dfrac{1}{2}\ln(x^2 + y^2)$, find u and v.

The flow is irrotational if ψ satisfies Laplace's equation which is given by

$$\frac{\partial^2 \psi}{\partial x^2} + \frac{\partial^2 \psi}{\partial y^2} = 0$$

Exercise **15(a) continued**

Solutions at end of book. Complete solutions available at www.palgrave.com/engineering/singh

Show that the flow is irrotational for

$$\psi = \frac{1}{2}\ln(x^2 + y^2)$$

Questions 7 to 9 are in the field of [**thermodynamics**].

(P is pressure, V is volume, T is temperature and R is the gas constant)

*7 The following terms are used in thermodynamics:

Isothermal compressibility $\kappa = -\dfrac{1}{V}\left(\dfrac{\partial V}{\partial P}\right)$

Thermal expansion coefficient

$$\beta = \frac{1}{V}\left(\frac{\partial V}{\partial T}\right)$$

Show that for the ideal gas equation, $PV = RT$, we have

$$\kappa = \frac{1}{P} \text{ and } \beta = \frac{1}{T}$$

8 An equation involving specific heats c_p, c_v is defined as

$$c_p - c_v = T\left(\frac{\partial V}{\partial T}\right)\left(\frac{\partial P}{\partial T}\right)$$

For the equation

$$\frac{RT}{P} = V + \frac{K}{RT}$$

(K is a constant)

find $c_p - c_v$.

*9 Van der Waals' equation is given by

$$P = \frac{RT}{V - b} - \frac{a}{V^2}$$

where a and b are constants. The critical isotherm occurs when

$$\frac{\partial P}{\partial V} = 0 \text{ and } \frac{\partial^2 P}{\partial V^2} = 0$$

A point which satisfies these two equations is called a 'critical point'. At the critical point the volume and temperature are V_c and T_c respectively. Find a and b in their simplest form for a critical point.

*10 [vibrations]   The displacement, $u(x, t)$, of a rod is a function of position x and time t:

$$u(x, t) = \left(A\sin\left(\frac{\omega}{\alpha}x\right) + B\cos\left(\frac{\omega}{\alpha}x\right)\right)$$
$$\times [C\sin(\omega t) + D\cos(\omega t)]$$

where $\alpha = \sqrt{\dfrac{E}{\rho}}$, ($E$ is modulus of elasticity, ρ is density of rod and ω is natural frequency of vibration). Show that $u(x, t)$ satisfies the wave equation:

$$\alpha^2 \frac{\partial^2 u}{\partial x^2} = \frac{\partial^2 u}{\partial t^2}$$

11 Find $\left(\dfrac{\partial^2 f}{\partial x^2}\right)\left(\dfrac{\partial^2 f}{\partial y^2}\right) - \left(\dfrac{\partial^2 f}{\partial x \partial y}\right)^2$ for the following functions:

a $f(x, y) = x^3 + y^3 - 3xy$

b $f(x, y) = 2xy + \dfrac{2000}{x} + \dfrac{2000}{y}$

SECTION B **Applications**

By the end of this section you will be able to:
▶ understand total differential and small change
▶ apply small change to engineering applications
▶ apply total differential formula to thermodynamics

B1 **Small changes**

Engineering components are subject to tolerance limits. For example, a resistor might be 10 kΩ ±5% where ±5% refers to the tolerance limit of the resistor. Also physical dimensions are prone to error measurements. In this section we examine the effects of such small changes.

Fig. 4

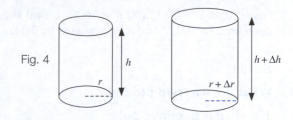

Consider a cylinder with volume V (Fig. 4).

Let Δr be a small change in the radius and Δh be a small change in the height, then the change in volume, ΔV, can be shown to be

$$\Delta V \approx \frac{\partial V}{\partial r}\Delta r + \frac{\partial V}{\partial h}\Delta h$$

Let's consider a general function of x and y, $f = f(x, y)$. Let Δx and Δy be small changes in x and y respectively. The change in f, Δf, is given by

$$\begin{aligned}
\Delta f &= f(x + \Delta x, y + \Delta y) - f(x, y)\\
&= f(x + \Delta x, y + \Delta y) - f(x, y + \Delta y) + f(x, y + \Delta y) - f(x, y)\\
&= \frac{f(x + \Delta x, y + \Delta y) - f(x, y + \Delta y)}{\Delta x}\Delta x + \frac{f(x, y + \Delta y) - f(x, y)}{\Delta y}\Delta y
\end{aligned}$$

$$\Delta f \approx \frac{\partial f}{\partial x}\Delta x + \frac{\partial f}{\partial y}\Delta y$$

Note the approximation sign. This is because

$$\frac{\partial f}{\partial x} = \lim_{\Delta x \to 0} \frac{f(x + \Delta x, y) - f(x, y)}{\Delta x}$$

$$\frac{\partial f}{\partial y} = \lim_{\Delta y \to 0} \frac{f(x, y + \Delta y) - f(x, y)}{\Delta y}$$

Hence the change in Δf is given by

15.8 $$\Delta f \approx \frac{\partial f}{\partial x}\Delta x + \frac{\partial f}{\partial y}\Delta y$$

B2 **Engineering applications**

Example 10 *electrical principles*

The power, p, consumed by a resistor of resistance R, is given by

$$p = \frac{E^2}{R}$$

Example 10 *continued*

Given that $E = 20$ V and $R = 1000\ \Omega$, find the approximate change in power if E is decreased by 0.5 V and R is increased by 50 Ω.

Solution

What are we trying to find?

The change in power, Δp.

Since p is a function of E and R:

$$\Delta p \approx \frac{\partial p}{\partial E}\Delta E + \frac{\partial p}{\partial R}\Delta R$$

From $p = \dfrac{E^2}{R}$ we have

$$\frac{\partial p}{\partial E} = \frac{2E}{R} \qquad\qquad \text{[Partially differentiating]}$$

Rewrite $p = \dfrac{E^2}{R} = E^2 R^{-1}$:

$$\frac{\partial p}{\partial R} = -E^2 R^{-2} = -\frac{E^2}{R^2} \qquad \text{[Partially differentiating]}$$

Substituting these, $\dfrac{\partial p}{\partial E} = \dfrac{2E}{R}$ and $\dfrac{\partial p}{\partial R} = -\dfrac{E^2}{R^2}$, into

$$\Delta p \approx \frac{\partial p}{\partial E}\Delta E + \frac{\partial p}{\partial R}\Delta R$$

gives

$$\Delta p \approx \frac{2E}{R}\Delta E - \frac{E^2}{R^2}\Delta R$$

We are given $\Delta E = -0.5$, with minus sign because E is decreased by 0.5 V. Also $\Delta R = 50$. Thus we have

$$
\begin{aligned}
\Delta p &\approx \frac{2E}{R}(-0.5) - \frac{E^2}{R^2}(50)\\
&= \frac{2\times 20}{1000}(-0.5) - \frac{20^2}{1000^2}(50) \qquad \left[\begin{array}{l}\text{Substituting}\\ E = 20 \text{ and } R = 1000\end{array}\right]\\
&= -0.04 = -40\times 10^{-3}
\end{aligned}
$$

The power is reduced by approximately 40 mW (milliwatts).

For the above example we can find the exact change in power by subtracting the original power, p, from the new power (new p) [the original values are $E = 20$ and $R = 1000$]:

Since E is reduced by 0.5 V and R is increased by 50 Ω, we have

$$\text{new } E = 20 - 0.5 = 19.5$$
$$\text{new } R = (1\times 10^3) + 50 = 1050$$

The new power is given by

$$\text{new } p = \frac{19.5^2}{1050} = 0.362$$

Before the change we had

$$p = \frac{20^2}{1000} = 0.4$$

The change in power is

$$(\text{new } p) - p = 0.362 - 0.4 = -0.038 \quad \text{(3 d.p.)}$$

or -38 mW. Hence the approximation of -40 mW is very close to the change of -38 mW.

? Why do we find an approximate change when we can evaluate the exact change?

Generally it is difficult or impossible to find the exact change because we are not given enough data. For the next example, we are only given the percentage error in measuring the physical dimensions, therefore we cannot evaluate the exact error in h. Go through the next example very carefully because it involves demanding algebra.

Example 11 *fluid mechanics*

The head loss, h, in a pipeline is given by

$$h = \frac{\lambda L V^2}{2gD}$$

where λ is a constant, L is the pipe length, V is the mean velocity of fluid, D is the diameter of the pipe and g is the acceleration due to gravity.

If the error in measuring L, V and D is $+0.1\%$, $+1.5\%$ and $+1\%$ respectively, find the approximate percentage error in h.

Solution

Note that h is a function of L, V and D because g and λ are constants. We can apply $\boxed{15.8}$ to functions of more than two variables. In this case we have

$$\Delta h \approx \frac{\partial h}{\partial L}\Delta L + \frac{\partial h}{\partial V}\Delta V + \frac{\partial h}{\partial D}\Delta D$$

We are given $h = \dfrac{\lambda L V^2}{2gD}$:

$$\frac{\partial h}{\partial L} = \frac{\lambda V^2}{2gD}, \quad \frac{\partial h}{\partial V} = \frac{2\lambda L V}{2gD} \qquad \text{[Partially differentiating]}$$

Rewriting h as

$$h = \frac{\lambda L V^2 D^{-1}}{2g}$$

Example 11 *continued*

$$\frac{\partial h}{\partial D} = -\frac{\lambda L V^2 D^{-2}}{2g} \qquad \text{[Partially differentiating]}$$

$$= -\frac{\lambda L V^2}{2g D^2}$$

Substituting these into Δh gives

$$\Delta h \approx \frac{\lambda V^2}{2gD}\Delta L + \frac{2\lambda L V}{2gD}\Delta V - \frac{\lambda L V^2}{2gD^2}\Delta D$$

How do we find ΔL, ΔV and ΔD?

Since the error in measuring L is $+0.1\%$, how do we write this in terms of a mathematical expression?

% means out of 100 therefore

$$0.1\% \text{ of } L = \frac{0.1}{100} L = 0.001L$$

Thus we have

$$\Delta L = 0.001L$$

Similarly we have

$$\Delta V = \frac{1.5}{100} V = 0.015V \text{ and } \Delta D = \frac{1}{100} D = 0.01D$$

Substituting these into

$$\Delta h \approx \frac{\lambda V^2}{2gD}\Delta L + \frac{2\lambda L V}{2gD}\Delta V - \frac{\lambda L V^2}{2gD^2}\Delta D$$

gives

$$\Delta h \approx \frac{\lambda V^2}{2gD}(0.001L) + \frac{2\lambda L V}{2gD}(0.015V) - \frac{\lambda L V^2}{2gD^2}(0.01D)$$

$$= \frac{\lambda V^2 L}{2gD}(0.001) + \frac{\lambda L V^2}{2gD}(2 \times 0.015) - \frac{\lambda L V^2}{2gD}(0.01)$$

$$= \frac{\lambda V^2 L}{2gD}[0.001 + (2 \times 0.015) - 0.01] \qquad \left[\text{Taking out } \frac{\lambda V^2 L}{2gD}\right]$$

$$= h(0.021) \qquad \left[\text{Because } h = \frac{\lambda V^2 L}{2gD}\right]$$

Hence $\Delta h \approx 0.021h$. The approximate percentage error is 2.1%.

In reality the error in measuring is more likely to be $\pm0.1\%$, $\pm1.5\%$ and $\pm1\%$ rather than $+0.1\%$, $+1.5\%$ and $+1\%$. With $\pm$ change the method of **Example 11** is likely to be wrong. The method of **Example 12** shows one way in which this problem can be overcome.

Example 12 *materials*

The coefficient of rigidity, η, of a wire of length L and diameter D is given by

$$\eta = \frac{KL}{D^4}$$

where K is a constant. Find the largest percentage error in η if errors of $\pm 0.3\%$ and $\pm 0.4\%$ occur in measuring L and D respectively.

Solution

We are given $\eta = \frac{KL}{D^4}$ where K is a constant. Partially differentiating:

$$\frac{\partial \eta}{\partial L} = \frac{K}{D^4} \quad \text{and} \quad \frac{\partial \eta}{\partial D} = -\frac{4KL}{D^5}$$

Putting these into

$$\begin{aligned}
\Delta \eta &\approx \frac{\partial \eta}{\partial L}\Delta L + \frac{\partial \eta}{\partial D}\Delta D \\
&= \frac{K}{D^4}\Delta L - \frac{4KL}{D^5}\Delta D \\
&= K\left(\frac{\Delta L}{D^4} - \frac{4L}{D^5}\Delta D\right) \qquad \text{[Taking out } K\text{]}
\end{aligned}$$

The percentage errors are given by

$$\Delta L = \pm 0.003L, \qquad \Delta D = \pm 0.004D$$

Substituting these into the above gives

$$\begin{aligned}
\Delta \eta &\approx K\left[\frac{\pm 0.003L}{D^4} - \frac{4L}{D^5}(\pm 0.004D)\right] \\
&= K\left[\frac{\pm 0.003L}{D^4} - \frac{4L}{D^4}(\pm 0.004)\right] \\
&= \frac{KL}{D^4}[\pm 0.003 - 4(\pm 0.004)] \qquad \left[\text{Taking out } \frac{L}{D^4}\right] \\
&= \eta[\pm 0.003 - 4(\pm 0.004)] \qquad \left[\text{Because } \eta = \frac{KL}{D^4}\right]
\end{aligned}$$

The largest error occurs when the signs are different, that is $+0.003$ and -0.004 or vice versa:

$$\begin{aligned}
\Delta \eta &\approx \eta[0.003 - 4(-0.004)] \\
&= \eta[0.019]
\end{aligned}$$

The largest error in η is 1.9%.

B3 **Total differential**

In the above we established

15.8 $\Delta f \approx \dfrac{\partial f}{\partial x}\Delta x + \dfrac{\partial f}{\partial y}\Delta y$

where f is a function of x and y, Δx and Δy are small but measurable changes in x and y respectively. By using the limiting process on   15.8   we have

15.9 $df = \dfrac{\partial f}{\partial x}\,dx + \dfrac{\partial f}{\partial y}\,dy$

where df is called the **total differential**.

The next example is pretty tough because you are expected to spot the transposition of the given equation to establish the result.

Example 13 *thermodynamics*

From the ideal gas equation

$$P = \frac{RT}{V}$$

show that

$$\frac{dP}{P} = \frac{dT}{T} - \frac{dV}{V}$$

where P, V, T and R represent pressure, volume, temperature and gas constant respectively.

Solution

From $P = \dfrac{RT}{V}$ we have P is a function of T and V (remember that R is a constant). So by   15.9

* $dP = \dfrac{\partial P}{\partial T}\,dT + \dfrac{\partial P}{\partial V}\,dV$

Use

$$P = \frac{RT}{V} = RTV^{-1}$$

Partially differentiating:

$$\frac{\partial P}{\partial T} = \frac{R}{V} \text{ and } \frac{\partial P}{\partial V} = -RTV^{-2} = -\frac{RT}{V^2}$$

Substituting these, $\dfrac{\partial P}{\partial T} = \dfrac{R}{V}$ and $\dfrac{\partial P}{\partial V} = -\dfrac{RT}{V^2}$, into   *   yields

Example 13 *continued*

$$dP = \frac{R}{V}\,dT - \frac{RT}{V^2}\,dV$$

By transposing the given equation, $P = \dfrac{RT}{V}$, we have

$$\frac{P}{T} = \frac{R}{V} \text{ and } \frac{P}{V} = \frac{RT}{V^2}$$

Putting these into $dP = \dfrac{R}{V}\,dT - \dfrac{RT}{V^2}\,dV$ gives

$$dP = \frac{P}{T}\,dT - \frac{P}{V}\,dV$$

$$= P\left(\frac{dT}{T} - \frac{dV}{V}\right) \qquad \text{[Taking out } P\text{]}$$

Dividing through by P gives the required result:

$$\frac{dP}{P} = \frac{dT}{T} - \frac{dV}{V}$$

SUMMARY

Let $f(x, y)$ be a function of x and y, and Δx, Δy be small changes in x and y respectively, then the change in f is approximately

15.8 $\Delta f \approx \dfrac{\partial f}{\partial x}\Delta x + \dfrac{\partial f}{\partial y}\Delta y$

We define the total differential, df, as

15.9 $df = \dfrac{\partial f}{\partial x}dx + \dfrac{\partial f}{\partial y}dy$

Exercise 15(b)

Solutions at end of book. Complete solutions available at www.palgrave.com/engineering/singh

1 [materials]   The moment of inertia, I, of a body is given by

$$I = \frac{1}{12}\,bd^3$$

where b and d are dimensions of the body. If $b = 5$ m, $d = 0.8$ m and the change in b is 5 cm and in d is 0.8 cm, find the approximate change in I. (The SI units of I are m⁴.)

2 [materials]   The modulus of rigidity, G, of a shaft of length L and radius R is given by

$$G = \frac{R^4\theta}{L}$$

where θ is the angle of twist. After heat treatment during a forging process, R is reduced by 0.5%, θ is increased by 2% and L is increased by 1.5%. Find the approximate percentage change in G caused by the treatment.

Exercise **15(b) continued**

Solutions at end of book. Complete solutions available at
www.palgrave.com/engineering/singh

3 〰 [*fluid mechanics*]   The rate of discharge, Q, of a liquid through an orifice (opening) of area A is given by

$$Q = 0.7A\sqrt{2gh}$$

where h is the depth of the centre of the orifice and g is the acceleration due to gravity.

Given that $A = (2 \times 10^{-3})$ m^2, $h = 1.75$ m and $g = 9.81$ m/s^2, find the approximate change in Q if A is increased by (0.2×10^{-3}) m^2 and h is decreased by 0.25 m. (The SI units of Q are m^3/s.)

4 ▦ [*electrical principles*]   The frequency, f, of an LC circuit is given by

$$f = \frac{1}{2\pi \sqrt{LC}}$$

where L represents an inductance and C a capacitance. If L is decreased by 1.5% and C is decreased by 0.5%, find the approximate percentage change in f.

5 ✈ [*aerodynamics*]   The lift, L, of a body in a fluid of density ρ is given by

$$L = \frac{C\rho V^2 A}{2}$$

where C is the lift coefficient, V is the free stream velocity and A is the area.

If C can be measured to within 1%, ρ to within 0.5%, V to within 0.6% and A to within 0.1%, find the largest percentage error in the value of L.

*6 〰 [*fluid mechanics*]   The rate of flow, Q, of a fluid through a pipe of diameter d and length L is given by

$$Q = \frac{p\pi d^4}{128\eta L}$$

where p is the pressure difference between the ends and η is the coefficient of viscosity.

If Q can be measured to within 0.5%, p to within 1.1%, L to within 0.3% and d to within 0.1%, calculate the maximum percentage error in the value of η.

7 ♨ [*thermodynamics*]   The ideal gas equation is given by

$$T = \frac{PV}{R}$$

where T is the temperature, V is the volume, P is the pressure and R is the gas constant. Show that

$$dT = \frac{VdP + PdV}{R}$$

SECTION C **Optimization**

By the end of this section you will be able to:
▶ find where stationary points occur
▶ distinguish between the three types of stationary points – maximum, minimum and saddle point
▶ identify the nature of stationary points
▶ find the solution to a practical optimization problem

C1 **Stationary points**

We have differentiated functions of two or more variables but what do the graphs of these functions look like?

We can use a computer algebra system or a graphical calculator to plot these functions. As discussed in **Section A** the graph of $z = f(x, y)$ is a surface. We can find analogies between the graphs of two independent variables, $z = f(x, y)$, and one independent variable, $y = f(x)$.

Compare the graphs of $y = x^2$ and the surface $z = x^2$ as shown in Fig. 5.

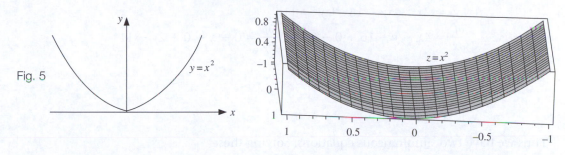
Fig. 5

Notice the similarities.

In MAPLE we have four choices for axes – boxed, framed, normal or none. In the plot for $z = f(x, y)$ the normal axes are difficult to visualize, hence the use of boxed axes. Figure 6 shows other functions of two variables plotted with the aid of MAPLE.

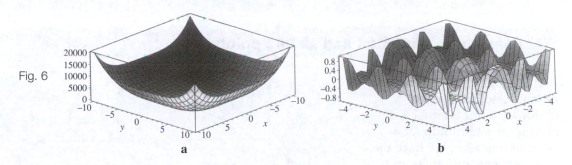
Fig. 6

a   b

Figure 6, part **a**, shows a portion of the surfaces of $f(x, y) = y^4 + x^4 - xy$ and Fig. 6, part **b**, shows $h(x, y) = \sin^2(y)\cos(x^2)$.

Let $f(x, y)$ be a function of two variables. The stationary points of $f(x, y)$ occur (Fig. 7) when

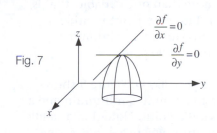
Fig. 7

15.10 $\dfrac{\partial f}{\partial x} = 0$ and $\dfrac{\partial f}{\partial y} = 0$

A stationary point is a point on the surface where **both** slopes in the x and y directions are zero. Remember that the stationary point in $y = f(x)$ is where

$$\dfrac{dy}{dx} = 0 \qquad [\text{Slope} = 0]$$

Example 14

Find where the stationary point(s) of

$$f(x, y) = 2x^2 + xy - 16x + y^2 - 11y$$

occur.

Solution

Differentiating $f = f(x, y)$ we have

$$f(x, y) = 2x^2 + xy - 16x + y^2 - 11y$$

$$\frac{\partial f}{\partial x} = 4x + y - 16 + 0 - 0, \qquad \frac{\partial f}{\partial y} = 0 + x - 0 + 2y - 11$$

Putting both these slopes to zero:

$$4x + y - 16 = 0$$
$$x + 2y - 11 = 0$$

Thus we have two simultaneous equations. Solving these

$$4x + y = 16$$
$$x + 2y = 11$$

gives $x = 3$ and $y = 4$. Hence the function $f(x, y)$ has a stationary point at $(3, 4)$.

C2 Maximum, minimum and saddle points

With the one independent variable case, $y = f(x)$, the stationary point is either a maximum, a minimum or a point of inflexion. We have the same concepts for functions of more than one variable, that is, a stationary point is either a maximum, a minimum or a saddle point.

Let $z = f(x, y)$. Figure 8 shows a maximum, a minimum and a saddle point. Note that a saddle point is where a curve in one direction is a maximum and the curve in the other direction is a minimum (Fig. 9, over the page).

Fig. 8

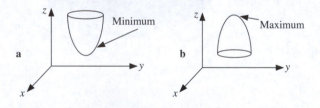

The 3-dimensional graph part **c** of Fig. 8 shows a saddle point at $(0, 0)$.

We can examine the graphs of these stationary points by using a computer algebra system, as the next example shows.

Fig. 9

Saddle point

Example 15

By using a computer algebra system (MAPLE) or a graphical calculator, plot the surfaces:

a $f(x, y) = y^2 - x^2$ for x between -5 and 5 and y between -10 and 10.
b $g(x, y) = e^{-x^2-y^2}$ for x between -3 and 3 and y between -3 and 3.
c $F(x,y) = xe^{-x^2-y^2}$ for x between 0 and 2 and y between -2 and 2.

Solution

a The MAPLE commands are

```
> f:=(x,y)->y^2-x^2;
```
$$f: = (x, y) \rightarrow y^2 - x^2$$

```
> plot3d(f,-5..5,-10..10);
```

Fig. 10a f

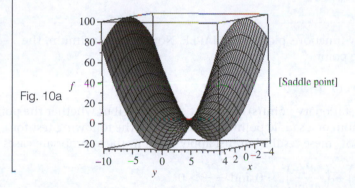

[Saddle point]

b

```
> g:=(x,y)->exp(-(x^2+y^2));
```
$$g: = (x, y) \rightarrow e^{(-x^2-y^2)}$$

```
> plot3d(g,-3..3,-3..3);
```

Fig. 10b g

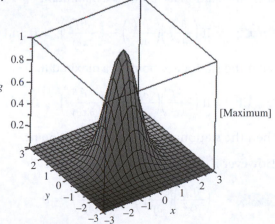

[Maximum]

Example 15 *continued*

c

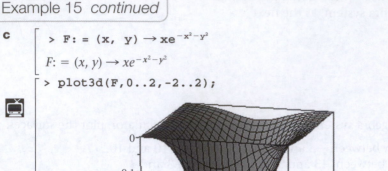

```
> F: = (x, y) → xe^(-x²-y²)
```

$$F: = (x, y) \rightarrow xe^{-x^2-y^2}$$

```
> plot3d(F,0..2,-2..2);
```

Fig. 10c F [Minimum]

Figures 10a, b and c show functions plotted in MAPLE. Notice the maximum, the minimum and the saddle point.

What are the limitations of 15.10 ?

 15.10 tells us where the stationary point(s) occur but does **not** tell us whether the points are a maximum, a minimum or a saddle point. We have to use the following test for $f = f(x, y)$. We will not show these results but just apply them in the appropriate cases.

 15.11   If $\left(\dfrac{\partial^2 f}{\partial x^2}\right)\left(\dfrac{\partial^2 f}{\partial y^2}\right) - \left(\dfrac{\partial^2 f}{\partial x \partial y}\right)^2 > 0$ and $\dfrac{\partial^2 f}{\partial x^2} > 0$

then the stationary point is a **minimum**.

 15.12   If $\left(\dfrac{\partial^2 f}{\partial x^2}\right)\left(\dfrac{\partial^2 f}{\partial y^2}\right) - \left(\dfrac{\partial^2 f}{\partial x \partial y}\right)^2 > 0$ and $\dfrac{\partial^2 f}{\partial x^2} < 0$

then the stationary point is a **maximum**.

 15.13   If $\left(\dfrac{\partial^2 f}{\partial x^2}\right)\left(\dfrac{\partial^2 f}{\partial y^2}\right) - \left(\dfrac{\partial^2 f}{\partial x \partial y}\right)^2 < 0$

then the stationary point is a **saddle point**.

However if

 15.14 $\left(\dfrac{\partial^2 f}{\partial x^2}\right)\left(\dfrac{\partial^2 f}{\partial y^2}\right) - \left(\dfrac{\partial^2 f}{\partial x \partial y}\right)^2 = 0$

at a stationary point, then no conclusion can be drawn.

In $\boxed{15.11}$ and $\boxed{15.12}$ we can replace $\dfrac{\partial^2 f}{\partial x^2} > 0$ with $\dfrac{\partial^2 f}{\partial y^2} > 0$ and $\dfrac{\partial^2 f}{\partial x^2} < 0$ with $\dfrac{\partial^2 f}{\partial y^2} < 0$

respectively.

$\boxed{15.11}$ to $\boxed{15.14}$ are called the second derivative tests.

Go through the next example very carefully. The difficulty can be in the algebra of solving simultaneous equations because these equations are **not** linear.

Example 16

Find where the stationary points of

$$f(x, y) = x^3 + y^3 - 3xy$$

occur and state their nature.

Solution

Let $f = f(x, y)$. First we find where the stationary points occur:

$$\frac{\partial f}{\partial x} = 3x^2 - 3y \text{ and } \frac{\partial f}{\partial y} = 3y^2 - 3x \qquad \text{[Partially differentiating]}$$

By equating these to zero we have the simultaneous equations

$\boxed{\dagger}$ $\qquad 3x^2 - 3y = 0$

$\boxed{\dagger\dagger}$ $\qquad 3y^2 - 3x = 0$

From $\boxed{\dagger}$ we have

$$3x^2 = 3y$$
$$x^2 = y$$

Substituting $y = x^2$ into equation $\boxed{\dagger\dagger}$ gives

$$3x^4 - 3x = 0 \qquad [\text{ Because } y^2 = (x^2)^2 = x^4]$$
$$3x(x^3 - 1) = 0 \qquad [\text{Taking out } 3x]$$
$$x = 0, \quad x = 1 \qquad [\text{Solving}]$$

If $x = 0$ then $y = 0$ and if $x = 1$ then $y = 1$ because $y = x^2$. Stationary points of f are at $(0, 0)$ and $(1, 1)$.

At $(0, 0)$ and at $(1, 1)$, do we have a maximum, a minimum or a saddle point?

We need to use $\boxed{15.11}$ to $\boxed{15.14}$. We have $f = x^3 + y^3 - 3xy$:

$$\frac{\partial f}{\partial x} = 3x^2 - 3y \qquad [\text{From above}]$$

$$\frac{\partial^2 f}{\partial x^2} = 6x - 0 = 6x \qquad [\text{2nd derivative}]$$

> **Example 16** *continued*
>
> $$\frac{\partial f}{\partial y} = 3y^2 - 3x \qquad \text{[From above]}$$
>
> $$\frac{\partial^2 f}{\partial y^2} = 6y - 0 = 6y \qquad \text{[2nd derivative]}$$
>
> $$\frac{\partial^2 f}{\partial x \partial y} = \frac{\partial}{\partial x}\left(\frac{\partial f}{\partial y}\right) \qquad \text{[Mixed derivative]}$$
>
> $$= \frac{\partial}{\partial x}\left(3y^2 - 3x\right) = 0 - 3 = -3$$
>
> Substituting all these into the second derivative formula 15.11 yields
>
> $$\left(\frac{\partial^2 f}{\partial x^2}\right)\left(\frac{\partial^2 f}{\partial y^2}\right) - \left(\frac{\partial^2 f}{\partial x \partial y}\right)^2 = (6x)(6y) - (-3)^2 \qquad \text{[Substituting]}$$
>
> $$= 36xy - 9$$
>
> At $(0, 0)$, which means $x = 0$ and $y = 0$, substituting these values gives
>
> $$36xy - 9 = 0 - 9 = -9 < 0 \qquad \text{[Negative]}$$
>
> By 15.13 , at $(0, 0)$ we have a saddle point.
>
> At $(1, 1)$, $x = 1$ and $y = 1$, we have
>
> $$36xy - 9 = (36 \times 1 \times 1) - 9 = 36 - 9 > 0$$
>
> $$\frac{\partial^2 f}{\partial x^2} = 6x = 6 \times 1 > 0 \qquad \text{[Positive]}$$
>
> By 15.11 , at $(1, 1)$ we have a minimum.

Make sure you find **all** the solutions to the simultaneous equations. A common mistake is to determine **one** solution and just test that. If you have non-linear simultaneous equations then there may be **more than** one solution.

C3 **Optimization problems**

We can tackle practical engineering problems involving the calculation of maxima and minima by finding values of variables which maximize or minimize a particular problem.

> **Example 17**
>
> Consider a closed rectangular box with dimensions x, y and z. If the fixed volume of the box is 1000 cm^3, find the dimensions so that it has a minimum total surface area. What conclusions can you draw from your result?

Example 17 *continued*

Solution

Figure 11 shows the box.

The volume $xyz = 1000$. By transposing, we can write   Fig. 11
this in terms of x and y only. We write the expressions
in terms of two independent variables because the
established formulae only apply to functions of two
variables. Dividing $xyz = 1000$ by xy gives

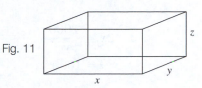

> ★ $z = \dfrac{1000}{xy}$

We first obtain the values of x and y which give minimum surface area and then we
substitute these into   ★   to find z.

The surface area, A, is given by

$$A = 2xy + 2yz + 2xz$$

We need to replace the z term so we can obtain a function in terms of x and y only.
Substituting $z = \dfrac{1000}{xy}$ into A gives

$$A = 2xy + 2yz + 2xz$$
$$= 2xy + 2y\left(\dfrac{1000}{xy}\right) + 2x\left(\dfrac{1000}{xy}\right)$$
$$= 2xy + \dfrac{2000}{x} + \dfrac{2000}{y} \qquad \text{[Cancelling]}$$
$$A = 2xy + 2000x^{-1} + 2000y^{-1}$$

At what values of x and y does A have a minimum surface area?

First we find where the stationary points occur by differentiating:

$$\dfrac{\partial A}{\partial x} = 2y - 2000x^{-2} + 0$$
$$\dfrac{\partial A}{\partial y} = 2x + 0 - 2000y^{-2}$$

By equating these to zero we have the simultaneous equations

$$2y - 2000x^{-2} = 0$$
$$2x - 2000y^{-2} = 0$$

From the first equation we have

$$2y = \dfrac{2000}{x^2}$$
$$x^2 y = 1000$$

Example 17 *continued*

From the second equation we have

$$2x = \frac{2000}{y^2}$$
$$xy^2 = 1000$$

Combining both equations yields

$$x^2 y = xy^2 \; (=1000)$$

Dividing through by xy gives

$$x = y$$

Substituting $x = y$ into $xy^2 = 1000$ we have

$$xx^2 = x^3 = 1000$$

Taking the cube root gives $x = 10$.

Since $x = y$ therefore $y = 10$.

Do these values $x = 10$ and $y = 10$ give the minimum surface area?

We need to check this by using tests 15.11 to 15.14 :

$$\frac{\partial A}{\partial x} = 2y - 2000x^{-2} \qquad [\text{From above}]$$

$$\frac{\partial^2 A}{\partial x^2} = 0 - 2(-2000)x^{-3} = 4000x^{-3} \qquad [\text{2nd derivative}]$$

Similarly we have

$$\frac{\partial A}{\partial y} = 2x - 2000y^{-2} \qquad [\text{From above}]$$

$$\frac{\partial^2 A}{\partial y^2} = 0 - 2(-2000)y^{-3} = 4000y^{-3} \qquad [\text{2nd derivative}]$$

$$\frac{\partial^2 A}{\partial x \partial y} = \frac{\partial}{\partial x}\left(\frac{\partial A}{\partial y}\right) \qquad [\text{Mixed derivative}]$$

$$= \frac{\partial}{\partial x}\left(2x - 2000y^{-2}\right) = 2 - 0 = 2$$

Example 17 *continued*

Putting these into formula 15.11 gives

$$\left(\frac{\partial^2 A}{\partial x^2}\right)\left(\frac{\partial^2 A}{\partial y^2}\right) - \left(\frac{\partial^2 A}{\partial x \partial y}\right)^2 = (4000x^{-3})(4000y^{-3}) - 2^2$$

Substituting $x = 10$ and $y = 10$ gives

$$(4000 \times 10^{-3})(4000 \times 10^{-3}) - 4 = 12 > 0 \qquad \text{[Positive]}$$

Hence the point $(10, 10)$ gives a maximum or a minimum.

We need to check for a minimum:

$$\frac{\partial^2 A}{\partial x^2} = 4000x^{-3} = 4000 \times 10^{-3} = 4 > 0 \qquad \text{[Positive]}$$

Hence by 15.11 , $x = 10$, $y = 10$ gives minimum surface area. To find z, use $z = \dfrac{1000}{xy}$:

$$z = \frac{1000}{xy} = \frac{1000}{10 \times 10} = 10$$

Conclusion – the minimum surface area of a rectangular box containing a fixed volume of 1000 cm^3 is a cube of 10 cm × 10 cm × 10 cm.

Note that even if you only have **one** solution, $x = 10$ and $y = 10$, you still need to confirm that this gives a minimum by using the second derivative tests.

SUMMARY

Stationary points of a function $f(x, y)$ occur when

15.10 $\dfrac{\partial f}{\partial x} = 0$ and $\dfrac{\partial f}{\partial y} = 0$

There are three types of stationary points – maximum, minimum and saddle point. We can decide what type of stationary point we have by using the second derivative tests:

15.11   If $\left(\dfrac{\partial^2 f}{\partial x^2}\right)\left(\dfrac{\partial^2 f}{\partial y^2}\right) - \left(\dfrac{\partial^2 f}{\partial x \partial y}\right)^2 > 0$ and $\dfrac{\partial^2 f}{\partial x^2} > 0$

then the stationary point is a **minimum**.

15.12   If $\left(\dfrac{\partial^2 f}{\partial x^2}\right)\left(\dfrac{\partial^2 f}{\partial y^2}\right) - \left(\dfrac{\partial^2 f}{\partial x \partial y}\right)^2 > 0$ and $\dfrac{\partial^2 f}{\partial x^2} < 0$

then the stationary point is a **maximum**.

15.13   If $\left(\dfrac{\partial^2 f}{\partial x^2}\right)\left(\dfrac{\partial^2 f}{\partial y^2}\right) - \left(\dfrac{\partial^2 f}{\partial x \partial y}\right)^2 < 0$

then the stationary point is a **saddle point**.

Exercise **15(c)**

Solutions at end of book. Complete solutions available at
www.palgrave.com/engineering/singh

1 Find where the stationary points of the following functions occur:

   a $f(x, y) = xy + \dfrac{1}{2}y^2 - 7y - 4x$

   b $f(x, y) = x^2 + y^2 + 4xy - 5x - 4y$

   c $f(x, y) = x^3 + y^2 + xy + 22y$

2 Find where the stationary points of the following functions occur and determine whether they are maxima, minima or saddle points:

   a $f(x, y) = x^2 + y^2 + 6xy - 10x - 14y$

   b $f(x, y) = x^3 + y^2 - x + y$

   c $f(x, y) = x^2 + y^3 + 4xy - 11x - 18y$

3 Find the dimensions of a rectangular tank, without a top, with minimum surface area and a fixed volume of 0.9 m^3. What can you conclude about your results?

4 An open rectangular tank has volume V. Find the dimensions of the tank for minimum surface area A.

5 An open-top rectangular storage tank has a volume of 9 m^3.

   i If the base of the tank has dimensions x and y then show that the total surface area A is given by

$$A = xy + \frac{18}{x} + \frac{18}{y}$$

   ii Find the dimensions of the tank for the surface area, A, to be a minimum.

*6 **i** The temperature, T, in a circular plate $x^2 + y^2 < 1$ is given by

$$T = x^2 + y^2 - x - y + 100$$

   Find the minimum temperature.

   ii If T is the same expression as part **i** then find where the stationary points occur if

$$x^2 + y^2 = 1$$

*7 A length of channel is made from a rectangular sheet of metal of length 9 m. The metal is bent as shown in Fig. 12. Find the values of x and θ for which the channel's capacity is a maximum.

Fig. 12

Examination questions **15**

Solutions at end of book. Complete solutions available at
www.palgrave.com/engineering/singh

1 An open-top rectangular cardboard box is to have a volume of 3 m^3. If A is the total area of cardboard and x and y are the length and breadth of the base show that

$$A = xy + \frac{6}{x} + \frac{6}{y}$$

Find the dimensions of the box which minimize the amount of cardboard required, giving each measurement to three decimal places. [NB: You must confirm that you have a minimum.]

University of Surrey, UK, 2005

2 Find and classify all the critical points of the function

$$f(x, y) = y^3 - x^3 + 3xy + 1$$

University of Manchester, UK, 2009

Examination questions **15 continued** Solutions at end of book. Complete solutions available at www.palgrave.com/engineering/singh

3 The cost, C, of constructing a box with sides of length x, y and $\dfrac{12}{xy}$ is given by

$$C = 4xy + \frac{72}{y} + \frac{48}{x} \quad x,\, y \neq 0$$

Find the values of x, y which minimize the cost.

Loughborough University, UK, 2008

4 The period T of small oscillations of a weight of mass m suspended on a (weightless) spring constant k is given by $T = 2\pi\sqrt{\dfrac{m}{k}}$ so that $k = 4\pi^2\dfrac{m}{T^2}$. This formula is used to measure the spring constant k. The mass m can only be measured to an accuracy of $\pm 1.5\%$ and the period T to an accuracy of $\pm 2\%$. Find the approximate maximum percentage error in the resulting measurement of k.

University of Aberdeen, UK, 2005

5 a A variable u is given as a function two variables x, y by

$$u = 3x^2 + 6x^2y - 4xy^2 - y^3$$

Find the partial derivatives $\dfrac{\partial u}{\partial x}$, $\dfrac{\partial u}{\partial y}$, $\dfrac{\partial^2 u}{\partial x^2}$, $\dfrac{\partial^2 u}{\partial y^2}$, $\dfrac{\partial^2 u}{\partial x \partial y}$.

b Show that the function $u = \sin(2x)\sin(4y)e^{-5t}$ of the three variables x, y and t is a solution of the partial differential equation:

$$\frac{\partial^2 u}{\partial x^2} + \frac{\partial^2 u}{\partial y^2} = c^2\,\frac{\partial u}{\partial t}$$

for some positive constant c, whose value should be calculated.

University of Aberdeen, UK, 2003

6 a Obtain the partial derivatives $\dfrac{\partial z}{\partial x}$, $\dfrac{\partial z}{\partial y}$, $\dfrac{\partial^2 z}{\partial x^2}$, $\dfrac{\partial^2 z}{\partial y \partial x}$ if

  i $z = x^3 + 5x^2y + 2y^3$

  ii $z = e^x \cos y$

b Find the value of p so that

$$z = e^{px}(x \cos y - y \sin y)$$

satisfies the equation $\dfrac{\partial^2 z}{\partial x^2} + \dfrac{\partial^2 z}{\partial y^2} = 0$.

c Find all the stationary values of

$$f(x, y) = 2x^3 + 6xy^2 - 3y^3 - 150x$$

and determine their nature.

Loughborough University, UK, 2006

Miscellaneous exercise **15** Solutions at end of book. Complete solutions available at www.palgrave.com/engineering/singh

1 [materials]   A beam of length L has a bending moment M given by

$$M = Kx + Px\left(1 - \frac{x}{L}\right)$$

where K is a constant, x is the distance along the beam from one end and P is the load. Determine $\dfrac{\partial M}{\partial P}$ and $\dfrac{\partial M}{\partial x}$.

2 [materials]   The bending moment M at a distance a of a beam of length L is given by

$$M = \frac{Kx}{L} + Pa\left(1 - \frac{x}{L}\right) - P(a - x)$$

(P is load and x is distance). Determine $\dfrac{\partial M}{\partial P}$ and $\dfrac{\partial M}{\partial x}$.

Miscellaneous exercise **15 continued**   Solutions at end of book. Complete solutions available at www.palgrave.com/engineering/singh

*3 [materials]   Given the stress function, $\Omega(x, y)$, as

$$\Omega(x, y) = Ax^3 y^2 + Bxy^2 \quad (A \neq 0, B \neq 0)$$

find

$$\frac{\partial^4\Omega}{\partial x^4} + 2\frac{\partial^4\Omega}{\partial x^2\partial y^2} + \frac{\partial^4\Omega}{\partial y^4} \; (= \nabla^4\Omega)$$

If $\nabla^4\Omega = 0$ then $\Omega(x, y)$ is called the Airy stress function. Test whether the above $\Omega(x, y)$ is an Airy stress function for $x \neq 0$.

4 [thermodynamics]   The pressure, P, of a gas is given by

$$P = \frac{RT}{V}$$

where T is the temperature, V is the volume and R is the gas constant. Find the approximate percentage change in P, if T is decreased by 1.3% and V is increased by 0.2%.

5 [electrical principles]   The resistance, R, of a wire of length L and diameter d is given by

$$R = \frac{cL}{d^2} \quad (c \text{ is a constant})$$

If the error in measuring L and d are 3% and 2% respectively, find the largest possible error in R.

6 [materials]   The axial stress, σ, is given by

$$\sigma = \frac{kr^2}{l^2} \quad (k = \text{constant})$$

where r is the radius of gyration and l is the length. Find the approximate percentage change in σ if r is increased by 1.3% and l is decreased by 0.6%.

*7 [fluid mechanics]   The fluid velocity, V, around a sphere of radius r is given by

$$V = V_0\left(1 + \frac{r^3}{x^3}\right)$$

The streamline acceleration, a, is defined by

$$a = V\frac{\partial V}{\partial x}$$

Show that

$$a = -3V_0^2\left(1 + \frac{r^3}{x^3}\right)\frac{r^3}{x^4}$$

At what value of x is $a = 0$?

*8 [structures]   A beam of length L (fixed at one end) and shear load S has load P given by

$$P = \frac{SA}{\alpha}\left[\frac{\sinh(\mu x)}{\mu\cosh(\mu L)} - x\right]$$

where α and μ are constants, A is the cross-sectional area and x is the length along the beam. The shear flow, q, is defined by

$$q = -\frac{1}{2}\frac{\partial P}{\partial x}$$

Show that

$$q = \frac{SA}{2\alpha}\left[1 - \frac{\cosh(\mu x)}{\cosh(\mu L)}\right]$$

*9 [structures]   A beam of length L has load P given by

$$P = A\cosh(\mu x) + B\sinh(\mu x) + \alpha$$

where A, B, μ and α are constants and x is the distance along the beam. Given that at $x = 0$, $P = 0$ and at $x = L$, $\dfrac{\partial P}{\partial x} = 0$, show that

$$P = \alpha\left[\tanh(\mu L)\sinh(\mu x) - \cosh(\mu x) + 1\right]$$

Miscellaneous exercise 15 continued   Solutions at end of book. Complete solutions available at www.palgrave.com/engineering/singh

10 [*materials*]   Radial, σ_r, tangential σ_θ and shearing $\tau_{r\theta}$ stresses are defined by

$$\sigma_r = \frac{1}{r}\left(\frac{\partial \Omega}{\partial r}\right) + \frac{1}{r^2}\left(\frac{\partial^2 \Omega}{\partial \theta^2}\right)$$

$$\sigma_\theta = \frac{\partial^2 \Omega}{\partial r^2}$$

$$\tau_{r\theta} = \frac{1}{r^2}\left(\frac{\partial \Omega}{\partial \theta}\right) - \frac{1}{r}\left(\frac{\partial^2 \Omega}{\partial r \partial \theta}\right)$$

For $\Omega = r^2\left[A\cos(2\theta) + B\sin(2\theta)\right]$

find σ_r, σ_θ and $\tau_{r\theta}$.

11 [*heat transfer*]   The one-dimensional heat equation is given by

$$\frac{\partial u}{\partial t} = k\frac{\partial^2 u}{\partial x^2}$$

where k is a constant, $u = u(x, t)$ is the temperature at time t and distance x of an insulated rod. Show that the following functions satisfy the one-dimensional heat equation:

a $u(x, t) = e^{-kt}\sin(x)$

b $u(x, t) = e^{-kt}\cos(x)$

c $u(x, t) = e^{-kt}\left[\cos(x) + \sin(x)\right]$

*d $u(x, t) = e^{-\eta^2 kt}\left[\cos(\eta x) + \sin(\eta x)\right]$
    where η is a constant.

*12 Figure 13 shows an open-top triangular container with a volume of 1 m³. Find the values of x, l and θ which give the minimum surface area A.

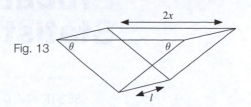
Fig. 13

[*Hint*: First find x, l and θ, which gives the stationary values of A, and then use mathematical software to confirm that these values do indeed give a minimum.]

You may use a computer algebra system for the remaining questions.

13 [*thermodynamics*]   The Dieterici equation is given by

$$P = \frac{RT}{V - b}e^{-a/RVT}$$

A critical point occurs at $\frac{\partial P}{\partial V} = 0$ and $\frac{\partial^2 P}{\partial V^2} = 0$.

Find an expression for a and b in terms of R, V and T at a critical point.

14 [*thermodynamics*]   The Redlich–Kwong equation is given by

$$P = \frac{RT}{V - b} - \frac{a}{V(V + b)T^{0.5}}$$

At a critical point find expressions for a and b in terms of R, V and T. (Critical point is defined in Question **13**.)

CHAPTER 16
Probability and Statistics

SECTION A **Data representation**

By the end of this section you will be able to:
- ▶ distinguish between continuous and discrete data
- ▶ construct frequency distributions
- ▶ draw a histogram and bar chart
- ▶ plot a frequency polygon

A1 **Discrete and continuous data**

The following are examples of discrete data:

- ▶ Number of people in a room
- ▶ Number of rejects on an assembly line
- ▶ Shoe size of children.

? **What do you think is the definition of 'discrete data'?**

Data which can only take **certain** values. The number of people in a room can only be 1, 2, 3, ... and not 1.23, 1.57, 10.11, ...

The following are examples of continuous data:

- ▶ Weights of people
- ▶ Output voltage of an analogue system
- ▶ Loads on a beam.

? **What do you think is the definition of 'continuous data'?**

Data which can take **any** values between two end points.

The weights of people can be 60.28 kg, 70.1 kg, ...

A2 **Frequency distribution**

? **What does the term 'frequency' mean?**

It is the number of times a particular value occurs in some data. The combination of particular values and their frequency is called a frequency distribution.

One way of representing the distribution of data is by a frequency table. This is a table that summarizes the data into some order.

Example 1

The number of rejects, in the last 30 days, from an assembly line has yielded the following data:

Example 1 *continued*

36	37	49	30	36	35
40	42	44	37	33	42
37	41	44	42	30	36
37	37	36	30	44	31
30	42	41	44	39	42

Construct a frequency distribution table.

Solution

Remember that frequency of a value is the number of times it occurs in the data. For example, there are 30 rejects on 4 days. We say the frequency of 30 is 4. We can summarize the above data as detailed in Table 1.

Number of rejects	Frequency f
30	4
31	1
33	1
35	1
36	4
37	5
39	1
40	1
41	2
42	5
44	4
49	1

TABLE 1

The representation of data is a lot clearer in the table.

? **We can check that all the data has been placed in the table. How?**

The sum of the frequencies should add up to 30 because there are 30 data values, that is $\sum f = 30$.   [Remember $\sum$ means 'sum of'.]. This is **no** guarantee that our frequency distribution is correct but it is a good guide.

? **This table is the frequency distribution for what sort of data?**

Discrete data (number of rejects)

We can use a similar idea to form a frequency distribution for continuous data. One way of representing continuous data is to group it into particular 'classes' or 'intervals', as the next example shows. This is particularly useful for a large amount of data.

Example 2a

The diameters, in mm, of 20 pipes are as follows:

40.6	40.7	40.9	41.0	41.1
41.4	41.5	41.7	41.2	41.2
41.9	41.3	41.4	41.6	41.8
41.6	41.2	40.5	40.5	41.9

Form a frequency distribution table, by grouping the data into five classes.

Solution

How do we form a frequency distribution for this data? We can group the data into classes, but classes of what size?

That depends on the data. The smallest value is 40.5 and the largest value is 41.9. If we use classes of size 0.3, then we will get five classes. Let's form the frequency distribution table for classes of size 0.3 (Table 2).

TABLE 2

Diameter d (mm)	Frequency
$40.45 \leq d < 40.75$	4
$40.75 \leq d < 41.05$	2
$41.05 \leq d < 41.35$	5
$41.35 \leq d < 41.65$	5
$41.65 \leq d < 41.95$	4

In this example 40.45, 40.75, 41.05, ... are called **class boundaries**.

These boundaries are normally chosen to lie between the observed values.

The **class width** is the difference between the upper class boundary and the lower class boundary.

Class width = (Upper class boundary) − (Lower class boundary)

For the above example:

Class width = 40.75 − 40.45
= 0.3

If we use classes of size 0.2, then we will need eight classes.

How many classes do we need if we use classes of size 0.1?

15 classes

This is an example of a frequency distribution with equal class width. A frequency distribution may have **unequal** class widths.

A3 **Histogram**

A histogram is a graphical representation of a frequency distribution.

Example 2b

Considering the data of **Example 2a**, draw a histogram for this data.

Solution

The frequency is plotted along the vertical axis and the grouped diameter of pipes along the horizontal axis.

Figure 1 shows a histogram of the data contained in Table 2.

The symbol, ℕ, used in Fig. 1 means that there is no data before the specified value, in this case 40.45.

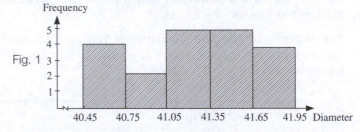

This is one of the simplest histograms to draw because it has equal class widths and so the height represents the frequency.

When either axis does **not** start at zero (in this case the horizontal axis starts at 40.45), it is normally abbreviated by omitting a section of the scale, indicated by ℕ.

In a histogram the **area** of the rectangle is proportional to the **frequency**.

Only in the case of equal class width do we have the height of each rectangle representing the frequency.

For histograms with unequal class widths we need to be careful, as Example 3a, below, shows.

Example 3a

The resistances of 100 resistors are given in Table 3a.

TABLE 3a	Resistance R (kΩ)	Frequency
	$1.1 \leq R < 1.2$	9
	$1.2 \leq R < 1.4$	20
	$1.4 \leq R < 1.5$	8
	$1.5 \leq R < 1.6$	12
	$1.6 \leq R < 1.8$	18
	$1.8 \leq R < 2.0$	12
	$2.0 \leq R < 2.3$	21

Draw a histogram for this data.

Example 3a *continued*

Solution

Remember that the frequency is proportional to the area. We need to choose a standard class width. By looking down the left-hand column of the table, we find the class widths are of sizes 0.1, 0.2, 0.1, 0.1, 0.2, 0.2 and 0.3. **Which class width would you choose to be standard?**

It really doesn't make much difference, but the most suitable seems to be 0.1 because there are three intervals with this width and it keeps the arithmetic easy, that is $0.1 \times 2 = 0.2$, $0.1 \times 3 = 0.3$.

If we choose our standard width = 0.1 then the second interval is **twice** the standard width and so we **halve** the frequency height. Similarly for the last interval, we take 1/3 of the frequency height (Table 3b). This new figure is known as the standard frequency.

TABLE 3b

Resistance $R(\mathrm{k\Omega})$	Class width (Standard width, SW)	Frequency	Standard frequency
$1.1 \leq R < 1.2$	0.1 (SW)	9	9
$1.2 \leq R < 1.4$	0.2 (2 × SW)	20	20/2 = 10
$1.4 \leq R < 1.5$	0.1 (SW)	8	8
$1.5 \leq R < 1.6$	0.1 (SW)	12	12
$1.6 \leq R < 1.8$	0.2 (2 × SW)	18	18/2 = 9
$1.8 \leq R < 2.0$	0.2 (2 × SW)	12	12/2 = 6
$2.0 \leq R < 2.3$	0.3 (3 × SW)	21	21/3 = 7

Plotting standard frequency against resistance values of Table 3b gives the histogram of Fig. 2a.

Again note that the data does not start at zero.

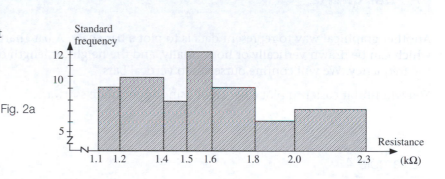

Fig. 2a

For unequal class widths the vertical axis gives the standard frequency. The frequency is evaluated by the area. Consider **Example 3a**: to evaluate the frequency from the histogram of the last interval, we multiply the height by 3. **Why?**

Because it is three times the standard width. Hence $3 \times 7 = 21$.

A4 **Frequency polygons**

Another graphical representation of a frequency distribution is a frequency polygon. There are two ways of constructing a frequency polygon:

1 Draw a histogram and join the midpoints of the tops of the rectangles.

2 Plot the standard frequency on the vertical axis against the midpoint of the interval.

Example 3b

Plot the frequency polygon for the data of **Example 3a**.

Solution

Which method should we use?
Method 1, because
we have already
drawn a histogram
for the data.

Figure 2b shows
the frequency
polygon of resistance
values.

Fig. 2b

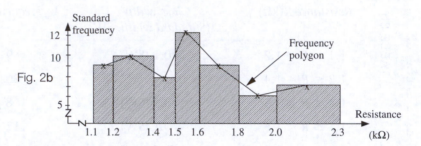

A5 **Bar charts**

Another graphical way to represent data is to plot a bar chart. A bar chart consists of bars which can be drawn vertically or horizontally, and the height or length of these bars gives the frequency. We will confine ourselves to vertical bars.

You will find it easier to plot a bar chart using appropriate software.

Example 4

Table 4 shows the number of new registrations with the Engineering Council at the end of each year. Draw a vertical bar chart to represent

a the number of CEng registrations against the year of entry

b the number of CEng, IEng and EngTech registrations against the year of entry.

Example 4 *continued*

TABLE 4	*Number of new registrations with the Engineering Council*			
	Year	*CEng*	*IEng*	*EngTech*
	2000	5096	1708	683
	2001	4932	1362	592
	2002	5180	789	574
	2003	4504	599	466
	2004	4518	484	758
	2005	5906	532	1580
	2006	5563	498	944
	2007	3489	586	839
	2008	3439	498	1343
	2009	3750	547	1314

Solution

a The number of CEng registrations is given in the second column. We plot a
series of vertical bars of the same width with the year plotted horizontally and the
number of CEng registrations (numbers in the second column of Table 4) vertically.
This is illustrated in Fig. 3a:

The number of new registrations for CEng

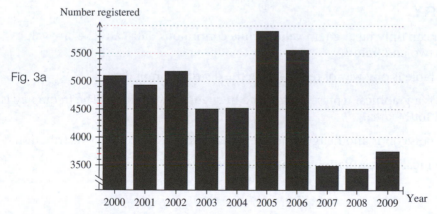

Fig. 3a

Note that the vertical axis starts at just above 3000 because all the entries in the CEng
column of Table 4 are above 3000. We could start at zero, but it would be more difficult
to visualize the difference between the various years.

Example 4 *continued*

b We can also plot three bars for each year showing each of the categories CEng, IEng and EngTech as illustrated in Fig. 3b:

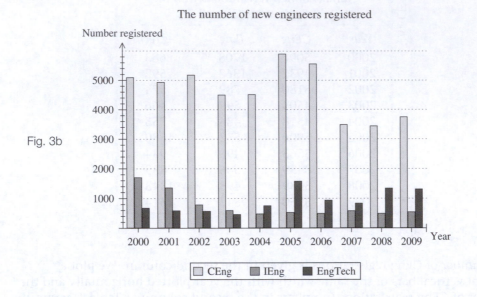

The number of new engineers registered

Fig. 3b

Note that this time the vertical axis starts at zero to enable the smaller quantities of IEng and EngTech registrations to be shown.

SUMMARY

Discrete data can only take certain values while continuous data can take any value between the two end points.

A frequency table is one way of representing the distribution of data.

A histogram is a graphical representation of a frequency distribution. The frequency is proportional to the area.

A frequency polygon is another graphical representation of a frequency distribution.

Another graphical representation of data is a bar chart.

Exercise 16(a)

Solutions at end of book. Complete solutions available at www.palgrave.com/engineering/singh

1 State whether the following are discrete or continuous data:
  a The weights of people
  b Marks in an examination
  c The number of motors in a batch
  d The lifetimes of light bulbs
  e The resistance value of a variable resistor.

Solutions at end of book. Complete solutions available at www.palgrave.com/engineering/singh

Exercise **16(a) continued**

2 The temperatures, to the nearest degree Celsius, for the last 30 days are as follows:

22	23	23	16	17	20
21	24	18	17	19	21
23	25	26	21	20	20
16	15	17	22	24	23
23	19	20	22	21	20

Construct a frequency distribution.

3 The heights, in m, of 40 students are shown below:

1.68 1.67 1.53 1.70 1.69 1.71 1.71 1.80
1.81 1.85 1.76 1.66 1.91 1.95 1.87 1.80
1.82 1.76 1.84 1.55 1.61 1.85 1.93 1.88
1.95 1.57 1.97 1.56 1.99 1.83 1.74 1.83
1.86 1.88 1.93 1.64 1.89 1.90 1.72 1.95

Construct a frequency distribution with an equal class width of 0.1.

4 Draw a histogram for question **3** with the same class width.

5 Draw a frequency polygon for question **3**.

6 The table below shows the time taken, in ms, for 105 op-amps to become fully operational:

Time taken t (ms)	Frequency
$10 \leq t < 20$	10
$20 \leq t < 30$	14
$30 \leq t < 40$	18
$40 \leq t < 45$	16
$45 \leq t < 50$	12
$50 \leq t < 55$	8
$55 \leq t < 70$	15
$70 \leq t < 100$	12

Draw a histogram to illustrate this data.

7 The following table shows the lifetime (in hours) of 88 light bulbs:

Lifetime L (hours)	Frequency
$150 \leq L < 200$	5
$200 \leq L < 250$	7
$250 \leq L < 300$	10
$300 \leq L < 350$	12
$350 \leq L < 400$	14
$400 \leq L < 450$	11
$450 \leq L < 500$	10
$500 \leq L < 550$	8
$550 \leq L < 600$	7
$600 \leq L < 650$	4

Draw a frequency polygon to illustrate this data.

8 The diameter of 100 piston rings are measured and recorded as follows:

Diameter d (mm)	Frequency
$86.00 \leq d < 86.10$	9
$86.10 \leq d < 86.15$	9
$86.15 \leq d < 86.25$	12
$86.25 \leq d < 86.40$	15
$86.40 \leq d < 86.45$	10
$86.45 \leq d < 86.50$	7
$86.50 \leq d < 86.70$	20
$86.70 \leq d < 87.00$	18

Draw a histogram and a frequency polygon for this data.

Exercise **16(a) continued**

Solutions at end of book. Complete solutions available at www.palgrave.com/engineering/singh

9 The table to the right shows the number of new registrations with the Engineering Council at the end of each year. Draw a vertical bar chart to represent the number of registrations of CEng, IEng and EngTech against each year on the same graph.

Comment on your graph and data.

Number of new registrations with the Engineering Council			
Year	CEng	IEng	EngTech
1984	3911	2391	1002
1985	5002	2774	1337
1986	5960	2682	1039
1987	6022	3066	1130
1988	5878	2281	1159
1989	4746	2335	1210
1990	9207	2559	1315
1991	5413	2634	1185
1992	5588	2128	1184
1993	6159	2050	1190
1994	5721	1556	1237
1995	5376	1433	1146
1996	5486	1579	1052
1997	5641	1595	903
1998	4792	1484	789
1999	5157	1562	916

SECTION B **Data summaries**

By the end of this section you will be able to:
► evaluate the mean
► understand what *standard deviation* means
► evaluate the standard deviation
► derive properties of mean and standard deviation
► evaluate the mean and standard deviation of data in a frequency distribution

B1 **Averages**

The sample mean, or average, of n numbers, $x_1, x_2, x_3, \ldots, x_n$, denoted by $\bar{x}$, is given by

16.1
$$\bar{x} = \frac{x_1 + x_2 + x_3 + \cdots + x_n}{n} = \frac{\sum\limits_{j=1}^{n} x_j}{n}$$

$$\left(= \frac{\text{sum of observations}}{\text{number of observations}} \right)$$

The notation $\sum\limits_{j=1}^{n} x_j$ means sum x_j from 1 to n, that is $x_1 + x_2 + x_3 + \cdots + x_n$.

Example 5

The lifetime (to the nearest hour) of 20 light bulbs is given by: 599, 601, 586, 604, 623, 624, 584, 613, 581, 600, 605, 592, 613, 617, 591, 599, 584, 616, 580 and 603.

a   **i** Without evaluating, estimate the mean from the data.
     ii Evaluate the mean lifetime of these light bulbs.

b   If only the ninth piece of data, 581, changes to 518, what will be the new mean?

Solution

a   **i** Since all the data is quite close to 600, a good estimate would be 600.

     ii Adding the data values and dividing by 20 gives

$$\text{Mean} = \frac{599 + 601 + 586 + \cdots + 603}{20} = 600.75$$

b   Remember that the mean is just an average.

With this change of data, is the new mean higher than 600.75 or lower?

Lower because we have entered a lower value, 518, rather than 581. The change can be evaluated by $\dfrac{581 - 518}{20} = 3.15$

The new mean = $600.75 - 3.15 = 597.6$

The easiest way to find the mean for the data of **Example 5** is to use a calculator. See the handbook of your calculator or the web for instructions.

Other averages to consider are the mode and the median. The mode is the most popular or most frequently occurring observation. The mode of

2, 2, 3, 4, 6, 6, 6, 7

is 6.

The median is the middle value of the data arranged in ascending (or descending) order. The median of

2, 5, 3, 6, 10, 4, 15

is evaluated by placing the data in order of magnitude

2, 3, 4, 5, 6, 10, 15

Hence the median is 5.

The median of the following eight pieces of data:

101, 137, 112, 67, 93, 12, 107, 152

is evaluated by first placing these data pieces into some order:

$$12, 67, 93, \underline{101,\ 107}, 112, 137, 152$$
$$\frac{101 + 107}{2} = 104$$

The mean of the two middle values, 101 and 107, is taken, thus

$$\text{median} = 104$$

? **Why use median as an average rather than mean?**

Example 6 illustrates why we sometimes use median as an average.

Example 6

Determine the daily mean time taken to get to work of the following data:

Name	Joe	Ben	Danny	Elly	Amy
Time taken to get to work (mins)	20	15	5	150	8

? The mean is $\bar{x} = \dfrac{20+15+5+150+8}{5} = 39.6$ minutes. Thus the mean time taken to get to work is approximately 40 minutes. This average is not particularly helpful. **Why?**

Because Elly's time of 150 minutes swamps the rest of the times. A better average would be the median which is the middle value of the ordered data 5, 8, 15, 20 and 150. Hence the median is 15 minutes.

Averages are often quoted in the media to refute or to highlight an injustice. For example, on average a woman earns less than a man. The connotations of this statement are that a woman gets paid less than a man for the same work. This is **not** correct because most of the high-paid work is taken up by men while a lot of part-time and low-paid employment is taken by women. Most of us have difficulty in interpreting statistical information, which should not be the case because most of statistics is common sense.

In an engineering experiment you often amass a lot of numbers, and you can use statistics to highlight important pieces of information from these, such as the mean or standard deviation (to be described below) of the data.

In this chapter we will however concentrate on the mean, as it is the most widely used average in statistics.

B2 **Standard deviation (s or SD)**

? The standard deviation is a 'measure of the spread' of data from the mean value. **How can we evaluate this?**

Consider the data 3, 4 and 5.

? **How can we evaluate the standard deviation for this data?**

Since we want to find out how far the data is away from the mean, we must first obtain the mean:

$$\text{mean } \bar{x} = \frac{3 + 4 + 5}{3} = 4$$

? **How can we evaluate how far each piece of data is from the mean?**

By calculating the difference (deviation from the mean):

$$3 - 4 = -1, \; 4 - 4 = 0 \text{ and } 5 - 4 = 1$$

To find a value for the spread we add these:

$$-1 + 0 + 1 = 0$$

By using this method we obtain a value for spread = 0.

? **What does this 0 signify?**

This signifies that **all** the data is centred on the mean and there is no deviation from the mean.

? As you can see this is **not** the case, since 3 and 5 are 1 away from the mean. **So how did we get an answer of zero?**

$$-1 \text{ and } 1 \text{ cancelled each other out in the addition}$$

? **How can we overcome this problem?**

We can square each difference from the mean, giving a positive or zero answer (squared deviation):

$$(-1)^2, \; 0^2 \text{ and } \; 1^2$$

So the spread can be measured by the sum of the squared deviations:

$$(-1)^2 + 0^2 + 1^2 = 2$$

Dividing this by the number of pieces of data gives us the 'measure of spread', called the **variance**:

$$\text{variance} = \frac{2}{3}$$

The **standard deviation** (s) is the square root of the variance. Hence

$$s = \sqrt{\frac{2}{3}}$$

Generally the variance of data $x_1, x_2, \ldots, x_n$ is given by

$$\boxed{16.2} \qquad \text{variance} = \frac{(x_1 - \bar{x})^2 + (x_2 - \bar{x})^2 + \cdots + (x_n - \bar{x})^2}{n}$$

$$\left[\text{variance} = \frac{\sum\limits_{j=1}^{n} (x_j - \bar{x})^2}{n} \right]$$

and the standard deviation, s, is given by the square root of the variance:

$$\boxed{16.3} \qquad s = \sqrt{\frac{(x_1 - \bar{x})^2 + (x_2 - \bar{x})^2 + \cdots + (x_n - \bar{x})^2}{n}}$$

$$\left[s = \sqrt{\frac{\sum\limits_{j=1}^{n} (x_j - \bar{x})^2}{n}} \right]$$

The standard deviation, $s = \sqrt{\text{variance}}$. A slightly easier formula to use in evaluating the variance is variance $= \dfrac{\sum x_j^2}{n} - (\bar{x})^2$.

By referring to formula $\boxed{16.2}$:

$x_j - \bar{x}$ is the deviation from the mean
$(x_j - \bar{x})^2$ is the squared deviation

The mean of the squared deviations gives us the **variance**. The square root of the variance is the **standard deviation**.

Remember that the standard deviation (SD) is a 'measure of spread' of data from the mean. **What does a small SD tell you about the data?**

That most of the data is close to the mean.

For a large SD, much of the data is far from the mean (Fig. 4). The standard deviation measures the spread of data from the mean.

Fig. 4

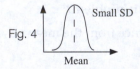

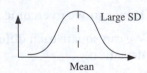

In other literature you might see the standard deviation and variance formulae with division by $n - 1$ rather than n, because the sample of data values has been taken from a population. In a nutshell, if we want to find the standard deviation of a given set of data then we divide by n, but if the data represents a sample from a population then we divide by $n - 1$. Throughout this book we will use the formula stated above: divide by n.

Example 7

The marks of an examination are given by the following data:

12, 27, 13, 21, 36, 56, 53, 55, 59, 83, 92, 75, 67, 80, 91, 99, 84 and 77.

Evaluate the standard deviation for this data.

> ### Example 7 continued

Solution

(?) **To use 16.3 , what do we need to evaluate first?**

$$\text{the mean } (\bar{x})$$

By 16.1 : $\bar{x} = \dfrac{12 + 27 + \cdots + 77}{18} = 60$

First find the variance, s^2, by 16.2 and then take the square root:

$$s^2 = \dfrac{(12 - 60)^2 + (27 - 60)^2 + \cdots + (77 - 60)^2}{18} = 742.444$$

Hence standard deviation, $s = \sqrt{742.444} = 27.25$ (2 d.p.). This is considered a large standard deviation, as you can observe that the exam marks are quite widespread from the mean of 60.

To find the standard deviation by hand is very tedious, as the above example shows. The easiest way to determine the standard deviation is to use your calculator.

> ### Example 8

Find the standard deviation of the following:

a 1, 2, 3, 4, 5 and 6   **b** 1001, 1002, 1003, 1004, 1005 and 1006
c 0.1, 0.2, 0.3, 0.4, 0.5 and 0.6   **d** 1.1, 1.2, 1.3, 1.4, 1.5 and 1.6

Solution

By calculator:

a $s = 1.71$ (2 d.p.)   **b** $s = 1.71$ (2 d.p.)   **c** $s = 0.17$ (2 d.p.)   **d** $s = 0.17$ (2 d.p.)

(?) **What do you notice about the results of Example 8?**

The standard deviation of **a** = standard deviation of **b**.

(?) **Are you surprised?**

This is not surprising since clearly, by looking at the data, we can see that the spread of data is the same. The data of **b** has only been shifted up by 1000.

(?) **What about c?**

$$\text{SD of } \mathbf{c} = \dfrac{\text{SD of } \mathbf{a}}{10}$$

In **Example 9** we will show that if the standard deviation of $x_1, x_2, \ldots, x_n$ is s then the standard deviation of $\dfrac{x_1}{k}, \dfrac{x_2}{k}, \dfrac{x_3}{k}, \ldots, \dfrac{x_n}{k}$ is $\dfrac{s}{|k|}$, where $k \neq 0$ is an arbitrary constant and $|k| = \sqrt{k^2}$.

Example 9

Let $y_i = \dfrac{x_i}{k}$ where $k \neq 0$ and $i = 1, 2, \ldots, n$. Show that:

i $\bar{y} = \dfrac{\bar{x}}{k}$ where $\bar{y}$ and $\bar{x}$ represent the mean values of y_i and x_i respectively.

ii $s_y = \dfrac{s_x}{|k|}$ where s_y and s_x represent the standard deviation (SD) of y_i and x_i respectively.

(Note: $|a| = \sqrt{a^2}$.)

Solution

i By **16.1** with $y_1 = \dfrac{x_1}{k}$, $y_2 = \dfrac{x_2}{k}, \ldots$ we have

$$\bar{y} = \frac{y_1 + y_2 + \cdots + y_n}{n}$$

$$= \frac{\dfrac{x_1}{k} + \dfrac{x_2}{k} + \cdots + \dfrac{x_n}{k}}{n} \qquad \left[\text{Substituting } y_j = \frac{x_j}{k}\right]$$

$$= \frac{1}{k}\left(\frac{x_1 + x_2 + \cdots + x_n}{n}\right) = \frac{1}{k} \underset{\text{by } 16.1}{\bar{x}} = \frac{\bar{x}}{k}$$

This is our required result.

ii By **16.2**:

$$s_y^2 = \frac{\displaystyle\sum_{j=1}^{n}(y_j - \bar{y})^2}{n} = \frac{\displaystyle\sum_{j=1}^{n}\left(\dfrac{x_j}{k} - \dfrac{\bar{x}}{k}\right)^2}{n} \qquad \left[\text{Substituting } \bar{y} = \frac{\bar{x}}{k}\right]$$

$$= \frac{1}{k^2} \frac{\displaystyle\sum_{j=1}^{n}(x_j - \bar{x})^2}{n} \qquad \left[\text{Taking out } \left(\frac{1}{k}\right)^2 = \frac{1}{k^2}\right]$$

$$s_y^2 = \frac{1}{k^2} \underset{\text{by } 16.2}{s_x^2}$$

$$s_y = \sqrt{\frac{1}{k^2}}\, s_x = \frac{s_x}{\sqrt{k^2}} = \frac{s_x}{|k|}$$

We have shown our required result.

Consider the data of **Example 8**. Let

$$x_1 = 1, \quad x_2 = 2, \quad x_3 = 3, \quad x_4 = 4, \quad x_5 = 5 \text{ and } x_6 = 6$$

Dividing these by 10 gives

$$y_1 = 0.1, \quad y_2 = 0.2, \quad y_3 = 0.3, \quad y_4 = 0.4, \quad y_5 = 0.5 \text{ and } y_6 = 0.6$$

The x mean is

$$\bar{x} = \frac{1 + 2 + 3 + 4 + 5 + 6}{6} = 3.5$$

By property **i** of **Example 9** we have the y mean, $\bar{y} = 3.5/10 = 0.35$.

The standard deviation of the x data is 1.71.

? **What is the standard deviation of the y data?**

Using property **ii** the standard deviation of the y data is $1.71/10 = 0.171$.

B3 Data in a frequency distribution

In this subsection we look at finding the standard deviation of data where each piece of data occurs one or more times and is therefore represented in a table. Consider the general frequency distribution (Table 5).

TABLE 5

Data	Frequency
x_1	f_1
x_2	f_2
x_3	f_3
...	...
...	...
...	...
x_n	f_n

The data in Table 5 says that x_1 occurs f_1 times, x_2 occurs f_2 times, etc.

The mean, $\bar{x}$, of this data is given by

16.4
$$\bar{x} = \frac{f_1 x_1 + f_2 x_2 + f_3 x_3 + \cdots + f_n x_n}{f_1 + f_2 + f_3 + \cdots + f_n} \quad \text{or} \quad \left[\bar{x} = \frac{\sum\limits_{j=1}^{n} f_j x_j}{\sum\limits_{j=1}^{n} f_j} \right]$$

and the variance, s^2, is given by

16.5
$$s^2 = \frac{f_1(x_1 - \bar{x})^2 + f_2(x_2 - \bar{x})^2 + f_3(x_3 - \bar{x})^2 + \cdots + f_n(x_n - \bar{x})^2}{f_1 + f_2 + f_3 + \cdots + f_n}$$

$$\text{or} \quad \left[s^2 = \frac{\sum\limits_{j=1}^{n} f_j(x_j - \bar{x})^2}{\sum\limits_{j=1}^{n} f_j} \right]$$

To find the standard deviation, s, we take the square root of the variance.

Example 10

The length of time (in years) that football players play for a particular club is shown in Table 6a below:

Length of time at club (in years) x	Frequency f
1	7
2	3
4	5
5	8
6	2
8	3

TABLE 6a

This table means that there were 7 players who played for this particular club for 1 year, 3 players played for the club for 2 years, 5 players played for the club for 4 years, etc.

Evaluate the standard deviation.

Solution

To find the standard deviation, what do we need to evaluate first?

the mean ($\bar{x}$)

By 16.4

$$\bar{x} = \frac{(1 \times 7) + (2 \times 3) + (4 \times 5) + \cdots + (8 \times 3)}{7 + 3 + 5 + \cdots + 3} = 3.893 \quad \text{(3 d.p.)}$$

This means that the mean number of years that a player spends at the club is 3.893 years. To find the standard deviation, we construct Table 6b.

Column A x	Column B $\bar{x}$	Column C $(x - \bar{x})^2$	Column D f	Col C × Col D $f(x - \bar{x})^2$
1	3.893	$(1 - 3.893)^2 = 8.369$	7	58.583
2	3.893	$(2 - 3.893)^2 = 3.583$	3	10.749
4	3.893	$(4 - 3.893)^2 = 0.011$	5	0.055
5	3.893	$(5 - 3.893)^2 = 1.225$	8	9.800
6	3.893	$(6 - 3.893)^2 = 4.440$	2	8.880
8	3.893	$(8 - 3.893)^2 = 16.869$	3	50.607
			$\sum f = 28$	$\sum f(x - \bar{x})^2 = 138.674$

TABLE 6b

Example 10 *continued*

By 16.5

$$s^2 = \frac{138.674}{28} = 4.953$$

Hence the standard deviation is found by taking the square root of both sides:
$s = \sqrt{4.953} = 2.23$ (2 d.p.).

As **Example 10** shows, the standard deviation calculation of data in a frequency distribution is very long and tedious. It's easier to use a calculator.

Try the next example on your calculator. It can be done by hand, but the numbers are not particularly easy to deal with.

Example 11

The age ranges of IEng registrations in 2009 are given in Table 7a below:

TABLE 7a	Age	19– 24	25– 29	30– 34	35– 39	40– 44	45– 49	50– 54	55– 59	60– 64	65– 69	70– 74	75– 79	80– 84	85– 89	90+
	Freq	5	166	434	1494	3062	4838	6177	6523	6481	2770	1598	953	586	247	46

Determine the mean age and standard deviation for IEng membership for the year 2009.

Solution

We first need to find the mid-value of each age range. This means we have (Table 7b):

TABLE 7b	Age	19– 24	25– 29	30– 34	35– 39	40– 44	45– 49	50– 54	55– 59	60– 64	65– 69	70– 74	75– 79	80– 84	85– 89	90+
	Mid-value x	21.5	27	32	37	42	47	52	57	62	67	72	77	82	87	95
	Freq f	5	166	434	1494	3062	4368	6177	6523	6481	2770	1598	953	586	247	46

Note that for the last entry, 90+, we used the upper limit of 100.

By entering the mid-values x and the frequency into our calculator we obtain

$$\text{Mean} = 55.75 \text{ and standard deviation} = 10.81$$

The mean age of the 2009 IEng membership is 55.75 years and the standard deviation is 10.81 years.

Example 12

Table 8a shows the maximum loads on various cables.

Max load L (kN)	Number of cables (frequency) f
$49.5 \le L < 54.5$	2
$54.5 \le L < 59.5$	3
$59.5 \le L < 64.5$	6
$64.5 \le L < 69.5$	10
$69.5 \le L < 74.5$	8
$74.5 \le L < 79.5$	11
$79.5 \le L < 84.5$	5
$84.5 \le L < 89.5$	1
$89.5 \le L < 94.5$	7
$94.5 \le L < 99.5$	3
$99.5 \le L < 104.5$	4

TABLE 8a

Determine the mean and standard deviation of this data.

Solution

? We have grouped continuous data. **To use formula 16.5 , what value of L do we consider?**

? **For $49.5 \le L < 54.5$ is $L = 49.5$ or 54.5 or …?** We take L to be the midpoint value, $L = 52$. Hence we have the solution shown in Table 8b.

Max load L (kN)	Midpoint L	Frequency f
$49.5 \le L < 54.5$	52	2
$54.5 \le L < 59.5$	57	3
$59.5 \le L < 64.5$	62	6
$64.5 \le L < 69.5$	67	10
$69.5 \le L < 74.5$	72	8
$74.5 \le L < 79.5$	77	11
$79.5 \le L < 84.5$	82	5
$84.5 \le L < 89.5$	87	1
$89.5 \le L < 94.5$	92	7
$94.5 \le L < 99.5$	97	3
$99.5 \le L < 104.5$	102	4

TABLE 8b

By using a calculator we get mean $\bar{L} = 76.33$ kN (2 d.p.) and standard deviation $s = 13.21$ kN (2 d.p.).

SUMMARY

The average mean, $\bar{x}$, is defined by

16.1   mean, $\bar{x} = \dfrac{x_1 + x_2 + x_3 + \cdots + x_n}{n}$

Standard deviation is a 'measure of spread' of data.

16.2   variance $= \dfrac{(x_1 - \bar{x})^2 + (x_2 - \bar{x})^2 + \cdots + (x_n - \bar{x})^2}{n}$

The standard deviation, $s = \sqrt{\text{variance}}$. For a frequency distribution, mean

16.4 $\bar{x} = \dfrac{f_1 x_1 + f_2 x_2 + f_3 x_3 + \cdots + f_n x_n}{f_1 + f_2 + f_3 + \cdots + f_n}$

Variance, s^2, is given by

16.5 $s^2 = \dfrac{f_1(x_1 - \bar{x})^2 + f_2(x_2 - \bar{x})^2 + f_3(x_3 - \bar{x})^2 + \cdots + f_n(x_n - \bar{x})^2}{f_1 + f_2 + f_3 + \cdots + f_n}$

Standard deviation is s – take the square root of s^2.

Exercise 16(b)

Solutions at end of book. Complete solutions available at www.palgrave.com/engineering/singh

1 The number of rejects on an assembly line are as follows:

12, 27, 34, 50, 5, 19, 39, 15, 8, 36, 29

Determine the median, mean and standard deviation for this data.

2 Determine the mean and standard deviation for the following data:
 a 5, 7, 12, 15, 37
 b 105, 107, 112, 115, 137
 c 0.5, 0.7, 1.2, 1.5, 3.7
 d 53, 73, 123, 153, 373
 What do you notice about your results?

[For questions 3 and 4: $\bar{x}$ and $\bar{y}$ represent x mean and y mean respectively. s_x and s_y represent the standard deviations for x and y respectively. Also $j = 1, 2, \ldots, n$.]

3 If $y_j = kx_j$ then show that $\bar{y} = k\bar{x}$ and $s_y = |k| s_x$.

4 If $y_j = kx_j + a$ then show that $\bar{y} = k\bar{x} + a$ and $s_y = |k| s_x$.

5 Find the mean and standard deviation of the following:
 i 27, 53, 61, 64, 89, 93, 98
 ii 270, 530, 610, 640, 890, 930, 980
 iii 0.27, 0.53, 0.61, 0.64, 0.89, 0.93, 0.98
 iv 32, 58, 66, 69, 94, 98, 103
 v 2800, 5400, 6200, 6500, 9000, 9400, 9900

6 The following data shows the mean critical load on various struts:

Critical load (kN)	150.0	157.0	159.0	162.0
Frequency	7	8	12	15
Critical load (kN)	163.5	166.0	168.3	
Frequency	9	4	2	

Evaluate the mean and standard deviation for this data.

Exercise **16(b) continued**

Solutions at end of book. Complete solutions available at www.palgrave.com/engineering/singh

7 A Goods-In department measures a sample of 50 resistors from a batch which claims that they are all 10 kΩ. The data is displayed in a frequency distribution as follows:

Resistance R (kΩ)	Frequency
$9.6 \leq R < 9.7$	1
$9.7 \leq R < 9.8$	2
$9.8 \leq R < 9.9$	5
$9.9 \leq R < 10.0$	17
$10.0 \leq R < 10.1$	18
$10.1 \leq R < 10.2$	5
$10.2 \leq R < 10.3$	1
$10.3 \leq R < 10.4$	1

i Estimate the mean, without evaluating it.
ii Do you think this data will have a large SD or a small SD?
iii Determine the actual mean and standard deviation.

8 Evaluate the mean and standard deviation for the following.
  a The tensile strength (MN/m²) of aluminium alloy is given by:

Tensile strength T (MN/m²)	Frequency
$320 \leq T < 350$	4
$350 \leq T < 380$	12
$380 \leq T < 410$	18
$410 \leq T < 440$	16
$440 \leq T < 470$	15
$470 \leq T < 500$	10
$500 \leq T < 530$	3
$530 \leq T < 560$	7

  b The Young modulus (GN/m²) for stainless steel is given by:

Young modulus (GN/m²)	193	194	195	196
Frequency	7	7	10	19
Young modulus (GN/m²)	197	198	199	200
Frequency	13	9	8	3

SECTION C **Probability rules**

By the end of this section you will be able to:
► understand the term *probability*
► understand some of the probability rules
► use these rules

There are many questions in life where we do not have precise answers. For example

1 forecasting weather
2 predicting the outcome of a football match between two evenly matched teams
3 predicting the lottery numbers.

In each of these cases there is an element of uncertainty. The mathematical technique used to describe this uncertainty is called **probability**. For example, insurance companies need to know the probability of an A320 aircraft crashing.

C1 **Introduction**

? **What does the term probability mean?**

Probability is a numerical value which measures the likelihood of an event A occurring. It is normally denoted by $P(A)$.

Let X be the event that a test rig conforms to specification. If it is known that 97% of the rigs comply with specification then we say that

$$P(X) = \frac{97}{100} = 0.97$$

This means the probability that X conforms to specification is 0.97.

The probability of any event lies between 0 and 1, that is $0 \le P(A) \le 1$. $P(A) = 0$ means that the event A **cannot** happen.

? **Can you think of any examples where $P(A) = 0$?**

A man being pregnant, the sun rising from the west, etc.

$P(A) = 1$ means that the event A **will** certainly happen.

? **Can you think of any examples where $P(A) = 1$?**

You will pass away at some time (unless science manages to conquer death), the sun will rise each day, etc.

If all possibilities are divided into n equally likely outcomes and the event A occurs for r out of the n outcomes, then the probability of A occurring, $P(A)$, is given by

$$P(A) = \frac{r}{n}$$

where $n \ge 2$ and $0 \le r \le n$.

? **What is the probability of obtaining a head on tossing a coin?**

There are **two** possible outcomes – heads or tails, hence $n = 2$. Since the event of getting a head can only occur **once**, therefore $r = 1$ and

$$P(\text{head}) = \frac{1}{2} \qquad \left[\text{The probability of a head is } \frac{1}{2} \right]$$

C2 **Notation**

Probability theory can be represented by using Venn diagrams. The sample space, S, is the representation of all the possible outcomes of an event. For example, if the event is throwing a six-sided die (Fig. 5 on the facing page) then

$$S = \{1, 2, 3, 4, 5, 6\}$$

If A is the event of getting an even number by throwing a die then $A = \{2, 4, 6\}$. The event, A, is called a subset of S.

? **What is the probability of getting an even number on throwing a die?**

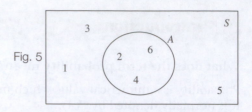
Fig. 5

There are three even numbers, 2, 4 and 6, and there are six possibilities, therefore

$$P(\text{even number}) = \frac{3}{6} = \frac{1}{2}$$

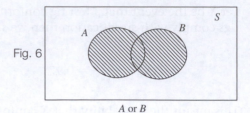
Fig. 6

Let B be the event of getting 5 or 6 on throwing a die, that is $B = \{5, 6\}$.

A or B

We first consider the general case with A and B being two different events. $A \cup B$ (read as 'A union B') are all the possible outcomes of A **or** B occurring (Fig. 6).

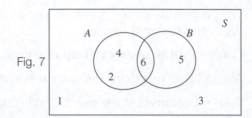
Fig. 7

For our example with $A = \{2, 4, 6\}$ and $B = \{5, 6\}$ as defined above we have the situation shown in Fig. 7.

Thus A or B has four possible outcomes, $\{2, 4, 5, 6\}$, and so

$$P(A \text{ or } B) = \frac{4}{6} = \frac{2}{3}$$

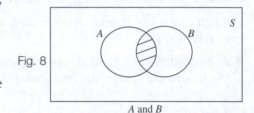
Fig. 8

A and B

$A \cap B$ (read as 'A intersection B') are all the possible outcomes of both A **and** B occurring (Fig. 8).

From our example with $A = \{2, 4, 6\}$ and $B = \{5, 6\}$ we consider Fig. 7. Since 6 is the **only** element in both A and B, the intersection, we have

$$P(A \text{ and } B) = \frac{1}{6}$$

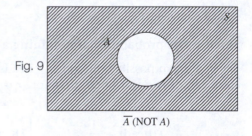
Fig. 9

$\overline{A}$ (read as 'not A') are all the possible outcomes of **not** A occurring (Fig. 9).

$\overline{A}$ (NOT A)

For our example with $A = \{2, 4, 6\}$ we have the situation shown in Fig. 10.

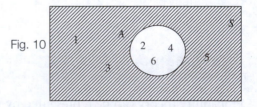
Fig. 10

Hence the event not A has three outcomes, $\{1, 3, 5\}$, therefore

$$P(\text{not } A) = \frac{3}{6} = \frac{1}{2}$$

C3 The *and* rule

If the occurrence of event A does not effect the occurrence of event B then A and B are said to be **independent** events.

If we toss a coin twice then the probability of getting a head on the second toss is 0.5, and is independent of what happened on the first toss. Thus each tossing of the coins is an independent event.

If A and B are independent events then

16.6 $\qquad P(A \text{ and } B) = P(A) \times P(B)$

The probability of getting two heads when tossing a coin twice is

$$P(\text{Head and Head}) = P(\text{Head}) \times P(\text{Head})$$
$$= \frac{1}{2} \times \frac{1}{2} = \frac{1}{4}$$

If events A and B are independent then so are **i** $\overline{A}$ and B, **ii** A and $\overline{B}$ and **iii** $\overline{A}$ and $\overline{B}$. Remember the notation: $\overline{A}$ means not A.

We can extend formula 16.6 for any sequence of independent events $E_1, E_2, \ldots, E_k$:

16.7 $\qquad P(E_1 \text{ and } E_2 \text{ and } E_3 \ldots \text{ and } E_k) = P(E_1) \times P(E_2) \times \cdots \times P(E_k)$

C4 The *not* rule

Either event A occurs or it does not occur, therefore

$$P(A) + P(\text{not } A) = 1$$

That is, the event A or not A covers **all** possibilities. Rearranging this gives

16.8 $\qquad P(\text{not } A) = 1 - P(A)$

Example 13

A component is subject to two independent tests A and B. The probability that a component passes test A is 0.7 and that it passes test B is 0.85. Determine the probability that a randomly chosen component:

  i does not pass test A
 ii does not pass test B
iii does not pass test A and does not pass test B
 iv passes both tests, A and B.

Solution

Let A and B denote the events of passing test A and passing test B respectively, then $P(A) = 0.7$ and $P(B) = 0.85$.

Example 13 *continued*

i P(a component does not pass test A) $= P(\text{not } A)$

$$\underset{\text{by } 16.8}{=} 1 - P(A)$$

$$= 1 - 0.7 = 0.3$$

ii Similarly $P(\text{not } B) = 1 - P(B) = 1 - 0.85 = 0.15$

iii P(a component [does not pass test A] and [does not pass test B])

$$= P([\text{not } A] \text{ and } [\text{not } B])$$

$$\underset{\text{by } 16.6}{=} P(\text{not } A) \times P(\text{not } B) = 0.3 \times 0.15 = 0.045$$

iv $P(A \text{ and } B) = P(A) \times P(B) = 0.7 \times 0.85 = 0.595$

Example 14

Let A, B and C denote three independent sections of a machine. The probabilities of each one working is given by

$$P(A) = 0.6, \quad P(B) = 0.55 \quad \text{and } P(C) = 0.8$$

Determine the probabilities that:

a all three sections are working
b none are working.

Solution

Let A, B and C denote the events of A, B and C working respectively.

a P(All three sections working) $= P(A \text{ and } B \text{ and } C)$

$$= P(A) \times P(B) \times P(C) \quad [\text{By } 16.7]$$

$$= 0.6 \times 0.55 \times 0.8$$

$$= 0.264$$

b P(None are working) $= P([\text{not } A] \text{ and } [\text{not } B] \text{ and } [\text{not } C])$

$$\underset{\text{by } 16.7}{=} P(\text{not } A) \times P(\text{not } B) \times P(\text{not } C)$$

$$\underset{\text{by } 16.8}{=} (1 - 0.6)(1 - 0.55)(1 - 0.8) = 0.036$$

16.6 $P(A \text{ and } B) = P(A) \times P(B)$

16.7 $P(E_1 \text{ and } E_2 \text{ and } E_3 \ldots \text{ and } E_k) = P(E_1) \times P(E_2) \times P(E_3) \ldots P(E_k)$

16.8 $P(\text{not } A) = 1 - P(A)$

Example 15

In a batch of 100 bearings, 10 are defective. If 5 are chosen at random (without replacement), what is the probability that all 5 are good?

Solution

The probability of the first bearing being good is $\dfrac{90}{100}$.

What is the probability of the second bearing being good?

Since we are not replacing the bearing there are now a total of 99 bearings, and we have already selected a good bearing in the first case and therefore there are only 89 good bearings remaining, hence

$$P(\text{Second bearing is good}) = \frac{89}{99}$$

$$P(2 \text{ good bearings}) = P([\text{1st good bearing}] \text{ and } [\text{2nd good bearing}])$$

$$= P(\text{1st good bearing}) \times P(\text{2nd good bearing})$$

$$= \frac{90}{100} \times \frac{89}{99}$$

We want to choose 5 good bearings, so repeating this process we have

$$P(5 \text{ good bearings}) = \frac{90}{100} \times \frac{89}{99} \times \frac{88}{98} \times \frac{87}{97} \times \frac{86}{96} = 0.584 \ (3 \text{ d.p.})$$

Example 16

An amplifier consists of two independent stages, A and B, connected in series (Fig. 11). The output of A is the input to B. The amplifier works if both stages A and B work. If the probability of A working is 0.72 and the probability of the amplifier working is 0.61, then find the probability that stage B works, given that stage A works.

Fig. 11

Solution

Let A and B denote stages A and B working respectively. Then we have $P(A) = 0.72$ and $P(A \text{ and } B) = 0.61$, because the probability of the amplifier working is 0.61. Since A and B are independent

$$P(A \text{ and } B) = P(A) \times P(B)$$

Rearranging gives

$$P(B) = \frac{P(A \text{ and } B)}{P(A)} = \frac{0.61}{0.72} = 0.847 \quad (3 \text{ d.p.})$$

C5 **The *or* rule**

For any events, A and B

16.9　　　$P(A \text{ or } B) = P(A) + P(B) - P(A \text{ and } B)$

A or B means A occurs or B occurs or both A and B occur.

We can justify **16.9** by the Venn diagrams shown in Fig. 12.

Fig. 12

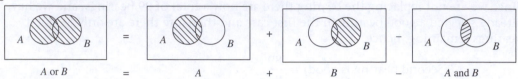

$A \text{ or } B$　　$=$　　　A　　　　$+$　　　　B　　　　$-$　　　$A \text{ and } B$

The intersection (A and B) is counted twice, and that is why we have to subtract the extra (A and B).

A and B are said to be mutually exclusive events if there is **no** element which is common between A and B (Fig. 13).

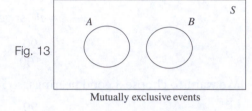

Fig. 13

Mutually exclusive events

If A and B are mutually exclusive events then $P(A \text{ and } B) = 0$ and

$$P(A \text{ or } B) = P(A) + P(B) - P(A \text{ and } B)$$
$$= P(A) + P(B) - 0$$
$$= P(A) + P(B)$$

16.10　　　$P(A \text{ or } B) = P(A) + P(B)$

? **What is the probability of getting a 3 or a 4 with a single throw of a die?**

Since 3 or 4 cannot occur together they are mutually exclusive events. (We cannot get a 3 and a 4 with a single throw of a die.) Thus

$$P(\text{getting a 3 or a 4}) = P(\text{getting a 3}) + P(\text{getting a 4})$$
$$= \frac{1}{6} + \frac{1}{6} = \frac{2}{6} = \frac{1}{3}$$

We can extend the formula **16.10** for any sequence of mutually exclusive events $E_1, E_2, \ldots, E_k$:

16.11　　　$P(E_1 \text{ or } E_2 \text{ or } E_3 \ldots \text{ or } E_k) = P(E_1) + P(E_2) + \cdots + P(E_k)$

If events A and B are independent then

16.12　　　$P(A \text{ or } B) = P(A) + P(B) - [P(A) \times P(B)]$

This is shown by **16.9** and **16.6**:

$$P(A \text{ or } B) = P(A) + P(B) - P(A \text{ and } B)$$
$$= P(A) + P(B) - \underbrace{P(A) \times P(B)}_{\text{by } \textbf{16.6}}$$

16.6　$P(A \text{ and } B) = P(A) \times P(B)$

Example 17

An electrical circuit has two switches A and B connected in parallel as shown in Fig. 14.

The probability of switch A working is 0.6 and of switch B working is 0.7. Find the probability that there is a connection between X and Y.

Solution

Let A denote the event that switch A works. Let B denote the event that switch B works. For a connection between X and Y we need A or B to work:

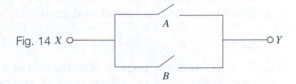

Fig. 14

$$P(A \text{ or } B) = P(A) + P(B) - P(A \text{ and } B)$$

$$\underset{\substack{\text{because } A \text{ and } B \\ \text{are independent}}}{=} P(A) + P(B) - [P(A) \times P(B)]$$

$$= 0.6 + 0.7 - [0.6 \times 0.7] = 0.88$$

C6 De Morgan's laws

In evaluating probabilities it is useful to employ the following laws from set theory. We assume these laws and apply them to find probabilities.

For any two sets A and B:

16.13
$$\overline{A \cup B} = \overline{A} \cap \overline{B}$$
$$\text{not}(A \text{ or } B) = (\text{not } A) \text{ and } (\text{not } B)$$

16.14
$$\overline{A \cap B} = \overline{A} \cup \overline{B}$$
$$\text{not}(A \text{ and } B) = (\text{not } A) \text{ or } (\text{not } B)$$

16.13 and 16.14 are known as de Morgan's laws.

In probability terms, where A and B are events we have

> The English mathematician, **Augustus de Morgan** (1806–71), was Professor of Mathematics at University College, London.

16.15
$$P(\text{not}[A \text{ or } B]) = P([\text{not } A] \text{ and } [\text{not } B])$$

and

16.16
$$P(\text{not}[A \text{ and } B]) = P([\text{not } A] \text{ or } [\text{not } B])$$

We can see how these rules work in practice below.

Example 18

An engineering module consists of coursework and an examination. It is known that 70% passed the examination, 85% passed coursework and 60% passed both. For a randomly chosen student, determine the probabilities that the student has

i passed the examination or coursework
ii failed both the coursework and examination.

Solution

Let E and C denote passes in examination and coursework respectively. We can visualize this example by Venn diagrams (Figs 15 and 16).

i P(passed the examination or coursework)

$$= P(E \text{ or } C)$$

$$\underset{\text{by } 16.9}{=} P(E) + P(C) - P(E \text{ and } C)$$

$$= 0.7 + 0.85 - 0.6$$
$$= 0.95$$

Fig. 15 E or C

ii P(failed both) $= P([\text{not } E] \text{ and } [\text{not } C])$

$$\underset{\text{by } 16.15}{=} P(\text{not } [E \text{ or } C])$$

$$\underset{\text{by } 16.8}{=} 1 - P(E \text{ or } C)$$

$$= 1 - 0.95 = 0.05$$

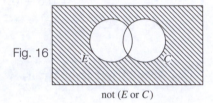

Fig. 16

not $(E \text{ or } C)$

which means there is a 5% chance of failing both, the coursework and the examination, which seems reasonable given the data.

C7 *At least* rule

If an event has two possible outcomes, then we call one of the outcomes a **success** and the other a **failure**. For example, the event of tossing a coin gives two possible outcomes – heads and tails. If we are asked to find the probability of obtaining at least n heads on tossing a coin, then we consider heads as success and tails as failure.

The probability of at least n successful outcomes can be calculated by

16.17 P(at least n successful outcomes)
$$= 1 - P(\text{less than } n \text{ successful outcomes})$$

Example 19

An inspector fails 5% of the motors from a batch. If two motors are chosen at random from the batch, evaluate the probabilities that

a both motors are good
b at least one motor is faulty.

Example 19 *continued*

Solution

$$P(\text{good motor}) = \frac{95}{100} = 0.95$$

a $P(\text{both motors good}) = P([\text{good motor}] \text{ and } [\text{good motor}])$

$$= P(\text{good motor}) \times P(\text{good motor})$$

$$= 0.95 \times 0.95 = 0.9025$$

b Let the selection of a faulty motor be a successful outcome, then

$$P(\text{at least one motor is faulty}) \underset{\text{by } 16.17}{=} 1 - P(\text{less than one motor is faulty})$$

$$= 1 - P(\text{both motors are good})$$

$$P(\text{at least one motor is faulty}) = 1 - \underset{\text{by part } \textbf{a}}{0.9025} = 0.0975$$

SUMMARY

If A and B are independent events then

16.6 $\qquad P(A \text{ and } B) = P(A) \times P(B)$

For any events A and B:

16.8 $\qquad P(\text{not } A) = 1 - P(A)$

16.9 $\qquad P(A \text{ or } B) = P(A) + P(B) - P(A \text{ and } B)$

If A and B are mutually exclusive events then

16.10 $\qquad P(A \text{ or } B) = P(A) + P(B)$

For independent events A and B:

16.12 $\qquad P(A \text{ or } B) = P(A) + P(B) - [P(A) \times P(B)]$

For any events A and B:

16.15 $\qquad P(\text{not } [A \text{ or } B]) = P([\text{not } A] \text{ and } [\text{not } B])$

16.16 $\qquad P(\text{not } [A \text{ and } B]) = P([\text{not } A] \text{ or } [\text{not } B])$

16.17 $\qquad P(\text{at least } n \text{ successful outcomes})$
$$= 1 - P(\text{less than } n \text{ successful outcomes})$$

Exercise **16(c)**

Solutions at end of book. Complete solutions available at www.palgrave.com/engineering/singh

1 In a large selection of electrical components, 10% of resistors are faulty, 5% of inductors are faulty and 3% of capacitors are faulty.
  a From a batch a resistor is chosen at random. Find the probability that it is not faulty.
  b By choosing randomly each of the three components, what is the probability that all three are faulty?

2 Three motors are chosen at random from a batch of which 2% are defective. Find the probability that all three are defective.

3 Two switches A and B are connected in parallel as shown in Fig. 17.

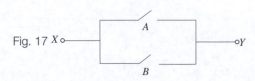

Fig. 17

Given that the probabilities of A and B functioning are

$$P(A) = 0.8 \text{ and } P(B) = 0.65$$

find the probability of a connection between X and Y.

4 The lifetimes, L, of 10,000 light bulbs produced by a certain manufacturer are as follows:

Lifetime L (hours)	Number of bulbs
$L \leq 200$	3000
$200 < L \leq 500$	3500
$500 < L \leq 1000$	2750
$L > 1000$	750

For a randomly selected bulb, find the probability that the bulb has

  i $L \leq 200$ hours
  ii $L \leq 500$ hours
  iii $L > 500$ hours.

5 A calculator manufacturer finds that 5% of the calculators do not have a memory button, 3% do not have a log button and 1% do not have log and memory buttons. For a randomly selected calculator, find the probability that the calculator

  a does not have a memory button or a log button
  b does have a memory button and a log button.

6 A communication network has two systems, A and B, connected in parallel and it only fails if both systems fail. The probability of A and B functioning properly is given by

$$P(A) = 0.88, \quad P(B) = 0.93$$

What is the probability that the communication network fails?

7 There are 50 components in a batch of which 4 are defective. Given that two components are randomly chosen, without replacement, and that the first component chosen is good, what is the probability that the second component is defective?

8 A toy is assembled from three different parts A, B and C. The probabilities of parts A, B and C being defective are given by:

$$P(A) = 0.15, P(B) = 0.25 \text{ and } P(C) = 0.02$$

Find the probability that a toy picked at random contains
  i no defective parts
  ii at least one defective part.

9 An electrical circuit consists of three switches A, B and C connected as shown in Fig. 18.

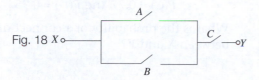

Fig. 18

The probability of A, B and C working are 0.7, 0.75 and 0.85 respectively. Find the probability that there is a connection between X and Y.

10 Three components A, B and C are connected in parallel as shown in Fig. 19.

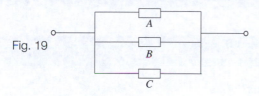

Fig. 19

A, B and C are independent.
The probability of each component working is as follows:

$$P(A) = 0.6, P(B) = 0.75 \text{ and } P(C) = 0.55$$

Find the probability that at least one component is working.

11 There are 60 components in a batch of which 3 are defective. Given that two components are randomly selected, without replacement, and that the first component is good, what is the probability that the second component is defective?

12 A box contains 100 fuses of which 5 are defective. If two fuses are chosen at random, without replacement, find the probabilities that
a both are defective
b none are defective
c only one is defective.

13 A system consists of two stages, A and B, in series (Fig. 20).

Fig. 20

The output of stage A acts as the input to stage B. Given that the probability of A working is 0.92 and the probability of the system working is 0.88, find the probability of B working given that A works.

14 5% of a batch of 120 components are faulty. If two components are chosen at random, without replacement, find the probability that the second component is faulty given that the first component is faulty.

15 A batch of 150 motors contains the following defects:

6 of type I defect
4 of type II defect
7 of type III defect.

From these defects we also have:

3 of type I and type II defects
2 of type II and type III defects
1 of type I, type II and type III defects.

One motor is chosen at random. If this motor has a defect of type II, find the probabilities that it also has

a a defect of type I
b a defect of type III
c defects of type I and type III.

16 A batch of 96 transformers contains 7 that are defective. If two transformers are selected at random, without replacement, find the probability that both are defective.

17 The reliability of an engine working to specification is 0.85. Two engines are tested at random. Find the probability that at least one of the engines is working to specification.

Exercise **16(c) continued**

Solutions at end of book. Complete solutions available at www.palgrave.com/engineering/singh

18 Switches A, B, C and D are connected as shown in Fig. 21.

Fig. 21

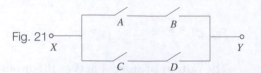

The probabilities of each switch working is given by

$$P(A) = 0.7, \; P(B) = 0.8,$$
$$P(C) = 0.77 \text{ and } P(D) = 0.95$$

What is the probability of a connection between X and Y?

SECTION D **Permutations and combinations**

By the end of this section you will be able to:
▶ evaluate permutations
▶ distinguish between permutations and combinations
▶ evaluate combinations

D1 **Permutations**

Do you remember what $n!$ (n factorial) means?

16.18 $n! = 1 \times 2 \times 3 \times 4 \times \cdots \times (n - 1) \times n$

For example, $6! = 1 \times 2 \times 3 \times 4 \times 5 \times 6 = 720$. Also we define $0! = 1$. We use $n!$ to count the number of ways of arranging (arrangements) n different items, as the next example shows.

Example 20

If car number plates are to contain the three letters A, B and C, how many different number plates can we form without repetition?

Solution

The first letter can be chosen in three ways – A, B or C. The second letter can be chosen in two ways and the third letter can be chosen in only one way. Hence the number of different plates $= 3 \times 2 \times 1 = 3! = 6$. The six different plates are ABC, ACB, BAC, BCA, CAB and CBA.

Doing the same example with four letters A, B, C and D gives us $4! = 24$ different plates.

The number of ways of arranging n different items is

16.19 $n!$

You can evaluate $n!$ on your calculator. Have a look at the instruction manual or the web.

If some of the n items are the same, then the number of arrangements reduces.
In **Example 20**, if we have the letters AAB, that is two letters out of 3 are the same, then the only arrangements would be AAB, ABA and BAA. This can be evaluated directly by

$$\frac{3!}{2!} = \frac{6}{2} = 3$$

In general, if there are n items of which p are the same then the number of arrangements is

16.20
$$\frac{n!}{p!}$$

A **permutation** is an arrangement where **order matters**. For example, AB is a different permutation from BA. The number of permutations of r items chosen from n different items is denoted by ${}^nP_r$ and given as

16.21
$${}^nP_r = \frac{n!}{(n-r)!}$$

Remember that **order does matter** for permutations.

Example 21

Consider A, B, C and D. If two letters are chosen, then find the number of all the possible permutations without repetition.

Solution

Permutations are AB, BA, AC, CA, AD, DA, BC, CB, BD, DB, CD and DC: 12 altogether.
Alternately if we use

16.21
$${}^nP_r = \frac{n!}{(n-r)!}$$

with $r = 2$ and $n = 4$ we obtain

$${}^4P_2 = \frac{4!}{2!} = 12$$

Example 22

The security code of an alarm consists of three single-digit numbers including zero followed by three letters, without any repetition. Evaluate how many different codes can by formed.

Solution

The number of permutations of choosing 3 numbers from 10 is given by

16.21
$${}^nP_r = \frac{n!}{(n-r)!}$$

Example 22 *continued*

With $r = 3$ and $n = 10$:

$$^{10}P_3 = \frac{10!}{(10-3)!} = \frac{10!}{7!}$$

The number of permutations of choosing 3 letters from 26 is given by

$$^{26}P_3 = \frac{26!}{23!}$$

Evaluating each of these on a calculator and multiplying gives

$$\text{Number of codes} = {}^{10}P_3 \times {}^{26}P_3 = 11\ 232\ 000$$

Thus more than 11 million different codes can be formed.

Most calculators will evaluate permutations directly.

D2 **Combinations**

Example 23

Consider the letters A, B, C and D. Without considering order, find the number of selections if

a 2 letters are chosen at random
b 3 letters are chosen at random.

Solution

a Selections are AB, AC, AD, BC, BD and CD; hence 6 selections altogether. Since order does not matter, AB is the same as BA and we count this as **one** selection. The number of selections can be evaluated by using the permutation formula 16.21 (where order does matter) and then halving it:

$$\frac{1}{2}{}^4P_2 = \frac{1}{2}\left(\frac{4!}{2!}\right) = \frac{4!}{2!2!} = 6 \quad \text{[Remember that } 2! = 2\text{]}$$

b For three letters chosen at random the selections are ABC, ACD, BCD and ABD; hence 4 selections. Remember that order does **not** matter: ABC, ACB, BAC, BCA, CAB and CBA all count as one selection rather than 6 (or 3!). We can use the permutation formula 16.21 and then divide by 3! (= 6).

$$\text{Number of selections } = \frac{1}{3!}{}^4P_3 = \frac{1}{3!}\left(\frac{4!}{(4-3)!}\right) = 4.$$

A combination is a selection of items where order does **not** matter. In the above example AB is the same combination as BA but AC is a different combination.

The number of combinations of r items chosen from n different items is denoted by ^nC_r and given by

$$\frac{1}{r!}\,^nP_r = \frac{1}{r!}\left(\underbrace{\frac{n!}{(n-r)!}}\right)$$
by 16.21

$$= \frac{1 \times n!}{r! \times (n-r)!} = \frac{n!}{r!(n-r)!}$$

Therefore

16.22 $$^nC_r = \frac{n!}{r!(n-r)!}$$

(^nC_r can be read as 'choosing r out of n'.)

Remember in combinations that order does **not** matter.

We can evaluate ^nC_r directly on a calculator. Evaluate $^{21}C_{15}$ by using your calculator. It should show 54264.

Example 24

In playing the lottery the object is to choose 6 numbers between 1 to 49. If these 6 numbers match the 6 numbers drawn from the machine (it does not matter about order), then you win the jackpot. What is the probability of winning this prize if you have bought 100 lottery tickets and have chosen different numbers on each ticket?

Solution

The number of ways of choosing 6 numbers from 49 is

$$^{49}C_6$$

because order does not matter. Since we have selected 100 different numbers, therefore

$$P(\text{winning the jackpot}) = \frac{100}{^{49}C_6} = \frac{100}{13\,983\,816}$$

Thus you have an approximately 100 in 14 million chance of winning the jackpot.

If you had brought only one lottery ticket, then there is an approximately one in 14 million chance of winning the jackpot.

16.21 $^nP_r = \dfrac{n!}{(n-r)!}$

Example 25

In a box there are 25 CD-ROM disks of which 4 are defective. If 3 disks are chosen at random, without replacement, from this box, what is the probability that

a all 3 disks are good?
b only 1 is defective?
c all 3 are defective?

Solution

The order of choosing disks is of no interest. So the number of ways of choosing 3 disks from 25 is given by $^{25}C_3$.

a For all 3 disks to be good, we choose 3 out of the 21 good disks, $^{21}C_3$. Hence

$$P(\text{all 3 good}) = \frac{^{21}C_3}{^{25}C_3} = 0.578 \quad (3 \text{ s.f.})$$

$$\left[\text{Alternatively } P(\text{all 3 good}) = \frac{21}{25} \times \frac{20}{24} \times \frac{19}{23} = 0.578 \right]$$

b Only 1 is defective and so 2 are good. We choose 1 out of the 4 defective disks, $^4C_1$, and 2 out of the 21 good disks, $^{21}C_2$. The probability of one defective is given by

$$P(\text{only one defective}) = \frac{^4C_1 \times {}^{21}C_2}{^{25}C_3} = 0.365 \quad (3 \text{ s.f.})$$

We multiply $^4C_1$ and $^{21}C_2$ because we require 1 defective disk **and** 2 good disks.

c For all 3 defective, we choose 3 out of the 4 defective disks, $^4C_3$:

$$P(\text{all 3 defective}) = \frac{^4C_3}{^{25}C_3} = 1.74 \times 10^{-3} \quad (3 \text{ s.f.})$$

SUMMARY

The number of permutations of r items chosen from n is given by

16.21 $$^nP_r = \frac{n!}{(n-r)!}$$

In permutations, order **does** matter.

The number of combinations of r items chosen from n is given by

16.22 $$^nC_r = \frac{n!}{r!(n-r)!}$$

In combinations, order does **not** matter.

Exercise **16(d)**

1 There are 60 light bulbs of which 10 are faulty. 7 bulbs are chosen at random, without replacement. Find the probability that 3 of these are faulty.

2 There are 60 fuses in a box of which 9 are defective. If 8 fuses are chosen at random, without replacement, then find the probability that

   i all 8 are good
   ii all 8 are defective
  iii at least 1 is defective.

3 In a game of lottery one has to select six different numbers between 1 and 40. If your six numbers match the six numbers drawn from the machine, then you win the jackpot (order does not matter).

a Find the probability of winning the jackpot if you have bought 150 lottery tickets.

b If each lottery ticket cost £1, then how much would you have to spend to obtain a probability of more than 0.5 of winning the jackpot?

4 A licence plate consists of 1 letter followed by 3 single digits and then 3 letters. Assuming all 26 letters and 10 numbers can be used, find the number of plates if

a no number or letter is duplicated
b numbers and letters can be duplicated.

SECTION E **Binomial distribution**

By the end of this section you will be able to:
► understand the term *discrete random variable*
► use the binomial distribution to find probabilities
► apply the binomial distribution formula

E1 **Discrete random variables**

Let $E_1, E_2, \ldots, E_k$ be mutually exclusive events, that is they have no elements in common. Then a random variable, X, gives each of these $E_1, E_2, \ldots, E_k$ a numerical value. The random variable is usually denoted by a capital letter. For example, let X represent the sum of dots on throwing two dice. We have

$X = 2$ is the numerical value for throwing (1, 1)
$X = 3$ is the numerical value for throwing (1, 2) or (2, 1)
$$\vdots$$
$X = 12$ is the numerical value for throwing (6, 6)

The word **random** is used because it is associated with an experiment and the value is determined by chance. The particular value a random variable takes is usually denoted by a small letter, $x, y, z, \ldots$.

The function which gives the probabilities of the random variable, X, is called the **probability density function** and denoted by $P(X = x)$.

In this and the next section we examine discrete random variables. A discrete random variable is a random variable, X, which can only take certain values $x_1, x_2, \ldots, x_n$ and the associated probabilities total 1. If

$$P(X = x_1) = p_1, P(X = x_2) = p_2, \ldots, P(X = x_n) = p_n$$

where $p_1, p_2, \ldots, p_n$ are probabilities then

16.23 $p_1 + p_2 + \cdots + p_n = 1$ [Remember all probabilities must total 1]

Example 26

The probability density function of a discrete random variable X is given by the formula

$$P(X = x) = kx^2 \quad \text{for } x = 0, 1, 2, 3$$

Evaluate the constant k.

Solution

Substituting the given values of x into $P(X = x) = kx^2$:

$$P(X = 0) = k(0^2) = 0$$
$$P(X = 1) = k(1^2) = k$$
$$P(X = 2) = k(2^2) = 4k$$
$$P(X = 3) = k(3^2) = 9k$$

Summing these results and equating to 1, because total probabilities = 1, gives

$$9k + 4k + k = 1$$
$$14k = 1 \text{ so } k = 1/14$$

E2 Binomial distribution

Example 27

On average, 7% of the motors produced by a machine are faulty. A random sample of 5 motors is selected from a large batch so that the probability remains constant. Determine the probabilities that

a 1 motor is faulty
b 2 motors are faulty
c 3 motors are faulty
d 4 motors are faulty
e all 5 motors are faulty
f no motor is faulty.

Do you notice any relationship between the number of faulty motors and their associated probabilities?

Example 27 *continued*

Solution

Let F and G denote faulty and good motors respectively.

$$P(F) = \frac{7}{100} = 0.07$$

Since 7% are faulty, therefore 93% must be good motors:

$$P(G) = \frac{93}{100} = 0.93$$

a $P(1 \text{ motor faulty out of } 5) = P(FGGGG)$

$$\begin{aligned} &= P(F) \times P(G) \times P(G) \times P(G) \times P(G) \\ &= 0.07 \times 0.93 \times 0.93 \times 0.93 \times 0.93 \\ &= 0.07 \times (0.93)^4 \\ &= 0.052 \end{aligned}$$

So the probability of getting 1 faulty motor among a sample of 5 is 0.052 – approximately a 1 in 20 chance of getting a faulty motor among 5.

? **Does this answer seem correct to you?**

No.

? **Is it too high or too low?**

Too low.

You would expect a higher chance of selecting 1 faulty motor among 5. This probability of 0.052 is the probability of getting the first motor faulty and the remaining 4 good. But we do **not** care about the order of the faulty motor, it can be the first motor or the second motor or the third motor, etc.

$$FGGGG \text{ or } GFGGG \text{ or } GGFGG \text{ or } \ldots$$

? **How many ways can we choose 1 faulty motor from a sample of 5?**

$^5C_1$

Hence the probability of a faulty motor is

$$\begin{aligned} P(\text{one motor faulty}) &= {}^5C_1 \times 0.052 \\ &= 5 \times 0.052 = 0.26 \end{aligned}$$

Alternatively we can evaluate this more accurately by

$$P(\text{one motor faulty}) = {}^5C_1 \times 0.07 \times (0.93)^4 = 0.262 \quad \text{(3 s.f.)}$$

Similarly working to an appropriate number of places we have:

b $P(2 \text{ motors faulty}) = {}^5C_2(0.07)^2(0.93)^3$

$$= 0.039$$

c $P(3 \text{ motors faulty}) = {}^5C_3(0.07)^3(0.93)^2$

$$= 0.00297$$

d $P(4 \text{ motors faulty}) = {}^5C_4(0.07)^4(0.93)$

$$= 1.116 \times 10^{-4}$$

> Example 27 *continued*
>
> **e** $P(5 \text{ motors faulty}) = {}^5C_5(0.07)^5(0.93)^0$
> $= 0.07^5 = 1.681 \times 10^{-6}$
> **f** $P(\text{no motors faulty}) = {}^5C_0(0.07)^0(0.93)^5$
> $= 0.93^5 = 0.696$
>
> Let X be the discrete random variable representing the number of faulty motors and x be the value X takes. Then
> $$P(X = x) = {}^5C_x(0.07)^x(0.93)^{5-x}$$

We can represent the probabilities of **Example 27** as in Table 9.

	Number of faulty motors	Probability
TABLE 9	0	0.696
	1	0.262
	2	0.039
	3	0.00297
	4	1.116×10^{-4}
	5	1.681×10^{-6}

This table gives the **probability distribution** for the number of faulty motors. We can denote the above probabilities as follows:
$$P(X = 0) = 0.696, P(X = 1) = 0.262, \ldots$$
where X is the discrete random variable.

? **What is the sum of the probabilities equal to?**

1.

Remember that a discrete random variable is a discrete variable whose associated probabilities all add up to 1.

For **Example 27**, you might also notice that all the terms are from the binomial expansion of
$$(0.93 + 0.07)^5$$
By 7.23 :
$$(0.93 + 0.07)^5 = \underbrace{0.93^5}_{P(X=0)} + \underbrace{{}^5C_1(0.93)^4(0.07)}_{P(X=1)} + \underbrace{{}^5C_2(0.93)^3(0.07)^2}_{P(X=2)}$$
$$+ \underbrace{{}^5C_3(0.93)^2(0.07)^3}_{P(X=3)} + \underbrace{{}^5C_4(0.93)(0.07)^4}_{P(X=4)} + \underbrace{0.07^5}_{P(X=5)}$$

The probability distribution shown in Table 9 is an example of a binomial distribution.

7.23 $(a + b)^n = a^n + {}^nC_1a^{n-1}b + \ldots + b^n$

? Why do we call this 'binomial'?

Because there are only two possible outcomes – a faulty motor or a good motor. Generally we denote two possible outcomes by success and failure. The probabilities associated with these outcomes are normally given by the letters p and q where p is the probability of success and $q = 1 - p$ is the probability of failure. For **Example 27**, a successful outcome is the selection of a faulty motor:

$$p = 0.07 \text{ and } q = 0.93$$

We always have $p + q = 1$.

It might seem confusing to associate success with the selection of a faulty motor, but if we nominate $p(\text{success})$ as the probability required by the question then it keeps each application of the binomial formula consistent. Since in **Example 27** we are interested in finding the probability of a faulty motor, therefore the selection of a faulty motor is considered as a success.

In probability theory the term 'trial' means a single execution of an experiment. Let n denote the number of independent trials ($n = 5$ for **Example 27**). Generally the probability of obtaining x successes in n independent trials is given by

16.24 $$P(X = x) = {}^nC_x p^x q^{n-x} = \frac{n!}{x!(n-x)!}p^x q^{n-x}$$

Remember that $q = 1 - p$ and that X is the random variable representing the number of successes in n trials. The probability distribution of 16.24 is called the **binomial distribution** and denoted by $B(n, p)$. Also note that

16.25 $$(q + p)^n = P(X = 0) + P(X = 1) + P(X = 2) + \cdots + P(X = n)$$
$$= 1$$

We can justify this by

$$(q + p)^n = \left(\underbrace{1 - p}_{= q} + p\right)^n = (1)^n = 1$$

E3 Applications

Example 28

From previous tests it is known that the probability of a resistor being out of tolerance is 0.15. From a large batch, a random sample of 20 resistors is chosen so that the probability remains constant. Evaluate the probabilities that

a exactly 7 will be out of tolerance
b at least 3 will be out of tolerance.

Example 28 *continued*

Solution

Let X be the random variable denoting the number of resistors out of tolerance. Consider 'resistor being out of tolerance' as success, so $p = 0.15$, $q = 1 - 0.15 = 0.85$ and $n = 20$. We need to use the binomial distribution formula:

16.24 $P(X = x) = {}^nC_x p^x q^{n-x}$

with these values:

  * $P(X = x) = {}^{20}C_x(0.15)^x(0.85)^{20-x}$

a Substituting $x = 7$ into   *   gives

$$P(X = 7) = {}^{20}C_7(0.15)^7(0.85)^{13} = 0.016$$

The probability of exactly 7 resistors being out of tolerance is 0.016 (3 d.p.).

b By 16.17:

P(at least 3 will be out of tolerance)
$= 1 - P$(less than 3 will be out of tolerance)
$= 1 - P$(0 or 1 or 2 will be out of tolerance)
† $= 1 - \big[P(X = 0) + P(X = 1) + P(X = 2)\big]$

Using   *   with values of x equal to 0, 1 and 2 respectively:

$$P(X = 0) = 0.85^{20} = 0.039 \qquad\qquad [{}^{20}C_0 = 1]$$
$$P(X = 1) = 20(0.15)(0.85)^{19} = 0.137 \quad [{}^{20}C_1 = 20]$$
$$P(X = 2) = {}^{20}C_2(0.15)^2(0.85)^{18} = 0.229$$

Substituting these into   †   gives

$$P\text{(at least 3 will be out of tolerance)} = 1 - (0.039 + 0.137 + 0.229)$$
$$= 0.595 \quad (3 \text{ d.p.})$$

Note that ${}^nC_n = 1, {}^nC_0 = 1$ and ${}^nC_1 = n$. For **Example 28**:
$${}^{20}C_{20} = 1, \ {}^{20}C_0 = 1 \text{ and } {}^{20}C_1 = 20$$

Example 29

The probability of passing an examination is 0.7. Out of 15 students, evaluate the probabilities that
a all 15 will pass
b none will pass
c at least 12 will pass.

16.17 P(at least n successes) $= 1 - P$(less than n successes)

Example 29 *continued*

Solution

Using the binomial distribution formula:

16.24 $P(X = x) = {}^nC_x p^x q^{n-x}$

with $p = 0.7$, $q = 0.3$, $n = 15$ and X the random variable which represents 'the number of students who pass':

★ $P(X = x) = {}^{15}C_x (0.7)^x (0.3)^{15-x}$

a With $x = 15$:

$$P(X = 15) = {}^{15}C_{15}(0.7)^{15}(0.3)^0$$
$$= 0.7^{15} = 4.748 \times 10^{-3} \quad (4 \text{ s.f.})$$

b Putting $x = 0$ into ★ :

$$P(\text{none will pass}) = P(X = 0) = 0.3^{15} = 1.435 \times 10^{-8} \quad (4 \text{ s.f.})$$

c $P(\text{at least 12 will pass}) = P(12 \text{ pass or 13 pass or 14 pass or 15 pass})$

† $= P(X = 12) + P(X = 13) + P(X = 14) + P(X = 15)$

Evaluating each of these probabilities by ★ gives

$$P(X = 12) = {}^{15}C_{12}(0.7)^{12}(0.3)^3 = 0.170$$
$$P(X = 13) = {}^{15}C_{13}(0.7)^{13}(0.3)^2 = 0.092$$
$$P(X = 14) = {}^{15}C_{14}(0.7)^{14}(0.3) = 0.031$$
$$P(X = 15) = (0.7)^{15} = 4.748 \times 10^{-3}$$

Substituting these into † gives

$$P(\text{at least 12 will pass}) = 0.170 + 0.092 + 0.031 + (4.748 \times 10^{-3})$$
$$= 0.298 \quad (3 \text{ s.f.})$$

SUMMARY

The sum of the probabilities, $p_1, p_2, \ldots, p_n$, that a discrete random variable takes is 1. That is

16.23 $p_1 + p_2 + \cdots + p_n = 1$

Binomial distribution is a probability distribution of a discrete random variable, X, such that

16.24 $P(X = x) = \dfrac{n!}{x!(n-x)!} p^x q^{n-x} \quad (\text{or } {}^nC_x p^x q^{n-x})$

where p is the probability of success, $q = 1 - p$ is the probability of failure and n is the number of independent trials. Also

16.25 $(q + p)^n = P(X = 0) + P(X = 1) + P(X = 2) + \cdots + P(X = n)$

$$= 1$$

Exercise **16(e)**

Solutions at end of book. Complete solutions available at www.palgrave.com/engineering/singh

1 10% of the resistors produced by a manufacturer are out of tolerance. If a sample of 5 resistors is selected at random, find the probability that

 a exactly 3 resistors are out of tolerance
 b 3 or fewer than 3 resistors are out of tolerance.

2 A component has a 1 in 25 chance of failing. Five components are chosen from a large batch so that the probability of failure remains constant. Determine the probability of

 a 1 component failing
 b more than 3 components failing.

3 The probability of an integrated circuit passing a quality test is 0.86. In a batch of 10 integrated circuits find the probabilities that

 i 2 will fail
 ii 8 will pass
 iii at least 2 will fail
 iv fewer than 2 will fail.

4 Determine the probability of having more boys than girls in a family of 7 children. Assume the probability of having a boy is 0.48.

5 The probability that a student passes a subject is 0.85. If the student takes 8 subjects, what is the probability that he passes more than 6 subjects?

6 a If X is the random variable which follows the binomial distribution with $n = 600$ and $p = 1 \times 10^{-3}$, evaluate (to 4 d.p.)

 i $P(X = 1)$
 ii $P(X = 2)$

 iii $P(X = 3)$

The Poisson distribution is described by

$$P(X = x) = \frac{e^{-np}(np)^x}{x!}$$

 b For this distribution, evaluate (consider the above values of n and p)
 i $P(X = 1)$ **ii** $P(X = 2)$ **iii** $P(X = 3)$
 c What do you notice about the results of **a** and **b**?

7 The probability of an England player scoring a penalty is 0.85. What is the probability that an England team will score more than 3 penalties out of 5?

8 The probability density function of a discrete random variable, X, is given by

$$P(X = x) = kx^4 \quad x = 0, 1, 2$$

Determine k.

9 The probability density function of a discrete random variable, X, is given by

$$P(X = x) = kx^3 \quad x = 0, 1, 2, 3$$

Evaluate the constant k.

10 Let X be the discrete random variable which follows the binomial distribution with parameters n and p. Show that

$$\sum_{x=1}^{n} xP(X = x) = np$$

SECTION F **Properties of discrete random variables**

By the end of this section you will be able to:
▶ find the mean and variance of a discrete random variable
▶ state the mean and standard deviation of the binomial distribution
▶ apply the Poisson distribution

F1 **Expected value**

Example 30

An experiment is carried out by throwing a die 100 times, and the results are recorded in Table 10.

TABLE 10	Score (x)	1	2	3	4	5	6
	Frequency	15	12	21	19	17	16

Determine the probability of each score and the mean score.

Solution

Let p_1, p_2, . . . , p_6 denote the probabilities for scores 1, 2,..., 6 respectively. Since a score of 1 appears 15 out of 100 times we have

$$p_1 = \frac{15}{100} = 0.15$$

Similarly

$$p_2 = \frac{12}{100} = 0.12,\ p_3 = \frac{21}{100} = 0.21,\ p_4 = \frac{19}{100} = 0.19,\ p_5 = \frac{17}{100} = 0.17,\ p_6 = \frac{16}{100} = 0.16$$

To find the mean we use our earlier formula given in **Section B:**

16.4
$$\overline{x} = \frac{f_1 x_1 + f_2 x_2 + f_3 x_3 + \cdots + f_n x_n}{f}$$

where $f = f_1 + f_2 + f_3 + \cdots + f_n$ = sum of frequencies. We can rewrite 16.4 as

$$\overline{x} = x_1\left(\frac{f_1}{f}\right) + x_2\left(\frac{f_2}{f}\right) + \cdots + x_n\left(\frac{f_n}{f}\right)$$

$$= x_1(p_1) + x_2(p_2) + \cdots + x_n(p_n) \quad \left[\text{Because } \frac{f_k}{f} = p_k\right]$$

Example 30 *continued*

Substituting $x_1 = 1$, $x_2 = 2, \cdots, x_6 = 6$ and $p_1 = 0.15$, $p_2 = 0.12, \cdots, p_6 = 0.16$ into
$\bar{x} = x_1(p_1) + x_2(p_2) + \cdots + x_6(p_6)$ gives

$$\bar{x} = (1 \times 0.15) + (2 \times 0.12) + (3 \times 0.21) + (4 \times 0.19) + (5 \times 0.17) + (6 \times 0.16)$$

$$= 3.59$$

We have evaluated the mean score of a die by an experimental approach. In this subsection we look at the mean from a theoretical viewpoint.

The expected value, $E(X)$, is the mean value of a random variable X and is normally denoted by μ, pronounced 'mew'. The symbol $\bar{x}$ is used for the mean of a sample and μ is used for the mean of a random variable.

The expected value, $E(X)$, of a discrete random variable X with values $x_1, x_2, \ldots, x_n$ is defined as

16.26 $E(X) = x_1\, P(X = x_1) + x_2\, P(X = x_2) + \cdots + x_n\, P(X = x_n)\quad [E(x) = \mu]$

Notice the similarity with the formula for the mean $\bar{x}$ above:

$$\bar{x} = x_1(p_1) + x_2(p_2) + \cdots + x_n(p_n)$$

Example 31

Evaluate $E(X)$ (or μ) for $P(X = x) = \dfrac{1}{6}$ for $x = 1, 2, 3, 4, 5$ and 6 where X represents the score on a die.

Solution

Writing the probability for each x gives Table 11.

TABLE 11	x	1	2	3	4	5	6
	$P(X = x)$	$\dfrac{1}{6}$	$\dfrac{1}{6}$	$\dfrac{1}{6}$	$\dfrac{1}{6}$	$\dfrac{1}{6}$	$\dfrac{1}{6}$

Multiplying each x with its associated probability and summing yields:

$$E(X) = \left(1 \times \frac{1}{6}\right) + \left(2 \times \frac{1}{6}\right) + \left(3 \times \frac{1}{6}\right) + \left(4 \times \frac{1}{6}\right) + \left(5 \times \frac{1}{6}\right) + \left(6 \times \frac{1}{6}\right) = 3.5$$

Since the probability distribution of Table 11 is symmetric, the mean value, $E(X)$ $(=\mu)$, is going to lie midway between 3 and 4, hence 3.5.

Notice the discrepancy between the mean, $\bar{x} = 3.59$, from **Example 30** and the expected value, $E(X) = 3.5$, from **Example 31** of the score on a die.

The value of $E(X)$ (or μ) is determined by the underlying theoretical probabilities but $\bar{x}$ is calculated in an experiment by using observed frequencies which are only approximately correct.

Next we apply this theory to the engineering field called Signal Processsing. In signal processing we want to get rid of interference and noise on the signal. Statistics and probability can quantify these unwanted disturbances. In particular, probability is used to understand the processes that generate signals. In signal processing you examine two signals and inquire whether the **variation** in the signal is down to noise or whether the underlying process is changing.

For example, you can generate a 100-point signal by tossing a coin 100 times and giving the signal 1 volt if the toss is a head and 0 volts if the toss results in a tail. In a 100-point signal the underlying theoretical probability gives a mean of exactly 0.5 volts because each outcome (heads and tails) is equally likely. However the mean of an actual 100-point signal is unlikely to be exactly 0.5 volts because each time a signal is generated the number of 1's and 0's is **different**.

The mean and standard deviation are **constant** with respect to the underlying probabilities that generated the signal but these statistics (mean and standard deviation) change each time for the actual signal.

 Example 32 *signal processing*

The probability distribution of a sampled signal is given in Table 12.

TABLE 12	x (volts)	1	2	3	4	5	6	7	8
	P(X = x)	0.10	0.12	0.15	0.20	0.17	0.13	0.07	0.06

Determine the mean voltage, μ (or $E(x)$).

Solution

As in the above example, we have

$$\mu = (1 \times 0.1) + (2 \times 0.12) + (3 \times 0.15) + (4 \times 0.2) + (5 \times 0.17)$$
$$+ (6 \times 0.13) + (7 \times 0.07) + (8 \times 0.06)$$

$$= 4.19 \text{ volts}$$

The mean voltage is $\mu = 4.19$ V (2 d.p.).

Generally we can write the formula 16.26 in shorthand notation as

16.27 $\qquad E(X) = \sum_{j=1}^{n} x_j \, P(X = x_j)$

The Greek symbol $\sum_{j=1}^{n}$ means sum from $j = 1$ to $j = n$.

The expected value of a function $g(X)$ of the discrete random variable X with particular values $x_1, x_2, \ldots, x_n$ is defined by

16.28 $\qquad E\big[g(X)\big] = g(x_1)P(X = x_1) + g(x_2)P(X = x_2) + \cdots + g(x_n)P(X = x_n)$

Notice the similarity with the formula for $E(X)$. This time we multiply the probabilities by the functions $g(x_1), g(x_2), \ldots, g(x_n)$. Formula 16.28 might seem difficult to comprehend, but the following example should make it more digestible.

Example 33

Consider the discrete random variable X with a probability density function

$$P(X = x_j) = \frac{x_j}{10} \text{ for } x_j = 1, 2, 3 \text{ and } 4$$

Determine $E(X^2)$.

Solution

Substituting the given values of x_j yields Table 13.

x_j	1	2	3	4
$P(X = x_j)$	$\dfrac{1}{10}$	$\dfrac{2}{10}$	$\dfrac{3}{10}$	$\dfrac{4}{10}$

TABLE 13

To find $E(X^2)$ we sum the product of x_j^2 and its associated probability:

$$E(X^2) = \left(1^2 \times \frac{1}{10}\right) + \left(2^2 \times \frac{2}{10}\right) + \left(3^2 \times \frac{3}{10}\right) + \left(4^2 \times \frac{4}{10}\right) = 10$$

The expected value function has the following properties for the random variables X and Y:

$$E(X + Y) = E(X) + E(Y)$$

The expected value of $X + Y$ is equal to the expected value of X plus the expected value of Y. Similarly

$$E(X - Y) = E(X) - E(Y)$$
$$E(kX) = kE(X) \text{ where } k \text{ is a constant}$$

The general formula is given by

16.29 $E\big[kf(X) + m\big] = kE\big[f(X)\big] + m$

where k and m are constants and f is any function of the random variable X. Formula 16.29 states that we can take out a constant, k, and the expected value of a constant, $E(m) = m$.

F2 **Variance**

The variance, σ^2, of a discrete random variable, X, is defined by

16.30 $\sigma^2 = E(X - \mu)^2$

where μ is the mean of X. (σ^2 is pronounced 'sigma squared'.)

Example 34

Show that

$$\sigma^2 = E(X^2) - \mu^2$$

where μ is the mean of X.

Solution

Applying 16.30 and expanding gives

$$
\begin{aligned}
\sigma^2 = E(X - \mu)^2 &= E(X^2 - 2\mu X + \mu^2) \quad [\text{Applying } (a - b)^2 = a^2 - 2ab + b^2]\\
&= E(X^2) - 2\mu E(X) + \mu^2 \\
&\quad\text{by } 16.29\\
&= E(X^2) - 2\mu\mu + \mu^2 \quad [\text{Remember that } \mu = E(X)]\\
&= E(X^2) - \mu^2 \quad [\text{Simplifying}]
\end{aligned}
$$

This is an important result:

16.31 $\qquad \sigma^2 = E(X^2) - \mu^2$

where

16.32 $\qquad E(X^2) = \sum_{\text{all } x} x^2 P(X = x)$

Example 35

Suppose that houses in a large village are sampled to ascertain the size of each family and we obtain the following probability distribution (Table 14):

x_j (size of family)	5	6	7	8
$P(X = x_j)$	0.17	0.3	0.35	0.18

TABLE 14

where X is the discrete random variable which represents the size of a family. Determine the mean size of the family μ and the variance σ^2.

Solution

Remember that $\mu = E(X)$ and so multiplying each x_j by the associated probability and adding together gives

$$
\begin{aligned}
\mu &= (5 \times 0.17) + (6 \times 0.3) + (7 \times 0.35) + (8 \times 0.18)\\
&= 6.54
\end{aligned}
$$

16.29 $E[kf(X) + m] = kE[f(X)] + m$ 　　　16.30 $\sigma^2 = E(X - \mu)^2$

Example 35 *continued*

?

Does this seem reasonable?

Since most of the probability is concentrated between families of size 6 and 7, a mean value of 6.54 seems reasonable.

Variance: $\sigma^2 = E(X^2) - 6.54^2$

Using Table 14 we have

$$E(X^2) = (5^2 \times 0.17) + (6^2 \times 0.3) + (7^2 \times 0.35) + (8^2 \times 0.18)$$
$$= 43.72$$

Substituting $E(X^2) = 43.72$ into $\sigma^2 = E(X^2) - 6.54^2$ gives

$$\sigma^2 = 43.72 - 6.54^2 = 0.948 \qquad \text{(3 d.p.)}$$

We can also use a calculator to evaluate the mean and variance of the distribution in Table 14.

F3 Poisson distribution

The Poisson distribution can be viewed as an approximation to the binomial distribution for large n and small p, that is a large sample size with small probability. It is also a useful distribution in its own right.

> The term Poisson comes from the French mathematician **Simeon-Denis Poisson** (1781–1840). He discovered the distribution towards the end of his life.

The probability $P(X = x)$ of the discrete random variable, X, is defined as

16.33 $$P(X = x) = \frac{e^{-\mu}\mu^x}{x!} \text{ for } x = 0, 1, 2, 3, \ldots$$

where μ is the mean of the distribution. In this case we say that the discrete random variable, X, follows the Poisson distribution.

Example 36

Let X be the number of failures of a system. If X follows a Poisson distribution with a mean of 2.5 failures per year, determine the probabilities of

a no failures in a year
b at most 2 failures in a year.

Example 36 *continued*

Solution

Using the Poisson distribution formula

16.33 $\qquad P(X = x) = \dfrac{e^{-\mu}\mu^x}{x!}$

with $\mu = 2.5$ gives

$$P(X = x) = \dfrac{e^{-2.5}2.5^x}{x!}$$

a $P(\text{no failures}) = P(X = 0) = \dfrac{e^{-2.5}(2.5)^0}{0!} = 0.082$ (3 d.p.)

b $P(\text{at most 2 failures}) = P(X = 0) + P(X = 1) + P(X = 2)$

$$= \dfrac{e^{-2.5}(2.5)^0}{0!} + \dfrac{e^{-2.5}(2.5)^1}{1!} + \dfrac{e^{-2.5}(2.5)^2}{2!}$$

$$= 0.544 \quad (3 \text{ d.p.})$$

Remember that $0! = 1$.

The mean value, μ, for the binomial distribution with parameters n and p is given by

16.34 $\qquad \mu = np$

You are asked to show this result in question **10** of **Exercise 16(e)**:

$$\mu = \sum xP(X = x)$$

where $x = 0, 1, 2, \ldots, n$ and $P(X = x) = {}^nC_x p^x q^{n-x}$ (remember that $q = 1 - p$).

The variance, σ^2, is given by

16.35 $\qquad \sigma^2 = npq$

By taking the square root of both sides we obtain the standard deviation, σ:

16.36 $\qquad \sigma = \sqrt{npq}$

Example 37

A manufacturer produces a batch of 1000 calculators. The probability that a calculator is defective is 1×10^{-3}. Evaluate the probability that the batch contains 3 defective calculators by

i binomial distribution   **ii** Poisson distribution.

Comment upon your results.

Solution

Since n is large (1000) and p is small (1×10^{-3}) we can use the Poisson approximation to the binomial.

Example 37 *continued*

i Let X be the number of defective calculators. Applying the binomial distribution

16.24 $P(X = x) = {}^nC_x p^x q^{n-x}$

with $n = 1000$, $p = 1 \times 10^{-3}$ and $q = 1 - (1 \times 10^{-3}) = 0.999$ gives

$$P(X = x) = {}^{1000}C_x (1 \times 10^{-3})^x (0.999)^{1000-x}$$

We want to find the probability of 3 defective calculators, therefore $x = 3$:

$$P(X = 3) = {}^{1000}C_3 (1 \times 10^{-3})^3 (0.999)^{997}$$
$$= 0.06128 \quad \text{(4 s.f.)}$$

ii Using the Poisson approximation, we will first need to find μ (mean value). **What is μ equal to in this case?**
By 16.34:

$$\mu = np$$
$$= 1000 \times 1 \times 10^{-3} = 1$$

Using

16.33 $P(X = x) = \dfrac{e^{-\mu}\mu^x}{x!}$

with $\mu = 1$ and $x = 3$ gives

$$P(X = 3) = \frac{e^{-1}1^3}{3!} = 0.06131 \quad \text{(4 s.f.)}$$

Note that the answers to **i** and **ii** agree to 3 s.f.

Example 37 illustrates that for small p and large n it is easier to use the Poisson distribution as an approximation to the binomial distribution. The mean and variance of the Poisson distribution are both equal to μ.

SUMMARY

The expected value, $E(X)$, is the mean value of a random variable, X, and is given by

16.26 $E(X) = x_1 P(X = x_1) + x_2 P(X = x_2) + \cdots + x_n P(X = x_n)$

The variance, σ^2, of a random variable, X, is given by

16.31 $\sigma^2 = E(X^2) - \mu^2$

where $\mu = E(X)$.

We say a discrete random variable, X, follows a Poisson distribution if

16.33 $P(X = x) = \dfrac{e^{-\mu}\mu^x}{x!} \qquad (x = 0, 1, 2, 3, \ldots)$

Exercise 16(f)

Solutions at end of book. Complete solutions available at www.palgrave.com/engineering/singh

1 The probability density function of a discrete random variable, X, is given by

$$P(X = x) = \frac{1}{5} \qquad x = 1, 2, 3, 4 \text{ and } 5$$

Find the expected value $E(X)$. What do you notice about your result?

2 Let X be the number of faulty items off a production line. In a sample of the items we obtain the following probability distribution:

x	1	2	3	4	5	6
$P(X = x)$	0.15	0.21	0.11	0.36	0.04	0.13

Determine $E(X^2)$.

3 🎛 [*signal processing*]   The probability distribution of a sampled signal is given by:

x (volts)	1	2	3	4	5
$P(X = x)$	0.1	0.2	0.4	0.2	0.1

Determine the mean, μ, and standard deviation, σ.

What do you notice about the mean value?

4 🎛 [*signal processing*]   The probability distribution of a sampled signal is given by:

x (volts)	1	2	3	4	5	6	7	8
$P(X = x)$	0.10	0.12	0.15	0.20	0.17	0.13	0.07	0.06

Determine the mean, μ, and standard deviation, σ.

5 🎛 [*signal processing*]   A random signal has the following probability distribution:

x (volts)	0	1	2	3	4
$P(X = x)$	0.35	0.33	0.19	0.12	0.01

Estimate a value for the mean.

Calculate the mean and standard deviation of the distribution.

6 The probability density function of a discrete random variable is given by

$$P(X = x) = kx^2 \qquad x = 1, 2, 3 \text{ and } 4$$

Determine
i k　　　　　　　**ii** $E(X)$
iii standard deviation　**iv** variance

7 The probability distribution of a discrete random variable is given by:

x	1	2	3	4
$P(X = x)$	0.1	0.3	0.2	0.4

Evaluate

i $E(X)$　**ii** $E(X^2)$　**iii** $E(X^2 + X)$
iv $E(5X^2 + 7X + 3)$

8 **i** Show that $E(k) = k$ where k is a constant.
ii Show 16.29 :

$$E[af(X) + b] = aE[f(X)] + b$$

9 Let X be the number of goals scored in a match. A survey of matches produces the following probability distribution:

x (number of goals scored)	0	1	2	3	4	5	6
$P(X = x)$	0.05	0.15	0.2	0.25	0.15	0.1	0.1

Determine the mean number of goals μ and standard deviation σ.

10 Assume the probability of having a girl or a boy is equally likely. Determine the expected number of girls in a family which has three children. Comment on your result.

Solutions at end of book. Complete solutions available at www.palgrave.com/engineering/singh

11 Suppose you roll a pair of dice and sum the total score on both dice. Draw a probability distribution table and predict the expected score. Evaluate the expected score and variance of this distribution.

12 Let X follow a Poisson distribution with a mean of 1.6. Find the following probabilities:

    a $P(X = 0)$ **b** $P(X = 1)$
    c $P(X \leq 2)$ **d** $P(X = 10)$

13 Consider a communication network where the number of errors follows the Poisson distribution. If the probability of an error in a bit is 1×10^{-3}, find the probability of

    a no errors in 4096 bits
    b more than 2 errors in 4096 bits.

SECTION G **Properties and applications of continuous random variables**

By the end of this section you will be able to:

▶ understand the term *continuous random variable*
▶ see the relationship between probability and its density function
▶ apply the properties of continuous random variables
▶ apply the above to reliability engineering
▶ find various functions of reliability engineering

This section is difficult compared to the previous sections. There are a number of functions related to reliability engineering which are used here. It is very easy to get confused between the various functions in this particular field.

G1 **Continuous random variables**

? **Can you think of any variables that are continuous?**

Output of an analogue system, mass, lifetime of a component, etc.

A continuous random variable, T, is a variable that can have any values between two end points, unlike discrete random variables which can only take certain values.

The probability of a discrete random variable is given by a sum.

The probability of a continuous random variable is given by an integral.

The probability density function of T, with a particular value t, is normally denoted by $f(t)$. The probability of T having values between a and b is given by the area under the probability density function, $f(t)$, as shown shaded in Fig. 22.

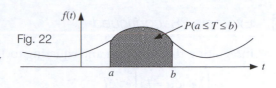

Fig. 22

The area can be evaluated by integrating:

16.37a $\qquad P(a \le T \le b) = \int_a^b f(t)\mathrm{d}t$

which means that the probability of T having values between a and b is the integral $\int_a^b f(t)\mathrm{d}t$. Consider the following example.

Let T represent the time that a component lasts, or the failure time of a component. Then $P(a \le T \le b)$ is the probability that the component will fail between (or at) the times a and b, and is evaluated by $\int_a^b f(t)\mathrm{d}t$.

The other property is

16.37b $\qquad P(-\infty < T < \infty) = \int_{-\infty}^{\infty} f(t)\mathrm{d}t = 1$

The probability of T having any value is 1. If T represents the failure time of a component then T can only have positive values and zero. Hence 16.37b becomes

16.37c $\qquad P(0 \le T < \infty) = \int_0^{\infty} f(t)\mathrm{d}t = 1$

This means that the component will eventually fail.

G2 Properties of continuous random variables

In this subsection we state some properties of continuous random variables. They are identical to similar properties of the discrete random variable from the previous section.

Example 38 *reliability engineering*

The probability density function of failure time (in years) of a component is given by

$$f(t) = ke^{-t}, \quad 0 \le t \le 3$$

i Determine k.
ii For this value of k find the probability that the failure time is at most 2 years.

Solution

i Let T be the continuous random variable. By formulae 16.37a and 16.37c we have

$$P(0 \le T \le 3) = \int_0^3 ke^{-t}\mathrm{d}t = 1$$

Example 38 *continued*

Integrating this function and equating to 1 gives

$$\int_0^3 ke^{-t}dt = k\int_0^3 e^{-t}dt \qquad\qquad \text{[Taking out the constant } k]$$

$$= -k\big[e^{-t}\big]_0^3 \qquad\qquad \left[\text{Integrating by } \int e^{kt}\,dt = \frac{e^{kt}}{k}\right]$$

$$\underset{\text{Substituting limits}}{=} -k\big[e^{-3}-e^0\big] = -k[0.05-1] = 0.95k = 1$$

Dividing by 0.95 yields $k = \dfrac{1}{0.95} = 1.05$.

ii We need to find $P(0 \le T \le 2)$ which is given by $P(0 \le T \le 2) = \displaystyle\int_0^2 ke^{-t}\,dt$

Substituting $k = 1.05$ into this formula yields

$$P(0 \le T \le 2) = \int_0^2 1.05e^{-t}dt = 1.05\int_0^2 e^{-t}dt$$

$$= -1.05\big[e^{-t}\big]_0^2 \qquad\qquad \text{[Integrating]}$$

$$\underset{\text{Substituting limits}}{=} -1.05\big[e^{-2}-e^0\big] = -1.05[e^{-2}-1] = 0.908$$

The probability that the failure time is at most 2 years is 0.908 (3 d.p.).

Let T be a continuous random variable with a probability density function given by $f(t)$. The expected value $E(T) = \mu$ of T is defined as

16.38 $\quad E(T) = \displaystyle\int_0^\infty tf(t)\ dt$

Example 39

The probability density function of a continuous random variable is given by

$$f(t) = \frac{t}{8}, \quad 0 \le t \le 4$$

Determine $E(T)$.

Solution

Using the above formula **16.38**, $E(T) = \displaystyle\int_0^\infty tf(t)\ dt$, with $f(t) = \dfrac{t}{8}$ and the limits 0 and 4 because the function is only valid between these values:

Example 39 *continued*

$$E(T) = \int_0^4 t\left(\frac{t}{8}\right)dt = \frac{1}{8}\int_0^4 (t^2)dt \qquad\qquad \left[\text{Taking out } \frac{1}{8}\right]$$

$$\equiv \frac{1}{8}\left[\frac{t^3}{3}\right]_0^4 = \frac{1}{24}[4^3 - 0^3] = \frac{64}{24} = \frac{8}{3}$$

Integrating

Hence $E(T) = 8/3$.

In general if $g(t)$ is a function of the continuous random variable T with a probability density function $f(t)$ then

$$E[g(T)] = \int_0^\infty g(t)\, f(t)\, dt$$

? **What is $E(T^2)$ equal to?**

$$E(T^2) = \int_0^\infty t^2 f(t)\, dt$$

By using the properties of integration we can show that

$$E[af(T) + bg(T)] = aE[f(t)] + bE[g(t)]$$

We can define the variance σ^2 of the continuous random variable T as

$$\sigma^2 = E(T-\mu)^2 \text{ where } \mu = E(T)$$

From this we can deduce the following:

16.39 $$\sigma^2 = E(T^2) - \mu^2$$

Example 40

Let T be the time taken in minutes to see a doctor. The surgery stipulates that no one will have to wait longer than 30 minutes. The probability density function is given by

$$f(t) = \frac{t}{450}, \quad 0 \le t \le 30$$

Determine

a the mean waiting time μ
b the standard deviation σ
c the probability that you will see a doctor within 10 minutes.

Solution

a The mean waiting time $\mu = E(T)$, which is defined above as

Example 40 *continued*

16.38 $E(T) = \int_{\sqrt 0}^{\infty} tf(t)dt$

Substituting the given $f(t)$ and the limits 0 and 30 into this formula yields

$$\mu = E(T) = \int_0^{30} t\left(\frac{t}{450}\right)dt = \frac{1}{450}\int_0^{30}(t^2)dt \qquad \left[\text{Taking out } \frac{1}{450}\right]$$

$$= \frac{1}{450}\left[\frac{t^3}{3}\right]_0^{30} = \frac{1}{450}\left[\frac{30^3 - 0^3}{3}\right] = 20$$

Thus the mean waiting time is 20 minutes.

b We first find the variance σ^2 and then take the square root to give the standard deviation. The variance was defined on the previous page as

16.39 $\sigma^2 = E(T^2) - \mu^2$

From part **a** we have $\mu = 20$, but we need to find $E(T^2) = \int_0^{\infty} t^2 f(t)\, dt$. Substituting

$f(t) = \dfrac{t}{450}$ and the limits 0 and 30 into this formula gives

$$E(T^2) = \int_0^{30} t^2\left(\frac{t}{450}\right)dt = \frac{1}{450}\int_0^{30}(t^3)dt$$

$$= \frac{1}{450}\left[\frac{t^4}{4}\right]_0^{30} = \frac{1}{450}\left[\frac{30^4 - 0^4}{4}\right] = 450$$

Substituting $E(T^2) = 450$ and $\mu = 20$ into $\sigma^2 = E(T^2) - \mu^2$ gives

$$\sigma^2 = E(T^2) - \mu^2 = 450 - 20^2 = 50$$

Taking the square root gives the standard deviation $\sigma = \sqrt{50} = 7.07$ (2 d.p.).

c The probability that you will see the doctor within 10 minutes is given by:

$$P(0 \le T \le 10) = \int_0^{10}\left(\frac{t}{450}\right)dt = \frac{1}{450}\int_0^{10} t\, dt$$

$$= \frac{1}{450}\left[\frac{t^2}{2}\right]_0^{10} = \frac{1}{900}[10^2 - 0^2] = \frac{1}{9}$$

This means there is a one in nine chance that you will see the doctor within ten minutes.

G3 **Reliability engineering**

Many of the engineering applications in this area lie in the field of reliability engineering.

Example 41 *reliability engineering*

The density function of failure time of a component is given by

$$f(t) = 0.3e^{-0.3t}$$

where t is in years. In reliability engineering, $f(t)$ is normally called the failure density function rather than the probability density function.

Find the probability that the component will fail within 5 years of operation.

Solution

Let T be the failure time of the component. Using

16.37a $P(a \le T \le b) = \int_a^b f(t)\mathrm{d}t$

with $a = 0$, $b = 5$ and $f(t) = 0.3e^{-0.3t}$ gives

$$P(0 \le T \le 5) = 0.3\int_0^5 e^{-0.3t}\mathrm{d}t$$

$$= 0.3\left[\frac{e^{-0.3t}}{-0.3}\right]_0^5 \qquad \left[\text{Integrating by } \int e^{kt}\mathrm{d}t = \frac{e^{kt}}{k}\right]$$

$$= -\left[e^{-0.3\times5} - e^0\right] \quad \text{[Substituting and cancelling 0.3's]}$$

$$= -\left[e^{-1.5} - 1\right] \quad \text{[Because } e^0 = 1]$$

Remember that taking a minus inside the brackets changes the signs:

$$P(0 \le T \le 5) = 1 - e^{-1.5} = 0.777 \quad \text{(3 d.p.)}$$

Hence there is more than a 77% chance that the component will fail within 5 years.

Example 42 *reliability engineering*

The failure density function, $f(t)$, for a particular set of components is given by

$$f(t) = \lambda e^{-\lambda t} \quad (t \text{ in years})$$

where $\lambda > 0$ is a constant. Show that the probability of a component failing within 10 years is given by

$$1 - e^{-10\lambda}$$

Example 42 *continued*

Solution

Using

16.37a $\qquad P(a \leq T \leq b) = \int_a^b f(t)\,dt$

with $a = 0$, $b = 10$ and $f(t) = \lambda e^{-\lambda t}$ yields

$$P(0 \leq T \leq 10) = \lambda \int_0^{10} e^{-\lambda t}\,dt \quad \text{[Taking out } \lambda\text{]}$$

$$= \lambda \left[\frac{e^{-\lambda t}}{-\lambda} \right]_0^{10} \quad \left[\text{Integrating by } \int e^{kt}\,dt \checkmark = \frac{e^{kt}}{k} \right]$$

$$= \frac{-\lambda}{\lambda} \left[e^{-\lambda t} \right]_0^{10} = -(e^{-10\lambda} - e^0) = 1 - e^{-10\lambda}$$

The cumulative distribution function, $F(t)$, of a continuous random variable, T, is defined as

16.40 $\qquad F(t) = P(T \leq t)$

For example, let T be the continuous random variable of failure time of a component. Then the cumulative distribution function, $F(t)$, is given by

16.41 $\qquad F(t) = P(0 \leq T \leq t)$

because T **cannot** have negative values. This distribution function in reliability engineering is called the **failure** distribution function. $F(t)$ gives the probability of a component failing within (or at) t years. We will use the word 'within' to mean $0 \leq T \leq t$ rather than $0 < T < t$.

For **Example 41**, we evaluated $F(5)$ – the probability of the component failing within 5 years.

Example 43 *reliability engineering*

The failure distribution function of a system is given by

$$F(t) = 1 - e^{-0.5t}$$

Determine the probability that the system will fail within 6 years.

Solution

Substituting $t = 6$ into our given function

$$F(t) = 1 - e^{-0.5t}$$

yields

$$F(6) = 1 - e^{-0.5 \times 6}$$

$$= 1 - e^{-3}$$

$$= 0.95 \quad \text{(2 d.p.)}$$

The probability that the system will fail within 6 years is 0.95 (2 d.p.).

? **What is the relationship between $f(t)$ and $F(t)$?**

Since $F(t)$ is the area under the graph of $f(x)$
between 0 and 1 (Fig. 23) we have

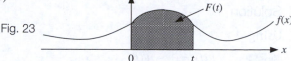

Fig. 23

16.42 $$F(t) = \int_0^t f(x)\,dx$$

Remember that $F(t)$ gives the probability of a component failing within t years.

Differentiating this we have

16.43 $$f(t) = \frac{dF}{dt}$$

G4 Reliability functions

In reliability engineering the continuous random variable, T, represents the failure time of a component. There are three particular functions which need to be defined:

1 Reliability function, $R(t)$, which gives the probability of a component lasting more than t years, that is does not fail within t years.

$$P(\text{does not fail within } t \text{ years}) = 1 - P(\text{fails within } t \text{ years})$$

$$= 1 - F(t)$$

where $F(t)$ is the failure distribution function. Hence

16.44 $$R(t) = 1 - F(t)$$

2 Hazard function, $h(t)$, is the failure rate of the component and is defined as

16.45 $$h(t) = \frac{f(t)}{R(t)}$$

Remember that $f(t)$ is the derivative of $F(t)$.
3 The Mean Time To Failure, $MTTF$, is defined as

16.46 $$MTTF = \int_0^\infty R(t)\,dt$$

Generally these functions are **only** defined over a strictly bounded domain.

Example 44 *reliability engineering*

Let the failure distribution function, $F(t)$, for a system be given by

$$F(t) = 1 - e^{-t/2}$$

Determine the reliability function, $R(t)$, for t equal to 5 years. This gives the probability that the system will last more than 5 years.

Example 44 *reliability engineering*

Solution

Using

16.44 $R(t) = 1 - F(t)$

with $F(t) = 1 - e^{-t/2}$, gives

$$R(t) = 1 - (1 - e^{-t/2})$$
$$= 1 - 1 + e^{-t/2}$$
$$= e^{-t/2}$$
$$R(5) = e^{-5/2} = 0.082 \quad (3 \text{ d.p.})$$

The probability that the system will last more than 5 years is 0.082.

Example 45 *reliability engineering*

The failure distribution function, $F(t)$, of a building system is given by

$$F(t) = 1 - e^{-t/100}$$

where t is in years. Determine

i the reliability function, $R(t)$
ii the hazard function, $h(t)$.

Solution

i By using $R(t) = 1 - F(t)$ we have

$$R(t) = 1 - (1 - e^{-t/100}) = 1 - 1 + e^{-t/100} = e^{-t/100}$$

ii $h(t) = \dfrac{f(t)}{R(t)}$. **What do we need to find first?**

$$f(t)$$

What is $f(t)$?

Using 16.43, it means we need to differentiate $F(t)$:

$$f(t) = \frac{d}{dt}\left(1 - e^{-t/100}\right) = 0 - \underbrace{\left(-\frac{1}{100}\right)e^{-t/100}}_{\text{by } 6.11} = \frac{1}{100}e^{-t/100}$$

Substituting these into $h(t) = \dfrac{f(t)}{R(t)}$ gives

6.11 $(e^{kt})' = ke^{kt}$ 16.43 $f(t) = dF/dt$

Example 45 *continued*

$$h(t) = \frac{\dfrac{1}{100}e^{-t/100}}{e^{-t/100}} = \frac{1}{100} \quad [\text{Cancelling } e^{-t/100}]$$

We can say the system has a constant failure rate because there are no variables in the answer.

 Example 46 *reliability engineering*

Determine the *MTTF* (Mean Time to Failure) of the failure density function

$$f(t) = 0.2 - 0.02t \quad 0 \le t \le 10$$

where t is in years.

Solution

What do we need to evaluate first?

$R(t)$ and $F(t)$ where these are defined for $0 \le t \le 10$.

This time we have $f(t)$ and need $F(t)$, therefore we need to integrate $f(t)$ to find $F(t)$. This is the reverse of the previous example.

By

16.42 $$F(t) = \int_0^t f(x)dx$$

with $f(x) = 0.2 - 0.02x$ (replacing t with x in the given function), we have

$$F(t) = \int_0^t (0.2 - 0.02x)dx$$

$$= \left[0.2x - \frac{0.02x^2}{2} \right]_0^t \quad [\text{Integrating}]$$

$$F(t) = 0.2t - 0.01t^2 \quad [\text{Substituting}]$$

Using $R(t) = 1 - F(t)$ with

$$F(t) = 0.2t - 0.01t^2$$

gives

$$R(t) = 1 - (0.2t - 0.01t^2)$$
$$= 1 - 0.2t + 0.01t^2$$

Next we use $MTTF = \int_0^\infty R(t)dt$ with $R(t) = 1 - 0.2t + 0.01t^2$ and the function is only valid for $0 \le t \le 10$:

$$MTTF = \int_0^\infty (1 - 0.2t + 0.01t^2)dt$$

$$= \int_0^{10} (1 - 0.2t + 0.01t^2)dt \quad [\text{Changing limits}]$$

Example 46 *continued*

$$= \left[t - \frac{0.2t^2}{2} + \frac{0.01t^3}{3} \right]_0^{10} \qquad \text{[Integrating]}$$

$$= 10 - \left(\frac{0.2 \times 10^2}{2} \right) + \left(\frac{0.01 \times 10^3}{3} \right) \quad \text{[Substituting]}$$

$$= \frac{10}{3} \text{ years} = 3\frac{1}{3} \text{ years}$$

The Mean Time to Failure, *MTTF*, is just over 3 years.

Example 47 *reliability engineering*

The failure density function for a set of components is given by

$$f(t) = 0.2(1 - 0.1t) \quad 0 \le t \le 10$$

where *t* is in years.

i Determine the reliability function, $R(t)$, and the hazard function, $h(t)$.
ii Find the probability that a component does not fail within 6 years.
iii In what time period will **all** the components fail?

Solution

i Note that by multiplying the brackets by 0.2 we have the same failure density function as **Example 46**:

$$f(t) = 0.2 - 0.02t \quad 0 \le t \le 10$$

Thus from this **Example 46** we have

$$R(t) = 1 - 0.2t + 0.01t^2$$

Substituting these into $h(t) = \dfrac{f(t)}{R(t)}$ gives the hazard function:

$$h(t) = \frac{0.2 - 0.02t}{1 - 0.2t + 0.01t^2} \quad 0 \le t \le 10$$

ii To find the probability that a component does not fail within 6 years we substitute $t = 6$ into $R(t) = 1 - 0.2t + 0.01t^2$:

$$P(\text{does not fail within 6 years}) = R(6)$$

$$= 1 - (0.2 \times 6) + (0.01 \times 6^2) = 0.16$$

iii The given density function is

$$f(t) = 0.2(1 - 0.1t) \quad 0 \le t \le 10$$

which says that a component will not last more than 10 years. So the fact that our limits are set at maximum 10 tells us that all components will fail within 10 years.

SUMMARY

The probability of a continuous random variable, T, having values between a and b is given by

16.37a $P(a \leq T \leq b) = \int_a^b f(t)\mathrm{d}t$

where $f(t)$ is the probability density function.

The expected value $E(T) = \mu$ is

16.38 $E(T) = \int_0^\infty tf(t)\,\mathrm{d}t$

The variance σ^2 is given by

16.39 $\sigma^2 = E(T^2) - \mu^2$

For the cumulative distribution function, $F(t)$, we have

16.40 $F(t) = P(T \leq t)$

In reliability engineering $f(t)$ and $F(t)$ are called the failure density function and the failure distribution function respectively.

The reliability function, $R(t)$, is defined by

16.44 $R(t) = 1 - F(t)$

The hazard function is defined as

16.45 $h(t) = \dfrac{f(t)}{R(t)}$

Mean Time To Failure, $MTTF$, is defined as

16.46 $MTTF = \int_0^\infty R(t)\mathrm{d}t$

Exercise 16(g)

Solutions at end of book. Complete solutions available at www.palgrave.com/engineering/singh

1 Let T (in minutes) be the continuous random variable which represents the waiting time for a bus. T has the probability density function given by

$$f(t) = \frac{t}{50}, 0 \leq t \leq 10$$

Determine

i mean μ **ii** $E(T^2)$ **iii** variance σ^2

iv the probability that waiting time is at most 5 minutes

v the probability that waiting time is more than 5 minutes.

2 Let T be the continuous random variable which represents the time (in years) it takes for a television set to break down. The probability density function of T is given by

$$f(t) = \frac{t^3}{2^{10}}, 0 \leq t \leq 8$$

Determine the probabilities that a television set breaks down

a within a year
b between the 1st and 2nd years
c after 6 years
d within 2 years or after 6 years.

Exercise **16(g) continued**

Solutions at end of book. Complete solutions available at www.palgrave.com/engineering/singh

3 The completion time T (in hours) for a certain task has the probability density function given by

$$f(t) = \frac{t}{48}\left(1 - \frac{t}{24}\right), \quad 0 \le t \le 12$$

a Compute mean time for completion μ.
b Compute $E(T^2)$.
c Compute variance σ^2.

⚙ **Questions 4 to 12 are in the field of** [*reliability engineering*].

4 Find the probability that a component will fail within 2 years for the following density functions:

a $f(t) = 0.5e^{-0.5t}$ **b** $f(t) = e^{-t}$

c $f(t) = 0.1e^{-0.1t}$

5 For the following failure distribution functions of a system determine the probabilities that the system will fail within 5 years:

a $F(t) = 1 - e^{-0.2t}$ **b** $F(t) = 1 - e^{-0.6t}$

c $F(t) = 1 - e^{-t/25}$

6 For the following failure distribution functions of a system determine the probabilities that the system will last more than 6 years:

a $F(t) = 1 - e^{-t/25}$ **b** $F(t) = 1 - e^{-t/2}$

c $F(t) = 1 - e^{-0.1t}$

7 The reliability function, $R(t)$, of a component is given by

$$R(t) = 1 - 0.04t^2 \quad 0 \le t \le 5$$

where t is in years.

i Evaluate the Mean Time To Failure, *MTTF*.
ii Find the hazard function, $h(t)$.

8 The hazard function

$$h(t) = 0.1t$$

Another definition of the reliability function is

$$R(t) = \exp\left[-\int_0^t h(x)\,dx\right]$$

Determine $R(t)$.

9

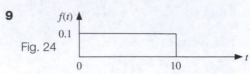

Fig. 24

Figure 24 shows a failure density function $f(t)$. Write $f(t)$ as an equation. Also determine the failure distribution function, $F(t)$, reliability function, $R(t)$, hazard function, $h(t)$, and *MTTF*.

10 Determine *MTTF* for the following failure density functions:

a $f(t) = 0.4\left(1 - \frac{t}{5}\right) \qquad 0 \le t \le 5$

b $f(t) = \frac{1}{3}e^{-t/3} \qquad t \ge 0$

where t is in years. (Remember that $\lim_{x \to \infty}(e^{-x}) = 0$.)

11 The median time to failure, t_{med}, is defined as

$$R(t_{\mathrm{med}}) = 0.5$$

where $R(t)$ is the reliability function. Determine t_{med} for

$$f(t) = 0.1e^{-0.1t} \quad (t \text{ is in years})$$

*12 For the reliability function, $R(t) = e^{-\lambda t}$, show that

$$MTTF = \frac{1}{\lambda} \text{ where } \lambda > 0$$

*13 A rectangular distribution is given by

$$f(t) = \frac{1}{b - a}, \, a \le t \le b. \text{ Show that}$$

i $\mu = \frac{a + b}{2}$ **ii** $\sigma = \frac{b - a}{\sqrt{12}}$

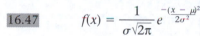

SECTION H **Normal distribution**

By the end of this section you will be able to:
▶ use the standard normal distribution to find probabilities
▶ examine examples which follow the normal distribution
▶ apply the standard normal distribution table to questions of probability

H1 Introduction

Suppose we measured the heights of 500 male adults.

❓ **What would the probability distribution curve look like?**

It would be symmetric about the mean because it seems equally likely that an individual will be 1 and 2 inches shorter or 1 and 2 inches taller than the mean.

❓ **What happens as you move further away from the mean?**

It is more likely that an individual is
1 or 2 inches shorter than the mean rather than
1 or 2 feet shorter than the mean. The
distribution might look like Fig. 25.

Fig. 25

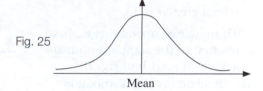

As you move further away from the mean, there
is less likelihood of finding an individual of that
height. There could be many curves which fit this description, but the one we examine in this section is called the normal distribution. The curve of Fig. 25 is a normal distribution which means it is **symmetric** about the mean and tails away at each extreme, and the probability density function, $f(x)$, is defined by

16.47 $$f(x) = \frac{1}{\sigma\sqrt{2\pi}} e^{-\frac{(x-\mu)^2}{2\sigma^2}}$$

Fig. 26

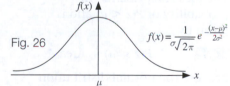

where μ is the mean and σ is the standard deviation.
Hence the graph of $f(x)$ is as shown in Fig. 26.

Try plotting $f(x)$ for different values of μ and σ on a graphical calculator. If you have MATLAB you can examine the normal distribution graph in one of the demonstrations. Try changing the values of the variables to see what happens to the graph.

H2 **Standard normal distribution**

Different μ and σ give different normal distribution curves.

Figure 27 on the page opposite shows normal distribution curves with the same mean, μ, but different standard deviations, $\sigma = 1$ and $\sigma = \sqrt{2}$.

Dealing with formula 16.47 is difficult. It is easier to use tables that are produced for the standard normal distribution which has mean = 0 and standard deviation = 1. We convert our normal distribution to the **standard** normal distribution by using the formula

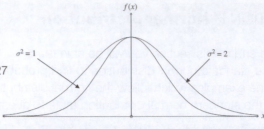

Fig. 27

$$16.48 \qquad z = \frac{x - \mu}{\sigma}$$

where z represents the number of standard deviations away from the mean and x is our continuous variable which follows a normal distribution.

We have carried out two operations in moving from **a** to **b** in Fig. 28:

1 From our given x we have subtracted μ so that the mean is shifted to 0.

Fig. 28

2 Then we have divided our result of **1** by the standard deviation, σ, which gives $\sigma = 1$.

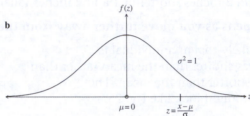

With these two operations we have converted to the standard normal distribution (part **b** of Fig. 28).

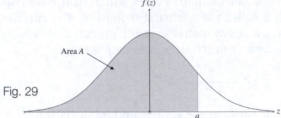

The standard normal distribution table (see **Appendix**, page 918) gives the area under the curve as shown in Fig. 29.

The area A represents the probability of $P(z < a)$ and is given by

Fig. 29

$$16.49 \qquad P(z < a) = \text{Area } A$$

because the normal distribution is a probability distribution.

Suppose we want to find the area corresponding to $z < 2.5$, we:

1 Look down the z (left-hand) column of the table of the **Appendix** on page 918 and find 2.5.

2 Look across at the 0 column (which is the second column) and we find .99379.

Hence the area under the normal distribution curve for $z < 2.5$ is 0.99379.

If we want to find the area corresponding to $z < 2.54$, then it is the same procedure as above but this time we look at the fourth column across from 2.5 and the area is 0.99446.

The total area covered under the curve is 1. Moreover $f(z)$ is symmetric about $z = 0$. Hence 0.5 of the area lies to the left of 0 and 0.5 to the right.

Example 48

Suppose that z follows the standard normal distribution. Calculate the following probabilities by using the table in the **Appendix** on page 918:

a $P(z < 0.5)$ **b** $P(z > 0.5)$ **c** $P(z < 1)$

Solution

We can illustrate each of these as the standard normal distribution.

a Using the table of the **Appendix** on page 918 with $z < 0.5$ gives 0.6915. Thus

$$P(z < 0.5) = 0.6915 \text{ (see Fig. 30)}$$

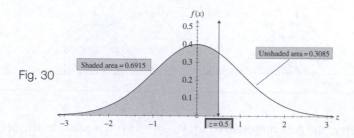

Fig. 30

Shaded area = 0.6915

Unshaded area = 0.3085

The value from the table gives the probability for $z < 0.5$.

b Since the total area under the curve is 1 we have

$$P(z < 0.5) + P(z > 0.5) = 1$$
$$P(z > 0.5) = 1 - P(z < 0.5) = 1 - 0.6915 = 0.3085.$$

This represents the unshaded area to the right of $z = 0.5$ in Fig. 30.

c Similarly we have Fig. 31. By using the table of the **Appendix** on page 918 we find $z = 1$ under the z column:

$$P(z < 1) = 0.8413$$

Remember that the area found from the table corresponds to the probability.

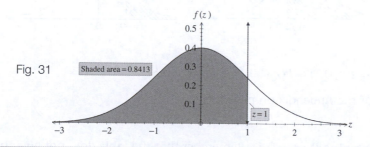

Fig. 31

Shaded area = 0.8413

Example 49

Suppose z follows the standard normal distribution. Evaluate the following probabilities:

a $P(z > -1)$ **b** $P(z < -1)$ **c** $P(-0.5 < z < 0.1)$

Solution

The table in the Appendix on page 918 *only* gives the area for positive values of z, so how are we going to evaluate $P(z > -1)$?

Since the normal distribution is symmetric we can examine $z < 1$.

a $z = -1$ where the minus sign means that the z value is below the mean (part **a** of Fig. 32).

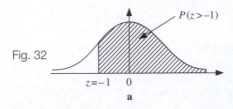
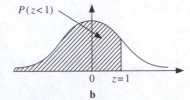

Fig. 32

Because of symmetry, the area represented by $P(z > -1)$ is the same as the area represented by $P(z < 1)$ (see parts **a** and **b** of Fig. 32). Thus by using the table we have

$$P(z > -1) = P(z < 1) = 0.8413$$

b $P(z < -1)$ is equal to the area on the left of $z = -1$ in part **a** of Fig. 32. By result **a** and knowing that the total area under the curve is 1, we have

$$P(z < -1) = 1 - 0.8413 = 0.1587$$

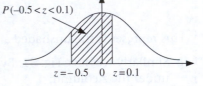

c See Fig. 33.

Fig. 33

We can find the shaded area of Fig. 33 by examining Fig. 34.

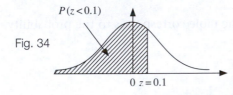

Fig. 34

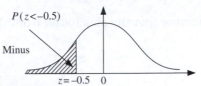

Thus

   $$P(-0.5 < z < 0.1) = P(z < 0.1) - P(z < -0.5)$$

By using the table of the **Appendix** on page 918

$$P(z < 0.1) = 0.5398$$

We also need to determine $P(z < -0.5)$. Remember that the table does not give the probability for negative z values, therefore

Example 49 *continued*

$$P(z < -0.5) = P(z > 0.5)$$
$$= 1 - P(z < 0.5)$$
$$= 1 - 0.6915 = 0.3085$$

Substituting these into ▮ * ▮ gives

$$P(-0.5 < z < 0.1) = P(z < 0.1) - P(z < -0.5)$$
$$= 0.5398 - 0.3085 = 0.2313$$

Example 50

The examination marks of an engineering module follow a normal distribution with mean $\mu = 40$ and the standard deviation $\sigma = 5$. If 1000 students took the examination and the pass mark is 35, how many students passed the examination?

Solution

This time we do not have the standard normal distribution as in the previous examples because the mean $\mu = 40$ and standard deviation $\sigma = 5$. We need to convert our data into the standard normal distribution so that we can use the table in the **Appendix** on page 918. **How?**

By using

16.48 $$z = \frac{x - \mu}{\sigma}$$

Our x value is the pass mark 35. Thus we have

$$z = \frac{x - 40}{5}$$
$$= \frac{35 - 40}{5}$$
$$z = -1$$

We therefore have the   Fig. 35
situation shown in Fig. 35.

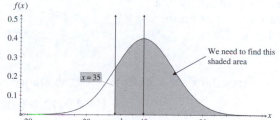

The area to the right of
35 in Fig. 35 represents
the students who have passed the examination. Using the table in the **Appendix** on page 918 with $z = 1$ gives 0.8413. We consider $z = 1$ rather than $z = -1$ because the normal distribution is symmetric, and the minus sign only indicates that z is to the left of the mean. The area above $z = -1$ is 0.8413.

The probability of a student passing the examination is 0.8413, therefore the number of students who passed the examination is

$$= 0.8413 \times 1000$$

$$= 841.3$$

841 candidates passed the examination.

Example 51

The resistance values of resistors in a batch follow a normal distribution with a mean of (1×10^3) Ω and a standard deviation of 100 Ω. Evaluate the probability of a resistor having a resistance of

a more than 1.2×10^3 Ω

b less than 0.8×10^3 Ω

c between 0.9×10^3 Ω and 1.1×10^3 Ω.

Solution

a Using

16.48 $$z = \frac{x - \mu}{\sigma}$$

with $\mu = 1 \times 10^3$ and $\sigma = 100$ we have

$$z = \frac{x - (1 \times 10^3)}{100}$$

$$= \frac{(1.2 \times 10^3) - (1 \times 10^3)}{100} \quad \text{[Substituting } x = 1.2 \times 10^3\text{]}$$

$$z = 2$$

The table in the **Appendix** on page 918 gives an area of 0.9772 for $z < 2$. Remember that the area corresponds to probability. Hence the probability of a resistance value of more than 1.2×10^3 Ω is

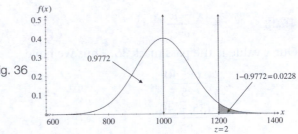

Fig. 36

$$P(\text{resistance} > 1.2 \text{ k}\Omega) = 1 - P(\text{resistance} < 1.2 \times 10^3 \text{ }\Omega)$$

$$= 1 - 0.9772 = 0.0228$$

See Fig. 36.

b 0.8×10^3 Ω is below the mean by the same amount as 1.2×10^3 Ω is above. For $x = 0.8 \times 10^3$ Ω, $z = -2$: since the normal distribution is symmetric, the probability of a resistance value less than 0.8×10^3 Ω is 0.0228. [Same as **a**.]

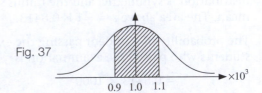

Fig. 37

c The probability of resistance values between 0.9×10^3 Ω and 1.1×10^3 Ω is represented in Fig. 37.

Example 51 *continued*

By 16.48 with $\mu = 1 \times 10^3$, $x = 1.1 \times 10^3$
and $\sigma = 100$ we have

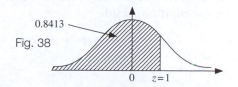

Fig. 38

$$z = \frac{(1.1 \times 10^3) - (1 \times 10^3)}{100} = 1$$

The table of the **Appendix** on page 918 gives
an area of 0.8413 for $z < 1$ (Fig. 38).

For $x = 0.9$:

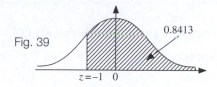

Fig. 39

$$z = \frac{(0.9 \times 10^3) - (1 \times 10^3)}{100} = -1$$

By symmetry for $z > -1$ we have Fig. 39.

Since the normal distribution is
symmetrical we can find the area between
$z = -1$ and $z = 0$ in Fig. 39 by subtracting
0.5 from 0.8413. So the area between
0.9 and 1.1 is as shown in Fig. 40.

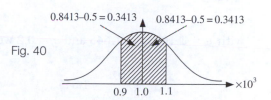

Fig. 40

The probability of a resistor having a value between $0.9 \times 10^3 \, \Omega$ and $1.1 \times 10^3 \, \Omega$ is

$$2 \times 0.3413 = 0.6826$$

Example 52

The lengths of certain components produced by a machine are normally distributed
with a mean of 1.107 m and standard deviation of 0.2 m. Find the values that constitute
the central 90% of the lengths.

Solution

Since the normal distribution is symmetric, each half
contains an area of 0.45, that is half of 0.9. We need
to find x_1 and x_2 of Fig. 41. The area less than the x_2
value is $0.5 + 0.45 = 0.95$.

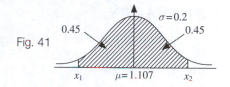

Fig. 41

Next we determine the corresponding z value from the table of the **Appendix** on page 918 for
0.95. Within the table we try to find the number 0.95. The closest number seems to be 0.9495.

z	0	1	2	3	4	5
•	•	•	•	•	•	•
•	•	•	•	•	•	•
1.6	•	•	•	•	0.9495	•

Example 52 *continued*

We need to add 0.0005 to this, 0.9495, to get 0.95, hence we look at the ADD columns on the right-hand side:

z	④	1 2 3 4 ⑤ 6 7 ADD
①.6	0.9495	· · · · 5

Therefore the z value for an area of 0.95 is 1.645.

Applying

16.48 $z = \dfrac{x - \mu}{\sigma}$

with $\mu = 1.107$, $z = 1.645$ and $\sigma = 0.2$ we have

$$\dfrac{x_2 - 1.107}{0.2} = 1.645$$
$$x_2 = 1.107 + (0.2 \times 1.645)$$
$$x_2 = 1.436$$

Similarly by symmetry:

$$\dfrac{x_1 - 1.107}{0.2} = -1.645$$
$$x_1 = 0.778$$

0.778 m to 1.436 m contains 90% of the central lengths.

SUMMARY

A normal distribution is a bell-shaped curve which is symmetric about the mean. We convert our normal distribution to a standard normal distribution so that we can use an established table to find the required probabilities. The conversion is carried out by

16.48 $z = \dfrac{x - \mu}{\sigma}$

where z is the number of standard deviations away from the mean.

Exercise **16(h)**

Solutions at end of book. Complete solutions available at www.palgrave.com/engineering/singh

1 Suppose that variable z follows the standard normal distribution. Calculate the following probabilities by using the table of the **Appendix** on page 918:

   a $P(z < 1.5)$

   b $P(z < 1.645)$

   c $P(z < 1.96)$

   d $P(z > 1.645)$

   e $P(z < -1.645)$

   f $P(z < -2)$

   g $P(-1.645 < z < 1.645)$

   h $P(-0.1 < z < 0.2)$

   i $P(-0.25 < z < 0.17)$

2 The lifetime of light bulbs follows a normal distribution with a mean of 500 hours and a standard deviation of 22 hours. Find the probability of a bulb lasting fewer than 540 hours.

3 The wearout times of a machine are normally distributed with a mean of 200 hours and a standard deviation of 10 hours. What is the probability that a machine has a wearout time of more than 220 hours?

4 The wearout times of a product have a normal distribution with a mean of 600 hours and a standard deviation of 20 hours. Determine the probability that such a product will have a wearout time of

   a more than 660 hours

   b fewer than 550 hours.

5 Fuses follow a normal distribution with a mean current of 2 amp and a standard deviation of 0.15 amp. Evaluate the proportion of fuses blowing with currents between 1.75 amp and 2.13 amp.

6 The heights of 500 students follow a normal distribution with a mean of 1.70 m and a standard deviation of 0.15 m. Determine the number of students who have a height

   a greater than 1.80 m

   b less than 1.50 m

   c between 1.65 m and 1.85 m.

7 The masses of students follow a normal distribution with a mean μ and a standard deviation of 1.2 kg. If 90% of the students have a mass of less than 70 kg, find the mean μ.

*8 The lengths of rods conform to a normal distribution with a mean of 0.5 m. If 95% of the lengths are less than 0.9 m, find the standard deviation σ.

9 The diameters of cylinders are normally distributed with a mean of 0.604 m and a standard deviation of 0.01 m. Find the values of diameters that contain the central 99% of the cylinders.

Examination questions **16**

Solutions at end of book. Complete solutions available at www.palgrave.com/engineering/singh

1 Find the mean and standard deviation of the following set of data

   2.35, 3.21, 1.85, 4.01, 2.73, 3.11, 2.16, 3.14

2 The lengths of a batch of steel rods are approximately normally distributed with mean of 2.8 m and standard deviation

of 0.13 m. Estimate the proportion of rods which are longer than 2.75 m or less than 2.95 m.

3 The speed of 50 cars passing a speed camera in a 30 mph zone is given on the next page:

Speed (mph)	20–23	23–26	26–29	29–32	32–35	35–45
Frequency	5	4	8	22	6	5

i Find the midpoint of each interval.
ii Find the mean of the distribution.
iii Find the standard deviation of the distribution.
iv If the speed follows a normal distribution with mean = 28 mph, and standard deviation = 1.25 mph, how many of the cars (in the sample of 50) would you expect to be driving between 26 and 29 mph.
v Comment on the sample.

Questions 1 to 3 are from University of Portsmouth, UK, 2007

4 A bag contains 12 balls, of which 5 are blue, 4 are red and 3 are yellow. Find the probability that if two balls are selected from the bag they will be the same colour.

5 The number of faults in a metre of copper wire has a Poisson distribution with average 4 faults. For a randomly selected metre of wire find the probability that there are more than three faults.

6 The weights of a certain type of chocolate bar are normally distributed with mean 50 g and standard deviation 2 g.

i Find the probability that a randomly selected bar weighs between 47 g and 52 g.
ii Bars are rejected if their weight lies outside the range 47–52 g. Find the probability that if ten bars are selected at random at least 3 will be rejected.
iii The probability that a bar weighs less than α grams is 0.8. Find the value of α.

Questions 4 to 6 are from University of Manchester, UK, 2008

7 a 3% of bolts made by a machine are defective. The bolts are packed in boxes of 50. Use a Poisson distribution to obtain the probability that the box contains at least two defective bolts.

b A random variable Z has a standard normal distribution, i.e. $Z \sim N(0, 1)$. Find the probability that it assumes a value

i less than 2.0
ii between -1.65 and -0.84.

c For electric light bulbs, the mean life is 1400 hours and the standard deviation is 200 hours. If the manufacturer guarantees a life of 1000 hours, what is the probability that a bulb chosen at random would fail to meet the guaranteed performance? (Use a normal distribution.)

d A chain is made of three links which are selected at random from a population of links. The strengths of the links are normally distributed with mean 500 units and standard deviation 10 units. Find the probability that the strength of the chain exceeds 490 units.

Loughborough University, UK, 2008

8 a In a certain batch of 20 valves, 2 are faulty. A sample of 4 valves is chosen at random from the batch. Find

i the probability that none of the valves in the sample are faulty
ii the probability that at least one of the valves in the sample is faulty.

Examination questions 16 continued

b Sampling and testing over a long period has shown that 3% of the components made on a particular production line are faulty. Write down the probability that a randomly chosen component is faulty.

A sample of 10 components is chosen at random. Assuming that the performance of each component is independent of the others in the sample, find

i the probability that all 10 of the components in the sample are in working order

ii the probability that exactly 9 of the components are in working order

iii the probability that at most one of the components is faulty.

University of Aberdeen, UK, 2004

Miscellaneous exercise 16

1 A manufacturer checks a sample of 50 capacitors from a batch which is labelled as 1 μF capacitors. The results are as follows:

1.091	1.020	0.901	0.905	0.943
1.010	1.047	0.984	1.084	1.093
1.061	0.953	0.961	0.999	1.089
1.064	0.951	0.947	1.042	1.021
1.019	1.003	0.910	1.067	0.932
1.089	1.086	0.988	1.031	1.022
0.954	1.002	1.014	0.998	1.035
1.065	1.057	0.996	1.072	1.080
1.008	1.013	1.007	0.998	1.069
1.012	1.006	0.997	1.098	1.003

Plot a histogram and a frequency polygon for this data.

2 The marks of students on an engineering module follow a normal distribution with mean of 45.5 and standard deviation of 5. Find the probability that a student selected at random doing this engineering module has

a fewer than 35 marks
b more than 50 marks
c between 50 and 60 marks.

3 The reliability of an engine working to specification is 0.9. If two engines are tested, what is the probability that at least one of them is working to specification?

4 An amplifier consists of two stages. Each stage has three distinct circuits, through which a signal can pass. Each stage is working if at least one of the circuits is activated by the signal. If the probability that a circuit is activated is 0.85, then determine the probability that

a the signal will pass the first stage
b the signal passes both stages.

5 Suppose there are n students in a room. What is the probability that no two of them have their birthday on the same day of the year for

i $n = 30$?
ii $n = 60$?

Are you surprised by your results?

(Assume there are 365 days in the year and a student is equally likely to have his/her birthday on any day of the year.)

6 In a multi-choice test there are 10 questions and 5 selections for each question. Only 1 selection is correct. If a student guesses all his/her answers, find the probability that the student will obtain

a all answers correct
b none correct
c exactly 4 correct.

7 The probability density function of a discrete random variable, X, is given by

$$P(X = x) = \frac{x^2}{k} \text{ for } x = 0, 1, 2, 3 \text{ and } 4$$

Determine

i k **ii** mean   **iii** standard deviation
iv variance.

8 A mechanical device has a wearout time distributed normally with a mean of 1100 hours and a standard deviation of 50 hours. Find the values that constitute the central 95% of the wearout times.

🔧 **The remaining questions, apart from 12, belong to [*reliability engineering*].**

9 For the reliability function

$$R(t) = \frac{k^2}{(k + t)^2} \quad t \geq 0$$

where k is a positive constant, show that the hazard function

$$h(t) = \frac{2}{k + t} \quad t \geq 0$$

10 A particular set of components has a constant failure rate, $h(t)$, given by

$$h(t) = \lambda \quad (\lambda \text{ is a constant})$$

By assuming the constant of integration to be zero, show that

$$R(t) = e^{-\lambda t}$$

11 The cumulative hazard function, $H(t)$, in reliability engineering is defined as

$$H(t) = -\int_0^t \frac{d}{dx} \Big[\ln[\, R(x)] \Big] dx$$

where $R(x)$ is the reliability function, $R(0) = 1$ and t is time. Show that

$$H(t) = -\ln[\, R(t)]$$

12 2% of the transformers in a batch are defective. Determine the largest sample size for the probability of no defectives to be > 0.8.

*13

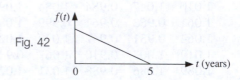

Fig. 42

Figure 42 shows a failure density function. What is the value of $f(t)$ at $t = 0$? Determine following functions of reliability engineering: failure distribution function, $F(t)$, reliability function, $R(t)$, and $MTTF$.

Sketch $F(t)$ and $R(t)$ on the same axes for $0 \leq t \leq 5$. What is the value of $F(t)$ and $R(t)$ for $t > 5$ and why?

14 The reliability function, $R(t)$, of a component is given by

$$R(t) = e^{-\lambda t}$$

where $\lambda > 0$ is a constant and t is in years.

Solutions at end of book. Complete solutions available at www.palgrave.com/engineering/singh

Miscellaneous exercise 16 continued

Plot, on the same axes, $R(t)$ against t for

a $\lambda = 1$ **b** $\lambda = 0.1$ **c** $\lambda = 0.01$

(You may use a computer algebra system or a graphical calculator.)

$R(t)$ represents the probability of the component surviving up to t years. What happens to $R(t)$ as λ gets smaller?

Note from your graphs that $R(0) = 1$ and $\lim_{t \to \infty} R(t) = 0$. Would you expect these results?

15 The failure density function, $f(t)$, for a particular set of components is given by

$$f(t) = \frac{4}{(1 + t)^5}$$

where t is in years. Determine the corresponding failure distribution function, $F(t)$, reliability function, $R(t)$, and the hazard function, $h(t)$. Determine the probability that the component lasts for more than 10 years.

By using a computer algebra system, or otherwise, plot $R(t)$ and $F(t)$ on the same axes for $t \geq 0$.

(You may use a computer algebra system or a graphical calculator for parts **a** and **b** of question **16**.)

16 The distribution function, $F(t)$, for a component is given by

$$F(t) = 4 - \frac{20}{5 + t/3} \quad 0 \leq t \leq 5$$

where t is in years.

a Determine the reliability function, $R(t)$, density function, $f(t)$, and the hazard function, $h(t)$.

b Plot the graphs of $F(t)$, $R(t)$, $f(t)$ and $h(t)$ on different axes for $0 \leq t \leq 5$.

c Determine the probabilities of a component lasting more than

  i 1 year **ii** 2 years **iii** 3 years
  iv 4 years **v** 5 years

d What does your answer to part **v** of **c** mean?

Solutions

For complete solutions see www.palgrave.com/engineering/singh

I(a) **1 a** > **b** < **c** < **d** ⩽ or ⩾ **e** <

2 a + **b** − **c** ÷ **d** × **e** −

3 a −5 **b** −2 **c** −2 **d** −22 **e** 0

4 a 3 **b** −12 **c** 12 **d** −3 **e** 3
   f −3 **g** −16

5 a −117 649 **b** 117 649 **c** −9 **d** −9
   e 1279 **f** 700

I(b) **1 a** 64 **b** 512 **c** 16 777 216 **d** 2025
   e 69 696 **f** 1 **g** −243

2 a ±7 **b** ±41 **c** 428 **d** 13 **e** 9
   f ±2 **g** 7 **h** ±2 **i** 3 **j** −3

3 a 8 **b** 13 **c** 36 **d** 6 **e** 10 **f** −10

4 a 20 **b** 108 **c** 2 **d** 3 **e** 2 **f** −2

I(c) **1 a** 30 **b** 42 **c** 54 **d** 42

2 a $2 \times 3^2 \times 5$ **b** $2^4 \times 3^2$ **c** 2×47
   d $3^2 \times 5 \times 11$

3 a 216 **b** 1728 **c** 2 178 000

I(d) **1** $3\frac{16}{113}$, 3, $2\frac{232}{323}$, $1\frac{169}{408}$, $9\frac{1}{7}$

2 a $\frac{1}{3}$ **b** $\frac{2}{5}$ **c** $\frac{18}{25}$ **d** $\frac{56}{75}$ **e** 1

3 a $\frac{16}{64} = \frac{1}{4}$ **b** $\frac{26}{65} = \frac{2}{5} \neq \frac{1}{5}$; not equivalent

   c $\frac{49}{89}$ and $\frac{4}{8}$ are not equivalent

4 a $\frac{10}{3}$ **b** $\frac{139}{69}$ **c** $\frac{987}{100}$

I(e) **1 a** $\frac{5}{6}$ **b** $\frac{13}{36}$ **c** $\frac{11}{60}$ **d** $\frac{12}{35}$ **e** $\frac{17}{12}$

2 a $\frac{1}{6}$ **b** $1\frac{23}{63}$ **c** $\frac{1}{145}$ **d** $\frac{61}{1271}$

3 a 1 **b** $\frac{47}{60}$ **c** $\frac{5}{84}$

4 a $\frac{1}{6}$ **b** $1\frac{59}{60}$ **c** 94

5 a $1\frac{1}{2}$ **b** $\frac{9}{20}$ **c** 94

6 a $1\frac{497}{1200}$ **b** $\frac{2}{97}$ **c** $\frac{1}{39780}$ **d** $\frac{1}{6360}$

   e $2\frac{1}{288}$ **f** $2\frac{1}{144}$ **g** $\frac{2870}{2871}$ **h** $\frac{2639}{2640}$

I(f) **1 a i** 1.618 **ii** 1.62 **b i** 4.6692 **ii** 4.67
   c i 2.503 **ii** 2.5 **d i** 0.374 **ii** 0.37

2 a i 1.62 **ii** 1.6 **b i** 2.9732 **ii** 3.0
   c i 0.1100 **ii** 0.1 **d i** 9.9 **ii** 10

3 a 1700 **b** 100 000 **c** 110 000 000 000

4 a 3.14, 2.72, 1.41 **b** 3.1, 2.7, 1.4

I(g) **1 a** 1.86×10^5 **b** 1.392×10^6 **c** 1.36×10^5
   d 3.439×10^{-8} **e** 9.51×10^{-8}
   f 9.29×10^{-3} **g** 2.58×10^{-5}
   h 1.496×10^7 **i** 2.7315×10^2 **j** 7.6×10^2

2 a 6 400 000 **b** 0.000 000 003 3
   c 0.000 072 92 **d** 300 000 000

3 a 12.75×10^{-3}, 12.75, 12.75×10^2, 12750
   b $3.14 \div 10^3$, 3.14×10^{-2}, 3.14×10^3

4 a 0.04 **b** 1.58 **c** 0.49

5 a 100 pF **b** 30 kΩ **c** 0.3 mA

6 a 8.536 kN **b** 75 MW **c** 200 GPa

7 a 3 m **b** 573×10^3 N **c** 25×10^6 J
   d 12×10^{-12} s **e** 25×10^{-3} W

8 a 3 **b** 3 **c** 10 000 **d** 30

I(h) **1** 111.6 km/h

2 9.81 m/s^2

3 765 miles per hour

4 1.206×10^{-3} g/cm^3

5 6.3 nm^4

6 0.465 m^2

7 a 13.4 m/s   **b** 22.22 m/s   **c** 2.99×10^8 m/s

8 2×10^6 N/cm^2

I(i) 1 a 135   **b** 144   **c** 441   **d** 175
   e 239   **f** 2311

2 a 2047   **b** -2 **c** 223/480

3 a 1   **b** 3   **c** 2 or 3   **d** -5 or 1
   e -5 or 5/14

4 a 461.81   **b** 83.78   **c** 523.60
   d 527.79   **e** 22.21

5 a 496   **b** 257

I(j) 1 a 10%   **b** $8\frac{1}{3}$%   **c** 87.5%   **d** $23\frac{1}{13}$%
   e 16.7%   **f** 258.3%

2 a $\dfrac{1}{25}$ **b** $\dfrac{9}{100}$ **c** $\dfrac{7}{40}$ **d** $\dfrac{1}{40}$

3 a 810 Ω to 990 Ω   **b** 1140 Ω to 1260 Ω
   c 26 730 Ω to 27 270 Ω   **d** 18 430 Ω to
   19 570 Ω   **e** 4 992 500 Ω to 5 007 500 Ω

4 114.6 km/h to 125.4 km/h

5 3.572 m

6 0.88%

7 4.97×10^{-3} m^3

8 8.6%

9 26.1 mA

I(k) 1 a 10:1   **b** 3:1   **c** 1:5:9   **d** 1:7:9:12
   e 4:5

2 a 8:15   **b** 42:29

3 a 4:5   **b** 13:18   **c** 1:2

4 0.128 m, 0.192 m and 0.32 m

5 Copper mass = 48 kg and zinc mass = 18 kg

6 Copper mass = 10.5 kg, zinc mass = 14 kg and nickel mass = 21 kg

7 Manufacturing = 180, building services = 270, vehicle = 450 and control = 360

MEI 1 1×10^{-9} m^3

2 a 125   **b** 10   **c** -2 **d** 2   **e** 13

3 a $\dfrac{19}{15}$ **b** $\dfrac{8}{15}$ **c** $\dfrac{10}{9}$

4 a i 3.142   **ii** 3.14   **b i** 3.16   **ii** 3.162
   c i 1.64   **ii** 1.645   **d i** 1 000 000
   ii 1 050 000
   e i 65 500   **ii** 66 000

5 a 1.79   **b** 3 or 4   **c** 8.19   **d** 118.08

6 a 1061.03   **b** 1×10^6 or 4×10^6

7 a 0.378 MV   **b** 10 μA   **c** 1.3 kΩ

8 a 95 Ω to 105 Ω   **b** 4.875 kΩ to 5.125 kΩ
   c 12.987 MΩ to 13.013 MΩ

9 6:43

10 $\dfrac{15}{4}$

11 15%

12 1.2 kW

13 a 1 333 333.33   **b** 407.45   **c** 335.12
   d 34 090.91   **e** 2.26

1(a) 1 $c = 25$

2 $I = 0.015$ A

3 $V_2 = 0.5 \times 10^{-4}$ m^3

4 $m = 0.094$ kg

5 $a = 1.6$ m/s^2

6 $\alpha = 6.8 \times 10^{-4}$/°C

7 $V = 11.4$ V

8 i $u = v - at$ **ii** $a = \dfrac{v - u}{t}$

9 $R = \dfrac{V}{I}$

10 $T = \dfrac{PV}{mR}$

11 $\omega = \dfrac{v}{r}\left(1 - \dfrac{S}{100}\right)$

1(b) **1** $V = \sqrt{PR}$

2 $P = \dfrac{\rho c^2}{\gamma}$

3 $A = \dfrac{2D}{\rho C v^2}$

4 $R = 73 \ \Omega$

5 $l = 62$ mm

6 $C = 1.4 \times 10^{-5}$ F $= 14$ µF

7 $D = \left(\dfrac{P}{2\pi k \rho n^3}\right)^{\frac{1}{5}}$

9 $r = \left(\dfrac{2T}{\pi \tau}\right)^{\frac{1}{3}}$

10 a $r = \dfrac{ER}{V} - R$ **b** $u = \sqrt{v^2 - 2as}$

c $K = \rho v^2 - \dfrac{4a}{3}$

d $r = \left(\dfrac{8vL\eta}{\pi Pt}\right)^{\frac{1}{4}}$ **e** $l = \dfrac{T^2 g}{4\pi^2}$

f $P = \dfrac{RT}{V - b} - \dfrac{a}{V^2}$

g $V_1 = \dfrac{W(n-1) + P_2 V_2}{P_1}$

h $r = \left(\dfrac{C}{P}\right)^{\frac{1}{n}}$ **i** $u = \dfrac{C^2 W^2 h}{f^2 D^2}$

1(c) **1 a** x^7 **b** $x^{\frac{7}{10}}$ **c** 1 **d** $\dfrac{1}{x^2}$ **e** $x^{\frac{11}{15}}$
f x^2 **g** x

2 a $(1 + y)^3$ **b** $(1 + x^2)^2$

c We *cannot* simplify this any further because under $\sqrt[3]{}$ we have different expressions

d $(x^2 + x + 1)^2$ **e** cannot simplify any further

3 0.734 m^3

1(d) **3 a** Incorrect   **b** Correct   **c** Correct
d Incorrect   **e** Incorrect

4 $ML^{-1}T^{-1}$

1(e) **1 a** $6x + 2$ **b** $-2x - 1$ **c** $-15y - 3$

d $3x^2 + 5x$ **e** $y - 4$ **f** $4x^2 - x$

2 a $\dfrac{wLx^3}{2EI} - \dfrac{wx^4}{2EI}$ **b** $\dfrac{wLx^3}{4EI} - \dfrac{3wx^4}{8EI}$

c $\dfrac{wL^2x^2}{16EI} - \dfrac{wx^4}{24EI}$

d $-\dfrac{wLx^3}{12EI} + \dfrac{wx^4}{24EI} + \dfrac{wL^3 x}{24EI}$

3 a $x^2 + 3x + 2$ **b** $6x^2 + 19x + 15$
c $4x^2 - 4x + 1$ **d** $4ab$ **e** $x^2y^2 + xy - x + 1$

5 a $R^2 - \omega^2 L^2$ **b** $\dfrac{1}{R^2} + \omega^2 C^2 - \dfrac{2C}{L} + \dfrac{1}{\omega^2 L^2}$

6 0 (zero)

1(f) **1 a** $4(x + y + z)$ **b** $8x(1 + y)$ **c** $2(x - 2y)$
d $x(3 - 2x)$ **e** $x(x - y)$ **f** $4x(4 + x)$
g $9x^2(1 - 3x)$

2 a $t\left(u + \dfrac{1}{2}at\right)$ **b** $\dfrac{m}{t}(v_2 - v_1)$ **c** $\rho A v_1(v_2 - v_1)$

3 $\pi r\left[r + (r^2 + h^2)^{\frac{1}{2}}\right]$

4 a $(x + 5)(x + 2)$ **b** $(x + 4)(x + 1)$
c $(x - 4)(x - 1)$ **d** $(x + 2)(x - 6)$
e $(2x - 1)(x + 1)$ **f** $(x - 4)(x + 1)$
g $(3x + 5)(7x - 2)$ **h** $(3x - 4)(2x + 3)$

5 a $(x - 1)^2$ **b** $(x + 1)^2$
c $(x - 6)(x + 6)$ **d** $(x - \sqrt{7})(x + \sqrt{7})$
e $(2x + 3)^2$

6 a $(Z - R)(Z + R)$ **b** $\left(\omega L - \dfrac{1}{\omega C}\right)\left(\omega L + \dfrac{1}{\omega C}\right)$

8 a $\dfrac{wx^2}{6EI}(3L - x)$ **b** $\dfrac{wx^3}{8EI}(2L - 3x)$

c $\dfrac{wx^2}{24EI}(x - L)^2$

1(g) **1 a** $x = \dfrac{1}{2}$ **b** $x = -2, x = -3$

c $x = 7, x = 3$ **d** $x = -\dfrac{1}{3}, x = \dfrac{5}{2}$

e $x = \dfrac{1}{5}, x = -3$ **f** $x = 1, x = -1$

g $x = 1$ **h** $x = -1$ **i** $x = -1, x = -2$

j $x = 1, x = 2$

2 $u = 5.31$ m/s

3 $x = \dfrac{L}{2}$ or $x = L$

For complete solutions see www.palgrave.com/engineering/singh

4 $t = 3.2$ s

5 $l = 12$ m, width $= 7$ m

6 $x = 5$

7 $x = 1.01$ m, $x = 0.25$ m

8 $t = 11.44$ s

9 $t = \dfrac{u \pm \sqrt{u^2 + 2gh_0}}{g}$

1(h) **1 a** $x = 1, y = 1$ **b** $x = 1, y = 1$
  c $x = 1, y = -1$ **d** $x = -29, y = -31$
  e $x = 3/4, y = 1/4$
  f $x = 3/4\pi, y = -1/4$

2 $a = 0.15, b = 35$ N

3 $a = 10$ m/s^2 and $u = 6.5$ m/s

4 $I_1 = 50$ mA and $I_2 = 73$ mA

5 $l_0 = 20$ m and $\alpha = 1 \times 10^{-4}$/°C

6 $R_1 = 1$ kΩ and $R_2 = 5$ kΩ

7 $F = K\rho Av^2$

EQ1 **1** $2x^2 - 9x + 10$

2 $a^{10/3}b^{19/10}$

3 -0.3551

4 $x = 5.5$ and $y = 1$

5 $x = 2$

6 $x = 0.1308, -6.7974$

7 $y^{-1/2}$

8 $x = \dfrac{11}{3}$

9 $x = -0.215, 1.549$

10 $m = \dfrac{M + 5t}{3t + 2}, m = 0.2584$

11 $x = \dfrac{108 + 3\sqrt{976}}{-8 + 2\sqrt{976}}, y = \dfrac{-4 + \sqrt{976}}{30}$

  and $x = \dfrac{108 - 3\sqrt{976}}{-8 - 2\sqrt{976}}, y = \dfrac{-4 - \sqrt{976}}{30}$

12 i $\dfrac{1}{36000}$ **ii** $2^{11}3^5$

13 0.8410

ME1 **1 a** $C = 1 - \left(\dfrac{u}{v}\right)^2$

4 41, 43, 47, 53, 61 and 71. All are prime numbers

5 a $f_0 = 2.25$ kHz   **b** $C = 25.3$ μF

7 $R = 1$ kΩ

8 $v = \dfrac{2D}{\rho A(C_D + kC_L^2)}$

9 $I = \dfrac{2E}{v_1^2 - v_2^2}$

10 $v_1 = \left(\dfrac{2(p_2 - p_1)}{\rho} + 2g(h_2 - h_1) + v_2^2\right)^{\frac{1}{2}}$

11 a $x = \dfrac{1}{5}$ **b** $x = 2$ **c** $x = 1, x = -2$

  d $x = \dfrac{1}{3}, x = -\dfrac{3}{2}$

12 $t = 1.43$ s

13 $u = 140$ m/s

14 $R = 3.42$ Ω

15 a $x = 5, x = 2$ **b** $x = 1, x = -1$

  c $x = \dfrac{1}{2}, x = 1$

  d $x = \dfrac{2}{5}, x = -\dfrac{1}{3}$ **e** $x - 3, x = 1$

17 $y = \dfrac{w}{36\,EI}(x^4 - 4Lx^3 + L^4)$

18 i $P = \dfrac{n^2\pi^2\,EI}{L^2}$ **ii** $P = 12 \times 10^6$ N

19 a $x = -0.38, x = -2.62$
  b $x = -0.59, x = -3.41$
  c $x = 0.29, x = -0.69$
  d $x = 0.28, x = -1.78$

21 $I_1 = 22$ mA and $I_2 = 70$ mA

24 a $x = 1, y = 1$ **b** $x = 58/5, y = -33/5$
  c $x = 1.77, y = 0.85$

26 $\zeta = 2\sqrt{mk}$

28 $\omega_1 = 20$ Hz, $\omega_2 = -20$ Hz, $\omega_3 = \sqrt{2}$ Hz

  and $\omega_4 = -\sqrt{2}$ Hz

31 $x = L, x = 0$

2(a) **1 a** $m = 1, c = 1$ **b** $m = 1, c = -1$

  c $m = 3, c = 5$ **d** $m = \dfrac{1}{2}, c = -7$

  e $m = 0, c = \pi$ **f** $m = 1, c = -\pi$

2(b) 1 a Acceleration = 0

b Total displacement = 301.5

2 $v = \begin{cases} \dfrac{1}{6}t & 0 \le t \le 60 \\ 10 & 60 < t \le 240 \\ -\dfrac{1}{12}t + 30 & 240 < t \le 360 \end{cases}$

3 $v = \begin{cases} 0.3t + 2 & 0 \le t \le 10 \\ 5 & 10 < t \le 20 \\ \dfrac{5}{3}(23 - t) & 20 < t \le 23 \end{cases}$

5 ii $t = 14$ s

6 a When $0 \le t \le 6$ and $14 \le t \le 20$

b Acceleration = −3.125

2(c) 5 Negative acceleration = deceleration, that is, the particle is slowing down

The initial deceleration is zero, and deceleration is continuously increasing

2(d) 1 a $(x - 2)^2 - 1$ **b** $(x + 4)^2 - 7$
c $(x - 3)^2 - 1$ **d** $(x - 5)^2 - 23$

e $25 - (x - 4)^2$ **f** $\left(x + \dfrac{7}{2}\right)^2 - \dfrac{45}{4}$

g $2\left(x + \dfrac{7}{4}\right)^2 - \dfrac{41}{8}$

2 a $x = 1, 3$ **b** $x = -4 - \sqrt{7}, 4 + \sqrt{7}$
c $x = 2, 4$ **d** $x = 5 \pm \sqrt{23}$

e $x = -1, 9$ **f** $x = \dfrac{-7 \pm \sqrt{45}}{2}$

g $x = \dfrac{-7 \pm \sqrt{41}}{4}$

3 Minimum voltage is −0.25 V

5 The maximum height is 36

6 The maximum height is 2041 m

7 Maximum value = $\dfrac{WL^2}{96EI}$

2(e) 8 The larger the k, the larger the velocity, v

9 As k increases, R decreases and as k goes to infinity, R goes to zero

10 As A increases, R decreases

13 a $R = 10$ **b** $R = 10 \times 10^3$ **c** $R = 1500$

Maximum power occurs at the value of the number inside the brackets

14 a $V = 30$ volts   **b** $V = 7.5$ volts

c $V = 5$ volts. Relationship is $V = E/2$

2(f) 1 a $x^6 + 6x^5 + 15x^4 + 20x^3 + 15x^2 + 6x + 1$

b $a^6 + 6a^5b + 15a^4b^2 + 20a^3b^3 + 15a^2b^4 + 6ab^5 + b^6$

c $1 - 6x + 15x^2 - 20x^3 + 15x^4 - 6x^5 + x^6$

2 a $x^4 + 20x^3 + 150x^2 + 500x + 625$

b $32 + 240x + 720x^2 + 1080x^3 + 810x^4 + 243x^5$

c $4096 - 18432x + 34560x^2 - 34560x^3 + 19440x^4 - 5832x^5 + 729x^6$

d $16x^4 - 32x^3y + 24x^2y^2 - 8xy^3 + y^4$

e $x^5 + 5x^3 + 10x + \dfrac{10}{x} + \dfrac{5}{x^3} + \dfrac{1}{x^5}$

f $x^4 - 4x^2 + 6 - \dfrac{4}{x^2} + \dfrac{1}{x^4}$

g $1 + 7x^2 + 21x^4 + 35x^6 + 35x^8 + 21x^{10} + 7x^{12} + x^{14}$

3 Amplitudes are 1, 9, 36, 84 and 126

4 $\dfrac{w^7}{16384} - \dfrac{7w^6x}{12288} + \dfrac{7w^5x^2}{3072} - \dfrac{35w^4x^3}{6912}$

$+ \dfrac{35w^3x^4}{5184} - \dfrac{7w^2x^5}{1296} + \dfrac{7wx^6}{2916} - \dfrac{x^7}{2187}$

EQ2 1 a −1.236, 3.236

2 a −1.43, 2.10

3 a $1 + 6x + 15x^2 + 20x^3 + 15x^4 + 6x^5 + x^6$, 1.772 (3 d.p.)

4 a $a^5 + 5a^4b + 10a^3b^2 + 10a^2b^3 + 5ab^4 + b^5$, $x^{10} + 5x^7 + 10x^4 + 10x + \dfrac{5}{x^2} + \dfrac{1}{x^5}$

5 $x = 0.184, 1.817$

6 a $(3x + 1)(3x - 5)$, $\left(-\dfrac{1}{3}, 0\right)$ and $\left(\dfrac{5}{3}, 0\right)$

b $9\left(x - \dfrac{2}{3}\right)^2 - 9, \left(\dfrac{2}{3}, -9\right)$

7 a $-1 + 4x - x^2$ **b** $x^2 + 4x + 3$

ME2 **6** $v = \begin{cases} t/2 & 0 \le t \le 20 \\ 2(t + 55)/15 & 20 < t \le 35 \end{cases}$

8 $s = 20t + 5$

10 As a increases y becomes less steep

11 $Q = \begin{cases} 0.4t & 0 < t \le 5 \\ 2 & t > 5 \end{cases}$

3(a) **1** 1, 4, 9, 1, 4, 9. Many to one

2 32, 212, 75.2

3 1, −1, 8, −8, 27, −27

4 −4.9, −48.36, −1

5 Not a function

6 a 0, 9.8, 19.6, 49   **b** $9.8t^2, 9.8t + 9.8$

7 a 5, 6.125, 14.375   **b** $t^3 + 2t^2 + t + 5$

8 $I^2 R$

9 $\dfrac{E}{2}, \dfrac{2E}{3}$

10 a $t = 0, t = 40.82$ **b** $380.4 + 180.4t - 4.9t^2$

3(b) **1 a** $f^{-1}(x) = \dfrac{x + 2}{7}$ **b** $f^{-1}(x) = \dfrac{1}{x}$ $(x \ne 0)$

c $f^{-1}(x) = 1 - x$ **d** $f^{-1}(x) = 5 - \dfrac{3}{x}$ $(x \ne 0)$

2 a $g^{-1}(t) = \dfrac{2t}{1 - t}$ $(t \ne 1)$

b $h^{-1}(x) = \dfrac{2x + 1}{x + 1}$ $(x \ne -1)$

c $f^{-1}(x) = \dfrac{x + 1}{2 - x}$ $(x \ne 2)$

3 ii $g^{-1}(t) = \dfrac{t + 1}{3t + 1}$

4 ii $g^{-1}(t) = \dfrac{t^3 + 3}{2 - t^3}$ $(2 - t^3 \ne 0)$

3(c) **2** The graphs are reflections of each other in the horizontal axis

4 The f graph is a stretch of the g graph (or vice versa)

3(d) **1 i** $2(x + 2)$ **ii** $2x^2 + 5x + 3$

iii $2x + 2$ **iv** $3x + 4$ **v** $-x - 2$

vi $\dfrac{x + 1}{2x + 3}$ $(2x + 3 \ne 0)$

2 i $4x + 9$ **ii** $8x + 21$ **iii** -3

3 i c

ii $ax^2 + (b + 2a)x + (a + b + c)$

iii $2ax + a + b$

4 i $x + 2$ **ii** $x^2 + 2x + 1$ **iii** $x^2 + 1$

5 i $x^2 - 1$ **ii** $x^2 - 1$ **iv** x **v** $x^2(x^2 - 2)$

6 i $\dfrac{3x - 6}{x}$ $(x \ne 0)$ **ii** x **iii** x

7 i x **ii** x

8 $f \circ f^{-1} = f^{-1} \circ f = x$

9 i $f \circ g$ or $g - 1$ **ii** $g \circ f$ **iii** $(g \circ f) + 7$

iv f or $g(f^2)$

10 $F(t) = \dfrac{t}{5}, R(t) = 1 - \dfrac{t}{5}$ and $h(t) = \dfrac{1}{5 - t}$

11 a $\dfrac{k}{s^2 + s + 0.01 k}$ **b** $\dfrac{1}{s + (k_1 + k_2)}$

c $\dfrac{1}{s + 2.3}$

12 $\dfrac{10s}{s^3 + 10s^2 + 21s + 10}$

3(e) **1 a** 1   **b** 0

2 10 m/s

3 a 6   **b** −3/2   **c** 2

4 2/3

5 0

6 1/k

7 0

8 2x

9 $3x^2$

3(f) **1** Same graph $|x - 1| = |1 - x|$

EQ3 **1** Domain and range are the set of all real numbers

2 a Reflected in the x axis, squashed horizontally by 3 and shifted up by 2

b $x = -3$

3 Domain is $x \geq -1$ and range is $f(x) \leq 2$

4 $8500

5 $-\dfrac{1}{3(x^2 - xh)}$

6 a 12   **b** $-\dfrac{13}{18}$

7 $f^{-1}(x) = \dfrac{3x + 5}{1 - 2x}$ provided $x \neq \dfrac{1}{2}$

8 a -5 **b** $\dfrac{1}{6}$ **c** -3

9 $fg(x) = -\dfrac{9 + 12x}{x^2}$,

$ff(x) = -x^4 + 8x^2 - 12$,

$g^{-1}(x) = \dfrac{3}{x - 2}$ provided $x \neq 2$

10 $f^{-1}(x) = -\sqrt{2 - x}$ provided $x \leq 2$

11 $p(t)$ goes from being positive to negative between $t = 0$ and $t = 4$

12 a 10   **b** 7/3   **c** 5/4

ME3 **1 i** 0   **ii** 5   **iii** $\dfrac{5x - 3}{2 - 3x}$ $(2 - 3x \neq 0)$

2 i $4.9t^2 + 9.8ht + 4.9h^2$ **ii** $9.8t + 4.9h$

iii $9.8 + 4.9h$ **iv** 9.8 **v** $9.8t$

3 i $3t^2 + 3th + h^2$ **ii** $3t^2$

7 $x = 5$, $x = 1$

8 i -2.625 **ii** 5.375

9 $f^{-1}(x) = \dfrac{1}{x}$ $(x \neq 0)$

10 $f^{-1}(x) = \dfrac{2x - 3}{5x - 2}$ $(5x - 2 \neq 0)$

11 $x = -2, -1$ and 2

12 ii 1   **iii** $2.718\,281\,828^x$

15 a $s = -3$, $s = -1$ **b** $s = 1.14$, $s = -6.14$

c no poles

16 $F(t) = \dfrac{t}{10}$, $R(t) = 1 - \dfrac{t}{10}$ and $h(t) = \dfrac{1}{10 - t}$

17 $R(t) = 1 - 0.3t + 0.015t^2$

18 $T(s) = 1000\,\dfrac{s}{500s^3 + 4000s + 8501s + 5000}$

4(a) **1** 0.57, 0.68 and 0.38

2 i $\dfrac{1}{2}$ **ii** $\dfrac{4}{\sqrt{3}}$ **iii** $\dfrac{3\sqrt{3}}{2}$ **iv** 1   **v** $\dfrac{\sqrt{2}+\sqrt{3}}{2}$

vi 1

3 AC $= 9.46$ m, span AB $= 17.16$ m

5 26.57°

6 4.62 m

7 42.1×10^3 mm² (3 s.f.)

8 a 60°   **b** 321.39 m   **c** 5.48 m   **d** 7.73 m

9 i $\cos(\theta) = \sqrt{\dfrac{2}{3}}$, $\tan(\theta) = \dfrac{1}{\sqrt{2}}$, $\operatorname{cosec}(\theta) = \sqrt{3}$,

$\sec(\theta) = \sqrt{\dfrac{3}{2}}$, $\cot(\theta) = \sqrt{2}$

ii $\sin(\theta) = \dfrac{\sqrt{5}}{3}$, $\tan(\theta) = \dfrac{\sqrt{5}}{2}$, $\operatorname{cosec}(\theta) = \dfrac{3}{\sqrt{5}}$,

$\sec(\theta) = \dfrac{3}{2}$, $\cot(\theta) = \dfrac{2}{\sqrt{5}}$

iii $\sin(\theta) = \dfrac{3}{5}$, $\cos(\theta) = \dfrac{4}{5}$, $\operatorname{cosec}(\theta) = \dfrac{5}{3}$,

$\sec(\theta) = \dfrac{5}{4}$, $\cot(\theta) = \dfrac{4}{3}$

10 1.242, 1.791 and 1.428

4(b) **1**

θ	0°	30°	60°	90°	120°	150°	180°
$y = \cos(\theta)$	1	0.866	0.5	0	-0.5	-0.866	-1

θ	210°	240°	270°	300°	330°	360°
$y = \cos(\theta)$	-0.866	-0.5	0	0.5	0.866	1

2

θ	0°	30°	60°	90°	120°	150°	180°
$y = \tan(\theta)$	0	0.577	1.732		-1.732	-0.577	0

θ	210°	240°	270°	300°	330°	360°
$y = \tan(\theta)$	0.577	1.732		-1.732	-0.577	0

4(c) **1 a** 45°, 135°   **b** 60°, 300°   **c** 60°, 240°
d 60°, 120°, 240°, 300°   **e** 60°, 300°
f 0°, 180°, 360°   **g** 25.38°, 154.62°

2 a $(180 \times n)° + 38.66°$

b $(360 \times n)° \pm 97.18°$

c $(180 \times n)° + \left[(-1)^n \times 12.71\right]°$

d $(60 \times n)° + \left[(-1)^n \times 15.35\right]°$

e No solution

f $(180 \times n)°, (180 \times n)° - 60°$

3 a 45°, 225°, 405°, 585°   **b** 90°, 270°

c 150.5°

d 58.31°, 148.20°, 238.31°, 328.20°

e 49.81°, 130.19°, 409.81°, 490.19°

f No solution

4 a 45°, 90°, 135°, 225°, 270°, 315°

b 0°, 15°, 165°, 180°, 360°

c 30°, 60°, 120°, 150°, 210°, 240°, 300°, 330°

4(d)
1 BC = 197 mm, AC = 135 mm

2 41 mm

3 110°

4 2.47 m

5 BC = 6.45 m, AC = 7.10 m

6 101°

7 2.6 m

8 77 mm

9 451 m

10 CD = 2.75 m, DE = 1.27 m

11 a $B = 56.44°$, $C = 93.56°$, AB = 11.98 or
$B = 123.56°$, $C = 26.44°$, AB = 5.34

b $A = 47.52°$, $B = 97.48°$, AC = 12.1 or
$A = 132.48°$, $B = 12.52°$, AC = 2.65

4(e)
1 **i** 2.15 rad   **ii** 0.23 rad   **iii** 2.2981 rad
iv 5.817 rad

2 **i** $\frac{\pi}{2}$ rad   **ii** $\frac{\pi}{6}$ rad   **iii** $\frac{11\pi}{6}$ rad   **iv** $\frac{\pi}{8}$ rad
v $\frac{\pi}{18}$ rad   **vi** $\frac{3\pi}{20}$ rad   **vii** $\frac{\pi}{60}$ rad
viii $\frac{4\pi}{5}$ rad

3 **i** 30°   **ii** 54°   **iii** 83.1°   **iv** $\left(\frac{10}{7}\right)°$
v $(1.15 \times 10^4)°$ **vi** 20°   **vii** $\left(\frac{140}{9}\right)°$
viii 19.5°

4 **i** 0.31   **ii** 0   **iii** −0.5

5 **i** $\frac{\pi}{2\sqrt{2}}$ **ii** 1   **iii** $\frac{1}{2}$

6 **i** 261.8 rad/s   **ii** 52.4 rad/s   **iii** 659.7 rad/s

4(f) In these solutions A = amplitude, T = period,
f = frequency

1 a $A = 10$, $T = \pi$, $f = 1/\pi$

b $A = 0.1$, $T = 20\pi$, $f = 1/(20\pi)$

c $A = H$, $T = 2$, $f = 0.5$

d $A = 1/\pi$, $T = 2/\pi$, $f = \pi/2$

e $A = 10000$, $T = 2\pi^3$, $f = \dfrac{1}{2\pi^3}$

f $A = \dfrac{1}{\pi e}$, $T = 2\pi^2 e$, $f = \dfrac{1}{2\pi^2 e}$

g $A = 220$, $T = 0.002$, $f = 500$

h $A = 1$, $T = \dfrac{\pi^2}{3}$, $f = \dfrac{3}{\pi^2}$

2 a $A = 4$, $T = 2\pi$, phase $= \pi/3$ rad lead

b $A = 1$, $T = 0.02$, phase $= 0.25$ rad lag

c $A = 20$, $T = 0.02$, phase $= \pi/5$ rad lead

d $A = 315$, $T = 2\pi/7$, phase $= 30°$ lead

e $A = 2$, $T = 1$, phase $= \pi/2$ rad lag

3 a 0   **b** $\pi/7$ lead   **c** $\pi/4$ lead   **d** 0.01 lag
e $\pi/2$ lag   **f** 10π lag   **g** π lag   **h** 1 lead

4 a $12\pi/175$ **b** 0.487   **c** $\pi/12$

5 a $2\cos(2t)$ **b** $7\sin(8t)$ **c** $10\sin(t - \pi)$

8 a $A = 5$ mA, $T = 16.67$ ms, $f = 60$ Hz,
phase $= 29.79°$ lead

b 2.48 mA for all values

c 2.79 ms   **d** 1.38 ms

4(g) **1** **i** $\dfrac{1}{2\sqrt{2}}\left(\sqrt{3} + 1\right)$ **ii** $\dfrac{\sqrt{3}}{2}$ **iii** $\dfrac{1}{2\sqrt{2}}\left(1 - \sqrt{3}\right)$
iv $-\dfrac{\sqrt{3}}{2}$ **v** $\dfrac{\sqrt{3} - 1}{\sqrt{3} + 1}$

4(i) **1** **i** $\sqrt{200}\cos(\omega t - 45°)$

ii $\sqrt{6}\cos(\omega t - 54.74°)$

iii $\sqrt{29}\cos(\omega t + 21.80°)$

iv $\sqrt{2}\cos(\omega t + 45°)$

v $2\cos(\omega t - 240°)$

vi $\sqrt{50}\cos(\omega t - 135°)$

3 Amplitude $= 2\sqrt{3}$ or $\sqrt{12}$, period $= 2\pi/3$

4 Amplitude $= 1.12$, period $= 1.99$ s, phase $= 26.57°$ lagging

5 i $x = 2\cos\left(10t - \dfrac{\pi}{6}\right)$, amplitude $= 2$

6 $r = 4\sqrt{2}\cos\left(2t - \dfrac{\pi}{4}\right)$, amplitude $= 4\sqrt{2}$ cm, period $= \pi$ s

EQ4 **1 b** Period is 4π, frequency is $\dfrac{1}{4\pi}$ and amplitude is 3

2 $\theta = (120n)° \pm 40°$

3 $2\cos(x - 150°)$

4 $\dfrac{1}{\sqrt{2}}\left[1 + \dfrac{1}{\sqrt{3}}\right]$

5 $\dfrac{7}{25}$

6 $t = \dfrac{\pi}{6}, \dfrac{5\pi}{6}, \dfrac{7\pi}{6}$ and $\dfrac{11\pi}{6}$

7 $-\dfrac{1}{\sqrt{2}}\dfrac{1}{2}\left[1 + \sqrt{3}\right]$

8 $\sqrt{\dfrac{1}{2}\left[1 - \dfrac{1}{\sqrt{5}}\right]}$

9 $\cos(\theta) = -\dfrac{4}{5}$, $\tan(\theta) = -\dfrac{3}{4}$, $\operatorname{cosec}(\theta) = \dfrac{5}{3}$, $\sec(\theta) = -\dfrac{5}{4}$ and $\cot(\theta) = -\dfrac{4}{3}$

10 Substitute $\operatorname{cosec}(x) = \dfrac{1}{\sin(x)}$

11 $x = 58.97°, 337.89°$

12 a Apply $\sin(A - B) = \sin(A)\cos(B) - \cos(A)\sin(B)$

  b Use $\sec(x) = \dfrac{1}{\cos(x)}$

13 $b = 28.28$

14 $\dfrac{1}{\sqrt{2}}\left[x - \sqrt{1 - x^2}\right]$

15 i Amplitude is 57, period time is 29.85 ms, frequency is 33.5 Hz, phase is 15° and phase time is 1.235 ms
  ii -14.65 volts **iii** 56.35 volts
  iv $t = 8.70$ ms **v** $t = 4.92$ ms

ME4 **7** 29.30, 26.48

8 58.41°, 48.19° and 73.4°

10 $x = 22.5°$

11 a $\pi/4, 3\pi/4, 5\pi/4, 7\pi/4$ **b** $\pi/6, 7\pi/6$

  c $\pi/6, 5\pi/6, 7\pi/6, 11\pi/6$

  d $\pi/8, 5\pi/8, 9\pi/8, 13\pi/8$

12 i Amplitude $= 20$, period $= 1/25$, frequency $= 25$

  ii 3.82×10^{-3}

14 20 s

15 0.75

17 i $128.17°, 231.83°$ **ii** no solution

  iii $128.17°, 231.83°$

18 a $\sqrt{3}$ **b** $1/\sqrt{2}$ **c** $\sqrt{3/2}$

19 $x = n\pi/3$ (n is a whole number)

5(a) **1** $V_2 = 0.14$ m^3

2 $V_2 = 0.081$ m^3

3 $V_2 = 0.046$ m^3

5(b) **1** 1520.26 N

2

x	-2	-1	0
$y = e^{-x}$	$e^2 = 7.39$	$e^1 = 2.72$	$e^0 = 1.00$
x	1	2	
$y = e^{-x}$	$e^{-1} = 0.37$	$e^{-2} = 0.14$	

3

v	0	0.1	0.2
i	0	6.39×10^{-7}	5.36×10^{-6}
v	0.3	0.4	0.5
i	4.02×10^{-5}	2.93×10^{-4}	2.2×10^{-3}
v	0.6	0.7	
i	0.0163	0.12	

4 a 10705 Pa **b** 8337 Pa **c** 5730 Pa

5 400°C, 390°C, 382°C, 361°C

6 $(0.5 \times 10^{-6})e^{-(2 \times 10^3)t}$

7 a $0.4(e^{-t/2 \times 10^{-6}} - e^{-t/1 \times 10^{-6}})$

   b $-0.05e^{-(10 \times 10^3)(t-1)}$

8 $5e^{-400t}(1 + 399e^{-600t} - 400e^{-1200t})$

10 $v = 20$ m/s

5(c) **1 a** 2.30 nep   **b** 4.61 nep   **c** -1.10 nep

   2 a 0.80 nep   **b** 1.50 nep

   3 60 dB

   4 0.316 W

   7 0.01 A

   9 a $2, \sqrt{2}$ **b** $10, \sqrt{10}$ **c** 100, 10

   10 0.105

   11 4.05

   12 a 2.926   **b** 0.342   **c** 0.946   **d** 1.057

   13 a 1.93   **b** 3.14   **c** 4.75

   17 a $\dfrac{1}{3}$ **b** ± 3.29 **c** -1 **d** 2.63

   19 $v = \dfrac{\eta T}{11600}\ln\left(\dfrac{i}{I_s} + 1\right), 0.62$ V

5(d) **1** $k = 3.14, b = 0$, circle

   2 $k = 10 \times 10^{-9}$

   3 iii $k = 43$

   4 $k = 0.075, n = -0.2$

   5 $k = 1.46, n = 0.676$

5(e) **1** 11013.23, -10.02, 11013.23, 1, 1, 0.46

   8 0.425, 0.470, 1.105

   9 18.39 m/s

   10 1.24 W

EQ5 **1** 2.13, 1.045 and 0.55

   2 a 37.3 W   **b** 83 W

   3 i $x = 4$ **ii** $x = \pm\sqrt{3.03}$ **iii** $x = 0.345$

   4 $p = 2.815V^2 + 9V$

   5 i $x = -1/3$ **ii** $x = 16$

   6 i $x = 1/3$ **ii** $x = 2$ **iii** $x = -2$ **iv** $x = 1/2$

   7 $x = 2$

8 a $2\log(A) + \dfrac{1}{2}\log(C) - 5\log(B)$

   b $x = 1.815$ **c i** 0   **ii** ± 2.39

9 i $I_s = 15.686$ **ii** $v = 4.375 \times 10^{-3}$

10 2

11 $-\tanh\left(\dfrac{1}{x}\right)$

12 a 2007   **b** 2012

13 a See **Example 21**   **b** Expand the right-hand side

14 a $\dfrac{13}{5}$ **b** $\dfrac{12}{13}$ **c** $\dfrac{312}{25} = 12.48$

15 $x = -1.227, 0.535$

ME5

   1 0.18 V

   4 63.2%, 86.5%, 95%, 98.2% and 99.3%

   5 i 1   **ii** 0.632, 0.865, 0.950, 0.982, 0.993

   iii 63.2%, 86.5%, 95%, 98.2%, 99.3%

   6 ii $v \to 10/k$ **iv** yes   **v** 10, 5, 2 (m/s)

   7 $n = -0.2$ and $k = 0.058$

   8 27.9 kN

   9 $\lim\limits_{t \to \infty} \theta = 300$

   12 ii $R(t) \to 0$ and $F(t) \to 1$

   14 Cable becomes more taut

   16 $t = \dfrac{\pi}{10}, \dfrac{\pi}{5}, \dfrac{3\pi}{10}, \dfrac{2\pi}{5}, \ldots, \dfrac{n\pi}{10}$

6(a)

   1 a 3   **b** -1 **c** $-\pi$ **d** 1000

   4 $v = 0$

   5 a 1   **b** 5   **c** $2x + 1$ **d** $-\dfrac{1}{x^2}$

   6 iii 455.625

6(b)

   1 a $3x^2$ **b** $3x^2(5x^2 + 1)$ **c** $\dfrac{1}{2}x^{-1/2} + \cos(x)$

   d $-\dfrac{1}{3}x^{-4/3} + e^x - \sin(x)$

   e $-\dfrac{3}{2}x^{-3/2} - \dfrac{5}{3}x^{-4/3}$

   2 i $75 - 3t^2$ **ii** 5 s

   3 $-\dfrac{1}{\rho^2}$

4 a $-3\sin(3x)$ **b** $\pi\cos(\pi x)$ **c** $11e^{11x}$

d $\dfrac{\pi}{2}\cos\left(\dfrac{\pi}{2}x\right)$ **e** $-\dfrac{\pi}{2}\sin\left(\dfrac{\pi}{2}x\right)$

5 $-2\pi fKN\cos(2\pi ft)$

6 a $2x\cos(x^2)$ **b** $e^{\sin(x)}\cos(x)$ **c** $2/x$

d $(3-6x^2)\sin(2x^3-3x)$ **e** $20x(x^2+1)^9$

f $(35x^6+12x^3)\sec^2(5x^7+3x^4)$

m $e^{\Sigma}[\sin(2\Sigma)+2\cos(2\Sigma)]$

n $-\sqrt{\dfrac{1}{(Z+1)\,(Z-1)^3}}$

o $e^{\Sigma}[(1+\Sigma)\sin(5\Sigma)+5\Sigma\cos(5\Sigma)]$

p $\dfrac{-4\delta}{\delta^4-1}$

4 $-e^{-kt}[k\cos(\omega t)+\omega\sin(\omega t)]$

6(c) 1 a $6x(x^2+2)^2$ **b** $2x\cos(x^2)$

c $-(2x+2)\sin(x^2+2x)$

d $\dfrac{3x^2+1}{x^3+x}$ **e** $(15x^4+3x^2)\cosh(3x^5+x^3)$

2 a $\cot(x)$ **b** $\tanh(x)$ **c** $-\sin(x)\sec^2[\cos(x)]$

d $2x(10)^{x^2}\ln(10)$ **e** $\dfrac{2\cos[\ln(x^2)]}{x}$

3 $(15\times10^{-3})e^{-(1\times10^3)t}$

4 $-\dfrac{E}{R}e^{-t/RC}$

5 $\dfrac{\eta Te^{-11600V/\eta T}}{11600I_s}$

6 $\dfrac{\alpha}{\eta}\,\tau I_s\,e^{\alpha v/\eta}$

7 $\dfrac{1}{2\sqrt{t}e^{\sqrt{t}}}$

8 $\dfrac{20}{3(5+t/3)^2}$

6(d) 1 $5(1-5t)e^{-5t}$

2 i $= -(1\times10^{-6})[999+1000t]e^{-1000t}$

3 a $\sin(t)+t\cos(t)$ **b** $\ln(x)+1$ **c** $e^{\theta}(1+\theta)$

d $e^u[\sin(u)+\cos(u)]$ **e** $\cos(2\alpha)$

f $\dfrac{1}{(1+q)^2}$ **g** $-\dfrac{1}{1+\sin(a)}$

h $\dfrac{e^z[\cos(z)+\sin(z)]}{\cos^2(z)}$ **i** $M^2(3\ln(M)+1)$

j $\dfrac{\beta[2\sin(\beta)-\beta\cos(\beta)]}{\sin^2(\beta)}$ **k** $\dfrac{1}{(1+h^2)^{3/2}}$

l $\ln[\sin(J)]+J\cot(J)$

6(e) 1 a $v=3t^2-20t+40,\ a=6t-20$

b $v=21t^2-100t+50,\ a=42t-100$

c $v=-10\sin(2t),\ a=-20\cos(2t)$

d $v=-30\pi\sin(10\pi t)+50\pi\cos(10\pi t),$

$a=-100\pi^2[3\cos(10\pi t)+5\sin(10\pi t)]$

2 i $v=6t^2-30t-36,\ a=12t-30$

ii $t=6$ s

3 $a=-9.8\,\text{m/s}^2$. Constant acceleration

4 $v=14.7$ m/s, $a=9.8\,\text{m/s}^2$

5 a $v=3-3\cos(3t),\ a=9\sin(3t)$

b $v=\dfrac{e^{-t/5}}{25},\ a=-\dfrac{e^{-t/5}}{125}$

c $v=(1-t)e^{-t},\ a=(t-2)e^{-t}$

d $v=\dfrac{6t}{1+3t^2},\ a=\dfrac{6-18t^2}{(1+3t^2)^2}$

e $v=\sqrt{5}\tanh(\sqrt{20}t),\ a=10\,\text{sech}^2(\sqrt{20}t)$

7 0

6(f) 1 a $-\tan(t)$ **b** $\dfrac{20-3t}{6t-2}$ **c** $\dfrac{3t^2}{2(t-3)}$ **d** $\dfrac{1}{2e^{t/2}}$

2 b $\dfrac{\cos(t)-t\sin(t)}{\sin(t)+t\cos(t)}$

3 ii $v_x=2(t-2),\ v_y=3t^2,$

$v=\sqrt{9t^4+4t^2-16t+16}$

4 ii $v_x=(t^2+2)^{1/2}.t,\ v_y=1$ **iii** $v=t^2+1$

5 i $v_x=16-24t^2,\ v_y=3t^2$

ii $v=\sqrt{256-768t^2+585t^4}$

6 $v_x = -r\sin(t)$, $v_y = r\cos(t)$,

$a_x = -r\cos(t)$, $a_y = -r\sin(t)$

7 i $v_x = 3t^2 - 2t$, $v_y = 10t - 3t^2$,

$a_x = 6t - 2$, $a_y = 10 - 6t$

ii $v = 11.31$ m/s, $a = 10.20$ m/s^2

6(g) 1 a $-\dfrac{x}{y}$ **b** $\dfrac{2 - 3x^2}{3y^2}$ **c** $-\dfrac{4x}{y}$ **d** $\dfrac{2 - x}{y - 3}$

2 a $\dfrac{y^2 - 2x}{2y - 2xy}$

b $-\dfrac{2[y + x\sin(x^2 + y)]}{2x + \sin(x^2 + y)}$ **c** $y\dfrac{2-x}{x(x-1)}$

d $-\dfrac{y}{x(2y + 1)}$ **e** $-[e^{-(x + y)}\cos(x) + 1]$

(The denominator is not zero in any of the
above cases)

3 i $\dfrac{4y - x^2}{y^2 - ax}$ **ii** $\dfrac{ay - x^2}{y^2 - ax}$, not defined

5 a $x^{\sin(x)}\left[\ln(x).\cos(x) + \dfrac{\sin(x)}{x}\right]$

b $(x^2 + 1)^{1/2}(3x + 1)^{1/3}\left[\dfrac{x}{x^2 + 1} + \dfrac{1}{3x + 1}\right]$

c $\dfrac{(x^2 + 1)^{1/2}(3x + 1)^{1/3}}{(x^4 - 2)^{1/5}}\left[\dfrac{x}{x^2 + 1}\right.$

$\left. + \dfrac{1}{3x + 1} - \dfrac{4x^3}{5(x^4 - 2)}\right]$

d $e^{(x^x)}x^x[\ln(x) + 1]$

e $x^{x^x + x - 1}[x[\ln(x)]^2 + x\ln(x) + 1]$

EQ6 1 i $-15x^{-4} - 7x^{-2} + 4x^5 - 9$

ii $2\left[\ln\left(\dfrac{2}{x}\right) - 1\right]$

iii $\dfrac{-4e^{-2x}[2\sin(5x) + 5\cos(5x)]}{\sin^2(5x)}$

iv $\dfrac{x[5 - 6x]}{\sqrt{1 + 5x^2 - 4x^3}}$

2 11.17

3 $\dfrac{e^x}{\sin(x)}[(x + 1)^2 - (x^2 + 1)\cot(x)]$

4 i $6x^5 - 15x^2 - 1$ **ii** $e^{2\theta}[2\cos(3\theta) - 3\sin(3\theta)]$

iii $\dfrac{x^2(1 - 2\ln(2x)) + 4}{x(x^2 + 4)^2}$ **iv** $2e^{\sin(2t)}\cos(2t)$

v $\dfrac{3x^2 - 1}{2\sqrt{x^3 - x}}$ **vi** $\dfrac{-2e^z}{(1 + e^z)^2}$

5 i $-\dfrac{\sqrt{2}}{3 - 2\sqrt{2}}$

ii $\dfrac{(x^2 + 1)^2(x + 7)^3}{2x - 3}\left[\dfrac{4x}{x^2 + 1} + \dfrac{3}{x + 7} - \dfrac{2}{2x - 3}\right]$

iii $-\dfrac{2x + y\cos(xy)}{2y + x\cos(xy)}$

6 $\dfrac{\cos[\ln(2x)]}{x}$

7 $-\dfrac{3e^{3x}}{e^{3x} + 1}$

8 Use a hyperbolic identity $\cosh^2(t) - \sinh^2(t) = 1$

9 a $\dfrac{df}{dx} = e^{\sin(x)}\cos(x)$, $\dfrac{dg}{dx} = 3x^2 + 2\sin(2x)$ and

$\dfrac{d}{dx}[f(x).g(x)]$

$= e^{\sin(x)}[(x^3 - \cos(2x) + 4\sin(x))\cos(x) + 3x^2]$

b $\dfrac{dy}{dx} = 0$, $\dfrac{d^2y}{dx^2} = -\dfrac{1}{9}$

10 i $\dfrac{2e^{x^2} + 4x^2e^{x^2} + 2y^2\sin(x) - 3yx^{1/2}}{4y\cos(x) + 2x^{2/3}}$

ii $P = (0, -1)$, $-\dfrac{\pi}{2} - 2$

11 $\dfrac{e}{1 - e}$

12 a $3^{x\ln(x)}\ln(3)[\ln(x) + 1]$

b $\dfrac{\ln(x^4) + 4}{2\sqrt{x}\ln(x^4)}$

c $(\ln(x))^{\cos(x) - 1}[x^{-1}\cos(x)$

$- \ln(x)\sin(x)\ln(\ln(x))]$

d $\dfrac{(4x - 1)^3}{(2x^2 - 1)^{3/2}(x + 1)^2}\left[\dfrac{12}{4x - 1} - \dfrac{6x}{2x^2 - 1} - \dfrac{2}{x + 1}\right]$

e $x\cos(x)e^{3x}[2 + x\tan(x) + 3x]$

ME6 1 0

4 $Q = 210t^2 + 3500$

5 $Q = (3.49 \times 10^3)e^{-3t}$

6 $\theta = \sin^{-1}\left(\dfrac{mg}{kl}\right)$, $\theta = \dfrac{\pi}{2}$

7 $x = \dfrac{mg}{k}$

8 $i = 0.036e^{-(2\times10^3)t} - 0.24e^{-(8\times10^3)t}$

10 $i = 0.0375\cos(75t)$

15 $v_x = 2t\sin(t) + t^2\cos(t)$,

$\qquad a_x = 2\sin(t) + 4t\cos(t) - t^2\sin(t)$

16 i $v_x = \cos(t) - t\sin(t)$,

$\qquad a_x = -2\sin(t) - t\cos(t)$

$\qquad v_y = \sin(t) + t\cos(t)$,

$\qquad a_y = 2\cos(t) - t\sin(t)$

$\quad$ **iii** $\dfrac{dy}{dx} = \dfrac{\sin(t) + t\cos(t)}{\cos(t) - t\sin(t)}$

17 $10^{(x^x)}x^x[\ln(x) + 1]\ln(10)$

7(a) **1 a** $(-3/2, -5/4)$

$\qquad$ **b** $(1,-2), (-1, 2)$

$\qquad$ **c** $(1, -2/3), (-1, 2/3)$

$\qquad$ **d** $(0,0), (5,625), (-5,625)$

$\quad$ **3** $(-5/3, 175/27)$ max, $(1, -3)$ min

$\quad$ **4** $(2, 8)$ pt of inflexion

$\quad$ **9 i** $p \to 0$ $\quad$ **ii** $0, 2$ $\quad$ **iii** -0.288 $\quad$ **iv** 0.0086

$\quad$ **10 i** $i \to 0$ $\quad$ **ii** e^{-1}

7(b) **1** $x = y = 25$ m

$\quad$ **2** $x = 60$ m, $y = 120$ m

$\quad$ **3** $C_L = \sqrt{\dfrac{C_D}{k}}$

$\quad$ **4** $r = h = 2/\pi^{1/3}$

$\quad$ **6** $x = L/2$, max $M = WL^2/8$

$\quad$ **7** $x = L$, max $y = WL^3/3EI$

$\quad$ **8** $WL^4/24EI$

$\quad$ **10 i** $\pi/3$ $\quad$ **ii** 7.8

7(c) **1** $(0,0)$ min, $(2,16)$ and $(-2,16)$ max

$\quad$ **2** $r = 25$

$\quad$ **5** $2\pi fk$

7(d) **1 i** 1 s, 3 s $\quad$ **ii** $v = 2t - 4$, $a = 2$ m/s^2

$\quad$ **2** $v = 25 - 9.8t$, $a = -9.8$ m/s^2,

$\qquad$ max height $= 31.9$ m

$\quad$ **3** 0 s, 2 s for $v = 0$ and 1 s for $a = 0$

4 $v = 9t^2 - 20t + 1$, $a = 18t - 20$

5 min velocity $= 0$ m/s

6 $v = 3(t^2 - 8t + 6)$

7 $\dfrac{ds}{dv} = -\dfrac{1}{v^2}$

8 $\omega = -248$ rad/s, $\alpha = 6$ rad/s^2

9 $\alpha = -\sin\left(2t + \dfrac{\pi}{6}\right)$, $t = \dfrac{5\pi}{12}$ s

7(e) **1 a** $y = 4x - 7$, $\;\; y = \dfrac{1}{4}(6 - x)$

$\qquad$ **b** $y = \dfrac{\pi}{2} - x$, $\;\; y = x - \dfrac{\pi}{2}$

$\qquad$ **c** $y = ex$, $\;\; y = e + e^{-1}(1 - x)$

$\qquad$ **d** $y = x - 1$, $\;\; y = 1 - x$

$\quad$ **2** $i = 7.68v - 8.19$

$\quad$ **3** $i = 541.34t + 1.19$

7(f) **1** $1 - \dfrac{x^2}{2!} + \dfrac{x^4}{4!} - \cdots$

$\quad$ **2** $x - \dfrac{x^2}{2} + \dfrac{x^3}{3} - \dfrac{x^4}{4} + \cdots$

$\quad$ **4** $\cos^2(x) = 1 - x^2 + \dfrac{x^4}{3} - \dfrac{2x^6}{45} + \cdots$

$\qquad$ $\sin^2(x) = x^2 - \dfrac{x^4}{3} + \dfrac{2x^6}{45} - \cdots$

$\quad$ **5 a** 1 $\quad$ **b** $1/2$

$\quad$ **6 a** $(x - 1) - \dfrac{(x - 1)^2}{2} + \dfrac{(x - 1)^3}{3} - \cdots$

$\qquad$ **b** $1 - (x - 1) + (x - 1)^2 - (x - 1)^3 + \cdots$

$\qquad$ **c** $e^3\left[1 + (x - 3) + \dfrac{(x - 3)^2}{2!} + \dfrac{(x - 3)^3}{3!} + \cdots\right]$

$\qquad$ **d** $\dfrac{1}{\sqrt{2}}\left[1 - \left(x - \dfrac{\pi}{4}\right) - \dfrac{1}{2!}\left(x - \dfrac{\pi}{4}\right)^2\right.$

$\qquad\qquad\qquad\qquad \left. + \dfrac{1}{3!}\left(x - \dfrac{\pi}{4}\right)^3 + \cdots\right]$

$\qquad$ **e** $1 + \dfrac{(x - e)}{e} - \dfrac{(x - e)^2}{2e^2} + \dfrac{(x - e)^3}{3e^3} - \cdots$

$\quad$ **7** $1 - \dfrac{1}{2}x^2 + \dfrac{x^4}{8} - \dfrac{x^6}{48} + \cdots$

$\quad$ **8** $x + \dfrac{x^3}{3} + \dfrac{x^5}{5} + \cdots$

7(g) **1 a** $1 + 20x + 190x^2 + 1140x^3 + \cdots$

$\qquad$ **b** $x^{15} + 15x^{14}y + 105x^{13}y^2 + 455x^{12}y^3 + \cdots$

$\qquad$ **c** $1024 - 5120Z + 11520Z^2 - 15360Z^3 + \cdots$

$\quad$ **2** $1 + Z + Z^2 + Z^3 + Z^4 + \cdots$

For complete solutions see www.palgrave.com/engineering/singh

3 $1 + \dfrac{W^2 x^2}{2T^2} - \dfrac{W^4 x^4}{8T^4} + \dfrac{W^6 x^6}{16T^6}$

where $-1 < \left(\dfrac{Wx}{T}\right)^2 < 1$

4 ii 2.6458

7(h) **1 a** $\displaystyle\sum_{n=1}^{\infty} \sqrt{n}$ **b** $\displaystyle\sum_{n=1}^{\infty} (2n)$ **c** $\displaystyle\sum_{n=1}^{\infty} (2n - 1)$

d $\displaystyle\sum_{n=1}^{\infty} \left[\dfrac{(-1)^{n+1}}{n}\right]$ **e** $\displaystyle\sum_{n=0}^{\infty} \left(\dfrac{1}{3}\right)^n$ **f** $\displaystyle\sum_{n=1}^{\infty} \left(\dfrac{2}{3}\right)^n$

2 a $\displaystyle\sum_{n=0}^{\infty} \left(\dfrac{1}{3}\right)^n = \dfrac{1}{2}$ **b** $\displaystyle\sum_{n=0}^{\infty} \left(\dfrac{1}{4}\right)^n = \dfrac{1}{3}$

c $\displaystyle\sum_{n=0}^{\infty} \left(\dfrac{1}{\pi}\right)^n = \dfrac{1}{\pi - 1}$ **d** $\displaystyle\sum_{n=0}^{\infty} \left(\dfrac{1}{m}\right)^n = \dfrac{1}{m - 1}$

3 a Converges, $\displaystyle\sum_{n=1}^{\infty} \left(\dfrac{1}{2^{2n-1}}\right) = \dfrac{2}{3}$ **b** Diverges

c Diverges **d** Converges, $\displaystyle\sum_{n=1}^{\infty} 10\left(\dfrac{1}{3}\right)^n = 5$

4 22.22 m (2 d.p.)

5 142.86 m (2 d.p.)

6 $A = 1$, the whole area is removed

7 a 1 **b** 1/3 **c** 1/9

8 £1111.11 (2 d.p.)

9 a Converges $\displaystyle\sum_{n=0}^{\infty} 8\left(\dfrac{1}{2}\right)^n = 16$

b Diverges, $\displaystyle\sum_{n=0}^{\infty} 3(2)^n$

c Converges, $\displaystyle\sum_{n=0}^{\infty} 16\left(\dfrac{3}{4}\right)^n = 64$

10 a $\displaystyle\sum_{n=1}^{\infty} \dfrac{1}{x^n} = \dfrac{1}{x - 1}$ **b** $\displaystyle\sum_{n=1}^{\infty} \left(\dfrac{x}{2}\right)^n = \dfrac{x}{2 - x}$

c $\displaystyle\sum_{n=1}^{\infty} \dfrac{1}{(1 + x)^n} = \dfrac{1}{x}$

d $\displaystyle\sum_{n=1}^{\infty} \dfrac{1}{(1 + x^2)^n} = \dfrac{1}{x^2}$

11 a Converges because $L = 0$

b Diverges because $L = +\infty$

c Diverges because $L = +\infty$

d Converges because $L = \dfrac{1}{2}$

e Converges because $L = \dfrac{1}{e}$

f Converges because $L = \dfrac{1}{3}$

g Converges because $L = 0$

h Diverges because $L = 3$

i Converges because $L = 0$

j Diverges because $L = \dfrac{11}{2}$

13 a i $L = \dfrac{2}{e} < 1$, series converges

ii Diverges because $L = \dfrac{3}{e} > 1$

b i $0 < x < e$ **ii** $x > e$

7(i) **1 a** -2.26 **b** -1.89 **c** -2.46

2 $-4.018, -8.967$ and -15.015

3 10.196 m/s

4 2.325 m

EQ7 **1 i** 30 m/s **ii** -6 m/s **iii** -12 m/s^2

2 a Maximum at $(-1, 3)$ and minimum at $(1, -1)$

b i 2 m **ii** 5 m/s **iii** 8 m/s^2

iv $t = 0$ or $t - 1$ s **v** 2/3 s

3 A square of edge 5 feet

4 2 in. by 3 in.

5 $1 + x + x^2 + \dfrac{1}{3} x^3 + \cdots$

6 i $1 + x - x^2 + \dfrac{5}{3} x^3 + \cdots$

ii $1 - 3x^2 + \dfrac{19}{6} x^4 + \cdots$

7 0.5049

8 a $y = 6x - 16$

b Maximum is 1 and it occurs at $x = -1$ and $x = +1$

9 $(1 - x^2)^{-1/2} = 1 + \dfrac{1}{2} x^2 + \dfrac{3}{8} x^4 + \dfrac{5}{16} x^6 + \cdots,$

1.005038

10 $2 - 2x + x^2 - \cdots$

11 $y = 2 - x$

12 $y = 3 - x$

13 Maximum at $r = a$ and max value is $4e^{-2}$. The graph goes through the origin

For complete solutions see www.palgrave.com/engineering/singh

14 a A circle of circumference of 100 cm but **no** square

b A square of length 20 cm and circle of circumference 20 cm

ME7 1 $x = 10$ m, $y = 5$ m

2 $x = 1/3$ m, $y = 2/3$ m

4 $x = 25/36$ m

5 $x = 2.16$ m, $y = 4.05 \times 10^{-3}$ m

6 $\theta = \pi/4$

7 $\theta = \pi/4$

8 $v = (t - 1)e^{-t}$, $a = (2 - t)e^{-t}$

11 $n = 7.58$

12 1

13 $y = x + \dfrac{2 - \pi}{4}$, $y = \dfrac{2 + \pi}{4} - x$

14 e^{-1}

15 $v = -5e^{-500t}$

16 $\dfrac{r}{2}$

18 i $x + \dfrac{x^3}{3!} + \dfrac{x^5}{5!}$

19 iii $\dfrac{\pi}{4}$

20 ii 0.446

21 $x = 2\sqrt{3}$ m and $y = 2$ m

22 $1 - \dfrac{v^2}{2c^2} - \dfrac{v^4}{8c^4} - \dfrac{v^6}{16c^6} - \cdots$

8(a) 1 a $\dfrac{x^2}{2} + C$ **b** $\dfrac{u^3}{3} + C$ **c** $\dfrac{z^4}{4} + C$

d $\dfrac{2t^{3/2}}{3} + C$ **e** $\dfrac{(\omega t)^2}{2} + C$ **f** $3t + C$

g $-t^{-1} + C$ **h** $2x^{1/2} + C$

2 vii 0.25

3 a 64   **b** 48.6   **c** 144   **d** $\dfrac{3}{2}$

8(b) 1 a $-\cos(t) + C$ **b** $\sin(t) + C$

c $\ln|\sec(t)| + C$ **d** $e^t + C$ **e** $\sinh(t) + C$

f $9.81t + C$ **g** $25x + C$ **h** $\dfrac{1}{2}x + C$

2 a $\sin(\omega t) + C$ **b** $\dfrac{10^x}{\ln(10)} + C$

c $\cosh(t) + C$ **d** $\ln|\sec(x) + \tan(x)| + C$

3 a $\dfrac{PV^{2.35}}{2.35} + C$ **b** $\dfrac{PV^{2.61}}{2.61} + C$

4 a $\dfrac{x^3}{3} + x^2 + C$ **b** $2x^{1/2} + \ln|\sec(x)| + C$

c $\tan(x) - 5e^x + x + C$

d $\dfrac{4x^{3/2}}{3} - \cos(x) - x + C$

5 ii $\dfrac{3 - 4x}{1 + x^2} + C$

8(c) 1 a $-\dfrac{\cos(7x + 1)}{7} + C$

b $\dfrac{\sin(7x + 1)}{7} + C$

2 a $\dfrac{\sin(\omega t)}{\omega} + C$ **b** $\dfrac{\sin(\omega t + \theta)}{\omega} + C$

3 a $-\dfrac{\cos(\omega t)}{\omega} + C$ **b** $-\cos(\omega t) + C$

4 a $\ln|x^2 - 1| + C$ **b** $\ln|x^3 - 3x^2 + 1| + C$

7 a $\dfrac{1}{7}\ln|7t - 1| + C$ **b** $\dfrac{1}{4}\ln|t^4 - 1| + C$

c $-\dfrac{1}{3}\ln|5 - t^3| + C$

8 a $\dfrac{e^{11x + 5}}{11} + C$ **b** $-\dfrac{e^{-2x + 1000}}{2} + C$

10 $v = u - gt$

12 i $v = 48 - 3t^2$ **ii** 4 seconds

13 $\dfrac{k^{1/\gamma}\gamma P^{\frac{\gamma - 1}{\gamma}}}{\gamma - 1} + D$ (D = constant of integration)

8(d) 1 $\dfrac{4}{3}$ units^2

2 $-\dfrac{4}{3}$ m

3 12.5 J

4 45.1 J

5 22.71 m

6 90.85 m

7 $P = 4.96$ kN, $R = 15.2$ kN m

8 1.3 kJ

9 8.1 minutes

10 9.9 kJ

15 $\dfrac{mr^2}{2}$

16 $\dfrac{49kr^{16/7}}{144}$

17 πuR^2

18 $\dfrac{V}{\omega T}\Big[\sin(\omega T + \theta) - \sin(\theta)\Big]$

19 $\dfrac{a}{\omega T}\Big[\sin(\omega T + \phi) - \sin(\theta)\Big]$

22 $\dfrac{\rho}{2\pi}\ln\!\left(\dfrac{b}{a}\right)$

23 $-(2 \times 10^3)\Big[e^{-(5\times10^3)t} - 1\Big]$

24 $\dfrac{Ir\sin(\beta)}{2(d^2 + r^2)}$

25 $\dfrac{I}{2\pi r}$

27 $(t^2 - 4t + 9)\exp\!\left[-\left(\dfrac{t^3}{3} - 2t^2 + 9t\right)\right]$

28 $(3 - t)\exp\!\left[\dfrac{t^2}{2} - 3t\right]$

29 $R(t) = 0.01t^2 - 0.2t + 1$,

$h(t) = \dfrac{0.02(10 - t)}{0.01t^2 - 0.2t + 1}$

30 $e^{-(4\times 10^{-3})t^{1/2}}$

32 $\dfrac{nC}{n - 1}\left[p_2^{\,1-\frac{1}{n}} - p_1^{\,1-\frac{1}{n}}\right]$

8(e) **1 a** $2\Big[e^x(x - 1)\Big] + C$

b $\sin(t) - t\cos(t) + C$

2 a $\dfrac{1}{9}\Big[3q\sin(3q) + \cos(3q)\Big] + C$

b $\dfrac{s^3}{9}(3\ln(s) - 1) + C$

3 $\dfrac{e^{2t}}{4}(2t - 1) + C$

6 $500(1 - e^{-t} - te^{-t})$

7 $50[t\sin(t) + \cos(t) - 1]^2$

8 161 A

8(f) **1 a** $\dfrac{1}{x + 1} + \dfrac{2}{x + 2}$ **b** $\dfrac{1}{t - 1} + \dfrac{1}{t + 1}$

c $\dfrac{3}{s - 1} - \dfrac{1}{s + 2}$ **d** $\dfrac{2}{2u + 1} - \dfrac{7}{u - 3}$

2 a $x - 1 + \dfrac{1}{x + 1}$

b $x^3 + x - 2 + \dfrac{1}{2}\left[\dfrac{3}{x + 1} - \dfrac{1}{x - 1}\right]$

3 a $\dfrac{2t + 1}{t^2 + t - 1} + \dfrac{2}{t - 1}$ **b** $\dfrac{1}{z - 1} + \dfrac{2}{(z - 1)^2}$

c $\dfrac{2x + 1}{x^2 + x + 1} + \dfrac{2x + 1}{(x^2 + x + 1)^2}$

8(g) **1 a** $2\ln|c + 2| + \ln|c + 1| + C$

b $\ln|\lambda^2 - 1| + C$ **c** $\ln\left|\dfrac{(a - 1)^3}{a + 2}\right| + C$

d $\ln\left|\dfrac{2y + 1}{(y - 3)^7}\right| + C$

2 a $\ln|p^2 + p - 1| + \ln(p - 1)^2 + C$

b $\ln|z - 1| + \dfrac{2}{1 - z} + C$

3 1.79

4 0.044

7 i $x - 3 + \dfrac{7}{x + 2}$ **ii** $\dfrac{x^2}{2} - 3x + 7\ln|x + 2| + C$

8(h) **1 a** $\dfrac{(b^2 - 3)^8}{16} + C$ **b** $\dfrac{(5s - 1)^{10}}{50} + C$

c $\dfrac{(a^3 - 2a^2 + 6)^5}{5} + C$ **d** $\dfrac{2(7q^3 - 5)^{3/2}}{3} + C$

e $\dfrac{2}{3}(p^2 - 3p)^{1/2} + C$

f $-\dfrac{1}{2}(\alpha^2 - 2\alpha + 10)^{-1} + C$

2 2 years

3 0.316

4 a $\dfrac{2}{3}$ (same answer as that in **Example 33**)

b 0

5 a $\dfrac{\sec^7(\beta)}{7} + C$ **b** $\dfrac{\tan^6(A)}{6} + C$

c $\cosec(A) - \dfrac{\cosec^3(A)}{3} + C$

16 $2e^{\sqrt{x}} + C$

17 $2e^{\sqrt{x}}[\sqrt{x} - 1] + C$

8(i)

2 $\dfrac{10}{\sqrt{2}}$

EQ8

1 a $\dfrac{3x^2}{2} + \cos(x) + C$ **b** 6.36 (2 d.p.)

2 a $\dfrac{5}{4}[2x\sin(2x) + \cos(2x)] + C$

b $\dfrac{5265}{8}$

3 $-10\cos(x) + C$

4 $\dfrac{[\ln(x)]^4}{4} + C$

5 i $16\dfrac{1}{2}$

ii $\dfrac{1}{\sqrt{2}}$

iii $\dfrac{1}{2}\ln(x^2 + 1) + C$

iv $-\theta\cos(\theta) + \sin(\theta) + C$

v $\ln\left|\dfrac{x-2}{x+1}\right| + C$

vi $\dfrac{\pi}{6}$

6 0.91

7 a $x^2 - 4x + 5\ln|x - 3| + C$

b $\ln|\cosec(x) - \cot(x)| + \cos(x) + C$

c $\dfrac{(3x-7)^{10}}{990}[30x + 7] + C$

d $x\sec(x) - \ln|\sec(x) + \tan(x)| + C$

8 a $\dfrac{15}{8}$ **b** $4[\ln(4) - 1]$

9 1.54

10 1

11 $\ln\left(\dfrac{4}{1.4}\right)$

12 i $e - 1$ **ii** $e - 5e^{-1}$

13 $-\dfrac{3}{\sqrt{2}}$

14 You will have to use $1 + \tan^2(\theta) = \sec^2(\theta)$

15 $2[\sqrt{x}\sin(\sqrt{x}) + \cos(\sqrt{x})] + C$

ME8 **1** $\dfrac{mv^2}{2}$

2 $\dfrac{EAl^2}{2L}$

3 $\dfrac{1}{2}k(x_2^2 - x_1^2)$

7 $\dfrac{x^2}{2} + \ln(x^2 - 4)^2 + C$

8 7.2 kJ/kg

9 $R\left[a(T_2 - T_1) + \dfrac{b}{2}(T_2^2 - T_1^2) + \dfrac{c}{3}(T_2^3 - T_1^3)\right.$

$\left. + \dfrac{d}{4}(T_2^4 - T_1^4) + \dfrac{e}{5}(T_2^5 - T_1^5)\right]$

10 $-\cot(x) - \tan(x) + C$

13 $\dfrac{C}{0.32}[V_1^{-0.32} - V_2^{-0.32}]$

14 i 0.245025   **ii** 0.25   **iii** 0.004975

15 $-\dfrac{k}{6}\left[\dfrac{r^3}{x^6} + \dfrac{2}{x^3}\right]$

18 $0.3e^{-(2\times10^3)t} - 0.125\,e^{-(8\times10^3)t} - 0.175$

19 i $\dfrac{20}{\pi}[1 - \cos(100\pi t)]$ **ii** $\dfrac{[1 - \cos(100\pi t)]^2}{\pi^2}$

21 π

23 $\dfrac{EID^2\pi^4}{64L^3}$

26 ii $\dfrac{\pi}{2}$

31 7.26×10^5 J/s

32 a 1.80×10^4 **b** 1.84×10^4
   c 2.30×10^4 **d** 2.93×10^4
   (units are kJ/kmol)

9(a) **1** 23 N s

2 a 0.743   **b** 1.147

3 i a 0.266   **b** 0.254

   ii 0.25   **iii a** -6.4% **b** -1.6%

4 53.91 m

5 69.65 m³/s

6 3.73 V

9(b)   **1** 0.999969, exact value = 1, % error = 0.0031%

2 0.681   **a i** 0.6797915375   **ii** 0.68597084676
  iii 0.6863708136   **iv** 0.6865841146
  b i 0.6838407449   **ii** 0.6863450655
  iii 0.6865041355   **iv** 0.6865883836

3 175.41 N s

4 17.62 N

5 0.3 A

6 9.58 J

7 149 m³/s

8 49.81 N

9 0.96×10^{-9}

10 0.159, 0.023, 0.001, 31.7×10^{-6}

11 a 2.553783362   **b** 2.553805688

9(c)   **1** 4 (units)²

2 Mean = 2 V, RMS = 2.31 V

3 2/π A

4 i 0 V   **ii** 0 V

7 Mean = 3.68 V, RMS = 4.10 V

8 2.38 mA

9 Mean = 1 V, RMS = 1.18 V, f = 1.18

10 0.75 V

11 3.86 kN

12 i 2/3   **ii** 2/3

9(d)   **1** $x = \dfrac{2.5}{4\pi^2}\,[\cos(2\pi t) - 1]$

2 $x = \dfrac{t^3}{6} + 2t + 3$

3 $a = 3.4$ m/s², $s = 3.6$ m

4 8.06 m/s

5 20.88 km

6 11.8 km (Simpson result)

8 2.02 s

9(e)   **1** 11.885 kJ

2 −21.592 kJ

3 −11.467 kJ

5 $\dfrac{P}{6EI}\,(3lx^2 - x^3)$

6 $\dfrac{w}{24EI}\,(4L^3x - x^4 - 3L^4)$

7 $\dfrac{w}{24EI}\left[L^4 - 4L^3x - (L - x)^4\right]$

9 0.946

10 i $1 - x^2 + \dfrac{x^4}{2!} - \dfrac{x^6}{3!} + \cdots$

   ii 0.743

EQ9   **1** 9/2

2 $3\pi - 4$

3 1.464 (3 d.p.)

4 $\dfrac{1}{2}[\cos(\pi) + 2(2\cos(\pi/2) + 3\cos(\pi/3))$
   $+ 4\cos(\pi/4)]$

5 0.78077 (5 d.p.)

6 $\ln(27) - 2$

7 0.3103 (4 d.p.)

8 a $s'(t) = -32t,\ s(t) = 16(16 - t^2)$

   b $t = 4$ s   **c** −128 ft/s

9 a $A = \displaystyle\int_{2}^{4} [-(x - 2)^2 - (4 - 2x)]\,dx$

   b 4/3

10 1.747, % error is 0.057%

11 a 12 **b** 5/2

12 $\dfrac{x^2}{2} - \dfrac{x^8}{16} + \dfrac{x^{14}}{336} - \cdots$

13 $b = 1$ and 2

14 $1 - \dfrac{1}{10} + \dfrac{1}{216} - \cdots$ and the fourth
   non-zero term is $\dfrac{1}{9360} < \dfrac{1}{1000} = 0.001$

ME9 **1** 14.02 kJ

2 430 J

3 930 kJ/kg

4 393 m

5 22.1 s

6 0.92 V

7 1.49 A

8 $\dfrac{3\pi}{2}$

9 Mean = 4.01 V, RMS = 4.19 V

10 b 200 m

11 15.5 kJ

12 $y = \dfrac{1}{EI}\left[\dfrac{2.5x^4}{12} - \dfrac{11x^3}{3} + \dfrac{38.5x^2}{2} - 30.625x\right]$

13 $\dfrac{22}{7} - \pi$

17 $\dfrac{9\pi^3}{8} - 2$

18 87 m

10(a) **1 a** $7 + j11$ **b** $-3 - j5$ **c** $-14 + j31$
  d $-14 + j31$ **e** $0.382 - j0.011$
  f $2.615 + j0.077$ **g** $9 + j14$ **h** $-5 + j12$
  i $-39 + j80$ **j** $-44 + j92$

2 $3 + j15, 3 - j15, -j3, -j, 0, \pi$ and e

3 $42 - j13$

4 $1, j, -j, 1, -j$

5 i $3 + j4, 2 + j11, -7 + j24, 0$ **ii** $2 + j$

6 $(0.0104 + j0.0055)$ S

7 a $a = 2, b = 3$ **b** $a = 0, b = 6$ **c** $a = 0, b = 2$
  d $a = 0.12, b = 0.16$ **e** $a = -3/2, b = 4$
  f $a = 1, b = 1$

8 a $20 - j15$ **b** $0.8 - j0.6$ **c** $5 - j$
  d $-1.7 - j0.1$

9 $Z_t = (0.756 + j0.805)\ \Omega$

10 $I = (0.385 - j1.923)$ A

11 a $z = -1 \pm j5$ **b** $z = 1 \pm j\sqrt{2}$

  c $z = \dfrac{7}{6} \pm j\dfrac{\sqrt{107}}{6}$

12 $(0.462 + j0.692)$ A

13 $I_s = (0.0843 + j0.0375)$ A, $I_p = (0.797 - j1.110)$ A

14 ii $\dfrac{R}{1 + \omega^2 C^2 R^2}, \dfrac{-\omega C R^2}{1 + \omega^2 C^2 R^2}$

10(b) **2 a** $\sqrt{2}\ \angle 45°$ **b** $\sqrt{2}\angle 135°$ **c** $2\angle(-30°)$
  d $2\angle(-120°)$ **e** $1\angle 90°$ **f** $x\angle 90°$
  g $3\angle 180°$ **h** $5\angle 0°$

3 a $1.73 + j$ **b** $4.6 + j1.95$ **c** $1.41 - j1.41$
  d $-0.90 - j1.07$

4 $|F| = 75.85$ N, $\arg(F) = -177.82°$

5 $F = 0$ N (equilibrium)

10(c) **1** $0.83\ \angle 51.57°$ A

2 $|z| = 0.2$ $\arg(z) = -26°$

3 a $\sqrt{3} + j$ **b** $\sqrt{3} + j$ **c** 240

5 a $6\angle(-45°)$ **b** $2\angle(-131°)$ **c** $15\angle 26°$

6 ii $|z| = 1, \dfrac{1}{z} = \dfrac{1}{3}\left(\sqrt{8} - j\right)$

8 Gain $= \dfrac{1000}{\sqrt{2}}$, phase $= -45°$

9 Gain $= 0.11$, phase $= -197.89°$

10 $I_1 = 1.27\angle(-29.93°)$ A, $I_2 = 0.94\angle(-8.13°)$ A

10(d) **1 a** $j2$ **b** $-2 - j2$ **c** -4
  d $9784704 - j2063872$ **e** $-76443 + j16124$
  f $12 - j316$

2 $1, j, -1$ and $-j$

3 -64

4 i $Z_0 = 1000\angle(-39.3°)\ \Omega$ **ii** $0.01005\angle 45°$ m^{-1}
  iii 884.2 m   **iv** 1.407×10^6 m/s

6 i Poles: $-4, 2 + j3.46$ and $2 - j3.46$
  ii Zeros: $\sqrt{2} + j\sqrt{2}, -\sqrt{2} + j\sqrt{2}, -\sqrt{2} - j\sqrt{2}$

    and $\sqrt{2} - j\sqrt{2}$

7 ii 0   **iii** 1

9 $-0.42 + j0.42$

10 i $1 \pm j\sqrt{3}$ **ii** $2^{1/3}\angle 20°, 2^{1/3}\angle 100°, 2^{1/3}\angle 140°,$
  $2^{1/3}\angle 220°, 2^{1/3}\angle 260°$ and $2^{1/3}\angle 340°$

For complete solutions see www.palgrave.com/engineering/singh

10(e) **1 a** $10e^{j\pi}$ **b** $10e^{j\pi}$ **c** $2e^{j\pi/2}$ **d** $\sqrt{2}e^{j\pi/4}$

   e $2e^{-j\pi/6}$

2 a je **b** $-e^{\pi}$ **c** $\sqrt{3}e^2 + je^2$ **d** $\dfrac{e^{1000}}{\sqrt{2}}(1-j)$

3 All parts are equal to 1

4 All parts are equal to -1

6 $2e^{-j\frac{5\pi}{6}}$

7 $5.45e^{-j2.76}$

8 a 0   **b** $e^t(6\cos(t) - 8\sin(t))$

10 $a = 0, b = 3$

EQ10 **1 i** $18 + j4$ **ii** $14 + j2$

2 a $10 - j5$ **b** $5^{1/4}\angle(13.28°)$ and $5^{1/4}\angle(193.28°)$

3 a $z_1 = \sqrt{2}\angle(60°)$, $z_2 = \sqrt{2}\angle(150°)$,
   $z_3 = \sqrt{2}\angle(240°)$ and $z_4 = \sqrt{2}\angle(330°)$ **b** $1\angle135°$

4 a $3 + 2i$, $3 - 2i$ **b** $\dfrac{1}{\sqrt{2}}(1 + i)$ **c** 2

5 a i $3 + 22i$ **ii** $10 - 11i$ **iii** $0.3 - 1.1i$
   iv $\sqrt{130}$ **v** -7 **vi** $3.6162\angle(3.1204°)$

   b $r_1 = 10^{1/6}\angle(-6.145°)$, $r_2 = 10^{1/6}\angle(113.855°)$

   and $r_3 = 10^{1/6}\angle(233.855°)$

6 $z_1 = 2\angle(30°)$, $z_2 = 2\angle(150°)$ and $z_3 = 2\angle(270°)$

7 $z_1 = 2\angle(36°)$, $z_2 = 2\angle(108°)$, $z_3 = 2\angle(180°)$,
   $z_4 = 2\angle(252°)$ and $z_5 = 2\angle(324°)$

8 a Cartesian: $-3\sqrt{2}(4 + i)$, $\dfrac{\sqrt{2}}{3}(1 - 4i)$,
   $0.8787 + 7.1213i$

   Polar: $3\sqrt{34}\angle(3.3866°)$, $\dfrac{\sqrt{34}}{3}\angle(-1.3258°)$,

   $7.1753\angle(1.448°)$

   b $1 + \sqrt{6}i$, $1 - \sqrt{6}i$

   c $2\angle\left(\dfrac{\pi}{3}\right)$, $1024\angle\left(\dfrac{4\pi}{3}\right)$, $z_1 = 2^{1/3}\angle\left(\dfrac{\pi}{9}\right)$,

   $z_2 = 2^{1/3}\angle\left(\dfrac{7\pi}{9}\right)$, $z_3 = 2^{1/3}\angle\left(\dfrac{13\pi}{9}\right)$

   d $\cos(3\theta) = \cos^3(\theta) - 3\cos(\theta)\sin^2(\theta)$

9 $\dfrac{9\pi}{10}$

10 a $\bar{z} = 2 + j$, $z + w = 5 + j$, $zw = 8 + j$,
   $\dfrac{1}{w} = \dfrac{3}{13} - \dfrac{2}{13}j$, $|z| = \sqrt{5}$

   b $4\angle\left(\dfrac{4\pi}{3}\right)$, $w_1 = 2\angle\left(\dfrac{2\pi}{3}\right)$ and $w_2 = 2\angle\left(\dfrac{5\pi}{3}\right)$

11 a $5 + j$, $-\dfrac{1}{2} + \dfrac{5}{2}j$ **b** $4\angle\left(-\dfrac{2\pi}{3}\right)$

   c $w_1 = \sqrt{5}\angle\left(\dfrac{\pi}{4}\right)$, $w_2 = \sqrt{5}\angle\left(\dfrac{3\pi}{4}\right)$, $w_3 = \sqrt{5}\angle\left(\dfrac{5\pi}{4}\right)$

   and $w_4 = \sqrt{5}\angle\left(\dfrac{7\pi}{4}\right)$

12 a $\sqrt{2}e^{-j\pi/4}$ **b** -1 **c** -1
   d $r_1 = 2\angle(0°)$, $r_2 = 2\angle(120°)$ and
   $r_3 = 2\angle(240°)$

ME10 **2 a** $j32$ **b** $2^6\left(-1 - j\sqrt{3}\right)$ **c** $\cos(4\theta) + j\sin(4\theta)$

3 a $x = \cos(\theta)$, $y = -\sin(\theta)$

   b $x = \dfrac{1}{2}$, $y = -\dfrac{\sin(\theta)}{2(1 + \cos(\theta))}$

4 a $1 + j$ **b** $\dfrac{1}{2}\left(\sqrt{3} + j\right)$

6 $F = 0$

7 $I_R = 12 - j12$, $I_Y = 11.94 + j22.46$,
   $I_B = -24.83 + j7.82$ and
   $I_N = -0.89 + j18.28$

10 $K = 25$

11 Gain $= 9.81$, phase $= -11.31°$

12 a Stable   **b** Critically stable

13 i $\sigma = -1$, $\omega = \sqrt{11}$ **ii** $\sqrt{12}$ rad/s   **iii** 0.29
   iv 0.45 s

14 $\omega = 1$: gain $= 0.995$ and phase $= -5.71°$;
   $\omega = 10$: gain $= 0.707$ and phase $= -45°$;
   $\omega = 100$: gain $= 0.0995$ and phase $= -84.289°$

16 $s_1 = 1\angle 0°$, $s_2 = 1\angle 60°$, $s_3 = 1\angle 120°$, $s_4 = 1\angle 180°$,
   $s_5 = 1\angle 240°$ and $s_6 = 1\angle 300°$

18 $\phi = \dfrac{3}{2}\left(x - \sqrt{3}y\right)$, $\psi = \dfrac{3}{2}\left(\sqrt{3}x + y\right)$

11(a) 1 a $\begin{pmatrix} 7 & 1 \\ 8 & 2 \end{pmatrix}$ **b** $\begin{pmatrix} 7 & 1 \\ 8 & 2 \end{pmatrix}$ **c** $\begin{pmatrix} 3 & 6 \\ 9 & -3 \end{pmatrix}$

d $\begin{pmatrix} 15 & 4 \\ 19 & 3 \end{pmatrix}$ **g** $\begin{pmatrix} 1 \\ -4 \end{pmatrix}$

h $\begin{pmatrix} -7 \\ -2 \end{pmatrix}$ **j** $\begin{pmatrix} 17 \\ -8 \end{pmatrix}$

e, f and **i** cannot be evaluated

2 $\begin{pmatrix} 0 & 1 & 4 & 3 & 0 & -3 & -4 & -1 \\ 0 & 1 & 2 & 6 & 7 & 6 & 2 & 1 \end{pmatrix}$

3 $A = \begin{pmatrix} 1 & 2 & 2 & 1 \\ 2 & 2 & 4 & 4 \end{pmatrix}$

a $\begin{pmatrix} -3 & -2 & -2 & -3 \\ 2 & 2 & 4 & 4 \end{pmatrix}$

b $\begin{pmatrix} 3 & 6 & 6 & 3 \\ 6 & 6 & 12 & 12 \end{pmatrix}$

c $\begin{pmatrix} 2 & 2 & 4 & 4 \\ -1 & -2 & -2 & -1 \end{pmatrix}$

d $\begin{pmatrix} -1 & -2 & -2 & -1 \\ 2 & 2 & 4 & 4 \end{pmatrix}$

e $\begin{pmatrix} 2 & 2 & 4 & 4 \\ 1 & 2 & 2 & 1 \end{pmatrix}$

f $\begin{pmatrix} -2 & -2 & -4 & -4 \\ -1 & -2 & -2 & -1 \end{pmatrix}$

4 $\begin{pmatrix} 0 & 0 \\ 0 & 0 \end{pmatrix}$

5 $\begin{pmatrix} 0 & 0 & 0 \\ 0 & 0 & 0 \\ 0 & 0 & 0 \end{pmatrix}$

11(b) 1 a $-8, -8$ **b** $-13, -13$ **c** $0, 0$

$\det A = \det B$

3 a i -9 **ii** 22 **iii** -198 **iv** -198

b i -1701.5 **ii** 624.3 **iii** -1062246.45

iv -1062246.45

c i -21.4 **ii** 79.27 **iii** -1696.38

iv -1696.38

$\det A \times \det B = \det(AB)$

4 a $\begin{pmatrix} 3 & -2 \\ -13 & 9 \end{pmatrix}$ **b** $\begin{pmatrix} 5 & -7 \\ -12 & 17 \end{pmatrix}$ **c** $\begin{pmatrix} 1 & 0 \\ 0 & 1 \end{pmatrix}$

5 a $\begin{pmatrix} -1/7 & 4/7 \\ 3/7 & -5/7 \end{pmatrix}$ **b** $\begin{pmatrix} -4/9 & 1/3 \\ 7/18 & -1/6 \end{pmatrix}$

c No inverse

6 $i_1 = -1.25$ A and $i_2 = 2.5$ A

7 $i_1 = \dfrac{47}{95}$ A and $i_2 = \dfrac{27}{95}$ A

8 a $x = 2$ and $y = 4$ **b** $x = -1$ and $y = 1$

c $x = \dfrac{1}{4}$ and $y = -\dfrac{1}{3}$

11(c) 1 a -117 **b** 1288 **c** -114

2 a $\begin{pmatrix} -2 & 1 \\ 3 & 5 \end{pmatrix}$ **b** $\begin{pmatrix} 1/3 & -3/7 \\ -2/5 & \pi \end{pmatrix}$

c $\begin{pmatrix} 7 & 2 & -3 \\ 3 & 6 & -3 \\ 4 & 1 & 1 \end{pmatrix}$ **d** $\begin{pmatrix} 1.17 & 9.39 & 2.11 \\ 1.36 & -1.45 & 5.20 \end{pmatrix}$

e $\begin{pmatrix} -7 & 3 & 8 & -4 \\ 2 & 6 & 3 & 6 \\ 1 & -4 & -3 & 7 \\ 5 & 7 & 5 & 0 \end{pmatrix}$

f $\begin{pmatrix} 1 & -9 & -4 & 9 \\ -3 & 4 & 2 & 19 \\ 7 & 6 & 8 & 11 \end{pmatrix}$

4 $t = 5 + j\sqrt{173}, t = 5 - j\sqrt{173}$

5 $C = \begin{pmatrix} 7 & 42 & -16 \\ -5 & -30 & 1 \\ -15 & -17 & 3 \end{pmatrix}$,

$C^T = \begin{pmatrix} 7 & -5 & -15 \\ 42 & -30 & -17 \\ -16 & 1 & 3 \end{pmatrix}$

6 $i_1 = -0.76$ A, $i_2 = -0.64$ A and $i_3 = 2.19$ A

7 a -42 **b** 44 **c** 13 **d** -85

8 a 0 **b** -509092.39

11(d) 1 a $x = 2, y = -4, z = -3$

b $x = 1, y = 2, z = 3$

c $x = 1/2, y = 1/4, z = 1/8$

2 $i_1 = 1.20$ mA, $i_2 = 0.28$ mA

3 $x = -6.2, y = 30, z = -10$

4 $\ddot{x}_1 = g/3, \ddot{x}_2 = -g/3, T = 4g/3$

5 $\ddot{x}_1 = -g/5, \ddot{x}_2 = g/5, T = 12g/5$

6 $I_1 = 2.56$ mA, $I_2 = 8.58 \times 10^{-5}$ A,

$I_3 = -0.615$ mA

11(e) 1 $x = 0, y = 0, z = 0$

2 Unique, $x = 1, y = 1$

3 Infinite, $x = a, y = -2a$

4 All are unique

5 a $\lambda = -4, \lambda = 2$ **b** $\lambda = 3 + j\sqrt{14}, 3 - j\sqrt{14}$

11(f) 1 a $-4, 7$ **b** $1, 3$ **c** $-3, 3$

2 a $\lambda = -4$, eigenvector $\begin{pmatrix} -3/11 \\ 1 \end{pmatrix}$;

$\lambda = 7$, eigenvector $\begin{pmatrix} 1 \\ 0 \end{pmatrix}$

b $\lambda = 1$, eigenvector $\begin{pmatrix} 1 \\ 2 \end{pmatrix}$; $\lambda = 3$, eigenvector $\begin{pmatrix} 1 \\ 1 \end{pmatrix}$

c $\lambda = -3$, eigenvector $\begin{pmatrix} -2 \\ 1 \end{pmatrix}$;

$\lambda = 3$, eigenvector $\begin{pmatrix} 1 \\ 1 \end{pmatrix}$

3 a $j, -j$ **b** $-1 + j2, -1 - j2$

c $1.50 + j3.57, 1.50 - j3.57$

4 $a\begin{pmatrix} -2/3 \\ 0 \\ 1 \end{pmatrix}$ corresponds to $\lambda = -5$ and $a\begin{pmatrix} 20 \\ 9 \\ 15 \end{pmatrix}$

corresponds to $\lambda = 4$

5 $\begin{pmatrix} 0 \\ 1 \\ 2 \end{pmatrix}$ corresponds to $\lambda_1 = 2$, $\begin{pmatrix} 0 \\ 1 \\ 1 \end{pmatrix}$ corresponds to

$\lambda_2 = 3$ and $\begin{pmatrix} 1 \\ -5 \\ -4 \end{pmatrix}$ corresponds to $\lambda_3 = 5$

6 1.40 s, 2.81 s

7 $0.68, 14.92, 3.80$

8 $0.21 + j5.73, 0.21 - j5.73,$

$-4.01 + j2.08, -4.01 - j2.08$

11(g) 1   i $\lambda_1 = 1, \mathbf{u} = \begin{pmatrix} 1 \\ 0 \end{pmatrix}$ and $\lambda_2 = 2, \mathbf{v} = \begin{pmatrix} 0 \\ 1 \end{pmatrix}$

ii $\mathbf{P} = \begin{pmatrix} 1 & 0 \\ 0 & 1 \end{pmatrix} = \mathbf{I}, \mathbf{D} = \begin{pmatrix} 1 & 0 \\ 0 & 2 \end{pmatrix} = \mathbf{A}$

b i $\lambda_1 = 0, \mathbf{u} = \begin{pmatrix} -1 \\ 1 \end{pmatrix}$ and $\lambda_2 = 2, \mathbf{v} = \begin{pmatrix} 1 \\ 1 \end{pmatrix}$

ii $\mathbf{P} = \begin{pmatrix} -1 & 1 \\ 1 & 1 \end{pmatrix}, \mathbf{D} = \begin{pmatrix} 0 & 0 \\ 0 & 2 \end{pmatrix}$

c i $\lambda_1 = 3, \mathbf{u} = \begin{pmatrix} 1 \\ -4 \end{pmatrix}$ and $\lambda_2 = 4, \mathbf{v} = \begin{pmatrix} 0 \\ 1 \end{pmatrix}$

ii $\mathbf{P} = \begin{pmatrix} 1 & 0 \\ -4 & 1 \end{pmatrix}, \mathbf{D} = \begin{pmatrix} 3 & 0 \\ 0 & 4 \end{pmatrix}$

d i $\lambda_1 = 1, \mathbf{u} = \begin{pmatrix} 2 \\ -1 \end{pmatrix}$ and $\lambda_2 = 4, \mathbf{v} = \begin{pmatrix} 1 \\ 1 \end{pmatrix}$

ii $\mathbf{P} = \begin{pmatrix} 2 & 1 \\ -1 & 1 \end{pmatrix}, \mathbf{D} = \begin{pmatrix} 1 & 0 \\ 0 & 4 \end{pmatrix}$

2 a $\begin{pmatrix} 1 & 0 \\ 0 & 32 \end{pmatrix}$ **b** $\begin{pmatrix} 16 & 16 \\ 16 & 16 \end{pmatrix}$

c $\begin{pmatrix} 243 & 0 \\ 3124 & 1024 \end{pmatrix}$ **d** $\begin{pmatrix} 342 & 682 \\ 341 & 683 \end{pmatrix}$

3 a i $\lambda_1 = 1, \mathbf{u} = \begin{pmatrix} 1 \\ 0 \\ 0 \end{pmatrix}, \lambda_2 = 2, \mathbf{v} = \begin{pmatrix} 0 \\ 1 \\ 0 \end{pmatrix}$

and $\lambda_3 = 3, \mathbf{w} = \begin{pmatrix} 0 \\ 0 \\ 1 \end{pmatrix}$

ii $\mathbf{P} = \mathbf{I}, \mathbf{D} = \begin{pmatrix} 1 & 0 & 0 \\ 0 & 2 & 0 \\ 0 & 0 & 3 \end{pmatrix}$

iii $\mathbf{A}^4 = \begin{pmatrix} 1 & 0 & 0 \\ 0 & 16 & 0 \\ 0 & 0 & 81 \end{pmatrix}$

b i $\lambda_1 = -1, \mathbf{u} = \begin{pmatrix} 1 \\ 0 \\ 0 \end{pmatrix}, \lambda_2 = 4, \mathbf{v} = \begin{pmatrix} 4 \\ 5 \\ 0 \end{pmatrix}$

and $\lambda_3 = 5, \mathbf{w} = \begin{pmatrix} 2 \\ 3 \\ 1 \end{pmatrix}$

ii $\mathbf{P} = \begin{pmatrix} 1 & 4 & 2 \\ 0 & 5 & 3 \\ 0 & 0 & 1 \end{pmatrix}, \mathbf{D} = \begin{pmatrix} -1 & 0 & 0 \\ 0 & 4 & 0 \\ 0 & 0 & 5 \end{pmatrix}$,

iii $\mathbf{A}^4 = \begin{pmatrix} 1 & 204 & 636 \\ 0 & 256 & 1107 \\ 0 & 0 & 625 \end{pmatrix}$

EQ11 1 i $-16\begin{pmatrix} 4 & 5 \\ 4 & 5 \end{pmatrix}$

ii $-4\begin{pmatrix} 15 & 17 \\ 19 & 21 \end{pmatrix}$

2 a $\mathbf{A}^2 = \mathbf{I}$ and $\mathbf{A}^3 = \dfrac{1}{7}\begin{pmatrix} 3 & -2 & -6 \\ -2 & 6 & -3 \\ -6 & -3 & -2 \end{pmatrix}$

b $\mathbf{A}^{-1} = \mathbf{A}$ and $\mathbf{A}^{2004} = \mathbf{I}$

3 $A^T = \begin{pmatrix} 1 & 1 & 1 \\ 0 & 1 & 2 \\ -1 & 1 & 3 \end{pmatrix}$, $CB = \begin{pmatrix} 0 & -2 \\ -1 & 1 \end{pmatrix}$,

$A^2 = \begin{pmatrix} 0 & -2 & -4 \\ 3 & 3 & 3 \\ 6 & 8 & 10 \end{pmatrix}$, **AB, B + C, A − B**

and BC^T is **not** valid.

4 a $A^2 = \begin{pmatrix} 7 & 10 \\ 15 & 22 \end{pmatrix}$, $B^2 = \begin{pmatrix} -1 & 0 \\ 0 & -1 \end{pmatrix}$,

$AB = \begin{pmatrix} -2 & 1 \\ -4 & 3 \end{pmatrix}$ and $BA = \begin{pmatrix} 3 & 4 \\ -1 & -2 \end{pmatrix}$

b $AB - BA = \begin{pmatrix} bg - cf & af + bh - be - df \\ ce + dg - ag - ch & cf - bg \end{pmatrix}$

5 i $\dfrac{2}{3^2}\begin{pmatrix} 1 & 1 \\ 1 & 1 \end{pmatrix}$

ii $\dfrac{2^2}{3^3}\begin{pmatrix} 1 & 1 \\ 1 & 1 \end{pmatrix}$

6 $A = \begin{pmatrix} 1 & 0 & -1 \\ 0 & 1 & -1 \\ 1 & 1 & 1 \end{pmatrix}$

7 $\det(A) = 165$

8 $A^{-1} = \begin{pmatrix} 1 & -1 & 2 \\ 1 & 2 & 0 \\ 0 & -1 & 1 \end{pmatrix}$

9 $\text{adj}A = \begin{pmatrix} 1 & 1 \\ 1 & 1 \end{pmatrix}$, $\det A = 0$ and

$\text{adj}A = \begin{pmatrix} 1 & 18 & -11 \\ 1 & 13 & -8 \\ 0 & -2 & 1 \end{pmatrix}$,

$A^{-1} = \begin{pmatrix} -1 & -18 & 11 \\ -1 & -13 & 8 \\ 0 & 2 & -1 \end{pmatrix}$

10 a $\lambda_1 = 1$ and $\lambda_2 = 3$

b $Q = \begin{pmatrix} 1 & 1 \\ 0 & 2 \end{pmatrix}$, $A^5 = \begin{pmatrix} 1 & 121 \\ 0 & 243 \end{pmatrix}$

11 Substituting the given eigenvalues and eigenvectors into $Au = \lambda u$

12 a $\lambda_1 = 2$, $u = \begin{pmatrix} 1 \\ -2 \end{pmatrix}$ and $\lambda_2 = 7$, $v = \begin{pmatrix} 2 \\ 1 \end{pmatrix}$

b $\vec{v}$ is an eigenvector belonging to an eigenvalue of 2 but $\vec{w}$ is **not** an eigenvector of A

13 a $\lambda_1 = 1$, $u = \begin{pmatrix} -1 \\ 1 \\ 1 \end{pmatrix}$ and $\lambda_2 = 2$, $v = \begin{pmatrix} 0 \\ 1 \\ 1 \end{pmatrix}$ and

$\lambda_3 = 0$, $w = \begin{pmatrix} 1 \\ -1 \\ 0 \end{pmatrix}$

b $\det(P) = \det\begin{pmatrix} -1 & 0 & 1 \\ 1 & 1 & -1 \\ 1 & 1 & 0 \end{pmatrix} = -1$ and because

$\det(P) \neq 0$ therefore matrix A is diagonalizable

ME11 **1 i** $\begin{pmatrix} 15 & 0 \\ 0 & 15 \end{pmatrix}$ **ii** $\begin{pmatrix} -42 & 0 \\ 0 & -42 \end{pmatrix}$ **ii** $\begin{pmatrix} ac & 0 \\ 0 & bd \end{pmatrix}$

2 $a = -2$, $b = 1$, $c = 3/2$, $d = -1/2$

3 $\begin{pmatrix} 4 & 0 \\ 0 & 4 \end{pmatrix}$, $\begin{pmatrix} 8 & 0 \\ 0 & 8 \end{pmatrix}$, $\begin{pmatrix} 1024 & 0 \\ 0 & 1024 \end{pmatrix}$

4 i $x = 5$, $y = -4$

ii $a = -6$, $b = 5$

iii $\begin{pmatrix} 5 & -6 \\ -4 & 5 \end{pmatrix}$

6 0

7 0

8 $I_1 = -68.45$ mA, $I_2 = -12.54$ mA,

$I_3 = 38.29$ mA

9 $i_1 = 2.19$ mA, $i_2 = 0.44$ mA, $i_3 = 1.28$ mA

10 $2\pi\sqrt{-\dfrac{2m}{k}}$, $\pi\sqrt{-\dfrac{2m}{k}}$

11 i $\lambda_1 = 4$, $\lambda_2 = 3$ **ii** For $\lambda_1 = 4$, $v = a\begin{pmatrix} 1 \\ 1 \end{pmatrix}$ and

for $\lambda_2 = 3$, $u = a\begin{pmatrix} 2 \\ 1 \end{pmatrix}$

12 $\dfrac{s + 1.4}{s^2 + 2s + 3}$

13 $\begin{pmatrix} \dfrac{2777}{2560} & -\dfrac{47}{7680} & \dfrac{1953}{2560} \end{pmatrix}$

14 a $\begin{pmatrix} -1804 & -2583 & -2727 \\ 12 & 5685 & 2699 \\ 1269 & 6387 & 1124 \end{pmatrix}$

b $\begin{pmatrix} -99.92 & 308.63 & 61.37 \\ 357.22 & -226.62 & 498.79 \\ 340.06 & -732.14 & 56.44 \end{pmatrix}$

For complete solutions see www.palgrave.com/engineering/singh

15

$$-0.2\left[\frac{(0.77 \times 10^9) + 515500s^3 - (0.29 \times 10^8)s^2 - (0.1 \times 10^9)s}{(1 \times 10^4)s^4 - (1.7 \times 10^4)s^3 - (2.54 \times 10^5)s^2 - (0.13 \times 10^7)s - (0.24 \times 10^8)}\right]$$

12(a)

1 13 kN, angle 22.26° below horizontal

2 8.66 kN horizontal and 5 kN vertical

3 8.49 kN horizontal and 8.49 kN vertical

4 3.94 kN horizontal and 9.19 kN vertical

5 $\frac{1}{2}\mathbf{b} - \mathbf{a}$

7 10.85 N horizontal

8 31.80 kN

9 24.27 kN at an angle of 45.54° anticlockwise from $\mathbf{F}_1$

10 23.18 kN at an angle of 17.76° clockwise from $\mathbf{F}_1$

12(b)

1 a 43.59 kN, angle 36.59° above the horizontal
   b 10.63 kN, angle 16.40° below the horizontal

2 181.96 kN

3 5.38 kN, angle 106.95° above the horizontal

4 21.13 kN, angle 167.04° below the horizontal

12(c)

1 a $-18\mathbf{i} + 14\mathbf{j} + 11\mathbf{k}$
   b $-176\mathbf{i} + 203\mathbf{j} + 46\mathbf{k}$

2 $7\mathbf{j} + 8\mathbf{k}$

3 $12\mathbf{i} - 7\mathbf{j} + 30\mathbf{k}$

4 $(45\mathbf{i} - 21\mathbf{j} + 15\mathbf{k})$ N

5 $(-32\mathbf{i} - 16\mathbf{j} + 4\mathbf{k})$ kN

6 $(0.88\mathbf{i} + 6.58\mathbf{j} + 9.15\mathbf{k})$ kN

7 $(254.645\mathbf{i} - 672.697\mathbf{j} - 1234.618\mathbf{k})$ kN, magnitude = 1428.86 kN

12(d)

1 a 23   **b** 0   **c** −211

2 59 kJ

3 12 J

4 6.03 J

5 i $x^2 + y^2 + z^2$ **ii** $x^2 + y^2 + z^2$

6 11.31°

7 i $\mathbf{v} = (3t^2 - 2t)\mathbf{i} + 4t^3\mathbf{j}$, $\mathbf{a} = (6t - 2)\mathbf{i} + 12t^2\mathbf{j}$
   ii 4.4°

12(e)

1 i $5\mathbf{i} + 26\mathbf{j} - 17\mathbf{k}$ **ii** $-5\mathbf{i} - 26\mathbf{j} + 17\mathbf{k}$

4 a $-11\mathbf{i} - 8\mathbf{j} + 14\mathbf{k}$ **b** $3\mathbf{i} - 8\mathbf{j} + 6\mathbf{k}$

5 $20\mathbf{i} + 4\mathbf{j} + 16\mathbf{k}$

EQ12

1 $-26\mathbf{i} - 12\mathbf{j} - 22\mathbf{k}$

2 a i 61 J   **ii** 45.85°

   b i $v = (3t^2 - 6t)\mathbf{i} + (12t - 2)\mathbf{j}$, $a = (6t - 6)\mathbf{i} + 12\mathbf{j}$

   ii $v = -2\mathbf{j}$ and $a = -6\mathbf{i} + 12\mathbf{j}$

   iii $t = 0$, $t = 2$

3 $43\mathbf{i} - 25\mathbf{j} + 2\mathbf{k}$

4 $\dfrac{(\mathbf{i} + 11\mathbf{j} + 7\mathbf{k})\mu q}{56\pi}$

5 i $-\mathbf{i} + \mathbf{j} + 3\mathbf{k}$ **ii** 0   **iii** $-7\mathbf{i} + 5\mathbf{j} - 4\mathbf{k}$
   iv -0.2304

6 i $\mathbf{j} - 11\mathbf{k}$ **ii** 1   **iii** $\mathbf{i} + 13\mathbf{j} - 2\mathbf{k}$ **iv** 0

   v 1.3335   **vi** $\dfrac{2}{\sqrt5}\mathbf{i} + \dfrac{1}{\sqrt5}\mathbf{k}$

   vii $\dfrac{1}{\sqrt{25.25}}(\mathbf{i} - 4.5\mathbf{j} - 2\mathbf{k})$

7 $\dfrac{1}{\sqrt{91}}$ and $-\dfrac{2}{\sqrt{210}}$

ME12

1 $\dfrac{5}{\sqrt{218}}\mathbf{i} + \dfrac{7}{\sqrt{218}}\mathbf{j} + \dfrac{12}{\sqrt{218}}\mathbf{k}$

2 i $\mathbf{i} + 2\mathbf{j} - 7\mathbf{k}$ **ii** $20\mathbf{i} + \mathbf{j} - 6\mathbf{k}$ **iii** 2   **iv** 2
   v 3   **vi** $29.4\mathbf{i} + 44.1\mathbf{j} - 14.7\mathbf{k}$ **vii** 20.9

3 90°

4 48.37°

5 10.84 kN at −55.46°

6 15.34 kN at 8.91°

7 $F_x = 23.77$ kN horizontal, $F_y = 31.55$ kN vertically down

8 a $63i + 69j + 48k$ **b** $86i + 225j + 111k$

9 35 J

10 468 J

11 a 13 J   **b** 149 J

12 $-3i + 6j - 21k$, 22.05 N m

13 $(29.72\,i - 4.95j - 24.77k)$ kN

15 $j(cb - ad)$

16 $-3, 3$

17 4

13(a) (C is the constant of integration)

1 $p = -\rho gz + C$

2 $p = -\dfrac{\rho\omega^2 r^2}{2} + C$

3 $y = 5x$

4 $y = e^x + C$

6 $v = 5t - 1.5t^2 + 8$, $s = 2.5t^2 - 0.5t^3 + 8t - 2.1$

13(b) **1** $y = e^{-x} - e^{-2x}$

2 $y = e^{2x} + Ce^x$

3 $y = \dfrac{x}{1 - x^2}$

5 $v = 1 - \dfrac{1}{t}$, $v \to 1$

6 iii $i \to \dfrac{E}{R}$

8 $i = (3.33 \times 10^{-3})\left[e^t - e^{-(3 \times 10^6)t}\right]$ A

13(c) **1 iv** $i = 0.368$ A

2 iv $v = 0.368E$

3 $v = E(1 - e^{-t/RC})$ V

4 ii $v = 15(1 - e^{-10t})$ V

iv 63.2%, 86.5% and 95.0% respectively of 15 V

13(d) **1 i** $\dfrac{dh}{dt} = -\left(34.79 \times 10^{-3}\right)h^{1/2}$

ii $h = \left[-(17.395 \times 10^{-3})t + 1.14\right]^2$

iii 1.09 min

---

2 i $\dfrac{dh}{dt} = -(6.27 \times 10^{-3})h^{1/2}$

ii $h = \left[-(3.13 \times 10^{-3})t + 1.265\right]^2$

iii 6.73 min

3 i $\dfrac{dh}{dt} = -(17.3 \times 10^{-3})h^{-3/2}$

ii $h = \left[-(43.25 \times 10^{-3})t + 15.59\right]^{2/5}$

iii 6 min

4 $h = \left[-\left(k\dfrac{g}{2}\right)t + 1\right]^2$

5 $\theta = 300 + 73e^{-(2.96 \times 10^{-3})t}$

6 $\theta = 300 + 100e^{kt}$

8 $\theta = 320 + 28e^{kt}$

13(e) **1** $y_1 = 1.1$, $y_2 = 1.231$ and $y_3 = 1.4025361$

2 At $x = 0.5$, $y = 1.8370655$

3 $y_1 = 0$, $y_2 = 1$, $y_3 = 3$, $y_4 = 6$ and $y_5 = 10$
Exact $y = 0$, 0.5, 2, 4.5, 8 and 12.5

4 $v(0.3) = 2.93262$

7 iv For $h = 0.25$, 83.85% and for $h = 0.005$, 1.82%

13(f) **1** $y_1 = 2.215$, $y_2 = 2.463075$

2 $y_3 = 1.436058$

3 $y(1.6) = 1.3759$

6 Euler's and improved Euler's methods have 4.286% and 1.062% errors respectively.

13(g) **1** $y_1 = 2.215513$, $y_2 = 2.464208$

2 $y_1 = 0.0627$, $y_2 = 0.1919$

3 $y(1.6) = 1.3774$

4 $v(2) = 9.1371$ m/s

EQ13 **1 i** $y = \sqrt{2x^3 + 1}$

ii $y = e^{2x} - e^x$

2 a $v = \dfrac{35}{1.5}\left(1 - e^{-\frac{t}{10}}\right)$ **b** 0.8961

3 10.44 a.m.

4 i $y = \dfrac{1}{3}\left(x^2 + \dfrac{8}{x}\right)$ **ii** $-x = y[\ln|y| + C]$

5 $y = -\dfrac{1}{2}\ln|-2(e^x + C)|$

6 $y = 2x + 2x^{1/2} - 1$

7 -12.25

8 $y = \cos(x) - 2\cos^2(x)$

9 $\dfrac{1}{4}\left[11\tan^2(t) + 9\right]$

10 $y = \dfrac{\sin(x) - x\cos(x) + C}{x^2}$

11 a 25 days   **b** No, because you will need to find $\ln|m - 3|$ which is impossible at $m = 3$

12 25.98 minutes

ME13 **2** $v = \dfrac{1}{0.1 - t}$

3 $v = \dfrac{1}{kt + 10}$ · Large k gives a large drop in velocity, v, with time t

9 i $v = 20(1 - e^{-0.1t})$ **ii** 20 m/s

10 $\dfrac{mg}{k}$

11 $\sqrt{\dfrac{mg}{k}}$

15 Runge–Kutta solution is closer to the exact solution than is the Euler's solution

18 iv 9 V
   v Smaller C is quicker to reach its final value of 9 V

19 iv 0.9 mA
   v Larger L takes longer for the current, i, to reach its final value of 0.9 mA

14(a) **1 a** $s = -4.9t^2$
   b $s = 1.63t^3$
   c $s = ut + \dfrac{1}{2}at^2$

2 a $y = Ae^{-3x} + Be^{-2x}$
   b $y = (A + Bx)e^{-2x}$
   c $y = e^x\left[A\cos(\sqrt{3}x) + B\sin(\sqrt{3}x)\right]$

3 a $y = Ae^{-2x} + Be^{-x}$
   b $y = Ae^{6x} + Be^x$
   c $y = (A + Bx)e^{3x}$
   d $y = e^{2x}[A\cos(x) + B\sin(x)]$
   e $y = (A + Bx)e^{4x}$

4 a $y = (A + Bx)e^{-3x}$

b $y = e^{-\frac{5}{2}x}\left[A\cos\left(\dfrac{5\sqrt{3}}{2}x\right) + B\sin\left(\dfrac{5\sqrt{3}}{2}x\right)\right]$

c $y = Ae^{m_1 x} + Be^{m_2 x}$ where $m_1 = \dfrac{-3 + \sqrt{41}}{2}$ and $m_2 = \dfrac{-3 - \sqrt{41}}{2}$

14(b) **1** $i = 2(e^{-2t} - e^{-5t})$

2 a $x = A\cos(3t) + B\sin(3t)$
   b $x = A\cos(4t) + B\sin(4t)$
   c $x = A\cos(\sqrt{2}t) + B\sin(\sqrt{2}t)$
   d $x = A\cos(\sqrt{5}t) + B\sin(\sqrt{5}t)$

7 $x = 5e^{-\omega t}(1 + \omega t)$

9 a $v = (A + Bt)e^{-t/2RC}$
   b $v = Ae^{m_1 t} + Be^{m_2 t}$ where

$$m_1 = \dfrac{1}{2RC}\left[-1 + \sqrt{\dfrac{L - 4R^2C}{L}}\right] \quad \text{and}$$

$$m_2 = \dfrac{1}{2RC}\left[-1 - \sqrt{\dfrac{L - 4R^2C}{L}}\right]$$

   c $v = e^{\alpha t}\left[A\cos(\beta t) + B\sin(\beta t)\right]$ where

$$\alpha = -\dfrac{1}{2RC}, \beta = \dfrac{1}{2RC}\sqrt{\left(\dfrac{4R^2C - L}{L}\right)}$$

10 $L = 0.294$ H

11 $\omega = \dfrac{1}{\sqrt{LC}}, \zeta = \dfrac{1}{2R}\sqrt{\dfrac{L}{C}}$

14(c) **1 a** -9 **b** $-x - 1$ **c** $3\cos(3x) - 11\sin(3x)$
   d $-\dfrac{1}{10}\left[3\cos(x) + \sin(x)\right]$ **e** $\dfrac{e^{3x}}{4}$

2 a $y = Ae^{-2x} + Be^{-x} + 3$
   b $y = Ae^{-x/3} + Be^x - 2x + 7$
   c $y = (A + Bx)e^{-x} + e^{5x}$
   d $y = Ae^{-4x} + Be^x + 3\cos(x) + 5\sin(x)$

3 $x = A\cos(\omega t) + B\sin(\omega t) + \dfrac{mg}{k}$ where

$$\omega = \sqrt{\dfrac{k}{m}}$$

4 $x = Ae^{2t} + Be^{-2t} - 0.125$

8 $i = Ae^{-(2.764\times10^4)t} + Be^{-(7.236\times10^4)t} + (50\times10^{-3})$

9 $y = Ae^{2x} + Be^{-x} - \frac{5}{3}xe^{-x}$

10 a $Y = Cxe^{-3x}$

　b $Y = x[a\cos(3x) + b\sin(3x)]$

14(d) 1 $i = -25e^{-3t} + 15e^{-5t} + 10 = 5(2 + 3e^{-5t} - 5e^{-3t})$

2 $i = 2.5e^{-3t} + 7.5e^{-t} - 10e^{-2t}$

3 $x = 2\cos(3t) + \sin(3t) - 2$

EQ14 1 $y = e^{-x}[\cos(x) + \sin(x)]$

2 i $y = 3\cos(4x) - \frac{1}{2}\sin(4x)$

　ii $y = Ae^{3x} + Be^{-5x} + \frac{2}{9}e^{4x}$

3 a $s = 2\cos(5t) + \sin(5t)$　　**b** $-\frac{3}{\sqrt{2}}$　　**c** 0

4 $y = 1.5\sin(x) + 0.001[2\cos(x) + x^2 - 2]$

5 $y = Ae^x + Be^{-2x} + 2x + 1,\ y = 3e^x + 2x + 1$

6 a $y = 2e^{-\frac{3}{2}x} + e^{-x}$

　b $y = A + Be^x + \frac{1}{10}[\cos(2x) - 2\sin(2x)]$

7 a is not a solution but **b** is

8 $y = Ae^x + Be^{-2x} + \frac{1}{10}e^{3x}$

9 $y = \frac{1}{2}[5\sin(x) + x\sin(x)]$

10 $y = A\cos(3t) + B\sin(3t) + t^2 - \frac{2}{9} - 2t\sin(3t)$

11 a $x = -\cos(3t) + \sin(3t) = \sqrt{2}\sin\left(3t - \frac{\pi}{4}\right)$,

　　amplitude $= \sqrt{2}$ and period $\frac{2\pi}{3}$

　b $x = e^t[A\cos(3t) + B\sin(3t)] + 2t + 1$

ME14 1 a $y = (A + Bx)e^{-4x}$　**b** $y = Ae^{3x} + Be^{-x}$

　c $y = Ae^{(3+\sqrt{2})x} + Be^{(3-\sqrt{2})x}$

2 a $y = (A + Bx)e^{-4x} + \frac{1}{2}$

　b $y = (A + Bx)e^{-4x} + \frac{1}{4}(2x - 1)$

　c $y = (A + Bx)e^{-4x} + \frac{1}{128}[8x^2 + 7]$

4 $T = Ae^{mx} + Be^{-mx}$ where $m = \sqrt{\frac{hP}{kA}}$

8 i $m^2 + \frac{1}{RC}m + \frac{1}{LC} = 0$

　ii $R_{cr} = \frac{1}{2}\sqrt{\frac{L}{C}}$

9 i $m^2 + \frac{R}{L}m + \frac{1}{LC} = 0$　　**ii** $R_{cr} = 2\sqrt{\frac{L}{C}}$

10 i $20\ \Omega$　　**ii** $i = (A + Bt)e^{-(10\times10^3)t}$

11 $v = e^{-2000t}[9\cos(4000t) + 4.5\sin(4000t)]$

18 $x = 2t + 0.776 - e^{-3t}\Big[0.195\sin\left(\sqrt{491}\,t\right)$
　　　$+ 0.776\cos\left(\sqrt{491}\,t\right)\Big]$

15(a) 1 a $\frac{\partial f}{\partial x} = 2x,\ \frac{\partial f}{\partial y} = 2y$

　b $\frac{\partial f}{\partial x} = \cos(x),\ \frac{\partial f}{\partial y} = -\sin(y)$

　c $\frac{\partial f}{\partial x} = 3x^2 + 3y,\ \frac{\partial f}{\partial y} = 3y^2 + 3x$

　d $\frac{\partial f}{\partial x} = \frac{1000}{x^2},\ \frac{\partial f}{\partial y} = 2$

2 $\frac{\partial M}{\partial P} = x,\ \frac{\partial M}{\partial x} = P + wx^2$

3 a $\frac{PL}{AE}$　**b** $\frac{3PL^3}{16EI}$

4 $\sigma_x = 12Ax^2y^2 + 2Bx^4,\ \sigma_y = 2Ay^4 + 12Bx^2y^2$

6 $u = \frac{y}{x^2 + y^2},\ v = \frac{-x}{x^2 + y^2}$

8 $c_p - c_v = T\left(\frac{R}{P} + \frac{K}{RT^2}\right)\left[\frac{RV + 2K/T}{(V + K/RT)^2}\right]$

9 $a = \frac{9RT_c V_c}{8},\ b = \frac{V_c}{3}$

11 a $36xy - 9$　**b** $\frac{16\times10^6}{x^3 y^3} - 4$

15(b) 1 $0.00853\ m^4$

2 Reduction of 1.5%

3 Increased by $(2.34\times10^{-4})\ m^3/s$

4 1%

For complete solutions see www.palgrave.com/engineering/singh

5 2.8%

6 2.3%

15(c) **1 a** (3, 4)   **b** (0.5, 1)   **c** $(-11/6, -121/12)$
and $(2, -12)$

2 a (2,1) saddle

b $\left(\dfrac{1}{\sqrt{3}}, -\dfrac{1}{2}\right)$ minimum and $\left(-\dfrac{1}{\sqrt{3}}, -\dfrac{1}{2}\right)$ saddle

c $\left(\dfrac{3}{2}, 2\right)$ minimum and $\left(\dfrac{25}{6}, \dfrac{2}{3}\right)$ saddle

3 $1.216 \times 1.216 \times 0.609$. One of the lengths is half of the other two

4 $(2V)^{1/3}, (2V)^{1/3}$ and $\left(\dfrac{V}{4}\right)^{1/3}$

5 $2.62 \times 2.62 \times 1.31$

6 i 99.5

ii $\left(\dfrac{1}{\sqrt{2}}, \dfrac{1}{\sqrt{2}}\right), \left(-\dfrac{1}{\sqrt{2}}, \dfrac{1}{\sqrt{2}}\right), \left(-\dfrac{1}{\sqrt{2}}, -\dfrac{1}{\sqrt{2}}\right)$

and $\left(\dfrac{1}{\sqrt{2}}, -\dfrac{1}{\sqrt{2}}\right)$

7 $x = 3$ m and $\theta = 60°$

EQ15 **1** $y = x = 1.817$ and $z = 0.909$

2 (0, 0) is a saddle point and $(-1, 1)$ is a minimum

3 $x = 2$ and $y = 3$

4 3.5%

5 a $\dfrac{\partial u}{\partial x} = 6x + 12xy - 4y^2, \dfrac{\partial^2 u}{\partial x^2} = 6 + 12y,$

$\dfrac{\partial u}{\partial y} = 6x^2 - 8xy - 3y^2, \dfrac{\partial^2 u}{\partial y^2} = -8x - 6y,$

$\dfrac{\partial^2 u}{\partial x \partial y} = 12x - 8y$

b $c = 2$

6 a i $\dfrac{\partial z}{\partial x} = 3x^2 + 10xy, \dfrac{\partial^2 z}{\partial^2 x} = 6x + 10y,$

$\dfrac{\partial z}{\partial y} = 5x^2 + 6y^2, \dfrac{\partial^2 z}{\partial y \partial x} = 10x$

ii $\dfrac{\partial z}{\partial x} = e^x \cos(y), \dfrac{\partial^2 z}{\partial x^2} = e^x \cos(y),$

$\dfrac{\partial z}{\partial y} = -e^x \sin(y), \dfrac{\partial^2 z}{\partial y \partial x} = -e^x \sin(y)$

b $p = 1$

c (5, 0) min, $(-5, 0)$ max, (3, 4) and $(-3, -4)$ saddle points

ME15 **1** $\dfrac{\partial M}{\partial P} = x - \dfrac{x^2}{L}, \quad \dfrac{\partial M}{\partial x} = K + \dfrac{P}{L}(L - 2x)$

2 $\dfrac{\partial M}{\partial P} = x - \dfrac{ax}{L}, \quad \dfrac{\partial M}{\partial x} = \dfrac{K - Pa}{L} + P$

3 $24Ax$. Ω is not an Airy stress function

4 1.5% decrease

5 7%

6 3.8% increase

7 $x = -r$

10 $\sigma_r = -2A\cos(2\theta) - 2B\sin(2\theta),$

$\sigma_\theta = 2A\cos(2\theta) + 2B\sin(2\theta)$ and

$\tau_{r0} = 2A\sin(2\theta) - 2B\cos(2\theta)$

12 $x = 2^{-1/6}$ m, $l = 2^{1/3}$ m and $\theta = 45°$

13 $a = 2RVT, b = V/2$

14 $a = 1.28RVT^{3/2}, b = 0.26V$

16(a) **1 a** Continuous   **b** Discrete   **c** Discrete
d Continuous   **e** Continuous

2

Temperature (°C)	15	16	17	18	19	20
Frequency	1	2	3	1	2	5

Temperature (°C)	21	22	23	24	25	26
Frequency	4	3	5	2	1	1

3 The shortest height is 1.53 m and the tallest is 2.00 m. Divide into 1.5–1.6, 1.6–1.7,....

Height h (m)	Frequency
$1.50 \le h < 1.60$	4
$1.60 \le h < 1.70$	6
$1.70 \le h < 1.80$	7
$1.80 \le h < 1.90$	14
$1.90 \le h < 2.00$	9

16(b) $\bar{x}$ = mean and s = standard deviation

1 median = 27, mean = 24.91, SD = 13.61

2 a $\bar{x} = 15.20$ and $s = 11.46$
b $\bar{x} = 115.20$ and $s = 11.46$

c $\bar{x} = 1.52$ and $s = 1.146$
d $\bar{x} = 155.00$ and $s = 114.6$

5 **i** $\bar{x} = 69.29$ and $s = 23.68$
ii $\bar{x} = 692.9$ and $s = 236.8$
iii $\bar{x} = 0.6929$ and $s = 0.2368$
iv $\bar{x} = 74.29$ and $s = 23.68$
v $\bar{x} = 7029$ and $s = 2368$

6 Mean $= 159.93$ kN, $s = 4.66$ kN

7 **i** Good estimate is 10 kΩ
ii Small SD because data is close to 10 kΩ
iii Mean $= 9.996$ kΩ, $s = 0.122$ kΩ

8 **a** $\bar{x} = 431.35$ MN/m^2 and $s = 55.62$ MN/m^2
b $\bar{x} = 196.29$ GN/m^2 and $s = 1.86$ GN/m^2

16(c) **1** **a** 9/10   **b** 3/20,000

2 8×10^{-6}

3 0.93

4 **i** 0.3   **ii** 0.65   **iii** 0.35

5 **a** 7/100   **b** 93/100

6 8.4×10^{-3}

7 $\dfrac{4}{49}$

8 **i** 0.62   **ii** 0.38

9 0.786

10 0.955

11 $\dfrac{3}{59}$

12 **a** $\dfrac{1}{495}$ **b** $\dfrac{893}{990}$ **c** $\dfrac{19}{198}$

13 0.957

14 $\dfrac{5}{119}$

15 **a** 3/4   **b** 1/2   **c** 1/4

16 $\dfrac{7}{1520}$

17 0.9775

18 0.88186

16(d) **1** 0.0716

2 **i** 0.249   **ii** 3.52×10^{-9} **iii** 0.751

3 **a** 3.908×10^{-5} **b** £1,919,190

4 **a** 258,336,000   **b** 456,976,000

16(e) **1** **a** 8.1×10^{-3} **b** 0.99954

2 **a** 0.170   **b** 1.239×10^{-5}

3 **i** 0.264   **ii** 0.264   **iii** 0.418   **iv** 0.582

4 0.456

5 0.6572

6 **a** **i** 0.3295   **ii** 0.0988   **iii** 0.0197
b **i** 0.3293   **ii** 0.0988   **iii** 0.0198
c The results are very similar

7 0.83521

8 $k = \dfrac{1}{17}$

9 $k = \dfrac{1}{36}$

16(f) **1** $E(X) = 3$, symmetric distribution therefore mean is 3

2 $E(X^2) = 13.42$

3 $\mu = 3$ V, $\sigma = 1.095$ V, symmetric distribution therefore mean is 3 V

4 $\mu = 4.19$ V, $\sigma = 1.94$ V

5 $\mu = 1.11$ V, $\sigma = 1.05$ V

6 **i** $k = \dfrac{1}{30}$ **ii** $E(X) = 3.33$ **iii** SD $= 0.83$
iv variance $= 0.69$

7 **i** 2.9   **ii** 9.5   **iii** 12.4   **iv** 70.8

9 $\mu = 3$, $\sigma = 1.643$

10 $\mu = 1.5$

11 $\mu = 7$, $\sigma^2 = 5.833$

12 **a** 0.20   **b** 0.32   **c** 0.78   **d** 6.12×10^{-6}

13 **a** 0.0166   **b** 0.776

16(g) **1** **i** 6.67   **ii** 50   **iii** 5.5111   **iv** 1/4   **v** 3/4

2 **a** 1/4096   **b** 15/4096   **c** 175/256   **d** 11/16

3 **a** $7\dfrac{1}{2}$ **b** $64\dfrac{4}{5}$ **c** $8\dfrac{11}{20}$

4 **a** 0.632   **b** 0.865   **c** 0.181

5 **a** 0.632   **b** 0.95   **c** 0.181

For complete solutions see www.palgrave.com/engineering/singh

6 a 0.787   **b** 0.050   **c** 0.549

7 i 10/3 years   **ii** $\dfrac{0.08t}{1 - 0.04t^2}$ $0 \le t \le 5$

8 $R(t) = e^{-0.05t^2}$

9 $f(t) = 0.1$, $F(t) = 0.1t$, $R(t) = 1 - 0.1t$,
$h(t) = \dfrac{0.1}{1 - 0.1t}$ all defined for $0 \le t \le 10$;
$MTTF = 5$ years

10 a 5/3 years   **b** 3 years

11 $t_{\text{med}} = 6.93$ years

16(h)   **1 a** 0.9332   **b** 0.95   **c** 0.975   **d** 0.05
   e 0.05   **f** 0.0228   **g** 0.9   **h** 0.1191
   i 0.1662

2 0.9655

3 0.0228

4 a 0.00135   **b** 0.00621

5 0.7592

6 a 126   **b** 45   **c** 235

7 $\mu = 68.463$ kg

8 $\sigma = 0.243$ m

9 0.57825 m to 0.62975 m

EQ16   **1** Mean = 2.82, SD = 0.648

2 Longer than 2.75 m is 0.65 (2 d.p.). Less than 2.95 m is 0.88 (2 d.p.).

3 ii and   **iii** Mean = 29.95, SD = 4.714
   iv 37

4 $\dfrac{19}{66}$

5 0.5665

6 i 0.7745   **ii** 0.4   **iii** $\alpha = 51.684$ g

7 a 0.4422   **b i** 0.9772   **ii** 0.151
   c 0.0228   **d** 0.5955

8 a i 0.6561   **ii** 0.3439
   b 0.03 **i** 0.7374   **ii** 0.2281   **iii** 0.9655

ME16 1

Capacitance (μF)	Frequency
0.900–0.930	3
0.930–0.960	6
0.960–0.990	3
0.990–1.020	16
1.020–1.050	7
1.050–1.080	7
1.080–1.110	8

2 a 0.0179   **b** 0.1841   **c** 0.18223

3 0.99

4 a 0.997   **b** 0.993

5 i 0.294   **ii** 0.006

6 a 1.024×10^{-7} **b** 0.1074   **c** 0.0881

7 i 30   **ii** 3.33   **iii** 0.83   **iv** 0.69

8 1002 hours to 1198 hours

12 Largest size is 11

13 0.4, $F(t) = 0.4t - 0.04t^2$,
$R(t) = 1 - 0.4t + 0.04t^2$, $h(t) = \dfrac{5}{3}$ years

15 $F(t) = 1 - \dfrac{1}{(1 + t)^4}$, $R(t) = \dfrac{1}{(1 + t)^4}$,
$h(t) = \dfrac{4}{1 + t}$, $\dfrac{1}{14641}$

16 a $F(t) = 4 - \dfrac{20}{5 + t/3}$, $R(t) = -3\left(\dfrac{t - 5}{15 + t}\right)$,
$f(t) = \dfrac{20}{3(5 + t/3)^2}$, $h(t) = \dfrac{20}{(t - 5)(15 + t)}$

   c i 3/4   **ii** 9/17   **iii** 1/3   **iv** 3/19   **v** 0
   d A component cannot last more than 5 years

(In **16** the functions are defined for $0 \le t \le 5$)

Appendix: Standard Normal Distribution Table

(Reproduced with permission. From Miller & Powell: *The Cambridge Elementary Mathematical Tables*, 2nd edn, 1979, Cambridge University Press.)

N(0,1)

Φ(z)

0   z

THE DISTRIBUTION FUNCTION Φ(z) OF THE NORMAL DISTRIBUTION N(0, 1)

ADD columns: 1 2 3 | 4 5 6 | 7 8 9

z	0	1	2	3	4	5	6	7	8	9	1	2	3	4	5	6	7	8	9
0.0	.5000	.5040	.5080	.5120	.5160	.5199	.5239	.5279	.5319	.5359	4	8	12	16	20	24	28	32	36
0.1	.5398	.5438	.5478	.5517	.5557	.5596	.5636	.5675	.5714	.5753	4	8	12	16	20	24	28	32	36
0.2	.5793	.5832	.5871	.5910	.5948	.5987	.6026	.6064	.6103	.6141	4	8	12	15	19	23	27	31	35
0.3	.6179	.6217	.6255	.6293	.6331	.6368	.6406	.6443	.6480	.6517	4	7	11	15	19	22	26	30	34
0.4	.6554	.6591	.6628	.6664	.6700	.6736	.6772	.6808	.6844	.6879	4	7	11	14	18	22	25	29	32
0.5	.6915	.6950	.6985	.7019	.7054	.7088	.7123	.7157	.7190	.7224	3	7	10	14	17	20	24	27	31
0.6	.7257	.7291	.7324	.7357	.7389	.7422	.7454	.7486	.7517	.7549	3	7	10	13	16	19	23	26	29
0.7	.7580	.7611	.7642	.7673	.7704	.7734	.7764	.7794	.7823	.7852	3	6	9	12	15	18	21	24	27
0.8	.7881	.7910	.7939	.7967	.7995	.8023	.8051	.8078	.8106	.8133	3	5	8	11	14	16	19	22	25
0.9	.8159	.8186	.8212	.8238	.8264	.8289	.8315	.8340	.8365	.8389	3	5	8	10	13	15	18	20	23
1.0	.8413	.8438	.8461	.8485	.8508	.8531	.8554	.8577	.8599	.8621	2	5	7	9	12	14	16	19	21
1.1	.8643	.8665	.8686	.8708	.8729	.8749	.8770	.8790	.8810	.8830	2	4	6	8	10	12	14	16	18
1.2	.8849	.8869	.8888	.8907	.8925	.8944	.8962	.8980	.8997	.9015	2	4	6	7	9	11	13	15	17
1.3	.9032	.9049	.9066	.9082	.9099	.9115	.9131	.9147	.9162	.9177	2	3	5	6	8	10	11	13	14
1.4	.9192	.9207	.9222	.9236	.9251	.9265	.9279	.9292	.9306	.9319	1	3	4	6	7	8	10	11	13
1.5	.9332	.9345	.9357	.9370	.9382	.9394	.9406	.9418	.9429	.9441	1	2	4	5	6	7	8	10	11
1.6	.9452	.9463	.9474	.9484	.9495	.9505	.9515	.9525	.9535	.9545	1	2	3	4	5	6	7	8	9
1.7	.9554	.9564	.9573	.9582	.9591	.9599	.9608	.9616	.9625	.9633	1	2	3	4	4	5	6	7	8
1.8	.9641	.9649	.9656	.9664	.9671	.9678	.9686	.9693	.9699	.9706	1	1	2	3	4	4	5	6	6
1.9	.9713	.9719	.9726	.9732	.9738	.9744	.9750	.9756	.9761	.9767	1	1	2	2	3	4	4	5	5
2.0	.9772	.9778	.9783	.9788	.9793	.9798	.9803	.9808	.9812	.9817	0	1	1	2	2	3	3	4	4
2.1	.9821	.9826	.9830	.9834	.9838	.9842	.9846	.9850	.9854	.9857	0	1	1	2	2	3	3	3	4
2.2	.9861	.9864	.9868	.9871	.9875	.9878	.9881	.9884	.9887	.9890	0	1	1	1	2	2	2	3	3
2.3	.9893	.9896	.9898								0	1	1	1	1	2	2	2	2
				.9901	.99036	.99061	.99086				3	5	8	10	13	15	18	20	23
								.99111	.99134	.99158	2	5	7	9	12	14	16	18	21
2.4	.99180	.99202	.99224	.99245	.99266						2	4	6	8	11	13	15	17	19
						.99286	.99305	.99324	.99343	.99361	2	4	6	7	9	11	13	15	17
2.5	.99379	.99396	.99413	.99430	.99446	.99461	.99477	.99492	.99506	.99520	2	3	5	6	8	9	11	12	14
2.6	.99534	.99547	.99560	.99573	.99585	.99598	.99609	.99621	.99632	.99643	1	2	3	5	6	7	8	9	10
2.7	.99653	.99664	.99674	.99683	.99693	.99702	.99711	.99720	.99728	.99736	1	2	3	4	5	6	7	8	9
2.8	.99744	.99752	.99760	.99767	.99774	.99781	.99788	.99795	.99801	.99807	1	1	2	3	4	4	5	6	6
2.9	.99813	.99819	.99825	.99831	.99836	.99841	.99846	.99851	.99856	.99861	0	1	1	2	2	3	3	4	4
3.0	.99865	.99869	.99874	.99878	.99882	.99886	.99889	.99893	.99896	.99900	0	1	1	2	2	3	3	4	4
3.1	$.9^3032$*	$.9^3065$	$.9^3096$								3	6	9	13	16	19	22	25	28
				$.9^3126$	$.9^3155$	$.9^3184$	$.9^3211$				3	6	8	11	14	17	20	22	25
								$.9^3238$	$.9^3264$	$.9^3289$	2	5	7	10	12	15	17	20	22
3.2	$.9^3313$	$.9^3336$	$.9^3359$	$.9^3381$	$.9^3402$						2	4	7	9	11	13	15	18	20
						$.9^3423$	$.9^3443$	$.9^3462$	$.9^3481$	$.9^3499$	2	4	6	8	9	11	13	15	17
3.3	$.9^3517$	$.9^3534$	$.9^3550$	$.9^3566$	$.9^3581$						2	3	5	6	8	10	11	13	14
						$.9^3596$	$.9^3610$	$.9^3624$	$.9^3638$	$.9^3651$	1	3	4	5	7	8	9	10	12
3.4	$.9^3663$	$.9^3675$	$.9^3687$	$.9^3698$	$.9^3709$	$.9^3720$	$.9^3730$	$.9^3740$	$.9^3749$	$.9^3758$	1	2	3	4	5	6	7	8	9
3.5	$.9^3767$	$.9^3776$	$.9^3784$	$.9^3792$	$.9^3800$	$.9^3807$	$.9^3815$	$.9^3822$	$.9^3828$	$.9^3835$	1	1	2	3	4	4	5	6	7
3.6	$.9^3841$	$.9^3847$	$.9^3853$	$.9^3858$	$.9^3864$	$.9^3869$	$.9^3874$	$.9^3879$	$.9^3883$	$.9^3888$	0	1	1	2	2	3	3	4	5
3.7	$.9^3892$	$.9^3896$	$.9^390$	$.9^404$	$.9^408$	$.9^412$	$.9^415$	$.9^418$	$.9^422$	$.9^4250$									
3.8	$.9^428$	$.9^431$	$.9^433$	$.9^436$	$.9^438$	$.9^441$	$.9^443$	$.9^446$	$.9^448$	$.9^4500$									
3.9	$.9^452$	$.9^454$	$.9^456$	$.9^458$	$.9^459$	$.9^461$	$.9^463$	$.9^464$	$.9^466$	$.9^4670$									

For negative values of z use Φ(z) = 1 − Φ(−z).

*$.9^3032$ signifies .999032, and similarly for other values quoted in this form.

Index